7th International Conference on 3D Radiation Dosimetry 2012

(IC3DDose)

Journal of Physics: Conference Series Volume 444

Sydney, Australia
4-8 December 2012

ISBN: 978-1-62748-676-7
ISSN: 1742-6588

Printed from e-media with permission by:

Curran Associates, Inc.
57 Morehouse Lane
Red Hook, NY 12571

CURRAN ASSOCIATES INC. proceedings.com

Some format issues inherent in the e-media version may also appear in this print version.

Printed by Curran Associates, Inc. (2013)

For permission requests, please contact the Institute of Physics
at the address below.

Institute of Physics
Dirac House, Temple Back
Bristol BS1 6BE UK

Phone: 44 1 17 929 7481
Fax: 44 1 17 920 0979

techtracking@iop.org

Additional copies of this publication are available from:

Curran Associates, Inc.
57 Morehouse Lane
Red Hook, NY 12571 USA
Phone: 845-758-0400
Fax: 845-758-2634
Email: curran@proceedings.com
Web: www.proceedings.com

Papers

Refresher

Invited Review

Clinical Challenges & Dosimetry Systems

New Dosimetry Systems

Polymer Gel Dosimetry

Radiochromic and Other Novel Dosimetry

EPID Dosimetry and Verification of Complex Treatments

IMRT and IGRT

2D versus 3D Dosimetry and 3D

Optical Dosimetry I

Dosimetry for Audit, SRS/SBRT and Small Fields

Optical Dosimetry II

3D Analysis and Ultrasound Evaluation of Gels

CBCT Dosimetry; Radiological Properties of Gels, X-ray CT, MRI

Motion, Brachytherapy, Heterogeneity and Miscellaneous

Poster

Preface

IC3DDose 2013, the 7th International Conference on 3D Radiation Dosimetry held in Sydney, Australia from 4–8 November 2012, grew out of the DosGel series, which began as DosGel99, the 1st International Workshop on Radiation Therapy Gel Dosimetry in Lexington, Kentucky. Since 1999 subsequent DoSGel conferences were held in Brisbane, Australia (2001), Ghent, Belgium (2004), Sherbrooke, Canada (2006) and Crete, Greece (2008). In 2010 the conference was held on Hilton Head Island, South Carolina and underwent a name-change to IC3DDose.

The aim of the first workshop was to bring together individuals, both researchers and users, with an interest in 3D radiation dosimetry techniques, with a mix of presentations from basic science to clinical applications, which has remained an objective for all of the meetings. One rationale of DosGel99 was stated as supporting the increasing clinical implementation of gel dosimetry, as the technique appeared, at that time, to be leaving the laboratories of gel dosimetry enthusiasts and entering clinical practice. Clearly by labelling the first workshop as the 1st, there was a vision of a continuing series, which has been fulfilled. On the other hand, the expectation of widespread clinical use of gel dosimetry has perhaps not been what was hoped for and anticipated. Nevertheless the rapidly increasing demand for advanced high-precision 3D radiotherapy technology and techniques has continued apace. The need for practical and accurate 3D dosimetry methods for development and quality assurance has only increased. By the 6th meeting, held in South Carolina in 2010, the Conference Scientific Committee recognised the wider developments in 3D systems and methods and decided to widen the scope, whilst keeping the same span from basic science to applications. This was signalled by a change of name from 'Dosgel' to 'IC3DDose', a name that has continued to this latest conference.

The conference objectives were:

- to enhance the quality and accuracy of radiation therapy treatment through improved clinical dosimetry
- to investigate and understand the dosimetric challenges of modern radiation treatments
- to provide a forum to discuss the latest research and developments in 3D and advanced radiation dosimetry
- to energise and diversify dosimetry research and clinical practice by encouraging interaction and synergy between advanced, 3D and semi-3D dosimetry techniques

We believe the conference program, with its excellent range of expert and specialist speakers, met these objectives.

Thanks are due to all invited speakers for their participation, to the Local Organising Committee members for all their hard work in making the conference happen, particularly the small core administrative support group, and to the range of academic, organisation and commercial sponsors who generously supported the meeting. The Scientific Committee members are also thanked for reviewing the submitted manuscripts and for assisting in the editorial process. Finally, all who travelled to Sydney, Australia for the meeting are acknowledged for choosing to attend and contribute to making this a successful conference.

Local Conference Organising Committee
David Thwaites (Conference Convener)
Clive Baldock
Leanne Price
Elizabeth Starkey
May Whitaker
Peter Greer
Lois Holloway
Phil Vial
Robin Hill

Conference Scientific Committee
Sven Back (Sweden)
Clive Baldock (Australia)
Cheng-Shie Wuu (USA)
Yves de Deene (Belgium)
Simon Doran (UK)
Geoffrey Ibbott (USA)
Andrew Jirasek (Canada)
Kevin Jordan (Canada)
Martin Lepage (Canada)
Mark Oldham (USA)
Evangelos Pappas (Greece)
John Schreiner (Canada)
David Thwaites (Australia)

David Thwaites
Director
Institute of Medical Physics
School of Physics
University of Sydney
NSW 2006
Australia

Clive Baldock
Executive Dean
Faculty of Science
Macquarie University
North Ryde
NSW 2109
Australia

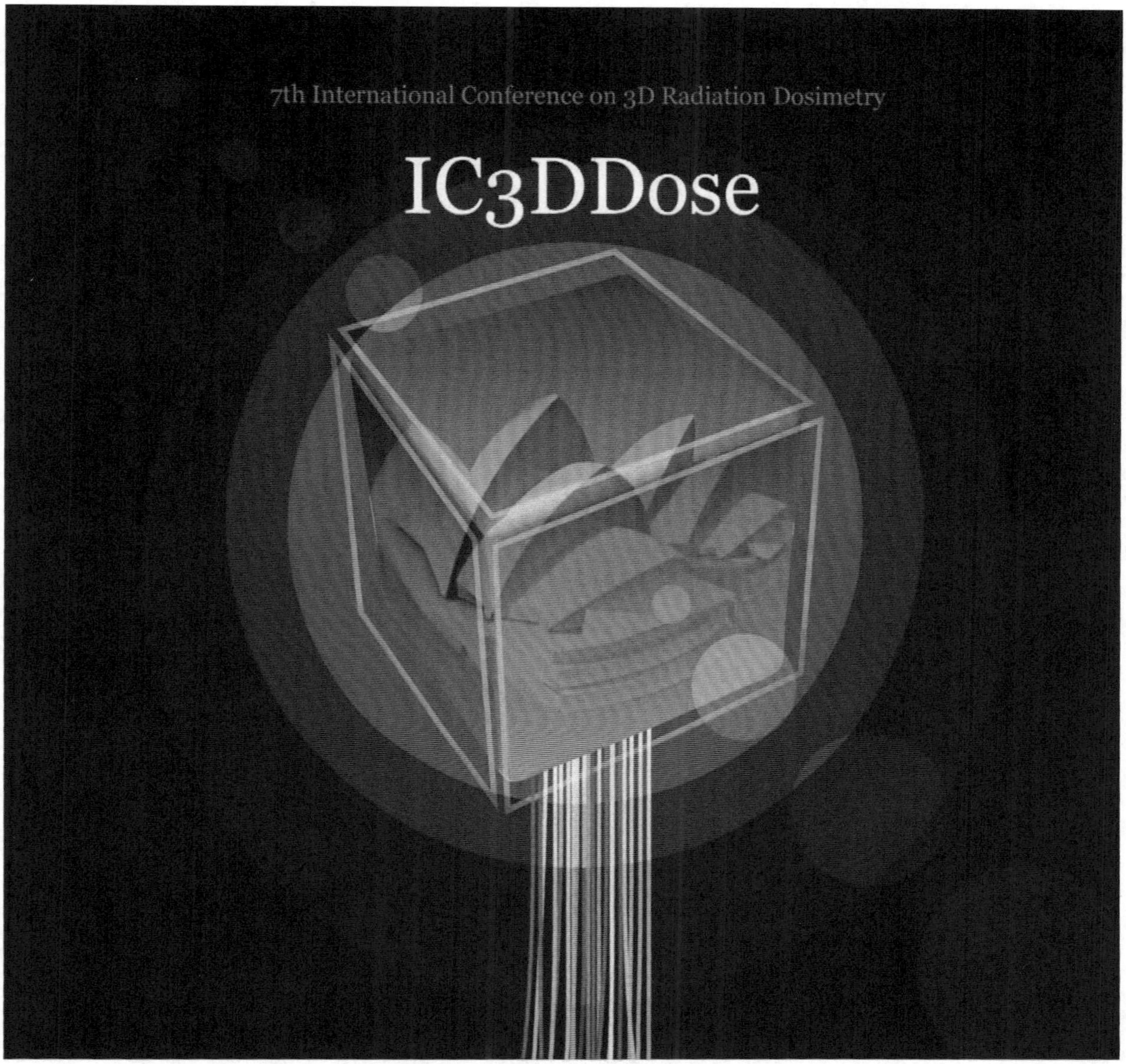

CONFERENCE PROGRAM

November 4 - 8 , 2012

Shangri-La Hotel, The Rocks, Sydney, Australia

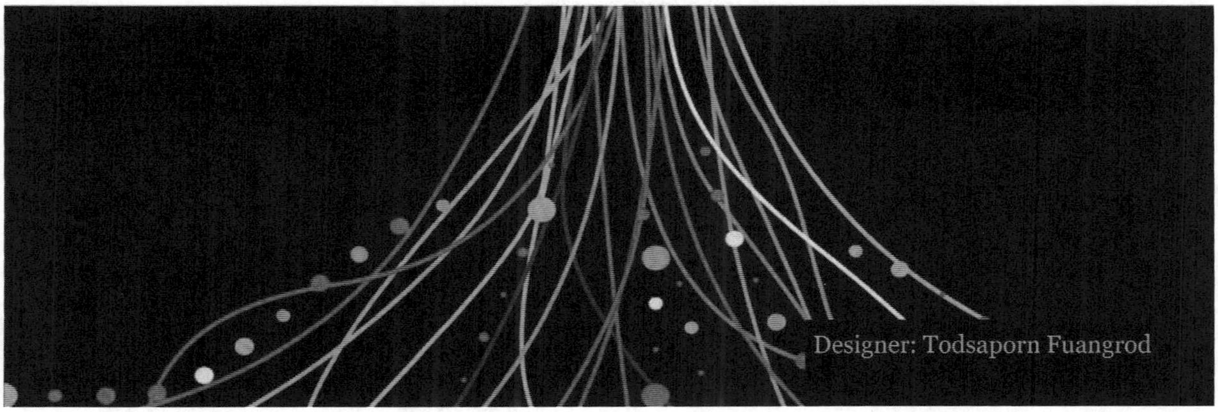

Thank you to our sponsors for their generous support

PLATINUM: ELEKTA

Sponsors of Ben Mijnheer and Jonathan Sykes; Conference dinner at MCA; Refresher program; USB drives for conference program; pens and notebooks

GOLD: SCANDIDOS

Sponsor of Stine Korreman; Welcome reception Harbour Cruise; Conference bags

SILVER: VARIAN

Sponsor of Peter Greer; and Lanyards

SILVER: NSW MINISTRY OF HEALTH

Sponsor of Thursday's invited review talks and the final horizon-scanning session; future directions and requirements of 3D radiotherapy dosimetry for advanced radiotherapy

SILVER: FACULTY OF SCIENCE, MACQUARIE UNIVERSITY AND INSTITUTE OF MEDICAL PHYSICS, UNIVERSITY OF SYDNEY

Sponsors of program booklet, proceedings and poster reception

BRONZE: CMS ALPHATECH AND SUN NUCLEAR

Sun Nuclear: Sponsor of Vladimir Feygelman
CMS Alphatech: drinks at conference dinner and poster reception

BRONZE: NUCLETRON

Jazz band at conference dinner and poster reception

BRONZE: CENTRE FOR MEDICAL RADIATION PHYSICS, UNIVERSITY OF WOLLONGONG

Sponsor of Peter Metcalfe

ITEM SPONSOR: NL-TEC

DJ for harbour cruise

ITEM SPONSOR: MEDTECH SOLUTIONS

Morning coffee sessions

ITEM SPONSOR: NSW/ACT BRANCH OF ACPSEM

Sponsor of Paul Keall

EXHIBITORS

Introduction and welcome

This meeting, the 7th International Conference on 3D Radiation Dosimetry, grew out of the DosGel series, which began as DosGel99, the 1st International Workshop on Radiation Therapy Gel Dosimetry in Lexington, Kentucky, proceedings of which are reproduced on your USB stick. The stated aim was to bring together individuals, both researchers and users, with an interest in three-dimensional radiation dosimetry techniques, with a mix of presentations from basic science up to clinical applications. One rationale was stated as supporting the increasing clinical implementation of gel dosimetry, as the technique appeared at that time to be leaving the laboratories of gel enthusiasts and entering clinical practice. Clearly by labelling the workshop as the 'first' there was a vision of a continuing series which has been fulfilled. On the other hand the expectation of widespread clinical use of gel dosimetry has perhaps not been as well-fulfilled. Nevertheless the rapidly advancing demands of advanced high-precision 3D radiotherapy technology and techniques has continued and quickened and the need for practical and accurate 3D dosimetry methods for development, verification, QA, etc has only increased with that. By the 6th meeting, held in South Carolina in 2010, the Organising Committee recognised the wider developments in 3D systems and methods and decided to widen the scope, whilst keeping the same span from basic science to applications. This was signalled by a change of name to the 6th International Conference on 3D Radiation Dosimetry, a trend that we have tried to nurture and develop further in this meeting. The objectives stay much the same:

- to enhance the quality and accuracy of radiation therapy treatment through improved clinical dosimetry.
- to investigate and understand the dosimetric challenges of modern radiation treatments.
- to provide a forum to discuss the latest research and developments in 3D and advanced radiation dosimetry.
- to energize and diversify dosimetry research and clinical practice by encouraging interaction and synergy between advanced, 3D, and semi-3D dosimetry techniques

We hope you find this program, with a great range of expert speakers in the field, meets these aims. We would like to thank, at the outset, all the invited speakers for agreeing to take part, the Scientific Committee members and the Local Organising Committee members for all their hard work in making the meeting happen, particularly our small core administrative support group, and the range of academic, organisation and commercial sponsors who have generously supported the meeting.

Finally we would like to thank all of you for choosing to come to this meeting. It is our great pleasure to welcome you to Sydney Australia and to the conference and we trust you will find the scientific sessions informative and stimulating and the city, the social events and the whole conference enjoyable and memorable. We will count this meeting a success if you renew old friendships and scientific collaborations and begin new ones.

Convenors

a b

(a)

David Thwaites, Director
Institute of Medical Physics
School of Physics, University of Sydney
david.thwaites@sydney.edu.au

(b)

Clive Baldock, Executive Dean
Faculty of Science
Macquarie University
clive.baldock@mq.edu.au

GENERAL INFORMATION

CONFERENCE ROOM

The conference will be held in the Cambridge Room each day.
This is located on Level 3 of the Shangri-La Hotel, 176 Cumberland Street, The Rocks, Sydney.

REGISTRATION DESK

The registration desk is located in the Foyer, Level 3. Opening hours are as follows:

Sunday	14:00 – 17:00
Monday	07:30 – 09:00, tea and lunch breaks
Tuesday	08:00 – 09:00, tea and lunch breaks
Wednesday	08:00 – 09:00, tea and lunch breaks
Thursday	08:00 – 0:900 only

SOCIAL PROGRAM

Sunday 4 November, 18:30 – 22:30
Sydney Harbour Cruise, Buffet dinner and drinks
Proudly sponsored by our Gold sponsor, Scandidos, together with NL-Tec
NOTE: Boarding 18:00-18.25 sharp. Don't be late! Two drop-offs at 21:30 and 22.30
Collection and drop off from Commissioner Steps Wharf at Circular Quay (on the quayside in front of the Museum of Contemporary Art)

Tuesday 6 November, 18:00 – 19:30
Happy Hour: Exhibition and Poster Viewing Drinks Reception
Proudly sponsored by our Silver sponsors, Faculty of Science, Macquarie University and the Institute of Medical Physics, University of Sydney and our Bronze sponsors, CMS Alphatech and Nucletron

Wednesday 7 November, 18:30 – 22:30
Conference Dinner, Harbourside Room, Museum of Contemporary Art
Proudly sponsored by our Platinum sponsor, Elekta
Bronze sponsors: CMS Alphatech, arrival drinks; Nucletron, jazz trio

PROGRAM

Monday 5 November 2012

8:15-8:30	Welcome & Introduction: C Baldock, D Thwaites
8:30-9.10	Refresher 1, chair: C Baldock K McAuley: *Gel fundamentals*
9:10-10:30	Session 1: Clinical Challenges & Dosimetry Systems Co-chairs: S Korreman, A Jirasek
9:10-9:40	Invited Review 1: S Korreman *The clinical need for 3D dosimetry and practical aspects of the use of current 3D systems in the clinic*
9:40-9:50	T J McCaw, J A Micka, L A DeWerd *Development and characterization of a three-dimensional film stack dosimeter*
9:50-10:00	S Arumugam, A Xing, G Goozee, L Holloway *Detecting VMAT delivery errors: A study on the sensitivity of the ArcCHECK-3D electronic dosimeter*
10:00-10:10	R Grant, G Ibbott, J Yang, J Adamovics, D Followill *Investigation of 3D dosimetry for an anthropomorphic spine phantom*
10:10-10:20	F Nordström, S af Wetterstedt, S Å J Bäck *4D dosimetry and its applications to pre-treatment quality control and real-time in vivo dosimetry of VMAT treatments*
10:20-10:30	H Norris, A Thomas, M Oldham *Evaluation of a clinically intuitive quality assurance method*
10:30-11:00	Tea/Coffee Break
11:00-12:30	Session 2: New Dosimetry Systems Co-chairs: P Metcalfe, S Doran
11:00-11.30	Invited Review 2: P Metcalfe *Developments in solid state dosimetry*
11:30-11:40	M Goulet, L Gingras, L Beaulieu, L Archambault *3D tomodosimetry using scintillating fibers: proof-of-concept*
11:40-11:50	T Gorjiara, R Hill, Z Kuncic, S Bosi, C Baldock *Water equivalence of micelle gels for x-ray beams*
11:50-12:00	M Alqathami, A Blencowe, U J Yeo, R Franich, S Doran, G Qiao, M Geso *Enhancement of radiation effects by bismuth oxide nanoparticles for kilovoltage x-ray beams: A dosimetric study using a novel multi-compartment 3D radiochromic dosimeter*
12:00-12:10	Y de Deene, J Vandecasteele, T Vercauteren *Low-density polymer gel dosimeters for 3D radiation dosimetry in the thoracic region: A preliminary study*
12:10-12:20	S B Ebenezer, M K Rafic, P Ravindran *Basic investigations on LCV micelle gel*
12:20-12:30	A Hiroki, S Yamashita, Y Sato, N Nagasawa, M Taguchi *New polymer gel dosimeters consisting of less toxic monomers with radiation-crosslinked gel matrix*
12:30-1:30	Lunch Break
1:30-3:00	Session 3: Polymer Gel Dosimetry Co-chairs: K Jordan, K McAuley
1:30-2:00	Invited Review 3: K Jordan *Review of recent advances in radiation dosimeters with optical readout*
2:00-2.10	T Juang, J Newton, M Niebanck, R Benning, J Adamovics, M Oldham *Customising PRESAGE® for diverse applications*
2:10-2:20	C H Yao, W T Hsu, S M Hsu, P Y L Ma, B T Hsieh, Y J Chang *NIPAM polymer gel dosimetry for IMRT four-field box irradiation using optical-CT scanner*
2:20-2:30	H Kawamura, K Shinoda, H Fuse, T Terunuma, K Miyamoto, T Sakae, A Matsumura *Comparison between polymer gel dosimetry and calculated dose with small field in stereotactic irradiation*
2:30-2:40	U J Yeo, M L Taylor, J R Supple, R L Smith, T Kron, R D Franich *Deformable gel dosimetry I: Application to external beam radiotherapy and brachytherapy*

PROGRAM

2:40-2:50	T Maeyama, N Fukunishi, K L Ishikawa, K Fukasaku, T Furuta, S Takagi, S Noda, R Himeno, S Fukuda *Response of aqueous dichromate and nanoclay dichromate gel dosimeters to carbon ion irradiation*
2:50-3:00	M Alqathami, J Adamovics, R Benning, A Blencowe *An investigation into ultra-sensitive substituted leucomalachite dye derivatives for use in the PRESAGE® dosimeter*
3:00-3:30	Tea/Coffee Break
3:30-5:10	Session 4: Radiochromic and other novel dosimetry Co-chairs: M Oldham, CS Wuu
3.30-4.00	Invited Review 4: M Oldham *3D dosimetry - systems and applications*
4:00-4:10	C Kurokawa, A Urushiyama *Measurements for dose distribution with a Photo-Stimulated Luminescence Dosimeter sheet*
4:10-4:20	P S Skyt, I Wahlstedt, E S Yates, L P Muren, J B B Petersen, P Balling *Exploring the dose response of radiochromic dosimeters*
4:20-4:30	J B Davies, S G Bosi, C Baldock *Photometric calibration of a radiochromic gel dosimeter*
4:30-4:40	P S Skyt, T S Kinnari, I Wahlstedt, E S Yates, L P Muren, J B B Petersen, P Balling *Diffusion properties of a radiochromic hydrogel dosimeter*
4:40-4:50	K Jordan *Non-diffusing photochromic gel for optical computed tomography phantoms*
4:50-5:00	J Davies, S Bosi, C Baldock *Verification of a non-diffusing gel dosimeter*
5:00-5:10	A Nasr, T Olding, L J Schreiner, K B McAuley Preliminary evaluation of diacetylene-based 3D micelle gel dosimeters
5:10-6:00	The Great Debate (moderated audience-participation debate on current issues in 3D dosimetry) Co-chairs: C Baldock, L J Schreiner

PROGRAM

Tuesday 6 November 2012

8:15-9:00	Refresher 2, chair: P Vial B McCurdy: *Dosimetry using EPIDs; systems, methods, applications*
9:00-10:30	Session 5: EPID dosimetry & verification of complex treatments Co-chairs: P Greer, B McCurdy
9:00-9:30	Invited Review 5: P Greer *3D EPID based dosimetry for pre-treatment VMAT verification*
9:30-9:40	H C Woodruff, P Greer *3D Dose reconstruction: Banding artefacts in cine mode EPID images during VMAT delivery*
9:40-9:50	M Tyler, P Vial, P Metcalfe, S Downes *Clinical validation of an in-house EPID dosimetry system for IMRT QA at the Prince of Wales Hospital*
9:50-10:00	C S Wuu, S Jelle Hoogcarspel, K Deh, W Y Hsu, J Admovics *3-D dose verification by cone-beam optical CT scanning of PRESAGE dosimeter*
10:00-10:10	S Arumugam, A Xing, P Vial, A Scotti, R Stirton, G Goozee, L Holloway *Impact of patient rotational errors on target and critical structure dose in IMRT: A 3D simulation study*
10:10-10:20	A Quinn, L Holloway, P Metcalfe *Image guidance during breast radiotherapy; a phantom dosimetry and radiation-induced second cancer risk study*
10.20-10.30	V Feygelman, D Opp, G Zhang, C Stevens, B Nelms *Experimental verification of the planned dose perturbation algorithm in an anthropomorphic phantom*
10:30-11:00	Tea/Coffee Break
11:00-12:30	Session 6: IMRT/IGRT Co-chairs: B Mijnheer, L Holloway
11:00-11:30	Invited Review 6: B Mijnheer *3D EPID-based in vivo dosimetry for IMRT and VMAT*
11:30-11:40	J Nilsson, A Karlsson Hauer, A Bäck *IMRT patient-specific QA using the Delta4 dosimetry system and evaluation based on ICRU 83 recommendations*
11:40-11:50	A Thomas, H Yan, M Oldham, T Juang, J Adamovics, F F Yin *The effect of motion on IMRT – looking at interplay with 3D measurements*
11:50-12:00	A Thomas, J O'Daniel, J Adamovics, G Ibbott, M Oldham *Comprehensive quality assurance for base of skull IMRT*
12:00-12:10	A Xing, S Arumugam, L Holloway, G Goozee *Validation of an IMRT beam model using a secondary treatment planning system as a 3D dosimeter*
12:10-12:20	T Olding, T Halsall, L J Schreiner, A Kerr *A Multi-Configurational Cylindrical Phantom Based Evaluation of Patient-Specific IMRT QA Tools*
12:20-12:30	J Vandecasteele, Y de Deene *Optical laser scanning of a leucodye micelle gel: preliminary results of a 3D dose verification of an IMRT treatment for a brain tumor*
12:30-1:30	Lunch Break

PROGRAM

1:30-3:00	Session 7: 2D versus 3D dosimetry; and 3D dosimetry for proton beams Co-chairs: V Feygelman, P Keall
1:30-2:00	Invited Review 7: V Feygelman *Dosimetric quality assurance of highly conformal external beam treatments: from 2D phantom comparisons to 4D patient dose reconstruction*
2:00-2:10	A Berg, M Wieland, J Naumann, O Jaekel *High resolution dosimetry in monoenergetic proton beam therapy on a normoxic polymer gel: the importance of high spatial resolution for reduced Bragg-Peak-quenching*
2:10-2:20	C S Wuu, Y Xu, X Qian, J Admovics, E Cascio, H M Lu *Dosimetric assessment of the PRESAGE dosimeter for a proton pencil beam*
2:20-2:30	S Park, C Jeong, B J Min, J Kwak, J Lee, S Cho, D Shin, Y Kyung Lim, S Yong Park, S B Lee *Feasibility Study of Proton-based Quality Assurance of Proton Range Compensator*
2:30-2:40	T Tominaga, S Hayashi, S Usui, H Kawamura, K Katahira *Application of a PAGAT/MgCl2 gel for dose measurements in a 150 MeV proton beam*
2:40-2:50	T Gorjiara, A Kacperek, Z Kuncic, C Baldock, S Doran *Preliminary characterization of PRESAGE® for 3D dosimetry of 62 MeV proton beam*
2:50-3:00	M Carroll, G Ibbott, R Grant, M Gillin, J Adamovics *Investigation of Photon and Proton Overlapping Fields in PRESAGE® Dosimeters*
3:00-3:30	Tea/Coffee Break
3:30-5:30	Session 8: Optical Dosimetry I Co-chairs: L Beaulieu, P Ravindran
3:30-4:00	Invited Review 8: L Beaulieu *Current status of scintillation dosimetry for megavoltage beams*
4:00-4:10	D S Kumar, E J J Samuel, Y Watanabe *Laser CT evaluation on normoxic PAGAT gel dosimeter*
4:10-4:20	P Naseri, D R McKenzie, P Liu, S Fleming, N Suchowerska *Array of square waveguides for scintillation dosimetry in external radiotherapy*
4:20-4:30	K Jordan, D Turnbull, J J Batista *Laser cone beam computed tomography scanner geometry for large volume 3D dosimetry*
4:30-4:40	L Florescu, G Ambartsoumian, C S Wuu *Effect of Refraction on Dose Reconstruction in Optical-CT Gel Dosimetry*
4:40-4:50	P B Ravindran, H M Thomas *Characterization of a cone beam optical scanner*
4:50-5:00	L Rankine, M Oldham *How effective can optical-CT 3D dosimetry be without refractive fluid matching?*
5:00-5:10	W G Campbell, A Jirasek, D Wells *Recent developments with a prototype fan-beam optical CT scanner*
5:10-5:20	Y J Chang, J Q Lin, B T Hsieh, C H Chen *A study on the reproducibility and spatial uniformity of N-isopropylacrylamide polymer gel dosimetry using a commercial 10X fast optical-computed tomography scanner*
5:20-5:30	M V Papoutsaki, E Pappas, A E Papadakis, C Varveris, J Damilakis, T G Maris *Dosimetric characteristics of N-vinylpyrrolidone based polymer gels: utilization depending on dose range*
5:30-5.40	A Cavan, J Meyer *A novel optical calorimetry dosimetry approach applied to an HDR Brachytherapy source*
6:00-7:30	Poster and exhibition reception

PROGRAM

Wednesday 7 November 2012

8:15-9:00	Refresher 3, chair: C S Wuu Y de Deene, S Doran, A Jirasek: *MRI, Optical and X-ray CT evaluation methods*
9:00-10:30	Session 9: Dosimetry for audit, SRS/SBRT and Small Fields Co-chairs: T Kron, J Warrington
9:00-9:30	Invited Review 9: T Kron *Dosimetry requirements for audit and clinical trials*
9:30-9:40	M Oldham, J Newton, L Rankine, J Adamovics, D Kirsch, S Das *How accurate is image guided radiation therapy (IGRT) delivered with a micro-irradiator?*
9:40-9:50	U J Yeo, M L Taylor, T Kron, D Pham, S Sivad, R D Franich *Potential dosimetric benefit of dose-warping based 4D planning compared to conventional 3D planning in liver stereotactic body radiotherapy (SBRT)*
9:50-10:00	M Niebanck, T Juang, J Newton, J Adamovics, Z Wang, M Oldham *Investigating the Reproducibility of a Complex Multifocal Radiosurgery Treatment*
10:00-10:10	T Olding, L Garcia, K Alexander, L J Schreiner, C Joshi *Stereotactic Body Radiation Therapy Delivery Validation*
10:10-10:20	S J Doran, A T Rahman, E Bräuer-Krisch, T Brochard, J Adamovics *Ultra-high resolution optical CT dosimetry for the visualisation of synchrotron microbeam therapy doses*
10:20-10:30	N Suchowerska, P Z Y Liu, A Ralston, P Naseri, P Abolfathi, J Lee, K Warrener, D R McKenzie *Scintillators for 3D and 4D dosimetry: current status and future potential for clinical translation*
10:30-11:00	Tea/Coffee Break
11:00-12:30	Session 10: Optical Dosimetry II Co-chairs: Y de Deene, N Suchowerska
11:00-11:30	Invited Review 10: Y de Deene *On the reliability of 3D gel dosimetry*
11:30-11:40	K Jordan *Optical computed tomography liquid calibration phantom*
11:40-11:50	W G Campbell, A Jirasek, D Wells *Revealing the impact of radiation-induced refractive index changes in polymer gel dosimeters*
11:50-12:00	D Ramm *Laser beam optical CT scanner for in-air gel readout: imaging artefacts*
12:00-12:10	H Svensson, P S Skyt, S Ceberg, S Doran, L P Muren, P Balling, J B B Petersen, S Å J Bäck *Feasibility study using MRI and two optical CT scanners for readout of polymer gel and Presage®*
12:10-12:20	T Juang, J Newton, S Das, J Adamovics, M Oldham *Preliminary investigation and application of a novel deformable PRESAGE® dosimeter*
12.20-12.30	K Jordan, A Walsh, M Peng, J Battista *Performance of latex balloons for optical computed tomography*
12:30-1:30	Lunch Break

PROGRAM

1:30-3:00	Session 11: 3D Analysis and ultrasound evaluation of gels Co-chairs: L J Schreiner, M Whitaker
1:30-2:00	Invited Review 11: L J Schreiner *Analysis/evaluation of 3D distributions and practical issues with gamma and chi comparisons*
2:00-2:10	C Ceberg *A note on the interpretation of the gamma evaluation index*
2:10-2:20	S Khoei, J V Trapp, C M Langton *Measurement of a PAGAT gel dosimeter by ultrasound computed tomography*
2:20-2:30	S Khoei, J V Trapp, C M Langton *Quantitative evaluation of polymer gel dosimeters by broadband ultrasound attenuation*
2:30-2:40	O Holmes, J Darko, T Olding, L J Schreiner *A potential modification of the γ-Evaluation: Mapping dose disagreements using γ-vector fields*
2:40-2:50	B Nelms, G Jarry, M Chan, C Hampton, Y Watanabe, V Feygelman *Real-world examples of sensitivity failures of the 3%/3mm pass rate metric and published action levels when used in IMRT/VMAT system commissioning*
2:50-3.00	M Al Sa'd, J Graham, G P Liney *A Software Tool for 3D Dose Verification and Analysis*
3:00-3:30	Tea/Coffee Break
3:30-5:40	Session 12: CBCT dosimetry; Radiological Properties of gels, X-ray CT, MRI Co-chairs: J Sykes, S Back
3:30-4:00	Invited Review 12: J Sykes *Dosimetry of CBCT: methods, doses and clinical consequences*
4:00-4:10	M V Papoutsaki, E Pappas, A E Papadakis, C Varveris, J Damilakis, T G Maris *Polymer gel dosimetry utilizing a 2D (SE) and a 2D (HASTE) multiple echo sequences*
4:10-4:20	T Fujibuchi, H Kawamura, K Yamanashi, A Hiroki, S Yamashita, M Taguchi, Y Sato, K Mimura, H Ushiba, T Okihara *Dose–response measurement in gel dosimeter using various imaging modalities*
4:20-4:30	T Gorjiara, Z Kuncic, J Adamovics, C Baldock *Monte Carlo water-equivalence study of two PRESAGE® formulations for proton beam dosimetry*
4:30-4:40	M B Kakakhel, T Kairn, J Kenny, J V Trapp *Reconstructing 3D x-ray CT images of polymer gel dosimeters using the zero-scan method*
4:40-4:50	H Johnston, M Hilts, J Carrick, A Jirasek *Characterization of the essential dosimetric properties of cosolvent-free polymer gel dosimeters: Recent progress in x-ray CT based normoxic polymer gel dosimetry*
4:50-5:00	K Vávrů, P Dvořák, J Tintěra, V Spěváček *Influence of preparation and calibration method of PAGAT dosimeter on TSE MR readout results*
5:00-5:10	S Hayashi, H Kawamura, S Usui, T Tominaga *Comparison of the influence of inorganic salts on the NMR dose sensitivity of polyacrylamide-based gel dosimeter*
5:10-5:20	Y R Huang, L L Hsieh, Y J Chang, B T Hsieh *The effect of chemical stability on the NIPAM gel dosimeter using 1H-NMR*
5:20-5:30	A Schmid, E Laistler, J Sieg, B Dymerska, M Wieland, J Naumann, O Jaekel, A Berg *Monomer consumption in MAGIC-type polymer gels in the Bragg-peak of proton beams observed by volume selective 1H MR-spectroscopy (MRS): proof of principle for high resolution MRS-methodology with a sensitive rf-detector*
6:30-10:30	Conference dinner at Museum of Contemporary Art

PROGRAM

Thursday 8 November 2012

8:15-8:50	Refresher 4:, chair: S Back D Thwaites: *Accuracy required and achievable in radiotherapy dosimetry*
8:50-10:50	Session 13: Motion, Brachytherapy, Heterogeneity, miscellaneous Co-chairs: A Jirasek, R Hill
8:50-9:00	B Nelms, V Feygelman *A novel method for 4D measurement-guided planned dose perturbation to estimate patient dose/DVH changes due to interplay*
9:00-9:10	S Ceberg, C Ceberg, M Falk, P M Rosenschöld, S ÅJ Bäck *Evaluation of breathing interplay effects during VMAT by using 3D gel measurements*
9:10-9:20	Anneli Edvardsson and Sofie Ceberg *The effect of systematic set-up deviations on the absorbed dose distribution for left-sided breast cancer treated with respiratory gating*
9:20-9:30	J Adamson, Y Yang, L Rankine, J Newton, J Adamovics, O Craciunescu, M Oldham *Towards comprehensive characterization of Cs-137 seeds using PRESAGE dosimetry with optical tomography*
9:30-9:40	A L Palmer, A Nisbet, D A Bradley *Semi-3D dosimetry of high dose rate brachytherapy using a novel Gafchromic EBT3 film-array water phantom*
9:40-9:50	J Poder, N Annabell, M Geso, M Alqathami, S Corde *ROPES eye plaque dosimetry: commissioning and verification of an ophthalmic brachytherapy treatment planning system*
9:50-10:00	N Govi, P Gueye, S Avery *Application of MAGAT polymer gel dosimetry in breast balloon brachytherapy*
10:00-10:10	J Vandecasteele, Y de Deene *Polymer gel dosimetry of an electron beam in the presence of a magnetic field*
10.10-10.20	K Penev, K Mequanint *The effect of mixed dopants on the Stability of Fricke gel dosimeters*
10:20-10:30	Y R Huang, Y J Chang, L L Hsieh, B W Yu, C H Chu, B T Hsieh *Preliminary dosimetry investigation of Tc-99m diagnostic radionuclide by NIPAM gel dosimeter*
10:30-10:40	U J Yeo, M L Taylor, J R Supple, R L Smith, T Krona, R D Franich *Deformable gel dosimetry II: Experimental validation of DIR-based dose-warping*
10:40-10:50	T Kairn, S B Crowe, J Kenny, J Mitchell, M Burke, D Schlect, J V Trapp *Dosimetric effects of a high-density spinal implant*
10:50-11:20	Tea/Coffee break
11:20-11:50	Invited Review 13, chair: D Thwaites P Keall *Dosimetry requirements for 4D radiotherapy*
11:50-12:40	Panel of all invited speakers Q&A, panel debate, future of dosimetry, requirements for developing applications Co-chairs: D Thwaites, C Baldock
12:40-13:00	Wrap up, closing remarks and goodbyes: C Baldock, D Thwaites
13:00-14:00	Lunch

IC3DDOSE 2012
COMMITTEE

David Thwaites (Conference Convenor)
Clive Baldock
Leanne Price
Elizabeth Starkey
May Whitaker
Peter Greer
Lois Holloway
Phil Vial
Robin Hill

IC3DDOSE 2012
INTERNATIONAL SCIENTIFIC COMMITTEE

Sven Back (Sweden)
Clive Baldock (Australia)
Cheng-Shie Wuu (USA)
Y de Deene (Belgium)
Simon Doran (UK)
Geoffrey Ibbott (USA)
Andrew Jirasek (Canada)
Kevin Jordan (Canada)
Martin Lepage (Canada)
Mark Oldham (USA)
Evangelos Pappas (Greece)
John Schreiner (Canada)
David Thwaites (Australia)

Journal of Physics: Conference Series **444** (2013) 012001 doi:10.1088/1742-6596/444/1/012001

Fundamentals of gel dosimeters

KB McAuley and AT Nasr
Chemical Engineering Dept., Queen's University, Kingston, Canada, K7L 3N6

Email: kim.mcauley@chee.queensu.ca

Abstract. Fundamental chemical and physical phenomena that occur in Fricke gel dosimeters, polymer gel dosimeters, micelle gel dosimeters and genipin gel dosimeters are discussed. Fricke gel dosimeters are effective even though their radiation sensitivity depends on oxygen concentration. Oxygen contamination can cause severe problems in polymer gel dosimeters, even when THPC is used. Oxygen leakage must be prevented between manufacturing and irradiation of polymer gels, and internal calibration methods should be used so that contamination problems can be detected. Micelle gel dosimeters are promising due to their favourable diffusion properties. The introduction of micelles to gel dosimetry may open up new areas of dosimetry research wherein a range of water-insoluble radiochromic materials can be explored as reporter molecules.

1. Introduction

This article describes some of the fundamental chemical and physical phenomena that occur in three-dimensional (3D) gel dosimeters used for detection and verification of dose distributions used in cancer radiotherapy. The relationships between these phenomena and dosimeter performance are described. First, Fricke gel dosimeters are discussed, followed by polymer gel dosimeters and two emerging types of gel dosimeters (i.e., micelle gel dosimeters and genipin gel dosimeters). Only gel dosimeters containing water and a gelling agent are considered. Other dosimeters, such as plastic PRESAGE™ dosimeters [1] are not discussed. This article focuses on recent work in gel dosimetry, which appeared in the literature since the previous 2010 IC3DDose conference.

2. Fricke gel dosimeters

Fricke or ferrous sulfate solutions, wherein the response to irradiation depends on the dose dependent transformation of ferrous (Fe^{2+}) ions into ferric (Fe^{3+}) ions, have been used to measure radiation doses for many years [2-4]. The initiation of modern 3D dosimetry is related to two important developments. The first development was the use of MRI to detect and quantify radiation-induced changes in Fricke solutions [5]. The second development was the spatial stabilization of dose information by dispersing Fricke solution throughout a gel matrix [6]. Unfortunatelly, the poor spatial stability of Fricke gels due to diffusion of Fe^{3+} ions constrains the permissible time between irradiation and measurements [7]. Limited success in reducing diffusion rates [8, 9] is obtained using different gelling agents (gelatin, agarose, sephadex and polyvinyl alcohol) and chelating agents such as xylenol orange, which induces colour changes that permit optical imaging [10-12].

Fricke gels are attractive for 3D dosimetry as they are easy to prepare, are radiologically tissue equivalent [13, 14] and give reproducible results [10]. However, like other popular gels used in dosimetry, Fricke gels are sensitive to conditions during preparation, irradiation and read-out (eg. impurities and temperature) [10]. The chemical yield (G value) of Fricke solutions is influenced by oxygen (i.e., G= 15.5 ±0.2 ions/100 eV for a dosimeter in equilbrium with air and G= 8.2±0.3 for an oxygen-free dosimeter) [16-19]. Although Fricke gel dosimeters may seem to be simple at first glance, Monte Carlo simulations that predict the fundamental behaviour of Fricke solutions (without a gelling agent) account for more than 60 chemical reactions to simulate interactions between radiation and oxygenated water. Eleven or more additional reactions are required to account for interactions with SO_4^{2-} and Fe^{2+} [19-22]. Additional reactions would be required to account for interactions with gelatin and a chelating agent. As a result, Fricke gel dosimeters are not really that simple. This inherent level of complexity should be kept in mind when evaluating other potential 3D dosimeters.

3. Polymer gel dosimeters

Polymer gel dosimeters are the most widely used 3D gel dosimeters [23]. They contain water and gelatin, along with monomers and crosslinkers that polymerize in response to free radicals generated by water radiolysis [10]. The amount of crosslinked polymer that forms and precipitates at each location in the gel depends on the local radiation dose and the local concentration of monomer and crosslinker [24, 25]. Formation of tightly crosslinked polymer particles (microgels) induces changes in the physical properties of the dosimeter that can be detected using several imaging techniques (e.g., MRI, optical CT, x-ray CT and ultrasound scans) [26-35]. The radiation dose distribution can then be estimated from the resulting 3D images and used to verify the treatment plan that was applied [10].

A variety of different monomers and crosslinkers have been tested for use in polymer gel dosimeters [10, 36]. Polyacrylamide gel (PAG) dosimeters are the most studied because they have fewer problems with dose-rate sensitivity and temperature sensitivity than other polymer gel dosimeters. PAG dosimeters consist of acrylamide (Aam) monomer and N,N'-methylene-bisacrylamide (Bis) crosslinker (see figure 1) dissolved in an aqueous gelatin matrix. The main reactions that occur during free radical copolymerization of acrylamide and bisacrylamide are shown via a cartoon in figure 2. Although linear polyacrylamide is water-soluble, crosslinked polyacrylamide precipitates. The precipitated polymer is held in position by the gelatin matrix, preserving spatial information via a more effective means than in Fricke gel dosimeters. Although the precipitated polymer molecules cannot readily diffuse, the unreacted monomers can easily diffuse through the gel during and after irradiation. Consequently, inaccurate dosimetry results can be obtained in situations wherein polymer radicals persist over long periods of time and are able to react with the diffusing monomer and crosslinker (e.g., in anoxic PAG gels where radicals can persist for longer than 12 hours [37] and in low-dose rate brachytherapy applications where radicals are generated continuously for many weeks [38]. Note that all current polymer gel dosimeters face this same problem with diffusion of reporter molecules.

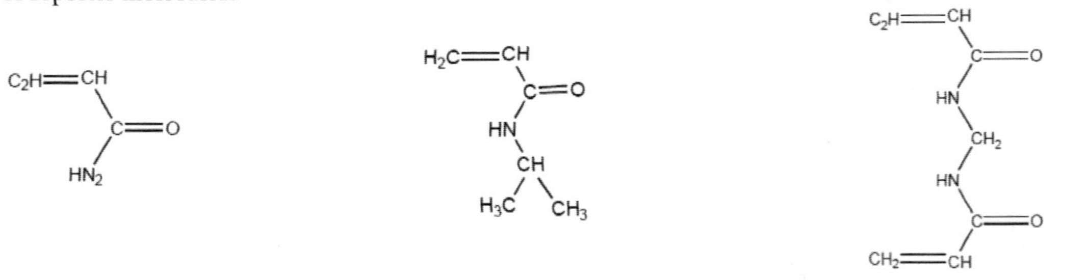

Acrylamide **NIPAM** **Bisacrylamide**

Figure 1: Chemical structure of acrylamide and n-isopropyl acrylamide (NIPAM) monomers and bisacrylamide crosslinker used in polymer gel dosimetry. Bisacrylamide is an effective crosslinker because two vinyl groups are available for polymerization.

Journal of Physics: Conference Series **444** (2013) 012001 doi:10.1088/1742-6596/444/1/012001

In some recent dosimetry studies, acrylamide (which is a severe neurotoxin and suspected carcinogen) has been replaced with n-isopropyl acrylamide (NIPAM), shown in figure 1, which has lower toxicity than acrylamide and is less likely to be ingested via inhalation due to its lower volatility and is less able to pass through human skin [29, 39-48]. Note that proper safety precautions (i.e., preparation in a fume hood and use of gloves and goggles) still need to be used when manufacturing these dosimeters. NIPAM-based dosimeters have similar reaction chemistry as the PAG chemistry in figure 2, but seem to be more susceptible to pre-polymerization, which can result in cloudy phantoms prior to irradiation [39, 42].

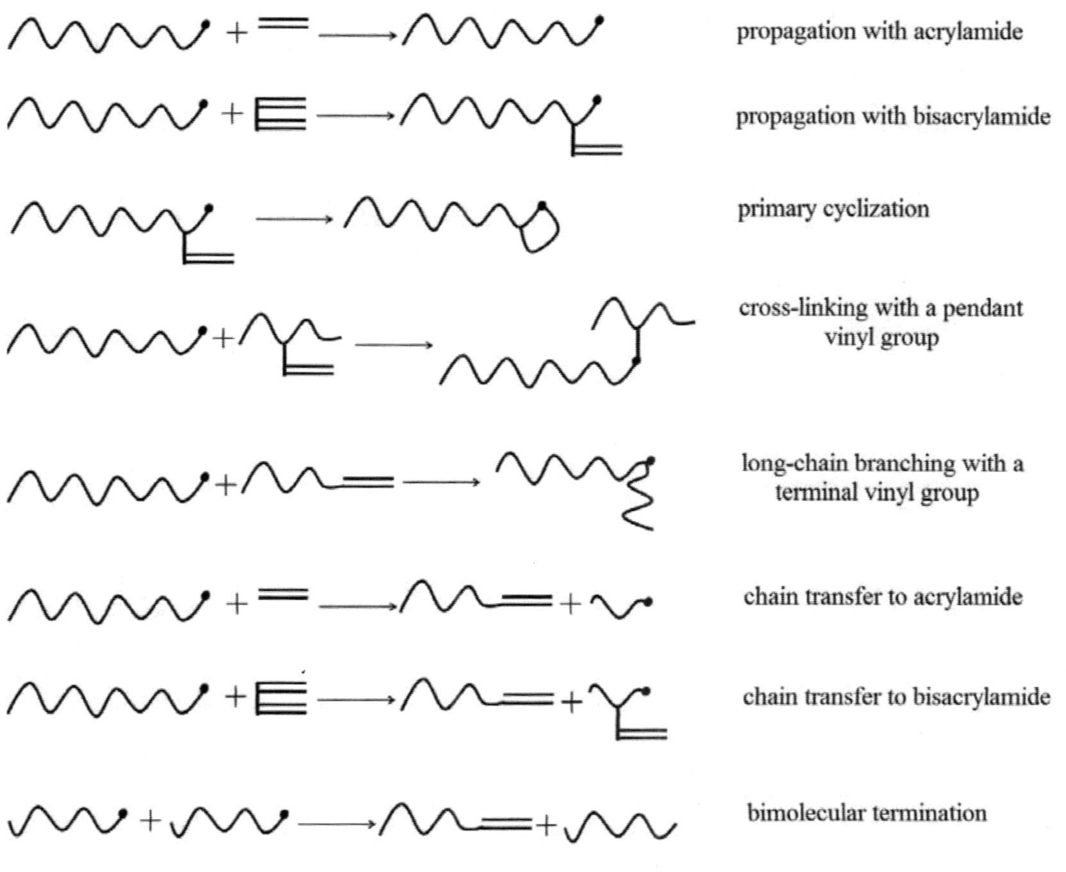

Figure 2: Main reactions in free radical copolymerization of acrylamide and bisacrylamide.

The sensitivity of PAG and NIPAM-based polymer gel dosimeters to radiation is directly related to %T, the total weight percent of monomer and crosslinker in the system [36, 37, 49], and to %C, the concentration of the crosslinker relative to the total monomer [37, 51]. Increases in both %T and crosslinker concentration tend to produce higher dose sensitivities, as measured by MR and x-ray CT [37, 40, 43, 49, 51-53]. The limited solubility of Bis crosslinker initially hampered the development of dosimeters with sufficiently high-dose sensitivities for accurate x-ray CT readout [54, 58]. Efforts to find effective crosslinkers with increased solubility were unsuccessful [59]. Recent NIPAM-based recipes with high %T have enabled additional bisacrylamide to be dissolved, resulting in polymer gel dosimeters with improved sensitivity and dose resolution for x-ray CT readout [41, 47, 48].

Measurement accuracy of PAG and NIPAM-based polymer gel dosimeters changes dramatically with %T and %C depending on the read-out technique used [60, 61]. For example, MRI techniques [62] can accurately measure dose distributions in PAG and NIPAM-based gels manufactured using traditional formulations that contains 6%T and 50 %C. However, cone beam optical imaging was complicated by light scattering when used to measure the dose distributions for such gels [26, 63, 64]. Reducing %T from 6 to 4 has improved the accuracy of cone beam optical measurements [26]. When using x-ray CT imaging, increasing %T to as high as 16 %, improves dose sensitivity and dose resolution [27]. It is clear that different recipes are suitable for use with different read-out techniques.

One of the major difficulties encountered by researchers when making polymer gel dosimeters is oxygen contamination [65-67]. Oxygen consumes primary free-radicals produced by water radiolysis and inhibits the growth of polymer radicals [68, 69]. Traditionally, PAG dosimeters were manufactured in oxygen-free glove boxes [67]. More recently, most polymer gel dosimeters are normoxic dosimeters that are manufactured in the presence of air and use an oxygen scavenger such as tetrakis hydroxymethyl phosphonium chloride (THPC) to remove oxygen that is initially dissolved in the gel solution [39, 70-73]. An additional benefit of using THPC in dosimeter recipes is that it helps to alleviate problems associated with long-lived radicals. Since THPC shortens the time period over which polymerization occurs, problems with edge enhancement are alleviated (i.e., there is a shorter time period when diffusing monomer and crosslinker can come in contact with polymer radicals). Unfortunately, THPC does not solve problems associated with oxygen that may leak into the phantom between manufacturing and irradiation [74, 75]. Minor oxygen leaks and interactions between this oxygen and THPC can cause severe dose inaccuracy [42, 76, 77]. Consequently, we recommend that internal calibration methods (e.g., based on depth dose) should be used instead of small calibration vials [29, 78-80]. Unexpected depth-dose behaviour will help to detect when oxygen has leaked into a phantom, so that results from contaminated phantoms can be discarded or used with caution.

4. Micelle gel dosimeters

Jordan and co-workers developed radiochromic micelle gel dosimeters for optical readout. The gel recipes consist of colourless leuco dyes (e.g., leuco malachite green or leuco crystal violet) emulsified in a hydrogel matrix using a surfactant [81, 82]. The leuco-dye molecules react with free radicals generated by water radiolysis, changing from colourless to deeply coloured as the radiation dose increases. Micelles are self-assembled aggregates of surfactant molecules that have both hydrophilic and hydrophobic parts. Above the critical micelle concentration (CMC), surfactant molecules orient themselves so that their hydrophobic parts repel away from surrounding water toward the centres of the micelles, leaving their hydrophilic parts in contact with water. The main purpose of using micelles in radiochromic micelle gel dosimeters is to emulsify the water-insoluble leuco-dye molecules within the hydrophobic core of the micelles to distribute the leuco dye throughout the 3D gel volume [82]. A second benefit is that micelles are significantly larger than individual leuco-dye molecules. As a result, the micelles, which contain the leuco dye, have low diffusivity within the gel matrix. Using emulsified leuco-dye molecules as reporter molecules results in improved spatial stability of dose information, compared with micelle-free optical dosimeters such as Fricke gel dosimeters and polymer gel dosimeters [81].

Current micelle gel dosimeters could benefit from further improvements as they are light sensitive and temperature sensitive during irradiation and tend to fade over time [82]. They also have relatively low dose sensitivity and may have significant dose-rate dependence [83]. One benefit of using micelles in 3D gel dosimeters is that, unlike traditional gel dosimeters, the reporter molecules do not need to be water soluble. In fact, very low or negligible water solubility will help to reduce diffusion and will improve spatial stability. Consequently, a range of new hydrophobic reporter molecules can be considered for use in 3D micelle gels. One type of water-insoluble reporter molecule that was recently studied is 10,12-pentacosadiynoic acid (PCDA) [81], which is the reporter molecule used in Gafchromic® films [85, 86]. PCDA changes colour in response to reaction with free radicals because it contains a diacetylene group (i.e., two carbon-carbon triple bonds separated by a carbon-carbon single bond). When diacetylenes oligomerize in response to free radicals, they produce intense colour changes due to conjugated double and triple bonds [86-88]. Unfortunately, PCDA and two other diacetylenes were shown to be unsuitable for micelle gel dosimeters because they did not oligomerize within the micelles. Nevertheless, micelle gel dosimeters merit further study and development due to their diffusion properties and the range of new reporter molecules that can be considered.

5. Genipin gel dosimeters

A recent gel dosimeter containing genipin, gelatin and water is currently being studied for radiotherapy dosimetry applications. Genipin is a natural cross-linker of many types of hydrogel polymers, including gelatin [89-92]. During the cross-linking reaction of gelatin with genipin in aqueous media, the mixture slowly changes from colourless to deep blue. The melting point of the resulting gel is much higher than genipin-free gelatin gels [89]. The transparent blue gels bleach in response to irradiation and the colour change can be optically quantified [90]. Stable 3D dose information can be obtained shortly after irradiation [89]. Genipin gels respond linearly to radiation doses between 100 and 1000 Gy [90]. Lowering the pH using sulfuric acid increases dose sensitivity of genipin gels to doses between 0 and 100 Gy [90] so that these gels may be promising for 3D radiotherapy dosimetry in future.

6. Conclusions

Fricke gels are used because they are relatively simple and reproducible, even though they have significant problems with spatial stability due to diffusion. The chemistry of Fricke gels is actually quite complicated and involves a large number of chemical reactions. In addition, their sensitivity depends on oxygen concentration. The effectiveness of oxygen-dependent Fricke gels should be kept in mind when searching for new and improved gel dosimeter recipes.

Oxygen sensitivity is an important problem in polymer gel dosimeters, even when they are produced using THPC as an oxygen scavenger. As a result, great care should be taken to prevent oxygen leaks during the time between gel manufacture and irradiation. Otherwise, misleading results could be obtained. The use of internal gel calibration (i.e., using depth-dose information) rather than small vials is recommended as a means of detecting when oxygen contamination has occurred. Polymer gel dosimeters containing NIPAM rather than acrylamide are being developed and used because they are safer to make and use. However, even though NIPAM is less toxic and less likely to be ingested by skin absorption or inhalation, proper chemical safety precautions still need to be taken when it is used. NIPAM gels with high %T enable additional crosslinker to be dissolved, so that improved dose sensitivity and resolution can be obtained for use with x-ray CT read out. Lower %T recipes are better for use with MRI and optical scanning; different recipes should be used with different imaging techniques.

Micelle gels show promise due to their favourable diffusion properties and the range of new water-insoluble reporter molecules that might be considered. Genipin gels also warrant further investigation and development.

7. References

[1] Gorjiara T *et al* 2011 *Med. Phys.* **38** 2265-74
[2] Fricke H and Morse S 1927 *Am. J. Roent. Radium Ther. Nucl. Med.* **18** 43
[3] Baldock C 2006 *J. Phys.: Conf. Ser.* **56** 14
[4] Baldock C 2009 *J. Phys.: Conf. Ser.* **164** 012002
[5] Gore J C *et al* 1984 *Phys. Med. Biol.* **29** 1189-97
[6] Appleby A *et al* 1986 *Med. Phys.* **14** 382-4
[7] Schreiner L J 2004 *J. Phys.: Conf. Ser.* **3** 9
[8] Harris P J *et al* 1996 *Phys. Med. Biol.* **41** 1745-53
[9] Baldock C *et al* 2001 *Australas. Phys. Eng. Sci. Med.* **24** 19-30

[10] Baldock C *et al* 2010 *Phys. Med. Biol.* **55** R1-63
[11] Hill B *et al* 2002 *Phys. Med. Biol.* **47** 4233-46
[12] Healy B J *et al* 2003 *Med. Phys.* **30** 2282-91
[13] Keall P and Baldock C 1999 *Australas. Phys. Eng. Sci. Med.* **22** 85-91
[14] Venning A J *et al* 2005 *Med. Phys.* **32** 1047-53
[15] Brown S *et al* 2008 *Appl. Radiat. Isot.* **66** 1970-74
[16] Allen A O 1961 *The radiation chemistry of water and aqueous solutions.* (Princeton: D. Van Nostrand Co.)
[17] Fricke H and Hart EJ 1966 *Chemical Dosimetry* in *Radiation Dosimetry* 2nd ed. Vol. II. FH Attix and WC Roesch (ed.) (Academic Press, New York)
[18] Katsumura Y *et al* 1988 *Radiat. Phys. Chem.* **32** 259-63
[19] Meesat R *et al* 2012 *Radiation Research* **177** 813-26
[20] Meesungnoen J *et al* 2001 *Radiat. Res.* **155** 269-78
[21] Autsavapromporn N *et al* 2007 *Can. J. Chem.* **85** 214-29
[22] Meesat R 2012 Ph.D. Thesis *Université de Sherbrooke* Québec, Canada
[23] Gustavsson H *et al* 2004 *Phys. Med. Biol.* **49** 227-41
[24] Lepage M *et al* 2001 *Phys. Med. Biol.* **46** 2827-39
[25] Lepage M *et al* 2001 *J. Appl. Polym. Sci.* **79** 1572-81
[26] Maryanski M J *et al* 1993 *Magn. Reson. Imaging* **11** 253-58
[27] Maryanski M J *et al* 1994 *Phys. Med. Biol.* **39** 1437-55
[28] Olding T *et al* 2011 *Journal of Medical Physics* **36** 3-14
[29] Jirasek A *et al* 2010 *Phys. Med. Biol.* **55** 5269-81
[30] Mather M L *et al* 2002 *Phys. Med. Biol.* **47** 1449-58
[31] Mather M L *et al* 2002 *Phys. Med. Biol.* **47** 4397-409
[32] Mather M L and Baldock C 2003 *Med. Phys.* **30** 2140-8
[33] Mather M L *et al Ultrasonics* **41** 551-9
[34] Mather M L *et al* 2003 *Phys. Med. Biol.* **48** N269-75
[35] Rintoul L *et al* 2003 *Appl. Spectrosc.* **57** 51-7
[36] Lepage M *et al* 2001 *Phys. Med. Biol.* **46** 2665-2680
[37] De Deene Y *et al* 2006 *Phys. Med. Biol.* **51** 653-73
[38] Nasr AT *et al* 2012 *Macromol. Theory Simul.* **21** 36-51
[39] Senden R J *et al* 2006 *Phys. Med. Biol.* **51** 3301-14
[40] Koeva V I *et al* 2009 *Phys. Med. Biol.* **54** 2779-90
[41] Chain J N M *et al* 2011 *Phys. Med. Biol.* **56** 2091-102
[42] Sedaghat M *et al* 2011 *Phys. Med. Biol.* **56** 6083-107
[43] Chang K Y *et al* 2011 *Nucl. Instrum. Methods Phys. Res. A* **652** 775-8
[44] Chang Y J *et al* 2011 *Nucl. Instrum. Methods Phys. Res. A* **652** 783-5
[45] Hsieh B T *et al* 2011 *J. Radioanal Nucl. Chem.* **290** 141-8
[46] Pak F *et al* 2012 *Med. Phys.* **39** 3737
[47] Jirasek A *et al* 2012 *Phys. Med. Biol.* **57** 3137-53
[48] Johnston H *et al* 2012 *Phys. Med. Biol.* **57** 3155-75
[49] Hilts M *et al* 2004 *Phys. Med. Biol.* **49** 2477-90
[50] Hurley C *et al* 2006 *Nucl. Instrum. Methods Phys. Res. A* **565** 801-11
[51] Maryanski M J *et al* 1997 *Phys. Med. Biol.* **42** 303-11
[52] Lepage M *et al* 2001 *Phys. Med. Biol.* **46** 1061-74
[53] Babic S and Schreiner L J 2006 *Phys. Med. Biol.* **51** 4171-87
[54] Trapp J V *et al* 2001 *Phys. Med. Biol.* **46** 2939-51
[55] Trapp J V *et al* 2002 *Phys. Med. Biol.* **47** 4247-58
[56] Trapp J V *et al* 2004 *Phys. Med. Biol.* **49** N139-46
[57] Hill B *et al* 2005 *Med. Phys.* **32** 1589-97
[58] Hill B *et al* 2005 *Br. J. Radiol.* **78** 623-30

[59] Koeva V I *et al* 2008 *Macromol. Symp.* **261** 157-66
[60] De Deene Y and Baldock C 2002 *Phys. Med. Biol.* **47** 3117-41
[61] Hurley C *et al* 2005 *Appl. Radiat. Isot.* **63** 443-56
[62] Murry P and Baldock C 2000 *Australas. Phys. Eng. Sci. Med.* **23** 44-51
[63] Bosi S *et al* 2007 *Phys. Med. Biol.* **52** 2893-903
[64] Bosi S G *et al* 2009 *Phys. Med. Biol.* **54** 275-83
[65] Lepage M *et al* 2002 *Phys. Med. Biol.* **47** 1881-90
[66] Hurley C *et al* 2003 *Phys. Med. Biol.* **48** 3043-58
[67] Venning A *et al* 2005 *Nucl. Instrum. Methods Phys. Res. A* **555** 396-402
[68] Baldock C *et al* 1998 *Phys. Med. Biol.* **43** 695-702
[69] Koeva V I *et al* 2009 *Macromol. Theory Simul.* **18** 495-510
[70] Jirasek A *et al* 2006 *Phys. Med. Biol.* **51** 1891-906
[71] De Deene Y *et al* 2002 *Phys. Med. Biol.* **47** 3441-63
[72] De Deene Y *et al* 2002 *Phys. Med. Biol.* **47** 2459-70
[73] Brindha S *et al* 2004 *Phys. Med. Biol.* **49** N353-61
[74] Michael G J *et al* 2000 *Phys. Med. Biol.* **45** N133-8
[75] Venning A J *et al* 2005 *Australas. Phys. Eng. Sci. Med.* **28** 105-10
[76] Sedaghat M *et al* 2011 *Phys. Med. Biol.* **56** 601-25
[77] Zehtabian M *et al* 2012 *Radiat. Meas.* **47** 139-44
[78] Chain J N M *et al* 2011 *Macromol. Theory Simul.* **20** 735-51
[79] Hilts M *et al* 2000 *Phys. Med. Biol.* **45** 2559-71
[80] Oldham M *et al* 1998 *Phys. Med. Biol.* **43** 2709
[81] Jordan K and Avvakumov N 2009 *Phys. Med. Biol.* **54** 73-6789
[82] Babic S *et al* 2009 *Phys. Med. Biol.* **54** 6791-808
[83] Vandecasteele J *et al* 2011 *Phys. Med. Biol.* **56** 627-51
[84] Nasr A T *et al* 2012 *Phys. Med. Biol.* (in press)
[85] Rad A N *et al* 1998 *Med. Phys.* **25** 2093-115
[86] Rink A *et al* 2008 *Med. Phys.* **35** 4545-55
[87] Ogawa K 1992 *Polym. Int.* **28** 25-33
[88] Kim J M *et al* 2005 *Macromolecules* **28** 9366-76
[89] Jordan K 2009 *J. Phys.: Conf. Ser.* **164** 012029
[90] Gorjiara T *et al* 2011 *Phys. Med. Biol.* **56** 4685-99
[91] Davies J B *et al* 2012 *Radiat. Phys. Chem.* **81** 1263-5
[92] Davies J B *et al* 2013 *Radiat. Phys. Chem.* **83** 19-27

Journal of Physics: Conference Series **444** (2013) 012002 doi:10.1088/1742-6596/444/1/012002

Dosimetry in radiotherapy using a-Si EPIDs: Systems, methods, and applications focusing on 3D patient dose estimation

BMC McCurdy
Medical Physics Department, CancerCare Manitoba, Winnipeg, MB, R3E 0V9, Canada. Department of Physics and Astronomy, University of Manitoba, Winnipeg, MB, R3T 2N2, Canada. Department of Radiology, University of Manitoba, Winnipeg, MB, R3T 2N2, Canada

E-mail: boyd.mccurdy@cancercare.mb.ca

Abstract. An overview is provided of the use of amorphous silicon electronic portal imaging devices (EPIDs) for dosimetric purposes in radiation therapy, focusing on 3D patient dose estimation. EPIDs were originally developed to provide on-treatment radiological imaging to assist with patient setup, but there has also been a natural interest in using them as dosimeters since they use the megavoltage therapy beam to form images. The current generation of clinically available EPID technology, amorphous-silicon (a-Si) flat panel imagers, possess many characteristics that make them much better suited to dosimetric applications than earlier EPID technologies. Features such as linearity with dose/dose rate, high spatial resolution, real-time capability, minimal optical glare, and digital operation combine with the convenience of a compact, retractable detector system directly mounted on the linear accelerator to provide a system that is well-suited to dosimetric applications. This review will discuss clinically available a-Si EPID systems, highlighting dosimetric characteristics and remaining limitations. Methods for using EPIDs in dosimetry applications will be discussed. Dosimetric applications using a-Si EPIDs to estimate three-dimensional dose in the patient during treatment will be overviewed. Clinics throughout the world are implementing increasingly complex treatments such as dynamic intensity modulated radiation therapy and volumetric modulated arc therapy, as well as specialized treatment techniques using large doses per fraction and short treatment courses (ie. hypofractionation and stereotactic radiosurgery). These factors drive the continued strong interest in using EPIDs as dosimeters for patient treatment verification.

1. Introduction

Radiation treatment delivery is becoming more complex, emphasized with the introduction of techniques such as dynamic intensity modulated radiation therapy (IMRT) and rotational IMRT. Rotational IMRT, commercially available as either RapidArc™ (Varian Medical Systems, Palo Alto, CA) or VMAT™ (Elekta AB, Stockholm, Sweden), is especially complex, delivering patient treatment while the gantry rotates, and involves simultaneous modulation of radiation aperture, dose rate, gantry speed, and potentially collimator and couch speeds. In addition to these new delivery technologies, there is increased clinical interest in aggressive treatment regimens that deliver the therapeutic dose in larger per-fraction doses and in fewer fractions compared to standard treatments. These techniques, including hypofractionation and stereotactic radiosurgery, place greater importance on the dosimetric accuracy of each individual treatment fraction delivered. In response to these developments, quality assurance bodies throughout the world acknowledge that patient treatment verification, preferably through *in vivo* patient dosimetry, is highly desirable for optimal patient safety during radiation treatment [1-4]. Electronic portal imaging devices (EPIDs) provide a means for this verification.

EPIDs were originally developed to improve the accuracy of patient setup by generating radiographs of the patient during treatment delivery. The image is formed by x-rays created by the megavoltage photon source in the linear accelerator (i.e. linac). Previously, several different technologies have been utilized to capture these images, including the scanning liquid ionization chamber array (commercialized as PortalVision by Varian Medical Systems), and camera-based systems viewing a phosphor scintillating screen. Camera-based systems used either analogue video cameras (Beamview^{Plus} by Siemens Healthcare) or CCD cameras (SRI-100 by Elekta AB). By today's standards, these first generation EPID designs produced somewhat poor quality radiological images and possessed some useful but very limited dosimetric characteristics. Beginning in the early 2000s and continuing throughout the decade, the current generation of amorphous-silicon (a-Si) EPID technology eventually became commercially available from three major linac manufacturers. Current a-Si EPID technology not only generates better quality images, but also possesses more useful dosimetric characteristics that will be discussed later in this review.

Most dosimetric applications have been investigated using the imagers in an 'integrating' acquisition mode where the image signal is integrated over an entire irradiation (i.e. during a treatment beam delivery) to produce a single image. However, recent studies have examined the dosimetric characteristics of the imagers operated in a 'movie' acquisition mode, where the image signal is read out at equal time intervals throughout an irradiation to produce a series of images. This is largely motivated by the time-dependent nature of rotational IMRT. Ultimately, the widespread availability of a-Si EPIDs have made them the subject of strong research activity into a broad range of dose applications including patient treatment verification.

There are a variety of ways to employ EPIDs as patient dosimeters, ranging from pre-treatment verification to full 3D dose estimation of the patient treatment from images gathered during treatment delivery. These methods have been described in detail by van Elmpt [5] and will only be briefly outlined in this review. Application examples for 3D patient dose estimation will be focused on here, since the rewards of daily 3D patient dose estimation are great: (1) mapping under- or over-estimates of patient dose onto the anatomy, thus providing clinically relevant feedback, (2) availability of cumulative dose over the entire treatment course, (3) improved treatment through adaptive radiotherapy approaches, and (4) production of a medical/legal record of delivered patient dose.

2. a-Si EPID systems and dosimetric characteristics

This section will review how a-Si EPID systems form images and also discuss system characteristics with respect to measuring dose.

2.1 a-Si EPID systems

The modern a-Si EPID is based on thin film semiconductor technology. The current generation of commercially available EPIDs use amorphous silicon (a-Si) deposited on a glass substrate to form arrays of photodiodes and field-effect transistors (FETs). An array of pixels is formed with each pixel consisting of a photodiode and an FET. Pixel pitches are currently 0.39×0.39 mm^2 (Varian) and 0.40×0.40 mm^2 (Elekta and Siemens), but with varying active areas of 40×30 cm^2 (Varian) and 41×41 cm^2 (Elekta and Siemens), corresponding to 1024×768 pixels and 1024×1024 pixels, respectively. The pixel arrays detect radiation indirectly, through the use of a scintillating screen (the standard phosphor used is gadolinium oxysulfide doped with terbium, Gd_2O_2S:Tb), which converts deposited radiation energy into optical photons that are then detected by the photodiodes. To improve detector efficiency, an additional sheet of metal is placed directly upstream of the scintillator, to help convert incident photons to electrons that then deposit dose in the scintillator. Typical metal layers in commercial EPIDs are 1 mm of copper (Varian and Elekta), although some users may place additional

Journal of Physics: Conference Series **444** (2013) 012002 doi:10.1088/1742-6596/444/1/012002

buildup on the detector especially if performing dosimetry with high-energy therapy beams above 6 MV. Antonuk *et al.* helped pioneer the development of the modern a-Si EPID systems for radiotherapy, and early publications by that group contain detailed discussions of the design and operation of the detector components for the interested reader [6-8].

2.2. Dosimetric characteristics of a-Si EPIDs

Desirable characteristics of any dosimeter include linearity (with dose and dose rate), reproducibility, high spatial resolution, no dead time, and real-time readout. Researchers have demonstrated many of these characteristics for a-Si EPIDs. For the PortalVision aS500/1000 (Varian) operated in integrating mode, Greer and Popescu's 2003 investigation [9] contains many useful results. A small dead time effect identified in that work was subsequently removed by the manufacturer through a software upgrade in the acquisition computer (from IAS2 to IAS3). Dosimetric properties of aS500/1000 EPIDs operated in movie mode were investigated by McCurdy and Greer [10]. For the iViewGT EPID system (Elekta), McDermott *et al* [11] and Winkler *et al* [12] document many dosimetric performance studies of the system operated in integrating mode. Long term reproducibility of the Varian system has been demonstrated to be <1% (all pixels) over a three year period [13] and of the Elekta system was demonstrated to be <0.5% (all pixels) over nearly two years [14]. The small pixel pitches of these systems (~0.4×0.4 mm^2) provide a high spatial resolution, vastly superior to most ionization chambers and also superior to computed tomography data sets where voxel sizes are typically about 1×1×2 mm^3.

Several investigators have observed image lag and image ghosting effects with a-Si EPIDs [11, 12, 15, 16]. Image lag is due to trapped charge in the photodiode which, when read out in subsequent frames, results in the EPID signal being offset. Image ghosting refers to the change in individual pixel gains due to the trapped charge modifying the electric field strength in the photodiode. These effects have been shown to result in a relative under response of 4-11%, and this may have a direct impact on dosimetric applications of EPIDs [17]. However, these issues are primarily limited to short irradiation times (ie. low number of monitor units) typically below those of routine clinical use, and also can be corrected for if desired [11, 12].

2.3. Challenges to using EPIDs for patient dosimetry

As useful for dosimetry as EPIDs have proven to be, there are still several challenges to applying them in clinical practice. First is the image ghosting/lag issue as described above. Second, the a-Si EPID is known to over respond to low energy photons (below about 0.5 MeV) as compared to a water equivalent detector [18, 19] due to the increased photoelectric effect in the copper/phosphor screen. Third, the a-Si EPID imaging units are fairly thin, cassette-like systems weighing a few kilograms, but this is enough material to cause a 'self-scatter' signal. That is, photon scatter generated within the EPID itself (not just primary photons from the linear accelerator head and/or scattered photons originating in the patient) will contribute to image signal [20]. Fourth, the optical photons scattering within the scintillator layer contribute to an optical glare effect (i.e. a blurring of the deposited dose image pattern). This effect is much smaller in a-Si EPIDs as compared to the previous generation of camera-based systems, and has been characterized [21]. Fifth, photon scatter will be generated by the treatment beam interacting with the patient, creating an additional, complex source of mostly lower energy photons incident on the EPID [19]. Sixth, for the Varian a-Si EPID, the imaging unit is mounted on a robotic arm that has been shown to contribute additional signal to the image from increased backscattered photons. The backscatter signal contribution is known to be asymmetrical, field size dependent, and field location dependent [22]. Seventh, for clinical use, the EPID systems are extended outwards from the main linear accelerator unit, into the treatment beam. When extended, the mounting systems are subject to gravitational forces which may cause mechanical flexion. This introduces small shifts in the EPID image location (intended position versus real position) as a function of gantry angle, of upwards of 1 mm (Varian) [23] and 4 mm (Elekta) [24]. Eighth, for some EPID dosimetry applications in rotational IMRT delivery, the gantry angle needs to be accurately known. It has been demonstrated [25] that gantry angle uncertainty associated with the acquired EPID images on Varian a-Si EPIDs typically exceeds the AAPM TG142 tolerance of 1° [26]. Ninth, for movie acquisition mode in Varian a-Si EPIDs, it has been shown that a small amount of image dose is missing from the total acquisition [10]. More recent work has shown this to be the result of incomplete image frame acquisition at the beginning of an irradiation [25].

The impact of these issues often depends on the specific EPID application. However, when using a-Si EPIDs for dosimetry, a variety of approaches have been shown to mitigate these challenges. Often empirical, correction-based techniques have been employed, which are attractive due to their simplicity. Others have shown that model-based approaches can be successful, trading off an increase in complexity to achieve robustness over a wide range of operating conditions.

3. Methods for EPID dosimetry

The extensive review of EPID dosimetry carried out in 2008 by van Elmpt *et al.* [5] provides a detailed cataloguing of the wide variety of methods found in the literature. The brief description that follows here is based on that work.

EPID dosimetry can be categorised as either 'pre-treatment verification', where the EPID measures treatment beams without the patient present, or 'treatment verification', where the EPID measures treatment beams during patient treatment. Further sub-categorisation can be applied depending on whether or not the treatment beams have passed through an attenuating medium before being measured by the EPID. If the treatment beams have not passed through an attenuating medium to reach the EPID, the method is referred to as 'non-transmission dosimetry' (or 'non-transit dosimetry'). Otherwise, the method is described as 'transmission dosimetry' (or 'transit dosimetry').

In general, dosimetry applications for EPIDs compare the dose estimated via an EPID measurement with an expected independently calculated dose. Therefore, a third layer of sub-categorisation can be employed, based on where the dose comparison is being made. There are two locations where the comparison can be made. The dose comparison may be made at the detector level (i.e. comparing a measured image to an expected image), typically in 0D (point comparison) or 2D (image comparison). Alternatively, the dose comparison can be made at the patient/phantom (i.e. inferring a dose in the patient/phantom from the measured EPID image and comparing to an expected dose in the patient/phantom). This comparison is usually made in 0D (point comparison), 2D (plane comparison), or 3D (volumetric comparison).

The special case where patient dose (0D, 2D, or 3D) is estimated from EPID measurements during patient treatment (verbosely described by the above nomenclature as an 'in-patient, transmission dosimetry, treatment verification method') is simply described as '*in vivo* patient dosimetry'.

Much work has been done investigating a-Si EPID dosimetric applications in all of these categories and the reader is referred to the van Elmpt *et al.* [5] review paper for a detailed summary of the EPID dosimetry literature up to 2008. The remainder of the discussion in this work will focus on 3D *in vivo* patient dosimetry applications of a-Si EPIDs.

4. Techniques for 3D *in vivo* patient dosimetry using EPIDs

Arguably the most valuable a-Si EPID dosimetric application is 3D *in vivo* patient dose reconstruction, where the delivered patient dose is estimated in 3D from measured EPID transmission images of the treatment beam(s). Even before the current generation of a-Si EPID technology, there was strong interest in using EPIDs for *in vivo* patient dose verification. Many groups investigated *in vivo* EPID dosimetry methods using earlier EPID technologies such as the scanning liquid ionization chamber design [27-29] or the camera-based designs [30-33]. A few of these techniques were modified and tailored for use with a-Si EPIDs. In particular the methods of Boellard *et al.* [27] and Louwe *et al.* [28] were adapted by Wendling *et al.* [34, 35], while the methods of Hansen *et al.* [32] and Spies *et al.* [33] were customized by Partridge *et al.* 2002 [36]. Other approaches estimating 3D patient dose include Jarry *et al.* [37] and Van Uytven *et al.* [38]. Although each of these efforts are technically distinct, they share the same general framework: (1) patient scatter is estimated and subtracted from the measured EPID image, resulting in an estimate of primary fluence entering the EPID, (2) the estimate of primary fluence is backprojected to a plane above the patient, and (3) the estimate of incident primary fluence is used as input to a patient dose calculation algorithm. The main techniques in the literature for utilizing a-Si EPIDs to recover 3D dose in the patient are summarized below.

The approach described by Partridge *et al.* [36] used a calibration technique to convert an EPID image into a map of patient thicknesses. A series of pre-generated, Monte Carlo kernels (as a function of thickness) were used in an iterative fashion to estimate the scattered radiation generated in the patient. Once the scattered radiation component was subtracted from the image, the remaining estimate of primary fluence was backprojected to a plane above the patient accounting for attenuation and the inverse square effect. This primary fluence estimate was then passed as input to a convolution/superposition style dose algorithm to calculate patient dose in 3D. One phantom was tested using a 5-field IMRT plan, demonstrating a 5.5% agreement in absolute dose.

Jarry *et al.* [37] estimated the dose signal in the EPID due to patient scatter by employing a Monte Carlo simulation. This scatter signal was subtracted from the measured EPID image to estimate the EPID dose due to the primary beam, which was assumed to be proportional to primary fluence in the EPID. The simulation also tracked photon energy spectra in small rectangular regions distributed throughout the image. Primary photons that were scored in the simulation were then back-projected through the patient CT data set towards an ideal point source (ie. an infinitesimal focal spot). This created a phase-space file above the patient that was then used as a radiation source to simulate the particle transport (in a forward fashion) through the CT data set. For several phantom test situations, including 3DCRT and IMRT fields, an absolute dose agreement of 5% was found.

Another approach has been recently proposed by van Uytven *et al.* [38]. This method used a detailed physical model to predict EPID images with the patient present. The model-based algorithm predicting a-Si EPID images was developed by Chytyk *et al.* [39-41], based on earlier work by that group [42-44]. The approach uses a detailed head fluence model of the linac, accounting for focal and extra-focal sources, multi-leaf collimator, and energy spectra. It also explicitly handles patient scatter (including energy spectra), energy response and self-scatter of the EPID, optical glare, and asymmetric backscatter from the support arm. The model is used as part of an iterative algorithm that removes the patient scatter from the EPID signal to estimate primary fluence entering the EPID, including energy spectra details. The iterative algorithm is similar to that of McNutt *et al.* [27], modified to take advantage of the information available in the extra-focal portion of the head fluence model. Once estimated, the primary fluence is back-projected above the patient, accounting for patient attenuation and the inverse square law, and then utilized by an in-house developed collapsed-cone convolution style patient dose calculation algorithm. Validation was performed on slab and anthropomorphic phantoms for simple square fields, dynamic IMRT, and rotational IMRT for several clinical treatment plans including prostate, lung, and head and neck sites. Agreement with the Eclipse treatment planning system (Varian) planned dose was observed to be within 2%/3 mm using absolute dose.

In 2009, Wendling *et al.* [34] extended to 3D that group's previous work reconstructing a 2D plane of dose within the patient [35]. A correction-based approach was taken, accounting for the dose response of the EPID as well as the self-scatter of the EPID detector. Scatter-to-primary ratios were measured for a range of homogeneous slab thicknesses and then used to estimate the scatter from the patient to the EPID. Patient attenuation and the inverse square law were handled. The spectral effects of beam hardening were accounted for through a depth-dependent attenuation correction factor. An in-house developed patient dose calculation algorithm was used, where measured depth-dependent patient scatter kernels were employed to estimate scatter dose contributions within the patient. Several clinically realistic treatments were tested including a prostate, a rectum, and a head and neck IMRT plan. Homogeneous phantoms were used with a 3%/3mm gamma comparison criteria, comparing high dose voxels (defined as voxels within the 50% isodose surface). The comparison demonstrated that the percentage of pixels in agreement (i.e. with gamma<1) was 99.9%, 98.1, and 96.6% for the prostate, rectum, and head and neck treatments, respectively. This work was recently validated for rotational IMRT [24] where tests included two prostate treatments, a lung SBRT, and a head and neck treatment. Tests run with an homogeneous phantom demonstrated differences in dose at isocentre of

-0.7%, -0.9%, +1.8%, and -4.7% respectively. In 2011, this work was further extended to make use of CT imaging data to calculate beam transmission [45]. This is attractive since it avoids the need to acquire a set of 'open field' EPID images (i.e. treatment field images but without any patient or phantom present), previously required by this method to generate a beam transmission factor image.

5. Clinical experience using 3D *in vivo* patient dosimetry with EPIDs

Of the *in vivo* techniques described in the previous section, there is currently only one that has been clinically implemented on a large scale. The methods developed at the Antoni van Leeuwenhoek Hospital of the Netherlands Cancer Institute [28, 29, 34, 35] have been used clinically since 2005. A retrospective analysis of nearly five years of clinical data collected was presented by Mans *et al.* [46]. The value of an *in vivo* EPID dosimetry program was clearly demonstrated, with 17 serious errors caught over 4337 patient plans analysed between January 2005 and August of 2009.

6. Summary

This paper presents a brief overview of a-Si EPIDs in the context of dosimetric applications focusing on 3D patient dosimetry. Clinically available a-Si EPID systems were discussed, highlighting dosimetric characteristics and remaining challenges to routine use. Available methods to use the EPID for dosimetry applications were outlined. A summary of the use of a-Si EPIDs during treatment to estimate 3D dose in the patient was presented.

Today's modern radiotherapy clinics are implementing more complex treatments such as dynamic IMRT and rotational IMRT, and there is increasing demand for specialized treatment techniques that incorporate large doses per fraction and short treatment courses (i.e. hypofractionation and stereotactic radiosurgery). These factors strongly motivate researchers to continue investigating EPIDs for *in vivo* patient treatment verification.

7. Acknowledgements

The author gratefully acknowledges the contributions of our research group team members, past and present, including Eric van Uytven, Tim van Beek, Pete McCowan, Ganiyu Asuni, Krista Chytyk, and Stephen Pistorius. Our Australian collaborators in Newcastle led by Peter Greer and including Pejman Rowshanfarzad, Mahsheed Sabet, Henry Woodruff, Todsaporn Fuangrod, Brian King, and Maura Monville, are also acknowledged for their many contributions.

8. References

[1] World Health Organization 2008 *Radiotherapy Risk Profile* (Geneva CH: World Health Organization)
[2] The Royal College of Radiologists, Society and College of Radiographers, Institute of Physics and Engineering in Medicine, National Patient Safety Agency, and the British Institute of Radiology 2008 *Towards Safer Radiotherapy* (London, UK: The Royal College of Radiologists)
[3] Derreumaux S *et al* 2008 *Radiat. Prot. Dosim.* **131** 130
[4] International Commission on Radiological Units 2010 *ICRU Report 83: Prescribing, Recording, and Reporting Photon-Beam Intensity-Modulated Radiation Therapy (IMRT)* (Oxford, UK: Oxford University Press)
[5] van Elmpt V *et al* 2008A *Radiother. Oncol.* **88** 289
[6] Antonuk L E *et al* 1992 *Med. Phys.* **19** 1455
[7] Antonuk L E *et al* 1995 *Radiographics* **15** 993
[8] Antonuk LE *et al* 1996 *Int. J. Radiat. Oncol. Biol. Phys.* **36** 661
[9] Greer P B and Popescu C C 2003 *Med. Phys.* **30** 1618
[10] McCurdy BMC and Greer PB 2009 *Med. Phys.* **36** 3028
[11] McDermott L N *et al* 2004 *Med. Phys.* **31** 285
[12] Winkler P *et al* 2005 *Med. Phys.* **32** 3095

[13] King B W *et al* 2011 *Australas. Phys. Eng. Sci. Med.* **34** 459
[14] Louwe R J *et al* 2004 *Med. Phys.* **31** 2989
[15] Siewerdsen J H and Jaffray D A 1999 *Med. Phys.* **26** 1624
[16] Partridge M *et al* 2002 *Nucl. Instrum. Methods Phys. Res. A* **484** 351
[17] McDermott L *et al* 2006 *Med. Phys.* **33** 2448
[18] Jaffray D A *et al* 1994 *Med. Phys.* **21** 45
[19] McCurdy B M C and Pistorius S 2000 *Med. Phys.* **27** 312
[20] McCurdy B M *et al* 2001 *Med. Phys.* **28** 911
[21] Kirkby C and Sloboda R 2005 *Med. Phys.* **32** 1115
[22] Rowshanfarzad P *et al* 2010 *Med. Phys* .**37** 2269
[23] Rowshanfarzad P *et al* 2012 *Med. Phys.* **39** 623
[24] Mans A *et al* 2010 *Radiother. Oncol.* **94** 181
[25] McCowan P M *et al* (*submitted to Phys. Med. Biol.*)
[26] Klein E E *et al* 2009 *Med. Phys.* **39** 4197
[27] McNutt T R *et al* 1996 *Med. Phys.* **23** 527
[28] Boellaard R *et al* 1998 *Int. J. Radiat. Oncol. Biol. Phys.* **41** 465.
[29] Louwe R J *et al* 2003 *Med. Phys.* **30** 2376
[30] Pasma K L *et al* 1998 *Phys. Med. Biol.* **43** 2047
[31] Pasma K L *et al* 1999 *Med. Phys.* **26** 2373
[32] Hansen V N *et al* 1997 *Med. Phys.* **24** 1477
[33] Spies L *et al* 2001 *Phys. Med. Biol.* **46** N203
[34] Wendling M *et al* 2006 *Med. Phys.* **33** 259
[35] Wendling M *et al* 2009 *Med. Phys.* **36** 3310
[36] Partridge M *et al* 2002 *Med. Phys.* **29** 1847
[37] Jarry G and Verhaegen F 2007 *Phys. Med. Biol.* **52** 2277
[38] van Uytven E *et al* 2012 *Med. Phys.* **39** 4634
[39] Chytyk K and McCurdy B M C 2006 *Med. Phys.* **33** 3333
[40] Chytyk K, McCurdy B M C 2009 *Med. Phys.***36** 1389
[41] Chytyk K *et al* (*submitted to Med. Phys.)*
[42] McCurdy B M C and Pistorius S 2000 *Med. Phys.* **27** 312
[43] McCurdy B M C and Pistorius S 2000 *Med. Phys.* **27** 2109
[44] McCurdy B M C *et al* 2001 *Med. Phys.* **28** 911
[45] Pecharromán-Gallego R *et al* 2011 *Med. Phys.* **38** 983
[46] Mans A *et al* 2010 *Med. Phys.* **37** 2638

Journal of Physics: Conference Series **444** (2013) 012003 doi:10.1088/1742-6596/444/1/012003

How to scan polymer gels with MRI?

Y De Deene
Department for Radiotherapy and Experimental Cancer Research, Ghent University, Belgium

E-mail: yves.dedeene@ugent.be

Abstract. The absorbed radiation dose fixated in a polymer gel dosimeter can be read out by several methods such as magnetic resonance imaging (MRI), optical CT, X-ray CT and ultrasound with MRI being the first method that was explored. Although MRI was considered as an elegant scanning technique, readily available in most hospitals, it was later found that using a non-optimized imaging protocol may result in unacceptable deviations in the obtained dose distribution. Although most medical physicists have an understanding of the basic principles of magnetic resonance imaging (MRI), the optimization of quantitative imaging sequences and protocols is often perceived as the work of MRI experts. In this paper, we aim at providing the reader with some easy guidelines in how to obtain reliable quantitative MRI maps.

1. Introduction

In the early days of the development of gel dosimetry, magnetic resonance imaging (MRI) was suggested as thé method to read-out gel dosimeters. The use of MRI as a non-destructive imaging method of a dosimeter gel was first proposed in 1984 by Gore *et al* [1] who showed that ferrous sulfate chemical dosimeters initially developed in 1927 [2] could be probed by nuclear magnetic relaxometry and hence by magnetic resonance imaging (MRI) [1]. Gore *et al* investigated the nuclear magnetic resonance (NMR) relaxation properties of irradiated Fricke or ferrous sulfate dosimetry solutions showing that radiation-induced changes, in which ferrous (Fe^{2+}) ions are converted to ferric (Fe^{3+}) ions, could be quantified using MRI and subsequently showed that Fricke dosimetry solutions dispersed throughout a gel matrix could be used to obtain three-dimensional (3D) spatial information using MRI. It was subsequently shown that irradiated Fricke-type gel dosimeters did not retain a spatially stable dose distribution due to ion diffusion within the irradiated dosimeters [3]. Fricke solutions with various gelling agents such as gelatin, agar, sephadex™ and polyvinyl alcohol (PVA) were investigated. Chelating agents to reduce diffusion in Fricke gels, such as xylenol orange (XO), had only limited success [4] and diffusion remained a significant problem in the advancement of gel dosimetry. Different models have been described that explain the mechanism of how the relaxation rates are affected by the paramagnetic substances [1, 5-7]. The spin-lattice relaxation rate ($R_1 = 1/T_1$) and the spin-spin relaxation rate ($R_2 = 1/T_2$) in Fricke gels is altered significantly upon irradiation. As R_1 of a non-irradiated Fricke gel dosimeter is small as compared to the R_2 of a non-irradiated Fricke gel dosimeter, the dynamic range of the Fricke gel dosimeter in relative terms is higher for R_1 than R_2. For this reason, R_1 mapping is preferred to R_2 mapping for Fricke gel dosimeters. Also R_1 maps (at least in the early days) can be obtained with a shorter acquisition time than R_2 maps which is of crucial importance in avoiding diffusion related blurring of the dose distribution.

Polymer systems for the use of radiation dosimetry were first proposed as early as 1954, where Alexander *et al* discussed the effects of ionizing radiation on polymethylmethacrylate [8]. In 1992, Kennan *et al* reported on NMR longitudinal relaxation studies performed on an irradiated aqueous solution of N,N'-methylene-bis-acrylamide (Bis) and agar, which showed that the relaxation rates increased with absorbed dose [9]. Polymer gel dosimeters are based on the conversion of co-monomers to polymer aggregates upon irradiation. This reaction alters the mobility of surrounding water molecules which also results in a change in R_1 and R_2 [10]. The dose-response of R_2 in gelatin based polymer gel dosimeters however is more pronounced than of R_1. To explain the effect that the radiation-induced polymerization has on the R_2 relaxation rate, a model of fast exchange [11] is adopted [12–14]. It is shown in later studies that not only the relaxation rate can be used as an imaging parameter but also other MR contrasts such as magnetization transfer [15–17] and chemical shift [18]. For a further explanation of these mechanisms, the reader is also referred to other papers in these proceedings. Different imaging sequences can be used to acquire quantitative images. These imaging sequences may differ in performance in terms of accuracy, precision and speed. It will be shown that for a specific imaging sequence, these three properties are interconnected.

The target figure of accuracy that is aimed in gel dosimetry for high-precision radiotherapy is about 3-5% of the maximum dose in regions of homogeneous dose and a spatial error of less than about 2-3 mm in regions of high-dose gradients. This figure of accuracy encompasses the overall dosimetry experiment. The problem in evaluating the overall accuracy of the dose maps obtained with gel dosimetry is that there is no "golden dosimetric standard" to compare with. The most reasonable strategy is to compare doses obtained with gel dosimetry with doses obtained by the most reliable dosimetry techniques that apply to a certain spatial dimension. For example, dose profiles of a single field (photons and electrons) can be compared with dose profiles obtained with an ionization chamber or diamond detector [19–20]. In two dimensions, gel dosimetry can be compared with film dosimetry [19, 21–22]. Dose distributions obtained with gel dosimetry have been compared with those calculated with treatment planning software [21, 23–29]. Errors that compromise the accuracy may occur at different stages of the dosimetry procedure [30].

At the stage of imaging the dosimeter, several imaging artifacts may cause errors in the final dose map. These errors may be classified in dose inaccuracies or in deformations of the dose maps. Studies of these different artifacts have resulted in different compensation strategies. The verification of the treatment plan can be seen as the major purpose of gel dosimetry in radiotherapy quality assurance. Besides the possibility of systematic errors, the dose maps will also contain stochastic noise. To minimize the stochastic noise in the images, the imaging sequence parameters should be optimized.

2. A step-by-step guide for MRI scanning of polymer gel dosimeters.

Although there is a vast amount of scientific literature available on different quantitative scanning methods and compensation of imaging artifacts, for the medical physicist who is planning to start using MRI as a readout technique for polymer gel dosimeters, the implementation of a reliable scanning protocol may seem very complex. Here is a step-by-step procedure that may guide you in implementing a quantitative MRI protocol to scan polymer gels.

2.1. Get yourself familiar with the basics behind magnetic resonance imaging (MRI)

Without studying the quantum mechanical description of nuclear magnetic resonance, it is essential that you familiarize yourself with the basic principles behind magnetic resonance imaging (MRI). Knowledge is the best medicine against beginners fear. There are some nice primers on the basics of MRI [31-35]. It may also be beneficial to allow yourself to play around with the parameters of a basic spin-echo sequence. It should be noted that in order to make the theory of NMR more accessible, some of these works have made compromises on the preciseness of the quantum mechanical model. More advanced textbooks on MRI physics and sequence development will give a more precise description of the quantum mechanical model [36-38].

2.2. *Choose a quantitative imaging sequence*

Theoretically, any kind of imaging sequence that generates an image in which the pixel intensity is related to absorbed dose in a monotonic fashion is a possible candidate. A dose map can be obtained by calibrating every pixel intensity to the absorbed dose by use of a set of calibration vials that have been irradiated with known doses. In general, due to the inhomogeneity of the radiofrequency (B_1-) field, contrast weighted images (T_1w, T_2w, MTw) suffer from severe image non-uniformity which results in poor accuracy. Eventually, this can be compensated by acquiring an additional B_1-field map and using the B_1-field map to compensate for the non-uniformity before calibration. However, other artifacts such as eddy currents and B_0-field inhomogeneity may also lead to image non-uniformity. A better approach to compensate for image non-uniformity is obtained by using quantitative parametric maps such as T_1-, T_2- and MTR-maps. The contrast mechanism (T_1, T_2, MTR) should be chosen on the basis of the kind of gel dosimeter that is scanned. In Fricke based gel dosimeters, both T_1 and T_2 are significantly affected upon irradiation and thus both T_1 and T_2 maps can be used. In polymer gel dosimeters T_2 is more affected than T_1 and thus T_2-maps and MTR-maps will yield the highest dose resolution. A comprehensive list of possible imaging sequences to acquire parametric maps is provided in [39] and listed in table 1. The parameter of interest is calculated from a set of images that are acquired where one imaging parameter is varied (see table 1).

Table 1: Overview of important quantitative MR imaging sequences for R_1 (= $1/T_1$), R_2 (= $1/T_2$) and magnetization transfer ratio (MTR) imaging. The variable that is changed to acquire different contrast weighted images is shown in the third column. If the variable is varied automatically within the sequence, the parameter is shown between squared brackets. The meaning of the availability is as follows: (*) The imaging sequence is not provided by the manufacturer. The sequence should be developed in house. (**) The sequence is available but significant changes to the imaging parameters are required. (***) The imaging sequence is readily available on all clinical MRI scanners.

Sequence type	Conditions	Variable	Post-processing	Availability	Spatial accuracy
1. *Quantitative R_1 imaging sequences ($R_1=1/T_1$)*					
Single spin-echo (SE)	TE short	TR (×2/×N)	Fit	***	Very good
Saturation recovery (SRGE/SRSE)	TR long, TE short	TM (×2/×N)	Fit	***	Good / Very good
Inversion recovery (IRGE/IRSE)	TR long, TE short	TI (×2/×N)	Fit	***	Good / Very good
Driven Equilibrium Single Pulse Observation of T_1 (DESPOT)	-	FA (×2/×N)	Fit	*	Good
Look-Locker (LL, TOMROP)	FA small	TI (×2/×N)	Fit	*	Good
Steady-State Free Precession (SSFP)	TR >> T_2	FA (×2/×N)	Anal. / Fit	**	Good
IR - Very fast acquisition (EPI, GRASE, HASTE)	TR long	TI (×2/×N)	Fit	**	Poor

2. Quantitative R_2 imaging sequences (R_2 = $1/T_2$)

Single spin-echo (SE)	TR long	TE (×2/×N)	Anal. / Fit	***	Very good
Fast spin-echo (FSE, TSE, RARE)	TR long	TE (×2/×N)	Anal. / Fit	***	Good
Multiple spin-echo (MSE, MC-SE)	TR long	[ΔTE (N)]	Fit	***	Very good
Steady-State Free Precession (SSFP)	TR << T_1 FA = 90°	[2 echoes]	Anal.	**	Good

3. Quantitative magnetization transfer (MT) imaging sequences

MT pulse prepared spin echo imaging sequence	TR long	MT pulse amplitude	Anal.	**	Very good
Pulsed MT steady state	TR short	MT pulse amplitude	Anal.	**	Good
Stimulated echo preparation	-	TM	Anal.	*	Very good

2.3. Develop image processing software for calculating parametric maps from a set of base images
The parametric maps are calculated from a set of base images with different image contrast. This is achieved by an adequate software program which is most often developed 'in house'. The importance of the choice of an appropriate cost-function to be minimized has been emphasized in [40]. It has been shown that the use of a chi-square minimization results in a more precise R_2 map as compared to a least-square minimization. In order to test the image processing software, a synthetic base image data set may be generated on which stochastic (or structural) noise may be added in order to test its performance.

2.4. Optimize the imaging sequence parameters for optimal precision
If the range of R_1, R_2 or MTR values is known in advance, the variable imaging parameters (column 3 in table 1) can be optimized to yield the highest achievable dose resolution. Tables of optimal scanning parameters for different R_2 ranges for a single-echo sequence and a multi-spin echo sequence are provided in [40] and [41] respectively. Theoretical derivations can be made for other imaging sequences using similar mathematical formalisms as provided in [40]. In most applications, a dose resolution $D_{\Delta\%}^{p=95\%}$ of at least 2% is required which results in a signal-to-noise ratio (SNR) in the dose map of approximately 138 ($D_{\Delta\%}^{p=95\%} = 1.96\sqrt{2}\,\sigma_D/(D_{max} - D_{min}) = 0.02$).

The dynamic range of the parametric map can be easily assessed using a series of standard contrast agent solution. The R_1 and R_2 are linearly proportional with the molar concentration of the contrast agent. To change both R_1 and R_2 of the test phantoms independently, a mixture of two contrast agents (e.g. Gd-DTPA and $FeSO_4$) or a mixture of a contrast agent and a gelling agent can be used.

2.5. Test the spatial and dosimetric accuracy of the imaging sequence

It has been well described that imaging artifacts may compromise the accuracy of the acquired dose maps [30, 42-45] both spatially as in dose. It is therefore vital that before a gel dosimeter is scanned, the performance of the imaging sequence in terms of accuracy is investigated. A simple test consists of scanning a 'blank' phantom (phantom with a non-irradiated gel) with similar spatial dimensions as the gel dosimeter phantom. It is important to leave the 'blank' phantom in the scanner room for at least 24 hours before scanning in order to equilibrate at the scanner room temperature. Quantitative parametric maps should be calculated from the 'blank' images and the signal homogeneity should be assessed. To isolate stochastic and structural deviations in the image, a voxel based analysis method can be used as described elsewhere [39]. The structural deviations should be below the tolerated error in the quantitative parametric MR maps. The tolerated error is obtained from the tolerable dose error as

$$\varepsilon_P = \frac{\partial P}{\partial D}\varepsilon_D \tag{1}$$

where P = R_1, R_2 or MTR and ε_P and ε_D are the tolerated error in the parameter P and in dose respectively.

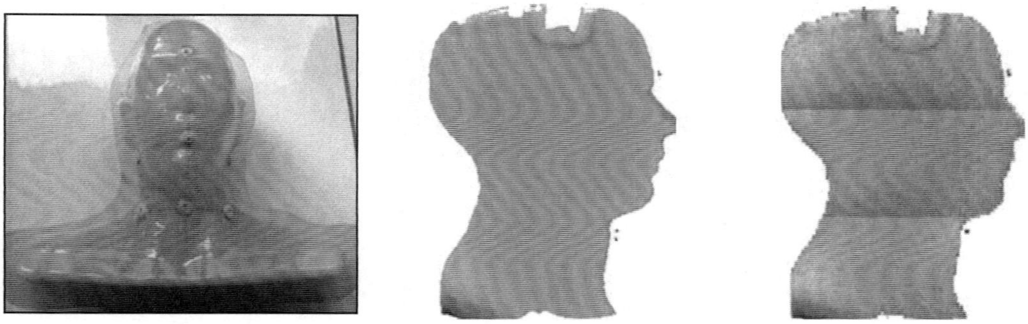

(a) **(b)** **(c)**

Figure 1: Antropomorphic 'blank' gel dosimetry phantom with fiducial markers (a) and corresponding sagital slice (b) A sagital reconstructed from a stack of 105 transverse slices (c) reveals the effect of temperature drift during scanning. Transverse scanning was performed in three interleaved blocks. The ring shaped cranial artifact is attributed to oxygen effects.

At this point, it is advisable to test the image uniformity for several RF coils and check if the addition of a water load improves the homogeneity. Also, it is recommended to perform this uniformity test with the same imaging parameters as in the actual dosimetry experiment, as some artifacts may be phantom related. Also temperature drift as a result of the RF energy from the imaging sequence is dependent on the number of imaging slices and number of acquisitions [44]. If temperature drift results in non-uniformity in the imaged volume, either the repetition time should be increased to decrease the overall specific absorption rate (SAR) or the sequence should be modified to obtain central k-space ordering [44]. Only if the error in the parametric map is below the tolerable error, it is recommended to proceed to the next step.

A geometric quality control QC phantom can be used to check the geometric accuracy [30,39]. It should be emphasized that geometric distortions originating from magnetic field distortions caused by susceptibility differences are phantom dependent. A magnetic field map can be acquired and used to correct the geometric image distortions [46-47].

If dose maps are to be acquired in different image orientations, it is also advisable to scan the 'blank' phantom with different slice orientations. Although at most modern MR scanners, eddy currents are minimized by actively shielded gradient coils, with some magnetic field gradient intensive imaging sequences, eddy currents may still render the dose-R2 response curve dependent on the slice orientation and other sequence parameters [42].

Journal of Physics: Conference Series **444** (2013) 012003 doi:10.1088/1742-6596/444/1/012003

2.6. Preparation of the gel dosimeter experiment

The gel dosimeter phantom and calibration vials are constructed. To minimize any systematic error introduced by the calibration vials, it is recommended to irradiate at least 15 to 20 calibration vials. The actual gel dosimeter phantom is provided with fiducial markers and scanned with CT or MRI to obtain the image data set on which the treatment is planned. The gel dosimeter phantom may be anthropomorphically shaped or may consist of a gel recipient inserted in an anthropomorphic cast. The isocenter on the treatment planning can be located with respect to the fiducial markers which facilitates the localization of the isocenter laser markers onto the antropomorphic phantom. Once the isocenter is determined, the isocenter laser lines are marked on the gel dosimeter phantom and the gel dosimeter phantom is irradiated according to the treatment plan. The calibration vials are irradiated with known doses of which the maximum dose is higher than the maximum dose in the treatment plan.

2.7. Transfer of the gel dosimeter to the MR scanner and scanning

Immediately after irradiation, the gel dosimeter phantom is transferred to the MR scanner room and left there for at least 24 hours in order to equilibrate at the MR scanner room temperature. The calibration vials are fixed onto the gel dosimeter phantom and scanned together with the gel dosimeter phantom. It may be helpful to make a special holder for the calibration vials.

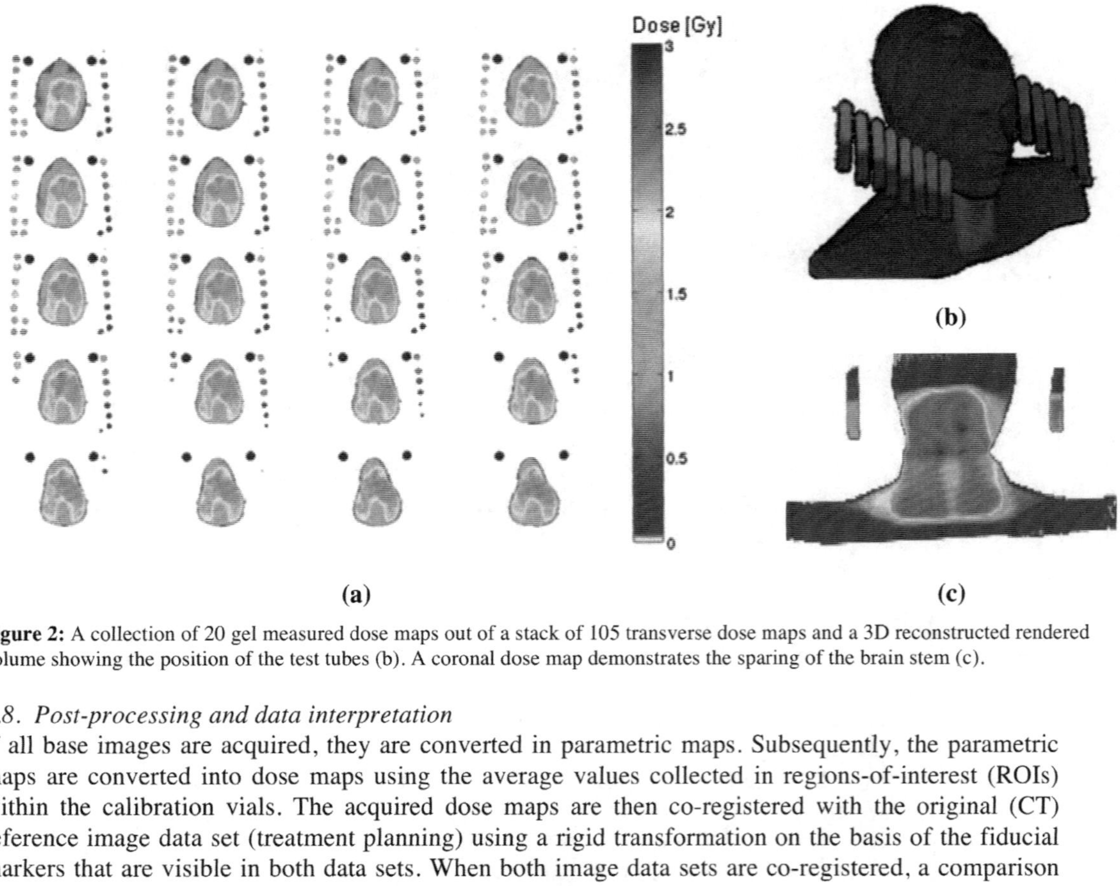

Figure 2: A collection of 20 gel measured dose maps out of a stack of 105 transverse dose maps and a 3D reconstructed rendered volume showing the position of the test tubes (b). A coronal dose map demonstrates the sparing of the brain stem (c).

2.8. Post-processing and data interpretation

If all base images are acquired, they are converted in parametric maps. Subsequently, the parametric maps are converted into dose maps using the average values collected in regions-of-interest (ROIs) within the calibration vials. The acquired dose maps are then co-registered with the original (CT) reference image data set (treatment planning) using a rigid transformation on the basis of the fiducial markers that are visible in both data sets. When both image data sets are co-registered, a comparison of the treatment planning dose grid and the gel measured dose distribution can be performed.

Based on clinically relevant dose tolerance and distance to agreement criteria, gamma-maps [48] can be created and dose volume histograms (DVHs) can be generated for both the treatment planning dose distribution and the gel measured dose distribution.

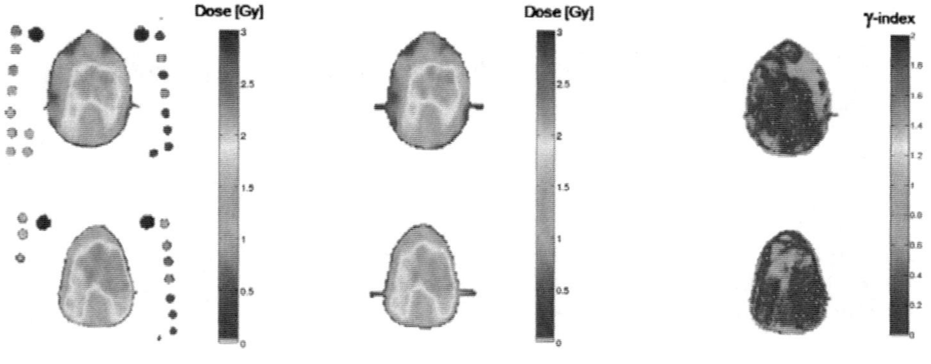

(a) **(b)** **(c)**

Figure 3: Gel measured (a), calculated (treatment planning) dose maps (b) and corresponding gamma plots (c).

3. Conclusion

In contrast to other imaging modalities, MR scanning of polymer gel dosimeters provides many degrees of freedom. Several quantitative MR properties (such as T_1, T_2 and MTR) can be imaged and several different MR sequences can be used to acquire these properties. Whatever property or sequence is used to generate a quantitative parametric MR image data set, the accuracy and precision should be assessed and optimized using 'blank' phantoms (i.e. phantoms similar to the actual gel dosimeter phantom but filled with a non-irradiated gel).

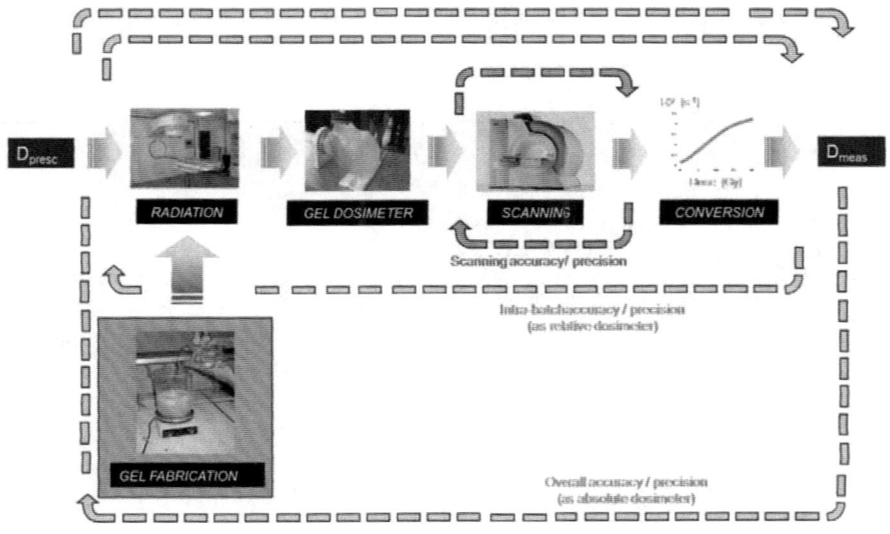

Figure 3: Gel dosimetry is performed in different steps. Errors can occur in each of the steps, leading to a decrease in the overall precision and accuracy. The accuracy and precision of each step can be determined independently.

The actual gel dosimetry experiment should be performed in a similar way as the patient treatment. The treatment planning should be performed on a reference image data set obtained from the gel dosimeter phantom and calibration vials should be irradiated with known doses. The gel measured dose distribution can be co-registered with the simulated (treatment planning) dose distribution on the basis of fiducial markers.

To obtain a picture of the overall accuracy of polymer gel dosimetry, also the fabrication and irradiation have to be included in the analysis (figure 3). This can be achieved through a reproducibility study of the complete gel dosimetry experiment from gel fabrication to dose distribution analysis with a well-characterized dose distribution (such as from a single square profile).

4. References

[1] Gore J C *et al* 1984 *Phys. Med. Biol.* **29** 1189-97
[2] Fricke H and Morse S 1927 *Am. J. Roentgenol. Radium Therapy Nucl. Med.* **18** 430-2
[3] Olsson L E *et al* 1989 *Phys. Med. Biol.* **34** 43-52
[4] Baldock C *et al* 2001 *Australas. Phys. Eng. Sci. Med.* **24** 19-30
[5] Podgorsak M B *et al* 1992 *Med. Phys.* **19** 87-95
[6] Duzenli C *et al* 1994 *Phys. Med. Biol.* **39** 1577-92
[7] Audet C and Schreiner L J 1997 *Med. Phys.* **24** 201-9
[8] Alexander P 1954 *Proceedings of the royal society of london series A - Mathematical and physical sciences* **223** 392
[9] Kennan R P *et al* 1996 *J. Magn. Reson. B* **110** 267-77
[10] Maryanski M J *et al* 1993 *Magn. Reson. Imaging* **11** 253-8
[11] Zimmerman J R and Brittin W E 1957 *J. Phys. C: Solid State Phys.* **61**
[12] Audet C 1995 *Ph.D. thesis, McGill University, Montreal, Quebec*
[13] De Deene Y *et al* 2000 *Phys. Med. Biol.* **45** 859-78
[14] Lepage M *et al* 2001 *Phys. Med. Biol.* **46** 1061-74
[15] Kennan R P *et al* 1996 *J. Magn. Reson. – series B* **110** 267-77
[16] Gochberg D F *et al* 2001 *Phys. Med. Biol.* **46** 799-811
[17] Lepage M *et al* 2002 *Phys. Med. Biol.* **47** 1881-90
[18] Murphy P S *et al* 2000 *Phys. Med. Biol.* **45** 835-45
[19] De Deene Y *et al* 1998 *Radiother. Oncol.* **48** 283-91
[20] Haraldsson P *et al* 2000 *Br. J. Radiol.* **73** 58-65
[21] De Deene Y *et al* 2000 *Magn. Reson. Med.* **43** 116-25
[22] Pappas E *et al* 2001 *Phys. Med. Biol.* **46** 783-97
[23] Ibbott G S *et al* 1997 *Int. J. Radiat. Oncol. Biol. Phys.* **38** 1097-103
[24] Oldham M *et al* 1998 *Phys. Med. Biol.* **43** 1113-32
[25] Meeks S L *et al* 1999 *Int. J. Radiat. Oncol. Biol. Phys.* **43** 1141-53
[26] Cosgrove V P *et al* 2000 *Phys. Med. Biol.* **45** 1195-210
[27] Ramm U *et al* 2000 *Phys. Med. Biol.* **45** N95-102
[28] Farajollahi A R *et al* 2000 *Phys. Med. Biol.* **45** N9-14
[29] Grebe G *et al* 2001 *Int. J. Radiat. Oncol. Biol. Phys.* **49** 1451-60
[30] De Deene Y 2006 *J. Phys: Conf Series* **56** 72-85
[31] Hornak JP 2007 The basics of MRI [web document] http://www.cis.rit.edu/htbooks/mri/
[32] Magnetic Resonance Technology – Information Portal 2008 MR-Tip [web document] http://www.mr-tip.com/
[33] Campus Medica 2007 e-MRI, Magnetic Resonance Imaging physics and technique course on the web [web document] http://www.e-mri.org/
[34] Kuperman V 2000 Magnetic resonance imaging – Physical principles and applications (Academic Press, San Diego, USA)
[35] McRobbie D W *et al* 2006 MRI from picture to proton (Cambridge University Press, Cambridge, UK)
[36] Slichter C P 1978 Principles of magnetic resonance (Springer-Verlag Berlin Heidelberg)
[37] Levitt M H 2002 Spin Dynamics: Basics of nuclear magnetic resonance (John Wiley and Sons, West Sussex, UK)
[38] Bernstein M A *et al* 2004 Handbook of MRI pulse sequences (Elsevier Academic Press, Burlington, Massachutes, USA)

[39] De Deene Y 2009 *J. Phys: Conf Series* **164** 012033, 1-42.
[40] De Deene Y *et al* 1998 *Signal Processing* **70** 85-101
[41] De Deene Y and Baldock C 2002 *Phys. Med. Biol.* **47** 3117-41
[42] De Deene Y *et al* 2000 *Phys. Med. Biol.* **45** 1807-23
[43] De Deene Y *et al* 2000 *Phys. Med. Biol.* **45** 1825-39
[44] De Deene Y and DeWagter C 2001 *Phys. Med. Biol.* **46** 2697-711
[45] De Deene Y *et al* 2001 *Phys. Med. Biol.* **46** 2801-25
[46] Park H W *et al* 1988 *Phys. Med. Biol.* **33** 339-49
[47] Cho Z H *et al* 1988 *Med. Phys.* **15** 7-11
[48] Low D A *et al* 1998 *Med. Phys.* **25** 656-61

How to perform an optical CT scan: an illustrated guide

SJ Doran
CRUK and EPSRC Cancer Imaging Centre, Institute of Cancer Research, Sutton, UK
Department of Physics, University of Surrey, Guildford, GU2 7RB, UK

E-mail: Simon.Doran@icr.ac.uk

Abstract. In this article, which accompanies one of the morning "refresher" sessions at the IC3D Dose 2012 conference, a workflow for optical CT scanning is described and the following steps in that workflow are illustrated: sample positioning, refractive index matching, the importance of apparatus cleanliness to the imaging process, the pre-scan and image reconstruction and post-processing.

1. Introduction

There is by now a considerable body of literature concerning the methodology for performing 3-D radiation dosimetry. As part of the IC3DDose 2010 "refresher course", I discussed in [1] two of the most widespread image acquisition techniques, Magnetic Resonance Imaging and Optical Computed Tomography (CT), and attempted to distil the existing research into a set of "hints and tips" to enable newcomers to the field to avoid common problems. This article is designed to be complementary to [1] and aims to provide a step-by-step guide to the process of acquiring an optical CT scan, illustrating the key procedures involved [2, 3].

2. Procedure for optical CT scanning

Figure 1 shows the complete workflow for the scanning process. A number of tasks are performed prior to the scanning experiment (during scanner commissioning, on receipt of a new sample batch and prior to the start of a set of imaging experiments), whilst others should ideally be performed for each sample that is scanned.

2.1 Sample positioning

An important requirement of the experiment as a whole is the ability to load a sample into the scanner at a consistent position and rotation angle. This turns out to be non-trivial; the various different scanners available commercially, and those developed in-house by research groups in this field, all use different systems for locating the sample correctly. A selection of these is shown in figure 2. As part of initial system commissioning and of on-going quality assurance procedures, it is advisable to investigate sample positioning, by repeatedly inserting the sample, rotating to a predefined angle, acquiring a projection image and then randomly resetting the position of the rotation stage before removing the sample. Forming subtraction images is an ideal way of revealing any differences in position of features in the projections.

2.2 Refractive index (RI) matching

Standard methods of optical CT, using either parallel-beam or cone-beam reconstruction make assumptions about the path of the light rays through the apparatus. Typically, the reconstruction algorithm does not take account of any deviations in the ray path brought about by refraction at the boundary between the sample and the surrounding medium. Any discontinuity in RI across this boundary leads to deviation of the light rays and “imperfect” data. This, in turn, gives rise to the well-known “wall artefact”. At a very low level, this artefact can be helpful, marking the sample boundary on images. However, as was shown in [4], if uncorrected, the artefact can lead to a significant “baseline roll” that extends much further into the sample than expected and seriously impacts on the quantitative nature of the scan. Polymer gels have refractive indices relatively near that of water and these can be accessed via mixtures of either glycerol (glycerine) or ethylene glycol with water. In each case, the additives are miscible in all proportions, and lead to RI ranges from 1.33 – 1.39 and 1.33 – 1.43 respectively. The exact RI value for any given sample will depend on a number of factors, including its composition, its temperature (hence the need for good equilibration with the bath of matching liquid) and, for polymer gels, the quantity of dose it has received.

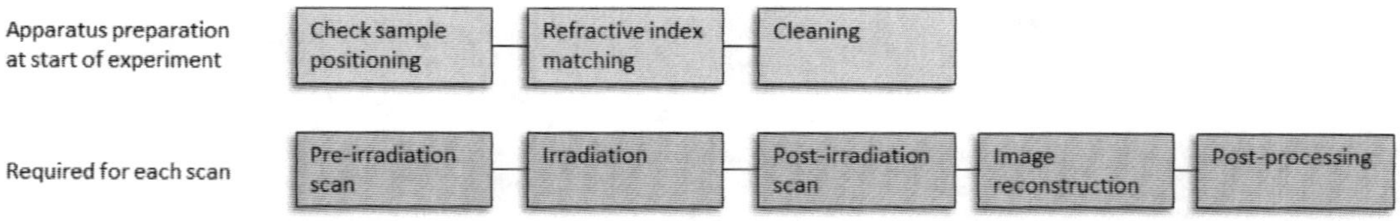

Figure 1: Procedure for performing an optical CT scan.

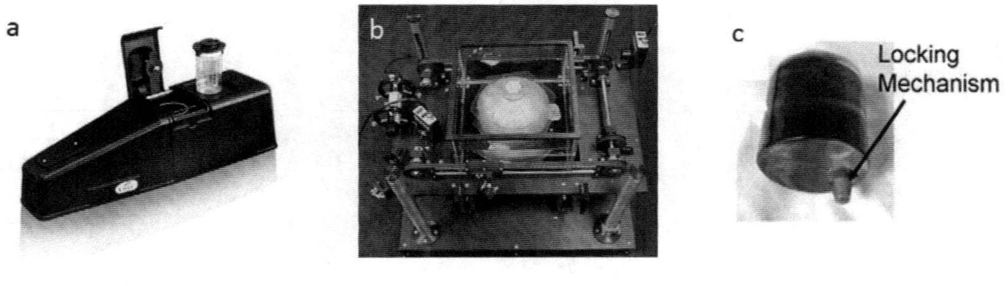

Figure 2: Different scanners have different ways of ensuring that the sample is positioned repeatably: (a) the Modus Vista uses a specially designed clamp to grip the top of the sample jar, with stepper motor providing the sample rotation from the top; (b) Octopus from MGS fixes the sample to a rotating platform; (c) Brady *et al.* use a sample that is moulded to fit in a unique place in the Duke University scanner — image from [2].

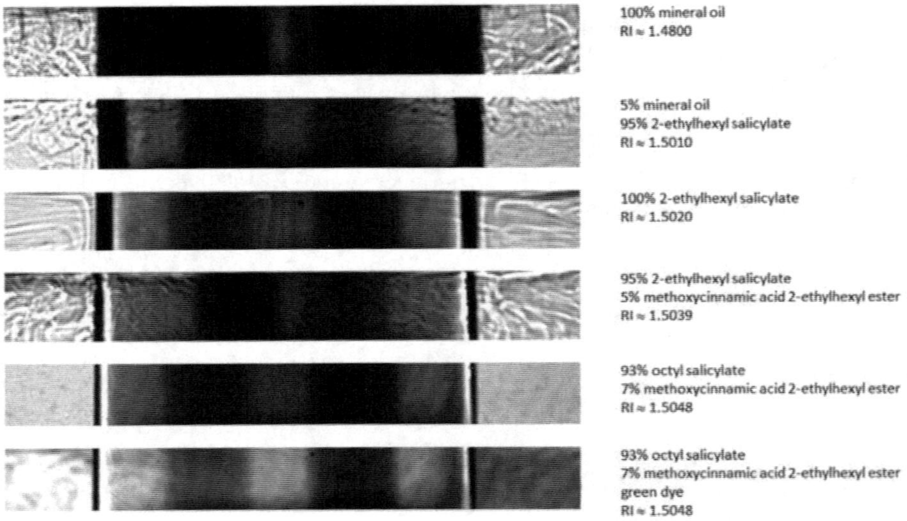

Figure 3: Stages in the process of matching the refractive index between the sample and the surrounding fluid. The final goal is to achieve images with the as little evidence of the sample-liquid boundary as possible. Note that this can be a time-consuming procedure, since mixing of the viscous fluids involved is slow. Only five iterations are shown here. It is not necessary for the mixture to come to a full equilibrium during each stage of the matching process, hence the swirling schlieren patterns of either unmixed or convecting fluid. Note: the absolute RI values are strongly dependent on temperature and so might vary from the values shown here, but the size of the differences between the RI values in successive rows is consistent.

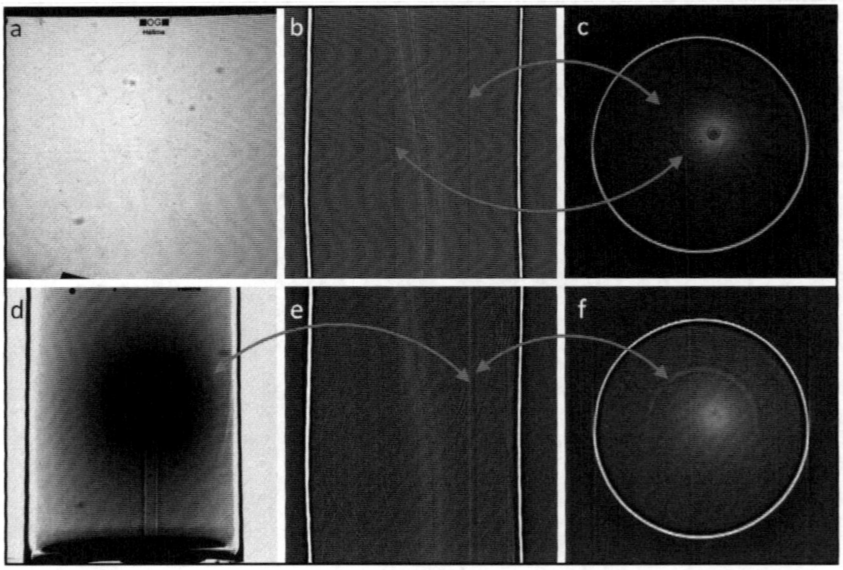

Figure 4: Illustration, in the context of a brachytherapy experiment, of how contamination within the imaging chain cause artefacts that severely impair the final dose maps. (a) Extreme example of a light field corrupted by dirt; zooming in on the image reveals both static features on one or more of the surfaces and dust motes within the matching liquid. The latter are seen very clearly on the accompanying video; (b) sinogram corrupted by vertical stripes corresponding to two small, intense features that do not vary in position as the sample rotates; (c) corresponding image, showing characteristic half-ring artefacts; (d) example projection dataset: the desired darkening from the dose deposited by the brachytherapy seed is well visualised and leads to a sinusoidal track in the sinogram (e), but the fixed imperfection leads to an unwanted vertical track in the sinogram and a half-ring artefact in the reconstructed image (f).

By contrast, the solid plastic radiochromic dosimeter PRESAGE® has a much higher RI of approximately 1.503 and matching can be achieved here with different combination of liquids. Previous research has used dibutyl and diallyl phthalates, but these a relatively toxic and must be handled with care, so we now prefer to use a mixture of 2-ethylhexyl salicylate (RI ≈ 1.502) and 4-methoxycinnamic acid 2-ethylhexyl ester (RI ≈ 1.54), supplemented by small quantities of mineral oil (RI ≈ 1.48) if the mixture RI needs to be reduced significantly. There remain inconveniences caused by the length of time it takes to mix the samples to achieve good matching throughout and the non-solubility in water of all of these chemicals, making disposal of used materials and cleaning of the apparatus an issue. Figure 3 follows the stages of a typical experiment to determine the optimal matching liquid composition. Note especially the exquisite (but unwanted) sensitivity to very small changes in refractive index, and the addition of a dye to the liquid, for the reasons outlined in [1].

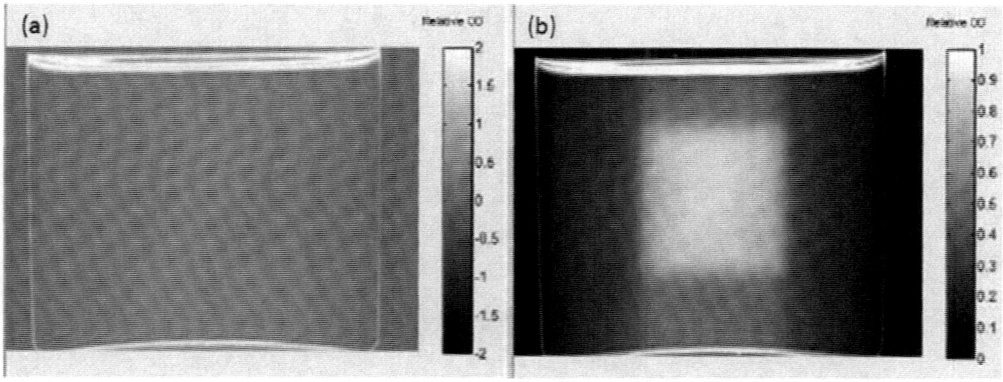

Figure 5: Example of reference pre-scan and corresponding post-irradiation scan — image from [3]

2.3. *Cleaning*

It is hard to overestimate the importance of maintaining the cleanliness of an optical CT scanner. Whilst many potential artefacts can be removed from the final reconstructed optical-CT images by judicious post-processing of the data, eliminating the problems at source will provide much higher quality input data into the system. Figure 4 shows an extreme example of the sort of problems that can occur and which were too severe to be entirely corrected by the pre-scan procedure below.

2.4. *Pre-scan*

Figure 5 shows an example of scans of a PRESAGE® sample before and after irradiation. The pre-scan serves three functions. Firstly, it allows us to calibrate the light-field provided by the scanner [5], which will typically not be entirely homogeneous, showing either a smooth variation or a set of isolated imperfections (Fig. 4a). Secondly, the pre-scan enables sample-specific features to be removed from the images. These fall into two categories: (i) Wall effects and baseline roll due to residual RI mismatch can either simply be subtracted, or can be modelled, potentially improving the SNR of the final image. (ii) Sample imperfections causing optical artefacts that are constant between the two scans can potentially be removed by subtracting the reference pre-scan. However, this voxel-by-voxel correction does require excellent reproducibility of sample positioning. The third function of the pre-scan is to provide the baseline map of the sample's optical density. Once this has been subtracted from the final post-irradiation data, what is left is a map of the *change* in optical density induced by the radiation and it is this quantity that is related to the absorbed dose.

The pre-scan procedure has several pitfalls. The most important is that it is difficult to manage on a shared system. The University of Surrey group has worked on several projects that would have involved long delays between pre- and post-irradiation scans (e.g., low-dose-rate brachytherapy, with an exposure time of several weeks and synchrotron irradiations in which several days' travel to the irradiation facility were required.) It is not always possible to guarantee that the equipment remains in the same state between pre- and post-scans. Typical issues are: changes in the incident light intensity (usually correctable); changes in field-of-view and focus (particularly challenging to account for when the system has manual lens adjustment); and changes in the refractive index matching liquid or additional contamination on the surfaces of the matching tank. Under these circumstances, we have found that it is useful to acquire a light-field scan (i.e., one with no sample present) to match each sample scan. In addition, it is also beneficial to acquire a scan with no light input to the camera (e.g., with lights off and lens cap on), in order to measure the so-called "dark-current image". This

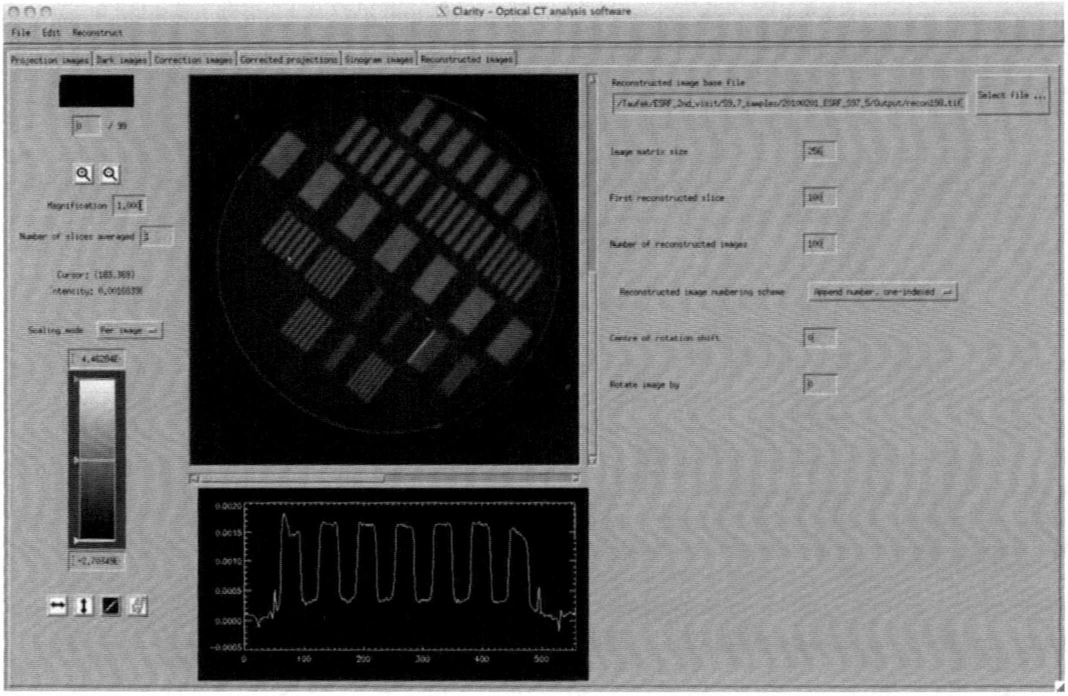

Figure 6: The *Clarity* application, written by the author at the University of Surrey, and running within the IDL (Interactive Data Language, Excelis Visual Information Solutions, Boulder, CO) data processing environment, is a typical example of the in-house image reconstruction solutions available.

information may be used to improve the reconstruction of the data [5]. Together, the light- and dark-field images allow one to perform a minimal set of corrections for the characteristics of the scanner, which provide a "basic image" that can be supplemented by further post-processing if sample-specific data become available. A further issue to be aware of is the need to monitor and correct for changes in the samples that are not related to radiation exposure. In another abstract to be presented at IC3DDose 2012 [6], we describe the effect of light on PRESAGE ® samples. However, these concerns apply equally, for example, to auto-oxidation of Fricke gels [7].

2.5. Irradiation and post-irradiation scan

For the reasons outlined above, sample irradiation and subsequent scanning should ideally be carried out as soon as possible after the pre-scan, and the post-irradiation scan should use identical parameters to the reference.

2.6. Image reconstruction and post-processing

These two areas overlap and different authors will have different opinions as to the extent to which image post-processing and enhancement are part of the reconstruction process itself. The major steps in the reconstruction are reforming the individual projections into a set of sinograms (Figs. 4b and 4e), filtering the sinograms, and performing back-projection, the mathematics of which is described in many popular texts, including [8]. Typical post-processing tasks might include artefact removal and image enhancement by median filtering. Finally, some form of calibration must be performed to convert the optical density values returned by the scan into dose.

There is currently no "standard" software for reconstructing 3-D dose maps using optical CT, each research group and manufacturer having developed their own processing environments. A typical example is shown in figure 6. It is the author's view that there would be considerable benefit for optical CT researchers to establish a working party to explore the creation of a standardised "community tool" that incorporates best practice and the most efficient reconstruction algorithms available. Such a tool could be designed to accept data from all the major groups and commercial scanners and would be complementary to specialist developments from individual groups. It would serve as a quotable "application of reference", against which other software could be compared.

3. Conclusion

Unlike the case of MRI, image acquisition using optical CT is still not entirely a "push button" technology, although manufacturers and groups in the field are moving steadily towards this position. This article has outlined the major steps of the workflow involved

4. References

[1] Doran SJ 2010 *J. Phys.: Conf. Ser.* **250** 012086
[2] Brady S L *et al* 2010 *Phys. Med. Biol.* **55** 2187
[3] Newton J *et al* 2010 *J. Phys.: Conf. Ser.* **250** 012078
[4] Abdul Rahman A T *et al* 2011 *Phys. Med. Biol.* **56** 4177-99
[5] Doran S J *et al* 2001 *Phys. Med. Biol.* **46** 3191-213
[6] Gorjiara T *et al* 2012 In *IC3Dose 2012 Proceedings* Sydney
[7] Babic S *et al* 2008 *Int. J. Radiat. Oncol. Biol. Phys.* **70** 1281-91
[8] Kak A C and Slaney M 1988 *Principles of Computerized Tomographic Imaging* (IEEE Press)

Considerations for x-ray CT polymer gel dosimetry

A Jirasek
Department of Physics and Astronomy, University of Victoria, Victoria, BC, Canada, V8W 3P6

E-mail: jirasek@uvic.ca

Abstract. This review paper addresses the basic considerations required when performing x-ray CT polymer gel dosimetry. The review focuses on recent developments pertaining to these basic considerations.

1. Introduction

X-ray CT remains an attractive modality for imaging irradiated polymer gel dosimeters (PGD). The system is robust, straightforward to implement, and x-ray CT scanners are readily available within radiotherapy clinics. While much of the research in the field is concentrated on MRI and optical CT imaging of irradiated PGD, x-ray CT imaging has been shown to be capable of providing images of irradiated PGD that are suitable for extracting useful dose information.

This overview is aimed primarily at highlighting developments within the field since the last review conducted in 2010 and earlier review work [1, 2]. As such, full details of early studies are not given here, but rather references are provided to these works. Some early results are provided for context.

2. General overview

The basic principle of x-ray CT PGD revolves around the fact that PGD undergo minute changes in density when irradiated with ionizing radiation. This was first demonstrated by Hilts *et al* in 2000 where they used an acrylamide based PGD irradiated with 6MV photons from a clinical linear accelerator [3]. The density change is attributed to a change in volume, as opposed to a change in mass, within the system. This volume change results from the radiation induced polymerization of monomer to crosslinker, along with a concomitant redistribution of water away from the formed polymer. The volume change has been estimated to be small (<1% for early acrylamide based systems) [1, 2]. While this is favourable in terms of maintaining the integrity of the dose distribution (i.e. radiation induced volume changes are not expected to lead to large material distortions), the small volume change has the negative effect of resulting in small changes in density and hence poor contrast in the CT images, as seen for example in Audet *et al* [4].

The images in the original work of Audet *et al* suffered from low contrast, and hence low dose resolution. A large effort since this first manuscript has been dedicated to improving the dose resolution of CT images acquired on irradiated PGD. However, this is not the only area of active research within the discipline. A large range of topics have been investigated and can be roughly categorized as outlined in table 1. Representative references are given to point the reader towards literature discussing the main topics outlined in the table. The remainder of the manuscript discusses recent work in some of these areas.

Journal of Physics: Conference Series **444** (2013) 012005 doi:10.1088/1742-6596/444/1/012005

Table 1: Recent research areas in CT PGD. Representative references are given, with concentration focused on works within the last two years.

Research Topic	**Reference**
Improving dose sensitivity (dose resolution)	[5-9]
Gel properties (stability, polymerization kinetics etc)	[5-8, 10, 11]
CT imaging parameter optimization	[12-14]
Post processing (filtering etc)	[15-19]
CT dose considerations	[15, 20, 21]
Clinical demonstrations	[4, 6]

3. Developments in x-ray CT PGD

3.1. Gel sensitivity

It is clear that the low dose resolution in images of irradiated PGD has been a key limiting factor in the technique gaining widespread usage. To address this issue a number of investigators have undertaken work with the aim of improving the sensitivity of PGD to radiation. Since the dose resolution of a system depends on both the slope of the dose response curve, as well as the uncertainty in the measurements, improving dose sensitivity should bring gains in dose resolution, if all other factors remain equal. Koeva *et al* undertook preliminary investigations in improving dose sensitivity of N-isopropyl-acrylamide (NIPAM) based PGD [7]. This initial study paved the path for later works, which included incorporating a cosolvent (isopropanol) into the formulation [22]. While isopropanol did significantly increase the dose sensitivity of NIPAM based PGDs due to the increased solubility limit of the crosslinker, the manufacturing technique and handling of the PGDs was complicated by the cosolvent. To overcome this, Chain *et al* proposed cosolvent-free NIPAM based PGD with increased sensitivity [5]. The increase in sensitivity was accomplished by recognizing that N-isopropyl-acrylamide is itself a cosolvent for the crosslinker, and hence by changing the relative fraction of monomer to crosslinker, an increased amount of crosslinker could be incorporated into the dosimeter and an increase in dose sensitivity could be realized. Figure 1 (left panel) illustrates the increase in dose sensitivity that is possible as the relative fractions of monomer and crosslinker are increased. Figure 1 (right panel) illustrates the resulting improvements in dose resolution with each formulation tested in figure 1 (left panel). It can be seen that PGD with high radiation dose sensitivity and high dose resolution can be manufactured for imaging with x-ray CT.

3.2. Gel properties

A number of researchers worked to establish the suitability of these new PGD formulations for robust x-ray CT PGD. Recently, Johnston *et al* undertook a comprehensive x-ray CT PGD analysis of Chain's formulation. Within this study they established the post radiation temporal stability, intra and inter-batch reproducibility, calibration robustness, and dose rate dependence of the new gel formulation. As an example, figure 2 (left panel) illustrates that post irradiation polymerization can occur for up to ~20 hrs post radiation, and that this polymerization can be detected with x-ray CT. Figure 2 (right panel) illustrated that the high sensitivity NIPAM gel does exhibit a dose rate dependence.

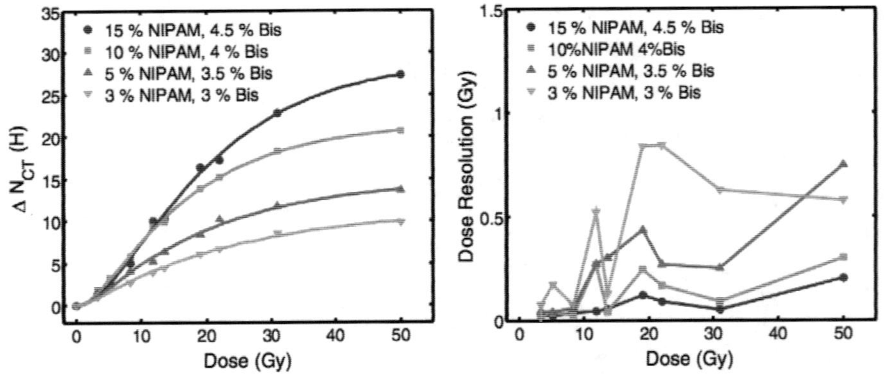

Figure 1: Left panel: Dose sensitivity and Right panel: resultant dose resolution in NIPAM based PGDs as a function of relative fractions of monomer and crosslinker. From [5].

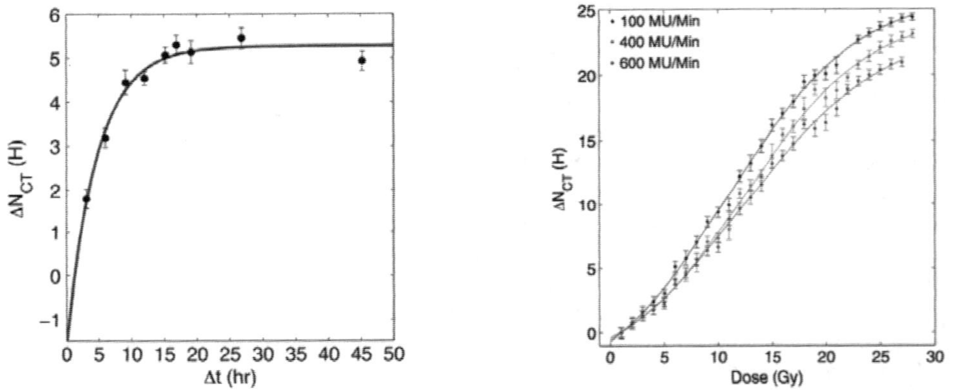

Figure 2: Left panel: Change in CT number as a function of time post irradiation. Polymerization continues for up to ~20 hrs post irradiation. Right panel: Dose rate dependence of the high sensitivity NIPAM PGD. From [6]

Interestingly, an opposite trend in dose rate dependence was observed in [11], although the concentration of monomer and crosslinker in their PGDs are different from that of Johnston. Chang *et al* also illustrate a potentially mild energy dependence of NIPAM based PGD, as shown in figure 3.

3.3. Imaging parameter optimization

A complete investigation of image acquisition parameter optimization was undertaken by Hilts [13] using an acrylamide based PGD and, as these results are now well established, they will not be discussed here. More recently, Ghavami *et al* undertook similar investigations with a NIPAM based PGD and reported similar results to that of Hilts *et al* [14]. These results are consistent with CT theory and are not, typically, dependent on the PG formulation.

One additional area that has been investigated recently is that of CT reconstruction technique on image quality in CT PGD. Hindmarsh used raw data from the CT scanner and investigated the effects of differing reconstruction techniques on resultant image quality [12]. Notably, the group moved away from filtered backprojection algorithms and investigated iterative techniques. As shown in figure 4, Hindmarsh *et al* demonstrate that the modulation transfer function can be significantly altered depending on the reconstruction technique employed.

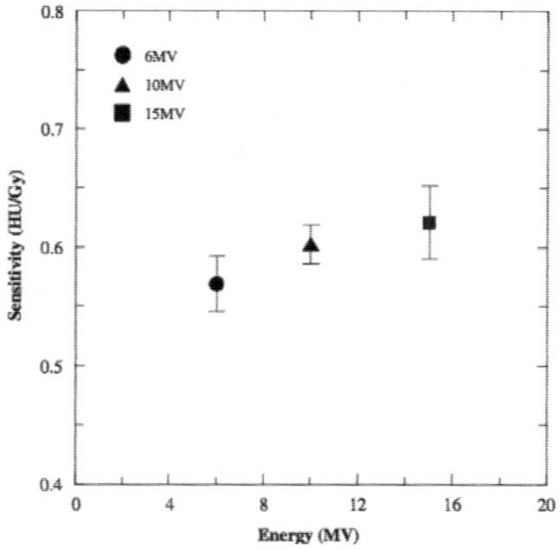

Figure 1: Sensitivity of NIPAM PG as a function of incident photon energy. From [11].

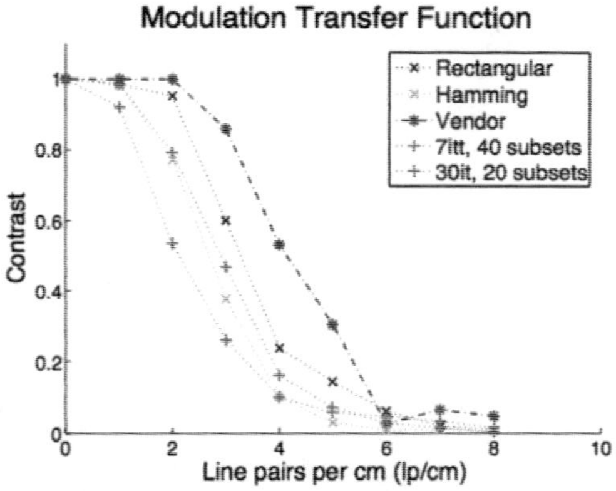

Figure 2: Effects of reconstruction technique on the modulation transfer function of the system. From [12].

3.4. Post processing

Due to the low contrast in CT PGD images, image filtering has traditionally been a key component in post acquisition processing [17-19, 23]. Recently, Jirasek et al identified a structural noise component, in addition to standard stochastic noise, within CT PGD images [16]. This structural noise is due to inherent gel structure and is typically several pixels in extent. Hence, standard stochastic noise filtering techniques have proven ineffective in filtering this structured component. A new class of filter was implemented to address the structural noise within CT PGD images. The Remnant Artefact Removal (RAR) technique is based on a class of signal removal techniques [24]. In the present implementation, RAR strips "signal", in this case considered to be the gel structure and unwanted remnant artefacts, from the image by utilizing fitting algorithms which discriminate between radiation induced large-scale dose structure (polymerization, ~>50 pixels in width) from inherent gel structure (~5 – 10 pixels in width). Figure 5 illustrates the efficacy of the Remnant Artefact Removal technique [16].

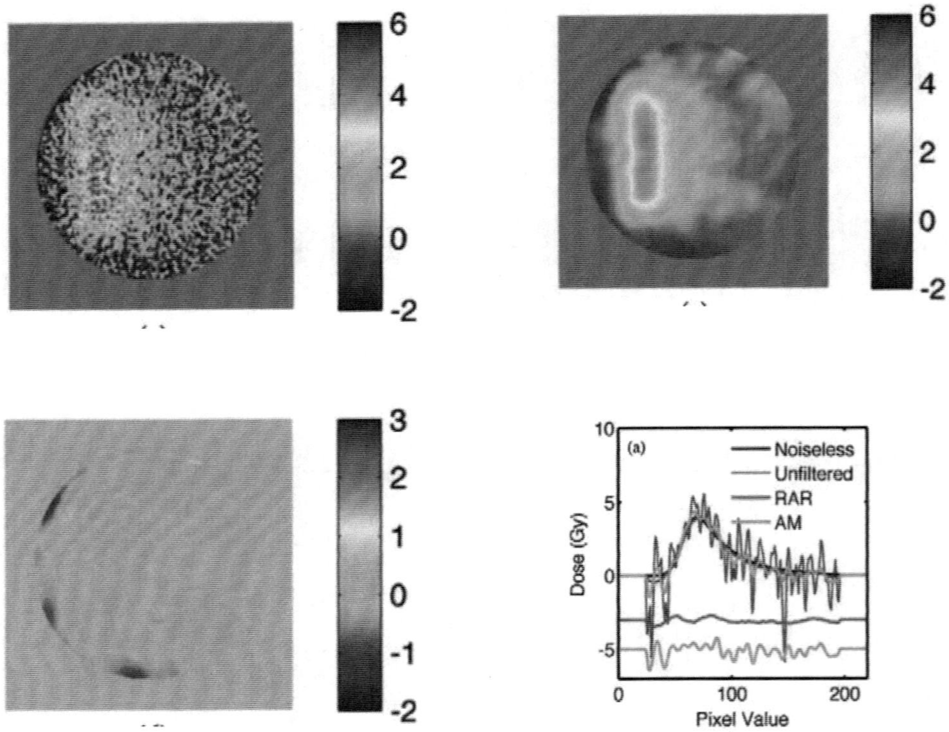

Figure 5: Top left panel: Synthetic wedge distribution with added stochastic and structural noise. Top right panel: Image of top left panel filtered with RAR. Bottom right panel: difference map between top right panel and known wedge distribution. Bottom right panel: Representative profile through wedge distribution illustrating unfiltered, RAR, and adaptive mean filtered results. Lower traces are difference profiles between filtered and unfiltered profiles. Colourbars and y-axis (bottom right panel) are in Gy. From [16].

3.5. CT dose considerations

Traditionally, image averaging has been used to improve the signal to noise ratio in CT PGD images [3]. However, limiting factors in this approach include time, CT tube load, and CT dose imparted to the PGD which can initiate further polymerization and hence distort the desired dose distribution [20]. Recently Kakahel *et al* introduced a "zero-scan" method which aims to improve the image quality of CT PGD images while minimizing the CT dose registered in the final image [15]. In their method they acquire a large number of images (16 – 360) and, on a pixel-by-pixel basis, fit a function (linear, exponential, etc) to the CT number vs scan number data. This allows for an interpolation back to a "zero-scan" CT number which contains very little CT dose. Furthermore, since the fit function incorporates data from a large number of scans, image noise is reduced in the zero-scan interpolated image. Figure 6 illustrates representative results of the zero-scan technique.

3.6. Clinical demonstrations

In general, very few CT PGD studies to date have focused on clinical demonstrations. Audet *et al* did so in an early study where they focused on a stereotactic radiosurgery application [4]. Recently, Johnston *et al* used a NIPAM based polymer gel, full gel calibration and RAR filtering to demonstrate clinical utility of CT PGD. They used a head phantom irradiated with calibration, and IMRT test irradiations. Generally, excellent agreement between planned and measured doses was observed and the work clearly demonstrated the potential of CT PGD for clinical applications. Figure 7 illustrates representative results from the work.

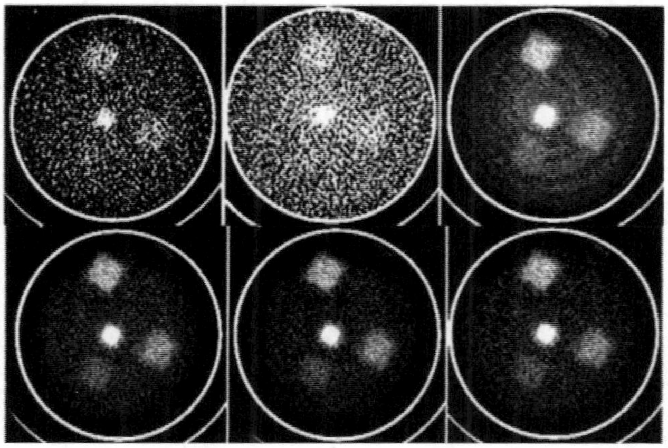

Figure 6: Top left: initial image. Top centre: 360th image in series. Note increased signal from polymerization due to CT dose. Top right: averaged image (360 averages). Bottom left: zero-scan interpolated image, exponential fit. Bottom centre: zero-scan interpolated image, linear fit. Bottom right: zero-scan interpolated image, quadratic fit. From [15].

4. Summary

Recent work has made excellent progress in moving CT PGD towards clinical utility. Efforts in gel recipe development and optimization, advanced filtering and artifact removal, CT dose consideration, and demonstrations of clinical potential have all worked to advance CT PGD towards clinical viability.

5. References

[1] Jirasek A 2010 *J. Phys.: Conf. Ser.* **250** 012070
[2] Jirasek A and Hilts M 2008 CT *J. Phys: Conf. Series* **164** 012038
[3] Hilts M *et al* 2000 *Phys. Med. Biol.* **45** 2559-71
[4] Audet C *et al* 2002 *J. Appl. Clin. Med. Phys.* **3** 110-8
[5] Chain J N M *et al* 2011 *Phys. Med. Biol.* **56** 2091-102
[6] Johnston H *et al* 2012 *Phys. Med. Biol.* **57** 3155-75
[7] Koeva V I *et al* 2009 *Phys. Med. Biol.* **54** 2779-90
[8] Jirasek A 2009 *Phys. Med. Biol.* **54** 907-18
[9] Hilts M *et al* 2004 *Phys. Med. Biol.* **49** 2477-90
[10] Sellakumar P and Samuel E J J 2010 *Radiat. Meas.* **45** 92-7
[11] Chang K Y *et al* 2011 *Nucl. Instrum. Methods Phys. Res. A* **652** 1775-8
[12] Hindmarsh J *et al* 2010 *J. Phys.: Conf. Ser.* **250** 012073
[13] Hilts M *et al* 2005 *Phys. Med. Biol.* **50** 1727-45
[14] Ghavami S-M *et al* 2010 *Reports of Practical Oncology and Radiotherapy* **15** 172–5
[15] Kakakhel M B *et al* 2011 *Med. Phys.* **38** 5130-5
[16] Jirasek A *et al* 2012 *Phys. Med. Biol.* **57** 3137-53
[17] Hilts M and Duzenli C 2004 I *Med. Phys.* **31** 39-49
[18] Jirasek A *et al* 2006 *Phys. Med. Biol.* **51** 2599-617
[19] Hilts M and Jirasek A 2008 *Med. Phys.* **35** 344-55
[20] Baxter P *et al* 2007 X-ray *Med. Phys.* **34** 1934-43
[21] Trapp J V *et al* 2002 *Phys. Med. Biol.* **47** 4247-58
[22] Jirasek A *et al* 2010 *Phys. Med. Biol.* **55** 5269–81
[23] Gopishankar N *et al* 2012 in *Modern Practices in Radiation Therapy* Ed. Gopishankar and Natanasabapathi (InTech) 327-38
[24] Schulze G *et al* 2005 *Appl. Spectrosc.* **59** 545–74

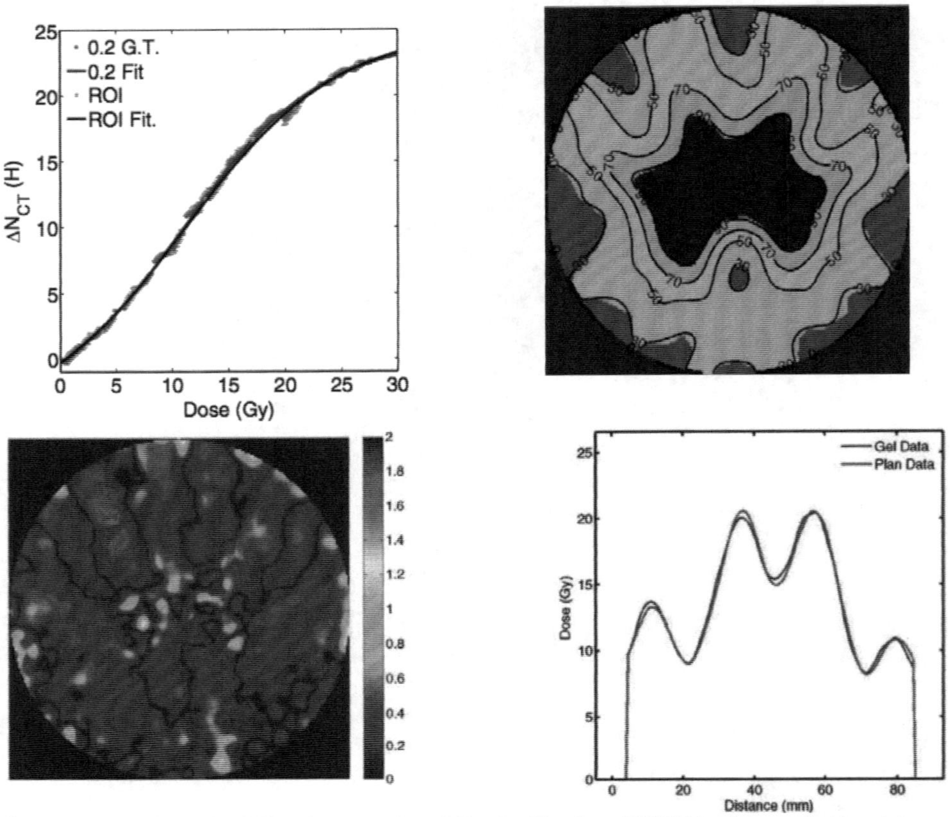

Figure 7: Top left panel: Calibration curve for a full gel calibration of NIPAM polymer gel. Top right panel: Gel colour map of binned isodose levels, compared with isodose lines from treatment planning calculations. Values shown are isodose percentages. Bottom left panel: Gamma map (3%, 3mm) between gel image and treatment plan (99.3% pass at isocentre slice, 93.4% pass through entire volume). Bottom right panel: Representative comparison between profile extracted from gel image and treatment plan. From [6].

Accuracy required and achievable in radiotherapy dosimetry: have modern technology and techniques changed our views?

David Thwaites
Institute of Medical Physics, School of Physics, University of Sydney, Australia

E-mail: david.thwaites@sydney.edu.au

Abstract. In this review of the accuracy required and achievable in radiotherapy dosimetry, older approaches and evidence-based estimates for 3DCRT have been reprised, summarising and drawing together the author's earlier evaluations where still relevant. Available evidence for IMRT uncertainties has been reviewed, selecting information from tolerances, QA, verification measurements, in vivo dosimetry and dose delivery audits, to consider whether achievable uncertainties increase or decrease for current advanced treatments and practice. Overall there is some evidence that they tend to increase, but that similar levels should be achievable. Thus it is concluded that those earlier estimates of achievable dosimetric accuracy are still applicable, despite the changes and advances in technology and techniques. The one exception is where there is significant lung involvement, where it is likely that uncertainties have now improved due to widespread use of more accurate heterogeneity models. Geometric uncertainties have improved with the wide availability of IGRT.

1. Introduction

There have been a range of published estimates of the accuracy required and achievable in radiotherapy dosimetry over at least the last 30+ years since ICRU report 24 [1] provided a summary in 1976. However most of these [see reviews in 2,3] have related to 3D CRT methods and hence before IMRT and the other rapidly advancing technologies and techniques in place today. Thus we are currently confronted with the critical need for accurate, practical dosimetry for commissioning, development, QA and both pre-treatment and in vivo verification for these technologies, for clinical practice and also for research; indeed dosimetry measurements, methods and resource requirements are recognized as a potential limiting step in the widespread clinical application of some technologies. The need has moved from 'simple' 1D and 2D dosimetry to requirements at all levels from 1D to 4D for IMRT, IGRT, ART, and 4D motion-managed RT [4]. The objective of all this development is to improve the balance of tumour control versus normal tissue complication incidence for curative treatments and to improve quality of life for patients receiving palliative treatments. However, these outcomes will not be safely achieved unless there are also matched improvements in accuracy in treatment delivery systems and appropriate dose measurement and modelling systems to support this [5]. There is considerable interest in this topic currently; with a forthcoming book from an IAEA expert group on accuracy and uncertainties in radiation therapy due to appear in 2013, lead by J. van Dyk.

This review is an update of the current author's series of evaluations of dosimetric accuracy for radiation therapy over the last 25 years [2,6-12] and repeats some text from those, where still relevant. It is limited to external beam megavoltage x-ray therapy and considers conventional and complex treatment methods, but essentially those that are applicable in the majority of departments. In general it considers curative treatments, as the treatment processes must be set up with the greatest accuracy to deliver those. In general, too, it considers best practice based on having comprehensive quality systems in place, including relevant QA for all steps and parts of the radiotherapy process [13-16]. This recognises that poor practice, in the extreme as evidenced in widely reported accident and error situations [17-18], can considerably worsen the accuracy achieved. Many of the estimates of accuracy have involved uncertainty analyses with best guesstimates of values. However, wherever possible here values are based on evidence, using optimum values as then representative of 'achievable' values and hence as the levels to be aimed for.

2. What is accuracy in radiotherapy dosimetry?

The term *accuracy* in radiotherapy is often used loosely. Firstly, absolute accuracy must be distinguished from precision or reproducibility. Overall uncertainty is a combination of type A uncertainties, assessable by statistical means (often loosely referred to in practice as random) and type B uncertainties, which must be assessed by other means (often loosely referred to in practice as systematic). Thus random uncertainties can be estimated from repeated independent observations and can be expressed as a standard deviation (sd), assuming a normal distribution. Systematic uncertainties on the other hand can only be estimated by an analysis of the process under consideration, assigning reasonable variations to parameters where the uncertainties are by definition not exactly quantifiable. They can be expressed as an effective sd, being taken as the estimate of the limits within which the correct value is expected to occur in around 70% of cases. Uncertainties of both types (sd or effective sd) can be combined in quadrature to provide an estimate of overall uncertainty [19-21]. As is well recognised, systematic errors contribute to absolute deviations from the 'correct' value in any situation whilst random errors are associated with precision. However the distinction can become blurred in certain circumstances. In radiotherapy, within one centre and one radiation modality, it is reproducibility of dosimetry that is critical. However in transferring experience from one centre to another or in intercomparing results between centres or between modalities, for example in clinical trials, some systematic uncertainties also become significant. In intercomparisons between a number of centres, some systematic uncertainties are effectively randomly distributed (eg beam calibration). On the other hand some systematic uncertainties are common to all participants and can be omitted from routine consideration eg. those associated with a given dosimetry protocol if all participants follow the same protocol, or those associated with basic physical data if the protocols involved are all based on the same consistent set. Such factors only require consideration when estimating absolute accuracy but play no part in estimations of dosimetric consistency, except where different modalities may be being compared. Thus a clear understanding is required of which uncertainties require inclusion in any given situation. However in practice it is often difficult to separate the two types [22].

As a further point, it should be noted that any stated values should also have their meaning defined. All values stated in the following, whether for individual parameters or for cumulative values where all involved uncertainties are combined in quadrature, are given as one standard uncertainty (or the sd for normal distributions, or effective sd for combined type A and type B uncertainties where normal distributions are assumed). To express the uncertainties in terms of confidence intervals, the standard uncertainty, u_c, is multiplied by the coverage factor, k. Where a normal distribution applies, k=2 provides a confidence interval of approximately 95% and k=3, of approximately 99%. Further information on uncertainties and confidence intervals can be obtained from a range of sources, eg [20-21]. It may be noted that various uncertainties in radiation treatment are not well represented by Gaussian distributions, but by one of a range of other possibilities, which can make analysis more problematic. In addition, for geometric uncertainties in radiation treatment, the combination of the two types of uncertainties is generally dealt with via margin recipes, assigning different importance to each type to reflect their impact [23]. Lastly, it may be noted that combination in quadrature (u^2_{total} =

$u^2_a + u^2_b + ...$) generally results in the larger uncertainties dominating the overall uncertainty, with consequences for consideration of where effort should best be applied to reduce uncertainties.

In summary, then, accuracy is a measure of how close a result is to the 'true value', whilst precision is a measure of the spread of independent determinations of the result where the latter is generally determined as the sd of the distribution of results.

3. Required accuracy in radiotherapy dose delivery based on the clinical evidence

3.1. Dosimetry

The clinical requirements on accuracy are based on evidence from dose response (dose effect) curves for tumour control probability (TCP) and for normal tissue complication probability (NTCP). These typically have sigmoidal shapes, with a threshold dose, relatively steep rises and saturation (100% effect) at high enough doses. Generally, in practical clinical situations the two relevant curves overlap along the dose axis [eg see 2], such that the dose to the tumour is limited by what can be tolerated by the most at-risk normal tissues. Radiotherapy optimisation and many advances in technology and techniques are aimed at improving this balance, ie maximising tumour control while maintaining tissue complications at an acceptable level. The steepness of the given TCP or NTCP curve versus dose defines the change in response expected for a given change in delivered dose. Thus uncertainties in delivered dose translate into either reductions in TCP from the optimised expected value, or increases in NTCP from the optimised expected value, both of which worsen the clinical outcome. The accuracy requirements are defined by the most critical (steepest) curves, observed for normal tissues or steeply responding tumours. Steepness data of clinically derived TCP and NTCP is reviewed in a number of publications, eg. [1,2,10,19,24] indicating wide variations in the slopes of the curves depending on tissue, effect, uncertainties in clinical dosimetry and in assessing outcomes, radiobiological variability, technique used, dose per fraction, etc. At the steepest parts of the dose-response curves, and for the steepest curves, 5% changes in dose can produce 10-20% changes in TCP and 20-30% changes in NTCP. Brahme [25-26] and others expressed these steepnesses in terms of the normalised dose gradient, γ, the percentage change in TCP or NTCP per 1% change in dose. This parameter will typically be greatest (curve steepest) in the 50% effect region, γ_{50}, but will vary depending on which part of the curve is of interest (generally lower than 50% for NTCP and hopefully greater than 50% for TCP; however it is often the value at 50% effect that is compared). The steepest clinical curves are for normal tissue effects, with γ_{50} values of up to 6 or 7% per 1% change in dose [26-27] and as a general requirement, it is necessary to base accuracy recommendations on the steeper dose-effect relationships encountered in routine clinical situations, as overall process accuracy must be established to meet these more demanding situations. It may be noted that whilst there have been new clinical data and many other studies of such information over recent years, including reviews such as the QUANTEC report/s for normal tissue effects (QUantitative Analysis of Normal Tissue Effects in the Clinic) [28], these general values are still applicable.

ICRU Report 24 [1] reviewed the more limited information available up to that time and considered that +/-5% accuracy was required in the delivery of absorbed dose to the target volume, but that in critical situations +/-2% may be required. However it was recognised that the latter figure was not generally realisable in then current practice. It was not made clear what uncertainty value the figures represented. Mijnheer et al [19] obtained a figure for required accuracy by considering normal tissue complications. They considered the steepness of dose-effect curves in terms of the percentage increase in absorbed dose to produce a change in the probability of normal tissue complications from 25% to 50%. A representative value of 7% was taken for this relative gradient and it was concluded that any transfer of clinical information from one centre to another will involve unacceptable risks of complications if overall uncertainty in absorbed dose is larger than this value. This was then assigned to the 2 sd level, resulting in a value of 3.5%, as one relative sd, as the general accuracy requirement on absorbed dose delivery. Brahme et al [26] considered the effects of variations in dose on tumour control for typical values, showing that the most critical loss in tumour control introduced by dosimetric inaccuracy is found at the highest level of tumour control probability. A general figure of 3% (relative sd) on the delivered absorbed dose to the patient was recommended as the tolerance level on accuracy in dose delivery, in order to keep variations in the probability of tumour control within acceptable limits. Thus overall a figure of 3% sd can be taken as a currently recommended general accuracy requirement, being considered as one relative sd, on the value of the dose delivered to the patient at the dose specification point. This implies there is a 95% probability that changes will be clinically observable for dose changes at twice this level, in situations described by the steeper dose-effect relationships. This is also consistent with more anecdotal evidence on clinical observations following inadvertent dose changes due to dosimetric errors [22].

Considering the different effects of systematic and random dose uncertainties on TCP and NTCP may give a further perspective. Overall, systematic uncertainties in the dose delivered will translate directly into changes in TCP and NTCP for the population of patients involved (or also as expected for an individual patient, but likely to be complicated by the effect of individual radio-sensitivity). For random uncertainties the effect is much more variable and depends on the steepness of the curve, the point of interest on the curve (where effects will now be greatest where the curvature of the curves is greatest), the level of uncertainty, etc. Generally, random uncertainties will smear effects and will reduce TCP and increase NTCP but in a less immediately clear way. It must be remembered too that some things randomly distributed between patients may be systematic for an individual patient. These considerations might indicate that tighter requirements are necessary on systematic uncertainties, of say better than 1 or 2%, with requirements still of around 3% on random uncertainties, to keep the changes in TCP and NTCP within acceptable tolerances (eg also within around 3-5%). Together these also imply that the overall required accuracy in absorbed dose distributions should be in the region of 3-5%. This may change for high-dose per fraction hypofractionated treatments, which would be significantly less forgiving of random uncertainties. Interestingly, these two sets of figures tie up with the levels of tolerance we typically set for pre-treatment or in-vivo verification of IMRT dose and dose distribution, yet we set these for typically 95% (of points or pixels) agreement, whether in phantom, 2D or 3D, or in vivo [29].

3.2. Geometry

In radiotherapy, geometric uncertainties translate to dosimetric uncertainties. Geometric uncertainties arise for a variety of reasons, including treatment machine specifications and tolerances, simulation and treatment set-up, patient or organ movement during treatment and changes of patient shape between fractions. Clinical information is limited concerning the effects of such variations on treatment outcome. Geometric miss of tumour/target will obviously decrease tumour control probability, whilst overlap of fields with adjacent normal structures, particularly critical organs, will be detrimental in terms of normal tissue complications. In general appropriate margins are defined around the target volume to allow for these uncertainties so it is difficult to find definitive data on the effect of inaccuracies. Conventional approaches to this have been to model the effects of overlap onto organs at risk or reduced coverage of target volume [e.g. 26], or to consider the various sources of uncertainty, combine them to give an overall value and then use the best practically achievable figures as the level to be recommended. On this basis the AAPM [30] arrived at a figure corresponding to 5 mm, one effective sd. Mijnheer et al [19] considered a wider set of data and recommended an accuracy of positioning of field edges and shielding edges of 4 mm, one sd, relative to expected anatomy. Even with this conventional approach, the requirements will be clearly more restrictive for stereotactic radiotherapy approaches, where geometric accuracy of around 1mm is required and again hypofractionated single or small number of fraction treatments are less forgiving of random geometric uncertainties.

However this approach no longer holds so clearly for newer technology and techniques, for at least three reasons; one is that geometric uncertainties now affect dose distribution within the target volume for IMRT and not just at the volume edges or interfaces to organs at risk; also IGRT, adaptive techniques and motion management techniques have provided the facility to reduce the uncertainties significantly as compared to conventional approaches; and lastly the desire to dose-escalate based on these techniques demands greater attention be paid to reducing margins on the boundary between PTV and OaR (PRV concepts), but also taking care not to compromise on TCP. Uncertainties now must also consider imaging system geometric calibration, QA and tolerances, and the performance of overall IGRT systems including image registration and corrective actions [31]. The BIR in a 2003 report [32] provides a useful summary and detailed discussion of geometric uncertainties in general and also for particular treatment sites, whilst a more recent UK multidisciplinary report covers in detail how to ensure geometric accuracy [33]. The evidence points to anatomical variations of the order of a few mm in conventional head and neck treatments to up to a few cm in pelvic and thoracic regions. This gives rise to typical uncertainties for conventional treatments [34], for example, of the order of 1-2, 4 and 4 mm systematic (Σ) for head and neck, pelvic and thoracic treatments respectively and 1-2, 4 and 6 mm for random (σ). Van Herk [34] considers these also for IGRT treatments and suggests uncertainties of the order of 1, 2.5 and 3 mm systematic for head and neck, pelvic and thoracic treatment respectively and <1, 1 and 6 mm for random. Applying his widely-used margin recipe ($m = 2.5\Sigma + 0.7\sigma$), intended to provide coverage of the CTV with the 95% isodose for 90% of patients) he suggests that margins reduce by factors between around 0.55 and 0.65 using IGRT, depending on treatment site and hence on likely motion, but this is likely to be very dependent on local implementation, treatment techniques and correction protocols and requires careful assessment in the local situation. He stressed the danger of over-confidence in the process and of reducing margins too much [35] and the potential loss of coverage of sub-microscopic spread that we probably fortuitously achieved before. In addition, of course, none of this addresses the biggest geometric uncertainty, ie. that of variations in the target volume delineation at the beginning of the process.

3.3. Summary

In summarising the currently available evidence, general recommendations on accuracy in radiotherapy are still the same as previously reviewed and concluded, ie. for +/- 3% (sd) on the absorbed dose delivered at the specification point (but expanded to include a tighter requirement on systematic uncertainty of ideally $\leq$ 1-2%); for +/- 3-5% (sd) on the dose at all other points in the target volume; and for 'a few' mm (sd) on geometric uncertainties. The latter figure for evidence-based geometric precision requirements ranges from sub-mm (sd) for the most critical stereotactic cranial treatments to between 2-4mm (sd) for other treatments, where the latter is dependent on the site and whether IGRT-based methods are being used. For geometric effects on dosimetry within target volumes for IMRT, the recommended tolerences may be considered to be those given for MLC performance in recommendations for IMRT system QA and verification, ie typically 1 mm or less (see section 4.2). Other than the IGRT-linked change, the overall recommendations are still quite closely in line with the previous analysis. All these are given as one sd and are general requirements for routine clinical practice. In certain cases, eg. palliative, higher values may be acceptable. However, it is usually impractical to have different standards of accuracy for different treatment systems, techniques and processes and so the general values must be widely applied. In some special cases smaller uncertainties may be demanded if very steep complication curves are involved or tight geometric tolerances are required, as in stereotactic treatments, and these will require additional resource, implementation, QA and verification. It must be appreciated that these figures are on final delivered dose to the patient at the end of all the contributing stages involved in the radiotherapy process. The dosimetry chain begins with primary dosimetry standards and the required basic physical data. It leads through dosemeter calibration, treatment beam calibration in reference conditions, relative dosimetry in all other conditions and the treatment planning process, including acquisition of patient data, modelling and specification of treatment volume and dose. It ends with the delivery of the prescribed, planned and accepted treatment to the patient under day-to-day conditions over the course of the treatment, as verified beforehand and/or during treatment by further imaging and dosimetry. The accuracy requirement on each part of the whole process must be significantly less than the overall recommendations to achieve the final values recommended. In general individual uncertainties are required to be of the order of 1% or 1 mm or less to achieve this.

4. Review of accuracy achievable in radiotherapy dosimetry

4.1. Estimates of overall dosimetric accuracy achievable from QA, in vivo dosimetry, audits and uncertainty analysis: conventional 3DCRT

Various attempts have been made to analyse the radiotherapy process to obtain cumulative uncertainties on delivered dose [6,7,18,19,22,25,26, 30]. Each major step, eg. absorbed dose to a reference point in water, measurement of relative doses and set up of the treatment planning systems, treatment planning of the individual patient and treatment delivery to the patient, has been broken down into sub-steps and best estimates of uncertainty have been assigned at each level for each contributing factor. The overall estimated cumulative uncertainties obtained have ranged from 2.5-8.5%, as one effective standard deviation (sd) [3]. This range in part reflects the judgements used by different authors, the number of separate sub-steps and parameters included and whether an optimal or a conservative approach was taken. A figure of 5% (sd) might be representative of these types of estimates, with smaller uncertainties for simpler treatments and larger for more complex.

More recently, uncertainty estimates based on harder evidence from intercomparisons, audits and *in vivo* dosimetry have been made from the author's own experience and programs [e.g. 2,9,11]. Table 1 reproduces a slightly modified summary [2] of these evaluations for external beam megavoltage x-ray treatments, following UK procedures and dosimetry protocols and utilising consistent results from

- the UK dosimetry intercomparisons and audits [36-37]
- the Scottish+ audit group of the UK radiotherapy dosimetry audit network [38], which includes intercomparisons in a number of phantoms simulating treatment sites and also some inter-centre *in vivo* intercomparisons
- analysis of QC records and specific experimental evaluation of uncertainties at many levels in the radiotherapy process in one department with a detailed quality system in place [9]
- systematic *in vivo* (diode) dosimetry in one department over more than 10 years [39,40].

The uncertainties are all given as effective sd values; therefore 95% confidence levels can be taken to be approximately twice these figures. It may be noted that the final cumulative patient values for a single centre are not very different to those for multi-centre intercomparisons. Line 1.2 , multi-centre, based on ion chamber intercomparisons around the UK, can be compared to TLD intercomparisons , eg as reported by the Radiological Physics Center (RPC) largely for N American centres, where the distribution of output calibration audit results for megavoltage photon beams has a standard deviation of 1.7% [41-42], which of course includes some contribution from the relative uncertainties of the TLD measurements themselves. Alternatively, the IAEA/WHO TLD programme has verified calibrations of beams in a wide range of centres, largely in hospitals in low and middle income countries. Overall the standard deviation of the whole set of results is 6.3%, but these are weighted by some rather large deviations. The percentage of results within the 5% acceptance limit is reported to be 96% after corrective actions. In general participants using consistent modern dosimetry protocols (eg TRS-398 [43] or similar) showed sd of less than 3% [44-45]. Thwaites and Williams [46] and ESTRO booklet 9 [47] summarise a range of dose intercomparisons for megavoltage photon beams in reference conditions and indicate generally that means are close to the expected value and sd are generally less than around 2% in developed countries following modern protocols. The initial results from the recently established Australian Clinical Dosimetry Service (ACDS) are in line with this, with the sd of the distribution of mailed dosimeter audit results being around 2% and ion chamber audit results around half that [48]. The two publications mentioned that summarise audits [46-47] also do so for audits in anthropomorphic or semi-anthropomorphic phantoms and show sd in the range of 1.3-3.5%, which can be compared to line 2 multi-centre uncertainty values.

Table 1: Optimal uncertainties (1 effective sd) in radiotherapy dosimetry based on experimental determinations. Beam calibration methods follow the UK system and dosimetry protocol (based on direct absorbed dose to water calibration factors for the whole beam quality range at the standards laboratory). It is assumed that uncertainties can be combined or separated in quadrature.

Source/step	*Single centre* **(different beams/times)**	*Multi-centre*
1. Dose at reference point in water phantom		
1.1. uncertainties quoted on calibration factors by the UK standards lab (NPL) are 0.7% (1 effective sd)	0.7%	0.7 %
1.2. variation in reference dose determination between beams and through time	0.5%	0.7-1%
1.3. combined	0.9%	1.0- 1.2%
2. Dose to phantoms representing various treatment sites (at a range of points within target volumes; given relative to reference dose)	0.8-1.8%	1.1-2.3%
3. Patient dose at specification point (based on estimates from *in vivo* dosimetry; for a wide range of treatment sites and techniques, given relative to reference dose)	1.5-3%	1.6 – 3.2 %
*+ where lung is significantly involved**	*(5%)**	*(5.1%)**
4. Estimated overall cumulative uncertainty on delivered patient dose at the specification point, including standards lab uncertainty	**1.7 – 3.1%**	**1.9 – 3.4%**
(+ where lung is significantly involved*	(5.1%)*	(5.2%)*

- see text for discussion of lung values

The overall cumulative uncertainties (Table 1) at the specification point are within, or close to, the recommended required values of 3% (1 sd), ie the clinical evidence-based requirements can be met on the experimental evidence available. The figures given are likely to be representative of fairly optimal situations in normal clinical practice. However overall uncertainties will be larger if any steps or sub-steps have larger uncertainties, whereas if special attention is given to minimising uncertainties for some special situations or treatments, values may be less. The only exception to the general values are those where there is significant involvement of lung. Here the *in vivo* measurements indicate that uncertainties increase, due both to motion and to the ability of planning systems to cope with such situations. This will depend on many factors, including the exact site, the beam directions, the planning algorithms, the lung correction approach, the amount of lung involved etc. However the lung-involved estimates were when advanced algorithms were not in such wide-spread use. It is likely now that these estimates would be reduced (better accuracy) if those measurements were re-done for clinical practice using advanced algorithms. There are a wide range of papers investigating the accuracy of older pencil-beam based models with simpler inhomogeneity corrections as compared to more modern convolution/superposition models using kernels to account for both photon and electron transport, or recently Monte Carlo-based approaches [49-51]. Generally the older algorithms can show deviations in situations involving significant lung tissue of up to 20%, whereas more modern

Journal of Physics: Conference Series **444** (2013) 012006 doi:10.1088/1742-6596/444/1/012006

algorithms show deviations of a few % maximum [49-51]. Thus this alone is likely to reduce the overall lung-involved uncertainty values to be much closer to those for other treatment sites.

The overall cumulative uncertainties were estimated then for patient delivered dose at the specification point. For other points within the target volume, additional uncertainties are present, which can be estimated by folding in uncertainties from multi-point measurements in phantom or from 2-D (or 3D) verification of fields and treatments (here for example, experience is rapidly growing from the development and implementation of IMRT verification methods). Optimal uncertainties in these points are likely to be similar to those at level 2 in the table, relative to the specification point, giving estimated cumulative uncertainties (1 sd) on the dose at other points in the target volume in the range of approximately 2 - 4.5%.

It may be noted that other realistic phantom intercomparisons have shown uncertainties which are in general agreement with Table 1, e.g. the START trial audit [52] had sd of 1.2 and 1.3 % in the results of multi-centre measurements at the specification point for 2D and 3D breast plans respectively. The RTO1 trial audit [53] showed a sd of 1.3% for multi-centre measurements at the specification point in a prostate phantom, comparing to locally planned values and an sd of 2% for measurements within the CTV and of 4% for all measurements, including some outside the target volumes.

4.2 *Estimates of overall dosimetric accuracy achievable for IMRT and more complex treatments*

There have been a range of recent publications and recommendations recognising the need for tighter tolerances and accuracy for IMRT, eg AAPM Task Group 106 [54] on beam data commissioning , Low et al [55] on dosimetry for IMRT, IPEM report 96's guidance for clinical implementation [56] and ESTRO booklet 9 [47] on verification. The presentations at the AAPM 2011 summer school on uncertainties in external beam radiation therapy [57] provide useful comprehensive data, eg Moran and Ritter re treatment delivery systems [58]. In general the recommendations are around 1mm, 1° and 1% for specific linear, angular and dosimetric parameters, but others [eg. 59] recommend tighter tolerances for IMRT for accurate dosimetry, eg on MLC leaf positioning. In addition, small field situations, which are more or less similar for small sub-fields in IMRT or VMAT, also have special considerations [60-61]. Comprehensive requirements for TPSs are discussed in a number of publications, notably IAEA [50], where performance is heavily dependent on the algorithm employed as well as imaging system/s accuracy as input and the range of other relevant parameters that can affect dose calculation accuracy, eg slice thickness, CT number accuracy, dose calculation grid, etc. It may be noted that ICRU 83 [62] recommends generic values of 3.5% for dose agreement between planned and measured (delivered) doses in low dose gradient areas and 3.5 mm for DTA in high dose gradient areas and that they change the prescription and reporting requirements for more complex 3D treatments to be dose-volume based, rather than point based.. There is also a growing set of literature around robust planning, able to cope with changes and uncertainties and also uncertainty-included planning and analysis, eg Jin *et al's.* [63-64] suggested dose uncertainty model for IMRT delivery. Their analysis for a series of prostate cases with varying inter- and intra-fractional motion indicated that consequent dose uncertainties for 95% of the CTV volume were from 1.3% to 2.9% and that prostate IMRT plans meeting similar plan objectives could have different dose uncertainties.

Therefore, it is unclear whether the increased complexity of IMRT will increase uncertainties, or the tighter tolerances on relevant parts of the process and greater effort put into QA and verification will decrease uncertainties and this may be a complex interplay of effects that might have different answers in different situations. Thus this short section considers a selection of the publications on QA, audit and in vivo dosimetry for IMRT to see whether the overall estimates of accuracy achievable may change when this is taken into account.

4.2.1. Phantom and pre-treatment verification measurements for IMRT

There is a wide range of recommendations on phantom measurements for departmental IMRT QA and verification processes [47, 54-57] and a large number of individual IMRT patients have had measurements made providing a wide body of information. These have generally used point dose measurements, often with 2-3% tolerance, to assess absolute dose delivery accuracy and also used 2D systems for verification of dose/fluence distributions [65-66]. In general, treatments are re-calculated onto a phantom , the treatment plan delivered and gamma index evaluation [67-68] carried out on a comparison between the expected dose distribution to be delivered (planning) and the measured actually-delivered dose distribution. The evaluation usually applies tolerances at the 3%/3mm level, for dose/distance-to-agreement, (usually in the range of 2%/2mm to 4%/4mm) and generally looks for 95-98% or so of points/pixels to lie within these tolerances, generally for the region lying at doses above some isodose level threshold, often the 20% isodose. These techniques have been extended into pseudo-3D tests of combined fields using appropriate 2D dosimeters in a phantom, or using one of the 3D systems or semi-3D systems [65-66, 69] now available, and delivering the whole treatment. A lot of effort has gone into this worldwide and has produced some feeling of confidence in local processes and procedures. However it is unclear exactly what statement these tolerances make on levels of accuracy achieved and there are papers which question the validity in particular of using the 2D verification approaches and whether in fact they do achieve what they intend [66,70]. Nevertheless if simplistically taken at face value, these are in-phantom measurements of the accuracy of delivery of IMRT and can be taken in the extreme to indicate that delivery is expected to be, and demonstrated to be, generally within 2-3% of absolute dose at a representative point and to within 3% and 3mm across the dose distribution, implying sd of 1-1.5% on dose to the representative point and at other dose points relative to this point and an overall sd in the range of 1.5-1.8%. This is well in line with the data given for single institutions in line 2 of Table 1. However, in fact it is not as simple as that and the results are highly dependent on differences in system, measurement resolution, analysis calculation and parameters, threshold, normalisation (global or local), tolerances set (in percentage pass rate, as well as dose or DTA values), actions taken, etc. [66-68,69]. In general it is too simplistic to reduce to a single gamma value and there should be at least an appreciation of where the differences lie and hence what their clinical significance is. Typically the 3 mm criterion on DTA may well imply that geometric uncertainties are within 1-1.8 mm, one sd (for pass rates of 100% to 90%), but linking overall dose uncertainties directly to the dose tolerance figure is a gross over-simplification and results in too low values of sd. This can be illustrated by considering the pass/fail rates if pushing the tolerances, eg 3%/3mm,3%/2mm, 3%/1mm, etc.; and/or by considering the dose and DTA information separately before combination. For example, by examining a range of 'passing' IMRT and VMAT deliveries using the Delta-4 3D verification system, it has been observed by the author that 97-100 % gamma index pass rates using tolerances of 3%/3mm have shown dose sd of 1.5% to 2.3% and geometric sd of 1-1.3 mm. Similarly 95% pass rates have shown 1.5-2.5% sd on dose and 1-1.5 mm sd on geometry, whilst 90% pass rates have shown 2-3% sd on dose and up to 1.9mm sd on geometry. It may be noted that these are using a 20% threshold and global normalisation.

A general warning must also be given that these numbers might not be directly exactly applicable to other systems and usage, but they do at least give an indication of the involved uncertainties and their ranges that may be linked to often applied tolerance levels and pass/fail values. Typically 95-100% pass rates are expected/observed using 3%/3mm, implying 1-1.5% uncertainty at a representative dose specification point and uncertainties at other points in the range of 1.8-2.9%. These can be compared to Table 1 (lines 1.3 and 2 combined) of 1.2-2.0 %. This analysis also raises questions of how robust are the tolerances used in gamma analysis at picking up the levels of errors expected [66-68, 70] and whether the two numerical tolerance values for dose and DTA should be the same; in fact it implies that the DTA value should be numerically less than the dose value (eg 3%/2mm)

4.2.2 In vivo dosimetry for IMRT

There are a range of approaches to in vivo dosimetry for IMRT, including modification of conventional 3DCRT methods using point dose measurements coupled with other information on the distribution delivery, 2D transmission systems (eg DAVID), 2D field-by-field transit dosimetry using EPIDs and with back-projection to a representative patient plane [29, 69, 71], and full 3D dose reconstruction for all fields using EPIDs [29, 72]. Here the latter will be considered, as the full 3D equivalent of isocentre dose reconstruction from entrance and exit dose measurements employed to obtain the uncertainty values in line 3 of Table 2. The NKI [29, 72] approach is one of the longest running systems of this type. For this, Mans et al have reported experience on over 4000 patients [29]. The report is primarily about the efficacy of the system at identifying gross errors (17/4333 patients verified over four years, ie 0.4%). However the tolerances for isocentre dose and for numbers of points within the 3%/3mm distribution gamma index requirements, coupled with the referenced preceding papers leading up to this, can also be used to indicate sd of dosimetric parameters, using a similar simplistic and extremes argument as employed above for results for delivered dose distributions in phantom. In addition some of these papers give specific dosimetric distribution sd data and there are also other similar sets of results and data from similar long-running systems, eg that of MAASTRO [73]. Overall this implies that the sd of differences between expected dose and delivered dose to the isocentre is in the range 1-1.6% from these studies and the overall sd for other points in the in vivo delivered dose distribution verifications are between around 2.5% and 3.3%, however now specifically for delivery of IMRT to patients. This can be compared with the values from Table 1 (lines 1.3 and 3 combined) of 1.7-3.2%, again reasonably well in line with the tabulated 3DCRT data. NKI has more recently extended this to VMAT verification with similar results, eg the mean difference between measured and planned isocentre dose is 0.2%, sd = 1.6% and with a spread of values from -3% to +5% [72]. Of course, such 3D in vivo dosimetry provides as near as practically possible a total end-to-end test of the overall process.

4.2.3. IMRT Audits and intercomparisons

A number of IMRT-related dose delivery audits have been carried out in Europe, some of which are summarised in ESTRO Booklet 9 [47], including the ESTRO-QUASIMODO audit simulating a prostate PTV and OaR optimized treatment [74]. These all show mean dose differences between delivered and expected doses to be close to 0 and sd to be between 1.5 and 3.9%. More recently, the UK IPEM QA group has carried out a relatively simple mailed audit of IMRT fields [75], using alanine for absolute dose and film for dose distribution, involving 57/62 of the centres in the country and evaluating distributions against 2%/2mm to 4%/4mm criteria. This showed point doses to have sd for phantom delivery between 1.5 and 2.5% and effective sd of distributions of 1-1.5%, ie overall cumulative uncertainties of between 1.8 and 3.4%. A more ambitious audit of rotational IMRT techniques has been piloted in 10 centres by the national Clinical Trials QA Group [76], following on from the successful PARSPORT clinical trial and QA audits [77], also showing results in line with the other IMRT audits. The pilot rotational audit demonstrated sd of 2% for points in the PTVand 2.9% when all points measured are taken into account.

The widest set of IMRT audits has been those conducted by the RPC to support advanced technology clinical trials, beginning in 2001. The RPC has six different phantoms representing different areas of the body and designed for end-to-end test audits of IMRT treatments of the head and neck, prostate and lung and stereotactic treatments to the brain, lung, spine, and liver [78-79]. They contain simulated targets and OaRs and can take point dosimeters and radiochromic film for distribution measurements in appropriate planes. RPC-measured doses are evaluated against centre-expected doses, using rather wide tolerances of +/- 7% for absolute PTV doses and +/- 4mm in higher dose gradient areas, eg between PTV and OaR, for head and neck, prostate and liver, and rather tighter tolerances for spine and lung audits (5%, 3-5 mm). For the head and neck audits, 70% of centres met the criteria on the first round, rising to around 80% on repeats. For the other audits the pass rates reported were in the range 50-85%, but some of these were for small numbers of centres. The largest group is for head and neck where over 750 institutions were audited. The distribution of differences between measured and expected doses to the PTVs showed means within 1-2 percent of zero and sd of around 5%. There was some evidence that smaller departments with poorer staff resources did worse. Summarising, Ibbott *et al.* [80] indicate that the studies suggest that the uncertainty in dose delivered to an anthropomorphic phantom with IMRT techniques is approximately 5% (sd).

Overall the results and indicated uncertainties from the European IMRT audits are in line with those for the multi-institution audits for 3DCRT, albeit showing a tendency to shift to rather higher sd. For the US studies, the uncertainties are significantly greater; this may partly reflect the stringency of the tests, but the tendency to correlate with the levels of staff resources in a given centre may say something about variable levels of QA in implementation of IMRT or in carrying out the audit irradiations.

4.3 Summary

Taking all this evidence into account, much of the information for IMRT as it pertains to uncertainties of dose delivery at the various levels to phantoms and patients is very similar to the evidence from 3DCRT, taking like-for-like situations. There is some evidence that the size of uncertainties achievable may shift upwards in more complex situations, but the range observed implies that similar uncertainties can be achieved. The widespread audits and in vivo reports on 3DCRT indicate that the levels of uncertainties reported here are representative of what it should be possible to generally achieve and the growing evidence from IMRT studies and the growing experience and expertise indicate that almost exactly the same levels of uncertainties ought to be achievable for IMRT.

5. Conclusion and observations

In this review of the accuracy required and achievable in radiotherapy dosimetry, the older approaches and evidence-based estimates for 3DCRT have been reprised, summarising and drawing together the author's earlier evaluations and repeating parts, where still relevant. Clinically-based accuracy requirements have not significantly changed, although a greater emphasis on high-precision delivery methods, including stereotactic radiotherapy, and on IMRT has focussed attention on reducing geometric uncertainties. At the same time the growing use of high-dose-per-fraction hypofractionated treatments, with steeper effective dose-response curves [81] may imply stricter dose and geometry requirements in these situations.

Available evidence for IMRT uncertainties has been reviewed, selecting information from tolerances, QA, verification measurements, in vivo dosimetry and dose delivery audits, to consider whether achievable uncertainties increase or decrease. Overall there is some evidence that they tend to increase, but that similar levels should be achievable (as evidenced by the lower uncertainty values in any of the spreads of values presented). Thus it is concluded that those earlier estimates of achievable dosimetric accuracy are still applicable, despite the advances in technology and techniques. The one exception is where significant amounts of lung are involved, where it is likely that uncertainties have now improved due to widespread implementation of more accurate heterogeneity modelling algorithms and are now likely to be more in line with the values presented for other treatment sites. The widespread availability of integrated IGRT systems has encouraged attention to geometric uncertainties and hence had a positive effect on what is achievable, which may further improve as motion management methods roll out more widely; all the evidence points to delivery to phantoms with sd of 1-1.5mm or so and delivery to patients in line with the figures in section 3.3, ie with sd of 'a few mm', depending on site and technique.

The accuracies and uncertainties presented here should be achievable, but require optimal approaches throughout, including comprehensive quality systems; attention to detail; safety, quality and accuracy cultures in radiotherapy departments [16]; and continuing vigilance. Practical, accurate, precise dosimeters and dosimetry systems are required to keep pace with the evolving complexity of technology and radiotherapy methods, for IMRT, small fields, 4D applications, etc. [82]. Increasingly, biological imaging and biologically-based adaptive radiotherapy (BART) [83] will support individualised personal medicine in radiation oncology, where knowledge of individual radiosensitivity will support individualised prescription and dose escalation. This will tend to stratify dose-response curves, leading to increased steepness of the relevant curves for specific patient sub-groups and hence fuelling the demands for even greater accuracy. Low dose effects are becoming more important requiring more accurate measurement and modelling of those. Dose escalation will require better quantitatively-supported decision criteria and the basis for trade-off, considerations, with uncertainty modelling included. Adaptive and 4D radiotherapy will require better more robust methods, not only of 3D and 4D dosimetry and deformable image registration, but also of deformable dose accumulation [84]. These techniques would be expected to improve dosimetric uncertainties in situations where changes and intra-fraction motion respectively are significant. Dose distribution evaluation criteria must evolve and be robust themselves; noting too that the measurement uncertainties are folded into the comparison of measurement versus expected doses, thereby affecting the tolerances themselves [68]. Similar considerations apply to clinical trial dosimetry [85-86], noting that greater uncertainties in dosimetry can detrimentally affect the outcome and conclusions of a trial, or mean that a greater number of patients are needed to give it the required statistical power.

Some final issues: are our QA tools fast enough, efficient enough, and integrated enough into delivery systems and processes to cost-effectively cope with the expansion of complexity of technology and techniques? It will become the norm for EPIDs to be used as dosimeters as well as imagers, maybe in combination with other integrated dosimetry systems, to routinely manage verification on-line in real time and hence time-and-cost-efficiently. Further effort is required by the manufacturers to provide usable practical tools to support this. The available informatics will also need development to further efficiently support increases in complexity.. Maintaining and improving dosimetric and geometric accuracy is the key to gaining improvements from advancing technology and techniques; outcomes may well be better from high-quality simpler techniques than poorly-controlled poor-accuracy advanced techniques; the optimum is to implement the newer techniques in a high quality, safety and accuracy environment to achieve both high precision and high accuracy for all patients. The more complex the treatments the greater the potential for problems, of course, so radiotherapy must be conducted within a consistent and sustainable quality framework. There is clear evidence that poor quality radiotherapy produces poorer outcomes [87] and lastly, both generally and as shown by that study [87], the community needs to address the inconsistencies in target volume delineation as one of the major remaining grey uncertainty areas in the radiotherapy process.

6. References

[1] Inernational Commission on Radiation Units and Measurements 1976 Determination of absorbed dose in a patient irradiated by beams of x or gamma rays in radiotherapy procedures *International Commission on Radiation Units and Measurements* ICRU Report 24 Bethesda, Maryland

[2] Mayles P, Thwaites DI (with Rosenwald J-C) 2007 Rationale and management of the quality system Chapter 37 *Handbook of Radiotherapy Physics: Theory and Practice* (eds. Mayles P, Nahum A and Rosenwald J-C) Taylor and Francis: London ISBN-13:978-0-7503-0860-1, 2007

[3] van Dyk J (Ed) 1999 The Modern Technology of Radiation Oncology: A Compendium for Medical Physicists and Radiation Oncologists *Medical Physics Publishing* Madison WI

[4] Thwaites DI and Malicki J 2011 Physics and technology in ESTRO and in Radiotherapy and Oncology: past, present and into the 4th dimension *Radiother.Oncol.* **100** 327-332

[5] Nystrom H and Thwaites D 2008 Physics and high-technology advances in radiotherapy: are they still worth it? *Radiother. Oncol.* **86** 1-3

[6] Thwaites DI 1989 Review and analysis of accuracy required and achievable in radiotherapy Accuracy in Radiotherapy *Phys. Med. Biol.* **34** 639

[7] Thwaites DI 1993 Accuracy requirements in radiotherapy Chapter 11, *A Practical Course in Radiotherapy Physics* Vol 2 (ed: Mayles. P, and Nahum, A E), The Institute of Cancer Research\The Royal Marsden Hospital, London

[8] Thwaites DI 1994 Uncertainties at the end point of the basic dosimetry chain *Measurement Assurance in Dosimetry* (ed: Svensson,H,) IAEA-SM-330 (IAEA, Vienna) 239-255

[9] Allahverdi M and Thwaites DI 1998 Achievable accuracy in radiotherapy dosimetry ESTRO 17 (Edinburgh, Sept 1998) *Radioth. Oncol.* **48** (suppl 1) S187

[10] Dobbs J and Thwaites DI 1998 Quality assurance and its conceptual framework Chapter 1 of *Physics Aspects of Quality Control in Radiotherapy* (eds: Mayles, P, et al) IPEM Report **81** (IPEM,York)

[11] Thwaites DI 2003 Verification and external audit of dose for conventional radiotherapy and the new technology: how well can we deliver radiotherapy doses? Current achievements in oncology Rep *Pract. Oncol. Radiother.* **8** (2) 225

[12] Thwaites DI 2006 Clinical dosimetry and QA for radiotherapy: developments in methods, applications and accuracy *Proc. Int. Conf. on quality assurance and new techniques in radiation medicine* IAEA: Vienna IAEA-CN-146 482-483

[13] Kutcher GJ, Coia L, Gillin M, Hanson WF, Leibel S, Morton RJ, Palta JR, Purdy JA, Reinstien LE and Svensson GK 1994 Comprehensive QA for radiation oncology: report of AAPM Radiation Therapy Committee Task Group 40 *Med. Phys.* **21** 581-618

[14] Thwaites DI, Scalliet P, Leer JWH and Overgaard J 1995 Quality assurance in radiotherapy (ESTRO report to the Commission of the European Union for the 'Europe against Cancer' programme) *Radioth. Oncol.* **35** 61-74

[15] Leer JWH, McKenzie AL, Scalliet P and Thwaites DI 1998 Practical Guidelines for the Implementation of a Quality System in Radiotherapy ESTRO Series *Clinical Physics for Radiotherapy* (ESTRO- Garant, Leuven)

[16] The Royal College of Radiologists, Society and College of Radiographers, Institute of Physics and Engineering in Medicine, National Patient Safety Agency, British Institute of Radiology 2008 Towards Safer Radiotherapy *The Royal College of Radiologists* London

[17] International Atomic Energy Agency 2000 Lessons Learned from Accidental Exposures in Radiotherapy *IAEA* Vienna

[18] International Commission on Radiological Protection (ICRP) 2000 Prevention of Accidental Exposures to Patients undergoing Radiation Therapy *ICRP* Publication **86** Pergamon

[19] Mijnheer BJ, Batterman JJ and Wambersie A 1987 What degree of accuracy is required and can be achieved in photon and neutron therapy? *Radiother.Oncol.* **8** 237-252

[20] International Organisation for Standardisation 1995 Guide to the Expression of Uncertainty in Measurement *International Organization for Standardization* Geneva

[21] Comite International des Poids et Mesures 2008 Evaluation of measurement data Guide to the expression of uncertainty in measurement *BIPM* Paris

[22] Dutreix A 1984 When and how can we improve precision in radiotherapy? *Radiother.Oncol.* **2** 275-292

[23] van Herk M 2004 Errors and margins in radiotherapy *Semin. Radiat. Oncol.* **14** 52-64

[24] International Commission on Radiation Units and Measurements 2006 *ICRU* Report 76 Measurement Quality Assurance for Ionizing Radiation Dosimetry Bethesda, Maryland, International Commission on Radiation Units and Measurements

[25] Brahme A 1984 Dosimetric precision requirements in radiation therapy *Acta. Radiol.Oncol.* **23** 379-391

[26] Brahme A, Ed. 1988 Accuracy requirements and quality assurance of external beam therapy with photons and electrons *Acta. Oncol. (*suppl 1) 5-76

[27] Bentzen SM 2005 Steepness of the radiation dose-response curve for dose-per-fraction escalation keeping the number of fractions fixed *Acta. Oncol.* **44** 825-828

[28] Bentzen SM, Constine LS, Deasy JO, Eisbruch A, Jackson A, Marks LB, Ten Haken RK and Yorke ED 2010 Quantitative Analyses of Normal Tissue Effects in the Clinic (QUANTEC): an introduction to the scientific issues *Int. J. Radiat. Oncol. Biol. Phys.* **76** S3-S9

[29] Mans A, Wedling M, McDermott LN, Sonke JJ, Tielenburg R, Vijlbrief R, Mijnheer B, van Herk M, Stroom JC 2010 Catching errors with in vivo EPID dosimetry *Med. Phys.* **37** 2638-2644

[30] American Association of Physicists in Medicine 1984 Physical aspects of quality assurance in radiation therapy *AAPM TG* Report **24**

[31] Korreman S, Rasch C, McNair H, Verellen D, Oelfke U, Maingon P, Mijnheer B and Khoo V 2010 The European Society of Therapeutic Radiology and Oncology-European Institute of Radiotherapy (ESTRO-EIR) report on 3D CT-based in-room image guidance systems: a practical and technical review and guide *Radiother.Oncol.* **94** 129-144

[32] British Institute of Radiology 2003 Geometric uncertainties in radiotherapy *BIR* London

[33] The Royal College of Radiologists, Society and College of Radiographers, Institute of Physics and Engineering in Medicine, National Patient Safety Agency, British Institute of Radiology 2008 On target: ensuring geometric accuracy in radiotherapy *The Royal College of Radiologists* London

[34] van Herk M 2011 Margins and Margin Recipes Chapter 8 (Palta J.R., Mackie T.R. Eds) Uncertainties in External Beam Radiation Therapy *Medical Physics Publishing* Madison WI

[35] Engels B, Soete G, Verellen D and Storme G 2009 Conformal arc radiotherapy for prostate cancer: increased biochemical failure in patients with distended rectum on the planning computed tomogram despite image guidance by implanted marker *Int. J. Radiat. Oncol. Biol. Phys.* **74** 388-91

[36] Thwaites DI, Williams JR, Aird EG, Klevenhagen SC, and Williams PC 1992 A dosimetric intercomparison of megavoltage photon beams in UK radiotherapy centres *Phys. Med. Biol.* **37** 445-461

[37] Nisbet A, Thwaites DI and Sheridan ME 1998 A dosimetric intercomparison of kilovoltage X-rays, megavoltage photons and electrons in the Republic of Ireland *Radiother. Oncol.* **48** 95-101

[38] Thwaites, DI 1996 External audit in radiotherapy dosimetry *Radiation Incidents* eds Faulkner K and Harrison RM (BIR, London 21-28; and, Thwaites, DI, Powley, S, Nisbet, A and Allahverdi, M, The United Kingdom's radiotherapy dosimetry audit network,in Standards and Codes of Practice for Medical Radiation Dosimetry, IAEA:Vienna (2003), IAEA-STI-PUB-1153, vol 2, 183-190

[39] Thwaites DI, Blyth C, Carruthers L, Elliott PA, Kidane G, McLeod A, Paolucci M and Stacey C 2003 Experience with in vivo diode dosimetry for verifying radiotherapy dose delivery: Practical implementation of cost-effective approaches, Standards and Codes of Practice for Medical Radiation Dosimetry *IAEA* Vienna IAEA-STI-PUB-1153 **2** 415-423

[40] Huyskens DP, Bogaert R, Verstraete J, Loof M, Nystrom H, Fiorino C, Broggi S, Jornet N, Ribas M, and Thwaites DI 2001 Practical Guidelines for the Implementation of in vivo Dosimetry with Diodes in External Radiotherapy with Photon Beams (Entrance Dose) ESTRO Booklet 5 *ESTRO* Brussels Belgium

[41] Aguirre JF, Alvarez P, Ibbott GS and Followill MD 2011 Testing, commissioning and validating an optically stimulated luminescence (OSL) dosimetry system for mailed dosimetry at the Radiological Physics Center *International Symposium on Standards, Applications and Quality Assurance in Medical Radiation Dosimetry (IDOS)* IAEA STI/PUB/1514 ISBN 978–92–0–116210–6 (IAEA: Vienna) **2** 411-421

[42] Aguirre JF, Tailor R, Ibbott G, Stovall M and Hanson W 2002 Thermoluminescence dosimetry as a tool for the remote verification of output for radiotherapy beams: 25 years of experience *Proceedings of the International Symposium on Standards and Codes of Practice in Medical*

Radiation Dosimetry IAEA-CN-96/82, International Atomic Energy Agency (IAEA) Vienna, Austria

[43] International Atomic Energy Agency 2000 Absorbed Dose Determination in External Beam Radiotherapy: An International Code of Practice for Dosimetry Based on Standards of Absorbed Dose to Water IAE*A* TRS-398 *International Atomic Energy Agency* Vienna

[44] Izewska J, Andreo P, Vatnitsky S, Shortt KR 2003 The IAEA/WHO TLD postal dose quality audits for radiotherapy: a perspective of dosimetry practices at hospitals in developing countries *Radiother. Oncol.* **69** 91-97

[45] Izewska J and Thwaites D 2002 IAEA Supported National Thermoluminescence Dosimetry Audit Networks for Radiotherapy Dosimetry *Proceedings of the International Symposium on Standards and Codes of Practice in Medical Radiation Dosimetry* IAEA-CN-96/137 International Atomic Energy Agency (IAEA) Vienna, Austria

[46] Thwaites DI and Williams JR 1994 Radiotherapy dosimetry intercomparisons *in Radiation Dose in Radiotherapy from Prescription to Delivery* (ed: Svensson H and Dutreix A) IAEA TECDOC 734 (IAEA, Vienna) 131-141

[47] Mijnheer B, Georg D (Eds.) ESTRO Booklet No. 9: Guidelines for the Verification of IMRT *ESTRO* Brussels, Belgium

[48] Williams I personal communication

[49] Morgan A, Knoos T, McNee S, Evans C and Thwaites DI Clinical implications of the implementation of advanced treatment planning algorithms for thoracic treatment *Radiother. Oncol.* **86** 48-54

[50] International Atomic Energy Agency 2004 Commissioning and quality assurance of computerized planning systems for radiation treatment of cancer IAEA TRS-430 *International Atomic Energy Agency*, Vienna

[51] Gershkevitsh E, Schmidt R, Velez G, Miller D, Korf E, Yip F, Wanwilairat S and Vatnitsky YS 2008 Dosimetric verification of radiotherapy treatment planning systems: results of IAEA pilot study *Radiother. Oncol.* **89** 338-346

[52] Venables K, Winfield EA, Aird EG and Hoskin PJ 2003 Three-dimensional distribution of radiation within the breast: an intercomparison of departments participating in the START trial of breast radiotherapy fractionation *Int. J. Radiat. Oncol. Biol. Phys.* **55** 271-279

[53] Moore R, Stanley S, Bidmead AM, Dearnaely DP, Mayles WPM and Warrington AP 2003 Quality assurance of dosimetry in UK MRC RTO1 trial of dose escalation using conformal radiotherapy in prostate cancer *Clin. Ocol.* **15** 58-59

[54] Das IJ, Cheng CW, Wats RJ, Ahnesjo A, Gibbons J, Li XA, Lowenstien J, Mitra RK, Simon WE and Zhu TC 2008 Accelerator beam data commissioning equipment and procedures: report of the TG-106 of the Therapy Physics Committee of the AAPM *Med. Phys.* **35** 4186-4215

[55] Low DA, Moran JM, Dempsey JF, Dong L and Oldham M 2011 Dosimetry tools and techniques for IMRT *Med. Phys.* **38** 1313-1338

[56] Institute of Physics and Enineering in Medicine 2008 Guidance for the clinical implementation of intensity modulated radiation therapy *IPEM* report 96 IPEM, York

[57] Palta JR and Mackie TR (Eds) 2011 Uncertainties in External Beam Radiation Therapy *Medical Physics Publishing* Madison WI

[58] Moran JM and Riter T 2011 Limits of precision and accuracy of radiation delivery systems Uncertainties in External Beam Radiation Therapy Chapter 10 (Palta JR and Mackie TR (Eds)) *Medical Physics Publishing* Madison WI

[59] Li JS, Lin T, Chen L, Price RA and Ma CM 2010 Uncertainties in IMRT dosimetry *Med. Phys.* **37** 2491-2500

[60] Alfonso R, Andreo P, Capote R, Huq MS, Kilby W, Kjall P, Mackie TR, Palmans H, Rosser K, Seuntjens J, Ullrich W and Vatnitsky S 2008 A new formalism for reference dosimetry of small and nonstandard fields *Med. Phys.* **35** 5179-5186

[61] Aspradakis MM, Byrne JP, Palmans H, Rosser K and Warrington AP 2010 Small Field MV Photon Dosimetry IPEM Report Number 103 *Institute of Physics and Engineering in Medicine* York UK
[62] International Commission on Radiation Units and Measurements 2010 ICRU Report 83: Prescribing, Recording, and Reporting Photon-Beam Intensity-Modulated Radiation Therapy (IMRT) Bethesda, Maryland *International Commission On Radiation Units and Measurements*
[63] Jin H, Palta JR, Kim YH and Kim S 2010 Application of a novel dose-uncertainty model for dose-uncertainty analysis in prostate intensity-modulated radiotherapy *Int. J. Radiat. Oncol. Biol. Phys.* **78** 920-928
[64] Jin H, Palta J, Suh HTS and Kim S 2008 A generalized a priori dose uncertainty model of IMRT delivery *Med. Phys.* **35** 982-996
[65] Korreman S 2013 Semi and virtual 3D dosimetry in clinical practice *7th Int. Conf. 3D Radiation Dosimetry* this issue
[66] Feygelman V and Nelms B 2013 Dosimetric quality assurance of highly conformal external beam treatments: from 2D phantom comparisons to 4D patient dose reconstructions *7th Int. Conf. 3D Radiation Dosimetry* this issue
[67] Low DA, Harms WB, Mutic S and Purdy JA 1998 A technique for the quantitative evaluation of dose distributions *Med. Phys.* **25** 656-661
[68] Schreiner LJ, Holmes O and Salomans G 2013 Analysis and evaluation of planned and delivered dose distributions: practical concerns with gamma and chi evaluations *7th Int. Conf. 3D Radiation Dosimetry* this issue
[69] McCurdy B 2013 Dosimetry in radiotherapy using a-Si EPIDs: systems, methods and applications, focussing on 3D patient dose estimation *7th Int. Conf. 3D Radiation Dosimetry* this issue
[70] Nelms BE, Zhen H and Tome WA 2011 Per-beam, planar IMRT QA passing rates do not predict clinically relevant patient dose errors *Med. Phys.* **38** 1037-44
[71] Greer P 2013 3D EPID-based dosimetry for pre-treatment verification of VMAT: methods and challenges *7th Int. Conf. 3D Radiation Dosimetry* this issue
[72] Mijnheer B, Olaciregui-Ruiz I, Rozendaal R, Sonke J-J, Spreeuw H, Tielenburg R, van Herk M, Vijlbrief R and Mans A 2013 3D EPID-based in vivo dosimetry for IMRT and VMAT *7th Int. Conf. 3D Radiation Dosimetry* this issue
[73] van Elmpt, W, McDermott, L, Nijsten, S, Wendling M, Lambin P, . Mijnheer, B 2008, "A literature review of electronic portal imaging for radiotherapy dosimetry," Radiother. Oncol. **88**, 289–309
[74] Gillis S, de Wagter C, Bohsung J, Perrin B, Williams P, Mijnheer BJ 2005 An inter-centre quality assurance network for IMRT verification: results of the ESTRO QUASIMODO project *Radiother. Oncol.* **76** 340-353
[75] Budgell G, Berresford J, Trainer M, Bradshaw E, Sharpe P and Williams P 2011 A national dosimetric audit of IMRT *Radiother. Oncol.* **99** 246–252
[76] Hussein M, Tsang Y, Thomas R, Gouldstone C, Maughan D, Snaith J, Bolton S and Clark CH 2012 Rotational radiotherapy in the UK - a pilot audit ESTRO 31 *Radiother. Oncol.* **103** (suppl 1) S30-31
[77] Clark CH, Hansen VN, Chantler H, Edward C, James HV, Webster G, Miles EA, Guerrero Urbano MT, Bhide SA, Bidmead AM and Nutting CM 2009 Dosimetry audit for a multi-centre IMRT head and neck trial *Radiother. Oncol.* Ireland **93**102-8
[78] Molineu A, Followill DS, Balter PA, Hanson WF, Gillin MT, Huq MS, Eisbruch A and Ibbott GS 2005 Design and implementation of an anthropomorphic quality assurance phantom for intensity-modulated radiation therapy for the Radiation Therapy Oncology Group *Int. J. Radiat. Oncol. Biol. Phys.* **63** 577-583

[79] Followill DS, Evans DR, Cherry C, Molineu A, Fisher G, Hanson WF and Ibbott GS 2007 Design, development, and implementation of the radiological physics center's pelvis and thorax anthropomorphic quality assurance phantoms *Med. Phys.* **34** 2070-2076

[80] Ibbott GS, Followill DS, Molineu HA, Lowenstein JR, Alvarez PE, Roll JE 2008 Challenges in credentialing institutions and participants in advanced technology multi-institutional clinical trials *Int. J. Radiat. Oncol. Biol. Phys.* **71** S71-S75

[81] Yorke E D, 2011 Dosimetric uncertainties and normal tissue tolerance, Ch.1, (Palta J.R., Mackie T.R.,eds), Uncertainties in External Beam Radiation Therapy, *Medical Physics Publishing, Madison, WI*

[82] Andreo P 2011 Accuracy Requirements in Medical Radiation Dosimetry International Symposium on Standards, Applications and Quality Assurance in Medical Radiation Dosimetry 1-23 IAEA, Vienna.

[83] Simpson B personal communication

[84] Keall PJ, Mageras GS, Balter JM, Emery RS, Forster KM, Jiang SB, Kapatoes JM, Low DA, Murphy MJ, Murray BR, Ramsey CR, van Herk MB, Vedam SS, Wong JW and Yorke E 2006 The management of respiratory motion in radiation oncology report of AAPM Task Group 76 Med. Phys. **33** 3874-3900

[85] Pettersen MN, Aird E and Olsen DR 2008 Quality assurance of dosimetry and the impact on sample size in randomized clinical trials *Radiother.Oncol.* **86** 195-199

[86] Kron T, Haworth A and Williams I 2013 Dosimetry for audits and clinical trials: challenges and requirements *7th Int. Conf. 3D Radiation Dosimetry* this issue

[87] Peters L, O'Sullivan B, Giralt J, Fitzgerald TJ, Trotti A, Bernier J, Bourhis J, Yuen K, Fisher R and Rischin D Critical Impact of Radiotherapy Protocol Compliance and Quality in the Treatment of Advanced Head and Neck Cancer: Results from TROG 02.02. *J. Clin. Oncol.* **28** 2996-3001

Semi- and virtual 3D dosimetry in clinical practice

SS Korreman
Department of Science, Systems and Models, Roskilde University
Department of Radiation Oncology, Rigshospitalet, University of Copenhagen
The Niels Bohr Institute, University of Copenhagen

Email: korreman@ruc.dk

Abstract. In this review, 3D dosimetry is divided in three categories; "true" 3D, semi-3D and virtual 3D. Virtual 3D involves the use of measurement arrays either before or after beam entry in the patient/phantom, whereas semi-3D involves use of measurement arrays in phantoms mimicking the patient. True 3D involves the measurement of dose in a volume mimicking the patient.There are different advantages and limitations of all three categories and of systems within these categories. Choice of measurement method in a given case depends on the aim of the measurement, and examples are given of verification measurements with various aims.

1. The role of 3D dosimetry in clinical quality assurance and technology development

The term 3D dosimetry is used to describe determination of delivered absorbed dose in a volume, typically as compared to the dose calculated in the treatment planning system. In the clinic, this type of dosimetric measurement is performed mostly in relation to patient specific quality assurance schemes.

1.1. Patient specific quality assurance

Patient specific QA is performed with the intention of ensuring that each treatment plan can be/is delivered such that the dose deposited in the patient is identical to the dose calculated and evaluated in the treatment planning system (TPS). The patient specific QA procedure must be performed using a measurement (dosimetry) system independent of the treatment planning and delivery systems, and should mimic as closely as possible the treatment situation.

Phantoms with embedded dose measurement arrays are therefore quite well suited for the purpose of these types of measurements when they are performed off-line. For patient specific QA, temporal resolution of detectors is not immediately relevant, as the property one really wants to measure is the accumulated dose to the patient for a specific treatment. Nevertheless, temporal resolution can be used to track the sources of any detected errors in a measurement.

The more complex a treatment plan is, the more warranted is the need for patient specific QA. Comprehensive commissioning and QA of all the subparts involved in the entire treatment chain is, however, by some regarded to render patient specific QA obsolete.

1.2. Commissioning

As part of the commissioning process of new equipment (hard- or soft-ware), it can be beneficial to run a number of actual patient plans and measure absorbed dose in a phantom in 3D. This tests the entire process of treatment planning to dose deposition end-to-end. Although a comprehensive programme of commissioning tests is planned, there may still be transfer and delivery issues which are not tested, particularly for the very complex cases of volumetric modulated arc treatments. Running a representative set of actual patient plans would give an indication as to whether or not such issues will arise in the clinical setting.

Additionally, propagation of small errors that are below tolerance levels during machine QA may in rare cases lead to significant deviations in actual treatment delivery.

1.3. Development and technology assessment

During development of a new treatment technique, it will often be essential to do occasional dose measurements to verify ability/accuracy of delivery with the new technique (debugging type action). This goes for development in delivery techniques as well as in planning techniques. Most often both high spatial and temporal resolution will be warranted for this purpose, and will often be more important than accumulated dose.

2. 3D dosimetry solutions – limitations and advantages

In this section, we will divide 3D dosimetry solutions into three categories; first category is true 3D dosimetry which coins at the moment only gel dosimetry, in the sense that this is the only measurement method which truly has the potential of presenting measured dose in a volume. The second category is termed semi-3D, and is characterized by measuring dose in a limited amount of points spread in the volume in which dose is delivered. The actual measurements can be said to constitute a sample of test points, but often calculation methods will be used to infer dose in the entire volume based on the few measurement points. The last category is termed virtual 3D, which is used to coin the feature that these systems do not in fact measure dose in a volume at all, but only infers dose by calculation from fluence type measurements performed either before or after beam entry.

2.1. "True" 3D dosimetry

Gel dosimetry is as close as we get to actually measuring absorbed dose in a volume. The focus of this review is commercially available 3D dosimetry and clinical use - gel dosimetry and development of this is described in detail in other papers in these proceedings, and we will hence defer from further description here. Instead only a few words will be said about limitations of gel dosimetry as compared to the more mature semi- and virtual 3D methods described in the sections below.

First of all, use of the term "true" dosimetry is of course misleading – a gel phantom is not the actual patient, and lacks the features of the patient (inhomogeneities), and most often also lacks the shape of the patient. The term true only refers to the 3D part in the sense that gel dosimetry can be used to measure dose in a volume rather than in sample points (see example in figure 1).

Also, the term "true" should not be mistaken to refer to the measurements as absolute dose measurements. In each gel phantom, the dose must be referenced to a benchmark measurement performed in the same gel batch and under controlled conditions (such as a simple depth dose).

The biggest limitation of gel dosimetry is, however, the time issue; it is a time consuming process to prepare a reliable gel phantom, and it is a time consuming and delayed process to read out dose after delivery. Read-out can only be performed after delivery using separate technology, and temporal resolution of measurement is not possible.

2.2. Semi-3D dosimetry

Several of the major vendors of measurement equipment offer phantoms with built-in measurement arrays – this includes the OCTAVIUS with 2D-ARRAY seven29 (PTW Freiburg GmbH), the MatriXXEvolution with MULTICube (IBA Dosimetry), the Delta4® (ScandiDos AB), the ArcCHECK (Sun Nuclear Corporation).

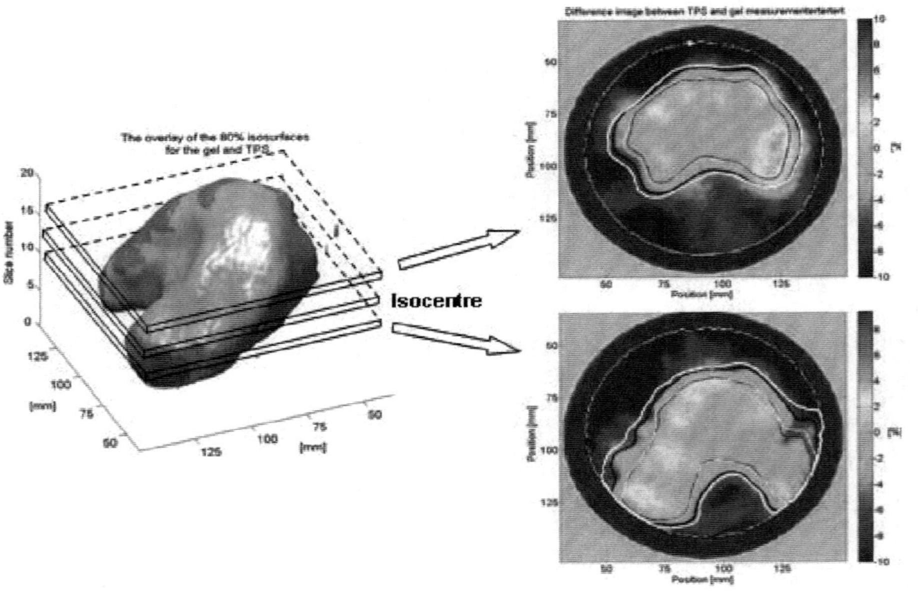

Figure 1: Example of a prostate treatment plan delivered with RapidArc treatment technique to a gel (as described in (1)). The 80% isodose contours are shown for the measurement and the treatment plan as calculated in the Eclipse treatment planning system, and with two slices of the phantom shown in more detail.

All of this equipment consists of one or more measurement array(s) - ion chambers or diodes – embedded in a solid phantom (see figure 2). This kind of dosimetry measurement is here termed semi-3D – dose is measured in a limited number of locations giving a direct test in these sample points only, while dose in the rest of the volume must be inferred from the limited measurements. The different systems described above have different advantages and limitations because they use different types of detectors in different configurations. Ion chambers are used in the OCTAVIUS and the MatriXXEvolution, whereas in ArcCheck and Delta4 diodes are used. The advantage of diodes is their small spatial extent, which limits the dose smearing effect occuring with larger detectors when dose gradients are present. However, they are not very closely spaced in the phantoms, and some modulation may be "lost" between measurement points.

Ion chambers have a limit with respect to spatial resolution given by their larger volume (even with the socalled "pixel" ion chambers used in the MatriXX array), which in effect smears out measurements, thereby loosing modulation and averaging dose gradients. On the other hand, ion chambers have the advantage of readily providing absolute dose.

Film can be used in the same way as described above by embedding the film in phantoms, and has the advantage of a much higher spatial resolution than an ion chamber or diode array. This is for instance relevant for volumetric modulated or IMRT plans, as these plans may contain a high degree of modulation, which can only be measured with high-resolution equipment.

Film, however, cannot provide temporal resolution of measurements, as opposed to all the other semi-3D equipment described here.

Temporal resolution of the measurement can be very useful for back-tracking the origin of potential errors which may be observed in the accumulated absorbed dose, specifically for advanced dynamic treatment delivery. If an error in the accumulated dose can be traced to stemming from one or more specific treatment delivery segments, it may be determined what the cause of the delivery error was. This information may in turn be used to make a decision on how to proceed in the specific case.

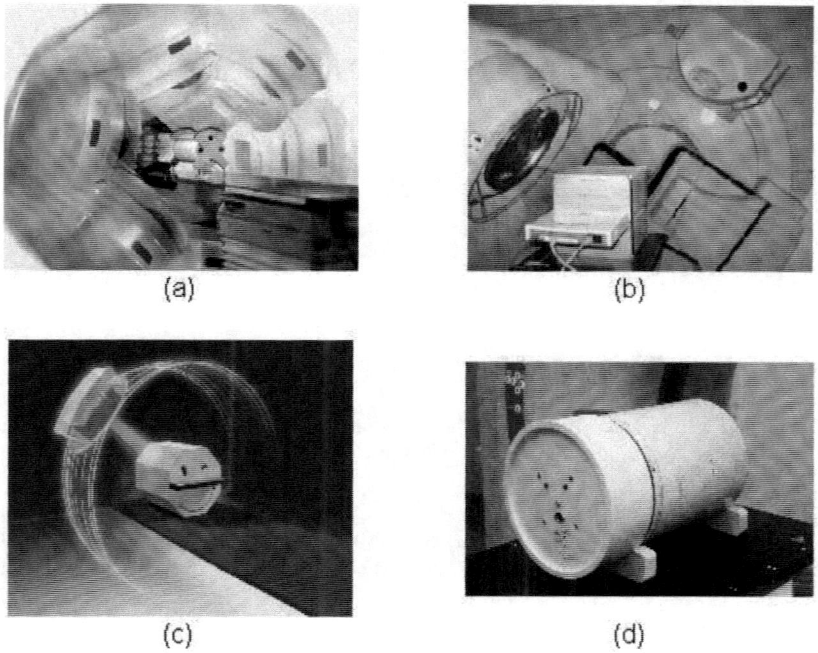

Figure 2: Four commercially available phantoms for semi3D dosimetry with embedded detector arrays. (a) Delta4® with two orthogonal diode arrays (courtesy of ScandiDos AB), (b) MatriXXEvolution two-dimensional ion chamber array with MULTICube (courtesy of IBA Dosimetry), (c) OCTAVIUS with 2D-ARRAY seven29 ion chamber array (courtesy of PTW Freiburg GmbH), and (d) ArcCHECK with a helical diode array cylindrical around the central axis (courtesy of Sun Nuclear Corporation).

2.3. Virtual 3D dosimetry

An alternative to phantoms such as described above, is to use the portal imager of the treatment machine that rotates with the gantry. The imager often has both very high spatial resolution and temporal resolution. However, rather than actually measuring accumulated absorbed dose delivered by the treatment machine, it measures beam fluence. Some mathematical modeling (e.g., back projection) is needed in order to convert this measurement to a representation of absorbed dose accumulated in a patient (or phantom) over the treatment delivery. There are several both commercial and non-commercial software available to perform this conversion (2, 3).

Portal dosimetry of this kind can be used for patient specific QA, and has the advantage of being useful with the patient both off and on the couch. With portal dosimetry, the response of the portal imager detector array can be accumulated over the treatment, either beaming through air, a phantom or the patient. In principle, this check is primarily a fluence check, and in this sense, it is a check that the beam delivery is performed correctly. If this technique is to be used to check the accumulated dose to the patient, a back projection calculation is needed, based on a model of the patient - the CT scan for treatment planning or possibly an online cone-beam CT scan of the patient.

Another virtual 3D approach is to attach a measurement array (or film) to the gantry, to allow it to rotate along with the gantry much as for the portal imager. This approach is currently available with two different scopes; attachment of a phantom with an embedded measurement array at isocentre distance, and attachment of a transmission ion chamber array close to the gantry head immediately after the beam exit (see figure 3). The latter has the advantage of being potentially usable during actual patient treatment for online dosimetry. Again, this requires a patient model (CT scan or cone-beam CT scan) for forward calculation of delivered dose.

Journal of Physics: Conference Series **444** (2013) 012007 doi:10.1088/1742-6596/444/1/012007

(a) (b)

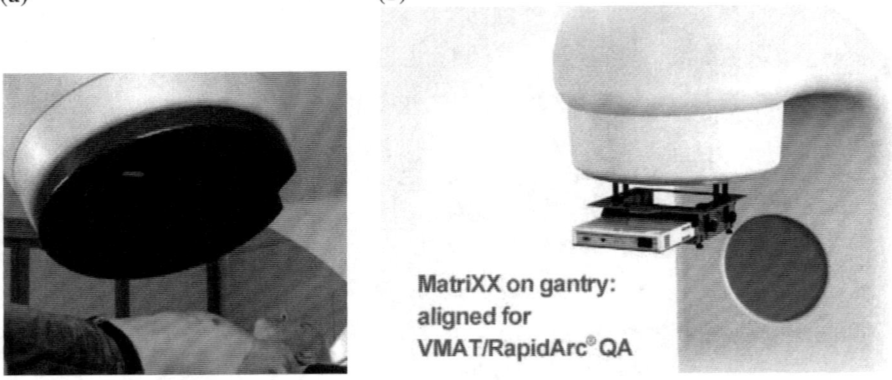

Figure 3: Two commercially available transmission ion chamber arrays for mounting on the gantry head. (a) Delta[4AT] (courtesy of ScandiDos AB), and (b) MatriXXEvolution (IBA Dosimetry, www.iba-dosimetry.com).

3. Focused use of 3D dosimetry

The choice of 3D dosimetry method and analysis method should be focused with regard to the aim of performing the measurement. In this section, considerations for appropriate choice of methods for a few examples of aims of dosimetry measurement will be given. First, a short description of common analysis methods for 3D dosimetry is given.

3.1. Analysis methods for 3D dosimetry measurements

The essential part of patient specific QA is to measure the actual dose deposited in a phantom mimicking the patient, or in the patient himself/herself. The measured dose is then compared with the dose calculated in the treatment planning system (as delivered to the phantom, not the patient). Comparison is most often performed using a gamma evaluation, in which both the dose difference and the geometric difference between measured and calculated dose is considered (4). In figure 4, an example is shown of measurement in a phantom with two orthogonal measurement arrays for a prostate volumetric modulated plan. Distributions are shown for dose differences at the measurement points, geometric distance to closest point of agreement, and for gamma values using 3 mm distance-to-agreement and 3 percent dose difference pass criteria. Appropriate pass criteria for gamma evaluation is an issue of debate, is dependent on parameters such as type of treatment, resolution of measurement equipment, and includes determination of the volume in which to perform the evaluation. Often a simple evaulation of gamma values will involve production of a frequency diagram of gamma values and a cut-off value for which a plan is considered overall to pass – for instance that no more than 5% of evaluation (measurement) points can have a gamma value above 1.

A limitation of gamma analysis is that there is no relation of the quantification of an error and specifically of the frequency distribution of errors to sensitivity of affected structures.

To complement gamma evaluation, dose-volume comparison is therefore occasionally also performed, and it is an additional feature in most commercial 3D dosimetry systems to provide this possibility. In short, a dose-volume analysis tool involves the inferred extrapolated (and interpolated) calculation of dose to the entire irradiated volume based on measurement points. Delineated structures are overlayed and dose-volume histograms are produced for the structures. These can then be compared to dose-volume histograms from the treatment planning system (for the phantom). A more sophisticated version of this feature performs the calculation of dose-volume histograms based on a CT scan of the patient and direct comparison can be performed to the dose-volume histograms of the treatment plan. As the measurements are performed in a phantom setup, this however requires a transformation of the measurements from the phantom to the patient.

(a)

(b)

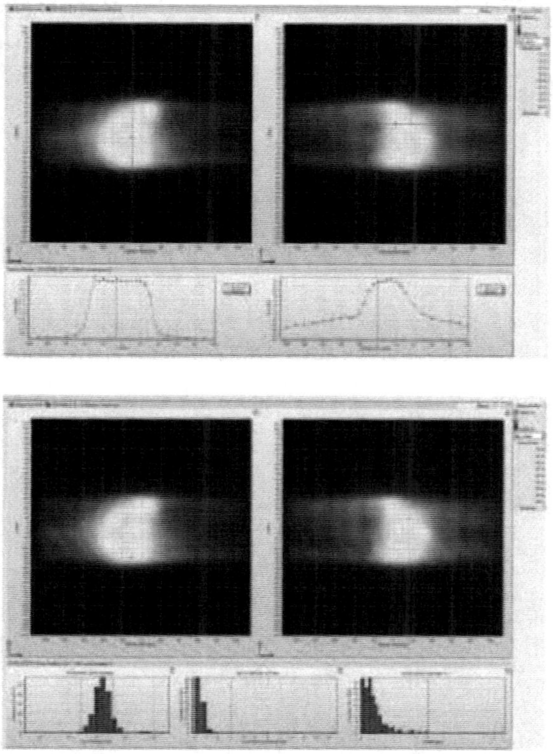

Figure 4: Data from measurements performed using the Delta4® phantom for a prostate RapidArc® plan. (a) The output of detectors in the two orthogonal arrays are shown in the top, below are two output profiles through the arrays are displayed (at the positions of the dotted lines in the detector views). The lines indicate dose calculated in the treatment planning system, and the dots indicate measured dose. (b) For the same measurement, the lower plots show distributions of point dose deviations, distances to agreement, and gamma values (with 3% dose deviation, 3 mm distance-to-agreement). Above, in the detector views, the points with gamma values above 1 are highlighted. Images in the figure courtesy of Lotte Fog, Department of Radiation Oncology, Rigshospitalet, Denmark.

3.2. Verification of treatment plans with high degree of modulation

High degree of modulation in a plan can often be seen in relation to intensity modulated delivery, and specifically in volumetric modulated arc therapy. For verification of such plans, large detectors and/or poor spatial resolution is not well suited, because the modulation will be smeared out and/or "fall between" measurement points. For delivery with dynamics of MLC and/or gantry, it is also of relevance to have the possibility of temporal resolution of the delivery during measurement.

For the case of volumetric modulated arc therapy, measurements should be performed with the planned rotation of the gantry. It is a common simplification of patient specific QA to collapse all delivery into gantry angle 0° for simplicity. This should not be done for volumetric modulated arc therapy, as the gantry rotation is a key component of the treatment – for the RapidArc® solution, the gantry position is for instance the driving force for the DMLC positioning. As a consequence of the gantry rotation being an essential part of arc therapy, it is also essential that equipment used in the quality assurance of these techniques is suited to handle gantry rotations. The equipment must therefore exhibit little or no angular dependence.

In light of the considerations stated above, a well suited dosimetric system is a virtual 3D system, using the portal imager. There is high spatial resolution with small detectors, high temporal resolution, and no angular dependence (the portal imager is inherently independent of gantry angle as the angle of incidence of the radiation does not change). The limitation is that there is no actual measurement of absorbed dose. If this is wanted, semi-3D dosimetry systems Delta4 or ArcCheck, both with small detectors, temporal resolution and little angular dependence, may be used. Limitations of these are that the spatial resolution is not highand high degree of modulation may fall between measurement points.

3.3. Verification of dynamics of 4D plan

Although 4D treatment using dynamic MLCs to track the respiratory movement of the target is not in clinical use yet, it is worth discussing this case as well, in relation to measurements performed in the scope of development of the 4D technique.

Special for this kind of treatment is that the target is moving, and it is particularly the ability of the treatment technique to track this movement. The measurement system must therefore emulate the target movement. This can be done using moving platforms on which to place the measurement phantom – platforms able to move in 3D, and programmable to perform motion mimicking actual patient target motion traces are preferable. Such a platform is for instance the HexaMotion® provided as an addition to the Delta4 phantom – another independent platform is the 4D phantom sold by Washington University.

The verification proces is not aimed at verifiying the plan itself, but to verify the ability to track motion. Temporal resolution is obviously necessary, and it is also necessary that spatial resolution is good enough to distinguish displacement of dose of the same magnitude (and less) of the phantom motion.

3.4. In-vivo dosimetry type QA

This type of measurements are performed with the aim of verifying the actually delivered to the patient during treatment. The classic in-vivo dosimetry measurement is performed as a point measurement on the patient surface using for instance a diode (5), but it is also a possibility to use virtual 3D dosimetry to provide a measurement on which (an estimate of) the dose in the patient can be inferred. In order to infer the actual delivered dose plausibly, back projection to a model of the patient must be performed – preferably to a model reflecting the actual patient setup and anatomy of the day, that is a conebeam CT scan.

This type of measurement requires use of either portal dosimetry or transmission chamber dosimetry – the advantage of the latter being that dose is measured (as transmission arrays are ion chambers), and the advantage of the former being the high spatial resolution of measurement points in the portal imager. The limitation of the latter is the poor spatial resolution which makes it less useful for clinical cases in which high precision is warranted.

An additional use of this type of virtual in-vivo dosimetry is that it can readily be used for plan-of-the-day adaptive treatment, in which the treatment plan is adapted to the daily anatomy of the patient based on conebeam CT, and the accumulated delivered dose of previous treatment fractions is incorporated. This is a concept which has been suggested in a number of publications (see for instance(6)), and has even been elaborated as a technique to account for respiratory motion (7). In the latter case, it is also relevant to use the temporal resolution of the measurements, which is available for all virtual 3D dosimetry systems.

3.5. High precision dose delivery verification

Examples of treatments in which it is relevant to measure dose with high accuracy are for instance cases that involve risk of dose spill into highly sensitive volumes in the patient, or small targets with high doses. In such cases, it may be necessary to take special precautions to ensure that dose is delivered according to plan in the sensitive volumes. It will usually be relevant to combine high spatial resolution with high accuracy in dose measurement. Virtual 3D dosimetry using the portal imager may not be well suited due to the lack of measurement of delivered dose (although the spatial resolution is good enough (8)), but on the other hand, virtual 3D dosimetry using transmission arrays are also not well suited due to the poor spatial resolution.

Semi 3D dosimetry generally do not show high spatial resolution, although the systems using diodes (Delta4 and ArcCheck) do have measurement arrays with small measurement volumes and closely spaced diodes. In the case of the Delta4 phantom, the phantom may be positioned and oriented such that the measurement arrays have their areas with highest spatial resolution in the sensitive volume.

For these kinds of measurements, it appears that true 3D dosimetry – gel measurements – is the only method that is really well suited. The limitation obviously is that this can only be done as pre-treatment verification, so that it may be even better to combine with in-vivo type measurements as described above.

4. Summary

In this review, 3D dosimetry is categorized as either "true" 3D, semi-3D or virtual 3D, with different advantages and limitations of all three categories. Virtual 3D involves the use of measurement arrays either before or after beam entry in the patient/phantom, whereas semi-3D involves use of measurement arrays in phantoms mimicking the patient. True 3D involves the measurement of dose in a volume mimicking the patient.

It is stressed that choice of the proper measurement method depends on the aim of the measurement. With different measurement systems and setups, different types of errors can be detected. A few examples have been given with different aims for verification measurements, and the limitations and advantages of different measurement systems for these examples have been elaborated.

Last, it should be said that the use of 3D measurements in general is important in quality assurance for radiotherapy, as should hopefully be evident from the review above. The types of errors detected when performing measurements cannot be completely predicted, even though the aim of the measurement is well defined. Often we will be surprised to find unexpected types of errors leading to the need for further and maybe other measurements – we meet the "green aliens" of dosimetry (expression gratefully borrowed from Lotte Fog, physicist af Rigshospitalet, Copenhagen). This is a good reason for performing measurements regularly, even when we believe we know our machines and treatment techniques well.

5. Acknowledgments

A grateful thanks for discussions and access to measurements to colleagues at Rigshospitalet Copenhagen, in particular physicists Lotte Fog and Bob Smulders, as well as co-workers in Lund and Malmö, in particular physicist Sofie Ceberg.

6. References

[1] Ceberg S *et al* 2010 *Phys. Med. Biol.* **55** 4885-98
[2] Mans A *et al* 2010 *Radiother. Oncol.* **94** 181-7
[3] Fogliata A *et al* 2011 *Br. J. Radiol.* **84** 534-45
[4] Low D A and Dempsey J F 2003 *Med. Phys.* **30** 2455-64
[5] Huyskens D P *et al* 2001 *Booklet no. 5 1st ed. ESTRO*
[6] Wright P *et al* 2009 *Phys. Med. Biol.* **54** 7121-33
[7] Webb S 2008 *Phys. Med. Biol.* **53** 1-21
[8] Fog L S *et al* 2011 *Phys. Med. Biol.* **56** 1853-63

Journal of Physics: Conference Series **444** (2013) 012008 doi:10.1088/1742-6596/444/1/012008

Review of four novel dosimeters developed for use in radiotherapy

P Metcalfe[1,2], A Quinn[1,2], K Loo[1], M Lerch[1], M Petasecca[1], J Wong[1], N Hardcastle[1,4], M Carolan[1,3], J McNamara[1,3] , D Cutajar[1], I Fuduli[1], A Espinoza[1], C Porumb[1] and A Rosenfeld[1]
[1]CMRP, University of Wollongong, NSW, Australia.
[2]Liverpool Cancer Therapy Centre, NSW, Australia
[3]Illawarra Cancer Care Centre, NSW, Australia
[4]Peter MacCallum Cancer Centre, Vic, Australia

E-mail: metcalfe@uow.edu.au

Abstract. Centre for Medical Radiation Physics (CMRP) is a research strength at the University of Wollongong, the main research theme of this centre is to develop prototype novel radiation dosimeters. Multiple detector systems have been developed by Prof Rosenfelds' group for various radiation detector applications. This paper focuses on four current detector systems being developed and studied at CMRP. Two silicon array detectors include the magic plate and dose magnifying glass (DMG), the primary focus of these two detectors is high spatial and temporal resolution dosimetry in intensity modulated radiation therapy (IMRT) beams. The third detector discussed is the MO*Skin*™ which is a high spatial resolution detector based on MOSFET technology, its primary role is *in vivo* dosimetry. The fourth detector system discussed is BrachyView, this is a high resolution dose viewing system based on Medipix detector technology.

1. Introduction

In recent years linear accelerator and brachytherapy systems utilised for radiotherapy have improved to enable the delivery of increasingly complex dose distributions that conform tightly to the cancer treatment target. These developments have not been accompanied by vendors updating and supplying sophisticated on-board dose monitoring systems. For example linear accelerator vendors still rely on dual ion chamber assemblys with monitor unit and symmetry feedback only. These chambers are placed well above the collimation system and multi-leaf collimators (MLCs) in a position where they provide no feedback about the complex modulated dose distribution that exits the MLC.

2. Methods

2.1. Commercial dosimetry array detectors

The requirement to produce dosimeters that can validate these complex modulated dose distributions has produced a myriad of pre vivo in phantom dosimetry test devices. These tests usually consist of a dosimetry phantom system. Some commercial examples include Delta4PT® which consists of two perpendicular diode arrays, ArcCHECK® which consists of a cylindrical diode array, and OCTAVIOUS® which houses a rotating array.

All these detector systems are placed in the treatment field before the first patient treatment fraction. The treatment plan is delivered to the dosimetry phantom system and the resulting dose map is compared with the radiotherapy treatment planning (RTP) computer dose map. In essence this is a pre-treatment test to ensure that the dose, segment and monitor unit information has been correctly transferred to the linac verification and MLC control system. These systems also test that the MLC positions are reasonably close to that planned.

Currently these methods do not account for any deviations in leaf consistency at each treatment fraction and hence the race is on to produce a system that can ascertain patient dose during each treatment fraction.

2.2. Magic Plate

The obvious positions to place this device are in the block tray (e.g. magic plate, Delta4 AV®) or behind the patient (EPID).The advantage of the EPID is it acquires the dose map after the beam has exited the patient and hence does not perturb the beam prior to the beam entering the patient. Most EPIDs have an amorphous silicon detector of very high spatial resolution. Some disadvantages are that the patient perturbs the beam. Another disadvantage is that most scintillation material currently used is of high atomic number so various corrections are required to account for beam energy off axis [1]. There is continuing research into the use of these detectors in direct mode[1] and perhaps low atomic number scintillators may be another option.

Dosimeters that are placed in the block tray seem a logical choice. The CMRP version of this concept consists of a series of epitaxial silicon diodes. The system is so named the Magic plate. Important characteristics of these detectors are that they do not reduce the beam transmission significantly (preferably <1%). Also their electron contamination contribution should not extensively alter the patient skin dose. Many of these detectors also consist of discrete detector elements, so they are constrained by the Nyquist criteria in terms of spatial resolution, in particular with regard to pitch between each detector array element. Placing these detectors in the block tray without full build up basically turns them into a fluence detector. While angular response becomes a small issue as the detector array is virtually perpendicular to the beam. The spectral properties of the beam off axis just beneath the MLC is not well characterised by Monte Carlo calculations.

The Magic Plate generation 1 prototype tested was a 2D array of 11 × 11 silicon diodes covering an area of 10 × 10 cm^2 (figure 1a). The diodes were mounted on a 0.5 mm thick Kapton substrate using the 'drop-in' technology. The physical size of a single diode is 1.5 × 1.5 × 0.425 mm^3. The Magic Plate was designed to be mounted on the linear accelerator head (accessory mount slot) in line with the radiation beam (figure 1b). It is designed to operate as a transmission detector measuring the 2D fluence map of the modulated radiation beam [2].

The diodes that was used in the Magic Plate was grown using the epitaxial-growth technique [3]. For the MP diodes, the p-epitaxial layer is a 50 μm thick p-Si grown on top of a 375 μm thick high resistivity p^+ substrate. The sensitive volume of the individual element defined by the n^+ region is 0.5 × 0.5 × 0.05 mm^3, whilst the detector pitch is 1 cm. The next prototype will have more detectors packed more closely. The number of detectors is limited by, packing density and readout capability (channels). Also detectors can be packed more closely if smaller fields such as Stereotactic radio-surgery (SRS) fields are to be studied. CMRP are currently working on a miniature SRS mobile tracking version of this detector.

2.3. Dose Magnifying Glass

The CMRP have developed a silicon strip detector that has very high 1D spatial resolution at very high temporal resolution The device is called a Dose Magnifying Glass (DMG). The 1st generation device had a ceramic base, however for the 2nd generation DMG is mounted on Kapton (Atomic number Z=6.6 and mass density= 1.45g/cm^3).The thickness of the Si wafer is 375 μm and the strip pitch is 200 μm. The silicon strips are 20 × 2000 μm^2 .The device has been found to be useful in measuring radiation fields that have rapid dose and time varying gradients.

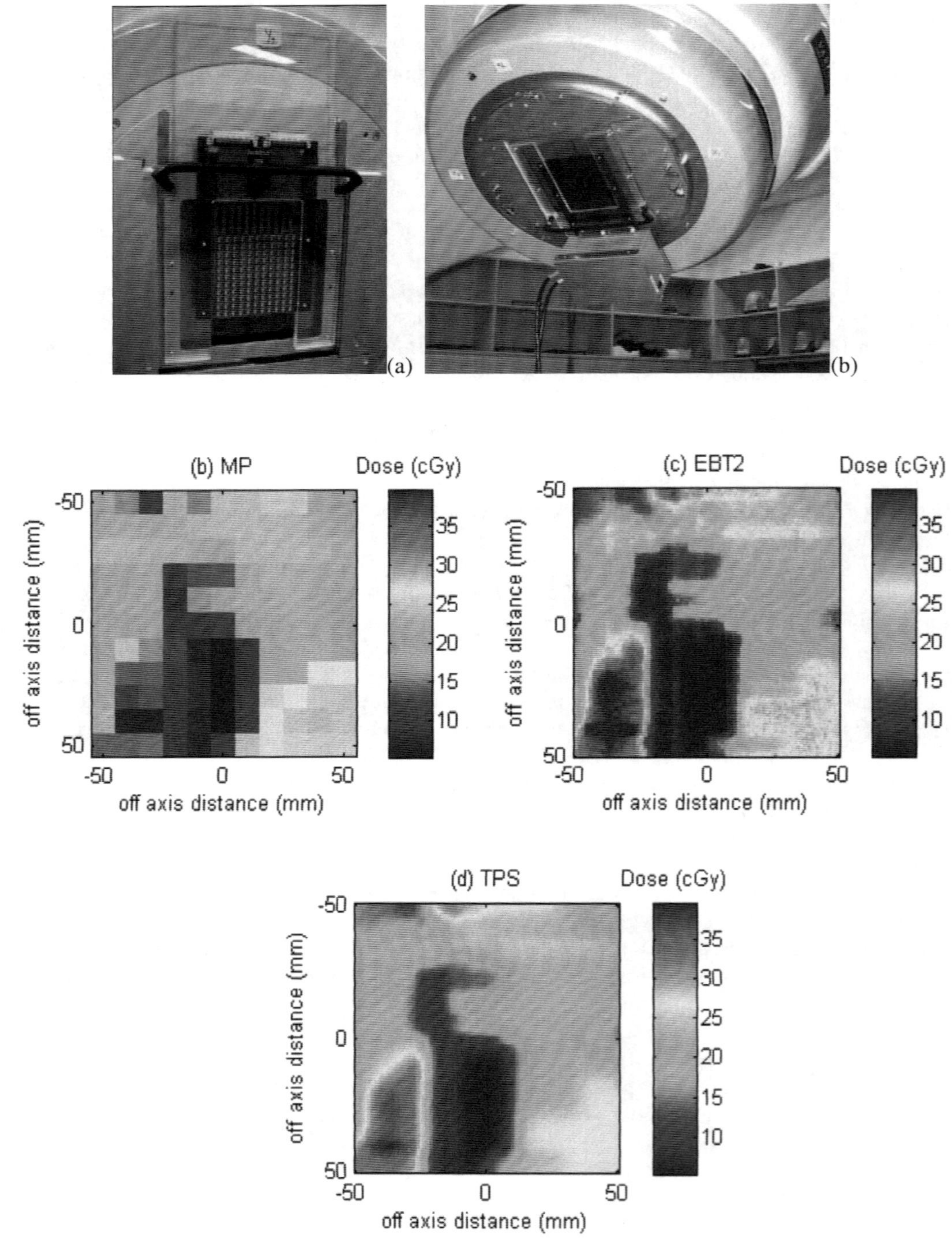

(c)

Figure 1: The CMRP Magic Plate Prototype 1(a) close up view, (b) mounted on linac head block tray, (c) intensity modulated radiotherapy (IMRT) dose maps, magic plate (left), Gaf film (middle), Pinnacle® RTP (right).

Journal of Physics: Conference Series **444** (2013) 012008 doi:10.1088/1742-6596/444/1/012008

We have reported the use of this device to track linac IMRT dose depositions - see Wong et al [4]. The device has also been reported in SRS small field dosimetry [5]. The DMG has also been used to measure Tomotherapy binary leaf speed [6].

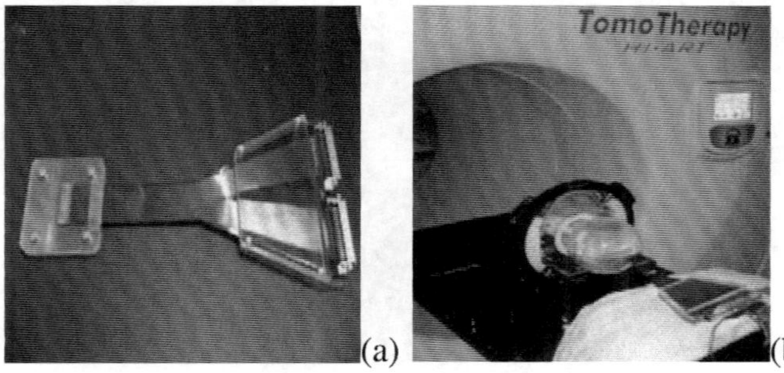

(a) (b)

Figure 2: The DMG with Kapton base. (a)The silicon strip detector is shown in the Perspex on left and the pig tail out to readout is shown on the right. (b) DMG placed in Lucy Phantom for TomoTherapy™ SRS measurements.

2.3. *MOSkin™ In Vivo Dosimeter*

The MO*Skin*™ is a metal oxide semiconductor field-effect transistor (MOSFET) - based radiation detector that has been developed to provide real time on-line *in vivo* dose measurements during radiation therapy. MOSFET-based radiation detectors use a single transistor to measure the radiation dose through the trapping of electrical charges within the transistor structures. Transistors are fabricated to be very small and to feature micron-sized radiation-sensitive structures; thus, these devices are ideal for measuring the radiation dose in areas of high-dose gradient, where the size of the detector may affect the reading. The MO*Skin*™ is mounted beneath a thin plastic cover with a uniform thickness that has been engineered to provide a depth of measurement of 0.07 mm, giving a skin dose equivalent depth when placed on a surface during radiation therapy [7].

The MO*Skin*™ detector is operated through a data collection unit, which can read out five devices at once in real time. Multiple data collection units may be used to operate more devices, and these may be connected to a computer for real time software-based analysis, see Figure 3.

The MO*Skin*™ detector has been successfully deployed to measure skin dose from conventional linear accelerators. Dose from megavoltage (MV) and kilovoltage (kV) image guidance devices has also been measured. The MO*Skin*™ device has a 2.5 mV cGy^{-1} sensitivity to MV X-rays, and an 8 mV cGy^{-1} sensitivity to kV X-rays (~100 kVp), the difference is attributed to the detectors enhanced sensitivity to the photoelectric effect. We have reported using these devices to assess skin dose during breast radiotherapy with immobilisation cast material [7]. Additionally the in-field and out-of-field skin dose has been measured for various image guidance and treatment scenarios involving breast tangents [9].

The MO*Skin*™ detector has been deployed for intra-cavity dosimetry during brachytherapy and external beam radiation therapy [10]. The design of the detector allows for the placement of the radiation-sensitive volume against the cavity wall, providing point-dose measurements on the cavity surface. Current work, which will be discussed, includes the continuing development of the devices with endo-rectal balloons (ERBs) in order to assess *in vivo* rectal wall dose during prostate external beam radiotherapy. Hardcastle *et al* showed that the MO*Skin*™ device may be integrated with an ERB to provide point-dose measurements from multiple locations of the rectal wall in real time during radiation therapy [11]. The ERB restricts motion of the rectum during irradiation and holds the dosimeters firmly against the cavity wall.

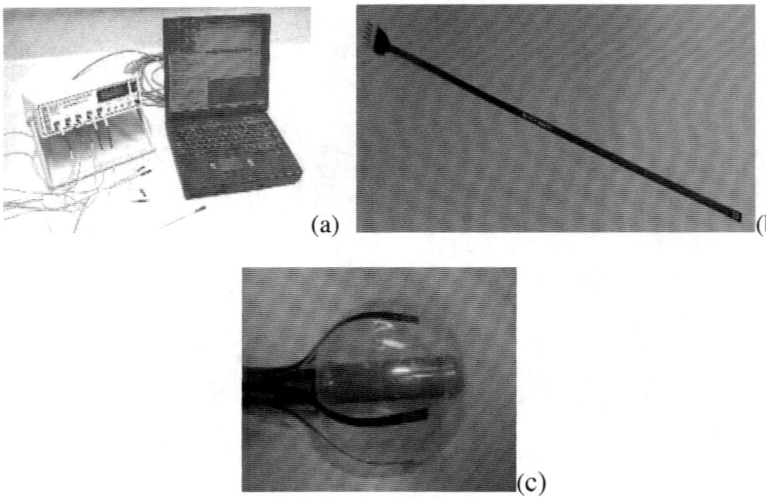
(a) (b) (c)

Figure 3: MO*Skin*™ Detector System: (a) read out device, (b) MO*Skin*™ detector on Kapton pigtail, and (c) multiple MO*Skins*™ mounted on a Endo Rectal Balloon (i.e. The Octo MOSkin prototype).

An *in vivo* dosimeter placed on the anterior rectal wall allows the clinician to measure the dose delivered to the section of the rectum receiving the highest dose. This is important where high doses are delivered to the residual prostate and anterior rectal wall, and would be particularly useful in the context of the limitations of dose calculation algorithms at cavity interfaces. Accurate knowledge of the rectal wall dose is needed to identify the relationship between the dose delivered and rectal toxicity. A secondary application of an *in vivo* anterior rectal wall dosimeter is target dose verification. The anterior rectal wall is generally contained within the planning target volume (PTV); therefore, any dosimeter placed on the anterior rectal wall can be used as a surrogate for the PTV dose at the posterior region of the volume. Recent articles have focused on implantable *in vivo* dosimeters for target dose verification [12-14]. It is less viable to implant these dosimeters in the post-prostatectomy setting, and is less invasive to use the MO*Skin*™ dosimeter mounted on an ERB. Furthermore, MO*Skins*™ are visible on CT. This ensures precise positioning of the dosimeter so that the dose can be compared with a dose-point calculation using a RTP computer. The application of the MO*Skin*™ to analyse the dose to the anterior rectal wall in the high-dose gradient region has been assessed [15]. The devices to be deployed in a proposed future study will be used in an eight-detector Octo-MO*Skin* configuration to track dose points around the rectal and anal wall interface.

2.4. BrachyView

The traditional pre-planning technique for I 125 Brachytherapy seed or I-192 implants relies primarily on transrectal ultrasound (TRUS) imaging and has limitations that may be overcome by intraoperative dynamic treatment planning (ITP).

An innovative pixellated detector integrated within the TRUS probe for real-time and in-vivo seed and anatomical imaging has been designed at CMRP. The dosimetry system, named BrachyView, is capable of providing accurate intraoperative planning and post-implant verification.

BrachyView incorporates three tiled Medipix2 detectors with a combined imaging area of 14x52mm^2 (256x768 pixels, each with size 55x55μm2), coupled to a multi-pinhole collimator as shown in figure 5. The Medipix design is a high resolution, highly pixellated silicon-based detector with properties ideally suited for obtaining radiation events from the implanted I-125 seeds, or in the case of HDR brachytherapy, Ir-192 implants. Three-dimensional reconstructed positions from multiple planar images are used to resolve the seed placement in real time.

Feasibility studies of the BrachyView system with simple seed arrangements show it can provide dynamic information about seed positioning. Further steps, including software development for imaging of multiple seeds in a realistic clinical scenario are underway. The eventual integration of Medipix in the TRUS probe will also open up the possibility of CT imaging for post-implant dosimetry, thereby streamlining the entire implant into a single-room procedure, where a single device has the capabilities to perform prostate volume studies, intraoperative planning, and post-implant dosimetry.

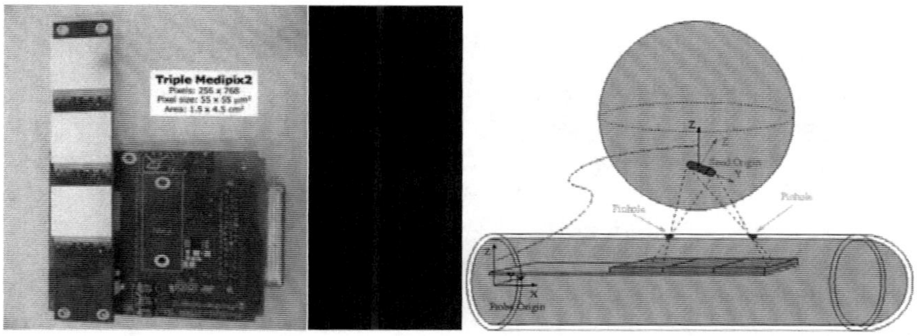

Figure 4: The BarchyView Dosimetry system (a) close up view of triple medipix detectors and electronics (b) Schematic of BrachyView Seed Origin verification

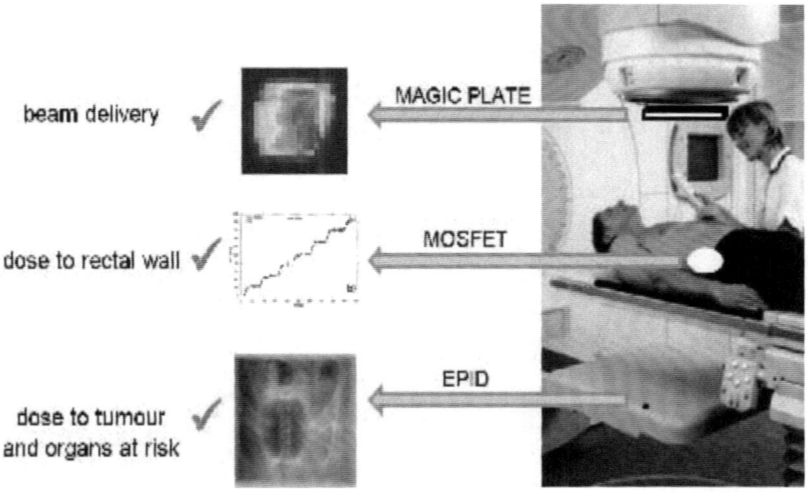

Figure 5: Concept Diagram showing how Magic plate, MO*Skin*™ and EPIDs may be used in future to validate IMRT dose delivery courtesy P. Vial (Private communication).

3. Summary

Four of the myriad of radiation dosimetry systems being developed at CMRP has been discussed. The common theme that develops during testing all these devices is that spatial, energy, and temporal resolution are important characteristics that need to be assessed for the successful implementation of any new dosimetry system.

Despite several established models, EPID-based transit dosimetry remains an untapped potential for a variety of reasons [16]. In recent years, the exciting possibilities of EPID *in vivo* dosimetry have been demonstrated with EPIDs combined with daily kV-CBCT to reconstruct the dose in three dimensions to verify hypofractionated rectal cancer treatment [17]. It would be an exciting prospect to combine the use of EPID with an incident fluence detector (e.g. magic plate) and an *in vivo* detector (e.g. MO*Skin*™) to give a detailed picture of dose at exit of the linac, dose at patient target and dose after exiting the patient.

Figure 5 illustrates a concept of how future patient treatments may be verified. A treatment plan may be compared with the reconstructed dose of the day and used in three-dimensional dose-comparison metrics to identify differences [18, 19].

Testing new detectors at CMRP is a bit like being a taste tester at a chocolate factory, some of the samples don't taste good to start with and may require some modifications to the recipe, but when a good prototype arrives that works well then that tastes real sweet.

4. References

[1] Greer P 2005 *Med. Phys.* **32** 3558-68
[2] Vial P *et al* 2008 *Med. Phys.* **38** 4362-72
[3] Sze S M 2001 *Semiconductor Devices: Physics and Technology* 2nd Edition (Wiley)
[4] Wong J H D *et al* 2010 *Med. Phys.* **37** 427-39
[5] Wong J H D *et al* 2011 *Med. Phys.* **38** 1226-38
[6] Wong J H D *et al* 2011 *Med. Phys.* **38** 2256-64
[7] Kwan I *et al* 2008 *Radiat. Meas.* **43** 929-32
[8] Kelly *et al* 2011 *Phys. Med. Biol.* **56** 1-13
[9] Quinn A *et al* 2011 *Med. Phys.* **38** 6222-7
[10] Kron T 2002 *Radiat. Prot. Dosimetry* **101** 445-8
[11] Hardcastle N 2010 *Phys. Med. Biol.* **55** 3859-71
[12] Beyer G P 2007 *Int. J. Rad. Onc. Biol Phys.* **69** 925-35
[13] Black R D *et al* 2005 *Int. J. Rad. Onc. Biol. Phys.* **63** 290-300
[14] Kry S F *et al* 2009 M *J. Appl. Clin. Med. Phys.* **10** 22-32
[15] Hardcastle N *et al* 2009 *Radiother. Oncol.* **92** 48-56
[16] van Elmpt W *et al* 2008 *Radiother. Oncol.* **88** 289-309
[17] McDermott L N *et al* 2008 *Radiother. Oncol.* **86** 35-42
[18] Bosi S G *et al* 2009 *Phys. Med. Biol.* **54** 275-283
[19] Hill R et al 2005 *Phys. Med. Biol.* **50** N331-44

Review of recent advances in non 3D dosimeters

K Jordan
London Regional Cancer Program, London Health Sciences Centre, Department of Medical Biophysics, Western University, London, ON, Canada

E-mail: kevin.jordan@lhsc.on.ca

Abstract. This review covers papers from the previous two years not being represented at this conference. Optically stimulated luminescence detectors, diamond detectors and radiochromic film are the major topics. An effort to link these papers to 3D dosimetry is included in this brief report. An overall theme of linear, water equivalent response is seen throughout the topics included.

1. Introduction

This review has a limited scope of reporting papers concerning dosimeters for radiotherapy that were published since the IC3DDose10 conference. It also attempts to complement research being presented at this meeting by highlighting dosimetry results that are not included at this meeting. The selection is intended to assist dosimeter researchers by providing a short list of papers outside their current focus. The review is broadly organized into point (0D) and planar (2D) dosimeters. Point detectors can be packaged as 1, 2, 3 and 4D dosimeters. Papers were selected from searches of PubMed, Web of Science and Summons databases with the keywords; dosimeter, ionizing radiation and radiation therapy. Papers dealing with doses greater than 50 Gy were excluded to limit material.

2. Miscellaneous

Valuable information related to radiotherapy dosimetry was published in recent review papers and more general data analysis papers. They are reported here for the benefit of readers interested in starting with a selected list of publications from the broad literature available.

2.1. Review papers

Small field dosimetry is challenging and provides an opportunity to learn the basic physical principles of dosimeter measurements. This information also provides the input to modeling with Monte Carlo calculations. The Institute of Physics and Engineering in Medicine report number 103, entitled Small field MV photon dosimetry, provides a wealth of information on this topic [1]. Dosimeter materials for 3D dosimetry can be evaluated using this report as a template. Often small field dosimetry is used to test the performance of an entire 3D dosimetry system. An even more challenging problem is accurate dosimetry for microbeam radiation therapy with x-ray beams generated by synchrotron sources where spatial resolution of tens of micrometers is required. A review of relevant high resolution dosimeters has been recently prepared, Bräuer-Krish et al [2].

Journal of Physics: Conference Series **444** (2013) 012009 doi:10.1088/1742-6596/444/1/012009

2.2. Analysis

Methods for comparison of large 3D dose distribution data sets have been limited since few 'bench mark' data sets are available. Many researchers are still quantifying performance of readout scanners and the dosimeter materials. However, experimental progress indicates that efforts to develop analysis methods are needed and dedicated sessions are now part of this conference program, see previous proceedings. It is anticipated that new analysis methods will stimulate design of 3D measurements and provide guidance on required characteristics of 3D dosimetry systems, for example, spatial and dose resolutions.

A standard comparison method involves spatially aligning a data set to the corresponding reference set and subtracting element by element. These dose difference arrays are easy to interpret and quantify. However, dose distributions are often normalized before comparison, eliminating absolute differences that may be important. For clinical comparisons, the 'gamma analysis' approach combines dose differences and distances to agreement is a practical method of deciding pass-fail treatment QA. Bak et al have developed a modified dose difference analysis which provides a weighting factor proportional to the local dose gradient. Results are similar to gamma analysis with the advantages of providing the sign and magnitude of the dose difference [3]. Measurement theory likely holds many useful solutions that could be adapted to 3D dosimetry. Comparison of data sets is a generic problem. Interaction with researchers in other disciplines will generate new ideas for analyzing results. An example of this cross discipline thinking is contained in a report examining Bland-Altman analysis for comparison of different radiation dosimeter performances as standards [4]. They compared different ion chambers for kV energy depth doses. Graphical analysis involves plotting dose differences against the mean dose and determining dependences. This analysis should generate ideas for comparisons of different 3D dosimetry systems.

Watanabe and Hayashi provide an excellent discussion of errors related to scaling doses for relative dosimetry [5]. This issue is very important when comparing data sets. In general, comparison of independently measured doses is much more useful. When differences are found outside the anticipated limits then investigations will reveal the causes of the discrepancies. Only after evaluating the dose differences should normalizations be performed. They have provided a practical example of errors introduced for dose analyses with radiochromic film and BANG3 polymer gel. The results emphasize that linear response in dosimeters is a valuable parameter for subsequent analysis.

3. Point dosimeters

Point detectors are common in radiotherapy dosimetry. They are often translated through watwer phantoms to generate dose profiles from ionizing radiation beams. Their sensitive volumes dictate the spatial resolution achievable. If their absorption and scatter coefficients are similar to water, they are more valuable since knowledge of the local beam energy is not required to obtain accurate dose to water measurements. OSL films can be considered as 0, 1 or 2D depending on the readout geometry. Bending the films allows dose on a non-planar surface to be recorded. This is sometimes considered a form of 3D dosimetry.

3.1. Optically stimulated luminescence dosimeters (OSL)

OSL dosimeters have been reviewed since DOSGEL06 for this meeting. Since then Al_2O_3:C packaged by Landauer Inc. as the NanoDot™ with the MicroStar reader has become the replacement technology for thermal luminescence dosimetry (TLD) systems. The ease of use, physical size and lower cost led to quick adaptation into radiotherapy clinics. This dosimeter also has the benefit of detailed studies over many decades of TLD and other OSL materials.

A pair of papers from the same authors have re-examined the angular dependence of the nanoDot dosimeter and found a 4% lower response when irradiated parallel to a 6 MV photon beam compared to normal incidence [6, 7]. The dosimeter was also modeled with Monte Carlo calculations and the effect was attributed to partial volume effects. The thin air pocket within the nanoDot casset was found to lower the angular dependence. The authors concluded that angular dependence should not be ignored when using these dosimeters for irradiation geometries that involve multiple beam angles. The second paper examines the response of the OSL's for measurements at beam locations away from the central axis where calibrations were performed. Generally, dosimeters are calibrated at or near central axis and at depths of charge particle equilibrium. In practice, these dosimeters are often used to measure dose just outside the beam edge or several centimeters from the beam edge where the energy spectrum is lower. In the case of patient treatment QA the OSL's may be at a location in a phantom that is sometimes in the beam and others outside with a range of incident angles. This inconsistency may lead to errors in measured dose. In the second paper, the authors examined out of field response and found doses up 30% lower than with ion chambers. This paper highlights the complexity of dosimetry and provides a clear example of how incorrect calibration procedures can lead to errors in measured dose. It can be cited when considering the overall importance of using energy independent dosimeters for complex 3D dosimetry.

OSL's can be further characterized following optical erasure of the previous signal. By this method dosimeters can be individually characterized when higher precision is required. In practice, the dosimeters can be grouped by sensitivity and then average group sensitivity is used for clinical dosimetry. There are now a few papers examining optical annealing effects, but there is no consensus on best protocol and or spectrum. Mrčela et al used both tungsten and UV filtered metal halide lamps, and found increased reproducibility in readings with optical annealing [8]. They found a reproducibility of 1% with annealing for repeated 1 Gy irradiations. My own experiments indicated blue light was more effective than green or yellow light. If the light source contains UV then higher readings are present following annealing. With LED's that peak near 450 nm we obtain background readings of 200-300 counts post annealing, compared to initial readings of a new dosimeter of ~50 counts. More results characterizing OSL's at diagnostic energy range have been reported [9]. The authors recommend optical bleaching to improve reproducibility, corrections for angular dependence and beam energy.

Many different materials have been reported for TLD dosimetry and the same process is likely to be repeated for OSL dosimeters. Dhabekar et al reported on the OSL performance of LiMgPo4:Tb,B phosphor [10]. The main advantage of this material is the large linear range from 10^{-3} to 10^{3} Gy. It has a dose sensitivity similar to Al_2O_3:C but a greater energy dependence below 100 keV. This report may stimulate development of similar phosphors. Jahn et al developed a 2D OSL reader with sub millimeter resolution for BeO films [11]. BeO has been studied as a dosimeter for several decades and its effective atomic number of 7.14 is close to tissue equivalence. A linear dose response from 10^{-3} to 10^{1} Gy was presented. Commercially available films with dimensions of 52x52x1 mm were used to record a dosimetric image. Further development of this technology should provide a very useful 2D dosimeter system.

3.2. Alanine powders

Alanine, lithium formate and lithium fluoride were compared for water equivalent energy response using Monte Carlo simulations [12]. Lithium formate with ESR readout provides the closest approximation to water response and exhibits sensitivity 5 times that of alanine and has a linear response from 0.2 to 1,000 Gy. These attributes make lithium formate an alternative dosimeter when accuracy is required from a small integrating dosimeter.

A study examining the feasibility of using alanine dosimeters in combination with Monte Carlo calculations for mixed neutron photon dosimetry indicates the method can provide consistent results for proposed boron neutron capture therapy [13].

3.3. Diamond diodes

Diamond detectors are of interest for water equivalent dosimetry since the atomic number is closer to water than silicon. Recently chemical vapour deposition has allowed commercial production of synthetic diamond with sizes suitable for radiation dosimetry. Lansley et at have compared samples of synthetic diamond films, 3x3x0.5 mm against a commercially available diamond detector from PTW

[14]. Overall performance was comparable. Spadaro et al examined response of detector and standard grade diamond films with two types of metal electrodes [15]. Synthetic diamond films have also been evaluated as thermal luminescence dosimeters and the performance is promising [16]. Agreement between 12 MeV electron depth dose and ionization chamber measured dose demonstrates these materials can be employed for accurate dosimetry. Repeatability of 4% maximum was reported. Single crystal synthetic diamond detectors have been compared against ion chambers and planning calculations for IMRT type dose distributions [17]. The results are comparable, indicating that diamond detectors may be useful for small field dosimetry. Independent studies are also reporting that commercial samples of CVC diamond films are suitable for radiotherapy dosimetry [18]. Larger crystals from other sources are also demonstrating acceptable performance [19]. Deposition of carbon nickel thin films by laser sputtering provided enhanced performance over both carbon and platinum electrodes, demonstrating importance of electrode quality on device performance [20].

3.4. Silicon diodes

Silicon diodes are commonly used in radiotherapy dosimetry because of their high sensitivity and potential sub millimeter spatial resolution. Drifts associated with accumulated dose and non-water equivalent energy response can be addressed by annealing with large doses and correct packaging of the detectors for specific applications. Silicon based dosimeters may become obsolete if synthetic diamond dosimeters become popular. However, silicon dosimeters have a large research literature to support continued development and are likely to be a less expensive technology for the near future. A thin 2D array of thin epitaxially grown diodes was characterized by Wong et al [21]. This technology could be used either in transmission mode or sandwiched into a water equivalent phantom. The sensitive volumes were 0.5x0.5x0.05 mm^3. Dose rate dependence, energy response and angular response were reported. Based on the reported data, thin diodes appear to be an improvement over current commercial diode dosimeters. Eklund and Ahnesjo have modeled silicon diode response with Monte Carlo simulations and have concluded, that thin active layers sandwiched between thicker inactive layers is a preferred geometry for dosimeters [22]. Cranmer-Sargison et al have evaluated the small field performance of the commercially available diode dosimeters and reported the reproducibility to be within 1.25% [23].

4. Planar dosimeters

Two dimensional sub millimeter resolution continues to be the strength of film dosimeters. Silver halide film is now obsolete as a commercial product due to replacement by CCD and CMOS diode arrays in digital imaging systems. However, high resolution, water-equivalent detector arrays with fast readout have yet to be commercialized. Currently the only commercial film dosimeter is the radiochromic product from Ashland Chemicals referred to as external beam therapy (EBT) film. Nearly all other types of dosimetry are compared with film dosimetry either for validation within a plane or as a cross reference at selected points. For example, 2D film dosimetry is commonly used to validate Monte Carlo simulations and selected planes within a 3D dose distribution measured or calculated. For this reason, the film dosimetry papers continue to dominate the literature in numbers. In fact, 50% of the papers in this review are related specifically to EBT film dosimetry. The EBT product continues to evolve. EBT is now obsolete, replaced by EBT2 and EBT3 films. Scanner technology continues to evolve as well. Commercial radiochromic film scanners have yet to be marketed and much work is involved in developing calibration protocols for flatbed document scanners. It is interesting that many of the issues associated with quantitative optical CT scanning are the same for radiochromic film scanning. Stray light is generally the limiting feature in obtaining the maximum information recorded in 2D and 3D dosimeter materials.

4.1. Radiochromic film

A review of radiochromic film dosimetry was recently prepared by Devic [24]. This paper serves as an efficient introduction to the field. The range of calibration strategies is summarized and provides a template for design of protocols for 3D dosimeters. The analogies between film and 3D dosimetry are useful to follow. Energy dependence and film uniformity are identified as current issues, but both can be solved with additional product development. A tremendous advantage of EBT2 film over, silver halide films is that it can be directly immersed in water for several minutes without measurable impact on dose response. This feature is documented in a paper by Arjomandy et al in which film strips are mounted in a water tank, collinear with photon, electron and proton beams and measured depth doses are compared with calibrated ion chamber results [25]. When high resolution is required, EBT2 film provides a "excellent secondary dosimeter". Stray light with flatbed document scanners is model dependent and requires careful evaluation to minimize impact of measured dose accuracy. Buston et al documented improvements that can be achieved for the Epson V700 model by placing additional white sheets of paper behind the film [26]. Generally, increasing the amount of scatter in an optical system degrades spatial resolution and dynamic range. A 1% uniformity was achieved over the entire scanning area.

Film uniformity has been reported by several authors as a problem with EBT2 film. While it is anticipated this is not a fundamental problem and improved manufacturing methods will make the uniformity within acceptable levels, users are left evaluating each film and developing protocols to minimize effect on measured dose. If each film is unique, then in principle, each pixel needs to be calibrated individually. Roozen et al have examined this issue and developed a pixel based calibration protocol that addresses the non-linear dose response of EBT2 film [27]. In this method a pre-irradiation is used to correct for non-uniformity film sensitivity (thickness) and record pixel sensitivity. Following irradiation and scanning of the dose distribution of interest, a second calibration irradiation is performed and used to correct the non-linear response of that pixel. This paper provides a clear example of the complexity of working with non-uniform, non-linear dosimeters and should be a warning to researchers working with 3D dosimeters. Mizuno et al evaluated EBT2 film homogeneity from different lots with an Epson GT-X970 scanner and recommended the each lot be evaluated before clinical use and the calibration curve be developed using net optical density of red channel only [28]. Another study by Aland et al reached the same conclusion with an Epson V700 scanner that incorporating the blue channel increased uncertainty [29]. EBT film dosimetry is actively being researched also by the staff of the manufacturer, Ashland Chemicals. Micke et al have developed several innovations in calibration protocols to maximize the accuracy of their product [30, 31]. They demonstrated an analysis method that uses the red, green and blue channels to correct for film non-uniformity pixel by pixel and maximize the dynamic range by merging dose responses for the three channels. They have also demonstrated calibration strategies using a 'master response' and individual sample responses to minimize the number of calibration points required to characterize a given film. With this approach accurate dosimetry within an hour of irradiation is demonstrated. This represents a significant achievement for making EBT2 practical for patient QA. The technique includes development of a formalism using a linear response function for the film-scanner system. A detailed comparison of different uniformity correction methods using the RGB channels demonstrated both improvements and degradations relative to the red channel netOD method [32]. Another detailed investigation found the RGB method proposed was equivalent to the single red channel approach with the advantage of not requiring a pre-irradiation film scan [33]. Mayer et al proposed another method involving ratios of the RGB images as an alternative to the "registered correction matrices" in the current analysis method of the manufacturer [34]. They concluded this method was simpler.

In order to achieve a high dynamic range in measuring a photon beam profile, Van den Heuvel et al, used three films and three central axis doses in order to use the scanner and film within optimum optical densities [35]. This technique, while useful for film dosimetry, may have less value in 3D dosimetry where the 3D stray light distribution is directly coupled the 3D dose image. However, the paper provides a clear demonstration of how to achieve the highest quality measurement possible within the limitations of the equipment available. EBT2 film with 40 Gy dose per fraction was demonstrated to have accuracy to less than 3% for 4 and 12 mm targets [36]. A very detailed film dosimetry protocol that achieves 2% accuracy was documented by Huet [37] and could provide technical insights for novice and experienced users of film and Epson scanners to review.

EBT film has been used to validate Monte Carlo calculations for 4-20 mm diameter, 6 MV photon beams [37], and identify dosimetric differences related to inter-leaf leakage, helical field junctioning and small field segments in IMRT deliveries [38]. Epson scanner comparisons reported similar dosimetry performance for 10000XL and V700 with V330 producing acceptable results [40]. An alternative to flatbed document scanners are devices specifically designed to digitize color film. Recently, immersion of film in refractive index matching fluid for scanning has been reported. This is completely analogous to the use of liquids for optical CT scanning of 3D dosimeters. Interference fringes, called Newton's rings, can be observed when scanning film in partial contact with a glass plate. EBT3 film was engineered to eliminate these fringes by placing small beads that the film face to increase air space and eliminate the optical interference. An alternative method involves suspending the film in air during transmission imaging. The performance of the Nikon Coolscan 9000 ED film scanner with EBT2 film is reported demonstrating high quality images for film sizes less than 6 x 20 cm^2 [41]. The spectral dependence of HS radiochromic film was re-evaluated and provides a useful reminder that both stray light and spectral effects contribute to the overall shape of dose response curves and that narrower spectral sources provide more linear responses [42]. Such information is directly applicable to design of 3D optical CT scanners and 2D film scanners optimized for readout of radiochromic films.

A recent paper by Devic et al describes fitting dose response curves to a linear function [43]. Results are presented using the green channel of EBT2 film irradiated to a maximum dose of 10 Gy. This approach allows accurate relative dosimetry to be performed without developing a multiple point calibration curve and first determining absolute dose. This innovation may have considerable value for analysis of 3D dosimeters with non-linear responses.

4.1.1. EBT2 versus Presage

Synchroton machines can generate high resolution dose distributions and are invaluable for studying the effects of microbeam radiotherapy. High resolution 3D dose distributions are being presented at this meeting using optical CT readout of Presage. An alternative approach involves sampling with a confocal microscope operating in fluorescence mode. Data reveals that Presage can record higher spatial resolution than EBT2 film due to diffusion within the latter [44]. This is also a reminder that luminescence signals need to be eliminated when measuring absorption from strongly attenuating samples, whether the samples are 2D films or 3D dosimeters. Inference from this data suggests films of Presage may be very useful 2D dosimeters.

Many formulations of Presage have been reported and exploration continues to provide enhanced sensitivities [45].

5. Scintillation dosimeters

Plastic fiber scintillators have been reviewed in previous DOSGEL reviews. The key advantages are water-equivalent response and typical volumes of 1 mm^3. They provide accurate dosimetry for small fields and with appropriate calibration protocols the coupling fiber 'stem effect' is minimized. In this program there is a review talk and several original presentations related to these dosimeters. For this reason, only other types of scintillation papers are reported in this review.

Analogous to larger 3D gel dosimeters, volumes of liquid scintillators can be used as both phantom and dosimeters. Imaging the visible scintillation light generated by dose deposition provides a real-time 3D (4D) dosimetry system. In principle, this process could also be performed with scintillation from appropriate gel materials, but liquids are likely better because of lower optical scatter and fewer chemicals to consider. The shapes are limited to the vessel that contains the liquid. By imaging at video frame rates, Archambault et al have demonstrated that spot scanned proton beamlets for intensity-modulated proton therapy can be verified [46]. Proton range, position and intensity were recorded with a 20x20x20 cm^3 liquid scintillator system with sub millimeter resolution.

Many studies have been devoted to developing methods to quantify the amount of light originating in the optical fiber connected to the plastic fiber dosimeter. This light is analogous to the stem effect with ion chambers. There have been a few reports in which time resolved detection has been employed to isolate the light originating in the dosimeter from the extraneous light from the coupling fiber. Many of the commercially available plastic scintillators emit light on the nanosecond time scale. The light from the fiber is due to several mechanisms and spans the range from picosecond to millisecond. By finding an appropriate combination of scintillator material and coupling fiber, it should be possible to effectively discriminate signals using time. Most medical linear accelerators have pulse trains that last several microseconds and a repetition rate of less than a kilohertz. This window between pulses can be used for accurate dosimetry provided dosimeters with suitable decay rates can be developed. Organic scintillators with lifetimes of 20 microseconds have been evaluated and the data indicates even longer lifetimes are required [47]. This approach may lead to new high performance dosimeter materials with applications spanning 0 to 4D dosimetry.

6. Conclusion

Radiotherapy dosimetry remains an active area of research as witnessed by the many topics listed here and those being presented at this meeting. Three-dimensional dosimetry [48-50] builds on the knowledge developed for point dosimeters and 2D films. Overall, the dosimetry trend continues to evolve towards energy-independent (water equivalent), linear dosimetry systems.

7. References

[1] Aspradakis M *et al* 2010 *IPEM* report number 103
[2] Bräuer-Krisch E *et al* 2010 *Amer. Inst. Phys. CP* **1266** 89-97
[3] Bak J *et al* 2012 *J. Appl. Clin. Med. Phys.* **12** 73-80
[4] Hill R *et al* 2010 *Radiat. Meas.* **45** 810-5
[5] Watanabe Y and Hayashi N 2012 *J. Appl. Clin. Med. Phys.* **13** 269-81
[6] Kerns J *et al* 2011 *Med. Phys.* **38** 3955-62
[7] Scarboro S *et al* 2012 *Phys. Med. Biol.* **57** 2505-15
[8] Mrčela I *et al* 2011 *Phys. Med. Biol.* **56** 6065-82
[9] Al-Senan R and Hatab Mustapha 2011 *Med. Phys.* **38** 4396-406
[10] Dhabekar B *et al* 2011 *Nucl. Instr. Methods. Phys. Res. B* **269** 1844-48
[11] Jahn A *et al* 2011 *Radiat. Meas.* **46** 1908-11
[12] Waldeland E and Malinen E 2011 *Radiat. Meas.* **46** 945-51
[13] Schmitz T *et al* 2011 *Acta. Oncol.* **50** 817-22
[14] Lansley S *et al* 2010 *Australas. Phys. Eng. Sci. Med.* **33** 301-6
[15] Spadaro S *et al* 2011 *Sens Actuators A Phys A* **171** 43-7
[16] Benabdesselam M *et al* 2011 *Diam. Relat. Mater.* **20** 520-2
[17] Tromson D *et al* 2010 *Diam. Relat. Mater.* **19** 1012-1016
[18] Betzel G *et al* 2012 *Physica Medica* **28** 144-52
[19] Ade N *et al* 2012 *Radiat. Phys. Chem.* **81** 232-9
[20] Abdel-Rahman M *et al* 2012 *Diam. Relat. Mater.* **22** 70-6
[21] Wong J *et al* 2012 *Med. Phys.* **39** 2544-58
[22] Eklund K and Ahnesjö 2010 *Phys. Med. Biol.* **55** 7411-23
[23] Cranmer-Sargison G *et al* 2011 *Radiother. Oncol.* **100** 429-35
[24] Devic S 2011 *Physica Medica* **27** 122-34
[25] Arjomandy B *et al* 2012 *Med. Phys.* **39** 912-21
[26] Butson E *Australas Phys. Eng. Sci. Med.* **34** 401-7
[27] Roozen K *et al* 2011 *Australas. Phys. Eng. Sci. Med.* **34** 281-9
[28] Mizuno H *et al* 2012 *J. Appl. Clin. Med. Phys.* **13** 198-205
[29] Aland T *et al* 2011 *Australas. Phys. Eng. Sci. Med.* **34** 251-60
[30] Micke A *et al* 2011 *Med. Phys.* **38** 2523-34

[31] Lewis D 2012 *Med. Phys.* **39** 6339-50
[32] McCaw T *et al* 2011 *Med. Phys.* **38** 5771-7
[33] van Hoof S *et al* 2012 *Phys. Med. Biol.* **57** 4353-68
[34] Mayer R *et al* 2012 *Med. Phys.* **39** 2147-55
[35] Van den Heuvel F *et al* 2011 *Med. Phys.* **38** 6443-8
[36] Hardcastle N *et al* 2011 *Med. Phys.* **38** 4081-5
[37] Huet C *et al* 2012 *Radiat. Meas.* **47** 40-49
[38] García-Garduño O *et al* 2010 *Radiother. Oncol.* **96** 250-3
[39] Kairn T *et al* 2011 *Australas. Phys. Eng. Sci. Med.* **34** 333-43
[40] Alnawaf H *et al* 2012 *J. Appl. Clin. Med. Phys.* **13** 314-21
[41] De Puysseleyr A *et al* 2012 *Phys. Med. Biol.* **57** 127-42
[42] Secerov B *et al* 2011 *Nucl. Instr. Methods Phys. Res. A* **633** 66-71
[43] Devic S *et al* 2012 *Med. Phys.* **39** 4850-7
[44] Annabell N *et al* 2012 *J. Synchrotron Rad.* **19** 332-9
[45] Alqathami M *et al* 2012 *Radiat. Phys. Chem.* **81** 867-73
[46] Archambault L *et al* 2012 *Med. Phys.* **39** 1239 46
[47] Beierholm A *et al* 2011 *Radiat. Meas.* **46** 1982-4
[48] Lepage M *et al* 2002 *Phys. Med. Biol.* **47** 1881-90
[49] Rintoul L et al 2003 Appl. Spectrosc. **57** 51-7
[50] Baldock C *et al* 2010 *Phys. Med. Biol.* **55** R1-63

Journal of Physics: Conference Series **444** (2013) 012010 doi:10.1088/1742-6596/444/1/012010

3D EPID based dosimetry for pre-treatment verification of VMAT – methods and challenges

PB Greer
Radiation Oncology Department, Calvary Mater Newcastle, Newcastle, NSW 2298, Australia. School of Mathematical and Physical Sciences, University of Newcastle, Newcastle, NSW 2310, Australia

E-mail: peter.greer@newcastle.edu.au

Abstract. This article presents an overview of pre-treatment verification of volumetric modulated arc therapy (VMAT) with electronic portal imaging devices (EPIDs). Challenges to VMAT verification with EPIDs are discussed including EPID sag/flex during rotation, acquisition using cine-mode imaging, image artefacts during VMAT and determining the gantry angle for each image. The major methods that have been proposed to verify VMAT with EPIDs are introduced including those using or adapting commercial software systems and non-commercial implementations. Both two-dimensional and three-dimensional methods are reviewed.

1. Introduction

It is recommended that more than 50% of cancer patients receive radiation therapy treatment [1]. Therefore this therapy represents a very important frontline treatment for cancer. With the aim to deliver higher doses of radiation to the tumour and lower doses to surrounding normal tissues, the delivery and planning approaches to radiation therapy are constantly evolving. The use of intensity modulated radiation therapy (IMRT) where the radiation beam fluence is non-uniform allows for greater shaping of dose distributions within the body and hence increased sparing of normal tissues. This has traditionally been delivered using several beams at fixed linear accelerator gantry angles. The non-uniform fluence is delivered by varying the beam aperture defined by the multileaf collimator (MLC). Because IMRT is a highly complex technique and the derived fluences are non-intuitive, the accuracy of delivery is usually verified by pre-treatment dosimetric measurements. This is mandated in some countries for every patient while in others it is optional (although extensively performed). The pre-treatment dosimetric verification comprises a comparison of a measured dose with the treatment planning system calculated dose. This can be performed on a beam-by-beam basis using two-dimensional (2D) measurement or on a whole-of-treatment basis using either a 2D measured dose-plane or less commonly a three-dimensional (3D) measurement.

Although the use of a continuously rotating gantry to delivery IMRT had been proposed for some time, the commercial implementation of this method did not occur until a new method for the derivation of the gantry motion, and multileaf apertures was derived by Karl Otto [2]. He termed this technique volumetric modulated arc therapy (VMAT). His method resulted in a highly efficient IMRT plan delivery and has therefore seen dramatic uptake in recent years. VMAT is commercially available as either RapidArc™ (Varian Medical Systems, Palo Alto, CA) or VMAT (Elekta AB, Stockholm, Sweden). The method allows for continuous variation of gantry speed, dose-rate and MLC aperture during delivery. The delivery is controlled by specifying control points at discrete gantry angles. For example for the Varian system two sets of control systems are used; 1) the fractional monitor units (MU) versus gantry angle, and 2) the MLC shape versus gantry angle. This highly complex delivery technique requires rigorous pre-treatment verification. Various existing dosimetry arrays have been utilised and in addition some new devices have been specifically designed for VMAT measurements. As with IMRT verification, the EPID represents a potentially very efficient method to verify VMAT, due to the very rapid setup time, immediate digital format, high resolution of the individual photodiodes, high level of reproducibility, and integration with the linear accelerator software. This review article outlines the major approaches that have been developed to date for the use of EPID for pre-treatment verification of VMAT and discusses some of the major challenges that must be overcome. The article is restricted in scope to verification of the recently developed VMAT technique of Otto (or similar) delivered on medical linear accelerators and recorded with two-dimensional EPID systems using non-transmission dosimetry (i.e. no phantom is present in the beam).

2. Challenges for pre-treatment dosimetric verification of VMAT with EPIDs

This section discusses some of the issues for using EPIDs for dosimetry that are specific to VMAT verification. More general issues for EPID dosimetry are covered elsewhere [3, 4].

2.1. EPID sag/flex during arc

When acquiring an integrated EPID image during gantry rotation the EPID will sag as a function of gantry angle. The effect of this is to displace the acquired frames from each other on the EPID detector with subsequent blurring of the integrated dose. Mans et al. [5] aligned single frames (typically one frame every 20 degrees) acquired during VMAT with the corresponding control point field outline. They applied the same shift to all neighbouring frames (over ±10 degrees). They quoted the magnitude of the EPID flex as 4 mm peak-to-peak at the EPID level. Rowshanfarzad et al. [6] measured the EPID sag for the Varian system as a function of gantry angle during arc rotation using a marker placed at isocentre and corrected for the effect of the marker displacement from the mean isocentre position. They found that the sag pattern is very reproducible but is accelerator dependent. The sag was largest in the inplane direction with an amplitude of approximately 1 mm (at the isocentre level) and smaller in the crossplane direction with an amplitude of less than 0.5 mm. One accelerator was found to have a much larger amplitude of nearly 5 mm in the inplane direction. They showed that the sag degrades Gamma results in the regions of steep dose gradients. To correct the sag, they acquired cine-mode images, correcting each image according to the previously derived sag map values at that gantry angle, and then recombined the images to form an integrated image. This approach was limited to shifting each image to the nearest pixel. Other authors have also measured the EPID sag [7, 8]. An entirely different approach was developed by Iori et al. who used an additional holding device which clamped to the accessory mount and held the EPID in position during rotation [9]. They were able to restrict the EPID sag to less than 1 mm with this device. The EPID in this case was mounted on an R-Arm which exhibits greater sag with gantry angle than the newer E-Arm.

2.2. Cine-mode imaging

For fixed beam IMRT, integrated mode EPID images are used for each beam, as the gantry angle can easily be verified visually. However for VMAT delivery, to fully verify accurate dose delivery requires that the dose is verified as a function of gantry angle. Therefore an integrated image acquired for the entire delivery would not fully validate the delivery as it compresses dose from all gantry angles into one image. This has prompted the use of cine-mode image acquisition where separate images are continuously acquired during the delivery, and these are associated with a gantry angle. For the Elekta system Mans et al. [5] modified in-house developed acquisition software to save every detector frame during imaging (~ 2.5 frames per second). For the Varian system the cine acquisition mode was not intended for dosimetry. Piermattei et al. [10] investigated cine-mode for the IAS2 system with a 0.6 Hz imaging rate. They used a low dose-rate of 100 MU/min and found good stability (within 1%, 2 SD) of the EPID signal for this mode using weekly measurements over a three month interval in both static and dynamic conformal arc modes. They also determined that signal reproducibility within each session and signal linearity were within 1% (2 SD). McCurdy and Greer [11] investigated the dosimetric performance of the IAS3 system cine-mode (7.5 Hz) compared to integrated mode. They found that the time-resolved signal agreed well with time-resolved ion-chamber measurements but that the mode under-responded to small dose levels. This is due to the two frames that contain partial signal at beam-on being discarded with this mode (Figure 1). Similarly the two partial frames at beam-off are also discarded. This was recently confirmed using a separate frame-grabber system to acquire all frames during cine-mode acquisition of a sliding window IMRT delivery. By comparing the first frame of the cine-mode acquired data with the separately acquired frames, the discarded frames were identified. These partial frames are included with integrated mode, (although recent investigations suggest that the very last partial frame in integrated mode is still discarded). Another problem with this type of imaging is that large numbers of images can be acquired for a single VMAT delivery. This can cause memory overflow problems during delivery.

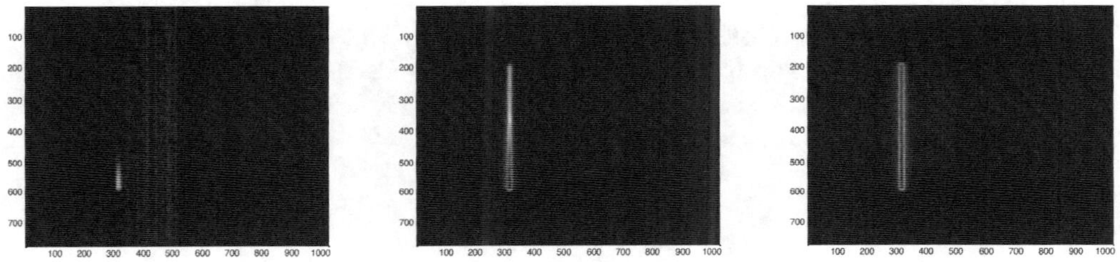

Figure 1: The first three frames acquired for a sliding window irradiation. The first two frames contain partial dose information as the beam-on occurred during the frame integration period. In the current Varian cine-mode implementation these first two frames (along with the final two partial frames that are read-out after beam-off) are discarded and some dose information is lost.

2.3. Image artefacts

For the Elekta system Mans et al. [5] reported image artefacts when they acquired individual frames using their in-house system and these were attributed to changes between discrete dose-rate levels during acquisition. This prevented them from determining the patient transmission factor using images acquired with and without the patient present, therefore they replaced this with a calculated transmission factor. For Varian systems VMAT is delivered using the pulse-drop servo to control the dose-rate (for fixed beam IMRT a pulse-length servo is used). This results in pulse-dropping artefacts in the frames of images acquired during VMAT delivery. These occur because different rows of the imager are read out at slightly different times. Therefore the time periods over which each row has integrated the dose are different. Each row can therefore integrate different numbers of beam pulses depending on when the pulses are dropped, with resulting different output signals. As the rows are read-out in batches, with each batch read-out very rapidly (virtually simultaneously), the result is banding artefacts where each batch of rows has integrated over a different number of pulses (Figure 2). These are not actual artefacts, the imager is accurately integrating the dose or pulses received, however these bands are problematic for dosimetry with cine-images as the non-uniform dose delivery is not included in any modelling or treatment planning system reference dose. The simplest method to reduce this is to average multiple frames for each cine image. However this has the disadvantage of averaging dose delivery over an increasing range of gantry motion. Further methods to account for these artefacts should be investigated.

2.4. Gantry angle for each cine-image

Methods are required to associate a gantry angle to each cine-image. Mans et al. [5] using their in-house acquisition software recorded the gantry angle via an iCom connection to the treatment machine and stored this with every recorded image frame. They investigated the lag of this gantry angle using a dedicated phantom with imbedded steel ball-bearings (BB's) to obtain the gantry angle from the image of the phantom. Their analysis found a lag of ~0.4 s or about one frame for the gantry angle. To compensate for this they used the average gantry angle of the acquired frame and the two successive frames. For a Varian C-Series linear accelerator the accuracy of the gantry angle stored in the DICOM image header was investigated by Ansbacher et al. [12] using a custom-designed "radiographic inclinometer" phantom. The phantom consisted of a pair of lead solder wires spiralled along the surface of a 20 cm long acrylic tube. As the gantry rotates the apparent intersection point of the wires moves proportionally along the axis of the tube. From the length of the intersection point the gantry angle range for each cine-image could also be measured. By comparing the imaged gantry angle and the header gantry angle they found that the header angle lagged the imaged angle for 3 frame averaged cine-images and from the number of repeated angle instances they suggested that the angle is updated no more than 2-3 times per second. The header angle accuracy was also investigated by McCowan et al. by comparison to dynalog file data, a separate encoder, and a similar radiographic inclinometer [13]. They found that the header angle can have errors of several degrees. Adamson and Wu [14] also determined the accuracy of the header angle as 1.0 ± 0.8 degrees although the reference angle source was not stated. At present to associate an accurate gantry angle with the EPID image requires the use of an external angle source which must be synchronised with the EPID images. The Varian TrueBeam™ accelerator uses an accurate encoder to populate the header with the gantry angle [D. Morf, Varian Medical Systems, private communication].

Figure 2: Example cine-mode frame of an open field illustrating the banding artefacts due to pulse-dropping during the acquisition.

3. Methods for pre-treatment dosimetric verification of VMAT with EPIDs- 2D

3.1. Commercial software systems

The Varian Portal Dosimetry™ system was developed for the pre-treatment verification of IMRT and has become widely used at many centres. The system has a model that uses details of the treatment plan to derive a prediction of the EPID response for the IMRT beam [15]. The model is very simple, a "fluence" matrix is first derived which represents the relative beam-on fraction. This is convolved with a triple-Gaussian kernel to model the dose-deposition in the EPID. A field-size dependent correction factor is applied which is derived by comparison of predicted and measured EPID response for a range of jaw-defined fields. The EPID response is calibrated to CU units by delivering a 10×10 cm^2 field of 100 monitor units to the EPID, and specifying a CU value. The prediction model then scales its data to CU so that the prediction and measurement can be compared in absolute CU units without normalisation. EPID images are acquired in integrated mode which collects image frames during the entire delivery and then stores an integrated CU value.

Iori et al. [9] have investigated the application of this system to VMAT verification for 23 fields. They stabilised the EPID detector using an ancillary holding device clamped to the accessory mount which holds the EPID in place during gantry rotation. As the prediction model required an IMRT plan, they used an "IMAT-SIM" converter to translate the VMAT delivery into a static gantry angle IMRT field plan. The Portal Dosimetry system was then able to predict the EPID image for this IMRT plan. The delivery at the linear accelerator was performed using both static gantry IMRT and VMAT. They found that the percentage of field-area that failed Gamma analysis with criteria of 3% and 3 mm was 3.0% ± 3.6% (mean ± 1 SD) for IMRT delivery and 3.4% ± 3.9% for the VMAT deliveries. A higher tolerance of 5% for the dose-difference was used for outer x-ray beam regions. The correlation between the fail rates for the IMRT and VMAT deliveries was within 1%.

The GLAaS algorithm converts measured EPID images to a dose-in-water plane [16]. These can then be compared to treatment planning system dose to water calculations for the IMRT beam. It does this by approximating the IMRT delivery as a series of N static segments and deriving correction factors that are applied to the EPID image based on an equivalent field size of each segment (termed equivalent window width Field ewwF). Empirically derived look-up tables are generated that relate EPID response to dose-in-water as a function of ewwF and depth in water. It also accounts for the difference in response of the EPID to open and MLC transmitted components of the delivered segment. Nicolini et al. [8] applied this method to VMAT plans delivered on Varian systems. They tested seven clinical-type deliveries using integrated images acquired over entire (360 degree) arcs. The EPID-derived dose-in-water was compared to Eclipse treatment planning system calculations where the VMAT delivery was collapsed onto an infinitesimally small gantry rotation, generating the integrated dose distribution for the entire arc on a single dose plane. Every control point was individually calculated and accumulated. They obtained Gamma pass rates with 3%, 3 mm criteria of 97.7 ± 1.2% (mean ± 1 SD) at 6 MV and 94.9 ± 1.3% at 18 MV. Some preliminary tests with shorter partial arcs of 6 or 12 degrees were also performed. For two plans, three consecutive partial-arcs were delivered and measured with the EPID. In these cases separate plans in Eclipse were generated encompassing the control-points corresponding to the arc. For the 6 degree arcs this comprised 4 control points (including start and stop points) with each control point representing 2 degrees of rotation. Tests of the short arcs were performed on a simplified VMAT implementation with constant dose-rate and gantry speed. Gamma analysis results were above 96.0% pass-rate with 3%, 3 mm criteria. In a much larger study from the same group a total of 275 patients with 375 arcs were examined [17]. The Gamma analysis results were 97.1 ± 2.4% (mean ± 1 SD) for a 3%, 3 mm criteria.

More recently Bailey et al. [18] investigated the EPID dosimetry algorithm EPIdose (Sun Nuclear Corporation, Melbourne, Fl, USA) for 26 VMAT deliveries. The EPIdose algorithm [19] converts the EPID image to a dose plane in water for direct comparison to the treatment planning system calculation. The method uses the MLC plan data and for each MLC segment shape derives a correction for the difference in EPID output factor to dose in water output factor. It also derives a correction for the difference in EPID response to open and MLC shielded portions of the segment. The EPID image is then convolved with a kernel to account for the difference in dose-deposition kernels of the EPID and water. Integrated mode EPID images were acquired for the deliveries, and the converted EPID data were compared to Eclipse calculated dose-planes. The Gamma evaluation results for 14 prostate fields was 98.2 ± 1.7% (mean ± 1 SD) with 3%, 3 mm criteria, and for 12 head and neck fields 95.3 ± 5.9%. Very similar results were obtained with a MapCheck device replacing the EPID measurements.

3.2. Other proposed methods

A non-dosimetric method has been developed by Bakhtiari et al [7]. In this method the MLC aperture is extracted from cine-mode EPID images acquired during VMAT delivery. These apertures are compared to the apertures defined in the MLC delivery file, which contains the MLC apertures for each of 177 control points (gantry angles). Using the gantry angle information in the EPID image header, the delivery file apertures were interpolated and then compared to the EPID measured aperture for that gantry angle.

4. Methods for pre-treatment dosimetric verification of VMAT with EPIDs- 3D

True 3D dosimetry with current EPIDs is not possible, as the EPIDs are inherently two-dimensional and are mounted on a support-arm. Three-dimensional dosimetry with EPIDs takes the approach of dose estimation (often termed reconstruction) in 3D within a virtual phantom or patient model calculated from the "in-air" EPID 2D measurement. A three-dimensional method to reconstruct dose in a patient model from EPID measured cine-mode transit dose images during treatment has been reported by Mans et al. [5] however this method is outside of the scope of this review. Ansbacher et al. [12, 20] have presented a method where a three-dimensional dose distribution is reconstructed in a virtual cylindrical phantom based on in-air cine-EPID measurements without any phantom present. This allows for direct comparison with an Eclipse dose calculation in a phantom with the same geometry. They compared a VMAT plan with the EPID reconstruction where every 10 EPID frames were combined into a single image before reconstructing the dose. Using a Chi 3-D comparison they found pass-rates of 84% for a high dose region (dose above 80% of the prescription) and 96% for a low dose region (dose 40-80% of prescription). By delivering the same plan as a conformal arc they were able to show that the Chi rates were markedly reduced to 55-57% with this method, however when these plans were "collapsed" onto one gantry angle the dose distributions were not able to be differentiated. This demonstrated the superior ability of gantry-angle resolved dosimetry to detect delivery errors for VMAT.

Another concept that has been recently reported in abstract form, has three-dimensional aspects in its approach [14]. The method used cine-EPID images acquired during delivery. The plan is modified so that the two most inferior MLC leaf pairs are open at all control points, and the collimator angle is zero or 90 degrees. A custom-built phantom is positioned so that it is visible on all images through the opened leaf pairs. This is used to measure the angular fraction through which each portal image is acquired. The expected values of delivered monitor units and the MLC positions at the beginning and end of each image are then derived from the angular increment. These are then used to derive a predicted detector response for this angular fraction using the Varian Portal Dosimetry prediction model outlined above. The two data sets are then separately stacked into a cube format with each image forming an x-y plane, with the z-coordinate representing the gantry angle. Each image has a leaf-motion axis and a cross-leaf axis. A projection through the data stack onto 3 planes is then formed by integrating along rays through the pixel data in the stack. The three data planes that are formed represent: leaf motion axis versus cross leaf axis; leaf motion axis versus gantry angle and cross leaf axis versus gantry angle. The 3 data planes for the measured and predicted data sets can then be compared.

5. Summary

VMAT is a highly complex delivery technique that requires careful and rigorous pre-treatment quality assurance. Ideally this quality assurance should verify the correct operation of the linear accelerator control systems for VMAT delivery, i.e. that the correct fractional MU is delivered between each gantry angle control-point and the correct MLC movement occurs between each control point. Current methods to perform pre-treatment VMAT quality assurance with EPIDs mostly do not fully realise this level of assurance, and they are predominantly 2D. Further research is required to provide more comprehensive and efficient pre-treatment quality assurance methods with EPIDs.

6. Acknowledgements

The author gratefully acknowledges the contributions of Pejman Rowshanfarzad, Brian King, Todsaporn Fuangrod, Henry Woodruff, and Michael Barnes at the Calvary Mater Newcastle. Boyd McCurdy and Peter McCowan at CancerCare Manitoba are also acknowledged for their many contributions. The author acknowledges the assistance of Daniel Morf and Varian Medical Systems iLab, Baden, Switzerland for assistance with technical queries on image acquisition and for the loan of the frame-grabber system used to investigate cine-mode acquisition.

7. References

[1] Delaney G *et al* 2005 *Cancer* **104** 1129-37
[2] Otto K 2008 *Med. Phys.* **35** 310-7
[3] van Elmpt W *et al* 2008 *Radiother. Oncol.* **88** 289-309
[4] Greer P B and Vial P 2011 *Concepts and Trends in Medical Radiation Dosimetry,* ed A B Rosenfeld, *et al.* pp 129-44
[5] Mans A *et al* 2010 3D *Radiother. Oncol.* **94** 181-7
[6] Rowshanfarzad P *et al* 2012 *Med. Phys.* **39** 623-35
[7] Bakhtiari M *et al* 2011 *Med. Phys.* **38** 1366-73
[8] Nicolini G *et al* 2008 *Radiat. Oncol.* **3**
[9] Iori M *et al* 2010 *Med. Phys.* **37** 377-90
[10] Piermattei A *et al* 2009 *Med Biol Eng Comput* **47** 425-33
[11] McCurdy B M C and Greer P B 2009 *Med. Phys.* **36** 3028-39
[12] Ansbacher W *et al* 2010 *J. Phys.: Conf. Ser.* **250** 108-11
[13] McCowan P M *et al* 2011 *Med. Phys.* **38** 3534
[14] Adamson J and Wu Q 2012 *Phys. Med. Biol.* **57** 6587-600
[15] Van Esch A *et al* 2004 *Radiother. Oncol.* **71** 223-34
[16] Nicolini G *et al* 2006 *Med. Phys.* **33** 2839-51
[17] Fogliata A *et al* 2011 *Br. J. Radiol.* **84** 534-45
[18] Bailey D W *et al* 2012 *J. Appl. Clin. Med. Phys.* **13** 82-99
[19] Nelms B E *et al* 2010 *J. Appl. Clin. Med. Phys.* **11** 140-57
[20] Ansbacher W *et al* 2006 *Med. Phys.***33** 3369-82

Journal of Physics: Conference Series **444** (2013) 012011 doi:10.1088/1742-6596/444/1/012011

3D EPID-based *in vivo* dosimetry for IMRT and VMAT

B Mijnheer, I Olaciregui-Ruiz, R Rozendaal, J-J Sonke, H Spreeuw, R Tielenburg, M van Herk, R Vijlbrief and A Mans
Department of Radiation Oncology, The Netherlands Cancer Institute - Antoni van Leeuwenhoek Hospital, Plesmanlaan 121, 1066 CX Amsterdam, The Netherlands

E-mail: b.mijnheer@nki.nl

Abstract. In this paper the various approaches of EPID-based *in vivo* IMRT and VMAT dose verification, and their clinical implementation, are described. It will be shown that EPID-based *in vivo* dosimetry plays an important role in the total chain of verification procedures in a radiotherapy department. EPID-based dosimetry, in combination with in-room imaging, is a fast and accurate tool for 3D *in vivo* verification of VMAT delivery. EPID-based *in vivo* dosimetry provides clinically more useful information and is less time consuming than patient-specific pre-treatment dose verification. In addition to accurate 3D dose verification, *in vivo* EPID-based dosimetry will also detect major errors in the dose received by individual patients, and provides a safety net for advanced treatments such as IMRT and VMAT.

1. Introduction

The implementation of IMRT and VMAT has increased the need for a high accuracy in the dose delivery to patients. For that purpose comprehensive quality assurance (QA) programs have been introduced to verify the correct functioning of all components in the radiotherapy treatment planning and delivery process. In addition to these QA programs of the separate components required for a patient treatment, often additional pre-treatment verification checks for individual patients are performed using a variety of phantoms in combination with ionisation chamber or diode arrays. With these different QA programmes in place, one may question the necessity for additional *in vivo* dose measurements during the actual treatment of an individual patient. The clinical use of *in vivo* dosimetry (IVD) in external beam radiotherapy has been addressed in a large number of studies, which were mainly related to the use of point detectors. Some review papers, for instance the recent IAEA Human Health Report Nr. 8 [1], identified a number of treatment errors by means of *in vivo* entrance and/or exit dosimetry during conventional radiotherapy. Also by means of EPID-based *in vivo* dosimetry a number of serious errors during 3D conformal radiotherapy [2] and IMRT delivery [3] were observed that could not have been detected by other QA checks using pre-treatment measurements. Although able to detect major errors, the main application of IVD is to assess all clinically relevant differences between planned and delivered dose. Another important aspect of IVD is that it also provides a record of the actual dose received by individual patients and fulfils legal requirements in some countries.

In this paper we will elucidate the current experience with EPID-based *in vivo* dosimetry for IMRT and VMAT dose verification, and the additional information that can be obtained compared to other patient-specific dose verification measurements.

2. EPID-based *in vivo* IMRT dose verification

2.1 IVD during IMRT

The use of EPID dosimetry has proliferated since several groups have demonstrated its unique possibilities for QA of IMRT, including IVD applications. Two approaches have been reported for using EPID-based *in vivo* dosimetry. In the first approach, a portal dose image with a patient in the beam at the position of the EPID is predicted using the planning CT data of that patient, which is then compared with a portal dose image measured with the EPID [*e.g.*, 4]. A limitation of this forward approach is that it is not always clear how dose differences in the plane of the EPID are related to dose differences in the patient. Several groups have therefore explored back-projection methods for the derivation of the patient dose distribution from a measured portal dose image. These models require the primary dose component at the position of the EPID, which is obtained by correcting the EPID response for the scattered component inside the EPID, and the radiation scattered from the patient/phantom. This primary radiation component is then back-projected to a point inside the patient/phantom and the scattered dose at that position is added (see figure 1).

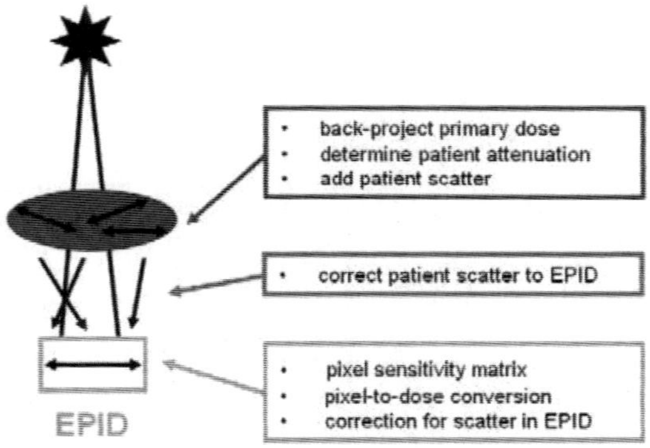

Figure 1: Schematic presentation of the various steps involved in the reconstruction of the dose distribution inside a patient/phantom from an EPID measurement using a back-projection model.

Nijsten *et al* [2] correlated the dose measured with an EPID on the central beam axis with dose values at 5 cm depth using a back-projection model having a (semi-) empirical relationship between these two quantities. Piermattei *et al* [5] reported a simple method for the *in vivo* determination of the midplane dose along the central beam axis using a transmitted signal measured by the central pixels of an a-Si type of EPID. Recently the first results of a national Italian project using this method have been published for photon beams generated by linacs of different manufacturers equipped with a-Si EPIDs [6]. The various back-projection models based on transit dosimetry have been discussed in a review article on EPID-based dosimetry by van Elmpt *et al* [7].

2.2 Clinical implementation of EPID-based in vivo dose verification of IMRT at NKI-AVL

In the Netherlands Cancer Institute-Antoni van Leeuwenhoek Hospital (NKI-AVL), back-projection algorithms have been implemented for the 2D and 3D dose verification of IMRT using a-Si EPIDs [8, 9]. For *in vivo* dosimetry, the reconstructed dose is generally compared for each IMRT field with the patient plan in a plane parallel to the EPID perpendicular to the beam intersecting the isocentre, *i.e.* by means of a multiple 2D approach. For some sites (*e.g.*, breast) the isocentre is not always very relevant for *in vivo* dose verification and a new reference point and other planes are chosen for the 2D γ-evaluation. As an example, the results of a verification of a 6-field IMRT plan for oesophagus cancer treatment are shown in figure 2 to illustrate how EPID-based *in vivo* dosimetry is clinically implemented. The comparison of the EPID-reconstructed and planned dose distribution is done by a 2D γ-evaluation method (3%/3mm dose-difference and distance-to-agreement) using the mean γ-

Journal of Physics: Conference Series **444** (2013) 012011 doi:10.1088/1742-6596/444/1/012011

value, the maximum 1% γ value (i.e. the 99th percentile of the gamma distribution) and the percentage of points with γ<=1 within the 20% isodose line of the planned maximum dose. In addition, the difference between the measured and predicted isocentre dose for each beam is provided, while the difference of the total dose is used as an alert criterion. The results of this analysis are given as an "EPID dosimetry report" for each patient. The yellow dot in the patient dosimetry report shown in figure 2 indicates that for fields 1, 2 and 6 at least one of the alert criteria is outside the tolerance level but still within the action level; *i.e.* a warning is given but the patient treatment may continue. The dose comparison is done versus the TPS dose calculation based on the anatomy in the planning CT scan. All curative patients at NKI-AVL are irradiated either with an IMRT or VMAT technique, and almost all are verified by means of EPID-based *in vivo* dosimetry during three fractions in the first week of their treatment.

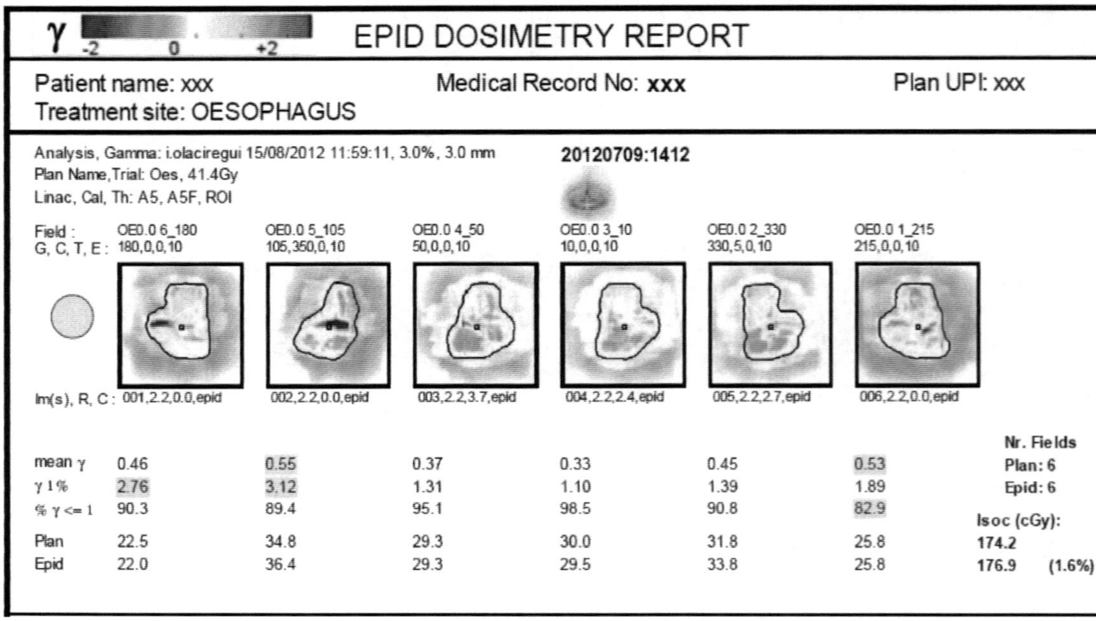

Figure 2: EPID dosimetry report showing the results of a 2D EPID-based *in vivo* dose verification of a 6-field IMRT treatment of an oesophagus cancer patient.

2.3 2D vs 3D IMRT dose verification

Verification of IMRT generally applies a 2D comparison of measured and planned dose distributions in a plane in a phantom, or patient, perpendicular to each IMRT beam, often using a γ-evaluation. 2D dose deviations in a single IMRT beam or segment may be diluted in the total 3D dose distribution and errors in the dose calculation or changes in the anatomy of a patient may no longer be detected. Furthermore identical criteria generally yield smaller γ-values in 3D dose verification than in 2D. These effects can clearly be seen by comparing figures 2 and 3.

Figure 3 shows the results of a 3D γ-evaluation of the total dose distribution, which has a green dot, i.e., all alert criteria are within tolerance level. Stricter γ-criteria for 3D dose verification, such as 2%/2 mm or 2%/1.5 mm, or adapting the tolerance/action levels, might therefore be necessary to trace the same number of deviations as observed when using 2D dose verification. For these reasons at NKI-AVL the multiple 2D approach is still used instead of a full 3D analysis. More research is needed to assess optimal values for alert criteria for 3D verification of IMRT and VMAT treatments of various treatment sites. This information is not only required for the analysis of 3D *in vivo* dose verification, but is also necessary for the evaluation of pre-treatment verification results. These studies are urgently needed because it has been shown by several groups that there exists a lack of correlation between passing rates based on γ-values during 2D verification of single IMRT fields, and clinically relevant dose differences, particularly with highly modulated fields [10-12].

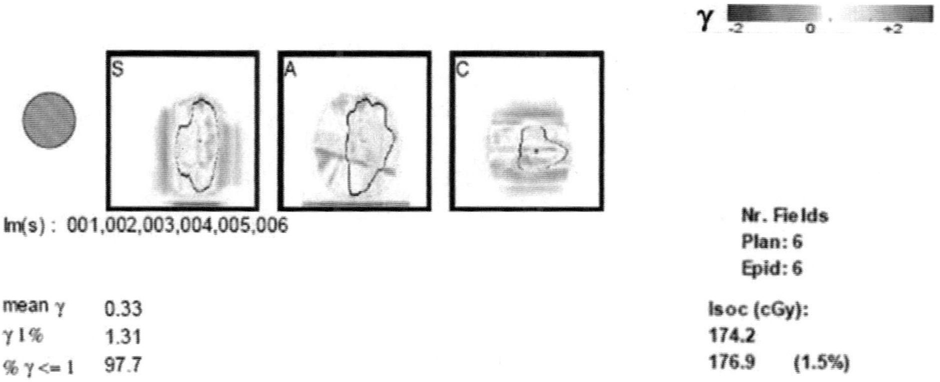

Figure 3: Results of a 3D EPID-based *in vivo* dose verification of the same 6-field IMRT treatment of oesophagus cancer as shown in figure 2.

3D EPID-based dose distributions provide the possibility to determine dose-volume histograms (DVHs) *in vivo*. Important clinical parameters can then be estimated from these DVHs, such as the median dose (D_{50}), the near-maximum dose (D_2) and the near-minimum dose (D_{98}) in the target volume as recommended in the new ICRU Report 83 [13]. In this way it is possible to estimate the effect of an under- or over-dosage in a small volume on the overall dose distribution in the target volume or organ at risk. However, knowledge of the anatomical location of this volume, as well as its size and dose level at which failure occurs, are important properties to evaluate IMRT QA results.

3. 3D EPID-based *in vivo* VMAT dose verification

3.1 Clinical implementation of EPID-based in vivo dose verification of VMAT at NKI-AVL

For VMAT verification it was necessary to modify our software to incorporate gantry-angle resolved image acquisition as well as an automatic correction for the EPID panel flex. Furthermore, the 3D back-projection model applied for IMRT verification [9] was adapted to include the transmission of the primary dose calculated from planning CT data instead of using measured transmission data. The total 3D dose distribution is reconstructed and compared with the planned dose distribution using a fast 3D gamma evaluation algorithm [14], with 3% of the prescribed dose and 3 mm as dose difference and distance-to-agreement criteria, respectively. Details of the adaptation of our back-projection EPID dosimetry method for accurate 3D dose verification of VMAT can be found elsewhere [15].

3.2 Examples of clinically relevant deviations

The results of the 3D verification of each arc of a clinical VMAT plan are also presented as an EPID dosimetry report, similar to that for 2D IMRT verification. Figure 4 shows a dosimetry report of the 3D *in vivo* verification of a dual arc VMAT treatment of a head-and-neck cancer patient, and figure 5 of a hypofractionated lung cancer patient VMAT treatment, respectively. The pictures illustrate the results of the 3D gamma evaluation in a sagittal, axial and coronal plane through the isocentre, while the data for the various gamma evaluation parameters within the 50% isodose surface, as well as the total dose at the isocentre, are also given. If the total dose is deviating more than 3.0% a warning is given (tolerance level exceeded), while for a deviation larger than 5% an error report is sent to a physicist (action level exceeded). Gamma evaluation statistics are reported with site-specific tolerance/action levels, which are for instance for the lung case: 0.50/1.00 for the mean gamma value, 2.0/4.0 for the maximum gamma value, and 85%/70% for the percentage of gamma values below one, respectively. The gamma evaluation data presented in figure 4 for the head-and-neck exceeded for both arcs the action level, while the measured dose at the isocentre was about 5% lower than the calculated value. Comparison of a cone-beam CT scan, made at the same day as the IVD measurement, and the planning CT scan showed that bolus was not present when making the planning CT scan. A new plan was generated with additional bolus material. Figure 5 shows that the results for the first arc of the hypofractionated lung case were OK but the data for the second arc were outside the action level. The cone-beam CT scan made after the second arc showed that the patient position was shifted about 1.5 cm between the two arcs. As a result of this observation, a cone-beam CT scan is now made routinely between the two arcs of hypofractionated VMAT treatments.

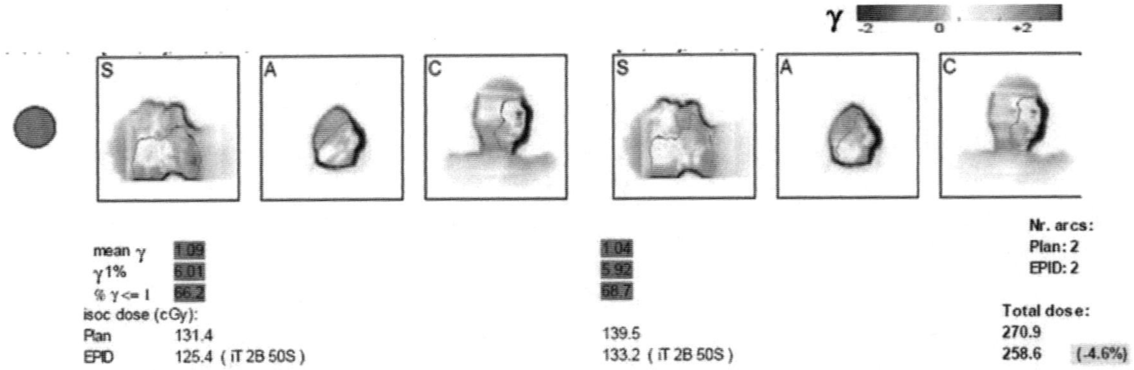

Figure 4: Results of a 3D EPID-based *in vivo* dose verification of a dual arc VMAT treatment of a head-and-neck cancer patient.

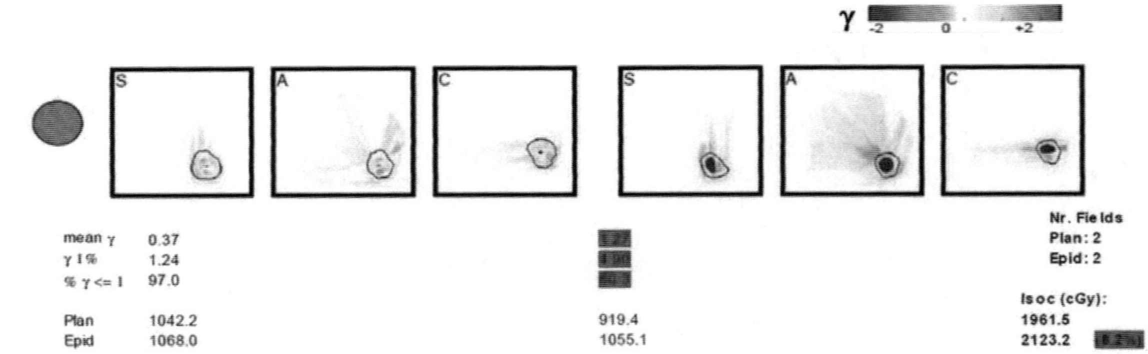

Figure 5: Results of a 3D EPID-based *in vivo* dose verification of a dual arc VMAT hypofractionated treatment of a lung cancer patient.

4. Discussion

4.1 Why 3D in vivo dose verification?

With respect to the verification of the patient position just before or during treatment, numerous in-room imaging tools are applied. Improved confidence in patient position during treatment does, however, not guarantee accurate dose delivery. It can, for example, not account for machine malfunctioning, data corruption, or changes in patient anatomy. The only way to directly verify that a patient received the correct dose during a given treatment fraction is to perform dose measurements during that fraction, *i.e.* by means of *in vivo* dosimetry. By performing IVD it will be possible to correct in most cases for dose errors in a timely manner. Performing one or a few IVD measurements, in combination with in-room imaging, is therefore a very practical approach in assessing the actually delivered patient dose over a series of fractions.

After the reports about recent radiation incidents it is likely that more pressure will be applied from safety authorities on the radiotherapy community to make IVD obligatory. These issues have been mentioned by a number of organizations [*e.g.*, 1] to answer the question "Why *in vivo* dose verification?" The use of a variety of point detectors for IVD has been described in detail in reports published by these organisations to verify conventional and three-dimensional conformal radiotherapy techniques. However, there is an urgent need to perform also IVD of IMRT and VMAT. EPID-based systems are becoming available that allow *in vivo* verification of the full 3D dose distribution during these types of patient treatment. Combining 3D *in vivo* dose measurements with 3D in-room imaging will serve together as a proof that also these more advanced treatment techniques are delivered as planned [16,17]. It may be expected that this novel approach of verifying both geometry and dosimetry in 3D during the same treatment session will be the start of a new way of looking at "Why 3D *in vivo* dose verification?".

4.2 Relation between in vivo dosimetry and other patient-specific dose verification methods

A drawback of applying *in vivo* dose verification maybe that part of the total dose has already been given before an error is detected, although this risk is reduced to a minimum when applying an online dose verification procedure. Certain types of error, *e.g.* in the dose calculation, can also be detected by means of phantom measurements before the start of a patient treatment. A large variety of instruments and methods is available for pre-treatment patient-specific dose verification. Pre-treatment dose measurements are generally performed in a homogeneous phantom, i.e. patient-specific anatomic information and tissue inhomogeneities are not taken into account, whereas these measurements are also often carried out under zero-degree gantry angle thus missing effects of incorrect gantry angle or gravity on dose delivery. Undoubtedly phantom measurements are an essential part of the commissioning process of a new IMRT or VMAT technique. It can however be argued whether patient-specific dose measurements are still useful after a certain number of patient treatments have been verified with pre-treatment phantom measurements considering the limitations of such an approach.

Pre-treatment dose verification is not capable of detecting a number of other errors due to, for instance, variation in patient anatomy or patient position, or the effect of immobilisation devices on the dose calculation and delivery. Furthermore, treatment parameters can (accidentally) be modified between pre-treatment verification and the first treatment. For these and other reasons, *in vivo* dosimetry is the only method capable of verifying the entire radiotherapy chain, including the actual patient treatment. Even if part of the total dose of treatments with multiple fractions has already been given before an error is detected by means of IVD, the remaining fractions can still be used to deliver the correct dose to the patient. The use of EPID-based IVD may in this respect be compared with the application of portal imaging; in both cases the actual treatment beam is used to verify during treatment the dose inside, or position of, the patient, respectively, applying online or offline decision protocols.

It should be noted that pre-treatment patient-specific verification measurements are generally performed by the physics team. These measurements should not be too cumbersome because the time required for this type of QA is proportional to the number of patients treated. By using IVD no additional treatment planning of a phantom irradiation or additional treatment time at the accelerator is necessary, while the measurements may be performed by radiation therapists. An additional advantage of IVD is therefore a reduction in workload compared to pre-treatment verification. Also an IVD measurement is often performed by the same people who irradiate the patient during the rest of their treatment, thus giving direct information about the situation during the actual patient treatment.

4.3 Final remarks

EPID-based IVD is until now performed in a routine way only in a limited number of institutions. The main reason that 2D and 3D *in vivo* dose verification is not yet applied on a large scale is that the dose reconstruction software only recently became commercially available, mainly for pre-treatment verification and not yet for IVD applications. Furthermore, software should include fully automated image acquisition and data analysis. Due to the large number of patients, there was an urgent need at NKI-AVL for automatic analysis of *in vivo* dose verification of IMRT and VMAT. Since August 2011 dose-reconstruction and γ -evaluation software runs automatically yielding a dosimetry report for inspection within minutes after treatment delivery without any manual intervention. In this way more time is available for medical physicists to investigate the reasons for differences between measured and planned dose calculations. For instance, checking the patiént and treatment data of that particular fraction, as well as additional phantom measurements will be needed to explain deviating *in vivo* dosimetry results. Automatic analysis may also allow a more frequent use of IVD during a series of fractions, for instance at all days in-room imaging is performed.

5. Conclusions

EPID-based *in vivo* dosimetry, in combination with in-room imaging, is a fast and accurate tool for 3D *in vivo* verification of IMRT and VMAT delivery. 3D *in vivo* EPID dosimetry provides clinically more useful information and is less time consuming than patient-specific pre-treatment dose verification. In addition to accurate 3D dose verification, *in vivo* EPID dosimetry will also detect major errors and provides a record of the dose received by individual patients.

6. References

[1] IAEA Human Health Report Nr 8 2012 Development of Procedures for in vivo Dosimetry in Radiotherapy *International Atomic Energy Agency* (Vienna: IAEA)
[2] Nijsten S M J J G *et al* 2007 *Radiother. Oncol.* **83** 65-75
[3] Mans A *et al* 2010 *Med. Phys.* **37** 2638-44
[4] van Zijtveld M *et al* 2009 *Med. Phys.* **36** 946-52
[5] Piermattei A 2006 *Med. Phys.* **33** 4414-22
[6] Piermattei A *et al* 2012 *Nucl. Instr. Meth. Phys. Res.* **B 274** 42–50
[7] van Elmpt W *et al* 2008 *Radiother. Oncol.* **88** 289-309
[8] Wendling M *et al* 2006 *Med. Phys.* **33** 259-73
[9] Wendling M *et al* 2009 *Med. Phys.* **36** 3310-21
[10] Kruse J J 2010 *Med. Phys.* **37** 2516-24
[11] Nelms B E *et al* 2011 *Med. Phy.* **38** 1037-44
[12] Wu C 2012 *Med. Phys.* **39** 3051-9
[13] ICRU Report 83 2010 *J. ICRU* **10** (1) (Oxford University Press)
[14] Wendling M *et al* 2007 *Med. Phys.* **34** 1647-54
[15] Mans A 2010 *Radiother. Oncol.* **94** 181-7
[16] McDermott L N 2008 *Radiother. Oncol.* **86** 35-42
[17] van Elmpt W *et al* 2009 *Int. J Radiat. Oncol. Biol. Phys.* **73** 1580-7

Dosimetric quality assurance of highly conformal external beam treatments: from 2D phantom comparisons to 4D patient dose reconstruction

V Feygelman[1] **and B Nelms**[2]
[1]Department of Radiation Oncology, Moffitt Cancer Center, Tampa, Florida 33612, USA
[2]Canis Lupus LLC, Merrimac, Wisconsin 53561, USA

E-mail: vladimir.feygelman@moffitt.org

Abstract. As IMRT technology continues to evolve, so do the dosimetric QA methods. A historical review of those is presented, starting with longstanding techniques such as film and ion chamber in a phantom and progressing towards 3D and 4D dose reconstruction in the patient. Regarding patient-specific QA, we envision that the currently prevalent limited comparison of dose distributions in the phantom by γ-analysis will be eventually replaced by clinically meaningful patient dose analyses with improved sensitivity and specificity. In a larger sense, we envision a future of QA built upon lessons from the rich history of "quality" as a science and philosophy. This future will aim to improve quality (and ultimately reduce cost) via advanced commissioning processes that succeed in detecting and rooting out systematic errors upstream of patient treatment, thus reducing our reliance on, and the resource burden associated with, per-beam/per-plan inspection.

1. Introduction

Following visionary work by Brahme [1, 2], inverse-planned radiotherapy became a widely clinically accepted technique. From the beginning, it was recognized that the complexities of planning and delivery of segmented beams in conjunction with intended sharp dose gradients would require, in addition to thorough commissioning, empirical dosimetric QA of each plan. This was certainly justified at first as many potential dosimetric pitfalls were subject of active investigation. These days, with IMRT being a staple treatment in the developed world, one could argue that the measurement of every plan is contrary to the principles of quality management and the physicists' time would be better spent on more thorough commissioning and credentialing of the IMRT delivery chain. We elaborate on this in the last section of the paper. However, given the high human cost of significant errors, the requirements of accreditation bodies, such as ACR/ASTRO in the United States [3], and relentless public scrutiny of "all things radiation", this is currently largely academic. Therefore, in the meantime it is imperative to continuously improve experimental IMRT dosimetry as it pertains both to system commissioning and individual patients. Realistically, only methods that are widely (i.e. commercially) available have the potential to significantly impact the clinical practice.

2. In-phantom measurements

The current requirement for IMRT program accreditation by the ACR/ASTRO [3] is "patient-specific end-to-end testing". This wording is somewhat important to those in the US, as it can be argued that it implies physical dose measurement in a phantom replacing the patient on the treatment table. This approach is intuitive to most medical physicists, as this is how all radiotherapy techniques, including IMRT, were historically developed.

2.1. Planar measurements

2.1.1. The humble beginnings – film and ion chamber

The ion chamber point dose measurements are still indispensable for system commissioning [4]. Since only a limited number of data points can be realistically acquired this way, film was originally employed as a planar dosimeter. It has a definite advantage in spatial resolution but also has some drawbacks as an absolute and even precise relative dosimeter, the most significant one being the dependence on the processing conditions. Although continuous use of either silver halide or radiochromic film for IMRT commissioning was advocated in no uncertain terms [5], the same author [6], followed by others [7], have demonstrated that most tasks can be successfully accomplished in the clinic with other tools, such as diode detectors, at least for the previously well-characterized MLCs. The only important parameter that appears to still need film measurements is the intra-leaf leakage. However, in our experience, these measurements are variable from leaf to leaf and noisy enough that a default value from the literature can be used just as well.

One of the practical problems with using silver halide film is the rapid disappearance of film processors from the hospital environment due to widespread switch to digital imaging. Even when a processor is still available, maintaining its stability adequate for dosimetric tasks without a sufficient daily volume of developed films presents an additional challenge. Radiochromic film is also not without its drawbacks, including polarization-induced angular dependence of optical density in the flatbed scanners and heating effects, among them [8]. In either case, using film is a time consuming two-step process, involving off-line scanning and analysis. The desire to obtain dose distributions for near-instantaneous online analysis led to the development of electronic 2D dosimetry arrays.

2.1.2. Planar electronic arrays

The most fundamental characteristics of the array are detector resolution (pixel size) and density (pixel pitch). For IMRT/VMAT QA, a detector $\leq$1 mm in size can be considered a point, and detector spacing between 5 and 2.5 mm is sufficient to faithfully represent any realistic dose profile [9, 10]. Widely used electronic arrays come with two types of detectors: diodes [11] or ion chambers [12, 13]. Diodes are small (~1 mm) but have energy-dependent response. The ion chambers are energy-independent but have larger active volumes leading to loss of resolution. All planar electronic arrays have angular sensitivity dependence [12, 14, 15]. For the composite IMRT or VMAT measurements it can be negated by either applying angular correction factors (easier done for fixed gantry delivery), or using a phantom that physically compensates for the angular dependence [14, 15]. Even with a perfectly calibrated planar array, when the beam is incident parallel to the detector plane, 2D modulation information degenerates into one dimension. Direct comparison of EPID-measured dose against the prediction can be considered a special case of high-resolution/high pixel density planar dosimeter that is always orthogonal to the beam. In its widely used commercial implementation (Varian Portal Dosimetry), the EPID measurement is compared not to the TPS calculations, but rather to a separate "portal dose predictor" algorithm. Strictly speaking, this violates the integrity of an end-to-end test, since the actual TPS dose distribution is not being evaluated [16, 17]. A great amount of work has been done to address the challenges of portal dosimetry stemming from the fact that a current EPID is fundamentally an imaging rather than a dose measuring device [16, 18], although first experiments on simultaneous imaging and water-equivalent dosimetry have been reported recently [19]. Recently, Gordon *et al* [20] have analyzed the limits of reliable fluence anomalies detection by pixel intensity deviations in the presence of geometrical registration errors and EPID noise. Two percent errors can be reliably detected on a minimum of 20 mm^2 area.

Kruse [21] has demonstrated that gamma analysis [22] passing rates with both an EPID and a chamber array, even with rather tight 2%/2mm dose-error/distance to agreement criteria, failed to reliably differentiate between what the author considered an acceptable and unacceptable plans. In addition, for electronic arrays (both 2D and 3D) with relatively low detector density (true of all ion chamber and diode arrays), the passing rate metric is subject to statistical uncertainties due to the sub-sampling, which can be significant as shown by Bailey *et al* [23].

2.1.3. *Quasi-3D array measurements and 3D dose reconstruction*

To counteract some shortcomings of the planar arrays, quasi-3D dosimeters were introduced as the next logical step. Two such devices, both based on diode detectors, are well characterized. The DELTA[4] [24-26] has two planar arrays arranged at a right angle and thus the 2D beam modulation information is preserved regardless of the gantry angle. The ARCCHECK [27-30] has a helical diode array and in the beam's eye view the detector arrangement is essentially invariant with the gantry angle. Certain performance differences stemming purely from the array geometry were pointed out [28].

The DELTA[4] detector spacing varies from 5 mm in the center to 10 mm on the periphery. The adjacent ARCCHECK diodes are always 10 mm apart, but the helical arrangement increases the effective detector density in the beam eye view. Three-dimensional dose in a cylindrical phantom can also be estimated via reconstruction using data inputs of a single planar array that rotates in synchrony with the gantry (OCTAVIUS 4D system). All these systems estimate full-density 3D dose in a *phantom* via various computational reconstruction strategies based on coarsely sampled measurements [24, 31], and these reconstructions (which might be considered "3D dose imaging") act as surrogates for the more elusive true 3D measurements.

2.1.4. *3D measurements*

The only current dosimeters that can directly measure 3D dose distributions are polymer gels, recently reviewed by Baldock *et al* [32]. They are fabricated from radiation-sensitive chemicals which upon irradiation polymerize as a function of dose. The degree of polymerization can be imaged by various methods. While polymer gels possess some desirable properties, their adoption in the clinic for routine QA and even system commissioning is limited by practical considerations well enumerated by Jursinic *et al* [14]. Feasibility of relative dosimetry with the optically scanned gel in the Radiologic Physics Center Head and Neck IMRT phantom was demonstrated by Sakhalkar *et al* [33]. While the strengthening of the IMRT credentialing process would be a welcome development, the requisite accurate calibration to dose, along with robustness to the stresses of physical shipping, remained to be demonstrated. An interesting application of a deformable dosimetric gel is direct validation of dose-warping algorithms [34].

3. In-patient 3D reconstruction

While reliable agreement between the calculated and measured/reconstructed dose in a phantom is the basis for the dosimetric commissioning of an IMRT system, its value for the meaningful patient-specific end-to-end testing is less clear. With the prevalent 3% *global* dose-error/3mm distance threshold criteria in particular [4, 35], γ-analysis passing rates for either per-beam single-plane [36] or quasi-3D [37] array geometries, had weak, and counter-intuitive, if any, correlation with the conventional clinical DVH metrics. Instances of the 3%/3mm passing rate metric's failure to detect systematic errors are numerous (see Refs [21, 36, 37] and our paper elsewhere in these proceedings). On the other hand, direct comparison of the planned and deliverable DVHs exhibits higher sensitivity and specificity and is expected to be more clinically meaningful and intuitive to both the physician and the physicist [37-39].

3.1. From fluence

A number of authors have suggested reconstructing the dose on the patient dataset from the measured modulated fluence [16]. Patient dose reconstruction from measured fluence can be broadly divided into two approaches: i) in-air measurements and ii) exit (transit) dosimetry, whereby the fluence modified by the patient is measured.

3.1.1. In-air fluence

For this method, in principle, any planar dosimeter, including film, positioned at any distance from the source can be used [40]. The dose on the patient planning CT dataset can be reconstructed from the dosimeter placed anywhere from the gantry head (for example, COMPASS device [41]) to ≥150 cm from the source (EPID [16]). The reconstruction method is a specialized dose calculation engine ranging in sophistication from pencil beam superposition [40] to Monte Carlo [38, 42, 43]. While the results comparing the dose reconstruction to film dosimetry on a thoracic phantom are encouraging for conventional and IMRT treatments [44], the most advanced and well-characterized EPID-based 3D dose reconstruction systems are not currently commercially available.

3.1.2. In vivo – exit fluence

The reported advantages of the EPID backprojection method is its relative simplicity compared to the Monte Carlo calculations [45] while mainatining the ability to catch errors related to the current state of the patient, which may differ from the time of the planning CT [46]. This method uses exit EPID dosimetry in conjunction with a backprojection reconstruction [47] or reversal of the convolution/superposition dose calculation in an extended phantom [48]. While sufficiently accurate with fixed gantry IMRT even in the presence of inhomogeneities, the backprojection reconstruction appears inadequate for the lung cancer VMAT treatments due to water-based scatter correction kernels [45]. Wendling *et al.* [45] reported a workaround, whereby a homogeneous dataset was substituted for the heterogeneous patient, which they termed *in aqua vivo* dosimetry. While improving the agreement between the planned and reconstructed doses, and allowing, for example, to detect the shifts in patient position, this appears to defeat the goal of comparing the planned and deliverable 3D dose distributions on the actual patient CT dataset.

3.2. From dose in phantom

3DVH software (Sun Nuclear) is capable of estimating the volumetric *patient* dose from the *phantom* measurements using perturbation methods, thus allowing direct DVH comparison while simultaneously fulfilling the true definition of the end-to-end test. The idea of the method is to use the dose distribution measured inside a QA phantom with a relatively low pixel density detector array to guide (correct when necessary) the TPS dose on the patient dataset, resulting in a high voxel-density patient dose grid [37, 49, 50]. Depending on the delivery technique, two types of phantoms can be used. With fixed-gantry IMRT, measurement-guided composite dose to the patient is reconstructed from the beam-by-beam measurements with a planar array (MAPCHECK) [50]. Dose reconstruction accuracy was verified in a homogeneous phantom with ion chamber and film [50] and using simulation in heterogeneous head/neck patients [37]. While introducing deliberate errors, Carrasco *et al* [39] have shown that while per-beam gamma analysis was prone to false positives/negatives, the 3DVH software correctly described the DVH of the plan which included the error.

For VMAT dosimetry, a cylindrical phantom (ARCCHECK) is a more appropriate choice. While the general approach is similar on the surface, this different detector geometry and the arc delivery require an entirely different method for measurement-guided dose reconstruction, called ARCCHECK Planned Dose Perturbation (ACPDP) [31]. ACPDP explicitly relies on the time-resolved nature of the ARCCHECK data, which have updates logged at 50 ms intervals. Along with the dose, the gantry angle determined by the virtual inclinometer [27] is also stored for each data update. At this point, both the low density (~ 10 mm) dose map on the cylindrical ARCCHECK surface and the gantry angle are known as a function of time. The DICOM RT Plan's beams' control points (CPs) are thus synchronized to absolute, corresponding delivery times, forming the basis of discretizing the delivery process into individual modulated sub-beams at ~2° intervals (which typically corresponds to resolution of 0.2-0.4 sec in terms of delivery time).

A relative 3D dose grid for each sub-beam is independently calculated by convolving a 3D impulse TERMA function throughout the phantom volume with the 3D scatter depth kernels. The next step is the position-dependent measurement-guided dose morphing and absolute scaling that converts full-density relative dose to absolute, using the relevant diode measurements as 3D spatial calibration data points. After that, a full-volume, high-density absolute dose grid is generated in the ARCCHECK phantom by summing all the component, time-resolved sub-beam dose grids. Drawing an analogy with a physical volumetric cylindrical dosimeter, this result could be thought of as a "virtual gel". A minor post-processing step fine-tunes, if necessary, the virtual 3D phantom dose by a global correction factor that best fits the virtual measurement to the composite (cumulative) doses per diode position. At this point, the high-resolution (sub-millimeter detectors), low density (~10 mm) cylindrical AC dose map is transformed into high voxel density (TPS resolution) volumetric dose grid on a solid cylindrical *phantom*.

The final step is to use this high density volumetric *phantom* dose grid to obtain measurement-driven estimate of the dose delivered to the *patient*. To that end, the voxel by voxel correction factors derived from the ratios of calculated (TPS) and reconstructed doses on the *phantom* are applied to the TPS dose distribution on the *patient* CT. This simple approach is based on perturbation theory and basic physics of Compton scatter, and is surprisingly accurate as verified by the direct ion chamber and film measurements in a homogeneous phantom [31] and the ion chamber and small optically stimulated luminescent dosimeters [51] in an anthropomorphic thoracic SBRT phantom. This 3D measurement-guided dose reconstruction system is commercially available.

4. In-patient 4D reconstruction

While 3D dose reconstruction on a patient is an improvement over the standard in-phantom QA techniques, the static approach does not take into account interplay between the organ motion and dynamic delivery. Quantifying any degradation of planned dose due to interplay prior to treatment, based on patient-specific target motion trajectories and patient-plan-specific delivery dynamics, would be of clear value in some cases, e.g. SBRT. Two emerging techniques attempt to address this issue.

4.1. From fluence

EPID-based 4D patient dose reconstruction framework was so far briefly described by Lin *et al* [42]. Both the tumor (fiducials) and MLC movements were continuously tracked by an EPID. The incident fluence distributions were sorted into their corresponding phases based on the tumor (fiducial) motion pattern detected in real time by the EPID and accumulated as the incident fluence map for each phase. Together with 4DCT, it was then used for Monte Carlo dose calculation using the EPID-detected MLC shapes. Deformable registration was performed to sum up the phase doses for SBRT treatment assessment.

4.2. From dose measurements in phantom

We have demonstrated by direct time-resolved measurements [31] that prototype 3DVH software can accurately estimate 4D VMAT dose in the patient based on the phantom measurements recorded every 50 ms and synchronized with the gantry movement through the virtual inclinometer [27]. The prototype relied on the TPS dose grids exported for each control point but this is mathematically not necessary. Instead, motion can be treated as another perturbation as described in a separate paper in these proceedings. With volumetric time-resolved dose, it is rather simple to accumulate the dose for any moving structure by propagating it through this 4D dose space either as a rigid or deformable object. This approach is currently being generalized for the fixed gantry IMRT delivery.

5. Looking Forward

The majority of the methods summarized in this work serve pre-treatment QA, with some aimed at per-fraction dosimetric verification. Each method has pros and cons in terms of accuracy, effectiveness, efficiency, and basic practicality. However, our summary thus far has been method-centric rather than system-centric, and now we will try to take a step back to see the bigger picture and ruminate on the future of dosimetric QA.

5.1. Modern radiation therapy as a high-tech manufacturing system

Deposition of a highly conformal dose to the patient is the ultimate output of a complex system, comparable to a high-tech manufacturing process involving many steps. At the very least, there are pre-treatment steps (such as imaging, contouring, planning, dosimetric QA) and per-fraction steps (such as patient setup, image guidance, and dose delivery), each one with its own sources of potential error and variation. This analogy to manufacturing allows us to compare our methods to those that helped revolutionize quality in the manufacturing sector starting in the post-WWII era.

Consider one of W.E. Deming's famous 14 Points [3] that states: "Cease dependence on inspection to achieve quality. Eliminate the need for inspection on a mass basis by building quality into the product in the first place." Deming recognized that searching for defects at the end of production (i.e. inspection) is too late in the process and, as a result, cost ineffective. One can draw a direct comparison to per-patient dosimetric QA, where detection, diagnosis, and mitigation of issues at the per-plan level clearly qualifies as "inspection on a mass basis" given the sheer number of radiation therapy plans. In these terms, "building quality into the product in the first place" is highly desirable, and would include such goals as optimizing accuracy of the TPS dose algorithms (by the vendors) and their implementation in the site-specific beam models (by the medical physicists), and maximizing accuracy and minimizing variation of the delivery system. As these are the goals of system commissioning, we can imagine that Deming might advise today's medical physicist to spend a lot more time commissioning the system in the first place, to detect and root out systematic errors by applying tight tolerances and stringent methods. To their part, vendors would strive for higher standards of accuracy and lower incidence of defects and imperfections. In addition, both vendors and clinicians would also employ continual improvement and abandon the concept of "meeting specifications."

An apt analogy is that of Aguayo's flywheel [52], wherein a flywheel is made of a rod with three heavy disks. Variation in the radial uniformity of those disks causes the spinning flywheel to wobble at high speeds, an undesirable result. Three strategies could be considered to reduce wobble: 1) inspect each disk, labelling heavy and light regions and using these to customize the mounting to offset the wobble; 2) assemble as usual and just inspect for wobble prior to shipment, rejecting those with wobble; or 3) work with the disk supplier (or fabrication process) to drive down the variation in the uniformity to the point that wobble rarely if ever occurs in the first place. Clearly, option #3 requires more time and resources up front, but is the wisest (and least costly) long term option compared to the very inefficient strategies #1 and 2. Today's conventions of dosimetric QA resemble #1 and #2 much more than #3.

5.2. Basic tenets of quality violated by common dosimetric QA conventions of today

Basic tenets of quality include: 1) build/design quality into a system, thus reducing reliance on inspection, 2) strive to meet tight tolerances with minimal variation, and 3) abandon the mindset of "meeting specs" and instead work to continually improve. These are largely violated by conventions of dosimetric QA. To start, insensitive metrics – such as the ubiquitous 3%/3mm γ-passing rate [4, 35] – allow many systematic errors to remain undetected [21, 36, 37]. Because system commissioning is in large part based on this metric, it hinders "building quality into the system", since both vendors and clinical physicists may erroneously believe their system is of higher quality than it is. This by definition violates the goal to strive for tight tolerances and inhibits driving down variation because that variation hides in deceptively high passing rates. Naturally, continual improvement is hamstrung as well, as meeting proposed action levels that are based on insensitive metrics [4] gives a false sense that there is nothing to improve. The relatively low performance of multi-institutional tests such as that run by the RPC [53], we believe, is at least in part a direct result of these violations of basic tenets of quality.

5.3. Re-designing commissioning goals

We suggest that the biggest improvements in dosimetric QA can be made by improving the commissioning process and standards. Figure 1 schematizes evolutionary changes in strategy that embrace the major tenets of quality. The noteworthy components of the commissioning process of the future are:

1. Identify specific, testable requirements of the system. Ideally, these would target individual potential errors or singular and testable components. Examples might include, small field PDD, output factors vs. field size and depth, leaf-end penumbra, etc. There will be many of these components and they may vary based on the delivery system (i.e. a conventional MLC-equipped C-arm linac may have a different subset of issues compared for example to a robot-mounted accelerator). The list of requirements should be comprehensive and centered around potential error sources.
2. Each testable requirement will have a test setup designed specifically for it. Clear instructions and, if applicable, common patient or phantom datasets are imperative in order to eliminate risk of variation of how the test is performed on a specific system.
3. Each testable requirement will have a custom method and metric(s) of analysis that are proven sensitive and preferably not binary (pass/fail) but rather a continuum of scores to assess even small variations.
4. When a test is performed and metric results generated, the result(s) will be added to a population database of identical systems. The site can compare itself against the population and see the high performance potential.
5. The population results then serve to drive benchmarking. System highest potentials are quantified and best practices can be shared. This can drive continual improvement not only of the individual system (i.e. improvement of their metric) but also of the population (i.e. reduction in variation and increase in mean performance level). Such metric distributions would also be valuable for regulatory agencies to use in site accreditation/credentialing.

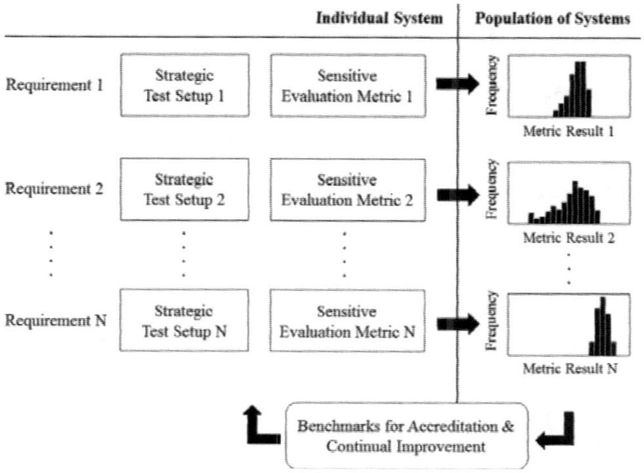

Figure 1: Schematic of an advanced commissioning process.

6. Conclusions

As IMRT technology continues to evolve, so do the dosimetric QA methods. In terms of patient-specific dose QA, we envision that the currently prevalent limited comparison of dose distributions in the *phantom* by γ-analysis will be eventually replaced by clinically meaningful 3D and 4D *patient* dose analysis. A number of approaches have been proposed, each with its own advantages and drawbacks, and it remains to be seen which ones will find their way into the wide spread clinical use. In terms of the bigger picture, we envision a future of QA built upon lessons from the rich history of "quality" as a science and philosophy. This future will aim to improve quality (and ultimately reduce cost) via advanced commissioning processes that succeed in detecting and rooting out systematic errors upstream of patient treatment. Safeguard checks against catastrophic failures such as data corruption during transfer from the TPS to the treatment management system can be easily automated and do not require physical measurements.

7. References

[1] Brahme A 1988 *Radiother. Oncol.* **12** 129-40
[2] Brahme A *et al* 1982 *Phys. Med. Biol.* **27** 1221-9
[3] ACR –ASTRO Practice guideline for intensity modulated radiation therapy (IMRT) [cited 6/18/2012] Available from: http://www.acr.org/~/media/ACR/Documents/PGTS/guidelines/IMRT.pdf
[4] Ezzell G A *et al* 2009 *Med. Phys.* **36** 5359-73
[5] Cadman PF 2011 *Med. Phys.* **38** 4464
[6] Cadman P *et al* 2005 *J. Appl. Clin. Med. Phys.* **6** 74-86
[7] Feygelman V *et al* 2010 *J. Appl. Clin. Med. Phys.* **11** 135-53
[8] Lynch B D *et al* 2006 *Med. Phys.* **33** 4551-6
[9] Dempsey J F *et al* 2005 *Med. Phys.* **32** 380-8
[10] Poppe B *et al* 2007 *Phys. Med. Biol.* **52** 2921-35
[11] Jursinic P A and Nelms B E 2003 *Med. Phys.* **30** 870-9
[12] Wolfsberger L D *et al* 2010 *J. Appl. Clin. Med. Phys.* **11** 241-51
[13] Poppe B *et al* 2006 *Med. Phys.* **33** 1005-15
[14] Jursinic P A, *et al* 2010 *Med. Phys.* **37** 2837-46
[15] Van Esch A *et al* 2007 *Med. Phys.* **34** 3825-37
[16] van Elmpt W *et al* 2008 *Radiother. Oncol.* **88** 289-309
[17] Bailey D W *et al* 2012 *J. Appl. Clin. Med. Phys.* **13** 82-99
[18] Nelms B E *et al* 2010 *J. Appl. Clin. Med. Phys.* **11** 140-57
[19] Vial P *et al* 2012 *Med. Phys.* **39** 4002
[20] Gordon J J *et al* 2012 *Med. Phys.* **39** 4959-75
[21] Kruse J J 2010 *Med. Phys.* **37** 2516-24
[22] Low D A *et al* 1998 *Med. Phys.* **25** 656-61
[23] Bailey D W *et al* 2011 *Med. Phys.* **38** 6053-64
[24] Feygelman V *et al* 2009 *J. Appl. Clin. Med. Phys.* **10** 64-78
[25] Sadagopan R *et al* 2009 *J. Appl. Clin. Med. Phys.* **10** 104-19
[26] Bedford J L *et al* 2009 *Phys. Med. Biol.* **54** N167-76
[27] Kozelka J *et al* 2011 *Med. Phys.* **38** 5021-32
[28] Feygelman V *et al* 2011 *J. Appl. Clin. Med. Phys.* **12** 146-68
[29] Letourneau D *et al* 2009 *Med. Phys.* **36** 1813-21
[30] Yan G *et al* 2010 *Med. Phys.* **37** 108-15
[31] Nelms B E *et al* 2012 *Med. Phys.* **39** 4228-38
[32] Baldock C *et al* 2010 *Phys. Med. Biol.* **55** R1-63
[33] Sakhalkar H *et al* 2009 *Med. Phys.* **36** 3371-7
[34] Yeo U J *et al* 2012 *Med. Phys.* **39** 2203-13
[35] Nelms B E *et al* 2007 *J. Appl. Clin. Med. Phys.* **8** 76-90

[36] Nelms B E *et al* 2011 *Med. Phys.* **38** 1037-44
[37] Zhen H *et al* 2011 *Med. Phys.* **38** 5477-89
[38] van Elmpt W *et al* 2008 *Radiother. Oncol.* **86** 86-92
[39] Carrasco P *et al* 2012 *Med. Phys.* **39** 5040-9
[40] Renner W D *et al* 2003 *Med. Phys.* **30** 2996-3005
[41] Boggula R *et al* 2010 *Phys. Med. Biol.* **55** 5619-33
[42] Lin M *et al* 2012 *Med. Phys.* **39** 3909
[43] van Elmpt W *et al* 2007 *Med. Phys.* **33** 2426-34
[44] van Elmpt W *et al* 2007 *Med. Phys.* **34** 2816-26
[45] Wendling M *et al* 2012 *Med. Phys.* **39** 367-77
[46] Mans A, *et al* 2010 *Med. Phys.* **37** 2638-44
[47] Wendling M *et al* 2009 *Med. Phys.* **36** 3310-21
[48] McNutt T R *et al* 1996 *Med. Phys.* **23** 1381-92
[49] Nelms B E and Simon WE. Radiation therapy plan dose perturbation system and method. US patent 7945022. 2011.
[50] Olch A J 2012 *Med. Phys.* **39** 81-6
[51] Jursinic P A 2007 *Med. Phys.* **34** 4594-604
[52] Aguayo R and Deming W E. *The American who taught the Japanese about quality*. NY: Simon and Schuster; 1991
[53] Ibbott G S *et al* 2008 *Int. J Radiat. Oncol. Biol. Phys.* **71** S71-5

Journal of Physics: Conference Series **444** (2013) 012013 doi:10.1088/1742-6596/444/1/012013

Current status of scintillation dosimetry for megavoltage beams

L Beaulieu[1,2i], M Goulet[1,2], L Archambault[1,2] and S Beddar[3]

[1]Département de Physique, de Génie Physique et d'Optique et Centre de recherche en Cancérologie de l'Université Laval, Université Laval, Québec, Québec G1K 7P4, Canada

[2]Département de Radio-Oncologie et Centre de Recherche du CHU de Québec, CHU de Québec, Québec, Québec G1R 2J6, Canada

[3]Department of Radiation Physics, The University of Texas MD Anderson Cancer Center, 1515 Holcombe Boulevard, Unit 94, Houston, Texas 77030

Email: beaulieu@phy.ulaval.ca

Abstract. Plastic scintillation dosimeters (PSD) have been a topic of a keen research interest for the past 20 years. Numerous PSD systems have been proposed, most times differentiating from the previous by a slight change in one or more components, such as the photodetector. However, a few major technological and engineering innovations have also been made. Over the past few years PSDs have been evaluated for small field dosimetry, in vivo dose measurements, building of arrays and much more. The present manuscript is intended to present the basic physics and properties of PSDs and its application over the past two decades.

1. Introduction

Detection of ionizing radiation using scintillating materials is one of the oldest techniques available to scientists. Key discoveries, such as x-rays by Roentgen, were made using a light-emitting screen. Various processes can lead to the emission of visible light following the interaction and excitation by ionizing radiation. The fast emitting component, usually in the range of 10^{-9} s, is called fluorescence. It constitutes the signal of interest. Other components such as phosphorescence or delayed fluorescence also produce visible light but at a much longer time scale. Therefore, efficient scintillation detectors for radiation detection usually make use of the fast emitting component. Also, two types of scintillation detectors are used on a regular basis and produce visible light through very different physical processes: organic scintillators and inorganic crystals. The reader is referred to the seminal books by Birk [1] or Knoll [2] for detailed reviews.

The applications of scintillators to medicine are numerous ranging from x-ray detection screens, gamma camera and PET scanners [3]. However, one of the first application to radiation dosimetry was reported in a manuscript entiltled "A new method of integral dose measurement with a plastic scintillator phantom" by Murai *et al* in 1963 [4]. In that paper, a plastic scintillator was chosen because is constituted a good approximation of soft tissues and the visible light emitted was shown to be proportional to the absorbed dose. However, one must wait for the publication of the two seminal papers by *Beddar et al* to see any significant interest for plastic scintillation detector as radiation dosimeters[5, 6].

The current manuscript presents a review of plastic scintillation dosimeters (PSDs), their challenges and their applications to photon and electron dosimetry.

2. Scintillation detector for radiation dosimetry

2.1. Basic physics properties

When ionizing radiation interacts with an organic scintillator, the energy deposited by the electrons generated in the medium excites orbital electrons in higher energy levels. This excitation process is very fast (10^{-15} s). Through internal conversion and vibrational relaxation, electrons decay to the S_{10} state within the 10^{-12} s. It is from the decay of S_{10} state to the S_{0x} that visible fluorescence light is emitted. The whole process is very fast and takes about 10^{-9} s. The gap between the S_1 and S_0 states is usually between 3 to 4 eV thus accounting from the blue-violet light usually seen from most scintillators. Furthermore, this band gap is large enough to ensure some temperature independence (room temperature translating to an energy of about 0.025eV). Finally, due to the differences in excitation and decay processes, the energy emitted as visible light produces an emission spectrum that is of lower energy than the excitation one. This is call the Stokes shift [2]. Thus organic scintillators are mostly transparent to their own emission.

It is important to note that a competing de-excitation mechanism is possible in which decays from the S_1 states to the triplet T_1 states occur followed by further decays from the T_1 to the S_0 states. This also results in visible light and is called delayed fluorescence and phosphorescence. This process is much less likely and is discarded in most dosimetry applications.

Plastic scintillators are composed of 97% of polyvinyltoluene (polystyrene in the case of plastic scintillating fibers) to which an organic fluor, the scintillating materials is added. These fluors are composed of aromatic chain with strong carbon double bonds. This is called a two component systems. To obtain emission of the green or orange light, a third component called a wavelength shifter is added. In addition, scintillating fibers also possesses a cladding that is usually made of PMMA, a material well known in radiation dosimetry.

Figure 1 shows some fundamental interaction properties of each of the materials described above, namely the mass energy absorption coefficients for photons and the stopping power for electrons. Both are normalized to water, the reference medium for radiation therapy dosimetry. Furthermore, relevant materials found in ion chambers and diodes are also shown. For materials composing scintillation detectors, it can be seen that for electrons, the stopping powers are almost independent of energy and within 2% of water over the full range covered by radiation therapy energies. Similarly, the mass energy absorption coefficients display a similar behavior for photons between 200 keV up to 3 MeV. This can be contrasted with silicon and even air. Thus from a basic interaction physics point of view, plastic scintillation detectors are water-equivalent and energy independent. This should translate to very little perturbation of the radiation fluence relative to the medium of reference (water); thus no correction factors should be needed for PSDs for most radiation therapy applications [7-9].

Finally, the physical density (1.032-1.06 g/cm^3), electron density (3.238-3.272 x 10^{23} e-/g) and elemental composition by weight forming the effective Z (7.7-8.5% H and 91.5-92.3% C) of the various components of a scintillator are all close to water.

2.2. Experimental validation

The various properties described above are all entrenched is a solid-state physics background. Yet, experimental validation under clinically realistic conditions is a critical step for adoption of PSDs. Beddar *et al* were the first to fully explored experimentally PSD properties for high-energy beam dosimetry. Namely in a series of two manuscripts, they have demonstrated the independence of PSDs to beam energy and dose rate, linearity of the light to dose, temperature independence, high spatial resolution and water-equivalence [5, 6]. Later on Archambault *et al* have shown that for small, point-like detectors PSDs also show no angular dependence if the Cerenkov effect is corrected for (to be discussed in details below) [10].

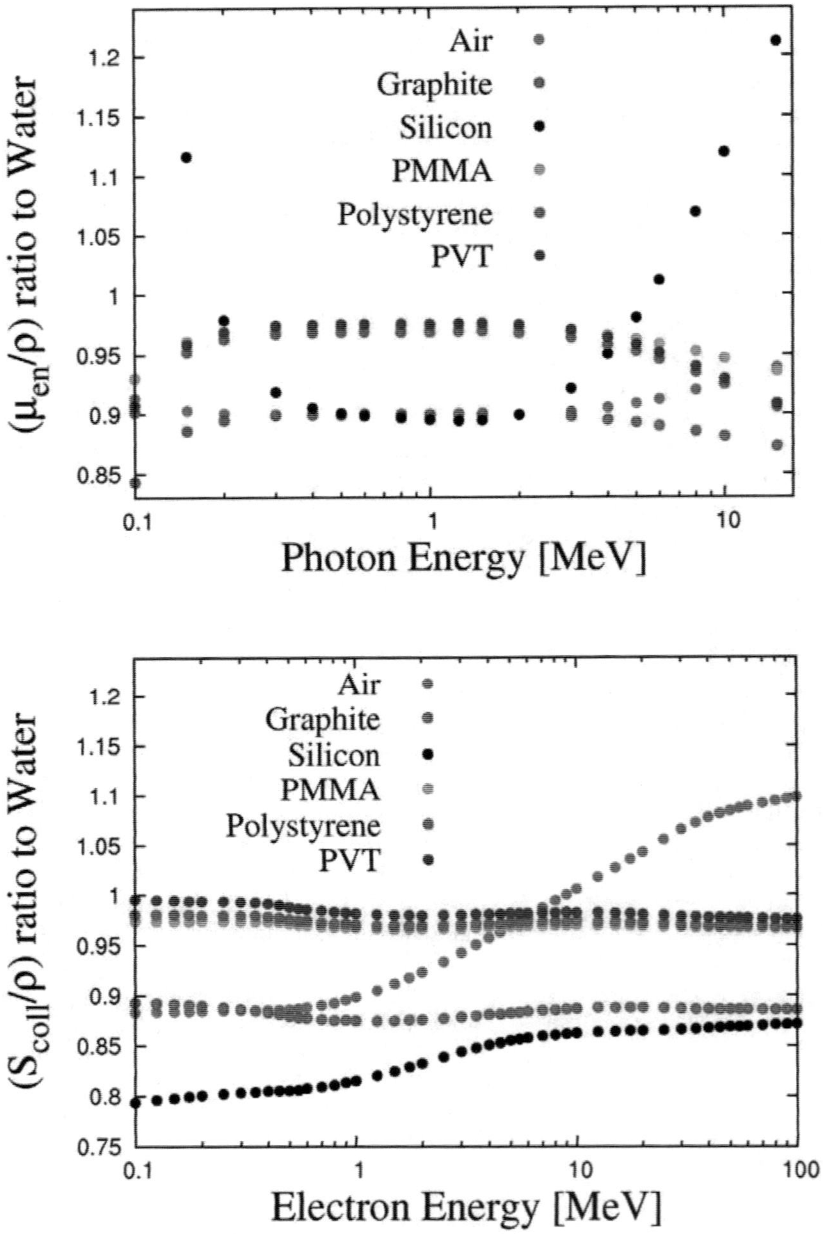

Figure 1: Mass energy absorption coefficients (upper panel) and electron mass stopping power (bottom panel) as a function of energy. Both are normalized to water and values were taken from the NIST website (www.NIST.org).

Validation of the same properties to clinical electron beams was detailed extensively by independent groups [11-13]. In particular Lacroix *et al* demonstrated that because PSDs measured directly the dose deposited by electrons without any correction factors, they can be used to extract ion chamber correction parameters [13].

While this is not the subject of this review, it is interesting to note that the measurements of absolute deposited dose for clinical proton beams with PSDs has not been thoroughly investigated at this time. Issue related to quenching effect are still being investigated in order to determine appropriate correction methods to correct for it [14].

2.3. Component of a PSD and design optimization

A PSD is made of three major building blocks: the scintillating probe (presented above), a light guide and a photodetector. This is called the optical chain. At all time, the linearity between the dose and light output depends not only of the scintillation properties but also of all component in the optical chain. Furthermore, each stage added in the path of the optical photons lead to a decrease of the light collection efficiency. First, the visible light produced in the scintillator must travel (through internal reflection) toward one of the exit face of the scintillator and be injected in the light guide (e.g. a clear optical fiber) (collection efficiency of about 5%). Interface between various components, e.g. scintillator to optical guide) is also a source a loss and the coupling efficiency is generally around 75-85% [15]. Light guide have intrinsic transport capability. Optical fibers, in particular the water-equivalent and flexible plastic types, are often used because of their enhanced light transport properties. Light attenuation in an optical fiber guide is generally less than 20% over a few meters. It is interesting to note that while most fiber-based PSDs use a cylindrical geometry, Yoo *et al* have shown that square fiber could be used as well [16]. The output of a light guide must be capture by a photodetector. Depending if the coupling to the photodetector is direct, passing through filters or a lens (or a combination thereof), the coupling efficiency can be as low as 5% and as high as 90%. Finally, the photodetector itself posseses an intrinsinc efficiency (quantum efficiency), which can vary from 20% to 90%. The overall efficiency over the complete optical chain is thus only of a few percent. Therefore, optimization of each component is critical [15, 17, 18].

Various photodetectors have been used over the years. In the 1990s, Beddar *et al* pioneered the use of a photomultiplier tube, while a similar design but using a photodiode was first validated by Létourneau *et al* [19]. Because of its simplicity, this configuration has been adopted by multiple groups ever since [12, 20, 21] The use of segmented RGB (red, green, blue) diode has been introduced in the early 2000s by Fontbonne *et al*, to implement a new Cerenkov subtraction feature [22]. RGB CCD camera has also been shown to be highly efficient for single point PSD and PSD arrays. Furthermore, high sensitivity electron multiply CCD was used by Archambault *et al* for fast (ms) in vivo dosimetry [1, 23]. Finally the used of PMT array was recently demonstrated by Liu et al [2, 24].

Overall the choice of plastic scintillator versus plastic scintillating fibers, cylindrical versus square fibers and of a particular photodetector should be dictated based on the specific application sought. In particular parameters such as light production (related to dose and dose rate), light collection, readout frequency and number of PSDs to read simultaneously (size of an array of PSDs) are all factors that need to be considered in the optimization of optical dosimeters, Lacroix *et al* have set-out a general framework enabling a quantitative evaluation [3, 25, 26]

3. The Cerenkov "problem"

Cerenkov is produced in a dielectric medium when electrons travel faster than light. This happen because the speed of light in the medium changes as c/n, where n is the index of refraction. For most plastics used in PSDs, the index n is between 1.48-1.60. This translates to electrons traveling faster than light in the medium for electron energy above 145-180 keV. Ongoing electrons polarize the medium with orbital electrons slightly perturbed; which results in dipole oscillations, then the relaxation process leads to the emission of visible light. Because that light travel slower than the exciting electrons, constructive interference occurs and the wave front will travel preferentially in the direction of the electrons [27] Cerenkov light spectrum is independent of the medium and peaks in the violet region. All other wavelengths in the visible spectrum are also present but the intensity varies as λ^{-3}.

Cerenkov light emission par mm of medium is factor of 20 to 100 lower than a scintillator. However the clear optical light guides can be meters long, with many cm exposed to ionizing radiation [28, 29]. If exposure was constant (always the same amount of fiber), Cerenkov would also be a constant and easily taken into account during the light to dose calibration. However, in radiation therapy field sizes, beam orientations and points of measurement all lead to variation in the amount of medium exposed. Therefore, the unwanted Cerenkov must be eliminated. In 2000 [19, 30], a detailed study of six types of dosimeters in small field photon beams concluded the following: "*If the problem with the relatively high cable signal can be solved, the plastic scintillator offers high resolution, near water equivalence and small directional dependence*".

The very first method proposed to eliminate this stem effect was the two fibers approach (figure 2). One clear fiber is used to collect the light emitted by the scintillator while a reference fiber is place side-by-side but without the scintillator [5, 28]. Assuming that both fibers see the same electron energy fluence in all irradiation condition, subtracting the signal of the reference fiber from the PSD results only in the scintillation light. This design has been replicated by multiple authors, with small variations, through out the years [12, 19, 31, 32]. This method worked perfectly for in-fields measurements in situation where no dose gradient is seen differentially but the two fibers, however, suffered from spatial limitation when both fibers see a different dose gradient.

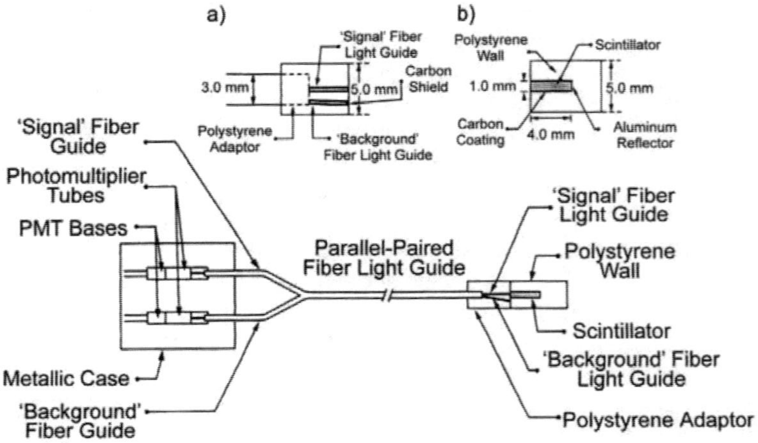

Figure 2: Diagram of the first prototype plastic scintillation detector, illustrating the two-fiber technique developed by Beddar *et al* from ref (5,6).

De Boer *et al* proposed another approach called optical filtering that takes advantage of the fact that the Cerenkov light is peaked in the violet region and that any scintillators emitting in a wavelength band slightly shifted will generate a signal that is much larger than the stem effect [33]. Thus an appropriate optical filter is selected, which removes wavelengths shorter than the scintillator emissions but let the higher wavelengths pass. This method was shown to remove about 50% of the stem signals over no removal [33]. Further optimization of this technique by Clift *et al* was able to reduce the stem effect by a factor of 2.4 [34]. The main advantage of this approach is to have a single probe and thus bypassing the limitation of the two-fibers method. However, because the Cerenkov light is present at all visible wavelengths, it is not as efficient.

A third approach, the temporal method, was proposed by Clift *et al* [35]. While plastic scintillators usually have very fast rise time (a few ns) and short decay constant (10-20 ns), this group uses a scintillator with a long decay constant (264 ns) and the fact that under pulsed beam, the stem effect shows a shorter decay time than the latter scintillator. Thus by timing the read-out of the scintillator relative to the beam pulse, essentially waiting for the stem effect to decayed-out after the beam pulse and measuring the scintillation light afterward. They were able to show that 99.9% of the stem effect was removed and doses as low as 0.2 Gy/min could be measured [35]. This method takes advantage of using a single probe, like the optical filtering approach. On the other hand, it also needs complex timing set-up using information about beam pulse from the accelerator. This further makes clinical implementation difficult and possibly accelerator model dependent.

In 2002, Fontbonne *et al* introduced what would become a leading method for stem removal: the chromatic technique [22]. This solution capitalizes on the differences between the optical spectra of plastic scintillation light and Cerenkov light. By measuring the light output of a PSD under two different wavelength filters, it is possible to decouple each phenomenon (scintillation and Cerenkov) and isolate the portion of the signal that is only proportional to the dose delivered to the sensitive volume of the detector. This methods has been shown to be very efficient while keeping the detection apparatus simple: filtering can be made by using external filters or even better on-chip RGB filter [10, 11, 22]. Archambault *et al* were the first to demonstrate in a systematic study that this method is superior to the approaches described above, in particular in the of dose gradients (small fields, brachytherapy, beam penumbra) [10]. Recently, Guillot et al have shown the calibration procedure for this technique is critical to ensure optimal subtraction in all measurement conditions [36].

Finally, Lambert *et al* proposed in 2006 a different approach for getting rid of the stem effect by using a hollow light-guide instead of a plastic optical fiber. Because no Cerenkov light is produced in such light-guide, there is no problem of Cerenkov [20]. Uses of this technique for small field applications have shown promises [12]. On the limitation side [37, 38], the hollow core fiber are rigid compared to plastic fiber which limits its use for in vivo applications and light transport is less efficient, showing attenuation between 5 to 10 times higher than plastic core fiber.

4. Modern Application

4.1. Small Field Dosimetry

Accurate measurement of small radiation fields is a well-known challenge for many dosimeters]. Lack of lateral charge-particle equilibrium [5, 6, 41], dose-averaging within the sensitive volume of dosimeters [2, 42] and differences between the composition of detectors and their surrounding media [5, 6, 43] all cause perturbations of the radiation field. Thus, discrepancies are seen between the measured dose and the actual dose that would be deposited in the medium in the absence of a detector [10, 43-45]. Semiconductor diode detectors are often used as a reference dosimeter for small field measurements due, in part, to their extremely small sensitive volume [11-13, 46]. Even with diodes, Monte Carlo simulations may nevertheless be required to accurately determine the dose delivered by small fields [13, 43, 46, 47]. Combining high precision measurement with detailed Monte Carlo simulation is not a trivial task and cannot be performed routinely in most clinics.

The use of small fields is becoming ubiquitous in radiation therapy and may be a source of errors if measurements are not conducted properly. Discrepancies have been observed between treatment planning systems and measurement in Intensity modulated Radiation Therapy (IMRT) that can be attributed to the presence of small fields [14, 41, 42, 48]. Techniques such as radiosurgery and stereotactic body radiation therapy (SBRT) rely heavily on small fields. Furthermore, specialized equipment such as the Cyberknife and Gamma-knife are only relying on small fields for delivering their treatments. To address such non-standard irradiation schemes a work-group has been established to better define how reference measurement should be made in non-standard conditions [15, 49].

In addition to several types of diodes, the current arsenal used for measuring small radiation includes small-volume ion chambers, films (radiographic and radiochromic), diamond detectors, liquid ion chambers and thermoluminescent dosimeter (TLD). The use of plastic scintillation detectors (PSDs) for small field dose measurement has been suggested nearly 15 years ago [16, 19, 50]. PSD are ideal for such measurement because they are water-equivalent and can be designed with small sensitive volume (i.e. PSDs can easily have dimensions smaller than a millimeter).

In recent years, PSDs have been increasingly used in radiosurgery and other small field techniques [12, 45, 51]. High precision/high accuracy measurements of MLC-shaped small fields [45] have shown output factors up to 15% higher than those measured with miniature ion chamber. This difference can be explained by the small size and water equivalence of PSDs. Thus proving that PSDs can be used to investigate possible discrepancies between TPS and measurements. To fully characterize the dose distribution produced by a given radiation beam, several profile and depth dose measurements are required. Such data acquisition can be simplified by the use of a detector array. However, given the high spatial resolution required for the characterization of a small field, most commercial arrays are inadequate for this task. Ion chamber arrays have individual detector widths of several millimeter and diode arrays cannot be packed much closer than 5 mm because of the surrounding electronics. PSDs do not suffer from such limitations, thus making PSD arrays an ideal tool for small-field characterization [16, 51].

Furthermore, because of their intrinsic water-equivalence, PSD arrays can be fully three-dimensional without the risk that detector element upstream affect the dose received by the detectors downstream. Films can also be used for 2D dose distribution measurements. However they are offline, integrating dosimeters while PSDs and other can be read in real-time [23], which provide useful information when delivering rotational treatments such as VMAT [52]. Another method for fully characterizing the dose distribution of a small radiation field could be to use the novel tomodosimetry concept [53]. Using tomodosimetry, it is possible to measure a 2D dose distribution with an arbitrarily high spatial resolution using only a small number of plastic scintillating fibers. With commercially available PSD and a clear demonstration of the potential of closely packed arrays [51] and tomodosimeters [53], it is likely that PSDs will become a reference for small field dosimetry.

5. In vivo dosimetry with PSDs

5.1. External beam radiotherapy

In addition to the dosimetric advantages PSDs have when compared to other detectors, they also can be easily incorporated with patient treatment applicators, inserted into Foley catheters or intraluminal interstitial needles as well as flexible catheters. The first study highlighting the use of PSDs for in vivo dosimetry and demonstrating their real-time monitoring capability of the dose rate delivery was demonstrated for external beam radiotherapy treatments by Archambault *et al* [23]. In that study, PSD systems designed into a modular detector patches, consisting of 5 closely packed sensors, were retrofitted to different rectal balloons and rectal inserts. Miniature PSD probes made out of green-emitting multi-clad scintillating fibers (0.5 mm diameter by 2mm length) were optically coupled to water-equivalent optical fibers were used to sample the dose reading using an electron multiplying (EM) CCD camera.

Doses of 200 cGy and 2cGy for all the PSDs were measured at a 0.15 s sampling rate and were accurate within 0.4% and 2.3% respectively. For an eight-beam intensity-modulated radiotherapy plan delivered to a prostate phantom with a rectal balloon inserted within it resulted in a very good agreement between the doses measured by the mounted PSDs and the treatment planning system. Overall, for all beams, the PSDs agreed with the treatment plan to within 0.5%. The main findings of the above study were: 1) a PSD multi-probe array could be read in real time with a CCD-based acquisition apparatus, 2) PSDs with high-spatial resolution (0.4 mm3) could be mounted on rectal balloons and used reliably for in vivo dose measurements, 3) the precision of the PSDs was 3% for measurements performed every 150 ms (1.5 cGy) and improved to 0.4% for a total dose of 200 cGy and 4) PSDs are capable of measuring dose with a high degree of accuracy.

5.2. Brachytherapy

The first PSD system built with the intention to be used for ^{192}Ir high dose rate (HDR) brachytherapy was developed by Arnfield *et al* in 1996 [54]. A plastic scintillator (BC408S, 1 mm diameter and 3 mm long) was attached to a 10 m silica fiber used a light guide and read by a photomultiplier tube. This system was never used for in vivo dosimetry however.

Also designed for HDR brachytherapy, a system has been described by Lambert *et al* (20) in 2006 and used clinically in a patient study in 2011[55]. The system called BrachyFOD is similar in design to the original of Beddar *et al* [5]. It uses a BC400 scintillator (0.5 mm diameter by 4 mm length) coupled to PMMA optical fiber guide (0.48 mm). The OPTIDOS Reader (PTW, Freiburg, Germany) composed of a photomultiplier and electrometer was used. 24 high-dose rate brachytherapy patients were enrolled in the study and dose to the urethra monitored. The PSD system underwent change after the first 14 patients in order to improve the positioning accuracy. A maximum dose departure of 9% from the expected values obtained by the treatment planning system was seen in the remaining 10 patients. This is inline with other brachytherapy in vivo studies using OSL [56] and Mosfet dosimeters [57, 58]. The authors of this study did not report seeing any temperature dependence for their in vivo measurement performed inside the urethra nor experiencing any stem effect when using these PSDs.

Another real-time in vivo PSDs for 192Ir HDR brachytherapy was developed by Therriault-Proulx *et al* [59] and validated in a phantom study [60]. The importance of removing stem effects for in vivo dosimetry using PSD in HDR brachytherapy treatments was assessed and was found to be required in order to be able to measure doses accurately at this energy range [59]. Therefore, a PSD system capable of stem removal was built using a tricolor (red-blue-green) photodiode connected to a dual-channel electrometer. The PSD system consisted of a 1.0 mm diameter x 3.0 mm long green scintillating fiber (BCF-60) coupled to a light-shielded 1 mm diameter 7 m long polymethylmethacrylate optical fiber (Eska Premier GH-4001). A clinically realistic Ir-192 HDR brachytherapy treatment plan was simulated by loading computed tomography images of an arbitrarily chosen patient with prostate cancer into the Oncentra version 3.2 Nucletron Treatment Planning System (TPS). The target volume (prostate) and surrounding organs at risk (i.e., bladder, urethra, and rectum) were contoured, and catheters were reconstructed in the software to reproduce the experimental setup. Simulations of treatment delivery were performed in a customized water phantom, which consisted of eleven catheters that were utilized as channels for dose delivery from the Ir-192 radioactive source, while two others were used to mimic dosimetry at the rectum wall and in the urethra using the above described PSD. The measured dose and dose rate data were compared to the expected values from the planning system. In-phantom dosimetry measurements of the treatment plan led to a ratio to the expected dose of 1.003±0.004 with the PSD at different positions in the urethra and 1.043±0.003 with the PSD inserted in the rectum. The verification of the dose delivered to the urethra within each catheter and at specific dwell positions led to average measured to expected ratios of 1.015±0.019 and 1.014±0.020, respectively. These values at the rectum wall were 1.059±0.045 within each catheter and 1.025±0.028 for specific dwell positions [60]. The study also demonstrated that the above-described system has the ability to detect positioning errors of the source and also allows for temporal verification of the treatment delivery.

5.3. Superficial therapy, radiology and interventional radiology

For low energy ionizing radiation, PSDs are not expected to retain their water-equivalence (see Fig 1, bottom panel). Furthermore, effect such as quenching can become important, displaying non-linearity between the deposited energy and light production [61, 62]. Jones and Hintenlang demonstrated experimentally that PSDs show a strong dependence to tube voltage between 40 and 120 kVp but for a given voltage they show linearity between light output and air kerma and a reproducibility of about 6% [63). Similar findings were made by Hyer *et al* in the context of dose measurements for CT scanner [64]. More recently, Lessard *et al* performed a systematic studies combining PSD measurements, cavity theory and Monte Carlo simulations to elucidate the energy dependence and possible quenching effect in this energy range [65]. A simple correction approach, using calculated x-ray tube spectra, resulted in absolute dose measurements within 2% of the expected values [65]. It is therefore quite likely that we will see more work in this area in the coming years.

6. 2D scintillation dosimetry

The water-equivalence property is of primary importance for applications beyond single-point dosimetry: it implies that the presence of a scintillating fiber in a water-equivalent phantom does not alter the original dose distribution in the phantom (i.e. the dose distribution in the absence of the detector).

This property makes scintillating fibers ideal to use in an array of detectors. The first PSD array encountered in the literature was made by Flühs *et al* for ophthalmic plaque dose measurements [31]. It uses a 1D array composed of 16 PSDs and uses the two-fiber method for handling the stem effect. Furthermore, this 1D array was mounted on a vertically translating stage also capable of translation. This gave the capability to map the dose in 3D space. Numerous investigators have developed experimental devices in both linear (1D) and planar (2D) configuration [16, 24, 51, 53, 66-71]. Among these, the highest density of detector was achieved by Gagnon *et al*. and Yoo *et al*., who achieve near 1 mm detector spacing [16, 51]. Their detectors were composed of two, crossed linear array of small length (< 5cm) and were developed for small field dosimetry. The highest number of detector in a single array was achieved by Guillot *et al*, with almost 800 scintillating fibers placed inside a 26×26 cm^2 detection plane [69, 71]. The resolution of this array was 10×10 mm^2 (except the two central profiles at 5 mm spacing) and was developed for IMRT and VMAT plan verification. As of today, no 3D experimental configuration of scintillating fiber arrays has been reported in the literature, although such arrangement would definitively be possible because of the water-equivalence of these detectors. However Kirov *et al* used liquid scintillator to characterize the dose distribution around brachytherapy sources [72] and Poenisch *et al* for high-energy photon beams [73].

Other investigators have opted for scintillator sheets instead of fibers to perform 2D dosimetry [74-76]. The scintillator sheets used ranged from 15x15 cm2 to 25x25 cm2 and were imaged with the combination of a CCD camera and a 45 degrees mirror. This methodology has the advantage of measuring 2D dose distribution in real-time with high resolution and without the discrete (i.e. incomplete) sampling performed by 2D arrays of scintillating fibers. 3D dosimetry has also been reported as possible using a vertical translation of the scintillator sheet [75].

Investigators have also looked for application of scintillating fibers in longer length (i.e. more than a few millimeters, as used in scintillating fiber arrays previously described) [68, 77]. Goulet *et al* developed a transmission detectors composed of 60 parallel, 27 cm long scintillating fibers for the on-line monitoring of MLC-driven radiotherapy treatment [68]. Collecting the scintillation light from both sides of each fiber and taking advantage of the important optical attenuation inside scintillating fibers, they were able to extract both the position and the fluence related parameters for each field under monitoring. These two independent parameters could then be evaluated and compared in real-time to their expected value (either theoretically of experimentally deduced). Using a similar fiber configuration, these same authors have developed a high resolution 2D dosimeter using tomographic principles [77]. Using a rotating array of 50 parallel fibers inside a water-equivalent phantom, they were able to measure the dose inside a 10 cm radius circle of detection with high spatial resolution (1×1 mm^2). This kind of methodology has the advantage of limiting the number of scintillating fibers needed to perform high-resolution dosimetry, and should find interesting applications in 3D dosimetry.

7. Conclusion

PSDs have gain in technological innovation and engineering refinement over the years. The major issue of the stem effect (Cerenkov and fluorescence) has been addressed and example clinical use shown in this documents. PSDs are poised to become reference dosimeters for small field dosimetry. Early results shows that PSDs can be used without need of correction for fields of 4-5 mm, such as encountered on the Cyberknife (Accuray, California) [78, 79]. New major developments are also expected based on the recent conference presentations demonstrating the possibility of having multiple scintillation probes on a single light guide [80, 81]. Finally, a point-like PSD system is now available commercially, enabling widespread clinical adoption [82].

8. References

[1] Birks J B 1964 *Theory and Practice of Scintillation Counting* Pergamon Press/Macmillan
[2] Knoll G F 2000 *Radiation detection and measurement* 3rd ed. John Wiley & Sons, Inc.
[3] Bushberg J T *et al* 2011 *The Essential Physics of Medical Imaging* Third Edition (Wilkins LW, ed.) Philadelphia, PA.
[4] Murai T *et al* 1964 *J. Radiat. Res.* **29** 23-34
[5] Beddar A S *et al* 1992 *Phys. Med. Biol.* **37** 1883-900
[6] Beddar A S *et al* 1992 *Phys. Med. Biol.* **37** 1901-13
[7] Beddar AS 2006 *Radiat. Meas.* **41** S124-33
[8] Beddar A S 1994 *Phys. Med. Biol.* **39** 253-63
[9] Cliftt M A *et al* 2000 *Phys. Med. Biol.* **45** 1885-95
[10] Archambault L *et al* 2006 *Med. Phys.* **33** 128
[11] Frelin A M *et al* 2005 *Med. Phys.* **32** 3000
[12] Lambert J *et al* 2010 *Phys. Med. Biol.* **55** 1115-26
[13] Lacroix F *et al* 2010 *Med. Phys.* **37** 4331-42
[14] Archambault L *et al* 2007 *Med. Phys.* **34** 2336
[15] Ayotte G *et al* 2006 *Med. Phys.* **33** 3519-25
[16] Yoo W J *et al* 2012 *Sensors* **12** 6404-14
[17] Beddar A S *et al* 2003 *Phys. Med. Biol.* **48** 1141-52
[18] Archambault L *et al* 2005 *Med. Phys.* **32** 2271-78
[19] Létourneau D *et al* 1999 *Med. Phys.* **26** 2555-61
[20] Lambert J *et al* 2006 *Phys. Med. Biol.* **51** 5505-16
[21] Beierholm A *et al* 2008 *Radiat. Meas.* **43** 898-903
[22] Fontbonne J M *et al* 2002 *IEEE Trans. Nucl. Sci.* **49** 2223-7
[23] Archambault L *et al* 2010 *Int. J. Radiat. Oncol. Biol. Phys.* **78** 280-7
[24] Liu P Z Y 2012 *Med. Phys.* **39** 1688-95
[25] Lacroix F *et al* 2009 *Med. Phys.* **36** 5214-20
[26] Lacroix F *et al* 2010 *Med. Phys.* **37** 412
[27] Attix FH 1986 *Introduction to radiological physics and radiation dosimetry* New York: Wiley-interscience
[28] Beddar A S *et al* 1992 *Phys. Med. Biol.* **37** 925-35
[29] Beddar A *et al* 2004 *Phys. Med. Biol.* **49** 783-90
[30] Westermark M *et al* 2000 *Phys. Med. Biol.* **45** 685-702
[31] Flühs D *et al* 1996 *Med. Phys.* **23** 304-427
[32] Ishikawa M *et al* 2009 *Phys. Med. Biol.* **54** 2079-92
[33] de Boert S F 1993 *Phys. Med. Biol.* **38** 945-58
[34] Clift M A *et al* 2000 *Phys. Med. Biol.* **45** 1165-82
[35] Clift M A *et al* 2002 *Phys. Med. Biol.* **47** 1421-33
[36] Guillot M *et al* 2011 *Med. Phys.* **38** 2140-50
[37] Liu P Z Y *et al* 2011 *Phys. Med. Biol.* **56** 5805-21
[38] Archambault L *et al* 2012 *Phys. Med. Biol.* **57** 3661-65
[39] Heydarian M *et al* 1996 *Phys. Med. Biol.* **41** 93-110
[40] Duggan D M and Coffey C W 1998 *Med. Dosim.* **23** 153-9
[41] Ding G X *et al* 2008 *Phys. Med. Biol.* **53** 3493-99
[42] Laub W U and Wong T 2003 *Med. Phys.* **30** 341-7
[43] Francescon P *et al* 2008 *Med. Phys.* **35** 504-13
[44] Sánchez-Doblado F *et al* 2007 *Int. J. Radiat. Oncol. Biol. Phys.* **68** 301-10
[45] Klein D M *et al* 2010 *Med. Phys.* **37** 5541
[46] Araki F 2006 M *Med. Phys.* **33** 2955-63
[47] Scott A J D 2008 *Med. Phys.* **35** 4671-84
[48] Sánchez-Doblado F *et al* 2007 *Physica medica* **23** 58–66

[49] Alfonso R *et al* 2008 *Med. Phys.* **35** 5179-86
[50] Beddar A S *et al* 2001 *IEEE Trans. Nucl. Sci.* **48** 924-8
[51] Gagnon J C *et al* 2012 *Med. Phys.* **39** 429
[52] Klein D *et al* 2012 *Radiat. Meas.* **47** 921-9
[53] Goulet M *et al* 2012 *Med. Phys.* **39** 4840
[54] Arnfield M R *et al* 1996 *IEEE Trans. Nucl. Sci.* **43** 2077-84
[55] Suchowerska N *et al* 2011 *Int. J. Radiat. Oncol. Biol. Phys.* **79** 609-15
[56] Andersen C E *et al* 2009 *Med. Phys.* **36** 5033-43
[57] Cygler J E *et al* 2006 *Radiother. Oncol.* **80** 296-301
[58] Bloemen-van Gurp E J *et al* 2009 *Int. J. Radiat. Oncol. Biol. Phys.* **73** 314-21
[59] Therriault-Proulx F *et al* 2011 *Med. Phys.* **38** 2176-9
[60] Therriault-Proulx F *et al* 2011 *Med. Phys.* **38** 2542
[61] Williamson J F *et al* 1999 *Phys. Med. Biol.* **44** 857-71
[62] Frelin A M *et al* 2008 *IEEE Trans. Nucl. Sci.* **55** 2749-56
[63] Jones A K and Hintenlang D 2008 *Radiat. Prot. Dosimetry* **132** 80-7
[64] Hyer D E *et al* 2009 *Med. Phys.* **36** 1711-6
[65] Lessard F *et al* 2012 *Med. Phys.* **39** 5308-16
[66] Archambault L *et al* 2007 *Med. Phys.* **34** 1583-92
[67] Lacroix F *et al* 2008 *Med. Phys.* **35** 3682
[68] Goulet M *et al* 2011 *Med. Phys.* **38** 1459-67
[69] Guillot M *et al* 2011 *Med. Phys.* **38** 6763-74
[70] Moon J *et al* 2012 *Appl. Radiat. Isot.* **70** 2627-30
[71] Guillot M *et al* 2010 *J. Phys.: Conf. Ser.* **250** 012006
[72] Kirov A S *et al* 2005 *Phys. Med. Biol.* **50** 3063-81
[73] Pönisch F *et al* 2009 M*ed. Phys.* **36** 1478
[74] Petric M P *et al* 2006 *Med. Phys.* **33** 96
[75] Frelin A M *et al* 2008 *Med. Phys.* **35** 1651
[76] Collomb-Patton V *et al* 2008 *Radiat. Prot. Dosimetry* **131** 100-9
[77] Goulet M, *et al* 2011 *Med. Phys.* **38** 3725
[78] Ralston A A *et al* 2012 *Phys. Med. Biol.* **57** 2587-602
[79] Morin J *et al* 2012 *Med. Phys.* (in press)
[80] Archambault L *et al* 2012 *Radiother. Oncol.* **103** S313-4
[81] Therriault-Proulx F *et al* 2012 *Radiother. Oncol.* **103** 8-9
[82] *Exradin W1, http://www.standardimaging.com/*

Dosimetry for audit and clinical trials: challenges and requirements

T Kron, A Haworth and I Williams
Peter MacCallum Cancer Centre, Department of Physical Sciences, University of Melbourne, Australia
Peter MacCallum Cancer Department, Australian Clinical Dosimetry Service, Melbourne, Australia

Email: tomas.kron@petermac.org

Abstract. Many important dosimetry audit networks for radiotherapy have their roots in clinical trial quality assurance (QA). In both scenarios it is essential to test two issues: does the treatment plan conform with the clinical requirements and is the plan a reasonable representation of what is actually delivered to a patient throughout their course of treatment. Part of a sound quality program would be an external audit of these issues with verification of the equivalence of plan and treatment typically referred to as a dosimetry audit. The increasing complexity of radiotherapy planning and delivery makes audits challenging. While verification of absolute dose delivered at a reference point was the standard of external dosimetry audits two decades ago this is often deemed inadequate for verification of treatment approaches such as Intensity Modulated Radiation Therapy (IMRT) and Volumetric Modulated Arc Therapy (VMAT). As such, most dosimetry audit networks have successfully introduced more complex tests of dose delivery using anthropomorphic phantoms that can be imaged, planned and treated as a patient would. The new challenge is to adapt this approach to ever more diversified radiotherapy procedures with image guided/adaptive radiotherapy, motion management and brachytherapy being the focus of current research.

1. Introduction

Radiotherapy is a locoregional cancer treatment modality that relies on accurate delivery of very high doses of radiation to a target whilst minimising dose to surrounding healthy structures. The required accuracy for dose delivery is typically quoted as being within +/- 5% of the prescribed dose [1, 2] while the spatial accuracy depends on the part of the body that is irradiated. It can vary from a few mm in the pelvis to less than 1mm when delivered to lesions in the brain. The consequences of not meeting these requirements can be very severe for the patient and radiotherapy departments are generally highly aware of the need for excellent quality assurance in the treatment planning and delivery process.

Most quality systems include the need for an external audit and many radiotherapy departments participate regularly in these audits [3, 4]. The largest two dosimetric audit networks at present are operated by the International Atomic Energy Agency (IAEA) [5, 6] and the American Radiological Physics Centre (RPC) in Houston [7, 8]. These two organizations also reflect the two key motivations for dosimetric audits: the IAEA provides a relatively simple audit of beam calibration that reduces the risk of misadministration while the RPC primarily supports co-operative clinical trials groups that use radiotherapy as part of clinical research [9]. More recently several other countries have established this type of service, and Australia has recently started to operate the Australian Clinical Dosimetry Service (ACDS) [10].

The objective of this paper is to describe some dosimetric audit systems and discuss their link to clinical practice and research. The second half of the manuscript is concerned with three examples for particularly challenging situations for radiotherapy audits: Image guided and adaptive radiotherapy (IGRT), Motion management and Brachytherapy.

2. Classification of dosimetric audits

Depending on the objective a range of different audit types are performed and several classification systems are used to describe them [11, 12]. In general, audits are classified according to their complexity. Table 1 lists the system recently introduced by the ACDS.

Commonly, a check of absolute dosimetry under reference conditions is referred to as a level I dosimetric audit [11, 12]. Figure 1 shows two such systems. The advantage of this type of audit is that they can be performed remotely by sending the dosimeters to participating centres where they are irradiated to a specified dose under clearly defined conditions [5, 6]. The detectors used for this type of audit are nearly exclusively luminescence detectors with thermoluminescence dosimeters (TLDs) [6, 13] being increasingly replaced by optically stimulated luminescence dosimeters (OSLDs) [14, 15] due to their faster readout and easier handling.

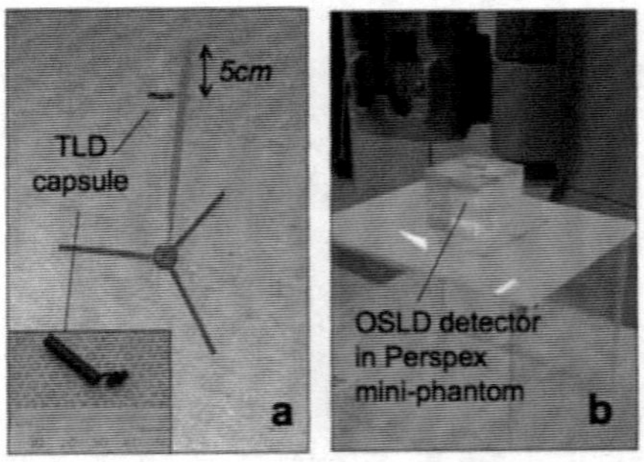

Figure 1: Set-up for verification of absolute photon beam calibration using a mailable dosimetry system. a) IAEA TLD capsule to be exposed in a water phantom at 5cm depth; b) RPC Perspex miniphantom with an OSL detector placed on a platform at 100cm SSD. The lack of full scatter conditions is accounted for using photon energy dependent correction factors.

However, a single dosimetric point dose measurements is not always sufficient to characterise a radiation beam adequately for radiotherapy. For example electron beams usually require verification of radiation quality and dose to interpret the readings appropriately. This is illustrated in figure 2 that shows the electron dosimetry module developed by the RPC. Two sets of three TLD capsules are placed at two depths in a Perspex phantom.

A tool for postal audits under non-reference conditions was recently introduced by the IAEA [12]. It is based on the jig shown in figure 1a and introduces a horizontal bar to allow placement of the TLD capsule off central axis. Also an extension of the vertical bar is available that places the reference depth at 10 cm which is more in line with recent calibration protocols [16, 17]. This would be a typical level II audit. An interesting future development may be level II audits based on two dimensional detector systems such as film [18] or detector arrays. The latter have mainly been developed for IMRT QA [19] but offer the advantage of acquiring large amounts of data in few exposures and as such, provide an easy means of characterising radiation beams.

Anthropomorphic phantoms are the primary tool used in level III dosimetric audits. Figure 3 shows an anthropomorphic phantom customised for dosimetric intercomparisons for clinical trials of prostate and rectal cancer radiotherapy [20, 21]. In a level III audit the whole chain of treatment from imaging to planning and delivery can be tested. While strictly speaking an anthropomorphic phantom is not required it makes the process as realistic as possible and typically motivates staff in participating centres.

Table 1: Dosimetric services offered by the Australian Clinical Dosimetry Service

Dosimetry level	ACDS	Detector type	Mode	System checked	Comments
Level I	Output under reference conditions	TLD, OSL	Remote	Every radiation beam	Identical to RPC audit
Level IB	Output under reference conditions	Ionisation chamber	On-site	Every radiation beam	Offered for new centres prior to opening
Level II	Dose distribution in physical phantoms	Detector array	Remote	Planning system	Can include inhomogeneity and allows clarification of level III findings
Level III	Anthropomorphic phantom end to end	Ion chamber, radiochromic film	On-site	Entire treatment chain	Treatment specific – most relevant for clinical trials

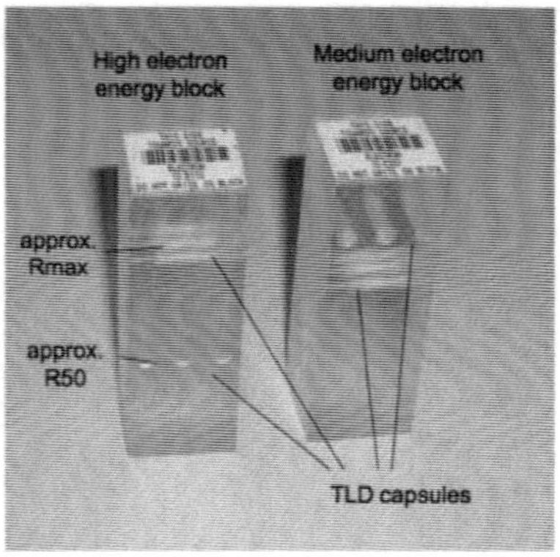

Figure 2: Perspex holders designed by the RPC for determination of electron energy (dosimeters approximately at depth of 50% dose, R50) and output (dosimeters approximately at depth of maximum dose). The TLD holders, which are exposed inserted into a small

Due to their weight and complexity anthropomorphic phantom based audits are often part of a site visit as was in the case of the phantom shown in figure 3. However, in principle anthropomorphic phantoms can also be mailed to users. The RPC has developed a suite of phantoms for this purpose [7, 8]. Most of them are plastic shells that can be filled with water after being sent to participating centres. The shells also contain dosimetric inserts with TLDs, OSLDs and/or radiochromic film that are shipped as integral part of the phantom or as separate module that must be inserted into the phantom prior to use [7, 8].

A variety of dosimeters have been used for dosimetric audits. While TLDs and OSLs are ideal for mailed dosimetry [5, 22], audits involving anthropomorphic phantoms typically also employ a two-dimensional dosimetry system such as radiochromic film [23, 24]. Ideally, one would like to take this one step further to three-dimensional systems such as gels [25] or presage [26]. Despite considerable interest, to date these systems are still largely experimental.

Finally, also *in vivo* dosimetry has been suggested as a tool for radiotherapy audits of various complexity [27, 28]. While possible in principle there are many additional uncertainties to account for, documentation and ethics need to be considered and many modern delivery techniques such as VMAT or IMRT would be difficult to assess. As such audits based on in vivo dosimetry are likely to stay more common in radiology where entrance dose is an important parameter in radiologic exposures that can be used for the establishment of diagnostic reference levels [29].

3. Auditing of modern radiotherapy delivery

As radiation delivery becomes more complex, auditing must test more aspects of the delivery to provide a reasonable assurance that a centre performs a particular treatment as per best practice. Anthropomorphic phantoms are well suited for this purpose and the RPC is performing a large number of audits using a variety of phantoms every year [7, 8, 24]. The results, which showed that a significant number of centres did not perform within tolerance, demonstrated in themselves that the audits are fulfilling an important role in identifying problems. To quote from Ibbott et al (2006): "The experience obtained through the irradiation of the phantoms by a number of institutions demonstrates that institutions vary significantly in their ability to deliver doses and dose distributions that agree with their own treatment plans" [8]. Repeat audits show in general that the results improve, a fact that has also been shown for the IAEA level I dosimetric audits [6].

An additional feature of complex audits is the need to be able to compare measured dose distributions with the ones planned for delivery. In the case of an audit it is not necessary to evaluate the quality of the dose distribution itself: the key point is if the dose predicted is identical to the dose given. In order to facilitate this, a clinical audit centre must have access to treatment planning or trials review software [30] that can display dose distributions from a variety of different planning systems.

In Australia, several audits of complex radiotherapy delivery have been performed. Figure 3c shows the dose distribution for a prostate cancer trial as assessed using radiochromic film [23]. The coronal sectioning of the phantom allows for easy film placement and provides a picture of the dose distribution. An ion chamber measurement in the centre of the phantom's prostate complements this dose distribution. To date, these have been typically part of clinical trials [23, 31]; however, the establishment of the ACDS is likely to lead to such services also being offered to any centre wishing to have an independent check of complex procedures.

4. Auditing for clinical trials

In the context of clinical trials an audit must not only support patient safety but also minimise the risk of not being able to answer the trial question. There have been several trials that have demonstrated that protocol compliance and quality assurance are very important for trial outcomes [32]. In the worst case, this can mask a significant difference between arms in a clinical trial. Conversely, good quality assurance can reduce the number of patients required in a clinical trial to answer the trial question [33].

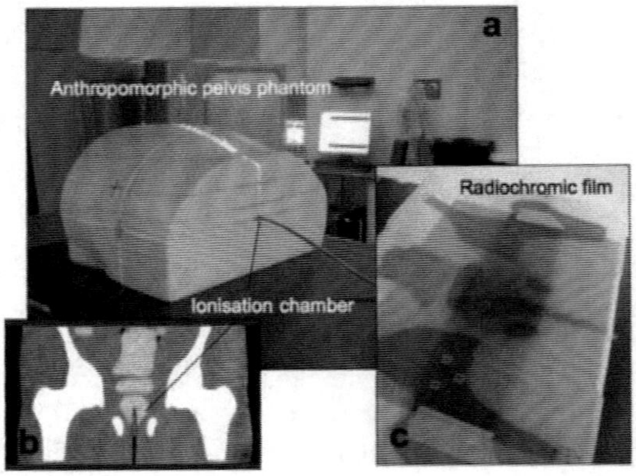

Figure 3: A pelvic anthropomorphic phantom 'Elvis' specially designed for dose assessment in two clinical trials of prostate and rectal cancer, respectively. Figure 3a shows a photo of the phantom while figure 3b shows a coronal reconstruction where the hole for an ionisation chamber can be seen. The phantom was built to mimic a real patient and is sectioned in coronal slices. Figure 3c shows a piece of radiochromic film which has just been exposed to an IMRT dose distribution for a prostate cancer delivery.

The level and complexity of the dosimetric audit therefore will vary depending on the importance of dosimetry for the trial question. If radiation is not part of the trial question (eg radiotherapy with or without administration of specific systemic treatment) possibly no external audit is required. However, if for example mucosal toxicity is a primary endpoint, it may be important to verify that the dose prediction by the planning system (or whatever other means is used to determine dose). In general though, audits for clinical trials need to go further than dosimetric audits as one needs to verify contouring and protocol compliance not just that what you see is what you get.

It has long been recognised that multicenter clinical trials not only test a new treatment approach but can also help to introduce new approaches and particularly technology safely [34-36]. This is something which is likely to be of increasing importance for advanced radiotherapy techniques and poses particular challenges as discussed in the next section.

5. Challenges for dosimetric audits

The increasing variety and complexity of radiotherapy makes audits increasingly challenging. In particular imaging plays an increasing role in the delivery and it is no longer sufficient to simply verify that the dose distribution is consistent with the planned distribution, additionally, an audit needs to assess that this dose distribution can be delivered to the correct part of the patient's anatomy. As such, comprehensive dosimetric audits need to consider the tools that allow determining how the dose delivery is optimised for each individual patient on every day of the treatment course.

In order to optimise resources many audit and clinical trials groups have commenced a risk management approach to auditing [31, 37]. One of the risk management approaches is failure mode and effect analysis (FMEA). The fundamental idea is not to perform activities such as audits according to a predetermined schedule but to analyse processes and give priority to a particular audit item depending on the likelihood of something going wrong and the severity of the consequences [38].

5.1. Image guided and adaptive radiotherapy

Image guided radiotherapy refers to the use of high quality imaging equipment in the treatment room prior or during the delivery of radiotherapy with the aim to ensure optimal treatment delivery [39, 40]. This can range from ultrasound to planar X-rays with or without fiducial markers, to cone beam CT and possibly even MRI [41, 42]. It is this wide variety of approaches and the need for decision making that make an audit of IGRT difficult. A few attempts have been made [43] and the American Radiation Therapy and Oncology Group (RTOG) has issues guidelines for the inclusion of IGRT into clinical trials (available from the RPC webpage: http://rpc.mdanderson.org/).

Even more complex is the assessment of adaptive radiotherapy where IGRT is used not only to position the patient but also to select or adjust a treatment plan. A recent multicenter clinical trial on adaptive radiotherapy of bladder cancer [44] has a program for credentialing of participating centres that specifically tests the ability of centres to perform the tasks required for treatment according to protocol. Figure 4 shows a Perspex body phantom used for credentialing of centres for adaptive radiotherapy which involves selection of a plan of the day selected based on volumetric imaging. The credentialing process tests the image guidance and decision making chain and involves inserting a target volume in a cylinder into the phantom the size and location of which is unknown to the operator. The task consists of imaging the phantom, identifying the cylinder and positioning it in the correct location based on reference images that are provided to participating centres [31].

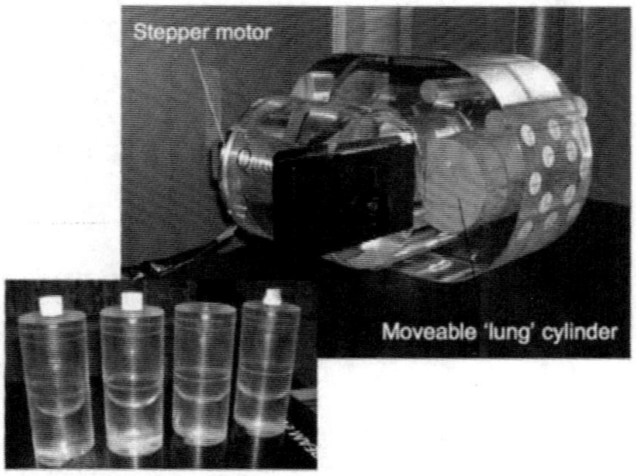

Figure 4: Perspex body phantom (Quasar, Modus Medical) with motion attachment. In the insert four custom built cylinders are shown that are used to test the ability of radiotherapy centres to identify a target and position it in the correct location.

5.2. Motion management

A different task for modern radiotherapy is motion management [45], be it in treatment planning or delivery. A number of groups have developed phantoms that have moveable parts and can be used for auditing. Of particular interest is the work of the RPC that has developed a lung and a liver phantom for auditing of stereotactic extracranial procedures. Like in other complex radiotherapy delivery techniques two-dimensional detector systems such as radiochromic film prove to be particularly useful [46].

5.3. Brachytherapy

The ICRP publication 97 reported on more than 500 brachytherapy events [47], including one death, that could have been prevented through rigorous quality assurance exercises, including independent audits. Many of the reported incidents were a result of human error and therefore brachytherapy audits ideally should include site visits that not only check dosimetry, but also check internal processes that may lead to random (patient specific) errors, or systematic errors that may affect all patients.

A small number of brachytherapy dosimetry audits have been conducted [48-52]. These audits vary in complexity, ranging from simple verification of the calibration of the local well chamber used to determine the reference air kerma rate of the ^{192}Ir source, to Level III audits involving measurement of dose in a phantom where the dwell positions and dwell times were calculated using the treatment planning computer. A small pilot study looking at the feasibility of conducting a brachytherapy audit was conducted in Australia. This study was based on the European audit described by Roue et al [52]. The Australian study was carried out in 7 centres using a small water phantom to confirm the dose predicted by the treatment planning system was consistent with measured dose using TLDs and 6-dwell positions. The study proved that such audits are feasible, and did detect one error at one treatment site. Recommendations for future studies, however, include the use of a phantom that more closely mimics a clinical situation, as this is more likely to detect errors that could occur in the clinic.

6. Outlook and conclusion

Audits are an essential part of quality radiotherapy. The complexity of modern radiotherapy makes a risk management approach essential as not all relevant aspects of a treatment delivery can be verified. By identifying the events that have the worst consequence and estimating how likely they are, it is possible to prioritise activities for dosimetric audits. It is likely that auditing in radiotherapy will go further along this road and include cost effectiveness, societal priorities and the desired outcome when deciding on new audits. It is also likely that dosimetric audits will not suffice to characterise practice and a more comprehensive audit is required. The IAEA has recently introduced such an audit specifically for radiotherapy [53-55]. However, whatever happens, given the high doses given in radiotherapy, there always will be an emphasis on safety: Primum non nocere.

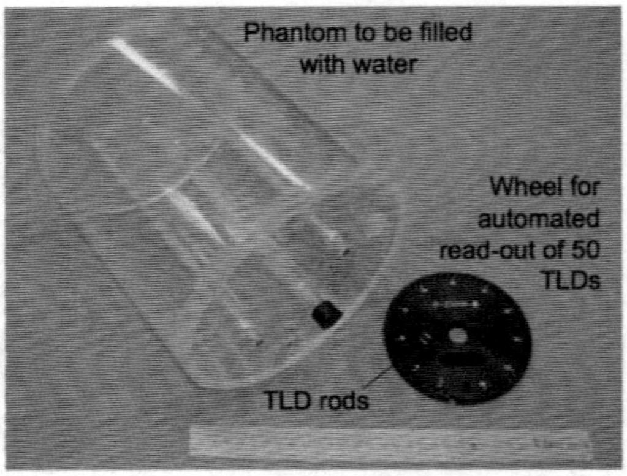

Figure 5: Brachytherapy intercomparison phantom. The phantom is mailable and can be filled with water at participating sites. TLDs are positioned in the centre hole and the brachytherapy source stepped through the three channels. Centres are asked to plan the phantom to deliver a certain dose.

7. References

[1] ICRU Report 24. In *ICRU reports* (International Commission on Radiological Units and Measurements, Bethesda, 1976)
[2] Mijnheer B J *et al* 1987 *Radiother. Oncol.* **8** 237-52
[3] Ferreira I H *et al* 2000 *Radiother. Oncol.* **55** 273-84
[4] Thwaites D *et al* 1995 *Radiother. Oncol.* **35** 61-73
[5] Izewska J *et al* 2003 *Radiother. Oncol.* **69** 91-7
[6] Izewska J *et al* 2002 *Radiat. Prot. Dosimetry.* **101** 387-92

[7] Followill D S *et al* 2007 *Med. Phys.* **34** 2070-6
[8] Ibbott G S *et al* 2006 *Technol. Cancer Res. Treat.* **5** 481-7
[9] Ibbott G *et al* 2008 *Int. J. Radiat. Oncol. Biol. Phys.* **71** S71-5
[10] Williams I *et al* 2012 The Australian Clinical Dosimetry Service: a commentary on the first 18 months *Australas. Phys. Eng. Sci. Med.* **35** 407-11
[11] Kron T *et al* 2002 *Int. J. Radiat. Oncol. Biol. Phys.* **52** 566-79
[12] Izewska J *et al* 2007 *Radiother. Oncol.* **84** 67-74
[13] Kron T 1995 *Australas. Phys. Eng. Sci. Med.* **18** 1-25
[14] Mrcela I *et al* 2011 *Phys. Med. Biol.* **56** 6065-82
[15] Reft C S 2009 *Med. Phys.* **36** 1690-9
[16] Almond P R *et al* 1999 *Med. Phys.* **26** 1847-70
[17] IAEA Technical report series N398 (International Atomic Energy Agency, Vienna, 2000)
[18] Novotny J *et al* 1997 *Phys. Med. Biol.* **42** 1277-88
[19] Low D A *et al* 2011 *Med. Phys.* **38** 1313-38
[20] Ebert M A *et al* 2011 *Med. Phys.* **38** 5167-75
[21] Harrison K M *et al* 2011 *Med. Phys.* **38** 5330-7
[22] Viamonte A *et al* 2008 *Med. Phys.* **35** 1261-6
[23] Healy B *et al* 2011 *Australas. Phys. Eng. Sci. Med.* **34** 195-202
[24] Molineu A *et al* 2005 *Int. J. Radiat. Oncol. Biol. Phys.* **63** 577-83
[25] Babic S *et al* 2008 *Int. J. Radiat. Oncol. Biol. Phys.* **70** 1281-91
[26] Sakhalkar H *et al* 2009 *J. Phys.: Conf. Ser.* **164** 12058
[27] Venables K *et al* 2004 *Radiother. Oncol.* **71** 303-10
[28] Kalef-Ezra J A *et al* 2002 *Radiat. Prot. Dosimetry* **101** 403-05
[29] Shrimpton P C *et al* 1986 *Br. J. Radiol.* **59** 749-58
[30] Ebert M A *et al* 2008 *Radiother. Oncol.* **86** 200-10
[31] Kron T *et al* 2012 *Radiother. Oncol.* **103** 293-8
[32] Peters L J *et al* 2011 *J. Clin. Oncol.* **28** 2996-3001
[33] Pettersen M N *et al* 2008 *Radiother. Oncol.* **86** 195-9
[34] Haworth A *et al* 2009 *Radiother. Oncol.* **90** 299-306
[35] Hurkmans C W *et al* 2009 *Radiat. Oncol.* **4** 1
[36] Kron T *et al* 2009 *J. Med. Imaging Radiat. Oncol.* **53** 412-8
[37] Dixon P and O'Sullivan B 2003 *Eur. J. Cancer* **39**, 423-9
[38] Huq MS *et al* 2008 *Int. J. Radiat. Oncol. Biol. Phys.* **71** S170-3
[39] Korreman S *et al* 2010 *Radiother. Oncol.* **94** 129-44
[40] Verellen D *et al* 2008 *Radiother. Oncol.* **86** 4-13
[41] van Herk M 2007 *Semin. Radiat. Oncol.* **17** 258-67
[42] Verellen D *et al* 2008 *Acta. Oncol.* **47** 1271-8
[43] Middleton M *et al* 2011 *Int. J. Radiat. Oncol. Biol. Phys.* **81** 1576-81
[44] Foroudi F *et al* 2011 *Int. J. Radiat. Oncol. Biol. Phys.* **81** 765-71
[45] Keall P J *et al* 2006 *Med. Phys.* **33** 3874-900
[46] Kron T *et al* 2011 *Radiat. Meas.* **46** 1920-3
[47] ICRP Publication 97 (Stockholm, 2005)
[48] Carlsson Tedgren A and Grindborg J E 2008 *Radiother. Oncol.* **86** 126-30
[49] de Almeida C E *et al* 2002 *Radiother. Oncol.* **63** 75-81
[50] Heeney C *et al* 2005 *Radiother. Oncol.* **74** 149-56
[51] Ochoa R *et al* 2007 *Radiother. Oncol.* **82** 222-8
[52] Roue A *et al* 2007 *Radiother. Oncol.* **83** 86-93
[53] IAEA. *Comprehensive Audits of Radiotherapy Practices: a tool for Quality Improvement - QUATRO*, (International Atomic Energy Agency, Vienna, 2007)
[54] Bosi S G *et al* 2009 *Phys. Med. Biol.* **54** 275-83
[55] Hill R et al 2005 *Phys. Med. Biol.* **50** N331-44

On the reliability of 3D gel dosimetry

Y De Deene[1,2] and J Vandecasteele[2]

[1]Institute of Medical Physics, School of Physics, University of Sydney

[2]Department for Radiotherapy and Experimental Cancer Research, Ghent University Hospital, De Pintelaan 185, 9000 Gent, Belgium

E-mail: Yves.DeDeene@UGent.be

Abstract. Gel dosimetry has a unique role to play in safeguarding conformal radiotherapy treatments as it covers the whole treatment chain and provides the radiation oncologist with the integrated dose distribution in 3D. A major obstacle that has hindered the wider dissemination of polymer gel dosimetry in radiotherapy centres is the lack of confidence in the reliability of the measured dose. Discrepancies in dose response of small versus large polymer gel dosimeters have been reported and although several hypothesis for these discrepancies have been postulated, the actual contribution of these error sources to the overall inaccuracy of the dose maps has not been determined. Several gel dosimetry research groups have chosen to use an internal calibration of gel dosimeters. In this study, the inter-and intra-batch reproducibility of the current state-of-the-art 3D gel dosimeters has been assessed. It is demonstrated that with a carefully designed scanning set-up, the overall accuracy that can be obtained with an independent calibration is well within 5% of all pixels.

1. Introduction

1.1. Gel dosimetry as a 3D dose verification method

Gel dosimetry provides a unique feature to display dose distributions occurring in clinical radiotherapy in three dimensions (3D) in humanoid shaped phantoms [1-3]. It has a unique role to play in safeguarding conformal radiotherapy treatments as it is the only dosimeter that covers the whole treatment chain and provides the radiation oncologist with the integrated dose distribution in 3D. However, some doubts with respect to the overall reliability have overshadowed the wider dissemination of polymer gel dosimetry in radiation centers. Significant differences in dose-R_2 response have been found between gel dosimeters that were poured in small calibration vials as compared to larger phantom recipients [4,5]. This finding severely compromises the use of an externally derived calibration curve to convert the measured R_2 maps to dose maps. Several possible reasons have been proposed as explanation for the observed discrepancies such as differences in cooling rate [6], changes in temperature during irradiation as result of the exothermal radiation induced polymerization reaction [7] and physicochemical changes related to an oxygen-antioxidant imbalance [8]. To circumvent the discrepancies in dose-R_2 response between small test tubes and larger volumetric anthropomorphic gel dosimeter phantoms many researchers have chosen to apply an internal calibration (i.e. the R_2 is calibrated to dose by using two points within the phantom as 'known' doses). Regardless the fact that with an internal calibration, the ability of gel dosimetry to safeguard absolute dose information of a treatment is lost, it should also be emphasized that there is a potential more severe danger with this strategy because the mentioned mechanisms may depend on radiation dose, dose rate, environmental conditions and size and shape of the phantom. There is therefore no evidence that the same mechanism for the observed discrepancies would not cause any dose read errors within the volumetric polymer gel dosimeter phantom itself.

The evaluation methods that have been applied by many gel dosimetry research groups to give evidence of the 'potential' of gel dosimetry are often based on a percentage number of pixels that pass gamma criteria between a gel dosimetry (GD) measured dose distribution and a treatment planning system (TPS) calculated dose distribution. This value obviously depends on the number of slices and the size (field-of-view) of the dose maps taken into consideration. Although it is recognized that gamma maps are a valuable tool in benchmarking a radiotherapy treatment plan, the use of a percentage value of passing pixels in benchmarking a dosimetry technique against a TPS calculated dose distribution can be misleading as proof of the reliability of the dosimetry technique. Moreover, it is not beyond discussion to what extend a TPS calculated dose distribution should be used as 'golden standard' to benchmark a new dosimeter. Of course, it is recognized that the intrinsic problem with the benchmarking of a 3D dosimeter is that there does not exist a 3D dosimetric 'golden standard' to compare with. The most reasonable strategy is to compare doses obtained with gel dosimetry with doses obtained by the most reliable dosimetry techniques that apply to a certain spatial dimension. For example, dose profiles of a single field (photons and electrons) can be compared with dose profiles obtained with an ionization chamber or diamond detector [9,10]. In two dimensions, gel dosimetry can be compared with film dosimetry [3,9,11]. Dose distributions obtained with gel dosimetry have been compared with those calculated with treatment planning software [3, 12-18]. Errors that compromise the accuracy may occur at different stages of the dosimetry procedure [19]. Given the uncertainties with respect to the dose-R_2 response, we argue that a more stringent criterion than gamma criteria should be applied for benchmarking 3D gel dosimetry.

1.2. Accuracy and precision

In this perspective it may be helpful to refresh the definition of accuracy and precision: *Accuracy* can be defined as the degree of conformity of a measured or calculated quantity (measurand) to its actual (true) value. *Precision* can be defined as the ability of a measurement to be consistently reproduced. Accuracy and precision can be defined statistically in terms of the deviation of the mean value from the reference value and the standard deviation of the mean value of many measurements respectively (figure 1). The measurement uncertainty of a single experiment thus comprises both systematic and random errors. Uncertainties have later been reclassified in type A and type B uncertainties where type A standard uncertainty is obtained from a probability density function derived from an observed frequency distribution, while type B standard uncertainty is obtained from an assumed probability density function that is bases on the degree of belief that an event will occur.

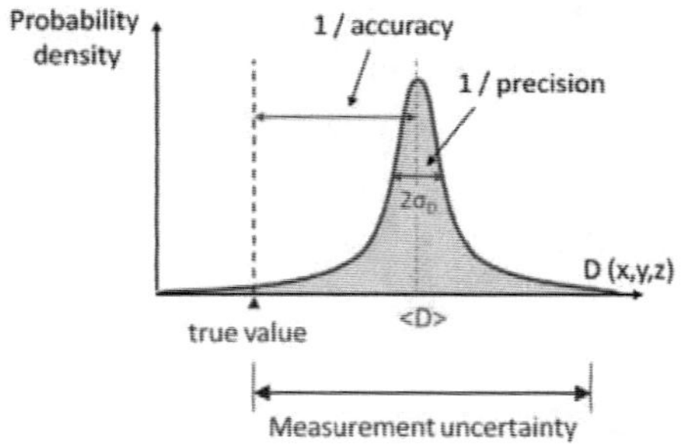

Figure 1. Accuracy and precision shown schematically in terms of probabilities of occurrence. Precision can be characterised in terms of the standard deviation of many measurements (experiments). The accuracy is characterized by the deviation of the mean of many measurements from the reference value.

In the case of gel dosimetry, uncertainty applies to both dose and space. In a gel dosimetry dose verification experiment (and any radiation treatment), the spatial and dosimetric dimensions are interwoven. It is theoretically impossible to extract both dosimetric and spatial errors from a measured spatial dose distribution (i.e. the result of a 3D gel dosimetry experiment). To comprise both spatial and dosimetric performance in one parameter, Low *et al* have introduced the gamma-index [20]. Other dose distribution comparisons are performed on the basis of maximum allowed dose differences [21].

To obtain a picture of the overall uncertainty of polymer gel dosimetry, also the fabrication and irradiation have to be included in the analysis. This can be achieved through a reproducibility study of the complete gel dosimetry experiment from gel fabrication to dose distribution analysis. An extensive reproducibility study, comprising the repetition of a gel dosimetry experiment 8 times, is proposed in this review and the contributions of different sources of inaccuracy are discussed. It is imperative to realize that in a clinical verification with gel dosimetry, consisting of a single experiment, the uncertainty of the dose measured in each voxel comprises both the systematic inaccuracy as the imprecision.

Gel dosimetry is performed in different steps and errors can occur at each stage (figure 2): (1) The polymer gel dosimeter is fabricated in a chemical laboratory and is stored until radiation. Any deviations in the chemical composition through inhomogeneous mixing, impurities in the recipients or oxygen leaks in the recipient can give rise to discrepancies between calibration tubes and volumetric phantoms [8]. Moreover, a difference in temperature course during storage may also affect the dose-R2 response [6]. Most of these deviations are compensated by using calibration phantoms that are fabricated from the same batch of gel. (2) The gel dosimeter phantom is then scanned with CT and the treatment planning is optimized (3) on the scanned set for a virtual planning target volume (PTV) and a set of critical organs. At this stage, just as with a patient, a reference coordinate system is allocated to the phantom by use of marker lines that are drawn on the phantom. In addition, stereotactic fixtures or fiducial markers [22] can be placed on the phantom that will later be visible on the MRI images or optical images and CT images in the case of optical or X-ray CT readout respectively for coregistration of the dose maps with the TPS. Positional setup errors are likely to result in deviations between the planned and the measured dose distribution. These deviations are not intrinsic to the gel dosimeter but are indicative for errors that may occur during the actual patient treatment. (4) Upon irradiation, a complex set of radiation-induced chemical reactions take place. On a molecular level, these reactions are probabilistic in nature.

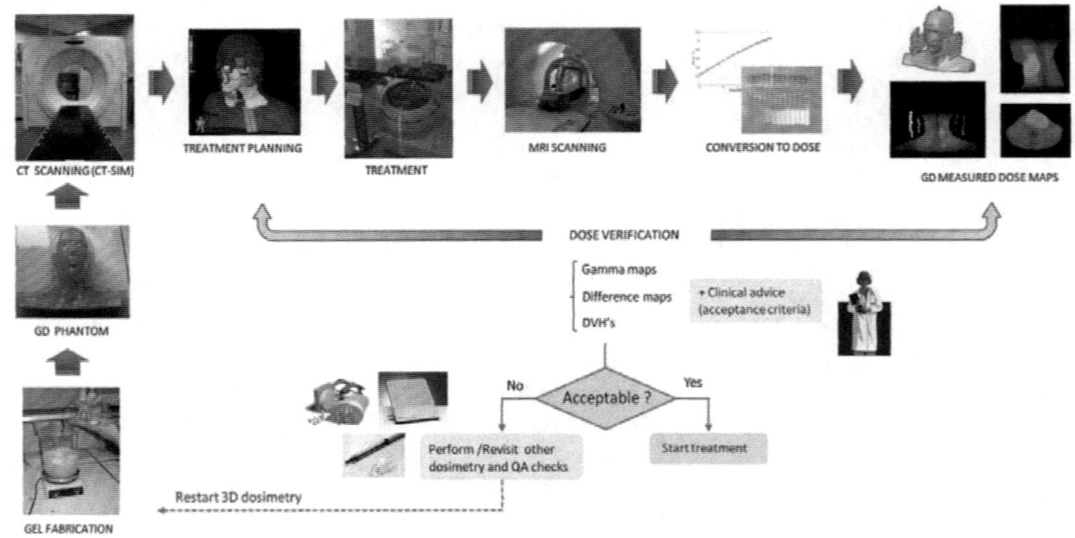

Figure 2. In treatment verification with gel dosimetry, the gel dosimeter phantom is taken through the entire treatment chain. Possible dosimetric or setup errors can occur at the different stages.

In most gel dosimetry applications, the voxel-size is several orders of magnitude bigger than the molecular size. As a result, this intrinsic radiochemical noise contribution can be easily neglected. Any positional set-up errors in treatment with respect to the planned treatment will affect the dose distribution. This error may also occur in patient treatment and is therefore not intrinsic to gel dosimetry (GD). In benchmarking GD however, the experimental set-up error should be minimized as much as possible.

Other sources of deviation in dose reading that may occur during irradiation of the gel dosimeter are related to the dependence of the irradiation response on the temperature of the 3D dosimeter during irradiation, on the dose rate and on the energy spectrum of the irradiation beam [23,24]. It is important to realize that differences in dose rate (and energy) will occur in the irradiated 3D volume even if the photon fluence rate (in terms of MU/min) of the individual beams is kept constant. Another important characteristic of the gel dosimeter is its temporal and spatial integrity [25]. These errors are inherent to the gel dosimeter and should be minimized in a good dosimeter as they compromise the specificity of the dosimeter. After irradiation, the gel dosimeter is scanned (5). During scanning, thermal noise will add to the measurements. In order to minimize the noise in the measured dose distribution the scanning parameters should be optimized [26,27]. The processing (fitting) of acquired data, may have a big influence on the amplification of the noise figure. For example, in quantitative R_2-NMR scanning it is found that a least-square fit will amplify the thermal noise in the base images to a larger extent than a chi-square (χ^2) based minimization. Imaging artifacts may result in systematic errors both dosimetric as spatially.

Table 1. Factors that determine the uncertainty in 3D gel dosimetry, classified in type A and type B. Factors indicated in red italic font also affect patient treatment.

	Type A (random error)		**Type B (systematic error)**	
	Dosimetric	Spatial	Dosimetric	Spatial
Chemical	• Variation in chemical concentration • Manufacturing temperature • Dose sensitivity		• Discrepancies between calibration vials and phantom • Chemical stability • Spatial integrity	• Volumetric contraction of the gel dosimeter
Radiation	• *Stochastic variations in the delivered dose* • Variations in the temperature during radiation • Reproducibility of calib. phantom-positioning • *Radiochemical noise*	• *Variations in the phantom position* • Spatial temperature variations in combination with temp. sensitive dose response	• Positioning error of the calibration phantom • Dose-rate dependent response • Energy-dependent response • Temperature dependence • Tissue equivalence • Recipient wall effects	• Phantom positioning error
Scanning	• Thermal noise	• Voxel size / shape (resolution)	• Voxel shape (bandwidth) • Imaging artefacts • Temperature during scanning	• Imaging artefacts

2. Type A uncertainty

The overall dosimetric precision is compromised by stochastic variations (random errors) that result from the several operations that take place in the dosimetry experiment. The first step in a gel dosimetry experiment is weighting the chemicals. Stochastic variations in the weighting will result in variations in the measured dose-related value (R_2, MT, OD) as the dose-response is determined by the chemical composition. It is found that other manufacturing conditions may also have an influence on the dose-response such as the temperature during fabrication. Stochastic variations in the controlled temperature between different experiments will therefore also lead to variations in the measured dose-related value. Also during irradiation there are different sources of stochastic variable contributions that determine the overall dosimetric precision such as variations in the dose delivery, variations in the temperature during irradiation and stochastic variations in the positioning of the calibration phantoms. Any form of scanning the gel dosimeter will introduce thermal detector noise. The noise contribution is determined by some scan parameters. Often, the scan parameters can be optimized in order to achieve an optimal figure of precision. In gel dosimetry, a calibration of the scanned parametric map (e.g. R_2 map) is needed to obtain a dose map. Any random errors and influence quantities that operate on the calibration samples will also affect the uncertainty of the dose map. Because the calibration vials and the volumetric gel dosimeter phantom are filled with gel having the same composition (same batch of gel), any random errors in weighting the chemicals will not result in a random variation in the final dose measurement. In addition, the stochastic variations in beam output during radiation of the calibration vials can also be assumed to be significantly less than 1% which is also supported by a reproducibility study where 20 (calibration) samples were irradiated with the same dose and under the same conditions. The resulting type A (random) uncertainty is therefore considered as only originating from thermal noise during scanning.

The concept of dose-resolution was introduced to evaluate *the intrinsic dosimetric precision* in terms of dose sensitivity and scanning signal-to-noise [44]. The dose resolution, written as D_Δ^p, is defined as the minimal detectable dose difference within a given level of confidence, p. The dose resolution is related to the standard deviation on dose σ_D by the equation

$$D_\Delta^p = \sqrt{2} \cdot k_p \cdot \sigma_D \qquad (1)$$

For a 95% confidence level, the dose resolution becomes $D_\Delta^p = 2.77\,\sigma_D$.

In most radiation dosimetry experiments, gel dosimeters are used in a relative manner in the sense that the dosimeter is exposed to the same treatment as the patient but with a different total radiation dose. The total dose delivered to the dosimeter is scaled to cover the active dose range of the dosimeter. In this context, it is preferable to use the concept of dose resolution relative to the operating dose range, here defined as *relative dose resolution* $D_{\Delta\%}^p$:

$$D_{\Delta\%}^p = \frac{D_\Delta^p}{(D_{max} - D_{min})} = \sqrt{2} \cdot k_p \frac{\sigma_D}{(D_{max} - D_{min})} \qquad (2)$$

If the dose maps are derived from quantitative NMR-R_2 maps, it can be shown that the relative dose resolution ($D_{\Delta\%}^p$) is equal to the relative R_2 resolution (${R_2}_{\Delta\%}^p$) which is defined in a similar way:

$$D_{\Delta\%}^p = \sqrt{2} \cdot k_p \frac{\sigma_D}{(D_{max} - D_{min})} = \sqrt{2} \cdot k_p \frac{\sigma_{R2}}{(R_{2max} - R_{2min})} = {R_2}_{\Delta\%}^p \qquad (3)$$

As mentioned earlier, the here defined dose-resolution does not include stochastic variations in chemical concentrations, in dose delivery or in the calibration procedure. For that reason, dose-resolution can be considered as an intrinsic lower limit of type A uncertainty.

It should be emphasized that the dose-resolution is not only related to the type of gel dosimeter but also to the scanning protocol [26,27]. In some publications, the concept of dose resolution has been used as the criteria to compare different types of gel dosimeters. This may be misleading as these studies report on dose resolutions obtained with sub-optimal scanning parameters. The concept of dose resolution however is very practical to optimize the intrinsic NMR sequence in terms of intrinsic precision [27]. In optimizing the NMR sequence, it is also important to take into account the number of slices that are required for the 3D dosimetry application as the optimization is also dependent on the number of slices.

Depending on the available scanning facilities three different optimization strategies can be followed:

1. _Only a single spin-echo sequence is available:_ It can be shown that if one has only a single spin-echo sequence a two-points method is preferred. In the two-points method, two differently T_2-weighted base images are acquired and the R_2 is derived algebraically from the signal intensity in the two images according to the equation:

$$R_2 = \frac{\ln S_1 - \ln S_2}{TE_2 - TE_1} \tag{4}$$

with S_i and TE_i the signal intensity and the echo time in image i respectively.
The relative dose resolution is then given by:

$$D^p_{\Delta\%} \cong \frac{\sqrt{2}k_p}{SNR_1 \cdot \Delta TE \cdot (R_{2,max} - R_{2,min})}\sqrt{\frac{1}{N_1 + N_2}\cdot\left(\frac{e^{2R_2\Delta TE}}{1-\psi_N} + \frac{1}{\psi_N}\right)} \tag{5}$$

SNR_1 is the signal-to-noise ratio in the first base image which is the image recorded with echo time TE_1. The echo time spacing $\Delta TE = TE_2 - TE_1$. Equation 5 also applies to the case where more acquisitions are taken in each image. The acquisition fraction is $\psi_N = \frac{N_1}{N_1+N_2}$ with N_1 the number of acquisitions (averages) of the first image (with TE_1) and N_2 the number of acquisitions (averages) of the second image (with TE_2). An optimal echo time spacing and acquisition fraction can be derived for a gel with R_2 values in a range $[R_{2,min}, R_{2,max}]$. Tables 1 and 2 in publication [27] give the optimal echo time spacing and the optimal acquisition fraction for the two-points method.

2. _A multiple spin-echo sequence is available with a fixed number of echoes:_ To acquire quantitative R_2 images, a multiple spin-echo sequence is highly preferred above a single spin-echo sequence because several differently T_2-weighted images are acquired within the same measurement time increasing the overall signal-to-noise ratio in the R_2-images. The quantitative R_2-image is then obtained by fitting a mono-exponential decay-function to the pixel intensities of corresponding pixels. On many scanners, a multiple spin-echo sequence is available on the scanner. Very often, the number of available echoes is 'hardcoded' in the sequence. In that case, it is advisable to optimize the echo time interval. The echo time interval is defined as the time interval between the first and the last echo in the multiple spin-echo acquisition window. The optimal echo time interval can be derived from a graph provided in figure 9 of reference [26]. It is also found that the optimization also depends on the fitting algorithm that is used to derive the R_2 value. For more than 7 echoes and using a χ^2-minimization fitting algorithm, the optimal echo time interval is approximately 2 times the T_2 value of the sample. To a good approximation this also applies to the median R_2 for a phantom containing a range of R_2 values.

3. *A multiple spin-echo sequence is available with an arbitrary number of echoes:* This is the most optimal situation from the perspective of optimization of the dosimetric precision when more slices are acquired. The reasoning behind this is based on the fact that within an optimal echo time recording period it is preferable to acquire as many echoes as possible. To optimise this sequence, the echo time spacing is taken as short as possible without introducing any artefacts. Then the number of echoes is optimized. Most often, there is an upper threshold on the available echoes in the multiple spin-echo sequence. When the recommended (optimal) number of echoes exceeds the number of available echoes, the echo time spacing is increased to cover the optimal echo time interval. In publication [27], table 5 gives the optimal number of echoes for a gel with R_2 values in a range [R_{2min}, R_{2max}]. The number of echoes is limited by the minimal inter-echo-time-spacing which is determined by machine related characteristics such as maximum gradient strength, sampling rate and SAR considerations. In practice, it is also advisable to check for imaging artifacts (uniformity, dose errors and geometrical distortions) while decreasing the inter-echo-time-spacing.

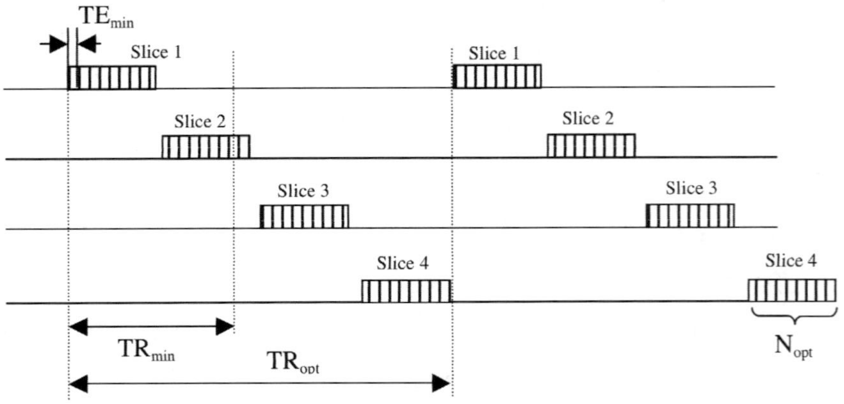

Figure 3. Multi-slice multi-echo sequence showing the optimal sequence parameters. The number of spin-echoes acquired for each slice is optimized.

Then, if more than one slice is acquired, the time interval between the end of the echo-time recording period and the repetition time can be used to scan another slice (figure 3). As from a certain number of slices, not all slices can be recorded within one repetition time. In that case, the repetition time can be extended to cover exactly the time needed to acquire all slices. The algorithm to determine the optimal echo time is described in [27].

3. Reproducibility study

3.1. Materials and Methods

In previous studies, the fundamental radiation properties of several gel dosimeters were studied and it was found that PAGAT gel dosimeters demonstrated the least dose rate and temperature dependence during irradiation [23,24] and dependence on cooling rate [6]. Therefore, we chose to use PAGAT gel dosimeters in this reproducibility study. The polymer gel dosimeter was fabricated according to a procedure as described elsewhere [23]. Both an intra-batch and inter-batch reproducibility study were performed (figure 4). In the intra-batch reproducibility study, a large quantity of polymer gel was fabricated and poured in eight spherical borosilicate flasks (250 ml each) and nineteen test tubes (calibration vials). In the inter-batch reproducibility study, a complete gel dosimetry experiment was performed eight times in which each time 0.6 l of gel was fabricated and poured in a spherical 250 ml borosilicate flask and corresponding nineteen test tubes. In both experimental designs, each of the eight spherical gel dosimeter phantoms was scanned together with its corresponding set of calibration vials.

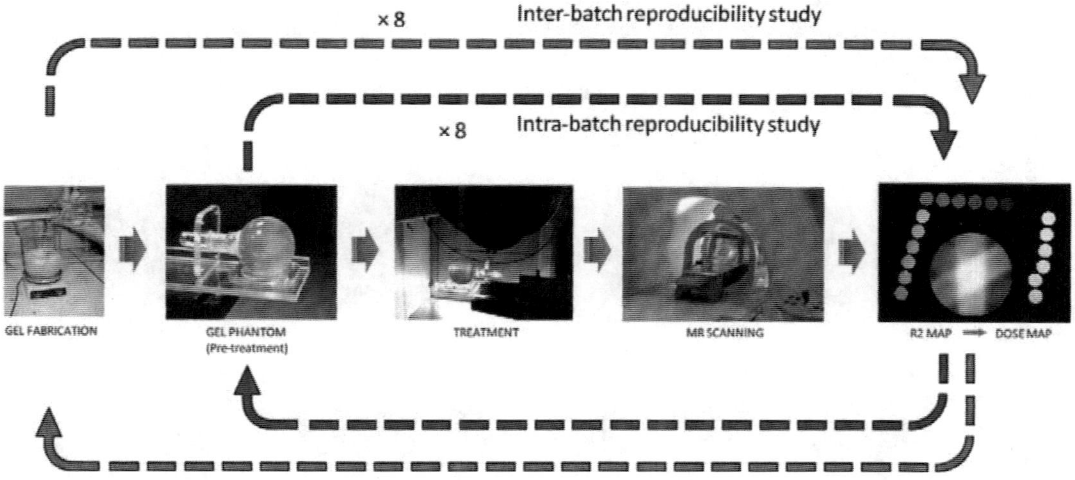

Figure 4. Inter- and Intra-batch reproducibility study of a gel dosimetry experiment.

After filling the recipients, the phantoms were stored in a large water reservoir (60 l) in order to ensure a similar temperature course in both the spherical phantoms and the calibration vials. A dedicated holder in PMMA was constructed to fixate the spherical phantoms during CT scanning, irradiation and MRI. CT images were taken from a spherical phantom which were transferred to the planning computer and a three beam coplanar irradiation was simulated using the treatment planning system Pinacle™ .

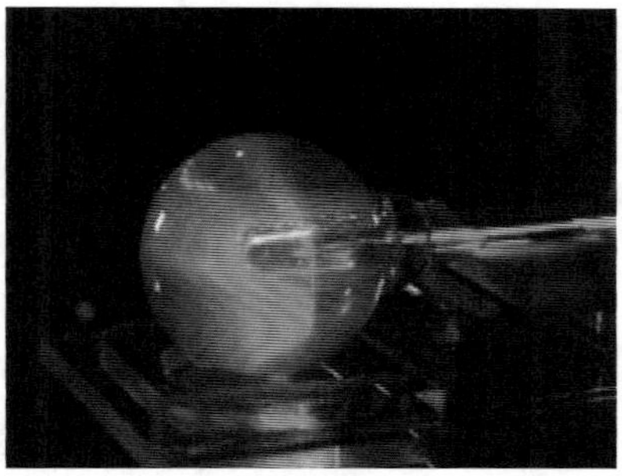

Figure 5. Reference measurement with an ionization chamber in the centre of the spherical phantom.

The irradiation treatment consisted of three coplanar beams with gantry angle at 180° (525 MU), 60° (150 MU) and -60° (375 MU) respectively. The calibration phantoms (test tubes) were irradiated at reference depth with a 10 cm × 10 cm photon beam at a variety of radiation doses to cover the calibration range. A reference measurement was performed in the centre of a similar spherical phantom filled with gelatine using a small volume ionization chamber (PTW semiflex 31010) (figure 5). The measured dose in the centre of the spherical phantom amounted to 11.28 Gy, corresponding with a dose difference of 0.4 % with the TPS.

Quantitative NMR spin-spin relaxation rate (R_2) maps were recorded from the spherical phantom surrounded by the corresponding calibration tubes with a 1.5 T MRI scanner (Avanto, Siemens) using a multiple spin-echo sequence with 32 echoes and echo spacing (ΔTE) 40 ms. The other scanning parameters were TR = 10 s, voxel size = 1 mm × 1mm × 5 mm, receiver bandwidth (BW) = 130 Hz/pixel and 10 slices, resulting in a total scan time of approximately 50 minutes. R_2 maps were constructed from 31 base images (ignoring the first base image) using a χ^2-minimization algorithm [26,27]. All gel dosimeter phantoms were scanned at different instances: 6h, 30h, 54h and 92h post-irradiation in the intra-batch reproducibility study and at 19h, 43h and 85h post-irradiation in the inter-batch reproducibility study. A detailed description of the full reproducibility study can be found in a publication which is currently in press [28].

3.2. Results and discussion

The dose distributions and deviations with respect to the TPS planned dose distribution (and ionization chamber) after calibration with their corresponding set of calibration vials were found to be very similar. Gel measured dose maps from the eight phantoms of the inter-batch reproducibility study are shown in figure 6. Although the dose maps look very similar, deviations become more apparent from dose profiles in comparison with a dose profile through the TPS (figure 7).

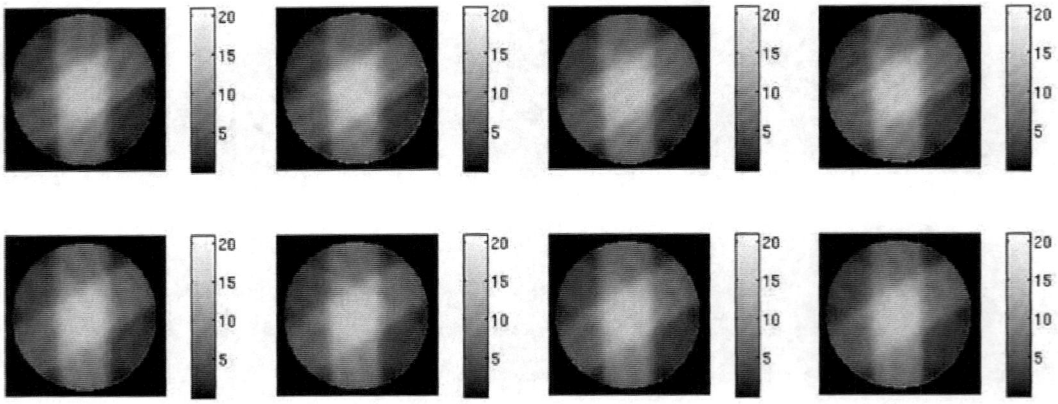

Figure 6. Gel measured dose maps through the isocentre of the eight different gel dosimeters of the inter-batch reproducibility study. The colour scale is in Gy.

Variations in R_2 in the order of 8% occur between the different gel dosimeters in the inter-batch reproducibility study (figure 7c). However, using the appropriate corresponding calibration curve for each gel dosimeter phantom, the actual average dose deviation between the different gel dosimeters is reduced to less than 3%. However, in a few cases at other scan times, larger deviations (in the order of 10%) were found for the same phantoms and a deviation with the ion chamber measured dose (and TPS calculated dose) of more than 10% is also apparent from figure 7d. The fact that a good precision was found in some scan sessions but not in all is indicative of an error source that changes after irradiation. It was later found that the major source of the irreproducible systematic dose error was attributed to changes in temperature during scanning between calibration vials and volumetric gel dosimeter phantoms (see section 4) that can be solved by the aid of active temperature stabilization in the MRI scanner (see section 5). It is obvious that by use of a renormalization using two internal dose points (in a low dose and high dose region) the measured dose distributions are matched (figure 7e). This renormalization procedure or relative scaling has been applied by most research groups. With this renormalized dose distribution, more than 97% of all points of the measured dose distribution fulfilled the gamma-criterion of 3% / 3mm.

Although in the intra-batch reproducibility study (results not shown), a better match was found in the initial R_2 distribution, the systematic and random dose errors were found to show a similar behaviour as these found in the inter-batch reproducibility study.

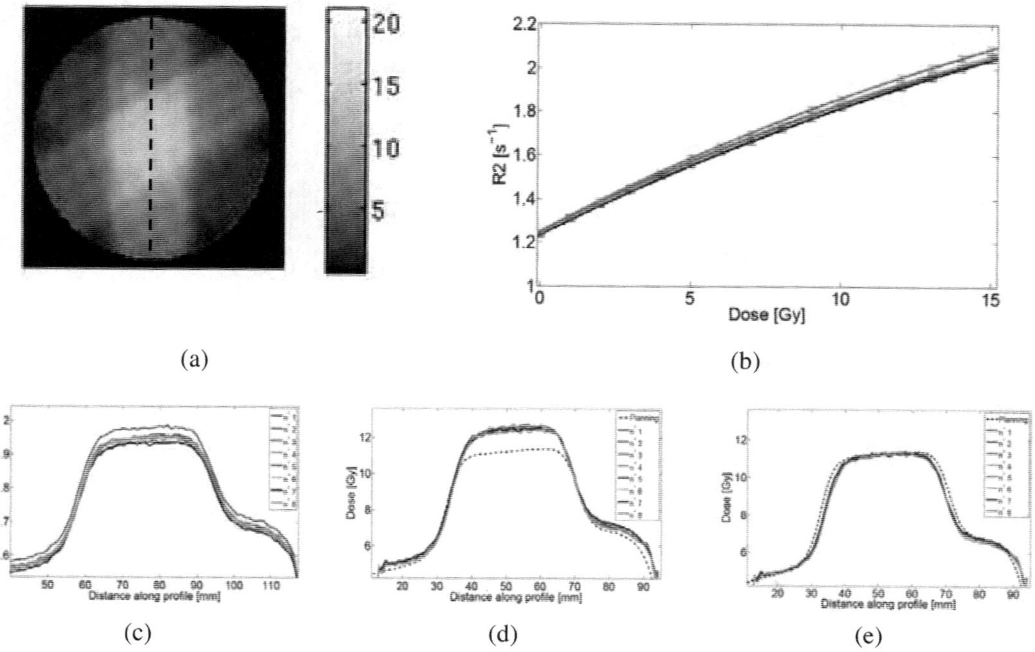

Figure 7. Dose map of one of the eight phantoms of the inter-batch reproducibility study (a) indicating the location of intensity profiles as shown in (c) - (e) for all eight phantoms. The corresponding dose-R_2 response curves for all eight phantoms are shown in (b). R2 profiles demonstrate some variation which after calibration with their corresponding dose-R2 calibration curves (b) disappear. However, a significant systematic deviation with the TPS planning profile (dotted line) remains. By applying a renormalization of the dose-R_2 calibration ('relative' internal calibration) the dose profiles match.

From this study, it was concluded that polymer gel dosimetry for these small sized volumetric gel dosimeter phantoms can be performed with a high dosimetric precision and accuracy by applying a renormalization of independent dose measurements using two internal reference points. In the verification of a treatment plan with gel dosimetry, it is necessary to verify the absolute dose in the reference points by use of an independent dosimetric point-measurement (ion chamber). For realistically sized humanoid shaped phantoms with volumes of more than 2 litre, a renormalization may not be sufficient because of internal temperature differences within the volumetric gel dosimeter phantom.

To assess the effect of various error sources on the dose reading, it is illustrative to consider the dose sensitivity. The dose sensitivity of a PAGAT gel dosimeter is in the order of 0.06 $s^{-1}\cdot Gy^{-1}$ (see figure 7b). With a realistic dose range of 15 Gy, a dose error of 1% with respect to the measured dose range amounts to 0.15 Gy or only $9\cdot10^{-3}$ s^{-1}. This places great demands on the accuracy and precision by which R_2 is measured.

4. Physico-chemical origins of dose error in polymer gel dosimeters

4.1. Gel fabrication and storage temperature

The influence of the temperature history during the storage period between fabrication and irradiation storage of the gel dosimeter on the dose response curve has been assessed for both PAGAT and MAGAT gel dosimeters [6]. After fabrication, different calibration phantoms were subjected to different temperature trajectories that are realistically encountered during cooling a gel in a refrigerator. The temperature trajectories were measured by use of fibre optic probes. It was found that the dose-R_2 response of MAGAT gel dosimeters was significantly more dependent on the storage temperature than that of PAGAT gel dosimeters.

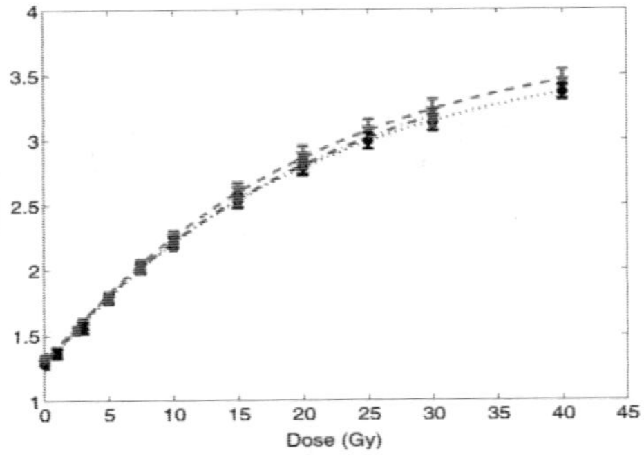

Figure 8. Dose-R_2 response plots of PAGAT gel dosimeters cooled down with different temperature courses [6].

To investigate the effect of a temperature gradient in a volumetric phantom after fabrication and before irradiation, a box-shaped Barex™ recipient was filled with PAGAT gel and a temperature gradient was induced in the phantom for 14 hours by exposing opposite sides of the phantom to different temperatures (5°C and 25°C respectively). The temperature gradient was measured by use of fibre optic temperature probes that were inserted in a similar blank phantom and in a similar experiment. More findings are detailed in a recent publication [29].

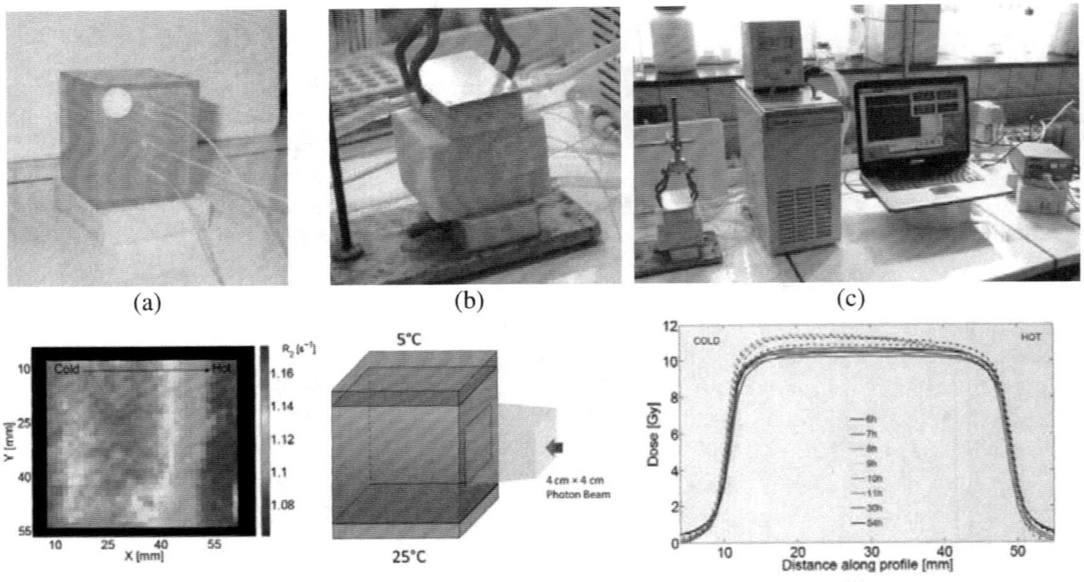

Figure 9. Experiment that illustrates the effects of temperature differences during storage (i.e. between fabrication and irradiation) on the measured dose distribution. A box-shaped gel phantom (a) is exposed to a temperature gradient during 14 hours before irradiation (b-c). An R_2 map of an unirradiated phantom shows a variation of 0.07 s^{-1} in the direction of the gradient (d). A phantom that is irradiated at a lateral side with a square beam also shows a negative dose gradient in the direction of the temperature gradient.

The R_2 of an unirradiated gel phantom that was exposed to a temperature gradient shows a negative non-linear response with increasing temperature (figure 9d). A dose overestimation is also found at the cold side of the phantom. However, this over-response decreases with time post-irradiation (figure 9f).

4.2. Irradiation temperature dependent dose-R_2 response

It was found that in some polymer gel dosimeters the temperature during irradiation has an influence on the dose-response [23]. This is likely due to a temperature dependent change in the diffusivity of the monomers in the gel matrix and a change in the chemical reaction kinetics. It is mainly the dose-R_2 sensitivity that is influenced by the temperature increase. The intercept is not affected (figure 10). The dependence on temperature during irradiation has also raised concerns on the influence of the reaction heat that is released during the exothermal radiation induced polymerization reaction [30].

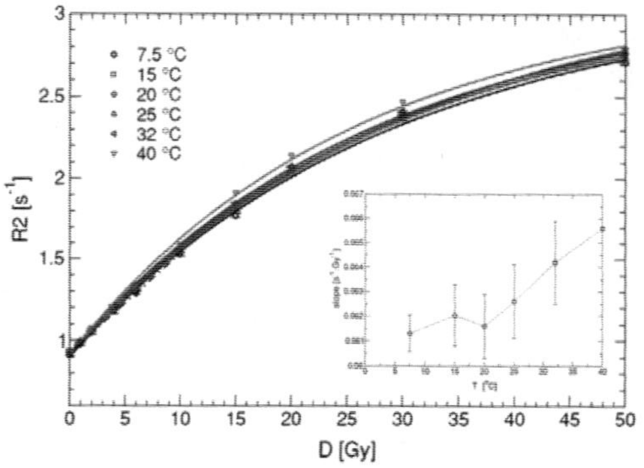

Figure 10. Dependence of the dose-R2 response on the temperature during irradiation. The inset shows the corresponding slope as a function of irradiation temperature [6].

In large volumetric phantoms, the temperature increase upon irradiation is not uniformly distributed within the phantom as a result of the non-uniform dose distribution. Moreover, after treatment the temperature course in a large volumetric phantom is also spatially non-uniform and depends on both the initial dose distribution and on the thermal characteristics of the phantom and boundaries. As a result, dose read errors may be expected. To investigate the effect of momentary temperature increases on the measured dose-R_2 response, calibration vials were irradiated under different imposed temperature trajectories [29].

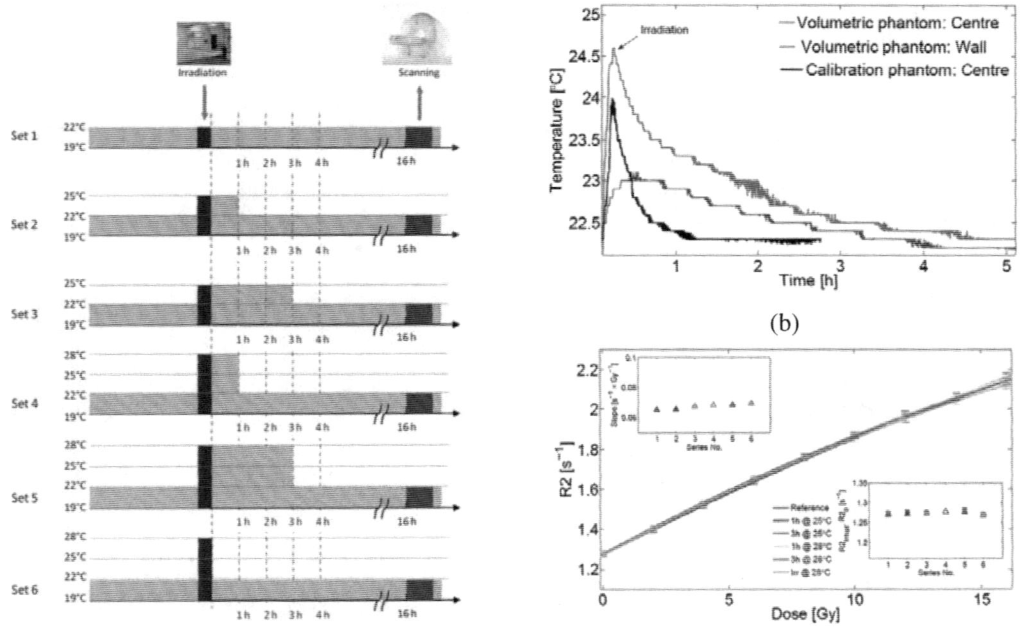

(a) (c)

Figure 11. Imposed temperature course on 6 different sets of calibration tubes (a) that mimic extreme conditions of exothermal heating at different locations in a volumetric phantom (b). The dose-R2 response curves of the six different sets are plotted in (c).

No significant effect of temperature changes during and immediately following irradiation could be detected (figure 11).

4.3. Chemical instabilities

Two kinds of chemical instabilities have been observed in polymer gel dosimeters [6, 31]. One affects the slope of the dose–R_2 response plot and is related to post-irradiation polymerization of the comonomer/polymer aggregates. It is observed that post-irradiation polymerization in a PAGAT gel dosimeter only lasts 12 hours after irradiation (figure 12b). The other instability affects the intercept of the dose–R_2 plot (figure 12c), lasts for up to 30 days and is related to the gelation process of gelatine. These mechanisms have later been confirmed through FT-Raman spectroscopy [32]. The chemical instability depends strongly on the polymer gel composition [6]. In order to minimize systematic errors it is advisable to scan the gel dosimeter phantom together with the corresponding calibration vials.

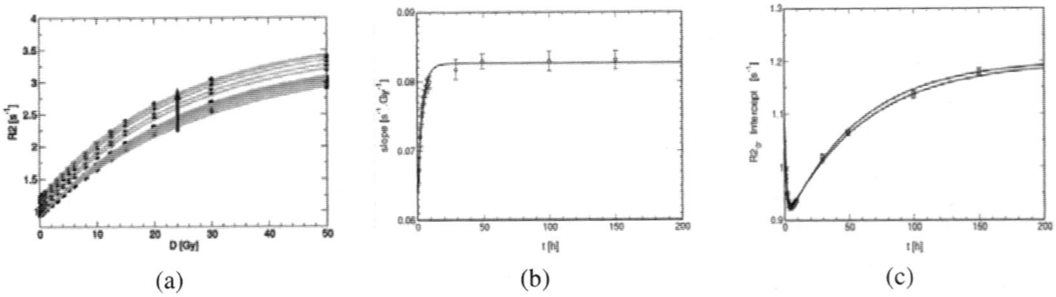

Figure 12. Dose–R_2 plots of a PAGAT gel dosimeter recorded at different post-irradiation times (a) and corresponding changes in slope (b) and intercept (c) [23].

4.4. Loss of spatial integrity

Loss of spatial integrity is a consequence of the diffusion of the active components or oxygen in the gel dosimeter. An extreme case is the Fricke gel in which the diffusion of ferric and ferrous ions results in a blurring of the measured dose distribution over time [33-35]. But also in polymer gel dosimeters can the diffusion of monomers at very high dose gradients during irradiation [36] and in the first hours after irradiation [25] have an influence on the measured dose distribution. Unreacted monomer diffusing from un unirradiated region may react with large polymer radicals in the irradiated region in the first hours after irradiation. A quantitative model is presented that links the loss of spatial integrity with the monomer diffusion and post-irradiation polymerization [37].

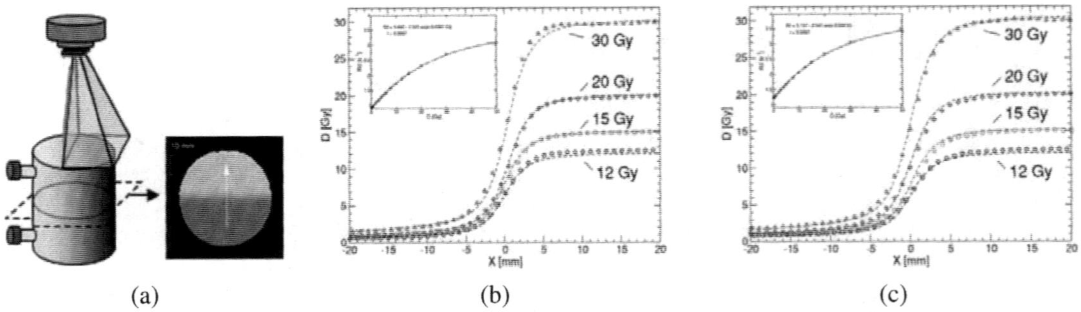

Figure 13. Dose penumbra at the edge of a half-blocked 10 cm × 10 cm square photon beam (a) measured with a PAGAT gel dosimeter acquired 10 hours post-irradiation (b) and 6 days post-irradiation (c).

No measured dose overestimation are found in PAGAT gel dosimeters for doses below 20 Gy (figure 13) while for MAGAT gel dosimeters dose overestimations have been seen near the penumbra region for doses of 10 Gy [23].

4.5. Dose rate dependent dose-R_2 response

As radiation induced polymerization reactions are regulated by complex reaction schemes involving initiation, propagation, cyclization, transfer and termination [38], it is not surprising that the dose-R_2 response of polymer gel dosimeters is dependent on dose rate [23]. Moreover, some reaction pathways are also diffusion controlled. The same mechanism is at the origin of the LET dependence of polymer gel dosimeters [39]. It has been found that this effect is more pronounced in MAGAT gel dosimeter as compared to PAGAT gel dosimeters [26] (figure 14). The influence of the dose rate dependent dose-R_2 response on the accuracy of dose verification with PGD should not be underestimated as it may lead to a depth dependent dose-response. A dose-rate dependency may be detrimental for a dosimeter. Nevertheless, the dose-rate dependence has not been documented for all gel dosimeter systems.

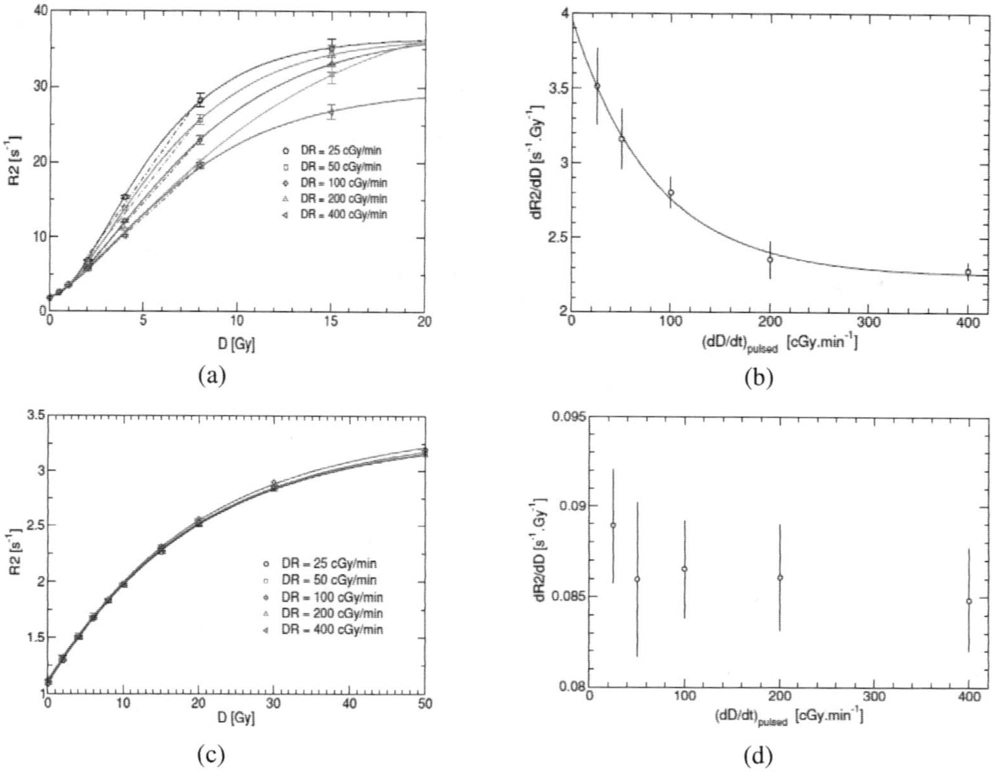

Figure 14. Dose-R_2 response curves for different dose rates for MAGAT (a) and PAGAT (c) gel dosimeters. The corresponding change in dose sensitivity of both gel dosimeters is recorded in (b) and (d) respectively.

4.6. Oxygen effects

With the introduction of 'normoxic' polymer gel dosimeters [40], using an anti-oxidant to scavenge oxygen from the gel, it was expected that problems related to oxygen infiltration in the gel would be solved [41]. Although the procedure of fabrication has been significantly simplified, the issue of oxygen infiltration after closing the dosimeter is still present [8,42]. This also restricts the use of different cast materials and restricts the storage time before irradiation [42] (figure 15). In PAGAT gel dosimeters a reduced dose-R_2 sensitivity is also found for increasing amounts of anti-oxidant [42].

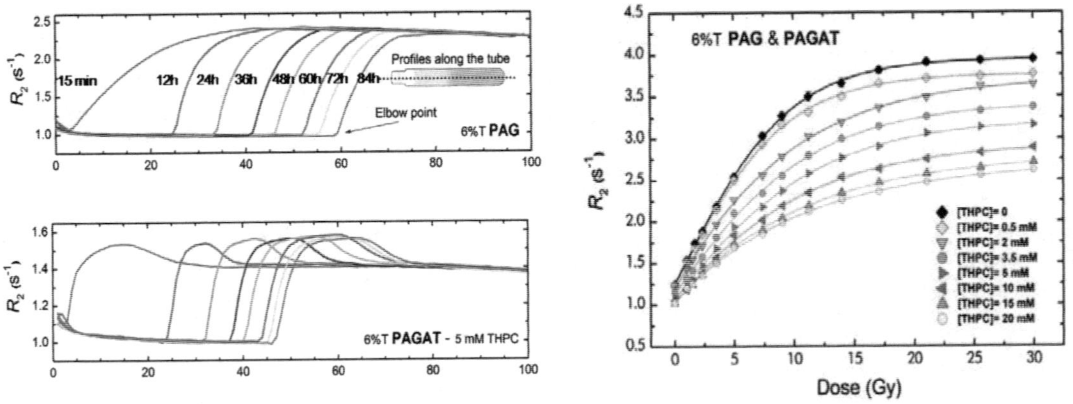

(a) (b)

Figure 15. Infiltration of oxygen from the top of the test tube (left to right) after irradiation in PAG and PAGAT gel dosimeters. Test tube samples have been irradiated to a dose of 5 Gy after the times indicated (a). The dose sensitivity of PAGAT gel dosimeters also depends on the amount of THPC. Adopted from Sedaghat et al [42].

Also, the effect of oxygen infiltration after irradiation has been investigated. No significant effect was detected for an oxygen redistribution 48 hours post-irradiation [29].

It is concluded that even with normoxic polymer gel dosimeters, cast materials should still be conscientiously chosen with Barex™ and glass as the optimal choice until now. The antioxidant should be well mixed in the gel and infiltration of oxygen from the inlet of the recipient should be avoided. No significant post-irradiation effect of oxygen could be established.

3.7. Recipient wall-effects

To avoid permeation of oxygen through the recipients wall, Barex® or glass are often used as phantom materials. It should be noted that some glass may contain heavy metals. These specific glass materials may result in a stronger attenuation of the incident beam and may also result in beam hardening. Some caution is therefore advised in selecting glass as cast materials.

It was shown by Monte Carlo simulations that the effect of a borosilicate glass wall and backscatter of a layer of air did not have a significant effect on the delivered dose in a test vial [43].

As differences in dose-R2 response were observed independently by many investigators, an experiment was designed in which a test tube was inserted in a volumetric phantom (figure 16).

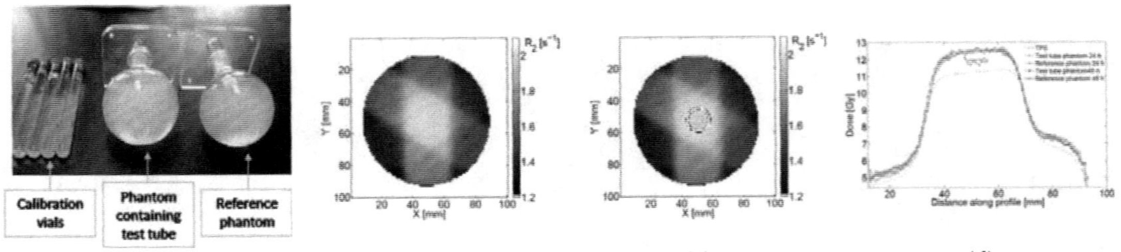

(a) (b) (c) (d)

Figure 16. Experiment in which a test tube with gel was inserted in a larger volumetric phantom (a). A small under response (c,d) was observed in the test tube as with respect to a reference phantom (b) which was attributed to a temperature difference during scanning of 0.3 °C.

A small difference in dose response was observed in the test tube as compared to a reference phantom. The observed effect has been found to be not reproducible and has been attributed to small temperature differences during scanning between the gel in the test tube and outside the tube in the volumetric phantom. All other MRI related effects including magnetic susceptibility differences, B_1 field inhomogeneities were excluded as possible explanations for the observed effect by additional measurements of the B_0 and B_1 magnetic field. Also the effect of the glass tube on the irradiation beam was excluded by independent pinpoint ion chamber measurements in the tube-containing and reference phantom. Repeated readouts with MRI demonstrated that the effect was also irreproducible in both an irradiated and a blank phantom.

5. Readout related sources of dose uncertainty in polymer gel dosimeters

Whatever imaging technique is applied to map a dose-effective contrast (R_1, R_2, MT, optical density, X-ray absorbance, ultrasound attenuation, dielectric constant, etc.) there are two essential criteria that should be accomplished in the imaging technique: First of all, the images should be accurate both in space as in dose (i.e. not contain systematic errors) and secondly, the stochastic deviations on each voxel should be as low as possible within a reasonable measurement time and for a certain image resolution. An interesting parameter to evaluate the stochastic deviations is the minimum detectable dose difference, defined as the dose resolution [44].

Imaging artefacts may compromise the dose accuracy in each voxel and can deform the image causing a spatial inaccuracy. Dosimetric imaging artefacts can be machine-related or object-related. Machine-related artefacts originate from imperfections in the scanning device while object-related artefacts originate from the dosimeter itself.

Table 2. Overview of important artifacts that may occur in MRI gel dosimetry classified by two criteria. The abbreviation as most often used in the literature is also given.

Geometrical distortions		Dose inaccuracies	
Machine related	Object related	Machine related	Object related
B_0 Field non-uniformity $B_0(x,y,z)$	Susceptibility differences $\Delta\chi(x,y,z)$	Eddy currents $\Delta B_0(t)$, $\Delta g_{x,y,z}(t)$	Temperature drift $T(t)$
Gradient non-linearity $g(x,y,z)$	Chemical shifts $\sigma(x,y,z)$	Stimulated echoes $M_{xy}(TE)$	Molecular self-diffusion $D(x,y,z)$
Eddy currents $\Delta B_0(t)$, $\Delta g_{x,y,z}(t)$		B_1 field non-uniformity $B_1(x,y,z)$	Standing waves and dielectric effects $B_1(x,y,z)$
		Slice profile $M_{xy}(z)$	Susceptibility differences $\Delta\chi(x,y,z)$

The most-important machine-related MRI artefacts are attributed to eddy currents, stimulated echoes, B_1 field inhomogeneity, imperfect slice profiles and standing waves. These machine-related artefacts may depend on the gel dosimeter shape and make it difficult to make general statements on the accuracy of the dosimeter. A larger phantom or a phantom with sharp edges may perform differently than a smaller cylindrical or spherical shaped phantom. Standing-waves can severely deteriorate the dose distribution in dosimeters with specific shapes and spatial dimensions but may be almost completely absent if the dosimeter phantom has a slightly different shape.

Object-related MRI artefacts are mainly attributed to a temperature drift during scanning or molecular self-diffusion. For a more detailed overview of different MRI artefacts and compensation strategies we refer also to previous communications [45,46].

In optical imaging, dosimetric artefacts are related to reflection and absorption by the recipient walls, off-axis positioning of the recipient, variation of the laser output and photo-detector and light-scattering by both impurities in the matching fluid, container and by the polymer [47-50].

4.1. Machine-related geometrical distortions

4.1.1. B_0 magnetic field inhomogeneities. Geometrical distortions may originate from a non-uniform magnetic field of the MR scanner. Important to note is that displacements and scaling of the whole phantom may occur both in the image plane as out of the image plane. The extend of the distortion is also dependent on the receive (pixel) bandwidth. The pixel bandwidth is inversely related to the readout time of each echo (frequency encoding window) in the pulse sequence. A shorter spin echo readout time will result in a larger pixel bandwidth and corresponding smaller distortion for a similar magnetic field non-uniformity. However this will go at the cost of signal-to-noise ratio, hence dose resolution. The magnetic field of the scanner is generally expressed by the manufacturer in parts-per-million (ppm) of the main magnetic field within a certain volume-of-interest. The displacement of a pixel can be easily calculated from equation 1:

$$\Delta x = \frac{\gamma \cdot \Delta B_0}{2\pi \cdot BW_{pix}} \cdot \left(\frac{FOV_x}{N_x}\right) \qquad (6)$$

where Δx is the shift in the frequency encoding direction in metric units, γ is the gyromagnetic constant ($\gamma = 2.6753\ 10^8\ s^{-1}\ T^{-1}$), ΔB_0 is the magnetic field inhomogeneity, BW_{pix} is the bandwidth per pixel ($BW_{pix} = 1/T_{ro}$), FOV_x is the field-of-view in the frequency encoding direction and N_x is the number of pixels in the frequency encoding direction. As an example, a 1 ppm field deviation on a 3T MRI scanner corresponds to a field inhomogeneity $\Delta B_0 = 3\ \mu T$ which results in a shift of approximately one pixel for a bandwidth per pixel of $BW_{pix} = 130\ Hz$ ($T_{ro} \cong 8\ ms$). Note that a higher spatial resolution (smaller pixel size: $\left(\frac{FOV_x}{N_x}\right) \downarrow$) will result in a smaller spatial shift in metric units.

Similarly, a shift in the phase encoding direction in pixel units is given by

$$\Delta y = \frac{\gamma}{2\pi} \cdot \Delta B_0 \cdot t_{ph} \cdot \left(\frac{FOV_y}{N_y}\right) \qquad (7)$$

where t_{ph} is the duration of the phase encoding gradient, FOV_y is the field-of-view in the phase encoding direction and N_y is the number of pixels in the phase encoding direction. The time of the phase encoding gradient is usually significantly smaller than the readout time (in the order of 1 ms) and the distortion can therefore be ignored in standard MRI sequences (in the case of R_2 mapping: a multiple spin-echo sequence). However, in the case of echo-planar imaging, the shift in phase-encoding direction is given by

$$\Delta y = \frac{\gamma}{2\pi} \cdot \Delta B_0 \cdot t_{esp} \cdot FOV_y \qquad (8)$$

where t_{esp} is the time between subsequent echoes (e.g. 1 ms) and FOV_y the field-of-view in the phase-encoding direction. For the same field homogeneity ($\Delta B_0 = 3\ \mu T$) and a field-of-view of 250 mm, this would result in an image distortion of more than 3 cm !

The shift in the slice selective direction is given by:

$$\Delta z = \frac{\Delta B_0}{G_z} \tag{9}$$

where G_z is the gradient strength of the slice selective imaging gradient. For the above example with a magnetic field inhomogeneity of 1 ppm and a slice selective magnetic field gradient strength of $G_z = 3\, mT/m$ the slice shift will also be 1 mm. Note that in the equations above the shifts and indices are taking in an orthogonal system of reference (i.e. x,y,z). Note that MRI allows the acquisition of oblique slices where the same equations apply but with the indices x,y,z replaced by the new coordinates.

To measure the magnitude of deformations from non-linear gradients and main magnetic field inhomogeneities several phantoms have been proposed [51-53]. In observing in-plane distortions a pin-cushion phantom is often used. To account for errors in the construction of the phantom, the phantom is first scanned by use of CT. By overlaying the images indicating the positions of the tubes of the MR images on the CT images a distortion map can be derived.

External (fiducial) markers on the phantom can also be used to allow image fusion with computer planning or other dosimetry techniques. These external markers may give a first indication of deformations or slice displacements. It is important to adequately chose the fiducial markers in order not to cause any misreading as a result of susceptibility differences or chemical shifts of the fiducial markers themselves. A scaling error of the phantom can also be detected by use image fusion/matching software.

4.1.2. Gradient non-linearity. The imaging gradients are expected to be linear. However, this only applies to the central part of the scanner. Distortions can result from any non-linearity in the magnetic field gradients and can be expected at the outer edges of a phantom with dimensions large as compared to the scanner. These distortions can be described by an infinitesimal spatial approach of equations (6)-(9). A distortion correction method for magnetic field and gradient non-linearity has been proposed [54]. Another method to correct for local magnetic field distortions is by a technique called 'view-angle-tilting' [55]. This method has the advantage that no post-processing is needed but has the disadvantage that the point-spread function is broadened which may lead to an erroneous interpretation of the spatial resolution.

4.1.3. Eddy currents. Apart from static deviations in the magnetic field and gradients, the magnetic field may also deviate from the theoretical expected shape due to time-varying magnetic field components. These time-dependent magnetic field deviations are caused by 'eddy currents'. The eddy currents are induced through switching of the imaging gradients giving way to an electromotive force that acts on the cryostat and metal casings. Several techniques to measure and analyze eddy currents in clinical scanners have been described [56-64]. The eddy current induced magnetic field is mostly described by a first order approximation composed of a global offset of the main magnetic field, $\Delta B_0(t)$ and a change in the magnetic field gradients, $\Delta g_i\ (i = x, y, z)$ [57]. The eddy current related spatial encoding errors are mainly due to eddy current fields that are present during frequency

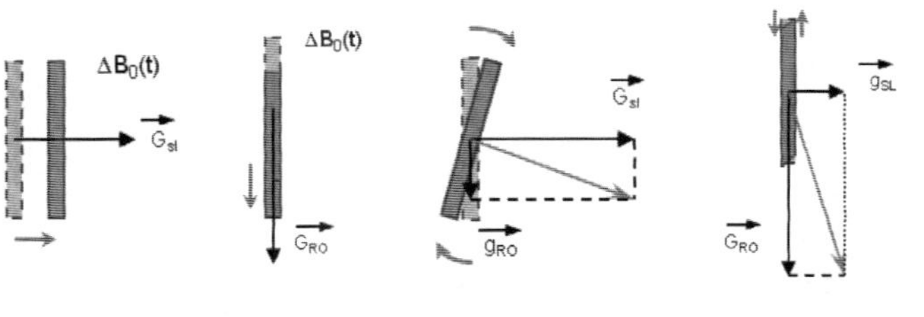

(a) (b) (c) (d)

Figure 17. Different slice displacements caused by eddy currents invoked by switching of slice selection and frequency encoding gradient. An offset in the main magnetic field during slice selection causes a slice shift in the direction of the slice selective gradient (a) while a main magnetic field offset during frequency encoding causes a shift in the direction of the frequency encoding gradient (b). A gradient offset induced by eddy currents of the frequency encoding gradient will cause a slice rotation in the direction of the freq encoding gradient (c) while a gradient offset induced by the slice selective gradient will cause a change in viewing direction (d).

encoding and slice selection. Depending on the moment in the sequence that eddy currents are present, they will cause slice shifts or slice tilting in different directions (figure 17). To measure slice displacements, a pyramidal phantom can be used [56].

4.2. Object related geometrical distortions

4.2.1. Magnetic susceptibility differences. In extending gel dosimetry to phantoms that include air cavities (for example to investigate effects of electronic disequilibrium), materials with different electron densities are inserted in the gel phantom. These materials very often have a different magnetic susceptibility (χ). This will result in susceptibility related distortions in the base images and in the final parametric images. Again, the distortion can be described with the equations (1)-(4). The distortions depend to a large extend on the position of the gel phantoms in the scanner. The magnetic field non-uniformity can be computed by numerically solving the Maxwell equations [65,66] or can be measured with MRI using a dedicated sequence [67]. Several compensation strategies have been developed to correct the image distortions [68-69]. An example of a susceptibility related artifact is shown in figure 18. This example was obtained by coincidence: In order to save some measurement time, two gel phantoms were placed in the scanner at the same time. The magnetic field and the frequency encoding direction are oriented upward in the images. It can be clearly seen that the phantoms are distorted near the interface between the phantoms. When the phantoms are oriented from left to right in the scanner the deformation disappears. Calculated magnetic field maps of both phantoms in an external magnetic field clearly demonstrate the much smaller field deviation when both phantoms are placed side-by-side perpendicular to the erternal magnetic field (figure 18c and d). Also changing the frequency and phase encoding direction will make the distortion disappear. Susceptibility induced deformations have been also observed when a guiding catheter for a brachy source is inserted inside a gel phantom [70]. In humanoid shaped phantoms, sharp boundaries may also result in susceptibility induced magnetic field distortions. These can be also compensated by placing the humanoid shaped phantom in a larger cylindrical recipient filled with paramagnetic contrast agent doped water. The contrast agent serves to lower the T_2 value beyond the first echo time in the multiple spin echo sequence in order to compensate artifacts from turbulent flow in the container.

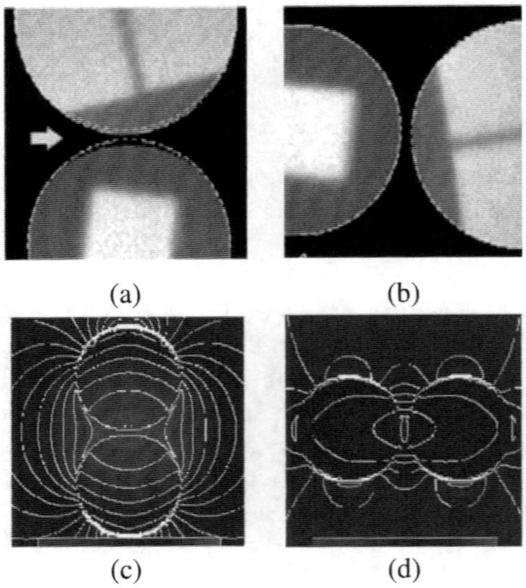

Figure 18. Two cylindrical gel phantoms positioned after each other in the direction of the external magnetic field of the scanner results in susceptibility related deformations (a) while if they are placed next to each other (b) no deformations are observed. The dashed lines are drawn on the image to show the actual boundary of the phantoms. Calculated magnetic field maps (c-d) demonstrate the significantly smaller effect in (b).

4.2.2. Chemical shifts. Also the inclusion of materials with a different chemical composition (organic liquids, fats) in the gel may result in a (chemical) shift of these objects in the frequency encoding direction.

4.3. Machine-related dose inaccuracies

4.3.1. Eddy-currents. Eddy-currents may also invoke dose inaccuracies. The magnetic fields induced by eddy currents experience a certain decay time. The succession of many imaging gradients in a multiple spin-echo sequence may lead to an increase in the magnetic field during the start of the sequence. This may lead to slice profile imperfections. It has been shown that eddy current effects can be minimized by playing out a gradient train before the start of the actual multi spin-echo acquisition [70]. This is attributed to a change in excitation history of stimulated echo-components. The result is a change in the measured T_2-decay curve. As the eddy currents (and especially the magnetic field offset) are dependent on the imaging direction, the disturbance of the excitation history of stimulated echoes is different and therefore also the measured R_2 values (figure 19). Before setting up an actual gel dosimetry experiment, it is important to investigate a possible dependence of the acquired R_2 values on imaging direction and other parameters. It is advisable to use a calibration plot that is derived from scans acquired with the same imaging parameters.

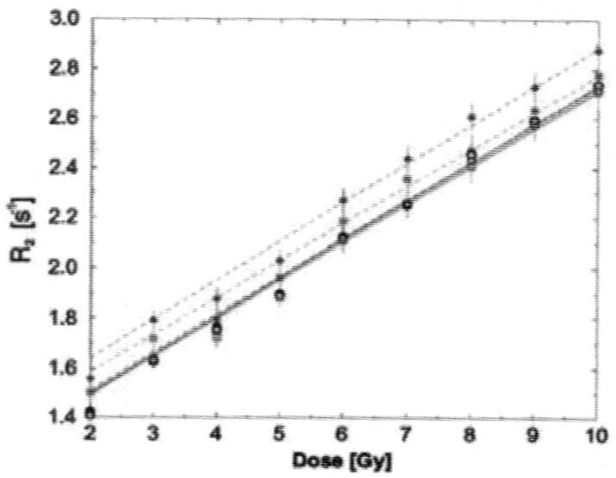

Figure 19. Dose-R_2 plots measured in three different orthogonal directions. The closed symbols (and dashed lines) are for measurements using an uncompensated standard multi spin echo sequence while the open symbols (and solid lines) are for measurements using a compensated imaging sequence [70].

4.3.2. Stimulated echoes. Stimulated echoes may occur in a multiple spin-echo sequence as a result of imperfect (finite) excitation and refocusing pulses. The finite excitation and refocusing pulses result in excitation profiles that deviate from a rectangular profile (figure 20). As a result, the spin magnetization pathways vary along the slice encoding direction [56,71]. It is typically found in a multiple spin echo sequence that the signal intensity in the first two spin-echo base images deviate from the exponential decay curve. It is advisable to remove the first two base images from the fitting. The spin-echoes in a multiple spin-echo sequence are separated at equidistant time intervals for the same reason.

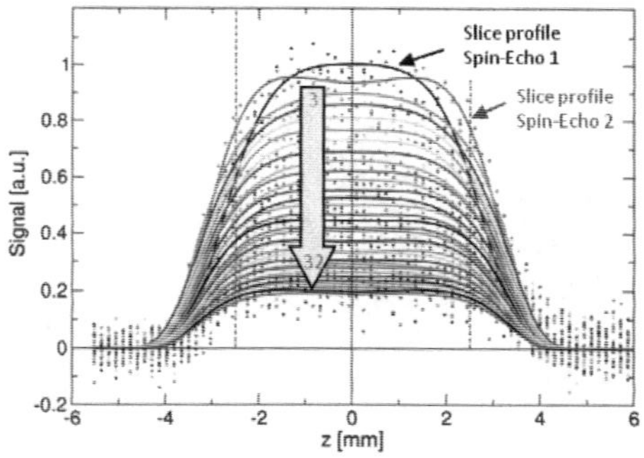

Figure 20. Measured slice profiles demonstrate deviations in the first two spin-echo images.

4.3.3. B_1 field non-uniformity. Different RF coils can be used to scan the gel phantom dosimeter. However, coils should be chosen very carefully as the radiofrequency field of the coils is only uniform within a limited region. With receive/transmit coils, an imperfect excitation may occur outside the homogeneous region. As a result, the spin magnetization history in the multiple spin-echo sequence may deviate and stimulated echoes will be created [72]. When scanned with a CP transmit/receive head coil, the R_2 map was found to be uniform only over an area of 120 cm in the centre of the coil, while the R_2 values (apparent dose values) decreased considerably near the edges of the coil (figure 21a).

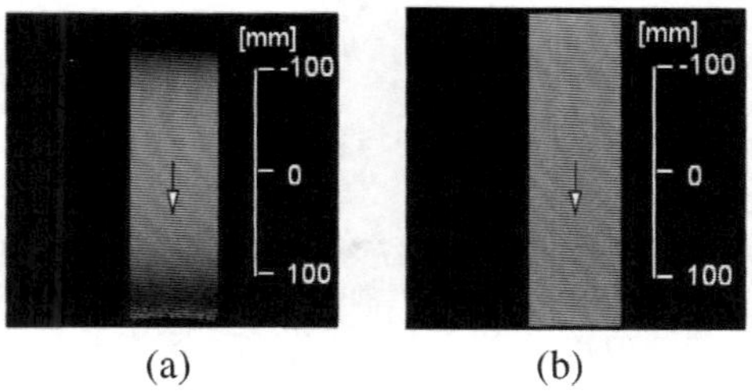

Figure 21. R2 maps of a homogeneous cylindrical gel phantom scanned with the head coil (a) and with the body coil (b).

The R_2 map of the homogeneous phantom is much more uniform when scanned with the body coil (figure 21b). However, measuring R_2 using the body coil as transmitter and receiver is at the cost of signal-to-noise. Modern MRI scanners are equipped with several receive only coils. When still transmitting with the body coil, a good compromise between homogeneity and signal-to-noise can be obtained but this should be used with caution. It is advisable to always scan a homogeneous ('blank') phantom to assess the R_2 homogeneity before scanning an irradiated gel phantom.

4.3.4. Slice profile imperfections and point-spread function. The effect of slice profile imperfections has been discussed in connection to eddy currents and B_1 field non-uniformity (figure 20). However, the slice profile also directly affects the spatial information of the dose distribution. As a result of an imperfect slice profile and imaging point-spread function, an outer-voxel contribution will be present in each voxel of the dose distribution. It is important to realize that the outer-voxel contribution depends on imaging parameters. The relative deviation from a rectangular block-shaped slice profile increases as the slice thickness decreases and the in-plane point-spread function (PSF) depends on both the T_2* and imaging parameters such as receiver bandwidth, image resolution and phase encoding gradient.

4.4. Object-related dose inaccuracies

4.4.1. Temperature drift. The dose-R_2 response of polymer gel dosimeters is scanning temperature dependent (figure 22). Any changes or non-uniformities in the temperature distribution during scanning will severely compromise the accuracy of the gel dosimeter phantom.

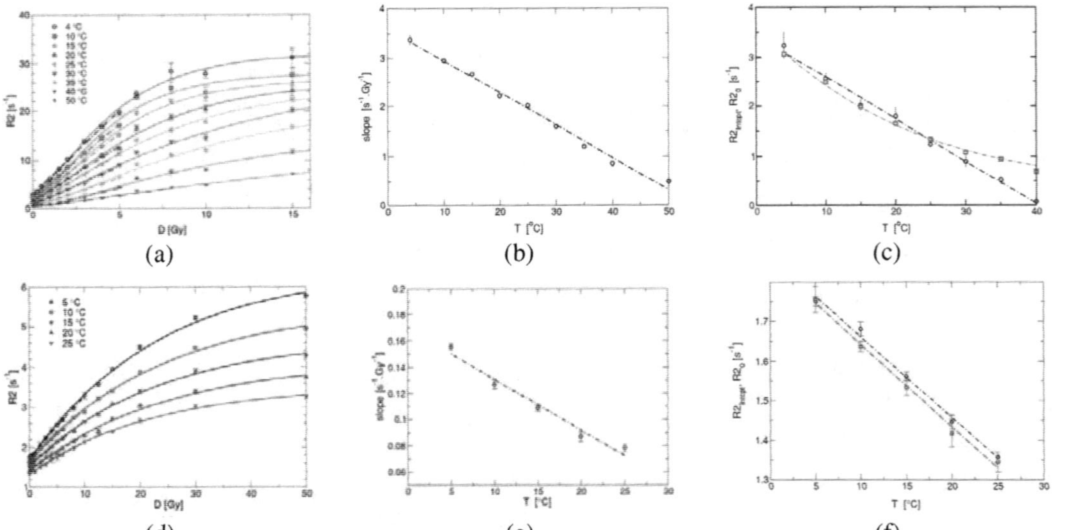

Figure 22. Dose-R_2 plot of MAGAT (a-c) and PAGAT (d-f) at different scanning temperatures. The slope (b,e) and intercept (c,f) are also plotted.

A systematic dose error will be induced when the calibration vials are at a different temperature than the volumetric GD phantom. With a mono-exponential saturation function as description of the Dose-R_2 response:

$$R_2 = R_{2,sat} - \Delta R_2 e^{-\alpha D} \qquad (10)$$

the dose error caused by a temperature deviation can be written as:

$$\frac{\partial D}{\partial T} = -\frac{1}{\alpha\,\Delta R_2}\left(\frac{\partial R_{2,sat}}{\partial T}e^{\alpha D} - \frac{\partial \Delta R_2}{\partial T} + \Delta R_2 D\frac{\partial \alpha}{\partial T}\right) \tag{11}$$

A temperature difference of only 1°C results in a dose error of 0.8 Gy or 8% at a dose of 10 Gy in a PAGAT gel (6 %T) [73]!

The temperature in the scanner room can easily fluctuate in the order of 1°C over a time span of a few hours (figure 23).

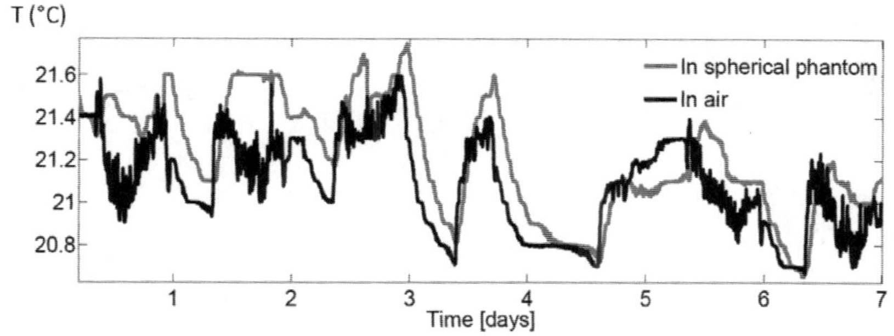

Figure 23. Temperature recorded in a scanner room inside a 250 ml spherical phantom (red line) and in air (black line) [73].

Temperature changes in the gel dosimeter may also occur as a result of the heat produced by the absorption of RF energy from the excitation and refocusing pulses. Especially in a 3D dose experiment where more than 30 slices are recorded, the total acquisition time can easily reach several hours and the specific-absorption-rate (SAR) can be substantial [74]. In figure 24, the temperature in the gel phantom during a long scanning experiment is shown.

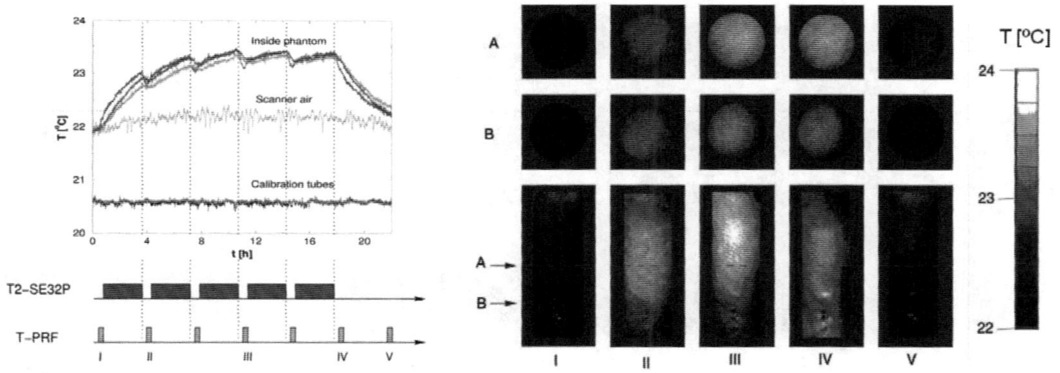

Figure 24. (a) Temperature drift during a long-term scanning experiment of a cylindrical gel phantom (length 30 cm × diam. 10 cm). The temperature is recorded by use of a fiber-optic thermometry system with 6 probes. The measurement history is also provided in a corresponding timetable (bottom). (b) Temperature images recorded with the proton resonance frequency (PRF) method at several time points during a multi-echo sequence as indicated in the timetable. The interval between two successive R_2 recordings is approx. 4 hours. The sequence was stopped after the recording of image IV.

Note that a temperature increase is observed during R_2 scanning but not during temperature mapping (FLASH-sequence) illustrating the high-energy transmittance during the multi-echo sequence. The imaging sequence can be made far less sensitive to temperature drift by using a centric k-space ordering scheme instead of a standard k-space linear ordering [74].

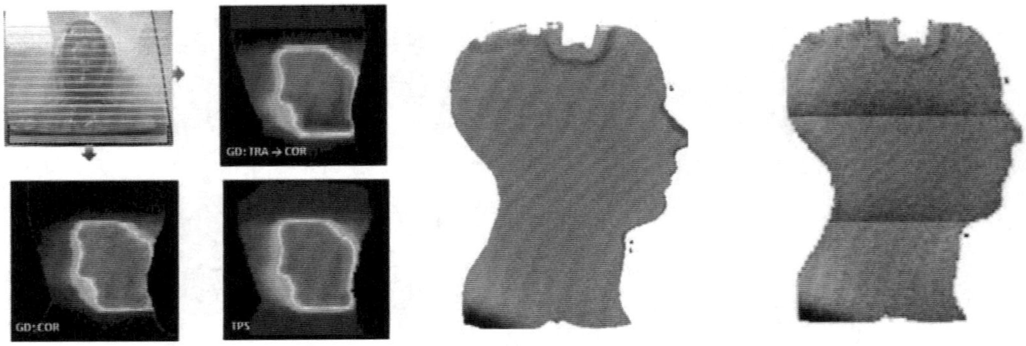

(a) **(b)** **(c)**

Figure 25. Dose deviations in a GD acquired dose distribution were attributed to temperature drift during scanning (a). The effect is clearly demonstrated by comparison of a sagital R_2 map reconstructed from a transverse set of R_2 maps (b) and a corresponding sagital R_2 map (c). Transverse scanning was performed in three interleaved blocks.

4.4.2. Molecular self-diffusion. A same set of calibration vials but scanned with a different resolution may give a different dose-R2 response. This is partially related to the effect of eddy currents that may be different for a different imaging resolution. However, also the molecular self-diffusion weighting by imaging gradients may have a significant effect especially in high-resolution images [75].

4.4.4. Magnetic susceptibility induced effects. As mentioned in section 4.2.1 will differences in magnetic susceptibility between gel, cast material and air create a non-uniform B_0 field. The non-uniform B_0-field can result in an off resonance effects that create signal voids in the base images. However, the spatial variations in signal are expected to scale with the same factor in all T_2 weighted base images when using a multiple spin echo sequence. As a result no significant effect in the resulting R2 maps will occur (figure 26).

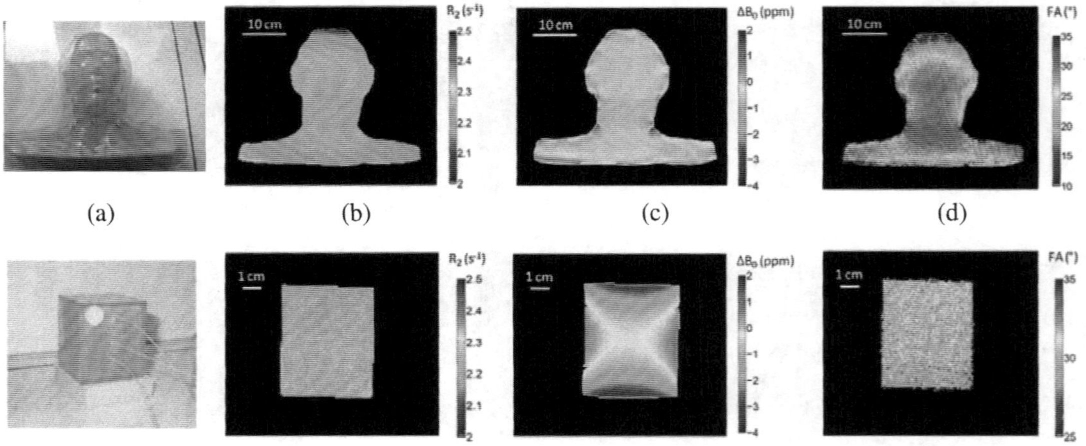

(a) (b) (c) (d)

(e) (f) (g) (h)

Figure 26. Two different blank gel phantoms (a,e) with corresponding R_2 maps (b,f), B_0 field maps (c,g) and B_1 field maps (d,h). No significant correlation between the B_0-field map and the R_2-map is observed. The non-uniformity in the cranial area of the humanoid shaped phantom is attributed to the B_1-field inhomogeneity (d).

However, caution is needed with other imaging sequences or other imaging contrasts such as magnetization transfer (MT) contrast that relies on the saturation of the off-resonance proton pool [76,77].

4.4.5. Standing waves and dielectric effects. The attenuation of high-frequency RF waves may be substantial and depends on the dielectric properties of the medium. As the carrier frequency of the RF pulses is proportional with the field strength, the effects may be more pronounced at 3T. When half the wavelength of the RF waves is a multiple of the size of the GD phantom, standing waves may occur [78] in the GD phantom and large B_1 field non-uniformities will be observed (figure 27). Changing the dielectric properties of the phantom by adding salt to the gel, will significantly change the wavelength of the RF waves in the GD phantom thereby improving the uniformity of the B_1 field and R_2 map. The uniformity of the R_2 map of the salt doped gel can be further improved by using a post-processing correction based on an acquired B_1-field map. The dielectric properties (electrical conductivity and relative dielectric permittivity) of the gel have been measured by use of a network analyser and theoretical calculations in a spherical phantom support the observations [73].

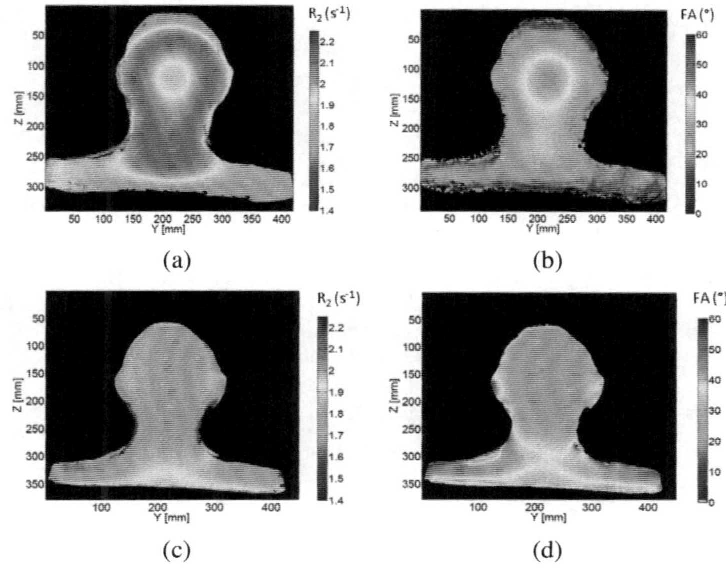

Figure 27. Coronal R2 maps (a,c) and corresponding B1 field maps (in units of effective flip angle) (b,d) of a conventional blank gel phantom (a,b) and of a salt doped (1% NaCl) gel phantom (c,d) scanned at 3T.

6. Relative contributions to the overall uncertainty in polymer gel dosimetry

In table III, the most important contributions to Type B uncertainties are summarized and some realistic values that are common in scanning humanoid shaped PAGAT gel dosimeter phantoms are given. The values listed only serve as typical examples and depend to a large extend on the type of scanner, the shape of the phantom and the coils.

From the overview of sources of uncertainty, it can be concluded that when gel dosimeters are handled under conditions of good practice (see also section 7), the physico-chemical mechanisms should not contribute to more than 2%. Note that for other gel dosimeters (such as MAGAT) where a significantly larger dose rate dependence is found, this figure of uncertainty may be exceeded. Scanner related sources of uncertainty may be kept well below 3 % by a proper scanning setup and/or by applying post-processing compensation methods such as for B_0- and B_1-field non-uniformity. A source of uncertainty that has a significant contribution to the overall uncertainty is related to temperature deviations and variations during scanning. These effects can be minimized by use of an imaging sequence with centric k-space reordering and/or an active temperature stabilization using thermal pads or a temperature stabilized water reservoir.

Table III. Contribution of different sources of uncertainty estimated on the basis of independent measurements, specifications and calculations for a typical PAGAT gel dosimeter scanned with a multiple spin echo sequence in a clinical MRI.

Mechanism	Realistic value	Estimated uncertainty
A. Physico-chemical mechanisms		
Storage temperature differences between calibration samples and volumetric gel dosimeter phantom		
• Cooling in fridge	15°C (5 h)	5 %
• Slow cooling	4 °C (5 h)	1 %
Irradiation temperature (passive temperature stabilization)	3 °C	1 %
Chemical instability (min. 10 h post-irradiation)		
• Calibration and volumetric phantom scanned subsequently (4 h): Absolute scanning time difference	5 h	2.5 %
• Calibration and gel dosimeter phantom scanned simultaneously: Absolute scanning time difference	1 h	0.5 %
Spatial integrity: $D_{max} < 20$ Gy	-	< 1%
Dose rate dependence:	-	< 1 %
Oxygen effect: In glass or Barex recipients	-	< 1 %
Recipient wall effects	-	< 1 %
B. Scanning related mechanisms: Spatial distortions		
B_0-field non-uniformities: (scanner dependent) for BWpp = 130 Hz		
• Within a spherical volume of 40 cm in a typical 1.5 T	$\Delta B_0 = 0.5$ ppm	0.25 pixel
• Within a spherical volume of 50 cm in a typical 1.5 T	$\Delta B_0 = 2$ ppm	1 pixel
Gradient non-linearity		
• Within a cylindrical volume (10 cm × 20 cm)	< 1 %	< 1 pixel
•		
Eddy currents		< 0.5 mm
Magnetic susceptibility differences Note: Spatial correction is possible	$\Delta B_0 = 3$ ppm	1.5 pixel
Chemical shifts	-	-
C. Scanning related mechanisms: Dose uncertainties		
Eddy currents		
• No EC compensation and using non-corresponding calibration set (obtained in another scan direction)	1 ppm	8 %
• With EC compensation	-	1 %
B_0 field non-uniformity		
• Within a spherical volume of 40 cm in a typical 1.5 T	$\Delta B_0 = 0.5$ ppm	< 1 %

• Within a spherical volume of 50 cm in a typical 1.5 T	ΔB_0 = 2 ppm	< 1 %
B_1 field non-uniformity (Strongly dependent on RF coil)	ΔB_1 = 10 % ΔB_1 = 20 % ΔB_1 = 30 % ΔB_1 = 40 % ΔB_1 = 50 %	1.5 % 5 % 8 % 13 % 20 %
Slice profile imperfections Depends on imaging parameters and dose distribution	-	-
Temperature fluctuations		
• Passive setup (without external temperature control)	1 °C	8 %
• Active temperature control in the scanner (thermal pads)	0.3 °C	< 3 %
Temperature drift		
• Linear k-space ordering	1 °C	8 %
• Centric k-space ordering	1 °C	< 1 %
Molecular self-diffusion (scanning direction dependence)	-	-
Standing waves and dielectric effects (see B_1 field non-uniformity)	-	-

7. An independent external calibration of gel dosimeters is possible

The temperature difference between calibration vials and volumetric GD phantom is a major factor of dose uncertainty (Type B). To better control the temperature during scanning, an active temperature controlled experimental setup was applied (figure 28b-c). The GD phantom and calibration vials are wrapped together in a thermal pad through which doped water is send to stabilize and homogenize the temperature in the scanner.

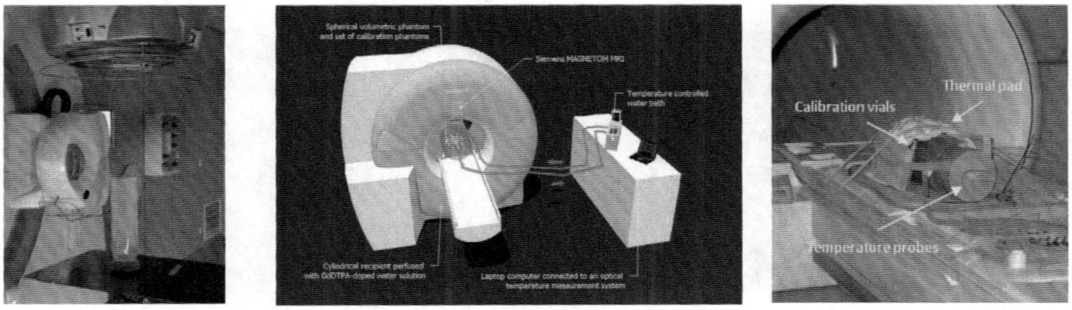

(a) (b) (c)

Figure 28. Experimental set-up for a temperature-controlled experiment. A 30 cm long volumetric phantom is irradiated with a square 4 cm × 4 cm photon beam (a) and scanned together with calibration tubes that are wrapped in a thermal pad that is connected to a circulating thermostatic water bath (c). The setup shown in (b) is similar to the temperature-controlled setup in (c) but where the spherical phantoms and calibration tubes of the earlier described reproducibility study were placed in a cylindrical recipient that is perfused with contrast agent doped water.

As an illustration of the feasibility of this experimental design we performed some preliminary tests on a 2.5 l cylindrical gel dosimeter phantom that has been irradiated with a 4 cm × 4 cm square photon beam (figure 28 and 29). A mono-exponential dose-R_2 response curve was derived from the calibration vials and used to convert the R2 maps to a dose map (figure 29b, left: R_2-map, right: dose-map). A depth-dose curve was extracted from a vertical profile through the dose-map (figure 29c). Within a homogeneous region of 180 mm in the longitudinal direction of the phantom, the dose-difference with an ionization chamber measured dose amounted to less than 5% with most of the points within 3% (see figure 29c and d). The non-uniform dose region beyond 180 cm of the phantom is attributed to a non-uniform B_1-field. In addition, at the outermost boundary of the phantom (beyond 250 cm from the entrance of the beam), a significant deformation can be observed as a result of a non-uniform B_0-field. In the results shown in figure 29, no compensation for the non-uniform B_0- or B_1-field has been performed.

In several experiments in the past we obtained a similar good agreement between gel dosimeter measured dose distributions and ion chamber measured dose, but in some experiments we found a (Type B) disagreement between the gel measured dose and the ion chamber (see also the tube experiment in figure 16d). We therefore have come to the conclusion that the irreproducibility in earlier experiments was attributed to a systematic temperature difference between calibration samples and the volumetric gel dosimeter phantom. Also, in a retake of the earlier experiment with spherical phantoms but now under temperature-controlled conditions, we found a decrease in dose deviation between gel measured dose and ion chamber measured dose in the centre of the tube to a value well below 3%.

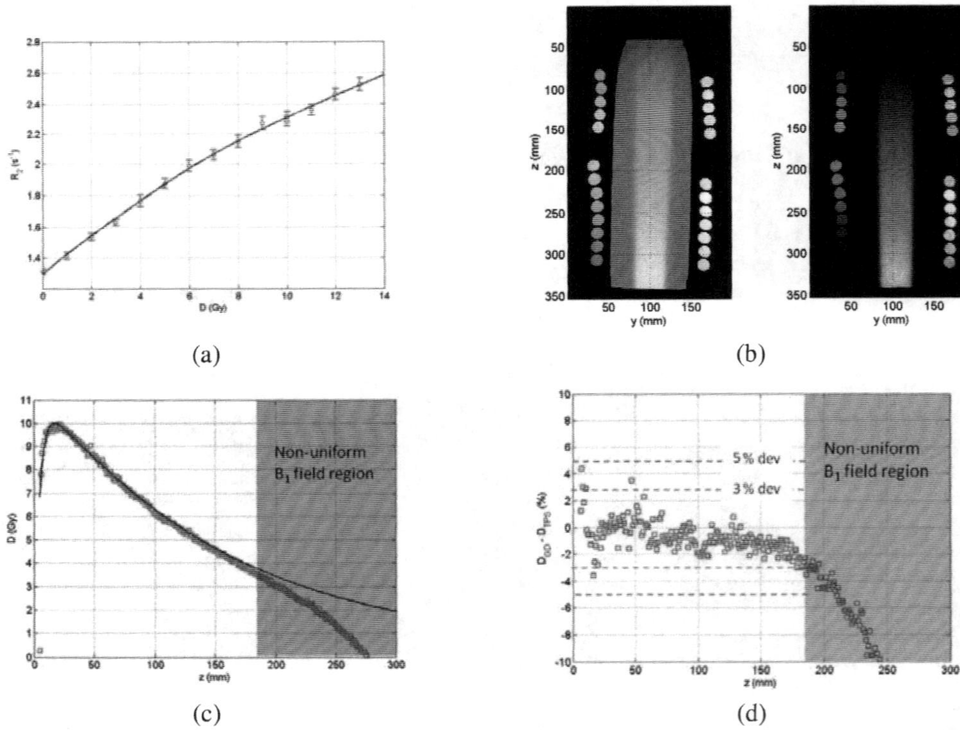

Figure 29. A temperature-controlled gel dosimetry experiment with absolute calibration demonstrates excellent agreement of a gel measured depth-dose curve with an ionization chamber measured depth-dose curve. The depth dose curve starts to deviate from the ionization chamber measurement beyond 180 mm as a result of B_1-field non-uniformity.

8. Quick guidelines for good-practice in polymer gel dosimetry with MRI readout

An extensive set of guidelines on how to scan polymer gel dosimeters with MRI has been discussed in previous conference proceedings [77]. Here we summarize it in the shape of a checklist.

1. Choose a polymer gel dosimeter type - Several polymer gel dosimeters have been discussed in the literature or are commercially available, but of only a few types all dosimetric properties have been reported (temperature dependence during irradiation, storage temperature dependence, scanning temperature dependence, dose rate dependence and energy and tissue equivalence). An important parameter is the dose rate dependence as there is no compensation for it. - If the dosimetric properties of the gel dosimeter have not been determined yet, it is advisable to do so.	
2. Choose an MR imaging sequence - There are several possible sequences and contrast mechanisms that can be used to scan the polymer gel dosimeter, but aside from the dose sensitivity of the imaged parameter, care has to be taken to image uniformity and image distortions. The most used imaging sequence is a multiple spin-echo sequence by which R_2 maps can be obtained. - To construct an R_2 map from the acquired base images, a pixel-wise fitting algorithm has to be implemented. From the perspective of dose resolution, it is advisable to use an algorithm that is based on the minimization of χ^2.	
3. Choose optimal sequence parameters for the polymer gel dosimeter - The sequence parameters (such as echo time spacing, number of echoes) can be optimized to yield the best dose-resolution in the parametric map. See the literature [26,27].	
4. Construct a cast for the humanoid shaped phantom. - The cast is preferably made from Barex, glass or any material with a very low oxygen permeability coefficient to minimize the infiltration of oxygen in the gel dosimeter.	
5. Fill the cast with a blank(non-irradiated) gel phantom. - Use a gel composition that has similar MRI properties as the gel to be used for the actual dosimetry (e.g. a gelatine gel doped with contrast agent and/or anti-fungicide). - Provide the phantom with fiducial markers that will be later visible on CT and MR images.	
6. Scan the blank phantom with CT and perform treatment planning on the scanned data set. - PTV and OARs are virtually created on the basis of realistic anatomy	
7. Scan the blank phantom and determine uniformity, distortion and signal-to-noise. - This is a vital step which is often overlooked. - The blank phantom should be scanned with similar parameters as will be used for scanning the actual gel dosimeter phantom.	

- During this investigative procedure, compensation strategies can be implemented.	
8. Fabricate polymer gel - Control temperature during fabrication and storage	
9. Perform the polymer gel dosimetry experiment - Dose prescription and beam data transported to Linac - Maximum dose scaled to limits determined by spatial integrity limit - Alignment of phantom on the basis of line markers, fiducial markers or additional immobilization device (in conformity with patient treatment) - Stabilize temperature in the gel dosimeter by use of thermal pads or circulating water reservoir before and during scanning. - (Acquire B_0- and/or B_1-field map for compensation of non-uniformities.)	
10. Compare the GD derived dose maps with TPS - Use gamma analysis carefully to avoid bias as a result of image noise (i.e.Use gel derived gel dose distribution as reference and TPS as evaluator)	

References

[1] Baldock C *et al* 2010 *Phys. Med. Biol.* **55** R1–63
[2] Vergote K *et al* 2004 *Phys. Med. Biol.* **49** 287–305
[3] De Deene Y *et al* 2000 *Magn. Reson. Med.* **43** 116-25
[4] Dumas E *et al* 2006 *J. Phys.: Conf. Ser.* **56** 239-41
[5] MacDougall N D *et al* 2008 *Br. J. Radiology* **81** 46-50
[6] De Deene Y *et al* 2007 *Phys. Med. Biol.* **52** 2719-28
[7] Cosgrove V P *et al* 2000 *Phys. Med. Biol.* **45** 1195-210
[8] Sedaghat M *et al* 2011 *Phys. Med. Biol.* **56** 6083-107
[9] De Deene Y *et al* 1998 *Radiother. Oncol.* **48** 283–91
[10] Haraldsson P *et al* 2000 *Br. J. Radiol.* **73** 58–65
[11] Pappas E *et al* 2001 *Phys. Med. Biol.* **46** 783–97
[12] Ibbott G S *et al* 1997 *Int. J. Radiat. Oncol. Biol. Phys.* **38** 1097–103
[13] Oldham M *et al* 1998 *Phys. Med. Biol.* **43** 1113–32
[14] Meeks S L *et al* 1999 *Int. J. Radiat. Oncol. Biol. Phys.* **43** 1141–53
[15] Cosgrove V P *et al* 2000 *Phys. Med. Biol.* **45** 1195–210
[16] Ramm U *et al* 2000 *Phys. Med. Biol.* **45** N95–102
[17] Farajollahi A R *et al* 2000 *Phys. Med. Biol.* **45** N9-N14
[18] Grebe G *et al* 2001 *Int. J. Radiat. Oncol. Biol. Phys.* **49** 1451–60
[19] De Deene Y 2006 *J. Phys.: Conf. Ser.* **56** 72-85
[20] Low D A *et al* 1998 *Med. Phys.* **25** 656-661
[21] Jiang S B *et al* 2006 *Phys. Med. Biol.* **51** 759-776
[22] Meeks S L *et al* 1999 *Int. J. Radiat. Oncology Biol. Phys.* **43** 1135-1141
[23] De Deene Y *et al* 2006 *Phys. Med. Biol.* **51** 653-673
[24] Karlsson A *et al* 2007 *Phys. Med. Biol.* **52** 4697-706
[25] De Deene Y *et al* 2002 *Phys. Med. Biol.* **47** 2459-2470
[26] De Deene Y *et al* 1998 *Signal Processing* **70** 85-101
[27] De Deene Y and Baldock C 2002 *Phys. Med. Biol.* **47** 3117-3141
[28] Vandecasteele J and De Deene Y 2013 *Phys. Med. Biol.* **58** 19-42

[29] Vandecasteele J and De Deene Y 2013 *Phys. Med. Biol.* **58** 43-61
[30] Salomons G J *et al* 2002 *Phys. Med. Biol.* **47** 1435-48
[31] De Deene Y *et al* 2000 *Phys. Med. Biol.* **45** 859-878
[32] Lepage M *et al* 2001 *Phys. Med. Biol.* **46** 2827-39
[33] Olsson L E *et al* 1990 *Phys. Med. Biol.* **35** 1623-1631
[34] Kron T *et al* 1997 *Magn. Reson. Imaging* **15** 211-221
[35] Harris P J *et al* 1996 *Phys. Med. Biol.* **41** 1745-1753
[36] De Deene Y *et al* 2001 *Phys. Med. Biol.* **46** 2801-2825
[37] Vergote K *et al* 2004 *Phys. Med. Biol.* **49** 4507-4522
[38] Fuxman A M *et al* 2005 *Chem. Eng. Science* **60** 1277-93
[39] Jirasek A and Duzenli C 2002 *Med. Phys.* **29** 569-77
[40] Fong P M *et al* 2001 *Phys. Med. Biol.* **46** 3105-13
[41] Hepworth S J *et al* 1999 *Phys. Med. Biol.* **44** 1875-84
[42] Sedaghat M *et al* 2011 *Phys. Med. Biol.* **56** 601-25
[43] Michael G J *et al* 2000 *Phys. Med. Biol.* **45** N133-N138
[44] Baldock C *et al* 2001 *Phys. Med. Biol.* **46** 449-60
[45] De Deene Y 2004 *J. Phys.: Conf. Series* **3** 87-114
[46] De Deene Y 2009 *J. Phys.: Conf. Ser.* **164** 012033
[47] Oldham M and Kim L 2004 *Med. Phys.* **31** 1093-1104
[48] Xy Y and Wuu C-S 2004 *Med. Phys.* **31** 3024-3033
[49] Doran S J and Krstajic N 2006 *J. Phys.: Conf. Ser.* **56** 45
[50] Doran S J 2010 *J. Phys.: Conf. Ser.* **250** 012086
[51] Price R R *et al* 1990 *Med. Phys.* **17** 287- 95
[52] Michiels J *et al* 1994 *Magn. Reson. Imaging* **12** 749-65
[53] Schad L R *et al* 1992 *Magn. Reson. Imaging* **10** 609-21
[54] Doran S J *et al* 2005 *Phys. Med. Biol.* **50** 1343-61
[55] Cho Z H *et al* 1988 *Med. Phys.* **15** 7-11
[56] De Deene Y *et al* 2000 *Phys. Med. Biol.* **45** 1807-23
[57] Jehenson P *et al* 1990 *J. Magn. Reson.* **90** 264-78
[58] Ahn C B and Cho Z H 1991 *IEEE Trans. Med. Imag.* **10** 47-52
[59] Wysong R E and Lowe I J 1991 *Magn. Reson. Med.* **17** 149-63
[60] Ahn C B and Cho Z H 1991 *Magn. Reson. Med.* **17** 149-63
[61] Liu Q *et al* 1994 *Magn. Reson. Med.* **31** 73-6
[62] Boesch C H *et al* 1991 *Magn. Reson. Med.* **20** 268-84
[63] Robertson S *et al* 1992 *Magn. Reson. Med.* **25** 158-66
[64] Schiff J *et al* 1994 *J. Magn. Reson. Series B* **104** 73-6
[65] Collins C M *et al* 2002 *Magn. Reson. Imaging* **20** 413-24
[66] Li S *et al* 1996 *Magn. Reson. Med.* **36** 705-14
[67] Park H W *et al* 1988 *Phys. Med. Biol.* **33** 339-49
[68] Crijns S P M *et al* 2011 *Phys. Med. Biol.* **56** 289-97
[69] Holland D *et al* 2010 *Neuroimage* **50** 175-83
[70] De Deene Y *et al* 2001 *Phys. Med. Biol.* **46** 2801-25
[71] Fransson A *et al* 1993 *Magn. Reson. Imaging* **11** 771-84
[72] De Deene Y *et al* 2000 *Phys. Med. Biol.* **45** 1825-39
[73] Vandecasteele J and De Deene Y 2013 *Phys. Med. Biol.* **58** 63-85
[74] De Deene Y and De Wagter C 2001 *Phys. Med. Biol.* **46** 2697-711
[75] Hurley C *et al* 2003 *Phys. Med. Biol.* **48** 3043-58
[76] Lepage M *et al* 2002 *Phys. Med. Biol.* **47** 1881-90.
[77] De Deene Y *et al* 2006 *Med. Phys.* **33** 2586-97.
[78] Tofts P S 1994 *J. Magn. Res. Ser. B* **104** 143-7
[79] De Deene Y 2010 *J. Phys.: Conf. Ser.* **250** 012015

Analysis and evaluation of planned and delivered dose distributions: practical concerns with γ- and χ- Evaluations

LJ Schreiner[1,2], O Holmes[2] and G Salomons[1,2]
[1]Cancer Centre of Southeastern Ontario at Kingston General Hospital, 25 King Street West, Kingston, ON, Canada, K7L5P9
[2]Department of Physics, Queen's University, Kingston, Ontario, Canada, K7L3N6

E-Mail: john.schreiner@krcc.on.ca

Abstract. One component of clinical treatment validation, for example in the commissioning of new radiotherapy techniques or in patient specific quality assurance, is the evaluation and verification of planned and delivered dose distributions. Gamma and related tests (such as the chi evaluation) have become standard clinical tools for such work. Both functions provide quantitative comparisons between dose distributions, combining dose difference and distance to agreement criteria. However, there are some practical considerations in their utilization that can compromise the integrity of the tests, and these are occasionally overlooked especially when the tests are too readily adopted from commercial software. In this paper we review the evaluation tools and describe some practical concerns. The intent is to provide users with some guidance so that their use of these evaluations will provide valid rapid analysis and visualization of the agreement between planned and delivered dose distributions.

1. Introduction

Improved treatment planning systems, dose delivery equipment (typically, linear accelerators) and therapeutic techniques have advanced modern radiation therapy by enabling clinicians to achieve more conformal high dose delivery to target volumes while sparing adjacent normal tissues. However, as discussed throughout these proceedings, these improvements have also increased significantly the requirement for dose delivery validation especially in the commissioning of treatment units and new treatment techniques and, in some settings, in the patient specific validation of dose delivery.

The measurement of the integrity of dose delivery during the commissioning of a new unit, or a new treatment technique, may involve the measurement of dose distributions in phantoms for test cases planned under well-defined conditions [1,2]. Such measurements can ensure correct performance and establish benchmark data for future quality assurance of the particular treatment protocol. The measurements often involve film or two or three dimensional (2D and 3D) diode or ion chamber array measurements on regular and anthropomorphic phantoms. These devices may also be used to test the delivered distributions for specific patients by exposing the devices in phantom to the same multileaf collimator leaf sequences, trajectories and MUs as planned for the treatment [3]. Patient treatment delivery validation may also be performed using dose reconstruction from exit beam measurements on Electronic Portal Imaging Devices (EPIDS) [4]. And finally whole treatment protocols may be regularly monitored in end to end test procedures using 2D and 3D dose delivery validation [5,6]. Gel and radiochromic 3D dosimetry has played a role in all these settings in select clinics.

2. Evolution of dose evaluation: introduction of dose and distance metrics

In all the situations above, the comparison of calculated dose distributions to physical measurement is a major task, since both the reference and the measured distributions usually consist of large 2D or 3D data sets [2,7,8,9]. Furthermore, the very nature of the comparison of dose distributions requires that the evaluation probe both dose and spatial domains [8]. An approach of past decades to accommodate dosimetric and spatial information by overlaying 2D dose contours in various planes for comparison was sufficient if the contours agreed, but the delivery was not easily evaluated if agreement failed. A better alternative for the comparison has been to calculate a dose difference map: a spatial representation of the numerical difference in the doses in the two distributions. There is a shortcoming with this technique in that the approach is inherently oversensitive in regions of high dose gradient where small spatial errors in either data set can lead to large dose differences between the measured and planned distributions [7,8,9]. Separate methods, such as the use of the distance to agreement (DTA) which reports the distance between the calculated point and the nearest point in a measured dose distribution with the same dose value [8], are needed to obtain useful comparisons in regions of high dose gradient. However, while DTA maps improve the dose comparisons in regions of high dose gradient, they tend to be overly sensitive in regions of uniform dose (low dose gradients) such as in the target volume.

A solution to these problems has been developed over the last 15 years through the introduction of various metrics combining agreement criteria for dose and distance in the distributions being compared [7,8,9]. Harms *et al.* [8] combined the complementary nature of the DTA and dose difference distributions in a software tool called the composite evaluation. In this evaluation, the dose difference and DTA distributions were independently evaluated as either meeting or exceeding predefined tolerances. Harms [8] used a dose difference criterion of 3 % (i.e., 3 % of the maximum dose) and a 3mm limit for the DTA, this was slightly different than the 3%/4mm tolerances suggested by Van Dyk for the evaluation of treatment planning systems [11]. The 3%/3mm criteria has been subsequently adopted in many other studies [9,12-18]. In Harms composite evaluation, the tolerance criteria were used to create two binary pass-or-fail distributions (corresponding to the DTA and dose difference) which were multiplied point by point to give a single binary distribution. The combined binary distribution contained a binary array (0's or 1's) indicating agreement or failure, respectively, within 3%/3mm. The final output of the evaluation showed regions which failed to meet both the dose and DTA criteria and, therefore, it did not suffer from increased sensitivity in regions of low or high dose gradient. But as it was only a binary map, the evaluation visualizing the outcome was not easily interpreted [8,9]. Also the approach heightened the impression of disagreement in regions of high dose gradient and was not representative of the significance of agreement in these regions.

3. Gamma dose evaluation

Low [7,9,12] formalized the approach of evaluating both dose and spatial characteristics of the two distributions being compared by the development of the gamma (γ) dose distribution method. A γ comparison is performed between two dose maps: one distribution is the 'reference plan' (typically from the treatment planning system, but see discussion below) and the other is the 'evaluated distribution', usually from a two or three dimensional dose measuring system. The reference distribution is treated as the true distribution, while the evaluated image is analyzed for its agreement with the reference as follows: every point in the reference image has a corresponding γ value; a measure of agreement at that location. Each possible point in the reference distribution can be coupled with any point in the evaluated image. For all pairs there exists a Γ, defined by the vector difference between the points.

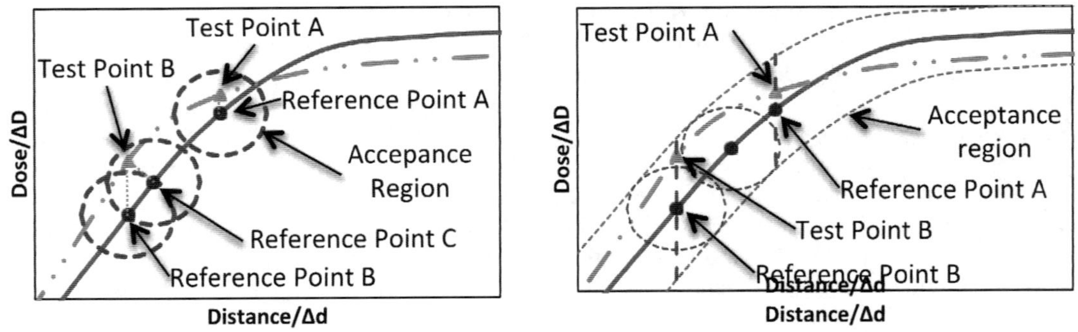

(a) (b)

Figure 1: a) An illustration of a 1D γ test. For each reference point (blue dot) there is a circular acceptance region (purple dashed line). The γ test for each reference point involves a search to find a point on the evaluation curve (a test point) that falls within the reference point's acceptance circle. The passing test point is not necessarily at the same physical resolution as the reference point (as shown there are only two test points on the evaluation curve, marked by green triangles). b) An illustration of a 1D χ test. For each reference point (blue dot) an extended dose limit criteria is calculated (purple dashed line) based on the dose gradient and distance tolerances at that point. The extended dose limits essentially indicate the upper and lower limits of the acceptance region surrounding the reference curve. The χ test for each reference point involves an interpolation of the test curve points to the positions of the reference points and a comparison of the interpolated test dose values to the extended dose limit criteria.

Tolerance criteria Δd_M and ΔD_M (e.g., 3mm and 3%) are used to normalize Γ along the distance and dose vector dimensions, correspondingly. The gamma index value, γ, is the smallest Γ that can be found considering the entire evaluated distribution. In 2D evaluation space $\Gamma^2 = 1$ describes a circle whose extent is defined by the tolerance criteria (see figure 1a), whereas in a 3D space the criteria define an ellipsoid. When $\gamma \leq 1$, the distributions agree within the stipulated tolerances. Conversely, when $\gamma > 1$, no point in the evaluated dose map can be found within the circle/ellipsoid and the dose distributions disagree at that location. As with the composite test [14], the γ function identifies if the the evaluated distribution passes ($\gamma \leq 1$) or fails ($\gamma > 1$) the comparison. Also, if the evaluated distribution fails, the value of γ indicates roughly the degree of failure relative to the dose and distance to criteria set in the evaluation [7]. For example, a $\gamma = 1.5$ fails by 50% which corresponds to a failure of 1.5% or 1.5mm for a 3%/3mm dose/DTA criteria; or 1% or 1mm for 2%/2mm criteria. If the vector nature of the gamma is evaluated [7,15,19] the test can give indications of whether the failure is primarily due to dose or distance to agreement failure.

4. Chi dose evaluation

Since it was introduced, the γ-tool has been refined, modified and evaluated by several authors [12-20]. One alternative to the γ function suggested was the chi (χ) evaluation proposed by Bakai *et. al* [14], which also provides a metric combining both dose and DTA criteria to evaluate the level of agreement between evaluation and reference dose distributions. However, in the χ approach the comparison between the reference and evaluated distributions is carried out differently than with the γ test. Instead of searching the test space for evaluation points that are closest to a given reference point, the χ test compares the reference dose to the test dose at the same point in space, but it scales the dose limit criteria according to dose gradient of the reference distribution.

The difference between γ and χ evaluations is illustrated in figure 1. Discrete points on the reference curve are systematically compared (see figure 1a) to discrete points on the evaluation curve. In the figure test point A on the evaluation curve is within the acceptance region for reference point A. The test point B is not within the acceptance region for reference point B, but then it is within the acceptance region for reference point C. This illustrates that the γ evaluation must search multiple test points to find a minimum Γ. When calculating gamma over a volume, this can become computationally expensive. The χ test takes a different approach. Instead of identifying an acceptance region about individual reference points, and determining if evaluation points fall within one of these regions, the χ test extends the dose tolerance based on the dose gradient and distance limit at that point. This is illustrated in figure 1b, where the χ dose limits, shown as the vertical dashed lines, have been extended beyond the dose-only tolerance (the circle) to reflect the limits of the "acceptance tube" surrounding the reference curve. In this way, the test points are each compared to the extended dose limits for that point with no need to search.

Beyond the fact that there is no need to search the test space, there are several other advantages to the χ test. First, the χ test returns a positive or negative value indicating whether a failure is an

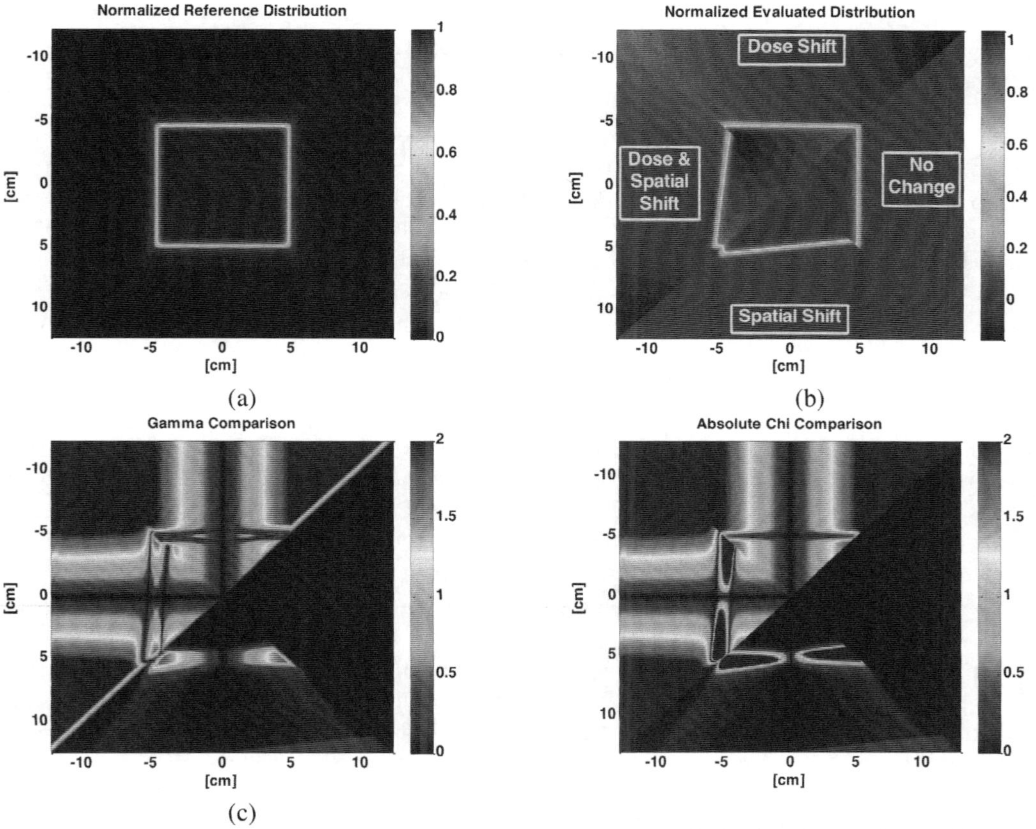

Figure 2: An illustration of the output from γ and χ comparisons between two test distributions that were first used by Low and Dempsey [13] to illustrated properties of γ evaluations. (a) The reference distribution is a normalized 10 cm by 10 cm square field originally from a 6MV irradiation. (b) The evaluated dose distribution is divided into quadrants along the diagonals, and each quadrant is perturbed from the original reference distribution in different ways as follows: the region at noon has the dose shifted by $0.12x$ % (x measured in mm), at three o'clock the distribution is not modified and so the evaluated and reference distributions correspond exactly at all points, in the quadrant at six o'clock , the points in the distribution have been shifted spatially by $0.12x$ mm (x measured in mm), and at nine o'clock, both shifts have been implemented (with the effect altered to depend on the y dimension). (c) Shows the γ comparison map calculated with the original γ-tool (reproduced with permission). (d) Shows distribution of the absolute value of the calculated χ [14] with an in-house χ-tool. (Note that the colour bar on the right presents the scale for the results of the dose comparisons, herein and in other figures below).

overdose or an under-dose (although this is not easily visualized in colour wash and so representations of the result of the χ evaluation, such as figure 2d, usually show absolute values.). In some situations, this allows one to more easily evaluate if a failure is clinically relevant (e.g., if it is an acceptable under-dosing in the region of an organ of risk). Another advantage with the χ evaluation is that one can assign different dose and distance limits in different regions of the dose distributions being compared and can also allow the dose criteria in the positive and negative directions to differ. This would enable one to apply tighter dose and/or distance criteria to the region around an organ of particular concern and looser criteria elsewhere.

The χ evaluation does have some limitations. Firstly, it requires that the points on the evaluation distribution are spatially matched to the reference points. This means that the data for the evaluation dose distribution may need to be interpolated, with the potential for some distortion. Secondly, the extended dose limits (i.e., establishing the acceptance regime outlined by the red curves in figure 1b) in regions where the dose gradient changes rapidly (where the second derivative is large). The result is that χ values near the shoulder portion of an evaluated curve may not be calculated accurately. In spite of these limitations, the χ test is a strong alterative to the traditional γ evaluation and it has recently been gaining broader clinical use.

5. Practical Considerations I: discretization and spatial resolution

While Low *et al.* presented a powerful theoretical concept for the evaluations based on continuous functions [9], real clinical comparisons are typically made between discretized representations of dose distributions often with the reference and evaluation dose data sampled at different spatial resolutions. The importance of resolution was first analysed by Depuydt *et al.* [12] in their clinical assessment of the γ evaluation. They were particularly concerned with overestimations of γ values caused by large grid spacing in the discrete dose distributions, particularly in regions of high dose gradient. To avoid overestimating they effectually reduced the continuous value of γ to a binary test equivalent to the composite evaluation [8]. Several other studies have been directed, in part, towards resolving this issue [14,17,18]. Low and Dempsey [13] noted that by re-sampling the distributions to 1x1mm2 grids, the error in γ associated with pixelization artifacts was reduced to less than 0.2, even in regions of high dose gradient. Others have minimized discretization artifacts by interpolating additional data points in the evaluated grid [17].

The effect of resolution on the resulting γ distributions is illustrated in figure 3, showing how the γ comparison maps and the observed pass rate (a common parameter used to roughly summarize the complex γ dose distribution comparison by reporting the percentage of points in the dose distribution that pass the specified γ criteria Δd_M and ΔD_M) change with spatial sampling (the bottom rows corresponding to different resolutions). When the film data (fixed at a resolution of 0.24 mm) are taken as the reference doses, and the Eclipse plan is set as the evaluation distribution, increasing the resolution of the evaluated distribution (from 2.5 to 0.24 mm) changes the pass rate from 80.9% to 91.3%. This result can be explained by the behaviour of the γ search. When the evaluated distribution has a coarse pixel size compared to the reference distribution, many reference pixels fall a significant distance from the nearest evaluated pixel. Thus, the γ value for many reference pixels reflect significant spatial misalignment purely as an artefact of the coarse evaluated resolution which is seen as the grid-lines in the low dose region of figure 3 (c). When the resolution of the evaluated distribution is increased to match that of the reference distribution, this spatial artefact is eliminated since each reference point has a directly corresponding pixel in the evaluated distribution. Increasing the evaluated resolution also provides each reference point with a greater range of dose values for comparison, further increasing the likelihood of finding a set of pixels which pass the gamma test. Note, however, that the change of resolution in the Eclipse plan does not significantly change the pass rate if the Eclipse plan is the reference distribution. This is discussed further below.

The requirement for interpolation adds a significant increased computational burden to γ evaluations [17] and the computation time grows as a third power of increasing grid resolution [18]. The computational speed of the algorithm can be enhanced by pre-calculating interpolation factors [17]. Another approach to reducing the evaluation time is to restrict searches around each reference point [15, 17, 18, 19]. For example, one can limit searches *a priori* so that points that "do not have a chance" of yielding the smallest value of Γ are eliminated: as soon as $\Delta d/\Delta d_M$ becomes larger than the smallest Γ found so far, the search is terminated. Others [18] have developed geometric approaches to determining the smallest Γ.

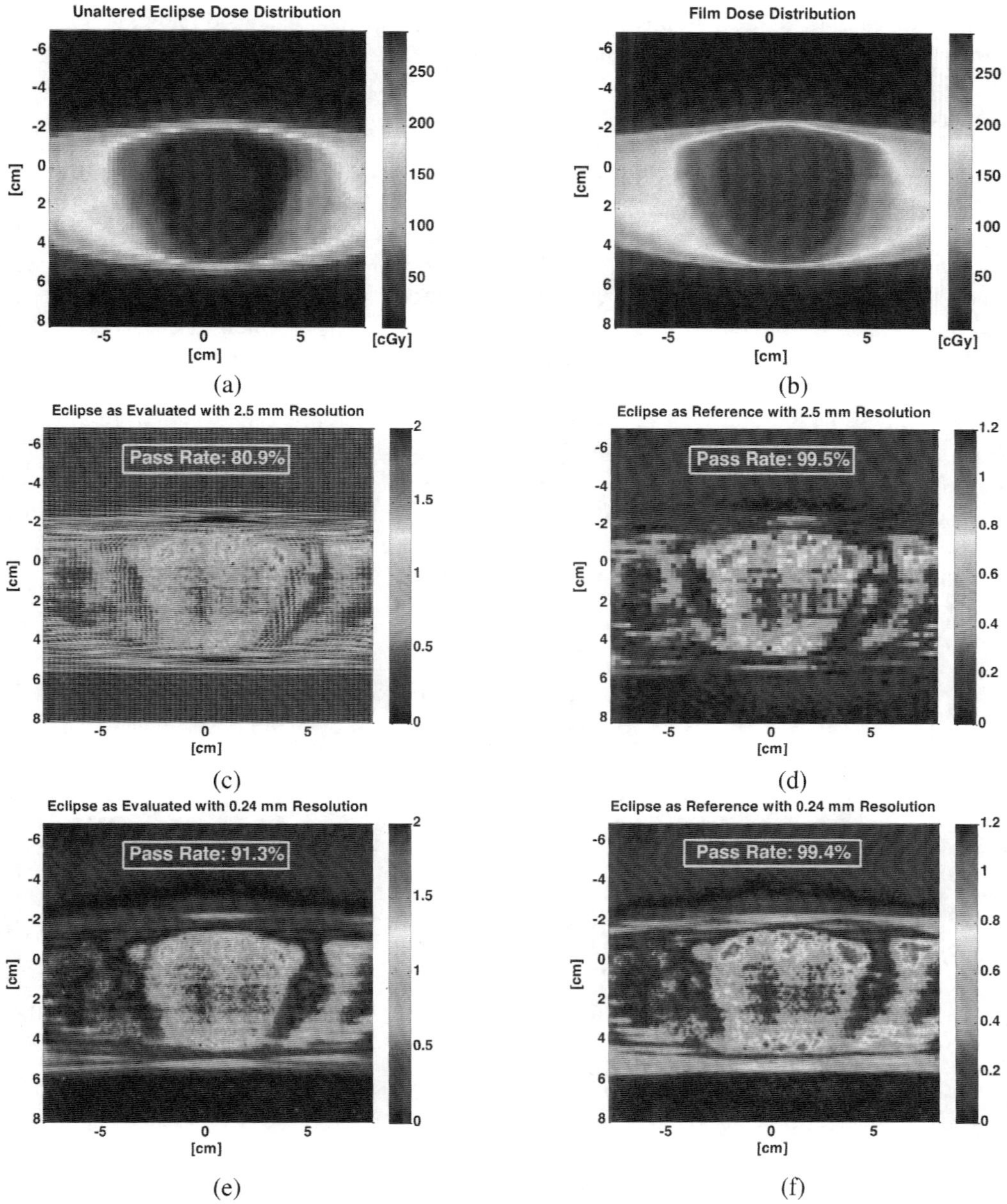

Figure 3: An illustration of the importance of two properties of the γ evaluation: i) the spatial resolution of the data sampling in the evaluated and reference dose maps, and ii) the status of the two distributions as reference or evaluated distributions. The top row shows (a) the calculated distribution for a prostate treatment plan from the Eclipse planning system and (b) a measurement of this dose distribution using radiochromic film at a 0.24 mm resolution. The middle row shows the γ comparison map from film data at the original 0.24 mm resolution and using the Eclipse data at the original 2.5 mm resolution as (c) the evaluated distribution and (d) the reference distribution. The bottom row shows the γ distribution from film data at the original 0.24 mm resolution and the Eclipse data interpolated to a 0.24 mm resolution as (e) the evaluated distribution and (f) the reference distribution.

6. Practical Considerations III: the role of reference and evaluation distributions

In the previous discussion of figure 3, it was noted that the results of the comparison of two dose distributions is affected by the designation of the distributions as reference or evaluated data sets. Neither γ evaluations, nor DTA map, are symmetric with respect to the distributions being compared. Before this is discussed in detail, it is important to note an important change of notation introduced by Low and Dempsey [13] to recognize that in clinical practice the distributions can both be calculated (e.g., output from two different planning systems), or from two separate measurements (e.g., on alternative dosimetric systems being evaluated for performance). For example, it may be necessary to perform comparisons between Monte Carlo computations and calculations made by commercial planning software, or between film measurements. So to avoid confusion, the terms 'reference' and 'evaluated' distributions were adopted by Low to replace 'measured' and 'calculated' distributions, respectively. This notation has been become widely adopted [15, 17, 18]. The convention is different than that used historically, for example, during the development of the DTA and dose difference tools in the evaluation of electron beam dose calculation software [10]. In the current usage it is usually understood that the dose comparison is used to measure the extent to which an evaluated distribution agrees with a reference distribution, which is treated as the *true* distribution (though, strictly speaking, obtaining a true distribution may not be possible).

That the designation of which distribution is set as the reference can affect the results of the γ comparison was illustrated previously in figure 3, where the role of the Eclipse dose distribution is switched in the bottom columns. This is also true for the χ comparisons (see figure 4). As noted above, part of the change in the γ comparisons can be related to differences in resolution. Noise in the data sets also contributes to the differences [7, 13]. In the examples shown in figures 3 and 4, the film dose distribution data are noisier and the Eclipse data smoother.

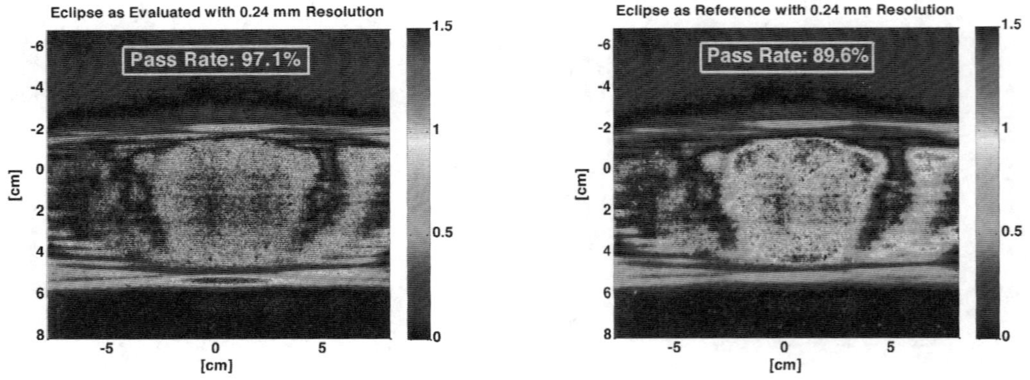

(a) (b)

Figure 4: An illustration of the importance of the status of the two distributions as reference or evaluated distributions for the χ evaluation comparing the Eclipse and film dose distributions shown in figure 3a and 3b, respectively. (a) the χ distribution map with the Eclipse data is the evaluated distribution. (b) the χ distribution with the Eclipse data is the reference distribution.

In figure 3 the side-by-side comparison maps (a)-(b) and (c)-(d) show substantial increases in pass rate when the noisy film data forms the evaluated dose distribution. This result stems from the search algorithm of the γ function which compares points in the reference distribution to all of the nearby points in the evaluated distribution. When the evaluated distribution contains noise, a greater range of dose values is made available for comparison with a given reference point. Therefore, the likelihood of finding a reference point that satisfies the dose and distance criteria increases. When the roles of the two distributions are reversed, the opposite is true, as extreme dose values from the film distribution are compared to a narrow range of nearby dose values in the smooth Eclipse distribution.

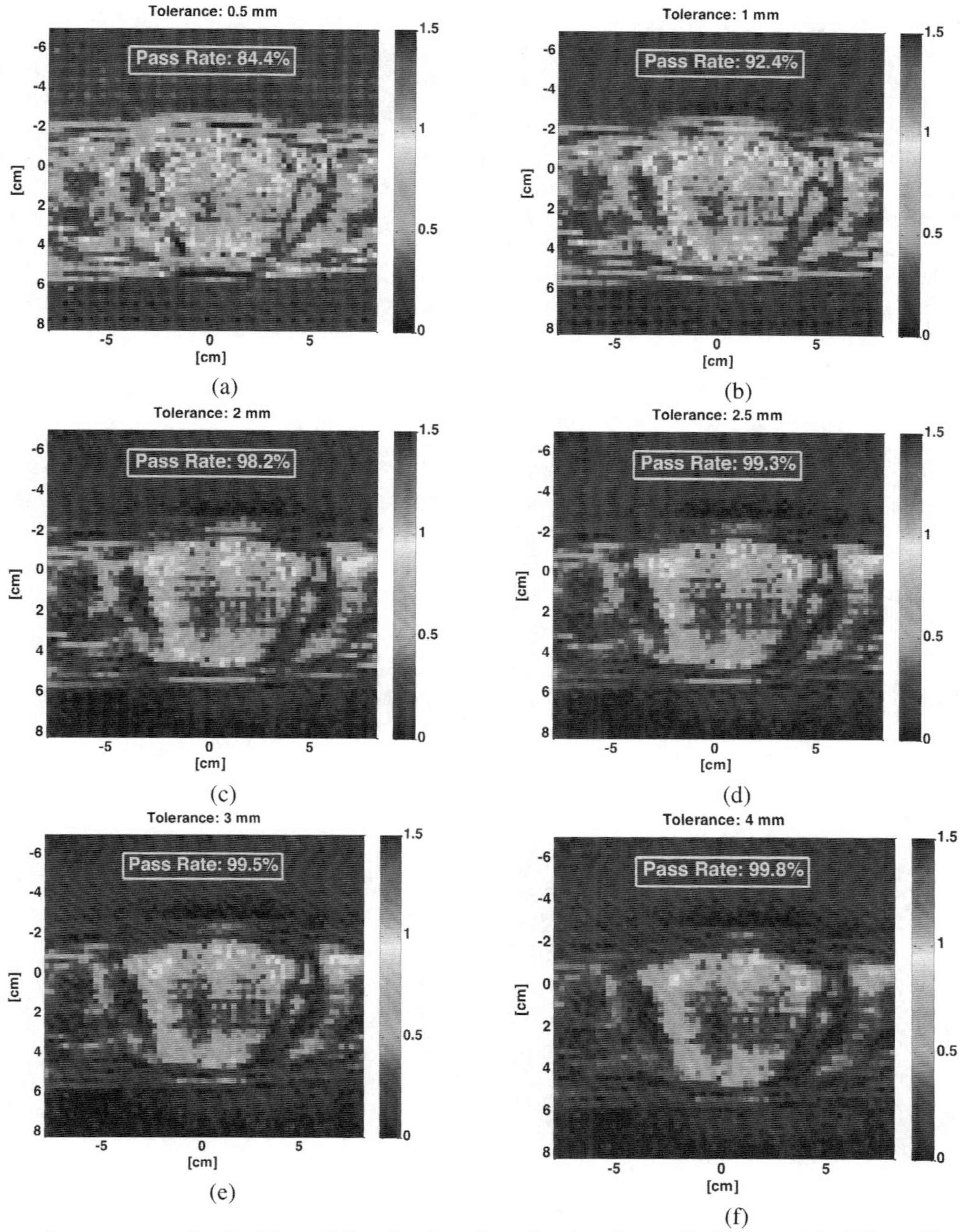

Figure 5: γ maps comparing the 2.5 mm Eclipse data from figure 3 as the reference distribution and the 0.24 mm film data as the evaluated distribution with the γ comparisons performed with different DTA criteria (labelled tolerance in each panel). Panes (a)-(f) display the results of altering the spatial DTA criterion of the γ test between 0.5 mm and 4 mm. Note that the agreement breaks down as DTA becomes smaller than the inherent resolution of the reference dose distribution.

The discrepancy in pass rate and appearance between the two χ distributions (in figure 4) highlights the effect of noise in the reference distribution. The noisy film measurement, when assigned to the role of reference distribution produces a more forgiving comparison. This is a result of the high gradient regions introduced by noise in the film measurement. The observed increase in gradient increases the acceptance region used when comparing reference and evaluated points, lowering the χ value at that point. The opposite is true when the smooth evaluated distribution serves as the reference distribution.

7. Other considerations and conclusions

Dose comparison is daily task for many medical physicists and γ and χ comparisons have become a common tool for treatment validation. Unfortunately, some of the characteristics of the dose distribution comparison tools are not well understood, and the evaluations can be misused. For example, the authors have seen manuscripts with γ dose distribution comparisons using DTA criteria smaller than the resolution of the data sets, which can lead to inappropriate γ maps (see figure 5). A common practice of reporting the agreement of planned and measured dose delivery (for example, in the acceptance of a particular patient's planned delivery) using only the percentage of points failing the γ evaluation as an assessment. This can reduce a useful evaluation (especially if mapped over structure sets from the planning process) to a single number which may not reflect the clinical relevance of the failures (for example, if the failure is under-dosing in a target area or over-dosing in an organ at risk).

Perhaps this report will encourage readers to take up further study and analysis of dose comparison techniques to more clearly set standards for use. More investigations to establish improved techniques (for example, by probing γ function behaviour with various dose and DTA acceptance criteria, Δd_M and ΔD_M [21]) and procedures for use would benefit the community.

8. Acknowledgements

Funding enabling our investigations of dose evaluation has been provided by the Canadian Institutes of Health Research (CIHR, FRN# 82914) and the Ontario Consortium for Adaptive Interventions in Oncology (OCAIRO) through the Ontario Research Excellence Program. We thank Dr. Daniel Low for providing test distribution approach from his work for us to use in our development, and Mr. Chris Jechel for his work generating the examples illustrating the nature of the dose evaluations.

9. References

[1] Yin F F *et al* 2010 *J. Phys.: Conf. Ser.* **250** 012002
[2] Schreiner L J 2006 *J. Phys.: Conf. Ser.* **56** 1
[3] O'Daniel J *et al* 2010 *J. Phys.: Conf. Ser.* **250** 012050
[4] Mijnheer B *et al* 2010 *J. Phys.: Conf. Ser.* **250** 012020
[5] Schreiner LJ 2009 *J. Phys.: Conf. Ser.* **164** 012001
[6] Schreiner L J *et al* 2011 Proc. Internat. Symp. Standards, Applications and Quality Assurance in Medical Radiation Dosimetry (IDOS) 2, 197-205 (IAEA Vienna, Austria)
[7] Low D 2010 *J. Phys.: Conf. Ser.* **250** 012071
[8] Harms W B *et al* 1998 *Med. Phys.* **25** 1830-6
[9] Low D A *et al* 1998 *Med. Phys.* **25** 656-61
[10] Mah E *et al* 1989 *Phys. Med. Biol.* **34** 1179-94
[11] Van Dyk J *et al* 1993 *Int. J. Radiat. Oncol. Biol. Phys.* **26** 261-73
[12] Depuydt T *et al* 2002 *Radiother. Oncol.* 309-19
[13] Low D A and Dempsey J F 2003 *Med. Phys.* **30** 2455-64
[14] Bakai A *et al* 2003 *Phys. Med. Biol.* **48** 3543-53
[15] Stock M *et al* 2005 *Phys. Med. Biol.* **50** 399-411
[16] Spezi E, and Lewis D G 2006 *Radiother. Oncol.* **79** 224-30
[17] Wendling M 2007 *Med. Phys.* **34** 1647-54
[18] Ju T *et al* 2008 *Med. Phys.* **35** 879-87
[19] Holmes O *et al* 2012 *J. Phys.: Conf. Ser.* (IC3DDose 2012 Proceedings)
[20] Clasie B M *et al* 2012 *Phys. Med. Biol.* **57** 6981-97
[21] Childress N L *et al* 2005 *Med. Phys.* **32** 838-50

Dosimetry of CBCT: methods, doses and clinical consequences

JR Sykes[1], R Lindsay[1], G Iball[1] and DI Thwaites[12]
[1]Medical Physics and Engineering, St James's Institute of Oncology, Leeds, UK
[2]Institute of Medical Physics, School of Physics, University of Sydney, Australia

E-mail: jonathan.sykes@leedsth.nhs.uk

Abstract. The use of Cone beam CT (CBCT) systems for Image Guided Radiotherapy is rapidly expanding in the developed world. With its use comes concern for the increased risks of additional radiation exposure. Quantification of the imaging dose is necessary in order to report, optimise and justify CBCT exposures. This article reviews the current methods of dose measurement and calculation including dose measurements in cylindrical phantoms, use of point dosimeters in anthropomorphic phantoms, calculation of dose using mathematical phantoms and calculation of individualised patient dose using Monte Carlo and model based techniques. Typical doses from commercial systems are reported and the clinical consequences, both risks and benefits, of using CBCT based IGRT reviewed briefly.

1. Introduction

Cone beam CT (CBCT) systems integrated into a radiotherapy treatment room (Linac) have become readily available within radiotherapy departments since Jaffray *et al* first published the concept in 1998 and becoming commercially available in 2005. The CBCT system enables a sequence of 2D radiographic projection images to be acquired from a kV source and flat panel detector imaging system as it rotates around the patient either before treatment or more recently during treatment [1]. These projection images can be reconstructed to produce a 3D "volumetric" image which is similar, but not of the same quality, as a conventional fan beam CT scanner. The reconstructed CBCT images can be used to correct patient position prior to treatment [2,3] or as a basis to adapt [4–6] the treatment plan to the changing anatomy of the patient during the course of their radiotherapy. This ability to visualize the target anatomy and nearby organs at risk at the point of treatment delivery and to subsequently make corrections either to the patient position or the dose delivered is often termed image guided radiotherapy (IGRT) [7]. The two main commercial radiotherapy CBCT systems from Varian and Elekta are the On-Board Imager (OBI) and Synergy system (XVI, X-ray volumetric imager) respectively.

The introduction of CBCT for IGRT proceeded initially, at least in some countries, with little regard for the extra radiation dose delivered by the imaging system, known as concomitant dose. The generalized justification for this was that the benefits of IGRT will outweigh the increased risks from the CBCT dose. However, more recently there has been increased interest in the concomitant imaging dose. This may be partly driven by a recent campaign aimed in particular at pediatric practice, to apply principles of ALARP – as low as reasonably practicable (Alliance for Radiation Safety in Pediatric Imaging, http://www.pedrad.org). There have also been a couple of publications that have made alarming statements about the magnitude of CBCT doses. For example Spezi *et al* found for 40 image fractions of a head and neck using the manufacturers image acquisition protocol that *"the mean concomitant dose in the beam-bordering region accumulated during the RT course from daily use of CBCT (assuming a total of 40 scans) could be in the range 1-2 Gy, with maximal doses in the range of 3-7 Gy."*[8]

With such doses there is rightly cause for concern e.g where the imaging dose increases the dose to the spinal cord for plans where intensity modulated radiotherapy (IMRT) has optimized the dose distribution to near cord tolerance. In such cases it is not easy to know whether the concomitant imaging dose is justified. However, in Stereotactic Ablative Radiotherapy (SABR) treatments of the lung and other anatomical sites such as prostate and spine, where the number of fractions is low (typically 3-8 fractions) and where the effect of geometric miss is much greater, the use of CBCT imaging is clearly justified. Also, for SABR, the total concomitant dose is low because of the reduced number of CBCT scans required. Justification of imaging dose, mandatory in UK legislation [9] and many other countries, is clearly required if CBCT is to be used safely in IGRT. In order to justify the IGRT dose, knowledge of the risks associated with the concomitant imaging, e.g risk of induction of secondary malignancy need to be balanced against the benefits, such as improved treatment outcome, whether that be increase in overall survival or local control or reduced toxicity and side effects both acute and long term. At the present time there is great uncertainty in the risks of secondary cancer induction from radiotherapy [10,11]. There is also very little published evidence on the benefits of IGRT protocols although these are likely to gradually emerge as results of clinical trials are published.

Of course justification of concomitant imaging dose is easier if it is lower and there are many publications which address the optimization of image acquisition, i.e reduction of imaging dose whilst maintaining sufficient image quality to perform the task of IGRT, which is, inherently, one of image registration [12]. The methods employed range from simply reducing the exposure, employing bow tie filters or "zonal" filters [13,14] reducing the number of projection angles used in reconstruction [12,15] and improving the reconstruction algorithms [16,17].

In order to minimize and justify the concomitant imaging dose we need to be able to measure it. The aim of this article is to review the methods employed and the progress to date in the measurement of concomitant imaging dose from CBCT.

2. Review of CBCT dose measurement methods

2.1. Measurement of CBCT dose in cylindrical phantoms

To understand dose measurement in CBCT it is worth taking a brief historical view of dose measurement in conventional fan beam CT (FBCT). The established method has been to measure the Computed Tomography Dose Index (CTDI) [18,19]. This is measured using a 100mm long cylindrical ionisation chamber in a cylindrical phantom. The $CTDI_{100}$, defined in equation (1), measures the dose to air over a volume which encapsulates the slice width plus the tails of the fan beam profile either side. This is performed for a single axial (i.e. not helical) rotation of the x-ray tube. For a narrow slice (<= 1cm) a 100mm long chamber is sufficient to capture enough of the profile without significant loss of accuracy. Capturing the entire profile, including the tails due to scatter, is equivalent to measuring the dose for a spiral acquisition,

$$\mathrm{CTDI}_{100} = \frac{1}{\mathrm{nT}} \int_{-50\mathrm{mm}}^{50\mathrm{mm}} \mathrm{D(z)dz} \tag{1}$$

where, n is the number of detector rows and T is the thickness of each row (mm).

Although there are some variants of the CTDI measurement, the $CTDI_{100}$ is generally accepted as a standard [20] and is the standard adopted by the IEC [21].

To estimate the average dose in the axial plane $CTDI_w$ equation (2) is calculated from the average of measurements at the centre and the periphery of the phantom.

$$CTDI_w = {}^1\!/_3\, CTDI_{center} + {}^2\!/_3\, CTDI_{periphery} \quad (2)$$

$CTDI_w$ alone only tells us what the average dose in a plane would be given a series of contiguous axial scans (stepped table feed) or helical scans with a pitch of unity. For a helical scan with pitch not equal to unity, volume CTDI ($CTDI_{vol}$) can be calculated using equation (3). The $CTDI_{vol}$ can be multiplied by the scan length to give the Dose Length Product (DLP) as in equation (4) in order to estimate the total absorbed dose, which can then be related to the stochastic effects of radiation exposure and used as a first order estimate in public health monitoring.

$$CTDI_{vol} = CTDI_w / Pitch \text{ (mGy)} \quad (3)$$

Where pitch is the table increment per revolution as a fraction of the detector width (nT)

$$DLP = CTDI_{vol} \times L, \text{ where L is the scan length (mGy.mm)} \quad (4)$$

The CTDI concept starts to break down with increasing cone angle for a number of reasons; 1) the dose profile along the central axis becomes non-uniform, 2) the tails of the dose profile extend well beyond the 10cm chamber length, 3) the phantom is not sufficient to capture the entire beam width and scatter and 4) the weightings of central and peripheral dose in $CTDI_w$ are not always a good estimate of the average dose across the volume.

In the publications arising soon after the introduction of CBCT in radiotherapy practise these problems were largely ignored [15, 22, 23] and researchers employed "CTDI like" techniques to measure the CBCT dose. They continued to use a 10cm chamber in a cylindrical PMMA phantom with the same diameters as the body and head CTDI phantoms (32cm and 16cm) respectively but lengthened often by placing two or more CTDI phantoms end to end. In recognition that this only measured an average of a central 10cm portion of the central axis, Amer *et al* gave this measurement the term Cone Beam Dose Index (CBDI). To account for the beam length of 25 cm being longer than the standard 15cm CTDI phantom Amer *et al* added a 15cm thicknesses of PMMA at each end of the phantom.

One method to ensure that the entire dose profile is acquired, at least for Cone angles typical of diagnostic CT scanners, is to measure the dose at the centre of a lengthened CTDI type phantom using a 300mm long cylindrical ion chamber [24].

An alternative is to measure the point dose using a 0.6 cm^3 Farmer type chamber ,as suggested by Fahrig *et al* [25], in cylindrical PMMA phantoms [26-28]. Song *et al* measured the dose at the centre of two CTDI type (body and head) phantoms placed end to end and gave this measurement the term CBCTDI [27]. Sykes *et al* used the same technique to measure the dose for both the Elekta Synergy and Varian OBI systems [28]. Typical doses for the two systems, taken from Sykes *et al* are given in Table 1.

The equivalence of measuring the dose at the centre of a helical scan of length, L $D_L(z = 0)$ and $CTDI_L$ the integral dose for a single static slice between –L/2 and L/2 was demonstrated by Dixon *et al* [29]. Furthermore, they demonstrated the equivalence of $D_L(z = 0)$ measured at the centre of a helical scan (scan length = L and pitch = 1) and a CBCT scan (aperture a) where a = L. This was confirmed experimentally by Mori *et al* who used a photodiode stepped through the beam to measure profiles of various beam widths [30]. They showed that dose profiles acquired along the central axis of both helical CT and CBCT scans are equivalent. An alternative to the stepping diode which would require many repeat CBCT scans is the use of a CT dose profiler (RTI Electronics AB, Mölndal, Sweden) as demonstrated by Palm *et al* [31].

Current recommendations for dose measurements published by the International Atomic Energy Authority (IAEA) and American Association of Medical Physics (AAPM) are divergent [32, 33].

The IAEA, have followed the IEC 60601-2-44 report [34] which is a pragmatic approach that can be performed with current dosimetry equipment. They note that even for a 10mm wide beam the $CTDI_{100}$, measured at the centre of the phantom, only collects 82% and 63% of the dose for the head and body phantoms respectively [35]. This illustrates that the $CTDI_{100}$ was never as accurate as one might desire with doses under-estimated for long scan lengths and over-estimated for short scan lengths. However, the $CTDI_{100}$ accuracy stayed constant for beam widths between 10mm and 40mm and only decreased significantly for beam widths greater than 40mm. The IAEA recommend a two tier approach to the $CTDI_{100}$ with measurement for beam widths of less than 40mm following the existing method but for those greater than 40mm they exploit equation (5) which states that the CTDI for a beam width greater than 40mm is related to the CTDI for a beam width less than 40mm by the ratio of the $CTDI_{free\text{-}in\text{-}air}$ at the two beam widths.

$$CTDI_{100,(N\times T)>40} = CTDI_{100,ref} \times \left(\frac{CTDI_{free-in-air,N\times T}}{CTDI_{free-in-air,ref}}\right)$$

where, $CTDI_{100,ref}$ is the $CTDI_{100}$ measured in a phantom for the reference beam of $(N\times T)_{ref}$ using an integration of 100mm, N = number of detector rows and T is the thickness of a single detector row and where $(N\times T)_{ref}$ is typically 20mm, $CTDI_{free-in-air,N\times T}$ is the $CTDI_{free-in-air}$ for a beam width of $N\times T$, and $CTDI_{free-in-air,ref}$ is the $CTDI_{free-in-air}$ for the reference beam width. (5)

Table 1: Cone Beam dose measurements (similar to $CTDI_w$) for standard imaging protocols on the Varian OBI and Elekta Synergy CBCT systems published in the UK Centre of Evidence Based Purchasing report [28]. Manufacturers recommended protocol settings may change over time based on the introduction of new technology or feedback from customers.

Varian OBI Imaging Protocol	Dose (mGy)	Elekta Synergy Imaging Protocol	Dose (mGy)
Low Dose Head	2.8	Low Dose Head	1.4
Standard Dose Head	5.6	Medium Dose Head	5.4
High Quality Head	27.8	High Dose Head	9.4
Pelvis	24.9	Pelvis M10	15.3
Pelvis Spotlight	20.2	Pelvis M15	12.5
		Pelvis M20	13.7

The measurement of $CTDI_{free-in-air,ref}$ is itself measured in two tiers with a single chamber (100mm length) position being used for beam widths less than 60mm and using two or three positions, each stepped by 100mm to cover beam widths larger than 60mm.

In the AAPM report [33], task group 111 present the theoretical underpinnings of measuring dose in axial or helical fan- or cone-beam CT with table translation and that in stationary-phantom cone beam CT. For axial or helical scanning they note that there is an equilibrium dose constant which is independent of the collimation or the pitch. The dose for any particular scan can then be determined as the product of the equilibrium dose constant and a factor (pnT/a) where p is the pitch, nT the total width of the detector i.e n rows of width T and a is the width of collimation. This considerably reduces the number of measurements that need to be made as long as the collimation, a, is known. The equilibrium dose constant is the dose measured at the central scan plane (z = 0) for a scan of length L_{eq} and with pitch p=a/nT. L_{eq} has to be sufficiently large that further increment of L does not significantly increase the measured dose.

Since the 10cm long chamber does not cover the entire beam profile, for wide angle CBCT the 10cm chamber can underestimate the dose by 2-5% [36]. This is one reason why the AAPM TG-111 methods are based on measuring the point dose (e.g using a Farmer type chamber) in a geometrical phantom (typically cylindrical) that is sufficiently long to provide full scatter conditions for the irradiated scan length.

AAPM TG-111 also introduces the concepts of integral dose (E_{tot}) and planar average equilibrium dose ($\overline{D_{eq}}$) and show that $E_{tot} = \rho \pi R^2 L \overline{D_{eq}}$ where R is the radius of the volume, L is the scan length and ρ is the mass density of the phantom. The integral dose,

"serves as a simplified indicator of patient risk: the presumption is that cancer risk increases the larger the dose and irradiation volume containing radiosensitive tissue."[33]

The practicalities of measuring $\overline{D_{eq}}$, the average dose over the scan plane are discussed briefly. Recognising that when $D_{eq}(r) = A + Br^2$ i.e $D_{eq}(r)$ has a parabolic form, then measurement of $D_{eq}(r)$ at two points such as the centre and periphery of the phantom as conventionally measured for $CTDI_{vol}$, would be reasonable. Note, however that this leads to $E_{tot} = \frac{1}{2} \cdot D(r=0) + \frac{1}{2} \cdot D(r = R-1)$ instead of the more commonly used formula $CTDI_{vol} = \frac{1}{3} \cdot CTDI(r=0) + \frac{2}{3} \cdot CTDI(r = R-1)$. Recognising that $D_{eq}(r)$ does not always follow a parabolic form AAPM TG-111 note that more detailed measurement or the use of Monte Carlo modelling might be required. This would be the case for the large fields of view of the Elekta Synergy and Varian OBI and also the medium field of view of the Elekta Synergy system where the detector panel is shifted laterally to extend the field of view. This creates a central cylinder which is exposed from all 360° whilst material in the remaining volume is only exposed from 180°.

AAPM TG-111 note that measuring the free-in-air dose equilibrium pitch product, is an important measurement to make at commissioning as it can be used for quality assurance purposes to assess constancy of exposure and can be used to infer the equilibrium dose measured in a phantom given a scanner with the same phantom factor (ratio of dose equilibrium in phantom to free-in-air dose equilibrium).

While AAPM TG-111 presents the theoretical underpinnings of the measurement of CBCT dose, it does not offer much in the way of standardisation of CBCT dose measurement. The report suggests but does not dictate the use of a Farmer type chamber and it discusses various phantom designs with various dimensions and cross-sectional shapes (circle or ellipse) and different materials but does not make any recommendations for a standard phantom and chamber as with the CTDI concept. The methods used by Islam *et al,* Song *et al* and Sykes *et al* [26-28] are in many ways closely aligned with TG-111.

2.2. Calculation and measurement of population based organ dose and effective dose from CBCT imaging

The $CTDI_{vol}$ measurement is an estimate of the average dose in the central axial plane of the scan and is typically calculated as one third of the central dose and two thirds of the peripheral dose. This is independent of the scan length and therefore does not relate to the total dose to the patient and any risk of radiation induced malignancy. A commonly used and very simple method to relate $CTDI_{vol}$ to total or examination dose is to multiply by the length of the scan. This is known as the dose length product (DLP).

For a more accurate assessment of radiation risk, the dose to individual organs and their respective organ sensitivities are needed. The effective dose (7), measured in units of Sieverts (Sv), is a summation of tissue equivalent doses (6) and tissue-specific weighting factors defined by ICRP 103 [37]. The effective dose can be related to radiation risk using, for example, data published on the Biological Effects of Ionizing Radiation (BEIR) by the National Academies Concerning Radiations Health Risks [38].

Equivalent dose for tissue/organ T (Sieverts)

$$H_T = \sum_R W_R \cdot D_{T,R} \tag{6}$$

where W_R is the weighting factor for radiation type R and $D_{T,R}$ is the absorbed dose for tissue T by radiation type R

Effective dose (Sieverts)

$$E = \sum_T W_T \cdot H_T \tag{7}$$

where, W_T is the weighting factor as given by ICRP 103 [37] and H_T is the equivalent dose for tissue or organ type T

One method of measuring organ dose in order to calculate effective dose is to use an anthropomorphic phantom for CBCT. This has been performed by a number of groups using small radiation dosimeters such as thermo-luminescent dosimeters (TLD) [15,22,23,31,36,39,40], fibre optic coupled water-equivalent plastic scintillators [41], silicon-photodiode dosimeters [42] and MOSFETs [43, 44].

An alternative method of estimating the effective dose is to use the ImPACT CT Patient Dose Calculator [45]. The ImPACT Patient Dose Calculator, designed originally for fan beam CT, uses a library of Monte Carlo calculated dose calculations [46] for organ doses in a humanoid mathematical phantom. The library covers numerous commercial CT scanners each characterised by the ratio of peripheral to central $CTDI_w$ and central to in-air $CTDI_w$ for both the head and body phantoms. To calculate the effective dose the operator selects the scanner type and then using a diagram of the humanoid phantom, the start and stop position of the scan required to include the organs covered in a typical scan. The software will provide individual organ doses with their weighting factors and equivalent doses as well as the total effective dose.

Ideally when using the ImPACT calculator the CT scanner for which the dose is to be calculated will be one of the scanners in the ImPACT library. If not, then the CT scanner can be matched to the closest one in the library using ImPACT factors derived from a linear combination of the ratios of the central and peripheral normalised $CTDI_{100}$ to $CTDI_{air}$. This method has been employed by several authors to match a CBCT scanner with fan beam CT scanners in the ImPACT library [22, 23]. Hyer *et al* compared the results of the ImPACT dose calculator with Monte Carlo measurements (described in 2.3. [47]. They measured in-air $CBDI_{100}$ and both central and peripheral $CBDI_{100}$ in the CTDI phantom to determine appropriate ImPACT factors for matching acquisition protocols, designed for the head, chest and pelvis with their respective acquisition settings e.g. fields of view on both the XVI and OBI CBCT systems, to a scanner in the ImPACT calculator. Organ doses from the ImPACT calculator were compared against previously published Monte Carlo calculated organ doses [41]. Hyer *et al* found that many organs agreed within 40% with generally better agreement for the pelvis scan however, some discrepancies of more than 100% were found. These differences were attributed to: differences in the anthropomorphic phantom and the mathematical phantom; the use of the ImPACT calculator for calculation of CBCT dose when it is designed for fan beam CT dose calculation; the inability of the ImPACT calculator to cope with partial scans and finally, a probable calculation mistake identified in the calculation of the Thymus dose. They conclude that the ImPACT dose calculator is not suitable for calculating CBCT dose.

Gu *et al* have modelled both kV- and MV-CBCT systems using MCPNX [48] and applied these models to calculate organ doses to the VIP-Man phantom that was developed from the National Library of Medicine's Visible Human Project [49]. They concluded that the effective dose for the Head and Neck and prostate was 8.53 mSv and 6.25 mSv respectively for a 125kVp kV-CBCT exposure of 1350 mAs.

2.3. Calculation of patient specific dose from CBCT imaging

The above methods of calculating dose from CBCT are based on calculating the dose to a typical patient represented by either the head or body CTDI phantoms or the humanoid model of the ImPACT calculator. In the diagnostic world of radiation protection it may be sufficient to relate the dose for CT scan protocols used by a particular hospital to the radiation risk for the purpose of justification and for reporting dose. However, in radiotherapy where many CBCT scans might be performed during the course of treatment it may be necessary to calculate the dose to specific critical organs to ensure that the combined treatment and concomitant imaging dose does not exceed the dose criteria specified in the design of the treatment plan. In such cases individualised patient dose calculations may be required. There are two main research groups who have developed Monte Carlo (MC) models of CBCT imaging systems Spezi *et al* have developed a model for the Elekta XVI CBCT scanner using EGSNRC/BEAMNRC and BEAMPP [8,50–52]. Ding *et al* have developed their MC model of the Varian OBI CBCT scanner using the EGSnrc/BEAMnrc code in their early work [53-55] and more recently with the BEAM/DOSXYZnrc code [56-58]. Chow *et al* have also modelled Elekta XVI using BEAMnrc and a Matlab based implementation of DOSXYZnrc [59].

Both Ding *et al* and Spezi *et al* have computed the dose on CT scans for a number of patients with a variety of anatomical locations and imaged using standard acquisition settings as supplied by the manufacturers [8, 60]. Both papers present doses to various organs of interest. A detailed analysis of their results is not presented here however, Table 2 summarises the dose for three anatomical sites. Note, the performance of the two CBCT systems should not be judged on the basis of the data reported in Table 2 since there is no reason to assume that the acquisition settings have been optimised to give the same trade-off between imaging dose and the quality of the image. This is evident in the head and neck doses where the dose for the XVI system is an order of magnitude less than the OBI system. Both authors present a number of interesting observations that are generally applicable to understanding patient dose from CBCT. Ding *et al* noted that the bone receives two to four times that of the soft tissue. Spezi *et al* showed that the bony structures consistently received a higher dose than the soft tissue with the mandible receiving three times the dose received by the surrounding soft tissues. This was caused by the increased mass-energy absorption coefficient due to the photoelectric interaction within the materials of higher atomic number. Spezi *et al* also showed that the addition of the bow-tie filter significantly reduced the dose by 22% in the Pelvis and 45% in the Chest. This was due primarily to the attenuation of the dose to the peripheral tissues but also the beam hardening affect. The bow-tie filter also reduces the scattered dose from the periphery of the patient to the imager which has the additional advantage of increasing image quality. Ding *et al* also calculated the dose for paediatric patients and observed that paediatric doses were of the order of two times that of an adult.

Table 2: Monte Carlo calculated patient doses (cGy) for three anatomical sites for the Elekta Synergy CBCT system and the Varian OBI CBCT system. Doses reported are for the body i.e. not to a specific organ.

	Pelvis/Abdomen	Head and Neck	Chest
Elekta XVI (Spezi *et al*) [8]	1.5 - 2.1	0.1 - 0.2	1.2 - 2.2
Varian OBI (Ding *et al*) [60]	1 - 5	3 - 9	2 - 9

Downes *et al* showed that the CBCT imaging dose has a left-right asymmetry due to the increased number of exposures at the start and stop gantry angle as the gantry rotation accelerates and decelerates at the beginning and end of each scan [52]. Unlike CT imaging where the patient is normally central in the CT scanner, in radiotherapy the isocentre is typically set to the centre of the target volume which may itself be offset from the centre of the patient's cross-section. Chow *et al* studied this effect and found for the pelvis phantom, variation in the mean dose of up to 20% for up to 10cm anterior-posterior shifts [61]. Dose variations for the chest and head and neck were typically between 7% and 17%.

One of the advantages of using Monte Carlo, apart from getting individualised dose measurements, is that dose distributions can be calculated for a range of situations including medium and large fields of view with offset collimation, partial arcs and offset isocentres.

An alternative to using Monte Carlo is to use a model based dose calculation similar to that used in planning the mega-voltage treatment. Aleai *et al* first showed that the then ADAC treatment planning system (TPS) , (now the Pinnacle TPS; Philips Medical Systems), can be adapted to calculate the dose at kilo-voltage energies by adjusting the photon spectrum and adding low dose scatter kernels to the collapsed cone superposition dose calculation algorithm [62-65].This method has since been enhanced by Ding *et al* who added a medium dependent correction (MDC) to account for the increased photo-electric cross section of bone which leads to the increased bone dose previously observed with Monte Carlo effects [66]. The MDC algorithm also successfully accounts for upstream and downstream effects. The results show a significant improvement compared to the calculation using the collapsed cone superposition model alone and agreement with Monte Carlo measurements is good especially upstream and downstream of the bone.

2.4. Combining dose with treatment planning dose

For radiotherapy the risk of concomitant imaging needs to be considered in the context of the risk of secondary cancer induction from radiotherapy treatment. In addition the dose to critical organs already receiving high doses from the treatment needs to be assessed to ensure the additional imaging dose does not exceed dose limits. The imaging dose needs to be considered both within the treated volume and also peripheral to the volume.

Qiu *et al* calculated dose using Monte Carlo for relatively large volume gynae IMRT treatments with field length of ~15cm and for CBCT scans of length ~24cm) [67]. They concentrated on modelling the in-field dose, discussing out-of-field dose only briefly. In-field doses for organs at risk were calculated using Organ Equivalent Doses calculated using Linear, Linear-Exponential and Plateau radiobiological models. The greatest increment in dose from imaging one CBCT per fraction was 2.5% for the bowel with the linear model but this reduced to 1.3% for the plateau model. For dose in the peripheral region, the CBCT dose was compared with the linac scatter and leakage doses. In the peripheral low dose regions, where there is low risk of secondary malignancies, the incremental dose from CBCT was found to be an order of magnitude less than the IMRT scatter dose and less than or equal to the linac leakage dose.

Chow *et al* concentrated on in-field dose and compared CBCT dose with the treatment dose for a prostate IMRT case [59]. The PTV dose rose by 0.6 Gy (0.8%) for a 78Gy/39# treatment which suggests the CBCT dose was ~1.5cGy scan. The femoral heads saw the largest increase in dose of ~2.5cGy (5%).

Perks *et al* measured the peripheral dose at the centre and on the surface of an anthropomorphic phantom [44]. They measured dose for a prostate IMRT treatment using MOSFETs and a kV-CBCT using TLDs. The dose from the IMRT dropped from the prescription dose of 2Gy (per fraction) down to 1cGy at 16 cm and 0.4cGy at 21 cm distances from the field edge. In comparison, the CBCT was 0.5cGy and 0.2cGy at the same positions respectively (7cm and 12cm from the imaged volume). They used an S20 collimator which arguably provides a longer field of view than necessary for prostate IGRT. The nominal dose per scan was 6 cGy which they acknowledge was twice that normally used in their clinic. To put this into context 6 cGy is four times the UK's diagnostic reference level for imaging the abdomen/pelvis and is arguably 3-5 times higher than necessary for adequate image quality for CBCT image guidance [68].

Harrison *et al* published two papers on the subject of combined treatment and imaging doses covering anatomical sites of; larynx, breast and prostate [69, 70]. They compared imaging dose from 2D-portal imaging and 3D-CT imaging with the treatment dose. While this work was not based on CBCT the differences between CT and CBCT doses are likely to be minimal so the work provides a good perspective on the relative impact of kV-imaging on the combined treatment and imaging dose. For the prostate they measured dose using TLDs in the Rando phantom. Neutron doses were also calculated for the 15MV beams. They calculated the dose to multiple organs both in-field and out-of field for combinations of a 37 fraction two phase prostate treatment with 26 CT images and 4 portal images. The excess relative risk was found to be < 0.1 for most organs with bone marrow showing the greatest increase in dose of up to 20%. They employed similar methods for the larynx and breast and concluded that the dose to critical organs increased by 5-20% with increases of up to 30% for bone surfaces and bone marrow. They noted that by far the larger component of dose to these organs was from scatter and leakage from the MV beam.

3. Clinical Consequences

3.1. Detrimental effects of radiation exposure

To date there has been no large scale epidemiologic studies of the cancer risks associated with CT scans. The evidence we have is derived from measurement and calculation of organ doses and applying organ-specific cancer incidence or mortality data derived from studies of atomic-bomb survivors [71]. The estimated attributable life time risk of death from cancer is ~0.01% increasing to 0.1% for exposures in early childhood [71]. However, these risks are calculated for the general population and not specifically for patients undergoing treatment with radiotherapy. Therefore the risk of imaging alone is small compared to the risk of treatment failure and other morbidities associated with radiotherapy treatment. It appears sensible that the risk from the imaging dose should be incorporated into the overall risk calculation including the treatment dose.

The primary risk of radiation exposure from radiotherapy, including any concomitant imaging, (excluding the risks of treatment failure and co-morbidities etc), is the induction of a secondary primary malignancy (SPM). As Tubiana notes in his review paper [11], these rarely occur before 10 years after treatment. However, with increased long-term survival rates the incidence of these malignancies is likely to increase. Tubiana found from cancer registries that the incidence could be as high as 20%. He also noted that SPM's tended to occur in tissues receiving more than 2Gy. Data derived from the US SEER cancer registry by Berrington de Gonzalez *et al* found that 9% of 5 year survivors developed a solid tumour and that the relative risk was highest for tissues that typically received more than 5Gy.

From the previous section we know that the imaging dose is typically small in comparison to the treatment dose. Nevertheless it adds to the radiation dose burden and contributes to the increased risk of SPM induction.

Very low doses are also associated with complications. Perks *et al* collated a number of such effects in their paper including: prolonged azoospermia at doses > 2.5Gy [72]; loss of ovarian function at doses < 2Gy [73] and hypothyroidism or thyroid nodules with median doses as low as 0.09 Sv [74]. Cataract formation can also occur with increasing likelihood for doses less than 2Gy [75].

3.2. Clinical Benefits.

As yet there is little published evidence on the clinical benefits of CBCT imaging. Chow *et al* calculated the NTCP increase from CBCT imaging during IMRT treatment of the prostate to be 0.5% although they recognised that the NTCP model used was relatively crude and did not take into account the RBE of kV imaging [59]. Nevertheless, they found that NTCP decreased by 3% when the CTV to PTV margin was reduced from 10mm to 5mm showing a net benefit of using CBCT imaging for every fraction of treatment. Kron *et al* showed that even when daily online IGRT (CBCT) was used with an adaptive strategy for bladder cancer, using plan libraries, integral dose to the whole irradiated volume and irradiated volume less the CTV, was less than the conventional treatment except in the case of some smaller volumes [76]. This is because on average the irradiated volume is smaller in the adaptive strategy than the irradiated volume that would be required to ensure the bladder is covered the majority of the time. Zelefsky *et al* compared cohorts of patients in which one group received prostate intensity modulated radiotherapy with IGRT and the other group received the same treatment but without IGRT [77]. They found that biochemical tumour control was significantly better for patients with high risk prostate carcinoma when IGRT was employed. In addition, late urinary toxicity was almost halved in the group with IGRT. While this study was performed using MV imaging and gold seed markers Moseley *et al* have demonstrated the equivalence of CBCT and gold seed marker/MV imaging based IGRT [78]. Recently, Bujold *et al* reviewed the literature and concluded that IGRT has enabled treatments such as hypo-fractionated stereotactic ablative radiotherapy of the lung, spine and liver [79]. They also conclude that,

"an improvement in relapse rate in prostate cancer, Hodgkin disease, and head and neck cancers using IGRT has been consistently reported."

and that,

"There is a suggestion that prostate and head and neck cancer patients might have lower toxicity with IGRT, especially when combined with other technical advances like IMRT."

4. Discussion

In this article various methods of calculating, estimating and measuring patient dose have been reviewed and the clinical consequences and benefits discussed briefly.

There is now a wealth of data measuring dose for the two main manufacturers of CBCT systems both using CBDI type techniques and using TLDs in anthropomorphic phantoms. Unfortunately, much of this data is difficult to relate back to protocols used on systems in the clinic as typically no reference dose is provided. Downes *et al* and Islam *et al* [52, 80] both suggest normalising doses to a reference dose measured in air. However, there is no standard employed by these authors. Such a standard would make interpretation of these results easier. In the absence of a standard reference dose, the results from each of these papers would have to be compared by normalising with the product of the exposure time and tube current.

Both IAEA and AAPM have produced recommendations on CTDI type measurements however there are foreseeable practical difficulties in implementing both and no published literature on the application of these methods to either the XVI or OBI CBCT systems. The AAPM-TG-111 report requires measuring the equilibrium dose from which the dose at any collimator setting can be determined. It is likely that this quantity, normalised to the dose equilibrium free in air, is transferable between machines of the same manufacturer/model. If so this could reduce the number of measurements required to provide dose calculations for all fields of view. The AAPM report discusses suitable phantoms but they are likely to be either excessively heavy due to their size or require filling with water, neither of which is ideal. It favours a small Farmer chamber which suits the typically large cone angles (scan length) of these CBCT systems which is good in that nearly all radiotherapy centres will have a Farmer chamber but unless they also have a superficial X-ray unit they are unlikely to have suitable calibration factors to provide a dose traceable to a national standard. The IAEA approach has the advantage of requiring fewer measurements, by reducing the number required in a phantom and also only requiring a single CTDI phantom by choosing a short length reference collimation. However, on the Elekta Synergy CBCT system the shortest collimation for the small field of view is ≈13cm which does not fit within the 10cm chamber. Additional collimators would need to be manufactured to make this work. An additional complication is the use of the offset detector for medium and large fields of view. It could be argued that a suitably calibrated 10cm chamber is more readily available than a Farmer chamber but this is only likely to be in big radiotherapy centres with Medical Physics departments supporting Diagnostic Radiology as well as Radiotherapy. Smaller centres are likely to prefer the use of a Farmer chamber even if this requires additional calibration at low energies. However, the small chamber is not large enough or small enough to measure the CTDI for a narrow collimation required as a reference in the IAEA method.

To follow good practise such as defined in the UK IRMER legislation re optimisation of patient dose, imaging doses should be reported, image acquisition protocols should be optimised and the use of CBCT for IGRT should be justified on an individual patient basis. But what accuracy of dose measurement do we need? For reporting, maybe a CTDI type measurement would suffice or perhaps calculation of individual organ doses and effective dose for a population of patients using either TLD measurements in an anthropomorphic phantom or using a dose calculator such as the ImPACT is required or do we need to measure these doses for an individual patient using MC or a suitable kV dose model? On the one hand, the uncertainty in the current risk estimates for secondary primary malignancy induction are so large, that a CTDI type measurement would seem sufficiently accurate. Additional corrections for patient size could be used to refine the dose estimate. There would still be some residual uncertainties such as the variation of the dose due to the location of the isocentre within the patient and dose gradients due to partial arcs. On the other hand if the additional risks of concomitant imaging are to be fully understand maybe an accurate dose calculation including individual patient organ doses is required so that eventually the risks can be correlated, with sufficient specificity, to the delivered dose.

For quality assurance purposes and optimisation of imaging dose then a basic CTDI type dose calculation is probably sufficient. However, standardisation of the measurement is important if published techniques are to be compared. There is still a great deal of work required to both improve CBCT technology and its image quality capabilities to minimise the radiation dose required and understand the level of image quality required for image guidance which in some cases can be performed accurately by automatic image registration algorithms with much less dose than would be desirable for a human operator to perform manual image registration [12].

Finally, it is clear from the literature that there are identifiable risks attributable to radiation dose and that increased imaging is likely to increase those risks. However there still exists great uncertainty in estimates of risk for secondary malignancies from radiotherapy mainly due to the lack of epidemiological data and detailed understanding of the underlying mechanisms for cancer induction [81]. These uncertainties if converted back to dose are in many cases likely to be greater than the total concomitant imaging dose. This makes the process of justification difficult and inexact with the danger that those involved with prescribing radiotherapy and associated IGRT protocols become complacent and dismiss the imaging dose without due consideration. Extra care is required for justification of CBCT imaging of paediatric patients. However, in many cases, the likely benefits of performing IGRT using CBCT, including reduced CTV-PTV margins and reduced geographic miss, are likely to outweigh the detriments.

5. References

[1] Nakagawa K *et al* 2008 *Int. J. Radiat. Oncol. Biol. Phys.* **72** S623

[2] McBain C A *et al* 2006 *Int. J. Radiat. Oncol. Biol. Phys.* **64** 625-34
[3] Oldham M *et al* 2005 Cone-beam-CT guided radiation therapy: A model for on-line application *Radiother. Oncol.* **75** 271-8
[4] Richter A *et al* 2008 *Radiation Oncology* **3** 42
[5] Elstrøm U V *et al* 2010 *Acta Oncologica* **49** 1101-8
[6] Burridge N *et al* 2006 *Int. J. Radiat. Oncol. Biol. Phys.* **66** 892-7
[7] Korreman S *et al* 2010 *Radiother. Oncol.* **94** 129-44
[8] Spezi E *et al* 2012 *Int. J. Radiat. Oncol. Biol. Phys.* **83** 419-26
[9] IRMER Health and Safety Executive UK 2000 *The Ionising Radiation (Medical Exposure) Regulations 2000*
[10] Berrington de Gonzalez A *et al* 2011 *The Lancet Oncology* **12** 353-60
[11] Tubiana M 2009 *Radiotherapy and Oncology: Journal of the European Society for Therapeutic Radiology and Oncology* **91** 4-15
[12] Sykes J R *et al* 2009 *Phys. Med. Biol.* **54** 7263-83
[13] Moore C J *et al* 2006 *Phys. Med. Biol.* **51** 2191-204
[14] Mail N *et al* 2009 *Med. Phys.* **36** 22-32
[15] Sykes J *et al* 2005 *Radiother. Oncol.* **77** 45-52
[16] Bian J *et al* 2010 *Phys. Med. Biol.* **55** 6575-99
[17] Wang J *et al* 2009 *Med. Phys.* **36** 252-60
[18] Hill B *et al* 2005 *Med. Phys.* **32** 1589-1597
[19] Hill B *et al* 2008 *Physica Medica* **24** 149-158
[20] Kim S *et al* 2011 *J. Appl. Clin. Med. Phys.* **12** 84-95
[21] IEC 2003 *IEC 60601-2-44, CT safety standard* (Edition 2, amendment 1)
[22] Amer A *et al* 2007 *Br. J. Radiol.* **80** 476-82
[23] Sawyer L J *et al* 2009 *Br. J. Radiol.* **82** 577-84
[24] Geleijns J *et al* 2009 *Phys. Med. Biol.* **54** 3141-59
[25] Fahrig R *et al* 2006 *Med. Phys.* **33** 4541
[26] Islam M K *et al* 2006 *Med. Phys.* **33** 1573
[27] Song W Y *et al* 2008 *Med. Phys.* **35** 480
[28] Sykes J *et al* 2010 Evaluation Report *NHS Centre for Evidence-based Purchasing (UK)*
[29] Dixon R L 2003 *Med. Phys.* **30** 1272
[30] Mori S *et al* 2005 *Med. Phys.* **32** 1061-9
[31] Palm A *et al* 2010 *J. Appl. Clin. Med. Phys.* **11** 3085
[32] IAEA 2011 IAEA Human Health Reports 5
[33] AAPM 2010 Report of AAPM Task Group 111
[34] International Electrotechnical Commission 2009 Medical electrical equipment - Part 2-44: Particular requirements for the basic safety and essential performance of X-ray equipment for computed tomography
[35] Boone J M 2007 *Med. Phys.* **34** 1364
[36] Osei E K *et al* 2009 *J. Radiol. Prot.* **29** 37-50
[37] International Commission on Radiation Protection 2007 The 2007 Recommendations of the International Commission on Radiation Protection.ICRP Publication 103 *Annals of ICRP* **37**
[38] Committee to Assess Health Risks from Exposure to Low Levels of Ionizing Radiation 2006 Health Risks from Exposure to Low Levels of Ionizing Radiation, VII Phase 22 vol VII Phase
[39] Stock M *et al* 2012 *Radiother. Oncol* **102** 355-63
[40] Wen N *et al* 2007 *Phys. Med. Biol.* **52** 2267-76
[41] Hyer D E *et al* 2010 *J. Appl. Clin. Med. Phys.* **11** 181–97
[42] Koyama S *et al* 2010 *Med. Phys.* **37** 4298
[43] Kim S *et al* 2010 *AJR Am. J. Roentgenol.* **194** 186–90
[44] Perks J R *et al* 2008 *Radiother. Oncol.* **89** 304–10

[45] ImPACT and Keat N 2011 ImPACT CT Patient Dosimetry Calculator 1.04
[46] Jones D G and Shrimpton P C 1991 NRPB-R250. Chilton, UK: *National Radiological Protection Board*
[47] Hyer D E and Hintenlang D E 2010 *Med. Phys.* **37** 4620
[48] Gu J *et al* 2008 Assessment of patient organ doses and effective doses using the VIP-Man adult male phantom for selected cone-beam CT imaging procedures during image guided radiation therapy *Radiat. Prot. Dosimetry* **131** 431-43
[49] Xu X G *et al* 2000 *Health Phys.* **78** 476-86
[50] Spezi E *et al* 2009 *Med. Phys.* **36** 127
[51] Spezi E *et al* 2011 *Med. Phys.* **38** 5254-63
[52] Downes P *et al* 2009 *Med. Phys.* **36** 4156
[53] Nelson W R *et al* 1985 The EGS4 Code System
[54] Rogers D W O *et al* 2011 BEAMnrc Users Manual
[55] Kawrakow I 2011 The EGSnrc Code System: Monte Carlo Simulation of Electron and Photon Transport 2001-11
[56] Ding G X *et al* 2007 *Phys. Med. Biol.* **52** 1595-615
[57] Ding G X *et al* 2008 *Med. Phys.* **35** 1135-44
[58] Ding G X and Coffey C W 2010 *Phys. Med. Biol.* **55** 5231-48
[59] Chow J C L *et al* 2008 *Med. Phys.* **35** 52
[60] Ding G X and Coffey C W 2009 *Int. J. Radiat. Oncol. Biol. Phys.* **73** 6107
[61] Chow J C L 2009 *Med. Phys.* **36** 3512
[62] Alaei P *et al* 1999 *Med. Phys.* **26** 1687-97
[63] Alaei P *et al* 2000 *Med. Phys.* **27** 2821-6
[64] Alaei P *et al* 2001 *Med. Phys.* **28** 194
[65] Alaei P *et al* 2010 *Med. Phys.* **37** 244
[66] Ding G X *et al* 2008 *Med. Phys.* **35** 5312-6
[67] Qiu Y *et al* 2012 *Radiother. Oncol.* **104** 257-62
[68] Sykes J R 2010 PhD *Thesis: Quantification of Geometric Uncertainties in Image Guided Radiotherapy* (University of Leeds)
[69] Harrison R M *et al* 2007 *Br. J. Radiol.* **80** 989-95
[70] Harrison R M *et al* 2006 *Br. J. Radiol.* **79** 487-96
[71] Brenner D J and Hall E J 2007 *N. Engl. J. Med.* **357** 2277-84
[72] Howell S and Shalet S 1998 *Endocrinol. Metab. Clin. North Am.* **27** 927-43
[73] Wallace W H B, *et al* 2003 *Hum. Reprod.* **18** 117-21
[74] Imaizumi M *et al* 2006 *J. Am. Med. Assoc.* **295** 1011-22
[75] Chodick G *et al* 2008 *Am. J. Epidemiol.* **168** 620-31
[76] Kron T *et al* 2010 *Radioth. Oncol.* **97** 485-7
[77] Zelefsky M J *et al* 2012 *Int. J. Radiat. Oncol. Biol. Phys.* **84** 125-9
[78] Moseley D J *et al* 2007 *Int. J. Radiat. Oncol. Biol. Phys.* **67** 942-53
[79] Bujold A *et al* 2012 *Semin. Radiat. Oncol.* **22** 50-61
[80] Islam M K *et al* 2006 *Med. Phys.* **33** 1573-82
[81] Schneider U 2011 *Genes* **2** 1033-49

Development and characterization of a three-dimensional film stack dosimeter

TJ McCaw, JA Micka and LA DeWerd
Department of Medical Physics, University of Wisconsin, Madison, WI 53705, USA

E-mail: mccaw@wisc.edu

Abstract. A three-dimensional (3D) film stack dosimeter (FSD) using Gafchromic® EBT2 film was characterized for use in external-beam radiotherapy. The FSD was found to have negligible energy dependence and orientation dependence less than 2% using Monte Carlo simulations. Percent-depth-dose measurements with the FSD aligned parallel and perpendicular to the beam axis agreed with Monte Carlo simulations within 2%. Measurements of a ^{60}Co slit field with the FSD and thermoluminescent dosimeters agreed with a gamma passing rate of 97.6% using 1.5%/1.5 mm criteria. Edge artifacts were smaller than 2 mm, minimally affecting the usable measurement volume. The FSD is water equivalent, energy independent and orientation independent within measurement uncertainty at ^{60}Co energies, providing a 3D dosimeter that can be analyzed with a desktop scanner.

1. Introduction

Advancements in conformal radiation therapy procedures have increased the need for three-dimensional (3D) dosimeters to verify dose calculation algorithms and treatment deliveries. Fricke gel and polymer gel dosimeters have been extensively investigated and have shown promise as 3D dosimeters [1-4]. However, the accuracy of Fricke gel dosimeters is limited by post-exposure diffusion [5-7], and the response of polymer gel dosimeters depends on the free-oxygen concentration during preparation [8]. Additionally, these dosimeters require specialized readout techniques, such as magnetic resonance imaging or optical computed tomography, and are subject to edge artifacts [9]. The development of a stable, easy-to-analyze 3D dosimeter is essential and could result in broader clinical implementation of 3D dosimetry. The purpose of this work was to develop and characterize a 3D radiochromic film stack dosimeter (FSD) that could be read out with a standard desktop scanner for use in external-beam radiotherapy. The energy dependence, orientation dependence and water equivalence of the FSD were investigated.

2. Materials and methods

2.1. Phantom design

Gafchromic® EBT2 film (International Specialty Products, Wayne, NJ) was used in this work. A phantom housing to maintain alignment of the films was fabricated out of Virtual Water™ (VW) (Med-Cal, Verona, WI). The phantom (figure 1) is a truncated sphere with a central cavity that holds 22 circular films, 3.8 cm in diameter, with a radial clearance of 60 μm. The films are separated by 1 mm thick VW spacers for a total stack height of 2.7 cm. Two semicircular tabs on the outer diameter of each film fix the azimuthal orientation of the stack. A cylindrical phantom housing (CPH) with a

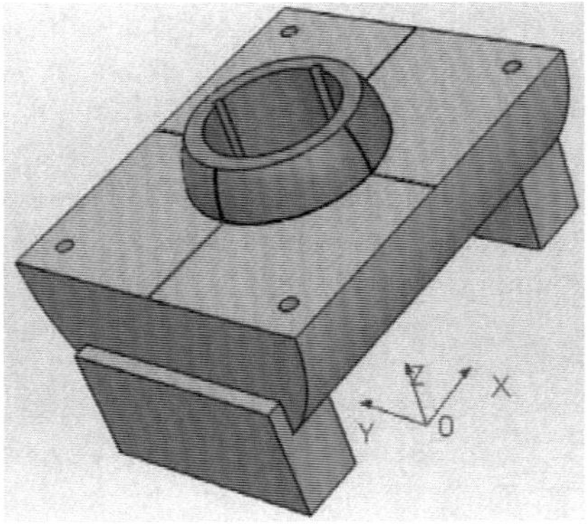

Figure 1: Cutaway schematic of the film stack dosimeter (FSD) phantom housing within the cylindrical phantom housing (CPH).

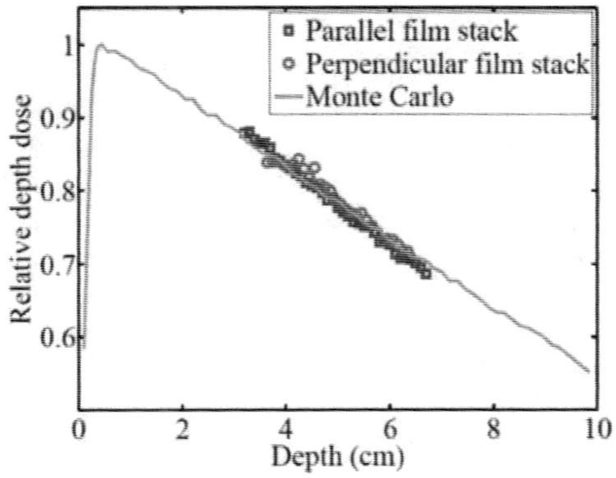

Figure 2: Percent-depth-dose profiles in a cylindrical water phantom simulated with MCNP5 and measured with the film stack dosimeter (FSD).

10 cm diameter and 12 cm length was also fabricated from VW. The CPH has a central cavity to accommodate interchangeable phantoms that house the FSD, thermoluminescent dosimeters (TLDs) and an Exradin A1SL ionization chamber (Standard Imaging, Middleton, WI), the latter two of which can be used to verify dose measurements made with the FSD.

2.2. Monte Carlo simulations

The response of EBT2 film to ionizing radiation is dependent on both the energy [10] and the relative orientation [11] of the incident radiation. The potential impact of these dependencies was explored with Monte Carlo (MC) simulations using MCNP5 v1.60. For all simulations and measurements, the coordinate axes were defined as shown in figure 1, with the beam axis parallel to the y axis. The simulations used a collimated photon point source with a ^{60}Co teletherapy spectrum.

The energy dependence was investigated by simulating the energy spectrum of the photon fluence through the film in the calibration and FSD geometries. The calibration geometry consisted of a (3x3) cm^2 piece of EBT2 film between two 5 cm thick VW slabs. The FSD was modeled at the center of the CPH with air gaps of 25 μm between films and VW spacers. An energy-binned cell flux tally, with bin widths of 25 keV, was used to determine the energy spectrum of the photon fluence through the calibration film and the top, middle and bottom films of the FSD. The orientation dependence of the FSD response was determined using an energy deposition tally to simulate the dose to a cylindrical volume parallel to the x axis in the central film. Simulations were completed with the source rotated about the central axis of the tally volume in 10° increments. The orientation dependence of film dosimeters is affected by the presence of air gaps around the film [11], so air gaps between films and spacers of 0, 25, 50 and 100 μm were considered.

The central-axis percent-depth-dose (PDD) profile in a water phantom from a (10x10) cm^2 ^{60}Co field at a source-to-axis distance (SAD) of 100 cm was simulated to compare the attenuation of the FSD with that of water. The size and orientation of the water phantom matched that of the CPH. The dose was tallied to cylindrical volumes at depth along the beam axis with radii of 0.5 mm and thicknesses of 0.1 mm. The tally volumes were spaced at intervals of 1.5 mm, decreasing to 0.5 mm near the depth of maximum dose.

2.3. Experimental verification

For experimental verification of the FSD, all exposures were performed with the University of Wisconsin Accredited Dosimetry Calibration Laboratory (UWADCL) Theratron 1000 ^{60}Co irradiator (Theratronics, Ontario, Canada). The UWADCL maintains a secondary absorbed dose-to-water standard, so use of the ^{60}Co irradiator reduced the uncertainty of the dose measurements used to characterize the FSD. The PDD profile of a (10x10) cm^2 field at a SAD of 100 cm was measured with the FSD oriented perpendicular and parallel to the beam axis for comparison with the orientation dependence simulations. The dose at a 5.6 cm depth in the CPH was measured using an A1SL ionization chamber to convert the MC-simulated PDD profile, normalized per source particle, to dose for comparison with the FSD measurements. The FSD was also used to measure the dose distribution from a 12 mm-wide slit field for comparison with a TLD phantom measurement. A slit collimator comprised of two 10 cm thick lead bricks was used. For each exposure, the respective dosimeter and phantom housing were placed within the CPH to ensure reproducible positioning.

2.3.1. Film calibration and analysis
The film response was calibrated and analyzed using the net optical density method described by McCaw *et al.* [12]. Sets of six (3x3) cm^2 pieces of EBT2 film were exposed to calibration doses of 0, 24.8, 36.6, 52.6, 76.2, 111, 160, 232, 337 and 489 cGy. Control films were used to account for background. All films were scanned individually in portrait orientation immediately prior to and one week after exposure using an Epson® Expression® 10000XL flatbed scanner (Epson America, Long Beach, CA) set to professional mode, 48-bit color and 72 dpi. The films were centered on the scan bed and lifted off of the glass surface using saturated radiographic films as masks. The film images were analyzed using MATLAB® (MathWorks, Natick, MA). Positioning of the films on the scan bed was confirmed to be reproducible within 1 pixel, so the films were compiled and registered without additional transformation to generate a 3D dose distribution.

2.3.2. TLD analysis.
LiF:Mg,Ti TLDs [Harshaw TLD-100 microcubes (Thermo Electron Corporation, Oakwood Village, OH)] were used to verify the FSD measurement of the ^{60}Co slit field. The TLD phantom holds 125 (1x1x1) mm^3 microcubes arranged in five crosses in each of five different slices parallel to the *x-z* plane. TLDs were annealed using the Cameron procedure [13] and stored for at least 24 h prior to exposure. The TLDs were read out with a Harshaw 5500 TLD automatic reader at least 24 h after exposure. The raw readings were corrected for background and non-linearity of the photomultiplier tube. Three sets of five TLDs were exposed to calibration doses of 5.00, 125 and 250 cGy. A linear fit was applied to the dose as a function of thermoluminescence.

3. Results
For energies below 500 keV, where the energy response of EBT2 film is greatest [10], the maximum difference in the simulated photon energy spectra for the calibration and FSD geometries was less than 0.5%. Overall, the greatest difference in the spectra for the two geometries was 1.4% at 1.175 MeV, where the energy response of EBT2 film is minimal [10]. For incident beam angles of 0° to 80° relative to the axis normal to the film plane, only the response in the 100 μm air gap geometry varied more than ±1%. For the parallel FSD geometry (i.e., 90° angle of incidence), the under-responses relative to the 0° orientation for air gaps of 0, 25, 50 and 100 μm were 0.9%, 1.4%, 1.7% and 2.5%, respectively. The air gaps in the FSD were measured to be approximately 20 μm. All simulations had an expanded ($k = 2$) Type A uncertainty of ±1%.

Because the PDD measurements with the FSD did not extend to the depth of maximum dose, the measurements were normalized to the maximum dose determined from the simulated PDD profile and the ionization chamber measurement. The measured and simulated PDD profiles are shown in figure 2. The FSD measurements agreed with the ionization chamber measurement within 1%. The depth dose values measured by the FSD in the perpendicular and parallel geometries agreed within 2%, verifying the results of the orientation-dependence simulations. Furthermore, the measured and simulated PDD profiles agreed within 2%, indicating the water equivalence of the FSD.

Select dose profiles at a depth of 4.1 cm in the CPH from FSD and TLD measurements of the ^{60}Co slit field are shown in figure 3. The dashed lines and error bars indicate the expanded ($k = 2$) overall uncertainty of the FSD (4.5%) and TLD (7.8%) measurements, respectively. The FSD and TLD data agreed within the measurement uncertainty. Additionally, the two dose distributions agreed with a

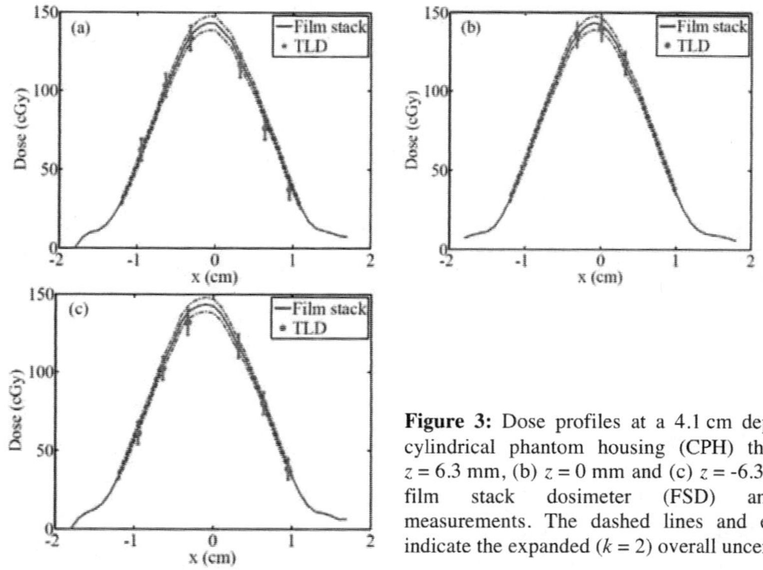

Figure 3: Dose profiles at a 4.1 cm depth in the cylindrical phantom housing (CPH) through (a) $z = 6.3$ mm, (b) $z = 0$ mm and (c) $z = -6.3$ mm from film stack dosimeter (FSD) and TLD measurements. The dashed lines and error bars indicate the expanded ($k = 2$) overall uncertainty.

gamma passing rate of 97.6% using 1.5%/1.5 mm criteria. Edge artifacts from the FSD measurements encroached less than 2 mm, permitting the use of smaller 3D detectors without loss of information.

4. Conclusions

A FSD using EBT2 film was developed and characterized. Within measurement uncertainty the FSD is water equivalent, energy independent and orientation independent at ^{60}Co energies. The measurement volume of the FSD is minimally affected by edge artifacts. Agreement with TLD and ionization chamber measurements verifies the accuracy of the FSD and analysis with a desktop scanner. The FSD is shown to be a valid 3D dosimeter for external-beam radiotherapy measurements. Future work will extend this investigation to higher-energy photon beams.

5. Acknowledgments

The authors would like to thank UWRCL and UWADCL customers, whose calibrations help support ongoing research at the UWMRRC.

6. References

[1] Kelly R G *et al* 1998 *Med. Phys.* **25** 1741-50
[2] Maryanski M J *et al* 1996 *Med. Phys.* **23** 699-705
[3] Baldock C *et al* 2010 *Phys. Med. Biol.* **55** R1-63
[4] Baldock C *et al* 1998 *Phys. Med. Biol.* **43** 695-702
[5] Pedersen T V *et al* 1997 *Phys. Med. Biol.* **42** 1575-85
[6] Harris P J *et al* 1996 *Phys. Med. Biol.* **41** 1745-1753
[7] Baldock C *et al Australas. Phys. Eng. Sci. Med.* **24** 19-30
[8] Maryanski M J *et al* 1994 *Phys. Med. Biol.* **39** 1437-55
[9] Oldham M and Kim L 2004 *Med. Phys.* **31** 1093–104
[10] Sutherland J G H and Rogers D W O 2010 *Med. Phys.* **37** 1110-6
[11] Suchowerska N *et al* 2001 *Phys. Med. Biol.* **46** 1391-7
[12] McCaw T J *et al* 2011 *Med. Phys.* **38** 5771-7
[13] Cameron J R *et al* 1964 *Health Phys.* **10** 25-9

Detecting VMAT delivery errors: A study on the sensitivity of the ArcCHECK-3D electronic dosimeter

S Arumugam[1], A Xing[1], G Goozee[1,2] and L Holloway[1,3,4]
[1]Liverpool and Macarthur cancer therapy centres and Ingham Institute, Liverpool Hospital, New South Wales, Sydney, Australia.
[2]University of New South Wales, Sydney, Australia.
[3]Institute of Medical Physics, School of Physics, University of Sydney, Sydney, New South Wales, Australia.
[4]Centre for Medical Radiation Physics, University of Wollongong, Wollongong, New South Wales, Australia.

E-mail: Sankar.Arumugam@sswahs.nsw.gov.au

Abstract. The sensitivity of the ArcCHECK 3D dosimeter in detecting VMAT delivery errors has been investigated. Dose and leaf positional errors of different magnitudes were introduced to whole arc and individual control points (CPs) of a simple open arc VMAT plan. The error introduced and error free plans were delivered and measured using the ArcCHECK device. The measured doses were compared against the treatment planning system calculated doses using gamma (γ) criteria with 2%/2mm and 3%/3mm tolerance levels. ArcCHECK effectively detected the dose errors resulting from MLC leaf positioning errors in limited CPs and Whole arc. For errors introduced to MU, ArcCHECK effectively detected the MU delivery errors in whole arc but not the MU errors introduced to CPs in integrated dose comparison.

1. Introduction

Modern radiotherapy techniques are highly complex in both planning and treatment delivery. The Volumetric Modulated Arc Therapy (VMAT) technique has been recently introduced and follows this trend [1, 2]. The complexity involved in the planning and delivery stages of VMAT necessitates the need for pretreatment plan specific dose verification using 3D dose measurements [3].

Gel dosimeters offer high resolution 3D dose measurements, but at present they are not in widespread clinical use due to various factors such as their high sensitivity to variations in chemical composition, limitations in readout methods, limited commercial availability and lack of user experience [4]. In recent years at least two electronic dosimeters, which enable semi 3D dose measurements, have been commercially introduced [5, 6]. ArcCHECK is one such device which uses semiconductor diode detectors arranged in a helical pattern in a cylindrical phantom. Feygelman et al [6] studied the characteristics of the ArcCHECK device extensively and reported its suitability and limitations for VMAT dose verification. However, to our knowledge, there is little evidence in the literature about the ability of the ArcCHECK to detect errors in a VMAT delivery. Yang et al studied the sensitivity of ArcCHECK in detecting systematic MLC leaf positioning error in Intensity Modulated Radiation Therapy (IMRT) and VMAT delivery by introducing errors to IMRT beams and VMAT arc [7] . In this work we have investigated the sensitivity of ArcCHECK in detecting MLC leaf positional and Monitor Units (MU) delivery errors in VMAT at both whole arc and individual control-point (CP) levels.

2. Materials and Methods

The Pinnacle Treatment Planning System (TPS), V9.0,(Philips Ltd, USA) was used to generate the VMAT plans considered in this study and described below. An Elekta-Synergy accelerator was used to deliver the VMAT plans using 6 MV photon beams and the ArcCHECK 3D dosimeter (Sun Nuclear Corporation, USA) was used for the dose verification. SNC Patient dose analysis software (Sun Nuclear Corporation, USA) was used to compare measured and planned dose matrices. The ArcCHECK dosimeter used in this study included the central cylindrical insert and all measurements were performed with the cylindrical insert in place. The absolute dose calibration procedure recommended by the manufacturer was followed to calibrate the device.

2.1. Simple VMAT plan

To study the sensitivity of ArcCHECK in detecting VMAT delivery errors, a treatment plan was generated on a digital CT representing ArcCHECK. The plan consisted of a single arc with gantry starting at 250^0 and ending at 110^0 and 111 CPs with 2^0 gantry spacing between them. The aperture size for each CP was set to 5 x 10 cm^2 and equal weights were assigned to all CPs. The ability of ArcCHECK to detect dosimetric errors was tested by introducing known errors of different magnitude to the planned VMAT arc in two steps:

In the first step, a range of dose errors 1%,-3%, 5%,-7% and -10% was introduced to the entire VMAT arc by changing the MU of the arc. Similarly a MLC leaf positional error was introduced by offsetting the MLC leaf banks by different magnitudes. The MLC leaves of both X1 and X2 banks were offset by 1 mm, -2 mm, 3 mm, -4 mm and 5 mm.

In the second step, dose and MLC leaf positional errors were introduced to two CPs of the arc. Two CPs at gantry angle 320^0 (Error ECP1) and 40^0 (ECP2) were chosen so that in the measurement of ECPI, the exit error would not be measured by ECP2. A dose error ranging in magnitude from 5% to 100 % was introduced to the weight of ECP1 and ECP2. Similarly an MLC error ranging in magnitude from 1 mm to 10 mm was introduced to the X1 and X2 leaf banks at ECP1 and ECP2 respectively. Dose and leaf position errors in the positive direction were introduced to ECP1 and in the negative direction were introduced to ECP2.

In both of the above mentioned steps the MLC leaf position and monitor unit weight errors were applied separately to the treatment field. The treatment field with no error and with the above mentioned errors was measured using ArcCHECK. The measured dose matrices were compared against the planned dose (with no error) to study the ability of ArcCHECK in detecting errors.

2.2. Dose matrix analysis

The calculated and measured dose matrices were analysed using gamma (γ) analysis [8] with a γ tolerance of 1. The γ analysis was performed with both 2%/2mm and 3%/3mm dose difference and distance to agreement tolerance levels. The percentage points passing γ criteria was analysed to assess the agreement between plan and measured dose matrices.

3. Results and Discussion

Table 1 shows the percentage points passing γ tolerance for the simple VMAT plan delivered with and without error. The ArcCHECK measurement showed good agreement with the calculated dose for the plan delivered without any error (96.5% and 100% of the points pass 2%/2mm and 3%/3mm tolerance levels respectively). For the plans delivered with MLC error to the whole arc, the pass rate decreased as the MLC error increased above tolerance level used in gamma analysis. This shows the ability of ArcCHECK to detect systematic errors in MLC leaf positions for VMAT delivery. Similarly the pass rate decreased as the magnitude of dose error introduced to the whole arc increased. However, the pass rate for delivery with positive dose errors was consistently reduced when compared to the negative dose errors although the magnitude of the negative errors was higher (table 1). This can be explained by a detailed analysis of the measured dose for the delivery without error. Figure 1 shows the dose profile for the planned and measured dose along the X axis and % difference between them. The dose measured at the beam entry points (distance between -205 mm and 205 mm in figure 1) varies from -1.6 % to 1.2 % whereas the dose measured at the exit points (distances from -325 mm to -205 mm and 205 mm to 325 mm in figure 1) is always higher, on average 0.8%, compared to the plan dose. Our Pinnacle beam model commissioning data showed that the percentage depth dose (PDD) calculated by Pinnacle for depths between 20 cm and 30 cm (ArcCHECK exit dose measurements are at 24 cm) agreed within ± 0.3% with ion chamber measurements in water. The high exit dose in the ArcCHECK measurement could be attributed to the difference in diode characteristics for the exit beam [9]. This inherent error in exit dose further increases the difference in measurements with positive error in dose delivery and thus decreased the pass rate of measurement points [10]. For the delivery with negative dose errors the opposite occurs thus increasing the pass rate of the measurement points.

Table 1: % points passing γ tolerance for 'simple VMAT plan' delivered with and without error

		% points passing γ			% points passing γ	
		γ Criteria			γ Criteria	
Error Scenario	Dose error (%)	2%/2 mm	3%/3 mm	MLC error (mm)	2%/2 mm	3%/3 mm
	No error	96.5	100	No error	96.5	100
	1	85.2	100	1	96.6	100
	-3	96.9	100	2	92.9	100
Whole arc	5	21.8	49.8	3	83.3	96.5
	-7	60.1	65.4	4	79.2	95.0
	10	13.5	19.7	5	70.6	87.6
	5	96.2	100	1	96.5	100
	7	95.5	100	2	96.4	100
Control point (CP)	10	95.6	100	3	95.5	99.6
	20	94.5	100	4	94.8	100
	40	94.1	100	5	95.2	100
	70	94.1	100	7	90.2	99.6
	100	94.0	100	10	90.8	99.3

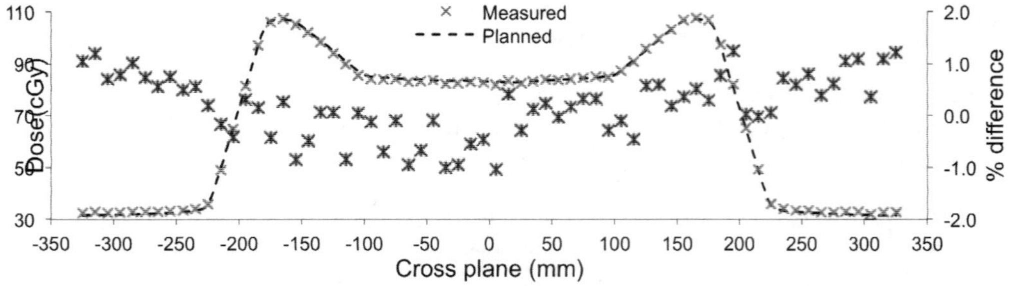

Figure1: Planned and measured dose profiles along the X direction for the simple plan delivered without error

For the plans with MLC errors introduced to CPs, the pass rate for the 2%/2 mm tolerance decreased only by 1.3 % between the plans delivered with no error and 5 mm error. As the error value increased further, larger changes (5.1% at 7 mm and 5.7 % at 10 mm) in the pass rate were observed (table 1). For the range of MU weight errors introduced to the CPs, only a very small change in the pass rate was observed. A decrease of 1.5% in the pass rate was observed between the plan with no error and the plan with a 100% weight error. Further, the pass rate for the 3%/3 mm tolerance was decreased only by 0.7% for the 10 mm error in MLC leaf position and no change in pass rate was observed for plans delivered with dose errors. The dose errors resulted from MLC leaf position and MU weight error are less than 3%, the tighter tolerance level detects these errors (table 1).

The delivery of clinical VMAT plans is far more complex than the simple plan investigated in this study. Further the accuracy of dose calculations performed by the TPS in VMAT calculations depends on the factors such as MLC leaf travel distance and gantry step size between CPs. Due to finite discretisation in arc calculations large MLC and gantry steps between CPs results in less accurate dose calculations [2, 6]. Feygelman et al [6] demonstrated the impact of the discretisation effect on the measurement plane of ArcCHECK and reported that ArcCHECK showed consistently low pass rate at the low tolerance levels compared to other detector systems. In this study we used 2^0 gantry spacing between CPs for the dose calculations of simple VMAT plan. In agreement with Feygelman et al we also observed that for measurement without delivery errors a high pass rate (100 %) was observed with 3%/3 mm tolerance and a relatively low pass rate (96.5 %) was observed with 2%/2 mm tolerance. In our study we simulated MLC leaf positions and MU delivery errors that likely to happen due to miscalibration and random misbehaviour of the treatment unit. ArcCHECK detected the MLC leaf position errors (table 1) but delivery errors resulting from errors in CP MU delivery was not detected by the device. One of the advantages of electronic 3D dosimeters over other integrated type 3D dosimeters is that electronic 3D dosimeters have high temporal resolution; enabling the verification of 3D doses measurements at individual CPs as well as integrated dose verification. In this study we investigated the ability of the ArcCHECK to detecting VMAT delivery errors using integrated dose measurements. In the next phase of this study we will investigate the ability of CP dose analysis to detect VMAT delivery errors and the impact of CP delivery errors on the dose delivered to clinical Regions of Interest (ROIs).

4. Conclusion

The ArcCHECK dosimeter was shown to effectively detect dose errors resulting from MLC leaf positioning errors. This included errors introduced to the whole arc and errors introduced only to two CPs of a simple single arc VMAT plan. For the simple single arc VMAT plan the ArcCHECK dosimeter only detected MU delivery errors occurring over the whole arc, not MU errors introduced to CPs using an integrated dose comparison. Future investigations will assess the effectiveness of individual CP dose analysis using ArcCHECK measurements for detecting these delivery errors and the impact of these errors on the dose to clinical ROIs.

5. References

[1] Otto K 2008 *Med. Phys.* **35** 310
[2] Webb S and McQuaid D 2009 *Phys. Med. Biol.* **54** 4345
[3] Ezzell G A *et al* 2003 *Med. Phys.* **30** 2089
[4] Baldock C *et al* 2010 *Phys. Med. Biol.* **55** R1-63
[5] Li JG *et al* 2009 *J. Appl. Clin. Med. Phys.* **10** 62
[6] Feygelman V *et al* 2011 *J. Appl. Clinl. Med. Phys.* **12** 146
[7] Yang W *et al* 2012 *Med. Phys.* **39** 3787
[8] Low D A and Dempsey J F 2003 *Med. Phys.* **30** 2455
[9] Jornet N *et al* 2004 *Med. Phys.* **31** 2534
[10] Hill R et al 2005 *Phys. Med. Biol.* **50** N331-44

Investigation of 3D dosimetry for an anthropomorphic spine phantom

R Grant[1,2], G Ibbott[1], J Yang[1], J Adamovics[3] and D Followill[1]
[1]Department of Radiation Physics, University of Texas MD Anderson Cancer Center, 1515 Holcombe Blvd., Houston, TX 77030 USA
[2]University of Texas at Houston Graduate School of Biomedical Sciences, Houston, TX 77030 USA
[3]Heuris Pharma, LLC, 412 Sunset Rd., Skillman, NJ, 08558 USA

E-mail: gibbott@mdanderson.org

Abstract. A new dosimetry insert for the Radiological Physics Center's spine phantom was designed to hold a specially molded dosimeter. The phantom was irradiated with the traditional insert loaded with radiochromic film and TLD, and then with the new 3D dosimetry insert. A comparison with the calculated dose distribution showed that PRESAGE® dosimeter, as well as the film and TLD system, agreed to within ±2mm. Further analysis of the 3D dosimeter, including a measured dose volume histogram, demonstrated the advantages of 3D dosimetry in a clinical environment.

1. Background

The ability of 3D dosimetry to obtain a large amount of dose information in a single irradiation can be advantageous in the clinic [1]. For areas in which dose margins must be closely monitored, such as spinal metastases, the availability of 3D dose information can allow the physicist to investigate high dose gradients and other regions of interest rather than be limited to measurements in selected planes which are most often restricted to the target center.

The Radiological Physics Center (RPC) has designed and implemented an anthropomorphic spine phantom that currently uses radiochromic film and thermoluminescent dosimeters (TLD) to evaluate treatments of spinal metastases [2]. The phantom contains a target located anterior to the vertebrae and avoidance structures simulating the spinal cord and esophagus. Two TLDs are located in the target. A piece of radiochromic film bisects the center of the target in the sagittal plane and a second film is located at the matching edge of the vertebra and the target in the coronal plane. These dosimeters allow for evaluation of the dose distribution along only these two planes. A 3D dosimeter encompassing the entire section of the spinal column allows for complete visual evaluation of the dose volume delivered and the ability to choose 2D planes for further analysis. For this study, we are using PRESAGE® [3], a solid tissue-equivalent [4, 5] radiochromic plastic and an optical-CT system for readout.

2. Methods

A second dosimetry insert for the phantom was created to hold a specially molded PRESAGE® dosimeter. The dosimeter matches the location of the TLD and film in the original insert and is shown in figure 1.

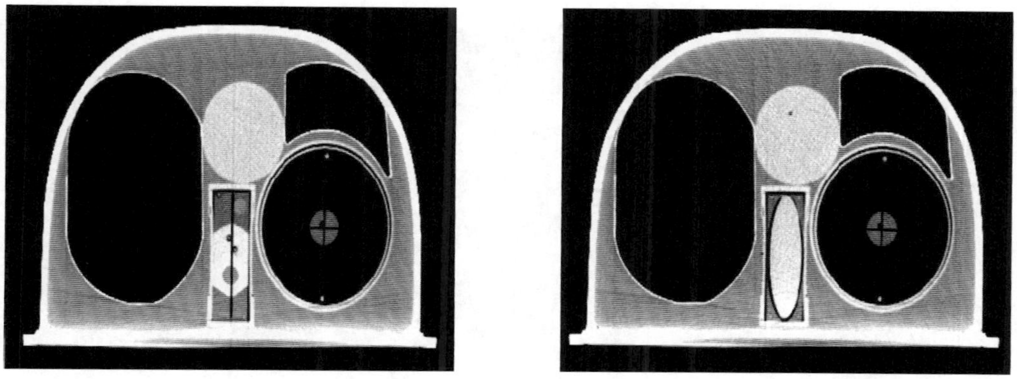

Figure 1. (A) Spine phantom with the original TLD and film insert. (B) Spine phantom with new PRESAGE® insert.

The phantom was CT imaged with each insert and the images were imported to the treatment planning system (TPS, "Pinnacle", Philips Radiation Oncology Systems, Fitchburg, WI). The target and the organs at risk were contoured. An intensity-modulated radiation therapy (IMRT) plan was created with a prescription of 6 Gy to 90% of the target with appropriate constraints to the normal structures.

The plan was delivered to the phantom twice; once with the TLD and film insert and once with the 3D dosimetry insert. The film was scanned with a CCD microdensitometer and while the PRESAGE® was analyzed with an optical-CT system, reconstructed to a 2 mm slice width.

The measured dose distributions were compared to the treatment plan calculated dose distribution using RPC in-house developed software or the Computational Environment for Radiotherapy Research (CERR) [6]. Film and PRESAGE® dose profiles were taken across several planes and compared for agreement. The distance to agreement (DTA) between the measured data and treatment plan, within the high dose gradient region, was quantified. The dose line profile through other areas of interest was also quantified. Dose volume histograms of structures within the PRESAGE® volume were calculated.

3. Results

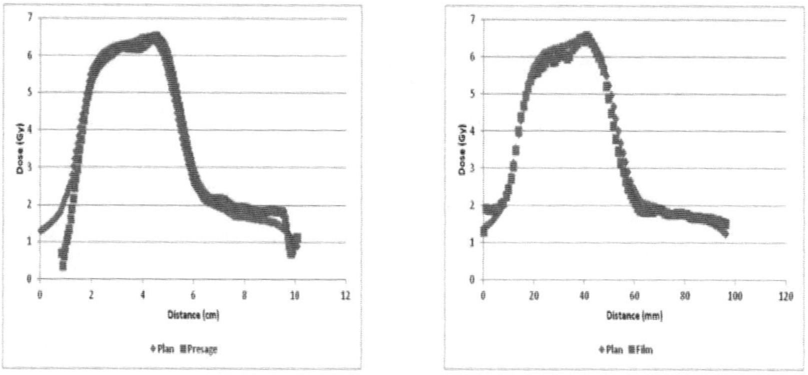

Figure 2: Dose Profiles comparing the treatment plan with PRESAGE® measurements and film measurements respectively.

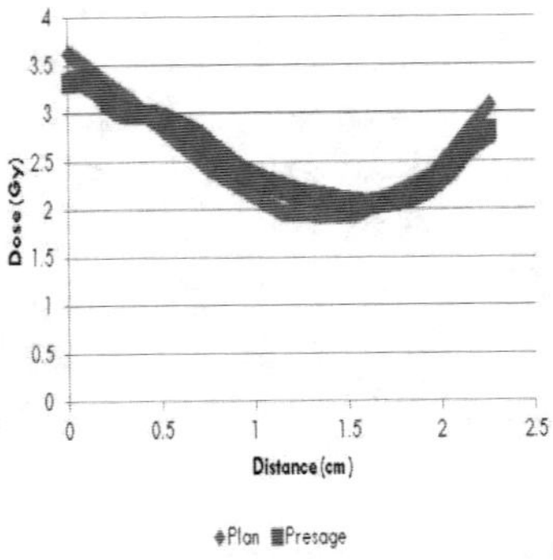

Figure 3: Dose Profile through the spinal cord.

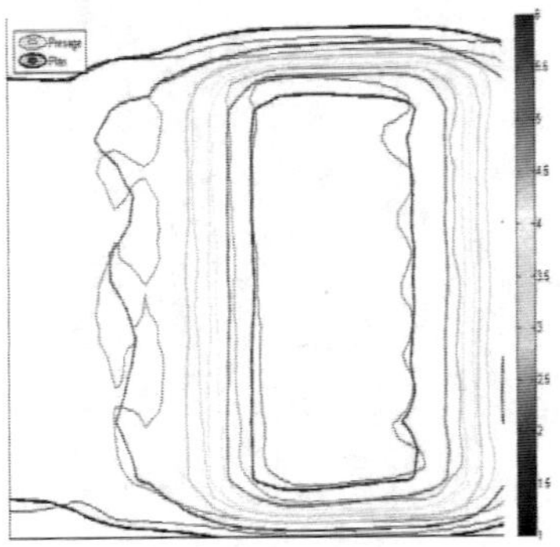

Figure 4: Isodose Plot through the sagittal plane

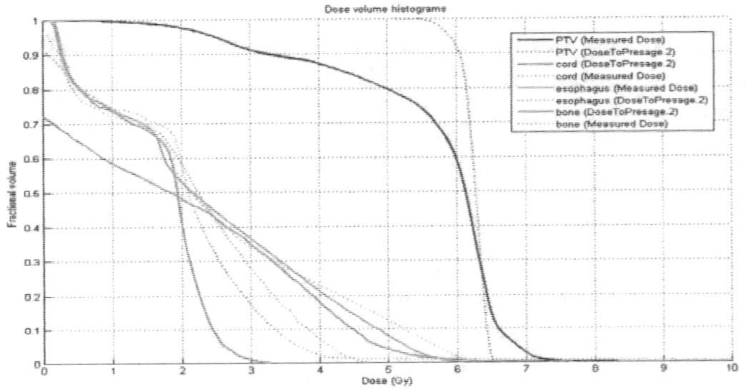

Figure 5: Measured and Calculated DVH for structures that fall inside the 3D dosimeter

The dose profiles show agreement within 2 mm in the anterior-posterior direction and within 1 mm in the superior-inferior direction using the PRESAGE®. The film agreed within 2 mm across all profiles. Representative dose profiles are shown above in figure 2.

Line profiles can also be taken across any area of interest as seen in figure 3 where a profile through the spinal cord shows agreement within 2 mm. Figure 4 is an isodose plot in the sagittal plane showing agreement with the treatment plan in a plane not covered by the film measurement. Figure 5 shows a dose volume histogram (DVH) comparing the calculated dose plan to a fully measured dose volume.

4. Discussion and Conclusions

The PRESAGE® dosimetry system shows great promise in phantom studies producing a measured volumetric dose set. Comparisons with film measurements confirm the 3D dosimeter is recording the expected dose values. Areas away from the traditional film planes were analyzed to demonstrate the comparatively large amount of measured data from a single irradiation. The major difference in dose coverage for the PTV from the DVH is attributed to the dosimeter not fully covering the PTV calculated dose. This measured DVH is only attainable with a volumetric dosimeter and can be advantageous in a clinical environment.

5. References

[1] Baldock C *et al* 2010 *Phys. Med. Biol.* **55** R1-R63
[2] Followill D S *et al* 2007 *Med. Phys*. **34** 2070-6
[3] Adamovics J and Maryanski M J 2006 *Radiat. Prot. Dosimetry* **120** 107-12
[4] Brown S *et al* 2008 *Appl. Radiat. Isot.* **66** 1970-74
[5] Gorjiara T *et al* 2011 *Med. Phys*. **38** 2265-74
[6] Deasy J O *et al* 2003 *Med. Phys*. **30** 979-85

4D dosimetry and its applications to pre-treatment quality control and real-time *in vivo* dosimetry of VMAT treatments

F Nordström, S af Wetterstedt and SÅJ Bäck
Department of Radiation Physics, Skåne University Hospital, Sweden
Department of Medical Radiation Physics, Lund University, Malmö, Sweden

E-mail: Fredrik.Nordstrom@med.lu.se

Abstract. In this study, a 4D dosimetry concept was developed. This concept included a method for calculation of 3D reference absorbed dose matrices at every control point of the delivery using a clinical treatment planning system (TPS). Further, the gamma evaluation method was extended to incorporate the 4th dimension of the TPS calculated dose distributions. The applications of the 4D dosimetry concept on pre-treatment quality control and real-time *in vivo* dosimetry were investigated.

1. Introduction

The rapid development and introduction of new treatment modalities (e.g. intensity modulated radiotherapy, IMRT, and volumetric modulated arc therapy, VMAT) has made it possible to deliver highly conformal, individually-shaped dose distributions. However, because of the rapid implementation of new technologies, there is widespread concern that current quality assurance programs do not provide adequate safeguards against treatment delivery errors [1].

Pre-treatment measurements have been widely employed as a part of routine patient-specific quality control (QC) of intensity-modulated treatments [2]. Because VMAT plan delivery involves continuous gantry movement, there are usually no plan-specific control measurements of anything other than the entire composite dose distribution and it is therefore possible that potential delivery errors at certain gantry angles are blurred. By analyzing VMAT delivery at a control point level, both systematic and plan-specific errors that are not visible in the composite plan might be identified.

Fast 4D dosimetry systems could also be used for real-time *in vivo* dosimetry. By allowing the *in vivo* system to interrupt the treatment if the measurements are outside pre-defined tolerance levels, serious overexposures to patients can be avoided. It is therefore possible that a real-time *in vivo* system could provide such confidence that pre-treatment quality control measurements can be reduced.

The aim of this study was to develop tools for measurement and evaluation of VMAT treatments on a control point level. Applications for both pre-treatment quality control and to real-time in-vivo dosimetry were investigated. The gamma evaluation method was extended to incorporate a 4th dimension (gantry angle interval) of the treatment planning system (TPS) calculated reference data.

2. Material and methods

The process of 4D dosimetry was divided into four different steps; extraction of 4D treatment planning system reference absorbed dose data, time resolved measurement, mapping of measured and calculated data, and comparison of measurements and calculations.

2.1. *Extraction of 4D TPS reference dose data*

In this study, a clinical version of the Eclipse TPS (Varian Medical Systems, Palo Alto, CA) was used for calculation of absorbed dose matrices in between each two consecutive control points (representing a small gantry angle interval) of VMAT treatment plans. Since the system only is able to calculate 3D dose matrices for each beam, the plan was exported to in-house developed MATLAB (MathWorks) software, where it was divided into a number of sub-arcs. The modified DICOM-RT plan files were imported back into Eclipse and 3D dose matrices were calculated using the Anisotropic Analytical Algorithm (version 10.0.0.28) for each imported beam (sub-arc).

2.2. *Time resolved measurement*

Two different detector systems were investigated in this study; the Delta[4] system (Scandidos AB, Sweden) and a prototype diode in-vivo dosimetry system integrated with the RayPilot real-time positioning system (Micropos Medical AB,). The Delta[4] system consists of a cylindrical polymethyl-methacrylate (PMMA) phantom with two orthogonal detector planes containing a total of 1069 p-Si diodes and has become a popular tool for pre-treatment QC of the total composite dose distribution of VMAT treatments. A research version of the Delta[4] software allowed for export of time stamped dosimetry readings. The RayPilot system is a clinical interstitial electromagnetic (EM) transponder system that can be used to monitor intrafractional prostate motion. A small p-i-n diode dosimeter was included in this device, which potentially could allow monitoring of target absorbed dose *in vivo* with little extra effort. The dosimetry readings from the two systems were analyzed using an application software developed in-house with support for 4D dose distributions.

2.3. *Mapping of measured and calculated data*

For VMAT plans, the gantry angle is directly related to the control point reached in the delivery. A gantry angle recording simultaneous with the cumulative detector reading can therefore be used to map the measured absorbed dose to the TPS calculated dose at the corresponding control point. Two different inclinometers (N62u, Nordic Transducer, Denmark and IK360, SIKO GmbH, Germany) were investigated for independent read-out of the gantry angle during delivery. Also, the use of gantry angle readings from DynaLog files, generated by the MLC computer, was investigated. The standard error of the mean in gantry angle readout was determined through the use of an alternating beam on/off pattern during continuous rotation for both a clockwise and a counterclockwise arc. The gantry angle was recorded simultaneous to read-outs of the beam's on/off status.

2.4. *Comparison of measurement and calculation*

The gamma evaluation method [3] was extended in order to incorporate the 4th dimension (gantry angle interval or time) of the reference data. In addition to the dose difference criterion (ΔD) and the distance-to-agreement criterion (Δd), a new time difference related criterion was introduced (Δt). For rotational treatments (e.g. VMAT), the gantry angle difference can be used as the Δt-criterion. The extended method (hyper-gamma, γ_H) simultaneously takes the dose difference, the distance in the Cartesian space and the time difference into account (figure 1) and is defined as:

$$\gamma_H(\vec{r}_m,t_m) = \min\{\Gamma_H(\vec{r}_m,t_m,\vec{r}_c,t_c)\}\forall(\vec{r}_c,t_c) \quad (1)$$

where

$$\Gamma_H(\vec{r}_m,t_m,\vec{r}_c,t_c) = \sqrt{\frac{\tau^2(t_m,t_c)}{\Delta t^2} + \frac{r^2(\vec{r}_m,\vec{r}_c)}{\Delta d^2} + \frac{\delta^2(\vec{r}_m,t_m,\vec{r}_c,t_c)}{\Delta D^2}} \quad (2)$$

As in the original gamma evaluation method, the evaluated point passes the test if $\gamma \leq 1$.

Journal of Physics: Conference Series **444** (2013) 012021 doi:10.1088/1742-6596/444/1/012021

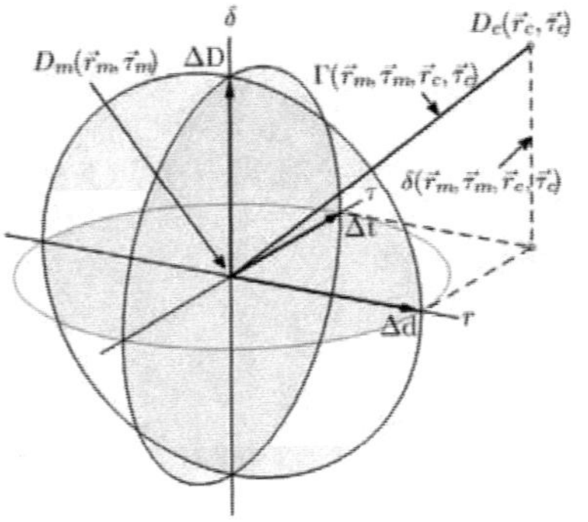

Figure 1: A graphical representation of the extended gamma evaluation method.

2.5. Clinical applications

The Delta[4] system was used to evaluate 20 clinical VMAT arcs on a control point level delivered using a Clinac iX (Varian Medical Systems, Palo Alto, CA). Each control point was analyzed using the hyper-gamma evaluation method with [3%/3mm/0.5°] and [2%/2mm/0.5°] criterion. The plans had been previously verified using the Delta[4] system and the traditional gamma evaluation method as a part of the clinical pre-treatment verification process. Comparisons between 3D and 4D evaluations were undertaken.

Measurements with the prototype EM-positioning and dosimetry system in an anthropomorphic phantom with an open water container were undertaken to evaluate the feasibility of using an implantable radiation dosimeter and EM localization system for real-time verification of VMAT of the prostate. Internal intrafractional motion was simulated using a programmable motion stage. The measured cumulative absorbed dosed were analyzed using the hyper-gamma method ([3%/3mm/1°]), as well as by dose difference comparisons between the measurements and the corresponding TPS calculated dose corrected for the recorded localization of the detector. The following intentional errors were introduced to investigate the detectability of errors using the proposed concept:

- Monitor units increased by 3% and 5%
- Collimator rotation offset by ±3° and ±5°
- Gantry angle offset by ±3° and ±5°
- MLC banks shifted by ±3 mm and ±5 mm
- MLC opening increased by 1 mm and 3mm
- MLC opening decreased by 1 mm and 3 mm with a minimum opening of 0.6 mm
- MLC fully retracted

3. Results and discussion

3.1. Uncertainties in gantry angle determination

The uncertainties associated with the gantry angle determination for two full arcs of different rotation directions were 0.20°, 0.13° and 0.083° (1 s.d.) for N62u, IK360 and DynaLog, respectively. The DynaLog files were associated with the smallest uncertainty, and were therefore used to group the dosimetry data from the Delta[4] measurements. However, the DynaLog data are not independent of the accelerator. Therefore, simultaneous IK360 measurements were used to verify the actual gantry angle readings. The DynaLog data cannot be read-out until the whole arc has been delivered; hence the N62u inclinometer was used for *in vivo* dosimetry.

3.2. Pre-treatment QC

The mean hyper-gamma pass rates (over all control points) ranged from 84.7 to 97.1% and 93.4 to 99.6% for the [2%/2 mm/0.5°] and [3%/3 mm/0.5°] criteria, respectively. The conventional gamma pass rates ranged between 97.7 and 100.0% for the [3%/3 mm] criteria. A higher hyper-gamma pass rate was generally observed for gantry intervals around zero degrees (gantry in upright position).

However, a significant difference (α=0.05) in average pass rate was found only for the first and last angle intervals.

A difficult task with time resolved dose distribution comparisons is to display the results in an intuitive way. In this study, the data were collapsed into a 2D view, displaying the hyper-gamma pass rate for each detector element (figure 2, left). Spatial areas with potential issues were identified from this view. Generally, similar patterns were recognized between the hyper-gamma pass rates and conventional gamma values. The detector elements with the lowest hyper-gamma pass rates were evaluated using dose difference profiles along the detector arrays as well as in control point intervals (figure 2, right). The positions of these detector elements in the beams-eye-views (BEV) for each control point were used to determine the cause of differences between TPS calculations and measurements. The main cause of deviations found for specific detector elements and control point intervals was the presence of complex MLC motion over the detector element (as seen from BEV).

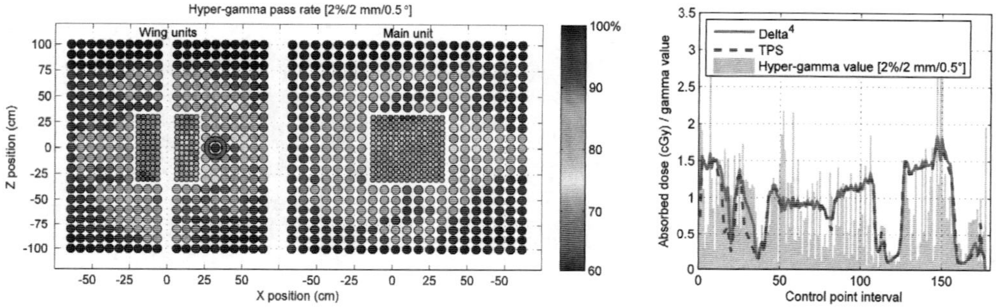

Figure 2: Hyper-gamma pass rate map [2%/2 mm/0.5°] (left) and a control point interval profile for the detector element (purple circle) with the lowest hyper gamma pass rate (right).

3.3. *In vivo dosimetry*

The maximum difference between all dosimeter readings and the corresponding TPS calculated cumulative absorbed doses for three consecutive deliveries with a static detector position ranged from -3.7% to -2.3% of the total TPS calculated absorbed dose and the hyper-gamma pass rate ranged from 94.9% to 100.0%. The introduction of intrafractional motion had a substantial impact on the measured absorbed dose for two of the trajectories investigated. The maximum dose differences for these two deliveries were -6.0% and 5.5%. However, using the position corrected TPS cumulative absorbed dose, the maximum dose differences were reduced to -3.2% and -2.8%, respectively. The absorbed dose difference exceeded the maximum absolute difference of the nonerroneous deliveries (3.7%) for three of the introduced errors and all trajectories. These errors were a 5% increase of MUs, MLC opening reduced by 3 mm, and a fully retracted MLC. The most severe treatment error (fully retracted MLC) was detected during an early stage of treatment. Serious radiation overexposure can be avoided in a clinical setting by allowing the *in vivo* system to interrupt treatment if the measurements are outside pre-defined tolerance levels. The hyper-gamma pass rate was reduced below 95% for 7 of the deliveries with introduced errors and a static detector position and it was also found to be sensitive to intrafractional motion (hyper-gamma pass rates for the two trajectories and for deliveries without errors introduced were reduced to 90% and 59%, respectively).

4. Conclusions

While the conventional gamma values provide an estimate of the accuracy of the entire delivery to a given point, the hyper-gamma and 4D analyzes can help identifying the actual cause of discrepancies between the calculated and measured absorbed dose. The beams-eye-view tool was valuable in order to investigate potential issues associated with specific control point intervals and detector elements.

It was found that if implantable EM-transponders are used for target localization, the addition of a dosimeter and an online dosimetric evaluation system provides a fast and resource-efficient *in vivo* dosimetry system.

5. References

[1] Saw C *et al* 2008 *J. Biomed Imaging Interv.* **4** e48
[2] Buonamici F B *et al* 2007 *Med. Phys.* **34** 1372-79
[3] Low D A *et al* 1998 *Med. Phys.* **25** 656-61

Evaluation of a clinically intuitive quality assurance method

H Norris, A Thomas and M Oldham
Duke University Medical Center, Durham, USA

E-mail: hannah.norris@duke.edu

Abstract. There is a pressing need for clinically intuitive quality assurance methods that report metrics of relevance to the likely impact on tumor control of normal tissue injury. This paper presents a preliminary investigation into the accuracy of a novel "transform method" which enables a clinically relevant analysis through dose-volume-histograms (DVHs) and dose overlays on the patient's CT data. The transform method was tested by inducing a series of known mechanical and delivery errors onto simulated 3D dosimetry measurements of six different head-and-neck IMRT treatment plans. Accuracy was then examined through the comparison of the transformed patient dose distributions and the known actual patient dose distributions through dose-volume histograms and normalized dose difference analysis. Through these metrics, the transform method was found to be highly accurate in predicting measured patient dose distributions for these types of errors.

1. Introduction

With the advent of new, more complex treatment methods in radiation therapy, the need for a 3D quality assurance method has become apparent [1, 2]. New devices have been introduced which make a "semi-3D" measurement with diode arrays; extensive studies have been performed evaluating the accuracy of these devices [3]. Oldham et al [4] recently reported a novel approach to clinically intuitive QA which yields clinical metrics from 3D dose measurements made in phantom. The method was retrospectively applied to 6 patients that received base-of-skull (BOS) IMRT treatment. Although good agreement was observed between the measured and planned DVHs in these cases, the uncertainty in the measured DVH curves was unknown. The present work characterizes and investigates this uncertainty intentionally introducing known errors to the same set of patients.

2. Methods

An assortment of mechanical, calibration and delivery errors were introduced into the treatment plan of each of the six patients. These included ±5° errors on the gantry, couch, and collimator for all beams, and 105% or 95% of the prescription dose. A total of 8 error-induced plans were therefore created for each patient. The *patient error dose distributions* corresponding to these errors were calculated using Eclipse Analytic Anisotropy Algorithm (AAA) [5] at 2.5 mm^3 isotropic resolution, with the heterogeneity correction applied. These dose distributions serve as the actual dose that would be delivered to the patient should these errors actually occur during delivery; they are the "gold standard" to which the transformed dose distributions will later be compared. Corresponding *phantom verification dose distributions* were calculated on an RPC-like anthropomorphic head phantom. These dose distributions act as the simulated 3D dosimetry measurements. The *transformed dose distribution* in the patient geometry is calculated with the transformation matrix, as described in Oldham et al. [4]. A schematic outline of the simulation rationale is shown in Figure 1.

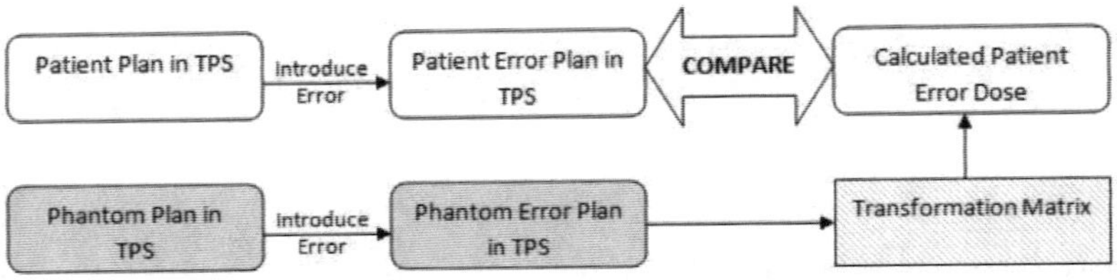

Figure 1: Schematic of methods; error-induced phantom dose distribution is used to calculate the transformed patient dose distribution, which is compared to the error-induced patient dose distribution. TPS refers to treatment-planning-system.

The accuracy of the transform method was evaluated by comparing the transformed dose distribution (upper right box in figure 1) with the gold standard, the patient error dose distribution. Clinically relevant metrics such as patient dose-volume histograms were obtained as well as 3D gamma passing rates and the Normalized Dose Distribution (NDD – a variation of gamma) [6].

3. Results

3.1. Normalized Dose Difference Based Metrics

The NDD values were calculated over both the entire patient volume for a global passing rate as well as for the PTV and a series of critical structures: the brainstem, the optic chiasm, the left and right eyes, and the left and right optic nerves. 2% dose difference and 2 mm distance to agreement were used. For the NDD metric, voxel values between -1 and +1 are passing.

Figure 2 shows an example of a trans-axial slice of patient case HN4 with induced error gantry +5° for all beams, overlaid with the actual dose distribution, the transformed dose distribution, and the NDD analysis of the two. The passing rate NDD values are 100% for all structures and 99.08% globally.

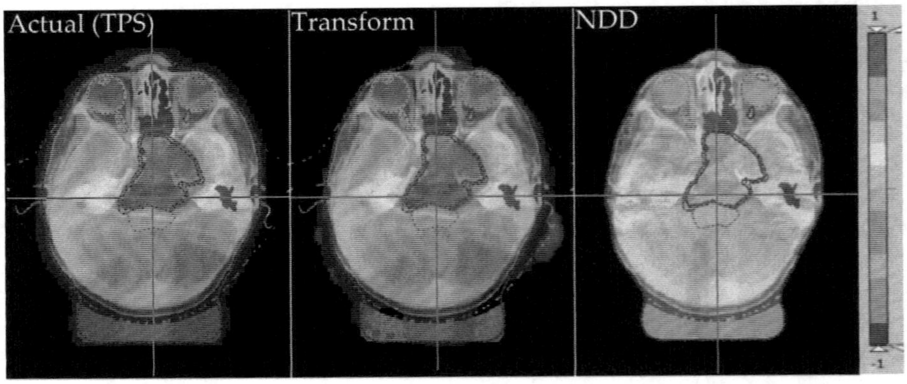

Figure 2: Actual (TPS) (left) and transformed (center) relative dose distributions with the 2%/2mm NDD analysis of the two (right), for patient HN4 with Gantry +5°.

Overall, NDD metrics comparing the transformed dose distribution with the actual dose showed very good agreement with the actual dose distributions when the NDD was performed with 2% dose difference, 2 mm distance-to-agreement criteria. The delivery errors where the dose normalization was changed by ±5% of the prescription dose had 100% pass rates for the global NDD as well as for each structure. The mechanical errors also showed very good passing rates for the same 2%/2mm criteria, especially for the collimator and couch errors, where the global passing rates were very high and the passing rates to the structures were at or near 100% for all patient plans. The gantry errors had the lowest NDD passing rates for the 2%/2mm criteria, but in general these passing rates were still very high (>98%).

3.2. Dose-Volume Histogram Based Metrics

With the overlaid transformed dose distribution on the patient CT set, a DVH can be calculated and compared to the actual DVH from the patient error dose distribution. This was done for all cases; the results were similar over all induced mechanical errors. Figure 3 shows an example of this for the prescription dose +5% on patient HN1 and the gantry rotated +5° on patient case HN4.

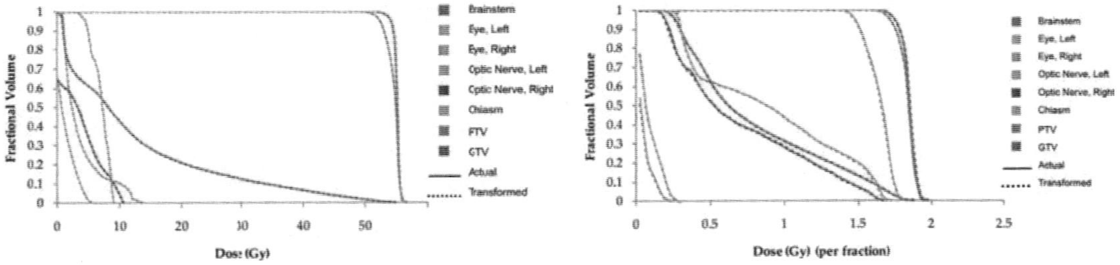

Figure 3: Best DVH agreement is seen for the dose delivery errors, as seen in patient HN1 with the prescription dose +5% (left); worst DVH agreement is seen for patient HN4 with the gantry+5°.

In general, the DVHs showed very good agreement between the transformed and the actual dose distributions. The delivery errors, where the plan normalization was increased or decreased by 5% of the planned prescription dose, showed indistinguishable DVH lines, indicating that the transform method produced perfect or near-perfect dose distributions. The mechanical errors, where the collimator, the couch, or the gantry were rotated ±5° from the planned values , also showed very good agreement in the DVH lines, although some slight deviations could be seen.

From the DVHs for all the plans, the minimum, maximum, and mean doses to the PTV were found for both the error-induced plan from the TPS and the transformed dose distribution. Correlations between the transformed dose values and error-induced planned dose values can be found by plotting the transformed dose value versus the error-induced planned dose value. Figure 4 shows this plot for the PTV as well as for the brainstem. These data show a high level of agreement between the transformed and the "gold standard" dose distributions.

For the PTV, the maximum dose has a linear fit given by y = 1.0004x, with $R^2 = 0.9998$; for the minimum dose, y = 0.9968x and $R^2 = 0.9995$; for the mean dose, the line of best fit is given by y = 0.9997x and $R^2 = 0.9999$. The line of best fit for the maximum dose to the brainstem is given by y = 0.9987x with $R^2 = 0.9999$. The line of best fit for the V_{20} data is given by y = 0.9999x with $R^2 = 0.9999$. These values again reflect a high level of accuracy between the actual and the transformed dose distributions.

4. Discussion

A high degree of accuracy was seen in the transform method, even when presented with relatively large mechanical and delivery errors. This is evidenced in the agreement of dose-volume histogram lines between the transformed and the actual dose distributions, as well as very high NDD passing rates under stringent criteria of 2% dose difference and 2 mm distance to agreement. This high level of accuracy in the transform method leads to the feasibility of performing clinically-relevant analysis of patient-specific quality assurance.

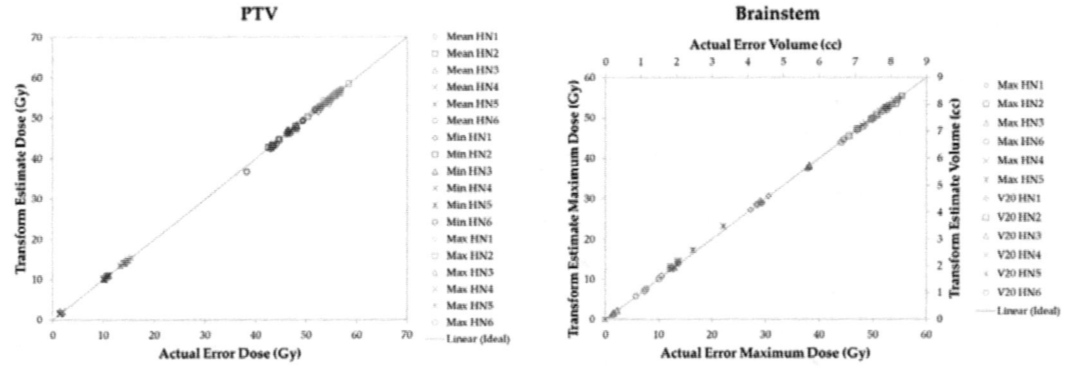

Figure 4: The maximum, minimum, and mean dose delivered to the PTV for all cases; transformed dose vs. error-induced TPS dose (left); maximum dose and V_{20} for the brainstem; transformed estimate dose vs. actual dose (right).

5. Conclusions

The transform method was shown to work well for the types of errors that were induced - rotating the gantry, couch, and collimator each ±5° and changing the prescription dose ±5%. These errors were rather large, certainly beyond what would be typically seen within the clinic, but the transform method continued to perform with a high level of accuracy. This level of accuracy can be attributed in part to the high resolution of 3D dosimetry.

The transform method is limited in clinical use by the size of the dosimeter. This can affect the transformed values for the organs at risk; if a PTV is large or if its volume is not completely contained within the dosimeter, then the accuracy of the transform method will be greatly reduced. In addition, this study is limited to head-and-neck/ base-of-skull IMRT treatments. Further investigation focus on the accuracy of the method for a wider range of scenarios, and other sites with greater sources of inhomogeneity.

6. Acknowledgements

This work was supported by NIH R01CA100835.

7. References

[1] Nelms B E *et al* 2011 *Med. Phys.* **38** 1037-44
[2] Baldock C *et al* 2010 *Phys. Med. Biol.* **55** R1-R63
[3] Feygelman V *et al* 2011 *J. Appl. Clin. Med. Phys.* **12** 3346
[4] Oldham M *et al* 2012 *Int. J. Radiat. Oncol. Biol. Phys.* (in press)
[5] Van Esch A *et al* 2006 *Med. Phys.* **33** 4130-48
[6] Jiang S B *et al* 2006 *Phys. Med. Biol.* **51** 759-76

3D tomodosimetry using scintillating fibers: proof-of-concept

M Goulet, L Gingras, L Beaulieu and L Archambault
Département de Radio-Oncologie et Centre de Recherche en Cancérologie, Hôtel-Dieu de Québec, 11 Côte du Palais, Québec, Québec, G1R 2J6, and Département de Physique, de Génie Physique et d'Optique, Université Laval, Québec, Québec, Canada

E-mail: beaulieu@phy.ulaval.ca

Abstract. We present a novel, high-resolution 3D dosimeter based on the tomographic acquisition of the dose using long scintillating fibers. This study aims to demonstrate the concept of the dosimeter with simulated acquisitions using the input dose from Pinnacle^3. The dosimeter is composed of concentric, cylindrical planes in which scintillating fibers are placed at different angles between the fiber and the cylindrical plane. Upon a complete rotation of the device around its central axis, the incident dose distribution on the cylindrical planes can be reconstructed using tomographic reconstruction algorithm. The 3D dose in the dosimeter can then be interpolated from the cylindrical 2D dose distributions. Using a simulated acquisition composed of two concentric cylindrical planes of 36 and 32 fibers each, we achieve below 1% local dose difference between the reconstructed 3D dose and the expected dose from Pinnacle^3 in the high dose, low gradient region of the volume encompassed inside the innermost cylindrical plane. The results show the potential of the method to perform high-resolution 3D dose measurements of both square and IMRT fields.

1. Introduction

Modern dose delivery techniques such as IMRT, VMAT and tomotherapy can produce dose distribution of very high conformity and complexity. These treatment modalities require a comprehensive, patient-specific quality assurance usually performed using 2D or 3D dosimetry of the incident beam [1]. Scintillating fibers (or plastic scintillation detectors) present numerous dosimetric advantages for volumetric (3D) dosimetry such as water-equivalence, which enable the 2D or 3D arrangement of detectors without perturbation of the incident beam [2, 3, 4, 5].

The main disadvantage of using punctual detectors to perform 3D dosimetry is the high number of such detectors required to cover a whole volume with a high density of detectors. To circumvent this issue, we developed the concept of tomodosimetry (the short for tomographic dosimetry), where scintillating fibers in the length of 10 to 20 cm are used to perform a tomographic acquisition of the incident dose distribution. A first application to 2D dosimetry has shown the potential of tomodosimetry to perform accurate, high-resolution dosimetry using a relatively little number of detectors [6].

In this work, we extend the concept of tomodosimetry to 3D dosimetry of external beam radiotherapy by performing the tomographic acquisition of the dose inside concentric, cylindrical plane of a phantom. The aim of this study is to evaluate the capacity of tomodosimetry to reconstruct 3D dose of both square fields and IMRT segments. To this end, light acquisition from each scintillating fiber was simulated using 3D input calculated with Pinnacle^3.

2. Methods and materials

2.1. Cylindrical tomography

All simulated fibers were confined into cylindrical planes (see figure 1) at a fixed distance from the axis of rotation of the cylinder volume. In order to perform a tomographic acquisition of the incident dose distribution, each fiber was placed at an angle Φ with respect to its cylindrical plane. As such, the sum of the light acquisition by a single scintillating fiber upon rotation of the cylinder volume around its central axis represents a projection of the dose deposited in the cylindrical plane at the angle Φ. Using many such scintillating fibers, multiple projections at different angles Φ can be acquired simultaneously with a single 360 degrees rotation of the cylinder.

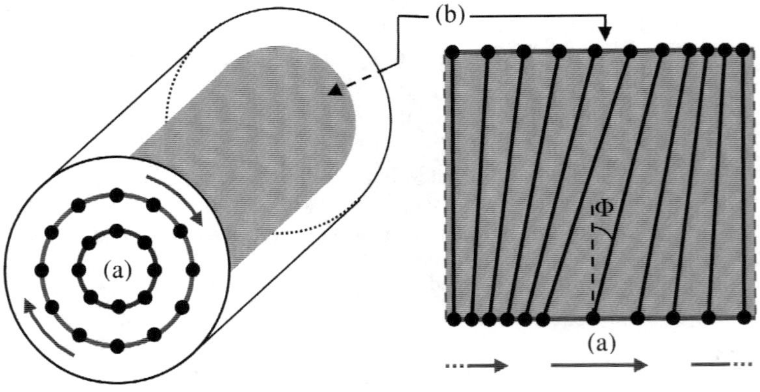

Figure 1: Scintillating fibers disposition in the phantom. Blue arrows indicate the direction of rotation. The radiation is expected to come from the figure top. Fibers are placed on the surface of one or more cylindrical volume (green, left). The angle Φ of each fiber on this surface is variable (right), allowing for tomographic acquisition of the dose following the rotation of the cylinder around its central axis.

2.2. Simulated acquisition and reconstruction

We simulated the tomographic dose acquisition using the 3D dose from Pinnacle[3] computed for a square field (10x10 cm^2) and an IMRT segment from a head-and-neck plan. We used two concentric cylindrical planes of radius 3.75 and 7.5 cm and length 20 cm. The outer cylinder plane was composed of 36, 1 mm diameter scintillating fibers oriented at an angle Φ (see figure 1) ranging from -45 to 45 degrees by steps of 5 degrees. The inner cylindrical plane was composed of 32 scintillating fibers oriented at an angle Φ ranging from 0 to 45 degrees by steps of 5 degrees. We further considered a scintillating fiber light acquisition in 5 degrees steps of the whole cylinder volume around its central axis (following the blue lines from figure 1), and reconstructed the 2D dose distribution at a resolution of 1x1 mm^2 on each cylindrical plane using an iterative reconstruction algorithm used in previous work [5]. The 3D dose was then interpolated at a resolution of 1x1x1 mm^3 in the cylindrical volume encompassed by one or both cylindrical plane, taking into account the $1/d^2$ decay of the dose (d = distance from the source).

3. Results

The reconstructed dose distribution on the inner cylindrical plane is presented in figure 2 for the IMRT segment. The angular position of 0 degree represents the top of the cylinder (i.e. the part closer to the incident beam), while the angular position of 180 degrees represents the bottom of the cylinder. Since the dose distribution is cylindrical, the left side (-90°) and the right side (270°) of the plane are coincident. In the same figure, the dose difference with respect to the input dose from Pinnacle[3] is shown for the same field. Namely, the dose reconstruction achieves a mean dose difference of 0.45%

of the maximum dose in the high dose ($D > 0.5D_{max}$), low gradient ($\Delta D < 0.01D_{max}$) region of the field. Similar accuracy is seen in the outer cylindrical plane for the same field (mean dose difference = 0.54%) and in both cylindrical planes of the square 10x10 cm^2 field (mean dose difference = 0.11% (inner plane) and 0.19% (outer plane)).

A summary of the 3D dose differences for the whole cylindrical volume is presented in Table 1 for both the square field and the IMRT segment. The mean dose difference in the high dose ($D > 0.75D_{isocentre}$), low gradient ($\Delta D < 0.01\ D_{isocentre}$) region of each field was better than 0.9% of the local dose in the volume located inside the inner cylinder, while the mean local dose difference was better than 1.2% in the volume comprised between the two cylindrical volumes. The 3D dose computed at the isocentre plane (i.e. the plane at 100 cm from the source) for the IMRT segment is shown in figure 3, along with the dose difference with respect to Pinnacle[3]. In this case the mean dose difference in the high dose ($D > 0.5D_{max}$), low gradient ($\Delta D < 0.01D_{max}$) region of the plane was of 0.76% of the maximum dose and the 3%/3mm gamma test pass rate was of 97.9% of the dose pixels above 10% of the maximum dose. Even better results are observed for the 10×10 cm^2 field, namely a mean dose difference in the high dose, low gradient region of the plane of 0.57% and a 3%/3mm gamma test pass rate of 100%.

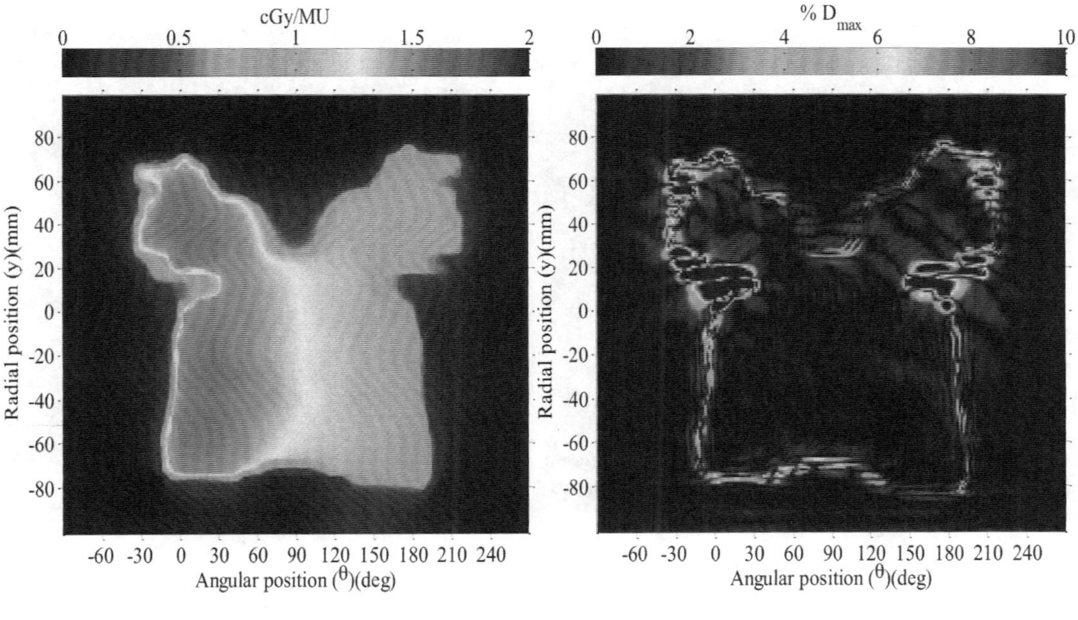

Figure 2: Left: 1x1 mm^2 dose reconstruction of the IMRT segment on the inner cylinder plane (radius = 3.75 cm). Right: dose difference with respect to the input dose from Pinnacle[3]. The reconstructed dose achieved a gamma test pass rate of 99.2% (3%/3mm, $D > 0.1D_{max}$) and a mean dose difference of 0.45% (high dose low gradient region of the field).

4. Discussion and conclusion

In this work, we have presented a proof-of-concept of a 3D dosimeter based on the tomographic acquisition of the incident dose on two concentric cylindrical planes. Simulation of light acquisition and dose reconstruction were performed and compared to the input dose of Pinnacle[3]. The obtained results demonstrate the potential of the proposed method to achieve 3D high-resolution dosimetry of both square and IMRT fields.

Table 1: Dose difference computed inside the cylindrical volume. The inner cylinder region refers to the volume encompassed inside the innermost cylindrical dose plane, while the outer cylinder region refers to the region between the two concentric cylindrical dose planes.

Field		$10x10\ cm^2$		IMRT segment	
3D region		Inner cylinder	Outer cylinder	Inner cylinder	Outer cylinder
Mean Absolute local dose difference (%)		0.54	1.12	0.84	1.01
Percentage of	1%	94.0	70.0	81.7	75.1
pixel under a local	2%	95.8	88.6	93.4	91.1
dose difference of	3%	96.5	96.3	96.1	96.4

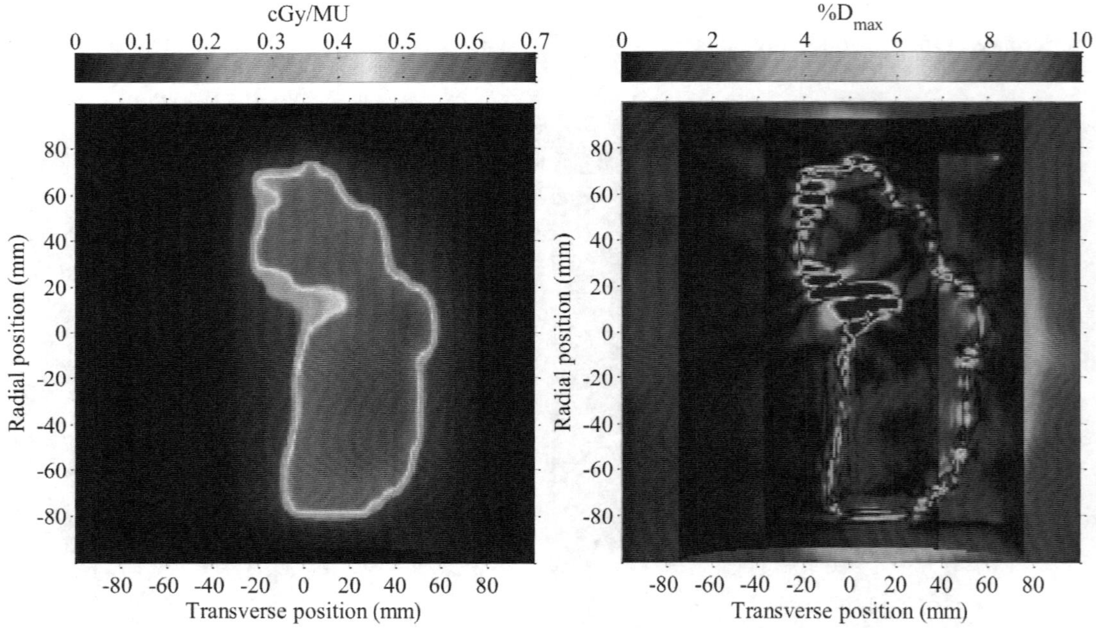

Figure 3: Left: 1x1 mm^2 3D dose reconstruction of the IMRT segment on the isocentre plane (source-to-plane distance = 100 cm). Right: dose difference with respect to the expected dose from Pinnacle3. The reconstructed dose achieved a gamma test pass rate of 97.9% (3%/3mm, $D > 0.1D_{max}$) and a mean dose difference of 0.76% (high dose low gradient region of the field).

5. References

[1] Low A *et al* 2011 *Med. Phys.* **38** 1313-38
[2] Beddar A S 2007 *Radiat. Meas.* **41** S124-33
[3] Guillot M *et al* 2011 *Med. Phys.* **38** 6763-74
[4] Guillot M *et al* 2010 *J. Phys. Conf. Ser.* **250** 012006
[5] Archambault L *et al* 2012 *Med. Phys.* **39** 1239-46
[6] Goulet M *et al* 2012 *Med. Phys.* (in press)

Journal of Physics: Conference Series **444** (2013) 012024 doi:10.1088/1742-6596/444/1/012024

Water equivalence of micelle gels for x-ray beams

T Gorjiara[1], R Hill[1,2], Z Kuncic[1], S Bosi[1,3] and C Baldock[4]
[1]Institute of Medical Physics, School of Physics, University of Sydney, NSW 2006, Australia
[2]Department of Radiation Oncology, Royal Prince Alfred Hospital, Camperdown, NSW 2050, Australia
[3]Department of Radiation Oncology, Prince of Wales Hospital, Randwick, NSW 2031, Australia
[4]Faculty of Science, Macquarie University, NSW 2109, Australia

E-mail: clive.baldock@mq.edu.au

Abstract. Micelle gel is a radiochromic hydrogel with the potential to be used as a three dimensional (3D) radiation dosimeter. Since an ideal dosimeter should present water equivalent properties, in this study the water equivalence of two formulations of micelle gel has been investigated by calculating electron density, effective atomic number, fractional interaction probabilities, mass attenuation coefficient. The depth doses for kilovoltage and megavoltage x-ray beams have also modelled using Monte Carlo code. Based on the results of this work, micelle gels can be considered as water equivalent dosimeters.

1. Introduction

Polymer gel dosimeters are manufactured from radiation sensitive chemicals, which upon irradiation polymerize as a function of the absorbed radiation dose [1]. These gel dosimeters which record the radiation dose distribution in three-dimensions (3D) have specific advantages when compared to one-dimensional dosimeters and two-dimensional dosimeters [2]. These 3D dosimeters are radiologically soft-tissue equivalent [3] with properties that may be modified depending on the application. The 3D radiation dose distribution in polymer gel dosimeters may be imaged using magnetic resonance imaging (MRI) [4, 5], optical-computerized tomography (optical-CT) [6, 7], x-ray CT [8, 9], ultrasound [10, 11, 12] or vibrational spectroscopy [13, 14].

The Micelle gel dosimeter is a radiochromic hydrogel containing leucodye which undergoes colour change upon radiation. Jordan *et al* used leuco Malachite Green (LMG) dye which converts to Malachite Green (MG^+) after exposing to ionizing radiation. This gel (will be referred as LMD1) also contains trichloroacetic, Triton X-100 and gelatine and has an absorption peak of 633 nm which is suitable to be readout by optical scanners [15]. Due to high dose rate dependency of the initial composition of leucodyes gels, a modified composition was suggested [16]. This new formulation of LMG (will be referred as LMD2) contains sodium dodecyl sulphate, chloroform, trichloroacetic acid, gelatine and LMG and shows good spatial stability and no energy dependency for megavoltage x-ray beams [17].

An ideal clinical dosimeter should exhibit water (or tissue) equivalent radiological properties and dose response. Water equivalence of polymer 3D dosimeters has been investigated previously [18-23]. In this study water equivalency and radiological properties of two different formulations of micelle

Journal of Physics: Conference Series **444** (2013) 012024 doi:10.1088/1742-6596/444/1/012024

gels have been investigated by calculating: electron density, effective atomic number, photon interaction probabilities and mass attenuation coefficient. We also determined depth doses in each of the dosimeters using Monte Carlo calculations for 50 kVp, 280 kVp and 6 MV x-ray beams.

2. Material and Methods

Previously published elemental weight fractions and mass densities of two micelle gel formulations (see Table 1) [17] were used for radiological properties and dose calculations.

Mayneord formula was used to calculate effective atomic number [24]. The fractional interaction probabilities for photoelectric absorption, Compton scatter and pair was calculated using the NIST XCOM cross sectional data for 1 keV - 20 MeV energy range [25]. The mass attenuation coefficients of the gels were also calculated using the elemental attenuation coefficients of the NIST database [26].

EGSnrc/BEAMnrc Monte Carlo package (Version 4, NRC, Ottawa, Canada) [27] was used to model Varian 21iXs linear accelerator (Varian Medical Systems, city, state, USA) and calculate the dose delivered to the micelle gels and water for 6 MV and 50 to 280 kVp x-ray beams. The simulation method has been discussed in details in the previously published articles by authors [20, 21].

Table 1: Elemental composition by fractional weight (w_k), density and calculated electron density and effective atomic number for the two micelle gel formulations and water.

Material	w_H	w_C	w_N	w_O	w_S	w_{Na}	w_{Cl}	ϱ (g.cm^{-3})	$ϱ_e$ (×10^{23} e.cm^{-3})	Effective atomic number
LMD1	0.1104	0.0229	0.0001	0.8649	-	-	0.0017	1.015	3.388	7.425
LMD2	0.1087	0.0379	0.0001	0.8414	0.0011	0.0016	0.0091	1.012	3.372	7.597
Water	0.1119	-	-	0.8881	-		-	1.000	3.343	7.417

3. Results

The calculated electron density and effective atomic number of the micelle gels dosimeter and water are presented in Table 1. Electron density of LMD1 and LMD2 is slightly higher than water by 1.4% and 0.9% discrepancy, respectively. The effective atomic number of LMD1 could be considered similar to water with 0.1% discrepancy while for LMD2 the difference from that of water is 2.4%.

Figure 1 shows the calculated fractional interaction probabilities for the micelle gels and water over the energy range 1 keV - 20 MeV. The maximum discrepancy in the fractional interaction probabilities of LMD1 from those of water are less than 2%. However, for LMD2, the discrepancies up to 17% and 8% for the photoelectric absorption and Compton scattering interaction probabilities are evident, respectively.

4. Conclusion

Our results demonstrate that both micelle gel formulations can be considered radiologically water equivalent for x-ray dosimetry. Micelle gels show more water equivalent radiological properties and dose response compare to PRESAGE® and polymer gel dosimeters such as MAGAT, MAGIC and PAGAT [18, 19, 23]. However, genipin gel still could be considered the most water equivalent polymer gel dosimeter [20, 28].

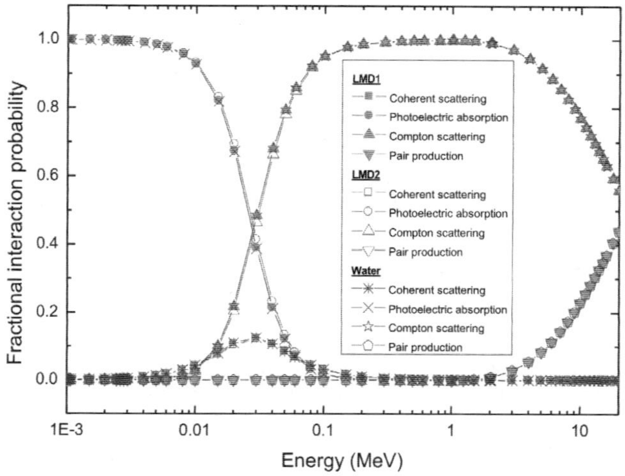

Figure1: Fractional interaction probabilities for the two micelle gel formulations and water.

The ratio of the mass attenuation coefficients of the micelle gels and water are plotted in figure 2 over the energy range 10 keV - 20 MeV. For LMD2 a peak is noticeable from 2 – 100 keV which is more than 8% higher than water. This can be attributed to the presence of high atomic number component; Cl, Na and S; in the gel. LMD1 has similar attenuation coefficient to water over the entire energy range.

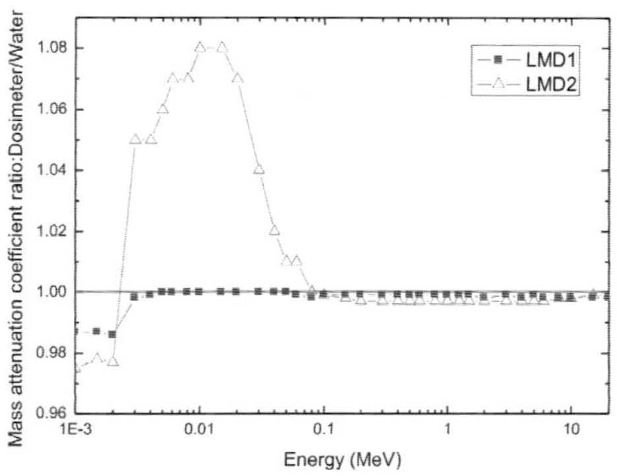

Figure 2: The mass attenuation coefficient ratio for two micelle gel formulations relative to water.

Figure 3 shows Monte Carlo calculated depth dose curves of the micelle gels and water for 50 kVp, 280 kVp and 6 MV x-ray beams. For 50 kVp, LMD1 has less than 0.5% difference from water while LMD2 differs from water by ≈ 2%. For 280 kVp, both gels could be considered water equivalent with less than 0.4% dose difference from water. In the buildup region of MV x-rays, delivered dose in both micelle gels is ≈ 1% higher than water due to existence of carbon and higher atomic number components. Consequently, after the buildup region, dose delivered in both gels is slightly less than water (by 0.5%).

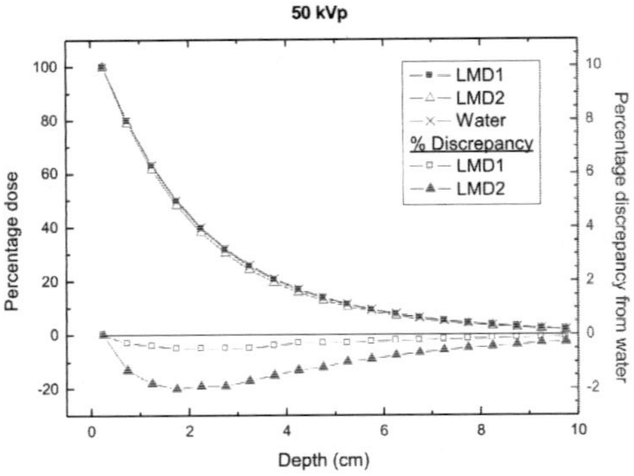

(a)

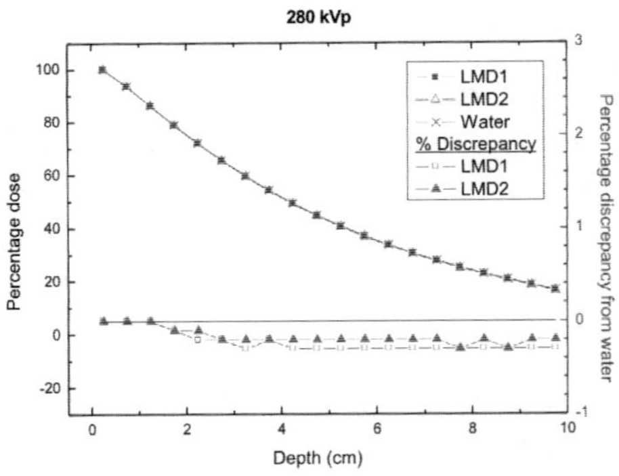

(b)

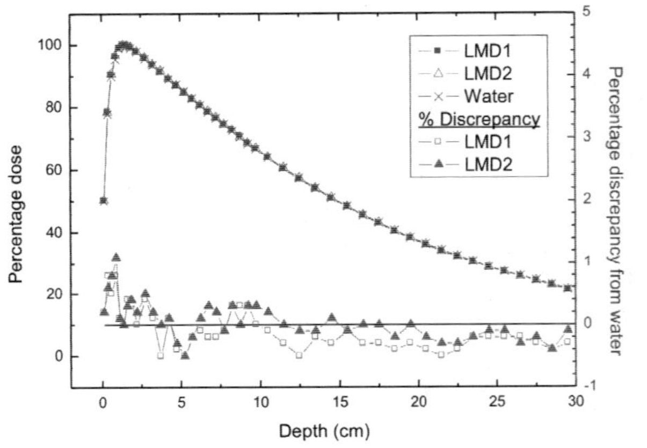

(c)

Figure 3: Monte Carlo percentage depth dose curves for (a) 50 kVp, (b) 280 kVp and (c) 6 MV x-ray beams for the two micelle gel formulations and water.

5. References

[1] Baldock C *et al* 1998 *Phys. Med. Biol.* **43** 695-702
[2] Baldock C *et al* 2010 *Phys. Med. Biol.* **55** R1-63
[3] Keall P and Baldock C 1999 *Australas. Phys. Eng. Sci. Med.* **22** 85-91
[4] Gustavsson H *et al* 2004 *Phys. Med. Biol.* **49** 227-41
[5] Lepage M *et al* 2002 *Phys. Med. Biol.* **47** 1881-90
[6] Bosi S *et al* 2007 *Phys. Med. Biol.* **52** 2893-903
[7] Bosi S G *et al* 2009 *Phys. Med. Biol.* **54** 275-83
[8] Trapp J V *et al* 2002 *Phys. Med. Biol.* **47** 4247-58
[9] Hill B *et al* 2005 *Br. J. Radiol.* **78** 623-30
[10] Mather M L and Baldock C 2003 *Med. Phys.* **30** 2140-8
[11] Mather M L *et al* 2002 *Phys. Med. Biol.* **47** 1449-58
[12] Mather M L *et al* 2003 *Ultrasonics* **41** 551-9
[13] Baldock C *et al* 1998 *Phys. Med. Biol.* **43** 3617-27
[14] Rintoul L *et al* 2003 *Appl. Spectrosc.* **57** 51-57

[15] Maryanski M J *et al* 1996 *Phys. Med. Biol.* **41** 2705-17
[16] Murphy P S *et al* 2000 *Phys. Med. Biol.* **45** 3213-23
[17] Vandecasteele J *et al* 2011 *Phys. Med. Biol.* **56** 627-51
[18] Brown S *et al* 2008 *Appl. Radiat. Isot.* **66** 1970-4
[19] Venning A J *et al* 2005 *Med. Phys.* **32** 1047-53
[20] Gorjiara T *et al* 2011 *Phys. Med. Biol.* **56** 4685-99
[21] Gorjiara T *et al* 2011 *Med. Phys.* **38** (4) 2265-74
[22] Gorjiara T *et al* 2010 *J. Phys.: Conf. Ser.* **250** 012036
[23] Gorjiara T *et al* 2010 *J. Phys.: Conf. Ser.* **250** 012053
[24] Khan F M 2010 *The physics of radiation therapy 4th ed. J. Pine (Philadelphia: Lippincott Williams & Wilkins)*
[25] Berger M J *et al* 2009 *XCOM: Photon Cross Sections Database. NIST Standard Reference Database 8 (XGAM) National Institute of Standards and Technology: Gaithersburg*
[26] Hubbell J H and Seltzer S M 2004 *Tables of X-Ray Mass Attenuation Coefficients and Mass Energy-Absorption Coefficients, National Institute of Standards and Technology: Gaithersburg*
[27] Rogers D W O 2006 *Phys. Med. Biol.* **51** R287-301
[28] Willowson K *et al 2008 Phys. Med. Biol.* **53** 3099-112

Enhancement of radiation effects by bismuth oxide nanoparticles for kilovoltage x-ray beams: A dosimetric study using a novel multi-compartment 3D radiochromic dosimeter

M Alqathami[1], A Blencowe[2], UJ Yeo[3], R Franich[3], S Doran[4,5], G Qiao[2] and M Geso[1]

[1]School of Medical Sciences, RMIT University, VIC 3083, Australia
[2]Department of Chemical & Biomolecular Engineering, The University of Melbourne, VIC 3052, Australia
[3]School of Applied Sciences and Health Innovations Research Institute, RMIT University, Melbourne, VIC 3001, Australia
[4]Department of Physics, University of Surrey, Guildford, Surrey, GU2 7XH, UK
[5]CRUK and EPSRC Cancer Imaging Centre, Institute of Cancer Research, Sutton, UK

E-mail: mamdooh.alqathami@rmit.edu.au

Abstract: The aim of this study is to present the first experimental validation and quantification of the dose enhancement capability of bismuth oxide nanoparticles (Bi_2O_3-Nps). A recently introduced multi-compartment 3D radiochromic dosimeter for measuring radiation dose enhancement produced from the interaction of X-rays with metal nanoparticles was employed to investigate the 3D spatial distribution of ionizing radiation dose deposition. Dose-enhancement factor for the dosimeters doped with Bi_2O_3-NPs was ~1.9 for both spectrophotometry and optical CT analyses. Our results suggest that bismuth-based nanomaterials are efficient dose enhancing agents and have great potential for application in clinical radiotherapy.

1. Introduction

High atomic number (Z) nanomaterials such as gold nanoparticles (AuNPs) have been demonstrated to induce dose enhancement, thus improving therapeutic efficacy. In particular, low-energy (kV) ionizing radiation in advanced radiotherapy. However, considering the extensive investigations involving radiation dose enhancement using AuNPs, no studies have been conducted to investigate the potential superiority of other high Z nanoparticles such as bismuth-based nanoparticles. Bismuth has a higher atomic number (Z = 83) than gold (Z = 79) and therefore, could potentially be utilized as a dose enhancing agent. The aim of this paper is to present experimental validation and quantification of the dose enhancement capability of bismuth oxide nanoparticles (Bi_2O_3-NPs) using a recently developed multi-compartment radiochromic dosimeter [1].

2. Materials and Methods

A multi-compartment radiochromic dosimeter, referred to as the Sensitivity Modulated Advanced Radiation Therapy (SMART) dosimeter was employed to measure the radiation dose enhancement produced from the interaction of X-rays with metal nanoparticles, and to investigate the 3D spatial distribution [2] of ionizing radiation dose deposition [1]. The SMART dosimeter had two compartments: an outer compartment with water-equivalent radiological properties and an inner compartment with the same radiological properties, however, doped with 0.5 mM Bi_2O_3-NPs (NanoArc®) (*ca.* 50 nm diameter). Methods for fabricating the SMART dosimeter, embedding nanoparticles in the polymeric matrix, irradiation and optical absorption measurements are discussed in detail elsewhere [1]. Doses were read out using a spectrophotometer and a cone-beam optical CT scanner (Vista™ Optical Scanner by Modus Medical Devices Inc.) [3-6].

3. Results and discussion

The SMART dosimeter employed in this study, complete with Bi_2O_3-NPs nanocomposite compartment, is shown in figure 1(a) after irradiation (10 Gy; 100 kV X-ray energy). The inner Bi_2O_3-NPs doped compartment of the SMART dosimeter showed a significantly higher radiation dose (> 80 %) compared with the surrounding control compartment (figure 1a). The significant increase in dose deposition in the area doped with Bi_2O_3-NPs is caused solely by the presence of bismuth, which is an effective radiation absorbing element.

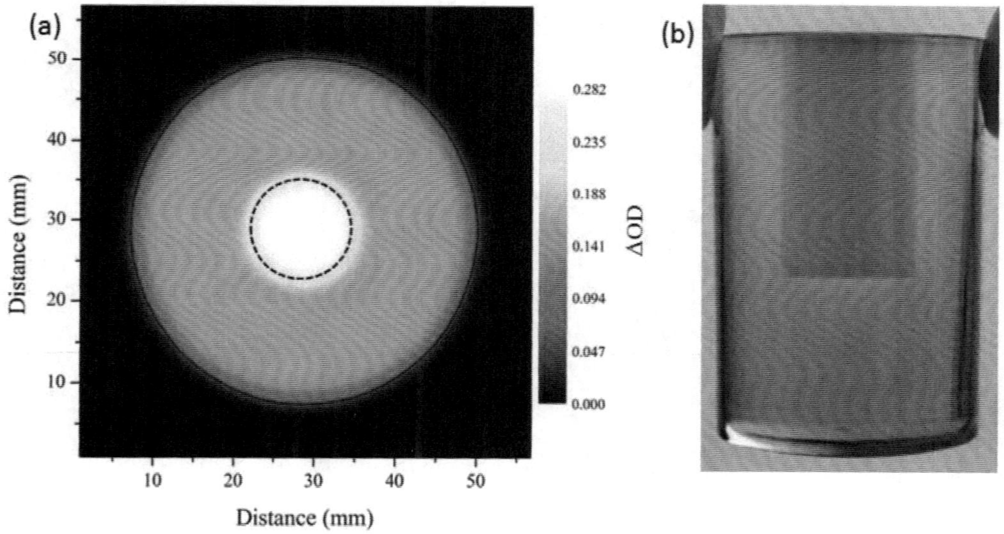

Figure 1: (a) Reconstructed radiation dose distribution a depth of 1 cm into the SMART dosimeter from the irradiated surface. The centre of the SMART dosimeter surrounded by a dashed line represents the area doped with Bi_2O_3-NPs. The scale in the colorbar is in ΔOD. (b) Photographs of the actual SMART dosimeter doped with 0.50 mM Bi_2O_3-NPs used in this study after exposure to 10 Gy using 100 kV energy.

When exposing the Bi_2O_3-NPs-doped compartment of the SMART dosimeter to radiation, photo-absorption by the Bi_2O_3-NPs causes an emission of low energy photoelectrons and Auger electrons leading to an enhancement in radiation dose deposition [7, 8]. This is clearly observed in the transverse plane (figure 1a) where a local boost in radiation dose can be seen in the compartment doped with Bi_2O_3-NPs. Therefore, the gain from utilizing Bi_2O_3-NPs in terms of a localized boost in radiation dose, relative to the dose to surrounding structures is clearly demonstrated by comparing the 3D dose distribution between the control and doped compartments given by the optical computed topography (CT) and converting the changes in optical density (ΔOD) to absolute dose. The boost in OD in the Bi_2O_3-NPs-doped compartment can also be seen when plotting OD line profiles through different transverse planes of the SMART dosimeter and depth dose curve (figure 2). From the optical CT measurement, the DEF was determined to be 1.94 (SD, 0.13) by taking the ratio for ΔOD at the centre of the Bi_2O_3-NPs doped compartment relative to the un-doped compartment throughout the effective depth of the dosimeter and taking the mean DEF value.

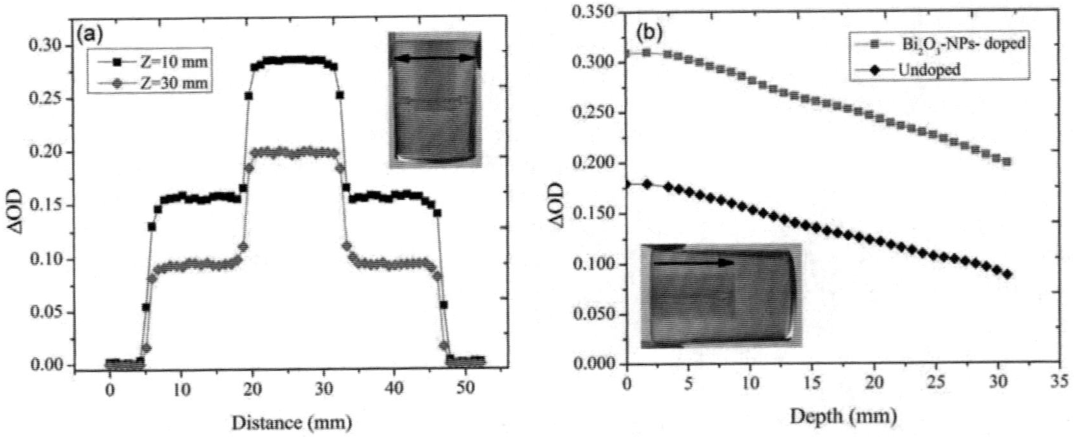

Figure 2: (a) ΔOD line profiles through the transverse plane at different depths of the SMART dosimeter. (b) Variation in ΔOD with depth through the undoped and Bi_2O_3-NPs-doped compartments of the SMART dosimeter. The actual SMART dosimeter is shown in both insets.

The absorption spectrum and measured changes in ΔOD versus the radiation absorbed dose at 633 nm for the undoped and Bi_2O_3-NPs-doped dosimeter cuvettes are shown in figure 3. Excellent correlation coefficient linearity ($R^2 > 0.999$) for the dose response was observed for both (with and without Bi_2O_3-NPs) over the applied radiation dose range. The dose enhancement factor (DEF) was defined as the ratio between the slope of the dosimeter doped with nanoparticles and the slope of un-doped dosimeter (control) [1]. The spectrophotometry analysis showed a DEF of 1.90 for the dosimeters doped with Bi_2O_3-NPs, which is in good agreement with the optical CT measurement (DEF = 1.94).

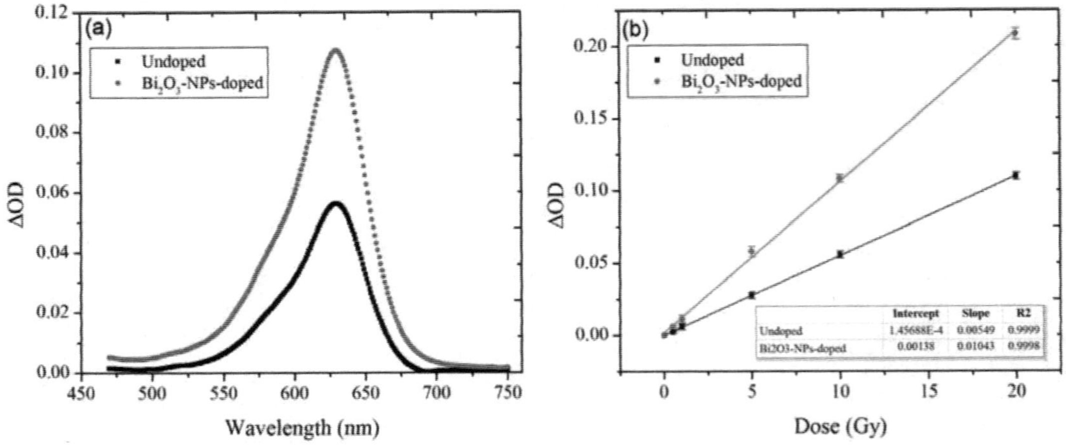

Figure 3: (a) A comparison between the absorption spectra of the undoped and Bi_2O_3-NPs-doped cuvettes after irradiation with 10 Gy using 100 kV energy. (b) **ΔOD** as a function of absorbed radiation dose for the undoped and Bi_2O_3-NPs-doped dosimeter cuvettes.

Journal of Physics: Conference Series **444** (2013) 012025 doi:10.1088/1742-6596/444/1/012025

4. Conclusion

In this study, we have experimentally demonstrated the efficiency of Bi_2O_3-NPs in radiation dose enhancement. Considering the fact that bismuth is not toxic, can be fashioned into nanomaterials and readily functionalized, and is one of the least expensive heavy metals, we believe further Monte Carlo, *in vitro*, and *in vivo* investigations involving bismuth-based nanoparticles are warranted to validate and fully understand dose enhancement achieved by Bismuth-based nanomaterials.

5. References

[1] Alqathami M *et al* 2012 *Int. J. Radiat. Oncol. Biol. Phys.* (in press)
[2] Baldock C *et al* 2010 *Phys. Med. Biol.* **55** R1-63
[3] Bosi S *et al* 2007 *Phys. Med. Biol.* **52** 2893-903
[4] Bosi S G *et al* 2009 *Phys. Med. Biol.* **54** 275-83
[5] Olding T *et al* 2010 *Phys. Med. Biol.* **55** 2819-40
[6] Olding T and Schreiner L J 2011 *Phys. Med. Biol.* **56** 1259-79
[7] Rahman W N 2009 *Nanomed., Nanotech. Biol. Med.* **5** 136-42
[8] Carter J D *et al* 2007 *J. Phys. Chem. B* **111** 11622-11625

Low-density polymer gel dosimeters for 3D radiation dosimetry in the thoracic region: A preliminary study

Yves De Deene[1,2], Jan Vandecasteele[1] and Tom Vercauteren[1]
[1]Department for Radiotherapy and Experimental Cancer Research, Ghent University Hospital, De Pintelaan 185, 9000 Gent, Belgium
[2]Institute of Medical Physics, School of Physics, University of Sydney

E-mail: Yves.DeDeene@UGent.Be

Abstract. Different low-density polymer gel dosimeters have been constructed that can be used to acquire the radiation dose distribution of IMRT treatments in the thoracic region. A heterogeneous phantom consisting of a low density polymer gel dosimeter sandwiched between two layers of soft tissue equivalent gel has been constructed. As a proof-of-principle, the phantom has been irradiated with a square 4 cm x 4 cm beam. The dose distribution is read out by use of both quantitative NMR spin-spin (R_2) and magnetization transfer (MT) imaging. Sources of error in the dose readout have been assessed and are discussed.

1. Introduction

Gel dosimetry provides a unique feature to display dose distributions occurring in clinical radiotherapy in three dimensions (3D) in humanoid shaped phantoms [1-2]. It has been shown that these polymer gel dosimeters (PGD), have electron densities and stopping powers very close to these of soft tissue [2] and are therefore considered as ideal 3D dosimeters for safeguarding the whole radiotherapy treatment chain.

In a previous study we have also demonstrated the potential to construct low density polymer gel dosimeters by beating the gel to a foam [3]. It was shown by micro-CT that the microstructure of these PGD foams had a very good resemblance to the microstructure of lung tissue. However, it was also shown that as a result of diffusive R_2 dispersion by the gas bubbles in the foam, quantitative R_2 mapping did not result in reliable dose maps. This was further complicated by practical difficulties in obtaining a homogeneous density distribution in the PGD foam. As an alternative to R_2 mapping, quantitative magnetization transfer (MT) imaging was suggested.

To obtain a more uniform density in the lung equivalent gel dosimeter, we now propose a different strategy by adding Styrofoam beads to a normoxic gel. Before adding gel to the Styrofoam beads, the beads are flushed with nitrogen gas. In addition, the suggested strategy lends itself better to the construction of antropomorphic gel dosimeters consisting of different density layers.

Low density (LD) PGDs could play an important role in 3D radiation dosimetry in the thoracic region (e.g. in the treatment of lung tumors). The construction of the LD PGDs described in this paper, makes it possible to create heterogeneous anthropomorphic 3D dosimeters. The significance of these heterogeneous LD PGDs is exemplified in radiation dosimetry in MRI-Linacs where the electron dose distribution is affected by the magnetic field and in the case of electronic disequilibrium [4].

Journal of Physics: Conference Series **444** (2013) 012026 doi:10.1088/1742-6596/444/1/012026

2. Methods and Materials

2.1. Low density polymer gel dosimeter construction

The polymer gel is composed of gelatin (300 Bloom, Type A) [8% (w/w)], methacrylic acid (MAc) [6% (w/w)], bis[tetrakis(hydroxymethyl)phosphonium]sulphate (THPS) [5 mm] and deionized water [approximately 86 % (w/w)]. In this case, Mac was chosen as the radiation active monomer because of its high sensitivity. A high sensitivity is desired to compensate the expected loss in SNR as a result of the lower density. First, gelatin is dissolved and allowed to swell in cold water for approximately 5 minutes. The gelatin mixture is heated to 45°C to bring it in the sol phase. Then, the mixture is allowed to cool down to 32°C after which the MAc is added slowly. In one type of gel, nitrogen is purged through the solution to expel the dissolved oxygen. In other types of gel, no additional nitrogen was bubbled through the solution. In mean time, nitrogen is purged through dry Styrofoam beads for approximately 30 minutes in a desiccator to evacuate the oxygen from the beads. Finally, THPS is added to the solution and is firmly mixed for approximately 3 minutes before pouring the solution in the phantom in which the Styrofoam beads were placed. A three layer phantom was constructed by placing the Styrofoam beads in a cylindrical cavity that was created by use of a nylon wire net (see figure 1). Calibration vials were also fabricated by placing Styrofoam beads in a series of test tubes before pouring the solution. The LD polymer gel had an electron density of 0.4 g/ml.

Figure 1. Three-layer density phantom with insert containing Styrofoam beads (a). The insert is composed of a nylon wire net that is clamped around Teflon rings that are suspended on glass bars (b). The composed gel phantom before irradiation (c) and after irradiation with a square 4 cm-by-4cm 6 MV photon beam (d). Volume rendered X-ray CT reconstruction of the phantom (e) and a volume rendered reconstruction of the segmented low-density insert with beam indicated (f).

2.2. Scanning low density polymer gel dosimeters
The gel dosimeter phantoms were MRI scanned with multi-spin echo sequences with 32 spin echoes and with different echo time spacings for R_2 mapping and with an in house built magnetization transfer (MT) imaging sequence with various MT pulse frequency offsets as described in previous work [4]. All scanning was performed on a Siemens 1.5T MRI scanner (Symphony).

2.2.1. Computer simulations of R_2 dispersion. The measured R_2 values in the low-density gel may be overestimated as a result of diffusive R_2 dispersion. This phenomenon has been encountered previously in hydrogel foams and can be turned in an advantage to estimate the bubble size of foam [6]. However, in the purpose of polymer gel dosimetry this is a disadvantage as it affects the R_2-dose response differently as in a water-equivalent gel dosimeter. Moreover, uncertainties in readout may arise at the boundary between low-density and unit-density polymer gel. To investigate the dependence of the diffusive R_2 dispersion on the size of the Styrofoam beads and on the echo time spacing, computational modelling was performed. The computational model of diffusive R_2 dispersion was validated by R_2 measurements on a benchtop NMR relaxometer (minispec 0.5T, Brüker) and on a 1.5T MRI scanner.

2.2.2. Magnetization transfer imaging. Magnetization transfer imaging was performed on a series of calibration tubes with low density (LD) and unit density (UD) gel irradiated to different doses in order to investigate the magnetization saturation as a function of frequency offset as an alternative to R_2 imaging. The MT data was fitted against a two-compartment model assuming a Gaussian line shape for the macromolecular proton pool [5].

3. Results and Discussion

3.1. Low density polymer gel dosimeter
The R_2 gel measured dose distribution in the three layer phantom is shown in figure 2. The R_2 maps were calibrated by using the calibration vials that can be seen on either sides of the phantom (figure 2a). A fair correspondence was found between the gel measured and treatment planning system (TPS) calculated dose distribution. However, a significant deviation with an overestimation in gel measured dose is apparent from the depth-dose profile (figure 2c) in the LD gel as compared to TPS calculated.

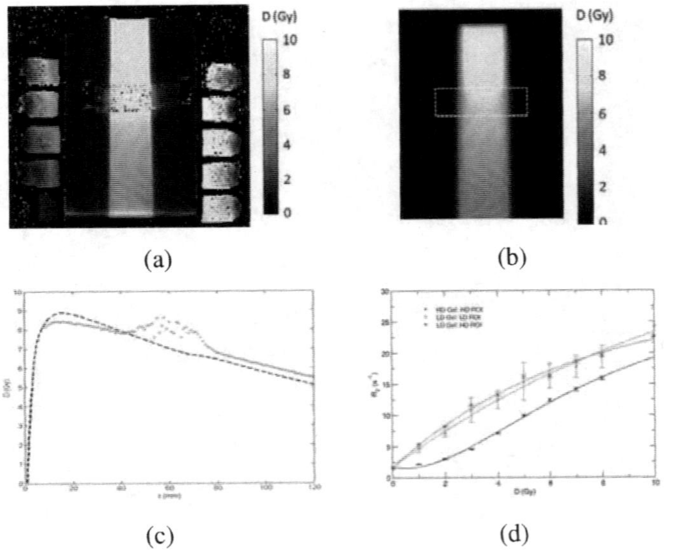

Figure 2. Gel measured (a) and TPS calculated (b) dose distribution in the three layer phantom. The depth-dose curve (c) shows a dose overestimation with the PGD in the LD region. The dose-R_2 response curve is different for UD and LD gel (d).

The deviation between the R_2 measured and TPS calculated dose distribution is attributed to: (1) A difference in oxygen distribution in the MAGAT gel formulation in the phantom and (2) diffusive R_2-dispersion caused by the internal magnetic field gradients that result from magnetic susceptibility differences between the Styrofoam beads and the gel fraction.

3.2. Diffusive R_2 dispersion

Diffusive R_2 dispersion originates from water molecules diffusing in the internal magnetic field heterogeneities will cause an additional nuclear spin dephasing that also depends on the inter-echo spacing of the multiple spin echo sequence [6].

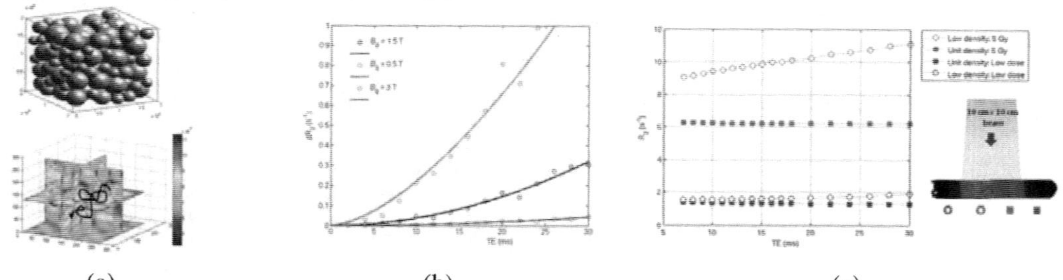

(a) (b) (c)

Figure 3. Simulations (a-b) and measurements (c) of the diffusive R_2 dispersion in the LD polymer gel dosimeter. The plot in (c) is obtained by scanning a long test tube containing a LD and a UD gel region and which has been irradiated with a lateral 10 cm x 10 cm field as indicated.

3.3. Magnetization transfer (MT) imaging

Magnetization as a function of saturation frequency offset shows that an optimal MTR sensitivity in both a UD PGD and a LD PGD is reached for a saturation pulse frequency offset of 600 Hz (figure 4).

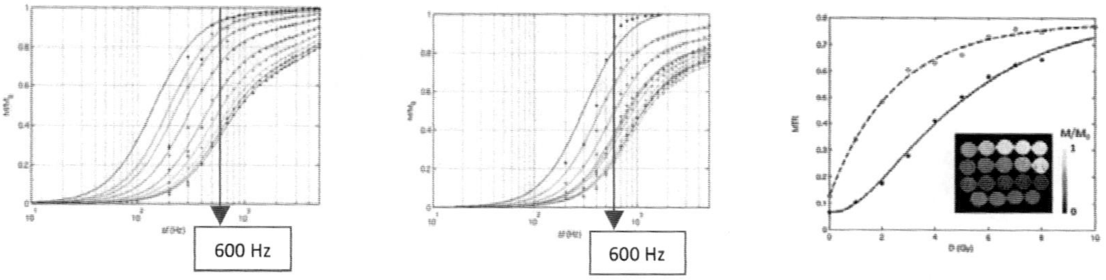

(a) (b) (c)

Figure 4. Magnetization as a function of saturation frequency offset for irradiated UD (a) and LD (b) gel samples. An optimum MTR sensitivity is found for $f_{off} = 600$ Hz. The corresponding MTR dose response shows a similar course as in R_2 for both UD (filled symbols) and LD (open symbols) gel.

4. Conclusions

Although a too significant discrepancy (approximately 10%) between the dose response in a UD and a LD PGD is found for clinical use, the error sources are well understood and there is room for improvement. There is great potential in using Styrofoam beads to lower the density of polymer gel dosimeters and for both R_2 and MT imaging.

5. References

[1] Vergote K *et al* 2004 *Phys. Med. Biol.* **49** 287-305
[2] Baldock C *et al* 2010 *Phys. Med. Biol.* **55** R1-63
[3] De Deene Y *et al* 2006 *Med. Phys.* **33** 2586-97
[4] Raaijmaker A J E *et al* 2008 *Phys. Med. Biol.* **53** 909-23
[5] Henkelman R M et al 1993 *Magn. Reson. Med.* **29** 759-66
[6] Baete S H *et al* 2008 *J. Magn. Reson.* **193** 286-96

Journal of Physics: Conference Series **444** (2013) 012027 doi:10.1088/1742-6596/444/1/012027

Basic investigations on LCV micelle gel

SB Ebenezer, MK Rafic, PB Ravindran
Department of Radiation Oncology, Christian Medical College, Vellore, India

E-mail: emailstoebi@gmail.com

Abstract. The aim of this study was to investigate the feasibility of using Leuco Crystal Violet (LCV) based micelle gel dosimeter as a quality assurance tool in radiotherapy applications. Basic properties such as absorption coefficient and diffusion of LCV gel phantom over time were evaluated. The gel formulation consisted of 25 mM Trichloroacetic acid, 1mM LCV, 4 mM Triton X-100, 4% gelatin by mass and distilled water. The advantages of using this gel are its tissue equivalence, easy and less preparation time, lower diffusion rate and it can be read with an optical scanner. We were able to reproduce some of the results of Babic et al. The peak absorption was found to be at 600 nm and hence a matrix of yellow LEDs was used as light source. The profiles obtained from projection images confirmed the diffusion of LCV gel after 6 hours of irradiation. Hence the LCV gel phantom should be read before 6 hours post irradiation to get accurate dose information as suggested previously.

1. Introduction

Verification of complex modern radiation therapy techniques such as 3-D CRT and IMRT need an integrating dosimeter that is tissue equivalent with good spatial resolution for dose verification [1-8]. In Fricke gel dosimetry, diffusion of the dose distribution [9, 10] after 2 hours creates inconvenience in the use of gel as a regular quality assurance tool.

Jordan et al developed a radiochromic hydrogel leuco crystal violet (LCV) micelle gel that has a low diffusion when compared to other gels [11]. An in-house optical cone beam CT (CBCT) scanner using yellow LED light source was used for evaluating the LCV gel dosimeter. In this study the absorption co-efficient and the diffusion of the LCV gel have been investigated.

2. Methods and Materials

2.1. Gel Preparation

The components used for the preparation of this gel was 4% w/w gelatin (Sigma, 300 bloom), 25 mM Trichloroacetic acid, (Sigma Aldrich), 1mM LCV, 4 mM Triton X-100 (Sigma) and triple distilled water. Gelatin was added to distilled water and heated to 30^{0}C with continuous stirring for 1 hour. Trichloroacetic acid, LCV and Triton X-100 were mixed together in distilled water and stirred at 45^{0}C for 30 minutes.

The two solutions were mixed after bringing the temperature of the second solution from 45^{0}C to 30^{0}C and stirred thoroughly for 10 minutes. The resulting pale-bluish violet coloured gel was poured into cylindrical phantoms (figure 1) (bulk gel of 1 Litre capacity) and cuvettes (figure 2) (4 ml volume). The gel phantoms were kept in the refrigerator at 4^{0}C until gelation.

2.2. Irradiation
The cylindrical gel phantom was irradiated to a dose of 30 Gy at a depth of 10 cm in the phantom using a 1.5 cm stereotactic cone attached to a Siemens Primus linear accelerator (Siemens, USA) that delivered 15 MV photons to the gel.

The cuvettes were kept in a water bath and irradiated with different doses on a Cobalt-60 teletherapy unit with lateral and backscatter conditions to find the attenuation data of the LCV gel.

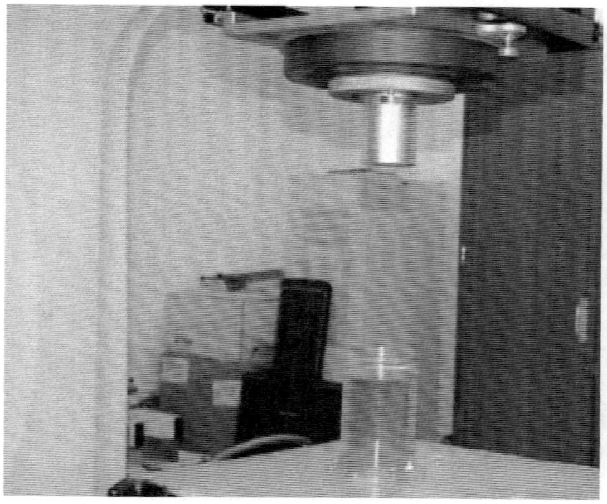

Figure 1: Irradiation setup of Bulk gel

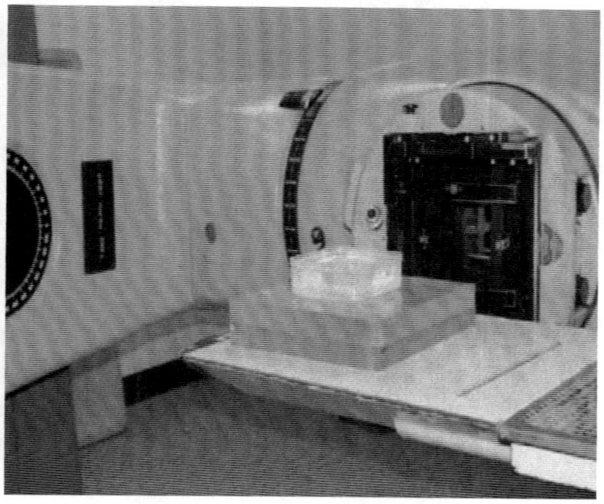

Figure 2: Irradiation setup of gel in cuvettes

2.3. Read out system
An in-house optical CBCT scanner was used for reading the dose matrix embedded in the gel phantoms. This scanner consisted of an aquarium of size 26 x 26 x 20 cm^3 with a turntable (rotated by a 4 pole stepper motor controlled with a computer) with angular graduations of 1 degree, a matrix of yellow LEDs as light source and a high resolution camera (Point Gray Fly Capture Camera) as a detector, The maximum resolution of the detector is 1024 x 768. 3D projection images (640 x 480) were obtained for 360 degrees. Reduction gears were used to ensure 1 degree rotations for every pulse sent to the stepper motor.

2.3.1. Optimal Readout wavelength determination.
For determining the optimal wavelength at which the LCV gel displays maximum absorption, the irradiated cuvettes with gel were read in a spectrophotometer (Schimadzu) for various wavelengths.

2.4. Gel Reconstruction
A pre and post irradiation scan of the phantom was carried out. The scans were performed with a time gap of 20 minutes, 1 hour, 2 hours, up to 6 hours 30 minutes and 7 hours post irradiation. Since the room temperature was at 32^0C, the water in the aquarium was maintained at 22^0C considering the melting point for LCV gel at 29^0C. The post irradiation scans were also performed at 12 and 19 hours.

The ratio of the pre and post scan data was obtained and the CT reconstruction of the optical attenuation was performed with 'ifanbeam' functions implemented in Matlab 7.5.0. Figure 3 shows the Matlab reconstructed dose distribution applying 'ifanbeam' function.

3. Results and Discussion
It was observed that the LCV gel under investigation had the peak absorption (figure 5) at a wavelength of 600 nm. The profiles obtained from the projection images of the bulk gel phantom (figure 6) showed no diffusion till 6 hours post irradiation and significant diffusion [11-13] when scanned at 6 hours 30 minutes post irradiation. However, the FWHM (Full width at Half Maximum)

measured from the profiles (figure 4) obtained from reconstructed data did not show any diffusion in the scans done till 7 hours post irradiation and a mild change after 7 hours.

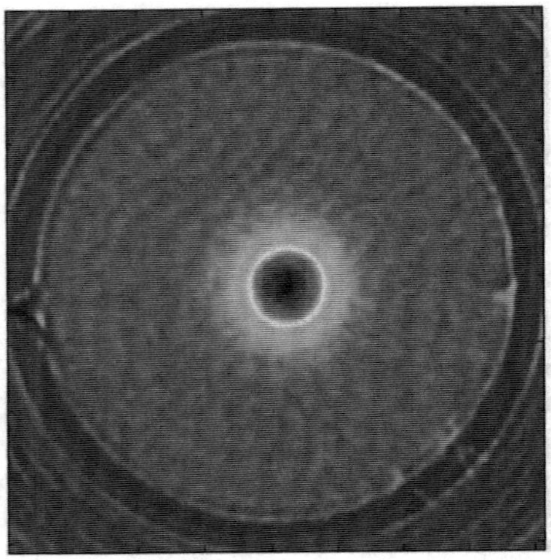

Figure 3: Reconstructed Dose

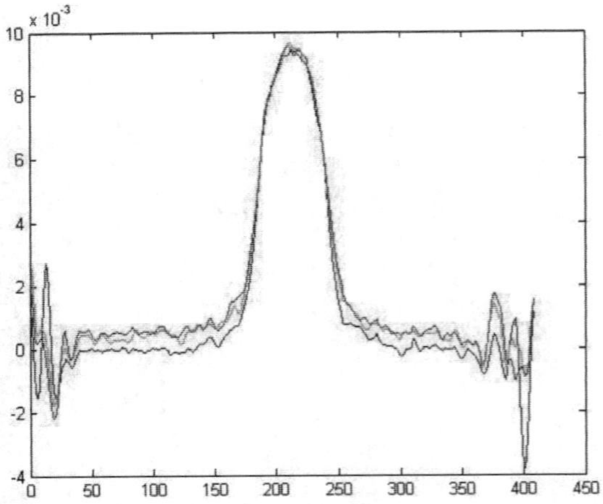

Figure 4: FWHM at 20 mins (blue), 6 hrs (green) and 6.5 hrs (red) post irradiation

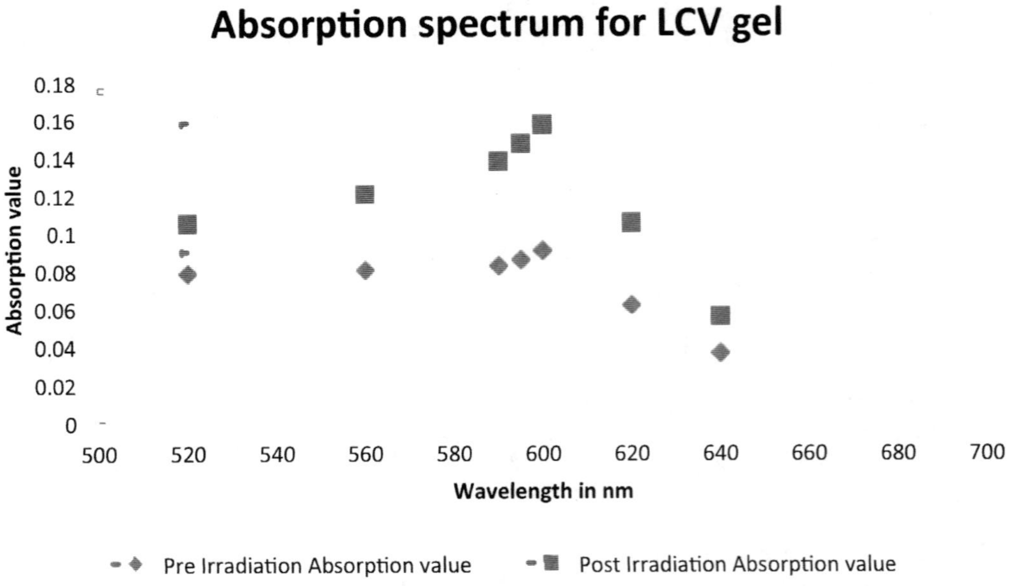

Figure 5: Plot showing the absorption spectrum of LCV gel irradiated to 20 Gy

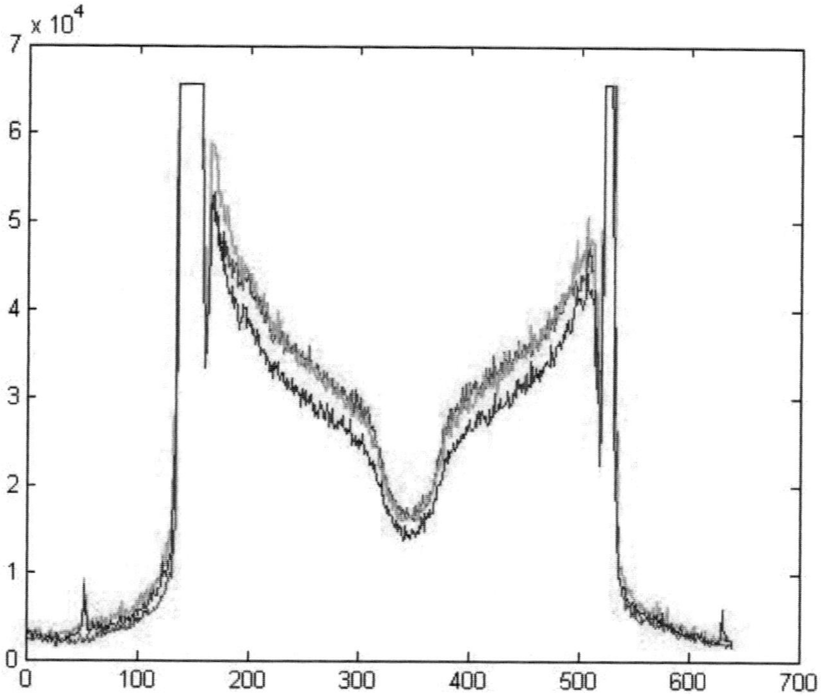

Figure 6: Profiles from projection images showing diffusion over time. 20 minutes (Red), 6 hours (Green) and 6.5 hours (Blue).

4. Conclusion

The feasibility of using the LCV gel with an optical CBCT scanner for radiotherapy dosimetry has been investigated. We were able to reproduce some previous published results [11-13]. Due to its lower diffusion rate compared to FX gel, it is concluded that the LCV gel scanned using optical CBCT scanner with a light source of wavelength 600 nm can be used as a 3D verification tool for high precision radiation therapy. As suggested by the previous authors [11-13] post irradiation scans should be performed before 6 hours to get accurate dose distribution. Care should be taken to keep the temperature low since this gel melts at 29^0C. Further work should involve making this gel more radiosensitive.

5. References

[1] Baldock C *et al* 2010 *Phys. Med. Biol.* **55** R1-63
[2] Hurley C *et al* 2005 *Appl. Radiat. Isot.* **63** 443-56
[3] Brindha S *et al* 2004 *Phys. Med. Biol.* **49** N353-61
[4] Bosi S G *et al* 2009 *Phys. Med. Biol.* **54** 275-83
[5] Lepage M *et al* 2002 *Phys. Med. Biol.* **47** 1881-1890
[6] Rintoul L *et al* 2003 *Appl. Spectrosc.* **57** 51-57
[7] Mather M L *et al* 2002 *Phys. Med. Biol.* **47** 4397-409
[8] Mather M L *et al* 2003 *Ultrasonics* **41** 551-9
[9] Harris P J *et al* 1996 *Phys. Med. Biol.* **41** 1745-53
[10] Baldock C *et al* 2001 *Australas. Phys. Eng. Sci. Med.* **24** 19-30
[11] Jordan K and Avvakumov N 2009 *Phys. Med. Biol.* **54** 141-53

[12] Babic S *et al* 2009 *Phys. Med. Biol.* **54** 6791-808
[13] Vandecasteele J *et al* 2011 *Phys. Med. Biol.* **56** 627-51

New polymer gel dosimeters consisting of less toxic monomers with radiation-crosslinked gel matrix

A Hiroki[1], S Yamashita[1], Y Sato[2], N Nagasawa[1] and M Taguchi[1]
[1]Environmental and Industrial Material Research Division, Quantum Beam Science Directorate, Japan Atomic Energy Agency, Takasaki, Gunma, 370-1292, Japan
[2]Chiyoda Technol Corporation, Higashi-ibaraki-gun, Ibaraki, 311-1313, Japan

E-mail: hiroki.akihiro@jaea.go.jp

Abstract. New polymer gel dosimeters consisting of less toxic methacrylate-type monomers such as 2-hydroxymethyl methacrylate (HEMA) and polyethylene glycol 400 dimethacrylate (9G) with hydroxypropyl cellulose (HPC) gel were prepared. The HPC gels were obtained by using a radiation-induced crosslinking technique to be applied in a matrix instead of a gelatin, which is conventionally used in earlier dosimeters, for the polymer gel dosimeters. The prepared polymer gel dosimeters showed cloudiness by exposing to ^{60}Co γ-ray, in which the cloudiness increased with the dose up to 10 Gy. At the same dose, the increase in the cloudiness appeared with increasing concentration of 9G. As a result of the absorbance measurement, it was found that the dose response depended on the composition ratio between HEMA and 9G.

1. Introduction

Polymer gel dosimeters composed of monomers and a gel matrix are drawing attention as a tool of the three-dimensional dose distribution measurements in the radiation therapy [1]. The polymer gel dosimeters are based on the following chemical behaviours. The monomers in the gel matrix are polymerized by the irradiation, and then the polymers aggregate to induce cloudiness in the gel matrix. Accordingly, an absorbance of the gel increases with the dose. Three-dimensional dose distributions are imaged finally by Optical CT [2], X-ray CT [3], MRI [4] or ultrasound [5].

Many polymer gel dosimeters have been developed [6, 7]. Currently, either acrylamide or methacrylic acid as a monomer and gelatin as a matrix are used in promising polymer gel dosimeters [8-11]. Although the gelatin gel is preparable conveniently, there is a possibility of the disappearance of the three-dimensional dose distribution because of the re-melting of the gel at the temperature over 35°C. On the other hand, since both acrylamide and methacrylic acid are harmful monomer, recipes for the polymer gel dosimeters consisting of less toxic monomers have been reported [12].

Polysaccharide derivatives such as carboxymethyl cellulose and carboxymethyl chitosan have been applied in industrial, medical, and agricultural fields because of safety and biodegradability. The polysaccharide derivatives usually undergo degradation by irradiation to a solid and a low-concentration aqueous solution. However, it was found that the gels of polysaccharide derivatives were obtained by γ-ray or electron beam irradiation to their highly concentrated aqueous solution without any crosslinker [13]. The gels prepared by a radiation-induced crosslinking technique show high water-absorption and thermal stability. Among the cellulose derivative gels, hydroxypropyl cellulose (HPC) gels are transparent materials [14].

In this work, we proposed new polymer gel dosimeters comprised of less toxic monomers and HPC gel prepared by radiation-induced crosslinking technique. Effects of the types of monomers and their composition ratios on the dose response of the polymer gel dosimeters were investigated.

2. Materials and methods

2.1. Preparation of HPC gels

HPC, which is the grade of 1000-5000 cP, was purchased from Wako Pure Chemical Industries, Ltd., Japan. Aqueous solution of 20 wt% HPC as a paste-state was sealed in polyethylene-nylon bags after pressing into a sheet of 1-mm thickness and degassing. The samples were irradiated to a dose of 10 to 70 kGy to obtain HPC gels. The irradiations were carried out using a Cockcroft-Walton type 2 MeV electron beam accelerator at the Takasaki Advanced Radiation Research Institute, Japan Atomic Energy Agency. The obtained HPC gels were immersed into an excess amount of distilled water to remove uncrosslinked HPC, and then vacuum-dried. Gel fraction of the obtained HPC gels was determined gravimetrically by measuring insoluble part after water extraction of sol. Degree of swelling (Sw) of the gels was calculated from weight ratio, $(W_s - W_d) / W_d$, where W_s is the weight of the swollen gel and W_d is the weight of the dried gel.

2.2. Preparation of polymer gel dosimeters

The dried HPC gels were immersed into aqueous monomer solutions consisting of monomers (5 g), tetrakis(hydroxymethyl)phosphonium chloride (THPC) (0.16 g), and ultrapure water (94.84 g). The monomer solutions with various composition ratios were prepared by using 2 of 14 kinds of monomers such as methacrylate- and acrylamide-types. The swollen HPG gels were vacuum-packed to avoid exposure to air in the storage. The prepared samples were stored at 5°C in a refrigerator until irradiation.

2.3. Irradiation

The prepared samples were irradiated up to 10 Gy by using a ^{60}Co γ-ray source at the Takasaki Advanced Radiation Research Institute, Japan Atomic Energy Agency. Dose rate was adjusted by varying the distance between the samples and source, which was in the range of 3.0 to 30 Gy/h. Dosimetry was preliminarily conducted with an ionization chamber. After irradiation, all samples were stored in the refrigerator.

2.4. Absorbance measurement

The absorbance of the polymer gel dosimeters were measured after 24 hours post-irradiation. The optical analysis was carried out using an ultraviolet and visible spectrophotometer (Hitachi High-Technologies Corporation, U-3310). The dose response of the polymer gel dosimeter was estimated from the absorbance at 660 nm as a function of dose.

3. Results and Discussion

Figure 1 shows the gel fraction and the degree of swelling (Sw) of HPC gels as the matrix of polymer gel dosimeters. The gel fraction of the HPC gel increased drastically up to 10 kGy to reach 75%, and then levelled off at about 90%. The Sw of the HPC gel swollen in pure water decreased from 26 to 8 at the dose of 10 to 70 kGy. Therefore, the HPC gel obtained at 10 kGy was selected as gel matrix for dosimeters because of the highest Sw. The Sw of the HPC gel swollen in each aqueous monomer solution exhibited in the range of 25 to 27. It was found that the Sw of the gel was independent of the type of monomers and their composition ratios. It would be worth noting that all of the swollen HPC gels were transparent.

Polymer gel dosimeters with 32 kinds of composition ratio were prepared by using 2 monomers which was selected from 14 monomers. Eighteen polymer gel dosimeters became cloudy by γ-irradiation at 10 Gy. It was found that 9 monomers can be utilized in the polymer gel dosimeter.

Among of them, methacrylate monomers such as 2-hydroxyethyl methacrylate (HEMA) and polyethylene glycol 400 dimethacrylate (9G) were efficient in terms of induction of the cloudiness of polymer gel dosimeter in comparison with acrylamide-type monomers such as acrylamide (AAm) and *N,N'*-methylene-bis-acrylamide (MBAAm). It is well known that polymer gel dosimeter comprised of AAm and MBAAm with gelatin, called PAG or PAGAT, became cloudy less than 10 Gy [15]. However, the polymer gel dosimeter with HPC gel matrix, which prepared by using the AAm/MBAAm aqueous solution, remained clear even at 10 Gy. This would be due to the effect of the matrix.

Figure 2 shows the photograph of the polymer gel dosimeters prepared by using the monomer solutions consisting of HEMA and 9G. The cloudiness of the polymer gel dosimeters increased with the dose. The polymer gel dosimeter comprised of HEMA and 9G at 3 and 2 wt%, respectively, exhibited higher cloudiness than that at 4 and 1 wt%. It was found that the polymer gel dosimeter with 3 wt% of HEMA and 2 wt% of 9G became cloudy at only 1 Gy. The absorbance measurement showed that the absorbance of the polymer gel dosimeters consisting of HEMA and 9G increased approximately linearly with an increase in the dose up to 10 Gy. The dose response that defined as the

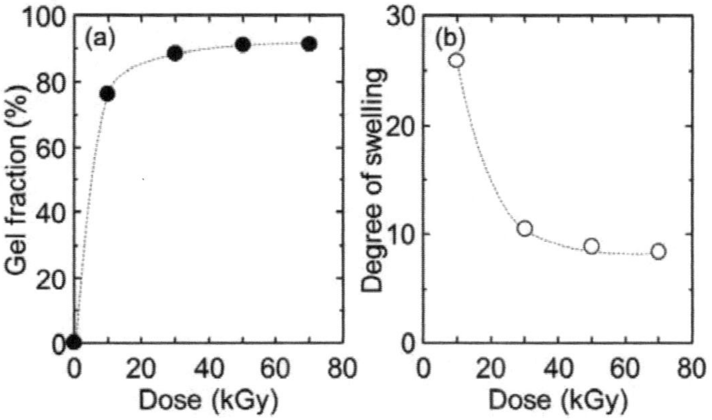

Figure 1: Gel fraction (a) and degree of swelling (b) of the HPC gels as a function of dose.

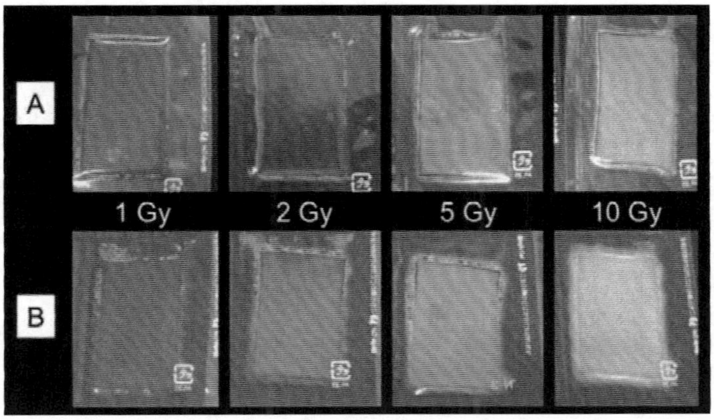

Figure 2: Photographs of the irradiated polymer gel dosimeters consisting of HEMA and 9G with HPC gel. The concentrations of HEMA and 9G are 4 wt% and 1 wt% (A), and 3 wt% and 2 wt% (B), respectively.

initial increment of absorbance per unit dose depended on the concentration of the 9G. It was found, therefore, that the dose response can be adjusted by changing the composition ratio.

4. Conclusion

New polymer gel dosimeters were prepared by using the radiation-crosslinked HPC gel instead of a conventional gelatin gel. The polymer gel dosimeters comprised of less toxic methacrylate-type monomers such as HEMA and 9G and HPC gel showed the cloudiness at only 1 Gy of ^{60}Co γ-irradiation. The absorbance of the polymer gel dosimeter at 660 nm increased approximately linearly with the dose up to 10 Gy. The dose response depended on the composition ratio between HEMA and 9G. It is expected that development of more sensitive polymer gel dosimeters can be achieved by selecting the kinds of monomers and optimizing their composition ratios.

5. Acknowledgement

This work was partially supported by JSPS KAKENHI (a Grant-in-Aid for Young Scientists (B)) Grant Number 23710073.

6. References

[1] Baldock C *et al* 2010 *Phys. Med. Biol.* **55** R1-63
[2] Bosi S *et al* 2007 *Phys. Med. Biol.* **52** 2893-903
[3] Hill B *et al* 2005 *Br. J. Radiol.* **78** 623-630
[4] Murry P and Baldock C 2000 *Australas. Phys. Eng. Sci. Med.* **23** 44-51
[5] Mather M L *et al* 2002 *Phys. Med. Biol.* **47** 1449-1458
[6] Lepage M *et al* 2002 *Phys. Med. Biol.* **47** 1881-1890
[7] Rintoul L *et al* 2003 *Appl. Spectrosc.* **57** 51-57
[8] Maryanski M J *et al* 1994 *Phys. Med. Biol.* **39** 1437-55
[9] Baldock C *et al* 1998 *Phys. Med. Biol.* **43** 695-702
[10] Hurley C *et al* 2005 *Appl. Radiat. Isot.* **63** 443-56
[11] Venning A *et al* 2005 *Phys. Med. Biol.* **50** 3875-88
[12] Senden R J *et al* 2006 *Phys. Med. Biol.* **51** 3301-14
[13] Yoshii F *et al* 2003 *Nucl. Instr. and Meth. B* **208** 320-4
[14] Hiroki A *et al* 2012 *JAEA-Review* **2011-043** 50
[15] Mariani M *et al* 2007 *Radiat. Phys. Chem.* **76** 1507-10

Journal of Physics: Conference Series **444** (2013) 012029 doi:10.1088/1742-6596/444/1/012029

Customising PRESAGE® for diverse applications

T Juang[1], J Newton[1], M Niebanck[1], R Benning[2], J Adamovics[2] and M Oldham[1]
[1]Duke University Medical Center, Durham, NC, USA
[2]Rider University, Lawerenceville, NJ, USA

E-mail: mark.oldham@duke.edu

Abstract. PRESAGE® is a solid radiochromic dosimeter consisting of a polyurethane matrix, a triarylmethane leuco dye, and a trihalomethane initiator. Varying the composition and/or relative amounts of these constituents can affect the dose sensitivity, post-irradiation stability, and physical properties of the dosimeter. This allows customisation of PRESAGE® to meet application-specific requirements, such as low sensitivity for high dose applications, stability for remote dosimetry, optical clearing for reusability, and tissue-like elasticity for deformable dosimetry. This study evaluates five hard, non-deformable PRESAGE® formulations and six deformable PRESAGE® formulations and characterizes them for dose sensitivity and stability. Results demonstrated sensitivities in the range of 0.0029 – 0.0467 ΔOD/(Gy·cm) for hard formulations and 0.0003 – 0.0056 ΔOD/(Gy·cm) for deformable formulations. Exceptional stability was seen in both standard and low sensitivity non-deformable formulations, with promising applications for remote dosimetry. Deformable formulations exhibited potential for reusability with strong post-irradiation optical clearing. Tensile compression testing of the deformable formulations showed elastic response consistent with soft tissues, with further testing required for direct comparison. These results demonstrate that PRESAGE® dosimeters have the flexibility to be adapted for a wide spectrum of clinical applications.

1. Introduction

PRESAGE® is a solid radiochromic dosimeter with the potential for use across a wide range of clinical applications [1-3]. In order to maximize PRESAGE®'s usefulness as a 3D dosimetry [4] tool, the dosimeter must be adaptable to meet application-specific dosimetry requirements. Examples of application-specific considerations are listed in table 1.

Table 1. Examples of Application-Specific Dosimetry Considerations

Application	Considerations
IMRT, VMAT	Tissue equivalent electron density
	Volume-independent dose response
	Un-irradiated transparency
Radiosurgery	High doses require low sensitivity
Remote dosimetry	Post-irradiation stability, low thermal effects
Brachytherapy	Tissue equivalence, high sensitivity, good stability
Deformable dosimetry	Elastic properties comparable to human tissues
Reusable dosimeters	Stable immediately post-RT, then optical clearing

The components of PRESAGE® are a transparent polyurethane matrix (approximately 90% of the PRESAGE® dosimeter by weight), a triarylmethane leuco dye, and a trihalomethane initiator. Formulations customized to specific applications can be made by varying the composition and/or relative amounts of these constituents. Physical characteristics (transparency, elasticity) of the PRESAGE® polyurethane matrix are dependent on the starting monomeric isocyanate and polyol, which are catalyzed by metal compounds. Recent studies have shown that the composition of these metal compounds and the trihalomethane initiator both influence dosimeter sensitivity [5, 6].

This study evaluates several formulations of PRESAGE®, including novel deformable formulations under initial stages of development, and characterizes them for a range of dose sensitivities and post-irradiation stability.

2. Methods

2.1. *PRESAGE® Chemistry and Synthesis*

The five formulations of the standard hard (Shore D Hardness 80) PRESAGE® (including two low sensitivity formulations specifically designed for high-dose applications) and six formulations of the deformable (Shore A Hardness 10-20) PRESAGE® under investigation are listed in detail in table 2. The leuco dye leucomalachite green (LMG) and its derivatives were synthesized by placing 1 equivalent of the corresponding aldedyde N,N-dimethylaniline (6 eq) and benzene (0.5M with respect to the aldehyde) in a flask fitted with a Dean-Stark trap and reflux condenser. p-Toluenesulfonic acid (0.1 eq) was added to the mixture and the solution refluxed. The reaction was monitored by thin layer chromatography until completion and worked up by diluting with benzene and washing with 10% sodium bicarbonate solution. The benzene and excess aniline were removed via azeotrope distillation with water. Flash column chromatography was performed to further purify the products. One LMG derivative (o-MeO-LMG DEA) was synthesized using N,N-diethylaniline.

Table 2. PRESAGE® formulations under evaluation.

	Formulation	Leuco Dye	Initiator Content (%)	Peak λ (nm)
Non-Deformable	SS1 (standard sensitivity)	2.0% LMG	0.75%	633
	SS2 (standard sensitivity)	2.0% LMG	0.50%	633
	SS3 (standard sensitivity)	1.7% o-MeO-LMG	0.40%	633
	LS1 (low sensitivity)	1.5% LMG	0.75%	633
	LS3 (low sensitivity)	1.0% o-MeO-LMG	0.75%	633
Deformable	D1	2.0% o-MeO-LMG	0.50%	633
	D2	2.0% o-MeO-LMG-DEA	0.50%	633
	D3	2.0% p-MeO-LMG	0.50%	620
	D4	2.0% dimethyl-LMG	0.50%	633
	D5	2.0% o-Cl-LMG	0.50%	647
	D6	2.0% o-Br-LMG	0.50%	639

2.2. *PRESAGE® Formulation Characterization*

2.2.1. *Sensitivity*

Sensitivity was determined from the change in optical density in sets of cuvettes containing each formulation under evaluation. Each set of cuvettes was irradiated on a 6 MV linac to known doses between 0 – 25 Gy for low sensitivity formulations and 0 – 8 Gy for all other formulations. Change in optical density was acquired using a Thermo Spectronic Genesys 20 spectrophotometer, and determined by subtracting the pre-irradiation absorption from the post-irradiation absorption for each cuvette at the peak absorption wavelength for each formulation. Cuvettes were stored in the dark following irradiation and tracked for up to 14 days to determine post-irradiation stability.

2.2.2. Tensile testing
The six deformable PRESAGE® formulations were subjected to tensile compression testing to evaluate the elasticity of the formulations. While Young's modulus values for elastic materials, such as soft biological tissues and deformable PRESAGE®, are highly dependent on the method of deformation [1], consistent tensile measurements allow an initial elasticity comparison between formulations. 10 mm x 10 mm x 25 mm samples of each formulation were subjected to compression by 80% (8 mm) at a rate of 2 mm/min using a Lloyd Model LRX Plus tensile tester.

3. Results

3.1. Sensitivity
The immediate post-irradiation sensitivities of each formulation are listed in table 3. All formulations exhibited a linear dose response. In the non-deformable formulations, use of o-MeO-LMG in the SS3 formulation increased the dosimeter dose sensitivity by approximately two-fold even with decreased initiator content. Deformable PRESAGE® formulations were found to exhibit sensitivities one order of magnitude lower than that of comparable formulations using the non-elastic polyurethane matrix.

Table 3. PRESAGE® sensitivity.

	Formulation	Sensitivity (ΔOD/(Gy·cm))		Formulation	Sensitivity (ΔOD/(Gy·cm))	
		Immediate			Immediate	2 days
Non-Deformable	SS1	0.0225	**Deformable**	D1	0.0012	--
	SS2	0.0213		D2	0.0014	--
	SS3	0.0467		D3	0.0056	0.0002
	LS1	0.0102		D4	0.0034	0.0028
	LS3	0.0029		D5	0.0005	--
				D6	0.0003	--

3.2. Post-irradiation Stability
Normalized post-irradiation sensitivities over time are shown in figure 1 for the five non-deformable PRESAGE® formulations. This figure illustrates the wide variety of response – from OD darkening (increasing sensitivity over time) to optical clearing (decreasing sensitivity over time) – achievable through modifications to the leuco dye and initiator. Exceptional post-irradiation stability (variation ≤1%) was exhibited by both standard sensitivity (SS3) and low sensitivity (LS1) formulations, suggesting promising applications in remote dosimetry for both standard and high dose applications.

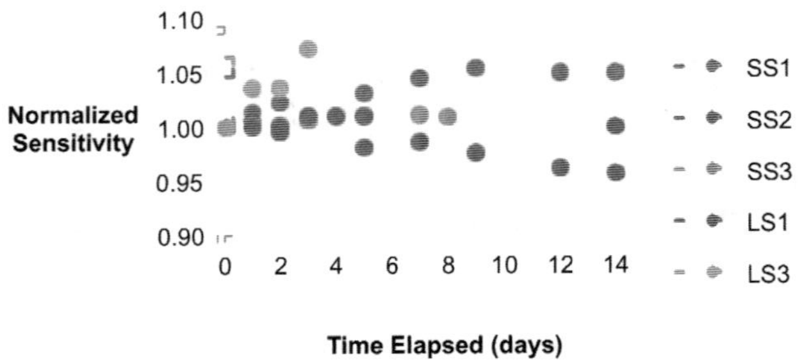

Figure 1: Normalized sensitivity over 14 days for all non-deformable formulations.

Sensitivity measurements were taken of the two most sensitive deformable formulations (D3 and D4) 2 days following irradiation as an initial assessment of post-irradiation characteristics (table 3). Both formulations exhibited decreases in sensitivity and ΔOD, suggesting potential reusability.

3.3. Elasticity of Deformable Formulations

Stress-strain curves for all deformable formulations are plotted in figure 2. The elastic response of all formulations show a stress-dependent Young's modulus (stress/strain) and increasing resistance to deformation with increasing applied stress, which is consistent with most soft biological tissues [7]. Further tensile testing is required to directly compare deformable PRESAGE® elasticity to that of tissues of interest.

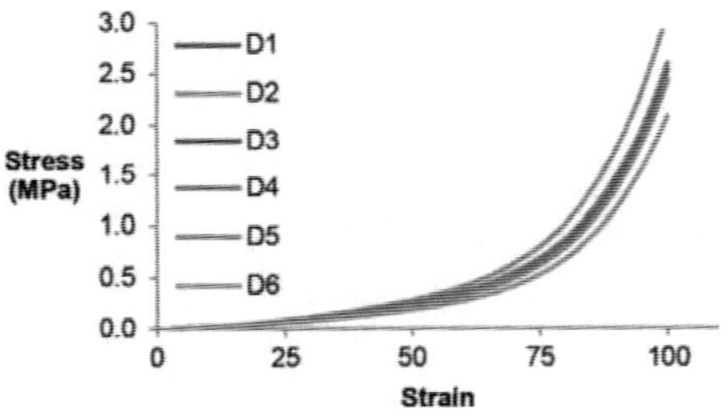

Figure 2: Stress-strain curves for deformable formulations determined through tensile compression. All samples were compressed by 80% (8 mm deflection). Young's modulus is represented by the slopes of the curves and is stress-dependent.

4. Conclusion

All PRESAGE® dosimeter formulations have in common a polyurethane matrix, a triarylmethane leuco dye, and a trihalomethane initiator which can be varied to adjust the properties of the dosimeter to suit application-specific dosimetry requirements. The most frequently used leuco dye has been LMG, but as shown above, formulations synthesized with the LMG derivative o-MeO-LMG are both more dose-sensitive per weight and more stable post-irradiation. Our results demonstrate that PRESAGE® dosimeters have the flexibility to be adapted for use in a wide spectrum of clinical applications, including low sensitivity for high dose applications, elasticity for deformable dose tracking, post-irradiation stability for remote dosimetry, and clearing for reusability.

5. Acknowledgements

This work was supported by NIH R01CA100835. We thank Patrick McGuire of the Duke University Mechanical Engineering and Materials Science Department for his help with tensile measurements.

6. References

[1] Brady S L *et al* 2010 *Phys. Med. Biol.* **55** 2187–201
[2] Clift C *et al* 2010 *Phys. Med. Biol.* **55** 1279–93
[3] Thomas A *et al* 2010 *Med. Phys.* **38** 4846
[4] Brown S *et al* 2008 *Appl. Radiat. Isot.* **66** 1970-4
[5] Gorjiara T *et al* 2011 *Med. Phys.* **38** 2265-74

[6] Baldock C *et al* 2010 *Phys. Med. Biol.* **55** R1-63
[7] Alqathami M *et al* 2012 *Radiat. Phys. Chem.* **81** 1688-95
[8] Alqathami M *et al* 2012 *Radiat. Phys. Chem.* **81** 867-73
[9] McKee C T *et al* 2011 *Tissue Eng.* **17** 155-64

NIPAM polymer gel dosimetry for IMRT four-field box irradiation using optical-CT scanner

CH Yao[1,2], WT Hsu[2,3], SM Hsu[1,2], PYL Ma[3], BT Hsieh[4] and YJ Chang[5,6]
[1]Department of Biomedical Imaging and Radiological Science, China Medical University, Taichung, Taiwan (R.O.C.)
[2]Graduate Institute of Clinical Medical Science, China Medical University, Taichung, Taiwan (R.O.C.)
[3]Department of Radiation Oncology, Puli Christian Hospital, Nan Tou, Taiwan (R.O.C.)
[4]Department of Medical Imaging and Radiological Science, Central Taiwan University of Science and Technology, Taichung, Taiwan (R.O.C.)
[5]Institute of Biomedical Engineering and Material Science, Central Taiwan University of Science and Technology, Taichung, Taiwan (R.O.C.)
[6]Department of Management Information Systems, Central Taiwan University of Science and Technology, Taichung, Taiwan (R.O.C.)

E-mail: ronchang@ctust.edu.tw

Abstract. The study assessed the dosimetric characteristics of the N-isopropylacrylamide (NIPAM) polymer gel dosimeter. Experiments on the intra-dosimeter consistency and reproducibility of NIPAM polymer gels were performed. A cylindrical NIPAM gel phantom measuring 10 cm (diameter) by 10 cm (height) by 3 mm (thickness) was irradiated using the four-field box treatment with a field size of 3 cm × 3 cm. A fast, optical computerized tomography scanner was used to scan the gel phantoms. The results showed that the dose profiles were consistent at various depths. The isodose lines agreed quantitatively with the calculated TPS dose and the measured NIPAM polymer gel dose within the 30 to 90 percentage isodose lines. In addition, the Gamma pass rates were determined to be 94.9%, 95.2%, and 95.7% at depths of 40 mm, 45 mm, and 50 mm, respectively, using 5% dose difference and 5 mm distance-to-agreement criteria. Using the same Gamma criteria, the Gamma pass rates were 95.1%, 95.3%, and 95.7% for the three replicated. The results indicated that the NIPAM polymer gel dosimeter was stable and reliable. The dosimetric characteristics highlighted the potential of NIPAM polymer gel dosimeter in radiotherapy.

1. Introduction

The rapid development of complex radiotherapy technologies, such as intensity-modulated radiation therapy (IMRT), cyber knife, tomotherapy, proton therapy, and brachytherapy, has made dose verification highly important [1-5]. One-dimensional and two-dimensional dose detectors are deemed insufficient for dose measurement in 3D complex distributions [6]. Polymer gel dosimetry is one of the most promising candidates in three-dimensional dosimetry for radiation therapy. However, oxygen (O_2) inhibition terminates the chemical polymerization reaction [7] Thus, antioxidants must be added to the gel compositions for gel preparation in the laboratory [8-11]. Fong et al. developed a nontoxic

MAGIC gel with less harmful gel monomers and compositions [8]. Senden et al. proposed the polymer gel, N-isopropylacrylamide (NIPAM) [11], which shows higher dose-response sensitivity and lower dose-rate dependence. In addition, Chang et al. revealed the long time temporal stability and high linearity of NIPAM polymer gel compositions [12]. In the present study, the NIPAM polymer gel recipe proposed by Chang et al. [12] was used in the IMRT four-field box irradiation to assess the internal uniformity and the reproducibility of dose response.

2. Materials and Methods

2.1. Gel preparation

The NIPAM polymer gel used in the study was composed of 5% gelatin, 5% NIPAM (97%, Sigma-Aldrich), 3% Bis (N,N'-methylene bisacrylamide), and 5 mM Tetrakis (hydroxymethyl) phosphonium chloride. The gel was formulated inside a fume hood under normal atmospheric conditions. Subsequently, the preparation process described by Senden and De Deene [8, 9] was performed. After completing the gel preparation, the gel was poured into cylindrical acrylic phantoms with the following dimensions: 10 cm (diameter), 10 cm (height), and 3 mm (wall thickness). The gel phantom was wrapped in aluminum foil to prevent photo-polymerization and then cooled to 7 °C in a refrigerator with water bath for 5 h until complete solidification was achieved. Three independent batches of gel phantoms manufactured at different lengths of time were prepared for the replicated experiments.

2.2. Treatment planning system and irradiation

A commissioned PROWESS treatment planning system was used to design the four-field box irradiation. The gel phantom was fixed using a vacuum bag with cross marks for alignment. The gel phantom was irradiated through IMRT (SIEMENS ONCOR Impression). The treatment field comprised four identical 3 cm × 3 cm square fields with gantries aligned at 0°, 90°, 180°, and 270°. The prescribed dose was 10 Gy at a beam-beam average dose of 2.5 Gy. The dose rate was 250 cGy, the focal point setting was at the center of the gel phantom, and the source surface distance (SSD) was 95 cm.

2.3. Scanning and data analysis

In this study, MGS OCTOPUS-10X Optical-CT scanner (MGS Research, Inc., Madison, CT, USA) was used as the read-out tool of the NIPAM polymer gel dosimeter. A single beam laser (780 nm wavelength, 30 mW) was used to scan the gel phantom, which was mounted on a turntable inside the scanning tank filled with a refractive-index matching liquid. Refractive index matching ensures that the rays transmitted by the gel propagate in a straight direction. After scanning, the image reconstruction program "reconQexp.m" written in MATLAB (The MathWorks, Natick, MA, USA) was used to reconstruct the projection data using the filtered back-projection technique. The final data presentation of the reconstructed image and the dose calibration were performed using the Java image-processing program ImageJ (version 1.43) developed by the National Institute of Health (NIH). Quantitative evaluation of dose distributions [13] was performed using gamma analysis [14, 15]. In this study, the measured NIPAM gel dose distributions were compared with the calculated treatment planning (TPS) dose distribution using 3% dose-difference and 5 mm distance-to-agreement (DTA) comparison criteria.

3. Results and Discussion

Figure 1 shows the line profiles of NIPAM polymer gel dosimeter at depths of 40 mm, 45 mm, and 50 mm for the post-irradiation time of 24 h. The maximum mean deviation is less than 3% in the center 4 cm region and the result indicates that intra-uniformity of gel is consistent in multiple slices.

Figure 2 shows the transverse view of the corresponding isodose lines at a depth of 40 mm. The calculated TPS dose (dashed line) and the measured NIPAM polymer gel dose (solid line) agree well within the 30 to 90 percentage isodose lines. However, the dose difference along the edge of the square field is larger than that in other regions. This erroneous phenomenon is probably a result of edge enhancement. The consistent results can be observed from the Gamma map (figure 3). The rejected regions are presented at the edge of the square field where high dose gradient are present.

Table 1 shows that the Gamma pass rates are 94.9%, 95.2%, and 95.7% at depths of 40 mm, 45 mm, and 50 mm, respectively, using 5% dose difference and 5 mm DTA criteria. Utilizing the same Gamma criteria, the Gamma pass rates for the three replicated experiments are 95.1%, 95.3%, and 95.7% (Table 2). Based on these values, the NIPAM polymer gel dosimeter is deemed very stable and reliable.

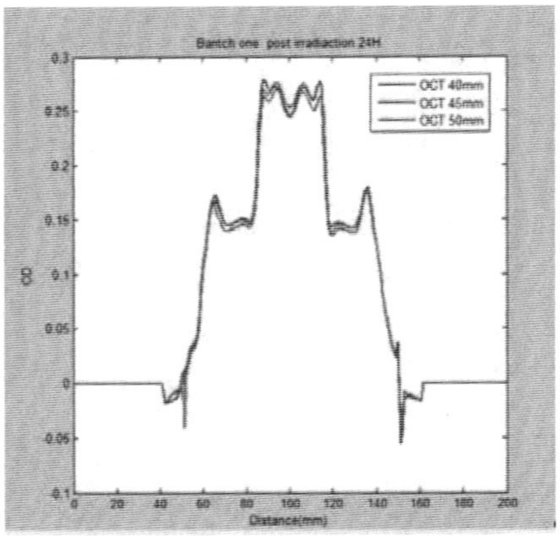

Figure 1: Line profiles at depths of 40, 45, and 50 mm for post-irradiation time of 24 h

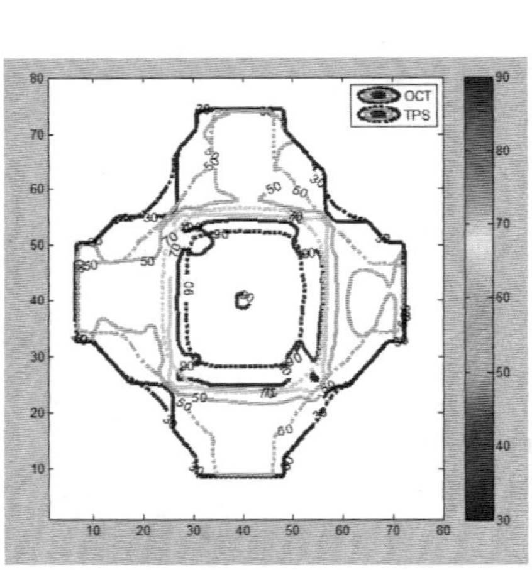

Figure 2: Isodose comparison between calculated TPS dose (dashed line) and measured NIPAM polymer gel dose (solid line)

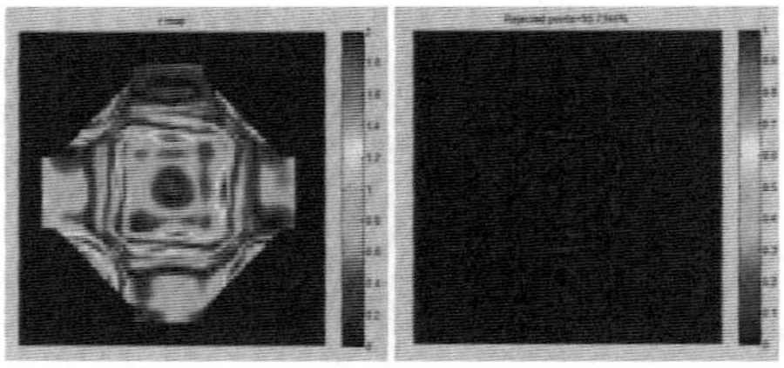

Figure 3: Representative 3D gamma comparison map of TPS and the measured NIPAM gel dose using 5% dose difference and 5 mm distance-to-agreement criteria

Table 1. Gamma index at various depths of NIPAM polymer gel

Interest in the area	$\gamma < 1$	$\gamma < 1$	$\gamma < 1$
Depth (mm)	40	45	50
Gamma criteria	5%/5 mm	5%/5 mm	5%/5 mm
Pass rate	94.9%	95.2%	95.7%

Table 2. Gamma indices of three batches of NIPAM polymer gel

Interest in the area	$\gamma < 1$	$\gamma < 1$	$\gamma < 1$
Batch	1	2	3
Gamma	5%/5 mm	5%/5 mm	5%/5 mm
Pass rate	95.1%	95.3%	95.7%

4. Conclusion

This detailed investigation of NIPAM polymer gel dosimeter provided promising results. The NIPAM polymer gel dosimeter showed stable and reliable results from intra-dosimeter comparison and replicated experiments. Likewise, the isodose lines agreed well for the 30 to 90 percentage ranges in transverse view. The dosimetric characteristics indicated that the NIPAM polymer gel dosimeter has great potential in radiotherapy.

5. References

[1] Baldock C *et al* 2010 *Phys. Med. Biol.* **55** R1-63
[2] Hurley C *et al* 2005 *Appl. Radiat. Isot.* **63** 443-56
[3] Gustavsson H *et al* 2004 *Phys. Med. Biol.* **49** 227-41
[4] Bosi S *et al* 2007 *Phys. Med. Biol.* **52** 2893-903
[5] Bosi S G *et al* 2009 *Phys. Med. Biol.* **54** 275-83
[6] Baldock C *et al* 1998 *Phys. Med. Biol.* **43** 695-702
[7] Lepage M *et al* 2001 *Phys. Med. Biol.* **46** 2827-839
[8] Fong P *et al* 2001 *Phys. Med. Biol.* **46** 3105-13
[9] De Deene Y *et al* 2002 *Phys. Med. Biol.* **47** 3441-63
[10] De Deene Y *et al* 2006 *Phys. Med. Biol.* **51** 653-73
[11] Senden R J *et al* 2006 *Phys. Med. Biol.* **51** 3301-14
[12] Chang YJ *et al* 2001 *Nucl. Instrum. Methods. Phys. Res. A* **652** 783-85
[13] Murry P and Baldock C 2000 *Australas. Phys. Eng. Sci. Med.* **23** 44-51
[14] Low D A *et al* 1998 *Med. Phys.* **25** 656-61
[15] Low D A and Dempsey J F 2003 *Med. Phys.* **30** 2455-64

Comparison between polymer gel dosimetry and calculated dose with small field in stereotactic irradiation

H Kawamura[1,2], K Shinoda[3], H Fuse[4], T Terunuma[2,4], K Miyamoto[3], T Sakae[2,4], and A Matsumura[2]
[1]Department of Radiological Sciences, Ibaraki Prefectural University of Health Sciences, 4669-2, Ami, Ami-machi, Inashiki-gun, Ibaraki, Japan
[2]Graduate School of Comprehensive Human Science, University of Tsukuba, 1-1-1, Tennodai, Tsukuba, Ibaraki, Japan
[3]Tsukuba Medical Center hospital, 1-3-1, Amakubo, Tsukuba, Ibaraki, Japan
[4]Proton Medical Research Center, University of Tsukuba, 2-1-1 Amakubo, Tsukuba, Ibaraki, Japan

E-mail: kawamura@ipu.ac.jp

Abstract. The purpose of this study is to investigate gel dosimetry for a small irradiation field in stereotactic radiotherapy. Treatment plans were generated by the Pinnacle^3 treatment plan system (TPS) for three different circular irradiated fields: 10 mm, 15 mm, and 20 mm. The polymer gels were irradiated to 6 Gy with 10-, 15-, and 20-mm-diameter collimators in 4 MV photon beams for stereotactic irradiation following TPS. Irradiated gels were evaluated with MRI at 1.5 T with R2 images. Firstly, the line profile of the irradiated center between TPS plan and the R2 image was compared. In the center profile at a dose calculated from the treatment plan, the full width at half maximum (FWHM) of 10-mm, 15-mm, and 20-mm collimators, were 13 mm, 19 mm, and 25 mm, respectively. In the center profile at R2 from the gel dosimetry, the FWHM were 13 mm, 20 mm, and 23 mm, respectively. Secondly, R2 images were converted to dosimetric maps to apply the gamma evaluation method. Comparison using gamma evaluation in the center of the irradiated plane between TPS plan and the dose map from the R2 image was performed. In gamma evaluation, when 3% and 3 mm criteria were used for comparison of the center plane of dose image from TPS and gel dosimetry, the pass ratio of the gamma criterion between calculated dose from the TPS and the dose map of irradiated gels in stereotactic irradiation was 98.6%. In comparison of the center profile and center plane, results of gel dosimetry were shown to have good agreement with the generated treatment plan dosimetric map for stereotactic irradiation.

1. Introduction

Polymer gel dosimetry that utilizes radiation-induced polymerization that is related to radiation dose, is a three-dimensional (3D) dosimetric tool for quality assurance [1-3][1]. Polymer gel dosimetry is desirable for radiation therapy for dose validation and to check whether the planned dose and areas are correctly irradiated [4]. It can be used to measure 3D doses in conformal radiotherapy, and it is efficient at measuring the 3D dose distribution in radiation therapy. Additionally, it has advantages of water and soft-tissue equivalent material dosimeters [5, 6][2, 3]. When a polymer gel is used as a phantom, it is possible to measure the radiation dose without disturbing the radiation fields and surroundings [7-9][4-6]. Accordingly, polymer gel dosimetry is a more suitable method than dosimetry using ion chambers for a small irradiated field.

Contemporary radiotherapy treatments include intensity modulated radiotherapy (IMRT) and stereotactic radiotherapy (SRT) applied to irradiate complicated targets and small targets. These treatment techniques need to employ 3D dosimetry to accurately deliver a treatment dose to a clinical target. In IMRT, in order to irradiate a clinical target more accurately following Treatment Planning System (TPS) for irradiation, the irradiated field is changed with a multi-leaf collimator and dose rate adjustment. However, it is difficult to evaluate accuracy with polymer gel dosimetry in IMRT. Because the polymer gel dosimetry is dependent on dose rate, it was also necessary to evaluate 3D dose distributions irradiated with a constant dose rate in the present study. In this way, polymer gel dosimetry could be used to improve IMRT irradiation.

Polymer gel dosimetry can also be used in small field irradiation. The purpose of the present study is to investigate gel dosimetry for a small irradiation field in stereotactic radiotherapy. Comparisons between polymer gel dosimetry and calculated dose from the TPS were performed in the center profile and the center plane. The gamma evaluation method was used for comparison in the center plane.

2. Materials and Methods

2.1 Preparation of polymer gel

A BANG-3 type polymer gel was prepared. The BANG kit was melted in a hot water bath at 55 degree, and 5 µM CuSO4 as modifying buffer and 1 mM ascorbic acid as antioxidant agent were added. Prepared gels were poured into polyethyleneterepthalate (PET) containers. The PET tubes were 30 mm in diameter and 70 mm in length for dose-R2 calibration. The PET containers were 50 mm, 50mm, and 70 mm in length for SRT irradiation (Fig. 1). The PET containers were filled with ascorbic acid solutions 2 days before gel preparation for the purpose of oxygen scavenging. The gels in PET containers were stored wrapped in aluminum foil in a refrigerator at 4 degree until irradiation.

2.2 Gel phantom design of stereotactic irradiation and treatment plan

Figure 1(a) shows a gel phantom for SRT irradiation. The phantom was a 200×140×125 mm Styrofoam container within the gel. The phantom was filled with water and air was turned out.

Treatment plans were 1.0-mm, 1.5-mm, and 2.0-mm simple circular irradiated fields generated by the Pinnacle^3 treatment plan radiotherapy system made by the Hitachi Medical Corporation. An X-ray CT device (CTS-20SPS, Shimadzu Corporation) was used for the stereotactic irradiation treatment plan. A non-helical scan was employed. The CT exposure conditions were as follows: a tube voltage of 120 kV, tube current of 130 mA, exposure time of 1.0 s and a slice thickness of 3.0 mm, an imaging field of view of 300×300 mm. The number of slices of the CT images was 30.

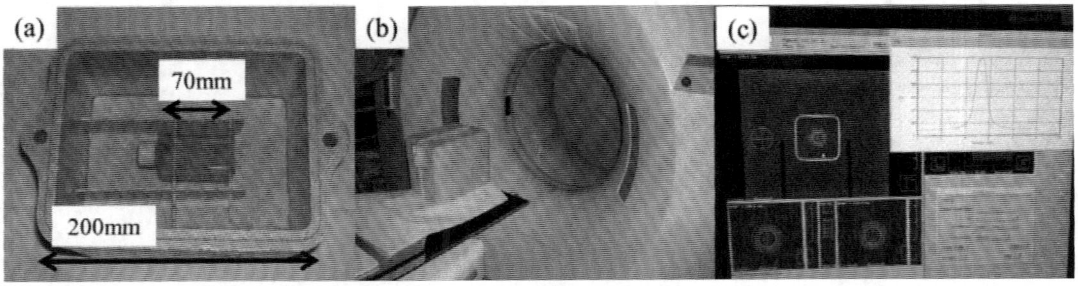

Figure 1: (a) left: Photograph of phantom including polymer gels; (b) middle: photograph of CT acquisition with phantom including gel acquisition treatment plan; (c) right: calculated dose map of treatment plan procedure from the acquired CT images. The red line shows the center profile.

2.3 Irradiation of prepared gels for calibration and for stereotactic irradiation

Prepared gels were irradiated with a photon beam at the Tsukuba Medical Center hospital using Linac (EXL-15DP, Mitsubishi Electric Corporation). The polymer gels in the PET tubes was irradiated with 4 MV photon beams (with no collimator) at the isocenter in a 300×300×300 mm^3 water tank for calibrations of dose versus R2 from 0 to 10 Gy in the beam axis. The phantom including the polymer for SRT was irradiated with 4 MV photon beams (with 10-, 15-, and 20-mm collimators, respectively) at the isocenter of a 300×300×300 mm^3 water tank to 6 Gy for SRT irradiation from the generated TPS plans in the beam axis. These gels after irradiation were stored in a refrigerator at 4 degree until MRI at 1.5 T.

2.4 MRI measurements of irradiated gels

MRI measurements were performed on a 1.5 Tesla TOSHIBA EXCELART Vantage scanner 1 day after irradiation. Irradiated gels were scanned with a quadrature (QD)Torso SPEEDER coil.

The gel was imaged using a multi spin echo pulse sequence (with a first effective echo time, TE1=30 ms, followed by TE2=60 ms). For each scan a repetition time of 4000 ms was used in order to reduce the influence of spin-lattice relaxation, with a 1 mm^2 resolution, and 5-mm-thick planes. The R2 images were calculated from two different echo time images using an in-house program.

2.5 Comparison in the center profile and in the center plane using gamma evaluation between gel measurements and calculated from the TPS

The R2 image in the center profile and the dosimetric map in the center plane of irradiated gels were compared with calculated data from TPS with three different fields. The R2 images were normalized in order to compare the calculated profile from TPS.

For comparison in the center plane, the R2 images of polymer gels were converted with a dosimetric map using R2 versus dose calibration curve at the center plane. The dosimetric map from irradiated gels and calculated dose map from TPS were compared in the central plane using gamma evaluation with DD system software (R-TECH INC) [10][7].

3. Results and Discussion

3.1 Calibration of dose response of polymer gel in photon beams and Linearity of R2 versus dose

The R2 calibration curve as a function of photon doses between 0 and 10 Gy was linear with regard to dose.The data points were obtained by averaging R2 values in the polymer gel. The fitted straight line had a slope of 1.99 ($Gy^{-1} \cdot s^{-1}$) and an intercept of 11.6 (s^{-1}). The coefficient of correlation was 0.996.

3.2 Comparison profile between gel measurements and TPS

Figure 2(a) shows the photograph of the non-irradiated polymer gel and SRT irradiated gels with 10-mm, 15-mm, and 20-mm collimators.

The comparison profiles in the center line between R2 from gels and calculated dose from TPS showed good agreement (Fig. 2(b-d)). In the center profile at the calculated dose from the treatment plan, the full width at half maximum (FWHM) of 10-mm, 15-mm, and 20-mm collimators were 13 mm, 19 mm, and 25 mm, respectively. In the center profile at R2 from the gel dosimetry, the FWHM of 10-mm, 15-mm, and 20-mm collimators were 13 mm, 20 mm, and 23 mm, respectively. The results between gel dosimetry and calculated dose in the central profile for SRT irradiation showed good agreement.

3.3 Comparison in the center plane between gel measurements and TPS using gamma evaluation

The pass-fail criterion for gamma evaluation method included a dose difference of 3% and a distance to agreement of 3 mm. In the case of SRT irradiation with a 10-mm collimator, the pass ratio of the gamma evaluation between calculated dose from the treatment plan and the dose map of the gel dosimetry in stereotactic irradiation was 98.6%. The results of gamma evaluation in the central plane showed good agreement between the dose map from gel dosimetry and calculated dose from TPS. A region with partial disagreement of the gamma evaluation method was seen with a low dose. This disagreement could have been caused by phantom size and uncertainly of the R2-dose conversion procedure. The isodose map shows the agreement with the high dose region in the center plane.

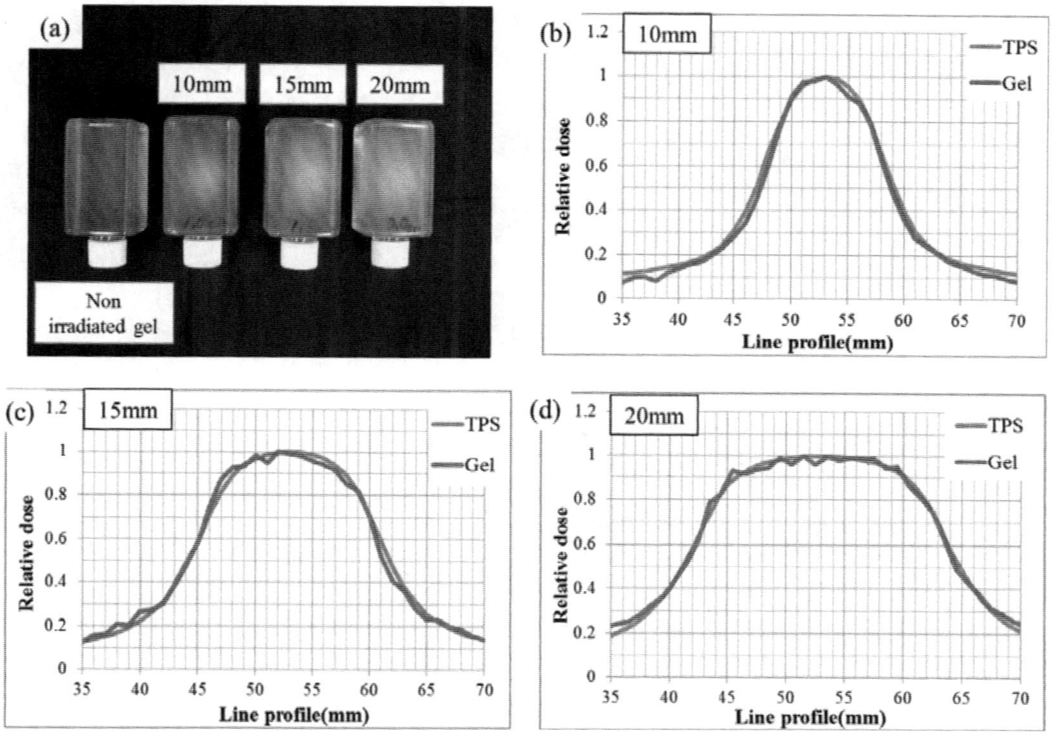

Figure 2: (a) upper left: photograph of irradiated polymer gels in stereotactic irradiation with 10-mm, 15-mm, and 20-mm collimators (left to right); (b) upper right: comparison between gel dosimetry and calculation from TPS in the center profile with SRS using 10-mm collimator; (c) lower left: comparison between gel dosimetry and calculation from TPS in the center profile with SRS using 15-mm collimator; (d) lower right: comparison between gel dosimetry and calculation from TPS in the center profile with SRS using 20-mm collimator.

4. Conclusions

This study showed good agreement between gel dosimetry and the generated treatment plan dosimetric map in the center of the profile and plane in stereotactic irradiation. In small field photon dosimetry, polymer gels can be an effective tool for comparison of the methods of gel dosimetry and generated calculated plans [11, 12].

5. References

[1] Baldock C *et al* 2010 *Phys. Med. Biol.* **55** R1-63
[2] Hill B *et al* 2005 *Br. J. Radiol.* **78** 623-30
[3] Murry P and Baldock C 2000 *Australas. Phys. Eng. Sci. Med.* **23** 44-51
[4] Lepage M *et al* 2002 *Phys. Med. Biol.* **47** 1881-90
[5] Keall P and Baldock C 1999 *Australas. Phys. Eng. Sci. Med.* **22** 85-91
[6] Venning A J *et al* 2005 *Med. Phys.* **32** 1047-53
[7] Baldock C *et al* 1998 *Phys. Med. Biol.* **43** 695-702

[8] Pappas E *et al* 2008 *Med. Phys.* **35** 4640-8
[9] Novotný J *et al* 2002 *Radiother Oncol.* **63** 223-30
[10] Low D A *et al* 1998 *Med. Phys.* **25** 656-61
[11] Rintoul L *et al* 2003 *Appl. Spectrosc.* **57** 51-7
[12] Lepage M *et al* 2002 *Phys. Med. Biol.* **47** 1881-90

Journal of Physics: Conference Series **444** (2013) 012032 doi:10.1088/1742-6596/444/1/012032

Deformable gel dosimetry I: application to external beam radiotherapy and brachytherapy

UJ Yeo[1], ML Taylor[1,2,3], JR Supple[1], RL Smith[1,3], T Kron[1,2] and RD Franich[1]
[1]School of Applied Sciences and Health Innovations Research Institute, RMIT University, Melbourne, VIC, Australia
[2]Physical Sciences, Peter MacCallum Cancer Centre, Melbourne, VIC, Australia
[3]William Buckland Radiotherapy Centre, The Alfred Hospital, Melbourne, VIC, Australia

E-mail: rick.franich@rmit.edu.au

Abstract. Inter- and intra-fractional variation in anatomic structures is a significant challenge in contemporary radiotherapy (RT). In this study, we describe the implementation of a novel deformable gel dosimetry system (dubbed 'DEFGEL') for application to external beam RT and brachytherapy experimental measurements. Complex / redistributed dose distributions due to applied deformations were readily observed and the discrepancies relative to a control case with an absence of deformation could be quantified. This work has obvious extensions to validation of deformable image registration algorithms, deformable dose calculation algorithms, and quality assurance of motion compensation strategies in RT.

1. Introduction

Temporal anatomic changes due to respiratory and other motion are a critical issue in contemporary radiotherapy (RT). The fact that many structures *deform* as a result of motion adds a further layer of complexity to the problem, both inter- and intra-fraction. Dose accumulation is a non-trivial practice in this context. Polymer gel dosimetry exhibits great utility in the verification of 3D spatial dose distributions of highly conformal treatment plans [1-4]. We have previously established a novel, tissue-equivalent 3D deformable gel dosimetric phantom, DEFGEL [5], for 4D deformable dosimetry. In this paper, we demonstrate the application of DEFGEL for experimental measurement of complex and redistributed dose in external beam radiotherapy (EBRT) and high dose rate (HDR) brachytherapy.

2. Materials and Methods

The DEFGEL dosimeter/phantom is comprised of a polymer gel in a latex membrane, moulded into a cylindrical geometry with 46 mm diameter [5]. A bilateral compression of 8 mm from each side (46 mm to 30 mm) was applied to generate a symmetric deformation of the DEFGEL using an acrylic compressor. DEFGEL phantoms were irradiated with photons (6 MV) from a Varian 21EX linear accelerator for EBRT, and with an Ir^{192} HDR brachytherapy source. For the EBRT delivery, a small field stereotactic plan of three dynamic arcs was adapted from a patient plan and 9 Gy was delivered to isocentre – half of the clinical prescription dose. Samples were irradiated in a water bath to remove dosimetric effects of surface curvature. For the HDR brachytherapy irradiation, a simple linear interval treatment plan was created using the Oncentra brachytherapy TPS to the CT images of DEFGEL with a standard catheter inserted. Five equally-weighted source dwell positions were planned with a 10 mm step size. In both treatment plan delivery schemes, doses were delivered to the DEFGEL in two different scenarios: (i) without deformation, (ii) with deformation (see figure 1(a) and 1(b)). Doses were read out using a cone-beam optical CT scanner (Vista™ Optical Scanner by Modus Medical Devices Inc.) [6-11]. The measured dose distributions delivered in the deformed state were then compared to those delivered in the undeformed state to quantify the dosimetric difference between the two scenarios.

3. Results and Discussion

Figure 1(c) depicts a comparison between the treatment delivered with and without deformation for EBRT, with the stereotactic field adapted from a patient plan. Figure 2 shows a comparison of the HDR brachytherapy treatment delivered with five dwells, with and without deformation. The impact of deformation on dose distributions is readily seen in the isodose contours and line profiles shown from the two aforementioned scenarios for both radiation delivery techniques. In both figures (figure 1 and figure 2), the deformation consisting of compression and release in the *y*-direction can be seen to elongate the field in that axis coupled with a contraction of the distribution in the orthogonal axes.

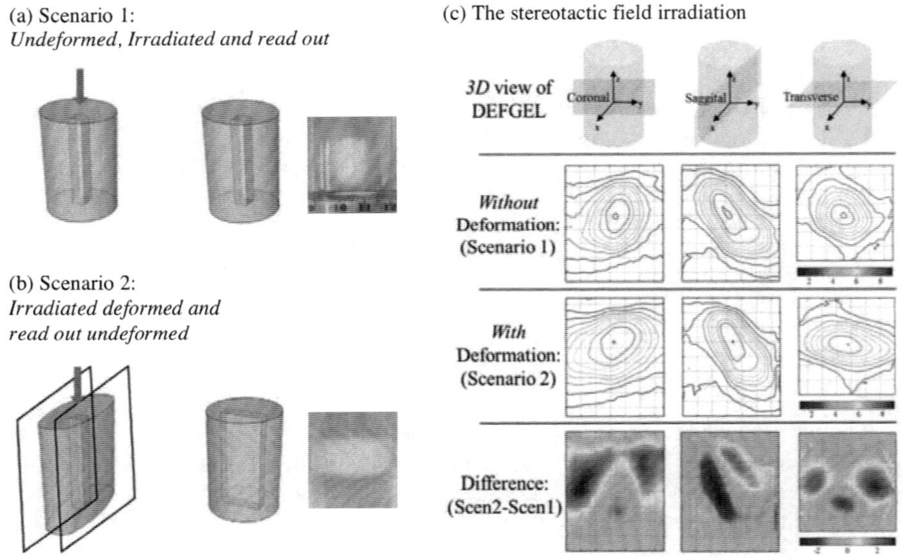

Figure 1: Irradiation schemes in two different scenarios: **(a)** Scenario 1 – irradiated and readout in undeformed state and **(b)** Scenario 2 – irradiated deformed, and read out in undeformed state. Photographs of measured dose distributions (1 × 2 cm^2 field) are also shown for each scenario. Dose distributions of three orthogonal planes for the stereotactic field irradiation are shown in **(c)**. The top row illustrates the three orthogonal planes in DEFGEL. The second, third, and fourth rows represent dose distributions of Scenario 1 and Scenario 2, and the difference thereof, respectively. The left, middle, and right columns correspond to the coronal, saggital, and transverse planes, respectively. All doses are in Gy and the grid spacing is 5 mm.

In figure 1(c), maximum doses were 9.17 Gy and 9.03 Gy for scenarios 1 and 2, respectively. Substantial dosimetric discrepancies up to approximately 3 Gy, or ~30% of the maximum dose delivered, are evident in all three planes.

For the Ir^{192} exposure in figure 2, it is plain from the results that there is a dosimetric difference between the planned and measured dose distributions when the dosimeter was in the deformed state during irradiation (Scenario 2). Similar to the external beam case, this figure shows elongation of the dose distribution in the direction of compression and relaxation, coupled with a contraction of the distribution in the orthogonal axis. This discrepancy increases with extent of deformation; note that each dwell position experienced different extents of deformation due to their varying positions relative to the centre of mass in the DEFGEL, as shown in figure 2(d). Consequently, the apparent dwell positions in the measured distributions appear different when comparing the deformed and undeformed dosimetry data. The planned length from the top to bottom dwell position was 40.0 mm (five dwells with a 10 mm step size), but the *measured* length between these maxima in the case of Scenario 2 was found to be 33.1mm (i.e. a 17.3 % reduction).

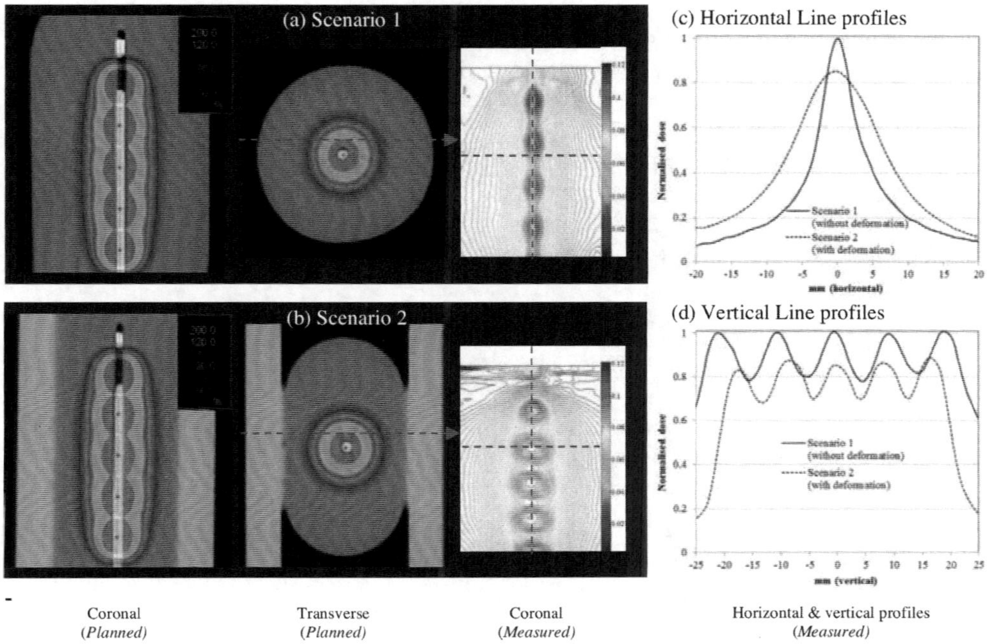

Figure 2: HDR brachytherapy irradiation of DEFGEL: 5 dwells of equal weighting and 10 mm spacing on a single catheter. **(a)** Scenario 1 and **(b)** Scenario 2: Planned dose distributions in coronal and transverse planes are presented in left and middle columns, respectively. The right column shows dose maps in planes indicated by the red dashed arrows (2.9 mm distance from the catheter) in the transverse plane (middle column). The dashed lines in the dose maps correspond to line profiles in the horizontal and vertical directions of the coronal planes shown in **(c)** and **(d)**. These figures show an elongated dose distribution in the direction of compression and relaxation, coupled with a contraction of the distribution in the orthogonal axis. The scale in the measured plane is in ΔOD.

For both treatment techniques, EBRT and HDR brachytherapy, the potential clinical implications of these results represent a decrease in dose to the planning target volume or increased dose to an organ at risk (or both). The use of DEFGEL for studying the effects of deformation on absorbed dose distributions is valid for comparable mass and density conserving deformations. Obvious anatomic examples would include prostate, breast, liver, etc. The relevant anatomical deformations would be those related to organ geometric changes induced by filling and emptying of the bladder, rectum, stomach, etc., as well as respiratory and cardiac motion during irradiation.

4. Conclusions

This methodology describes the use of a tissue-equivalent, 3D dose-integrating deformable phantom that yields dosimetric information for both external beam and HDR brachytherapy. It is demonstrated that the system is potentially capable of reproducibly emulating the physical deformation of an organ, and therefore can be used to evaluate absorbed doses to deformable targets and to validate deformation algorithms.

5. Acknowledgements
This work is supported by an RMIT University Research & Innovation Emerging Researcher Industry Award (Dr R Franich, Dr Taylor). Mr Smith is supported by Cancer Australia Grant 616614, funded by the Radiation Oncology Section, Australian Government Department of Health and Ageing.

6. References
[1] Baldock C *et al* 2010 *Phys. Med. Biol.* **55** R1-63
[2] Lopatiuk-Tirpak O *et al* 2008 *Med. Phys.* **35** 3847-59
[3] De Deene Y *et al* 2002 *Phys. Med. Biol.* **47** 2459-70
[4] Hurley C *et al* 2006 *Nucl. Instrum. Methods Phys. Res. A* **565** 801-11
[5] Yeo U J *et al* 2012 *Med. Phys.* **39** 2203-13
[6] Islam K *et al* 2003 *Med. Phys.* **30** 2159
[7] Bosi S *et al* 2007 *Phys. Med. Biol.* **52** 2893-903
[8] Bosi S G *et al* 2009 *Phys. Med. Biol.* **54** 275-83
[9] Bosi S G *et al* 2009 *Appl. Opt.* **48** 2427-34
[10] Olding T *et al* 2010 *Phys. Med. Biol.* **55** 2819-40
[11] Olding T and Schreiner L J 2011 *Phys. Med. Biol.* **56** 1259-79

Response of aqueous dichromate and nanoclay dichromate gel dosimeters to carbon ion irradiation

T Maeyama[1], N Fukunishi[1], KL Ishikawa[2,3,4], K Fukasaku[2,5], T Furuta[7], S Takagi[2,6], S Noda[2,7], R Himeno[2,7] and S Fukuda[8]
[1]RIKEN/Nishina Center for Accelerator-Based Science, Wako, Japan
[2]RIKEN/Computational Science Research Program, Wako, Japan
[3]The University of Tokyo/Department of Nuclear Engineering and Management, Graduate School of Engineering, Tokyo, Japan
[4]The University of Tokyo/Photon Science Center, Graduate School of Engineering, Tokyo, Japan
[5]Himon'ya Hospital/ Department of Neurosurgery, Tokyo, Japan
[6]The University of Tokyo/Department of Mechanical Engineering, Graduate School of Engineering, Tokyo, Japan
[7]RIKEN/Advanced Center for Computing and Communication, Wako, Japan
[8]National Institute of Radiological Sciences/ Research Center for Charged Particle Therapy, Chiba, Japan

E-mail: maeyama@riken.jp

Abstract. We have recently reported the significant reduction of radiation product diffusion by the incorporation of clay nanoparticles into dichromate gel (DCG) dosimeters. In this work, we investigate the influence of the nanoclay addition and gelation on the MRI R_1 ($1/T_1$) image response of the dichromate dosimeter to the therapeutic carbon ion beam ($^{12}C^{6+}$ 290 MeV/u). The MRI R_1 distribution in the aqueous dichromate solution well reproduces physical dose-depth distribution with a high linear-energy-transfer (LET) efficiency. The nanoclay DCG dosimeters, on the other hand, exhibit composition-dependent LET efficiency degradation, while a sharp Bragg peak can still be detected. These results indicate that the nanocomposite gel addition may induce change in the radiation-induced reaction mechanism.

1. Introduction

Gel dosimetry [1, 2] is a suitable method for the validation of complex 3D dose distributions [3] encountered in contemporary radiotherapy including scanning-beam carbon ion radiotherapy [4]. However, implementation of polymer gels for carbon ion radiotherapy dosimetry has exhibited dose quenching around the Bragg peak [5-7]. Although the other gel dosimeter based on liquid chemical dosimeter such as Fricke and ceric sulfate gel dosimter suffer from diffusion produced by irradiation [8, 9], it is expected that the dose response of the ceric sulfate gel shows relative insensitivity to LET changes [10, 11]. We have recently succeeded [12] in suppressing the diffusion in DCG dosimeters by adding clay nanoparticles, exploiting the property of the dichromate ion to adsorb to clay [13] and nanocomposite gel [14]. A liquid dichromate dosimeter based on radiation-induced reduction of $Cr_2O_7^{2-}$ is stoichiometrically represented with the radiation-chemical yield of water radiolysis.

$$G(\mathrm{Cr}^{3+}) = 2\times G(\mathrm{Cr}_2O_7{}^{2-}) = 1/3[G(\mathrm{H}) + 2G(\mathrm{H}_2\mathrm{O}_2) - G(\mathrm{OH})] \quad (1)$$

The dose response of the dichromate gel dosimeter is expected to be unaffected by LET changes, since the OH and H yields cancel each other [15]. In this work, we investigate the influence of the nanoclay addition and gelation on the response of the dichromate dosimeter to the therapeutic carbon ion beam ($^{12}C^{6+}$ 290 MeV/u).

2. Material and methods

The aqueous dichromate solution prepared base on reference [16]. The nanoclay used here is a synthetic hectorite called Laponite XLG (Rockwood). Details of the chemical compositions of gel or liquid dosimeters are summarized in table 1. Beam of $^{12}C^{6+}$ 290-MeV/u were provided from the Biological irradiation port in the Heavy Ion Medical Accelerator in Chiba (HIMAC) at National Institute of Radiological Science (NIRS). The dose-depth distribution were measured by ionization chamber using binary filter system consisting of plates made out of PMMA with different thicknesses corresponding to water thicknesses [17]. Liquid samples were irradiated in an array of 1-cm cuvettes at an entrance surface dose of 450 Gy with 56.33mmH_2O PMMA filter. The prepared gels were enclosed in Pyrex cylinder (Iwaki Glass Co., 9827TST: bottom thickness 2.1mm, diameter 32mm, length 230mm). Irradiations were undertaken from the bottom of Pyrex cylinder at an entrance surface dose of 470 Gy. The MR images were measured at 1.5-tesla (Intera Achieva Nova Dual, Philips Medical Systems, Best, The Netherlands) using Q-body coil. Calculation of the relaxation rates ($R_1 = 1/T_1$) was accomplished by using a “turbo-mixed” sequence (TR = 2260 ms, TE = 19 ms, TI = 500ms, ETL = 12, pixel spacing = 0.75 mm, NSA = 4).

Table 1: Chemical composition of dichromate gel and liquid

	$Na_2Cr_2O_7$ mM	$HClO_4$ mM	AGAR w/v %	LaponiteXLG w/v%
Gel1	1	1	2.5	1
Gel2	1	25	2.5	1
Gel3	10	25	2.5	1
Liquid	2 mM $K_2Cr_2O_7$	50 mM $HClO_4$	0.1 mM $AgClO_4$	

3. Results and discussion

The MRI R_1 distribution in the aqueous dichromate solution well reproduces physical dose-depth distribution with a high LET efficiency (figure 1). Since the product under irradiation in a 1-cm cuvette was homogeneous, we compared the total dose calculated from 1-cm length convolution of dose-depth distribution measured by ionization chamber and δR_1 values. In the dose distribution, the dose ratio of the Bragg peak dose to the entrance surface dose is approximately 3, while the δR_1 ratio of the value at the Bragg peak to the value at the entrance surface is approximately 2.5. The LET efficiency ε defined the following equation is approximately 0.8 at the Bragg Peak. From this result, we found that the aqueous dichromate solution dosimeter shows weak LET dependence.

$$\varepsilon = \left[\frac{\delta R_1}{dose}\right] \text{at the Bragg peak} \;/\; \left[\frac{\delta R_1}{dose}\right] \text{at the entrance surface} \quad (2)$$

The δR_1 distributions of the nanoclay DCG are shown in figure 2 with dose-depth distribution. Sine initial energy of ion was degraded by 2.1 mm Pyrex glass at entrance surface, the δR_1 distributions were shown with adding 3.82mmH_2O depth. These results indicate that the dose response of nanoclay DCG is highly sensitive to compositional variation. Specifically, the sensitivity of the nanoclay DCG increases with increasing concentrations of perchloric acid or dichromate. We explicitly investigated concentration variations up to 500-mM dichromate. Gels containing high concentrations of dichromate ion or perchloric acid take a long time to fix. As a result, inhomogeneous MRI images were obtained (not shown). In the dose-depth distribution, Dose at the Bragg peak indicates magnitude approximately 5 times higher than the surface dose. The δR_1 at the Bragg peak, in contrast, indicates magnitude approximately 1.5 times higher than the surface. We found that the nanoclay DCG response decreased with increasing LET. Especially, the LET efficiency was approximately 0.3 at the Bragg peak.

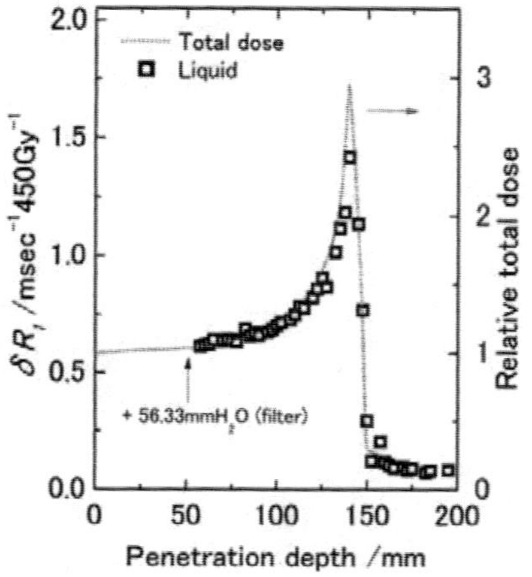

Figure 1: The δR_1 value of the aqueous dichromate solution after irradiation with 450 Gy of $^{12}C^{6+}$ 290MeV/u. The symbols indicate measured values (MRI). The dashed line indicates calculated total dose in 1cm cuvvete from the 1-cm length convolution of the dose in figure 2.

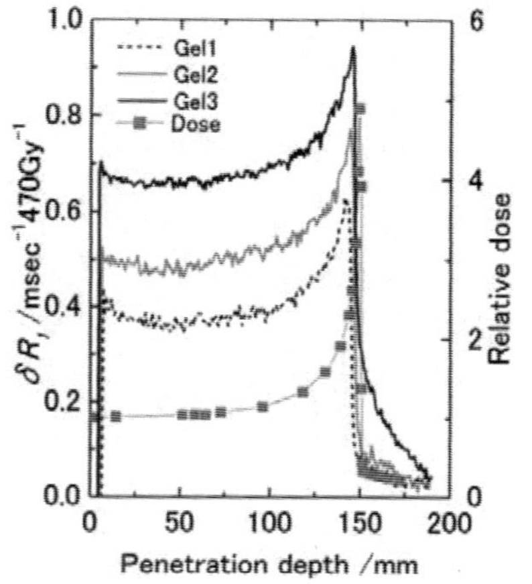

Figure 2: The δR_1 distribution of nanoclay DCG after irradiation with 470 Gy of $^{12}C^{6+}$ 290 MeV/u in the case of Gel 1, 2, 3. The symbol-line show dose measured.

4. Conclusion

We have investigated the LET efficiency of the nanoclay DCG and aqueous dichromate solution using carbon ion irradiation ($^{12}C^{6+}$ 290 MeV/u). The aqueous dichromate dosimeter exhibits an LET efficiency (0.8) at the Bragg peak superior to that of the aqueous Fricke and coumarin solutions (< 0.5) [18, 19]. The results for the nanoclay DCG dosimeters indicate that the composition significantly influences the response and that the LET efficiency is degraded compared with that of the aqueous dichromate dosimeter.

5. Acknowledgments

This research was supported by Research and Development of the Next-Generation Integrated Simulation of Living Matter, which is a part of the development and use of the Next-Generation Supercomputer Project of MEXT, Japan. It was also partly supported by RIKEN President's Discretionary Fund (Strategic Programs for R&D), the special postdoctoral researcher program of RIKEN and MEXT (JSPS) KAKENHI (24760725). A part of this study was performed as the Research Project with Heavy Ions at NIRS-HIMAC.

6. References

[1] Lepage M *et al* 2001 *Phys. Med. Biol.* **46** 2827-39
[1] Rintoul L *et al* 2003 *Appl. Spectrosc.* **57** 51-7
[3] Baldock C *et al* 2010 *Phys. Med. Biol.* **55** R1-63
[4] Haberer T *et al* 1993 *Nucl. Instrum. Meth. A* **330** 296-305
[5] Kantemiris I *et al* 2009 *J. Phys.: Conf. Ser.* **164** 012055

[6] Ramm U *et al* 2004 *Radiother. Oncol.* **73** S99-101
[7] Ramm U *et al* 2000 *Phys. Med. Biol.* **45** N95-102
[8] Baldock C *et al* 2001 *Australas. Phys. Eng. Sci. Med.* **24** 19-30
[9] Harris P J *et al* 1996 *Phys. Med. Biol.* **41** 1745-53
[10] Schreiner L J 2004 *J. Phys.: Conf. Ser.* **3** 9-21
[11] Bäck S A J *et al* 2002 *Appl. Radiat. Isotopes* **56** 895-99
[12] Maeyama T *et al* 2012 *IFMBE Proc.* **39** 1183-6
[13] Kumagai Y *et al* 2011 *Radiat. Phys. Chem.* **80** 876-83
[14] Wang Q *et al* 2010 *Nature* **463** 339-43
[15] Barr N F and Schuler R H 1959 *J. Phys. Chem.* **63** 808-12
[16] Al-Sheikhly M *et al* 1988 *Int J Rad Appl Instrum* **32** 545-51
[17] Yamashita S *et al* 2008 *Radiat. Phys. Chem.* **77** 439-46
[18] Bhat N N *et al* 2003 *Radiat. Phys. Chem.* **68** 909-16
[19] Maeyama T *et al* 2011 *Radiat Phys Chem.* **80** 1352-7

An investigation into ultra-sensitive substituted leucomalachite dye derivatives for use in the PRESAGE® dosimeter

M Alqathami[1], J Adamovics[2], R Benning[2] and A Blencowe[3]
[1]School of Medical Sciences, RMIT University, VIC 3083, Australia
[2]Department of Chemistry and Biology, Rider University, Lawrenceville, NJ 08648, USA
[3]Department of Chemical & Biomolecular Engineering, The University of Melbourne, VIC 3052, Australia

E-mail: mamdooh.alqathami@rmit.edu.au

Abstract. A comparison between a commercial leucomalachite green (LMG) dye and three newly synthesised derivatives incorporated into the PRESAGE® dosimeter was carried out to determine their effect on the sensitivity and post-response photofading of the dosimeter. For all of the substituted LMG derivatives (with either methoxy, chlorine or bromine substituents), the sensitivity of the resulting dosimeters to radiation dose increased significantly and was dependent on the type of LMG derivative used, with the bromine substituted derivative showing the highest sensitivity increase (450%) followed by chlorine and methoxy substituted derivatives (340%, 200 %, respectively) relative to commercial LMG. All LMG dyes investigated showed similar post-response photofading characteristics except the methoxy-LMG derivative, which showed a slight improvement in post-response photo-retention.

1. Introduction

The PRESAGE® dosimeter is a unique 3D radiochromic dosimeter with great potential for clinical applications [1, 2]. Unlike polymer gel dosimeters [3], the PRESAGE® dosimeter is a clear solid polyurethane matrix containing leucomalachite green (LMG) dye as a reporter component and halogenated carbons as a radical source [4, 5]. Free radicals generated from the radiolysis of the halo-carbon bond upon irradiation oxidize the LMG dye leading to a change in absorbance [6]. The change in absorbance is linear with respect to the absorbed radiation dose [5]. Some of the attractive features and advantages of the PRESAGE® dosimeter over gel dosimeters include its lack of sensitivity to oxygen and diffusion [5]. Furthermore, the radiological tissue-equivalent [7, 8] PRESAGE® dosimeter is solid, easily handleable, can be fashioned into any shape and requires no supporting container. As a part of a series of PRESAGE® dosimeter component-specific optimization studies [4, 9], the aim of this study was to investigate several new LMG dye derivatives. The LMG dye is one of the critical components of previous PRESAGE® dosimeter formulations, however, the commercial LMG dye currently suffers from a number of potential limitations mainly related to its sensitivity to radiation dose and post-response photofading. The new LMG derivatives were specifically developed to overcome these limitations.

2. Materials and Methods

2.1. PRESAGE® fabrication
The chemical agents used in this study were: Polyurethane resin (Crystal Clear 200, Smooth-On, Easton, PA USA), which are supplied in two parts (Part A and Part B); The radical initiator, tetrabromoethane, and commercial LMG dye (Sigma-Aldrich, St Louis, MO); The newly synthesised LMG derivatives, referred to as MeO-LMG, Cl-LMG and Br-LMG (Heuris Pharma, LLC (Skillman, NJ)). In order to directly compare between the LMG dye and its derivatives, the concentration of the radical initiator and dye were kept constant throughout and only the type of leuco dye was varied. The dosimeters were prepared in poly(methyl methacrylate) spectrophotometer cuvettes with a wall thickness of 1 mm and internal dimensions of 1 × 1 × 4.5 cm^3 [10]. Detailed description of the fabrication process is described elsewhere [4].

2.2. Absorbance change measurements
Absorbance change measurements were acquired using a dual-beam Perkin Elmer Lambda 25 UV-VIS spectrophotometer (Perkin Elmer, Waltham, MA, USA). The absorption spectrum over the visible wavelength region (470–750 nm) with 1 nm intervals was initially used to determine the absorption maxima of the PRESAGE® dosimeters with LMG and its derivatives. The absorption of each PRESAGE® cuvette was measured pre- and post-irradiation. To investigate the influence of the different leuco dyes used on post-response photofading, absorption acquisitions were conducted at different time intervals (1, 3, 6, 24, 48, 92, 120, and 168 hours post-irradiation). All PRESAGE® cuvettes were kept in a cold (*ca.* -18 °C) and dark environment pre- and post-irradiation to avoid accidental exposure to ultraviolet or visible light.

2.3. Irradiation
The PRESAGE® dosimeter cuvettes were irradiated using a 6 MV medical linear accelerator (Elekta Synergy, Crawley, UK), using a dose rate of 5 Gy/min and various radiation doses (0, 0.5, 1, 5, 10, 20 and 30 Gy). The field size was set to 10 × 10 cm. The cuvettes were sandwiched between two solid water phantoms with the top one measuring 1.5 cm in thickness. The FSD was set to 100 cm from the solid water surface.

3. Results and Discussion
The molecular structure of the commercial LMG dye and its derivatives employed in this study are shown in figure 1. Upon irradiation the color formation occurs when the methine proton (C-H) is cleaved leading to the formation of a colored cation. In the three new derivatives, one of the aryl protons of LMG was substituted with either methoxy (MeO), chlorine (Cl) or bromine (Br) groups (figure 1).

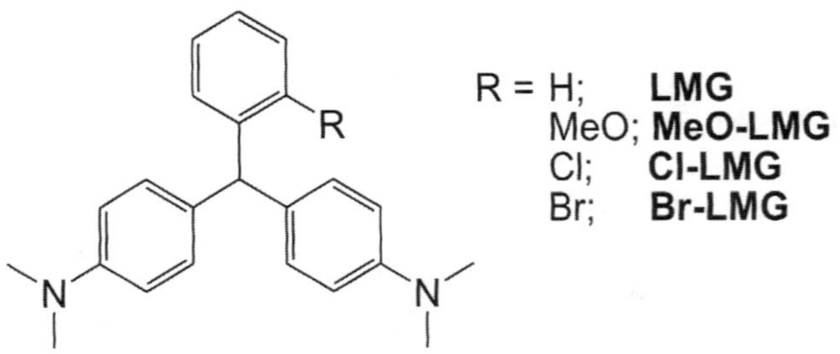

Figure 1: Molecular structure of the leuco dyes investigated. R = the substituted part of the parent LMG.

The absorption spectra of the LMG dye and its three derivatives after oxidisation are shown in figure 2. For all leuco dyes, the absorption maximum (λ_{max}) was found to peak at *ca.* 633 nm (red region of the spectrum), which is a typical visible absorption λ_{max} of the oxidized form of LMG (malachite green). This suggests that the new LMG derivatives could be used with current optical tomography imaging systems such as Vista™ (Modus Medical Devices Inc.) with optimum sensitivity.

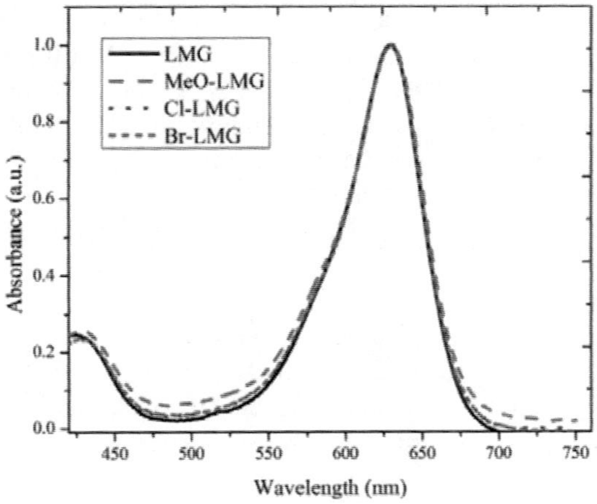

Figure 2: Normalized absorption spectra of the PRESAGE® dosimeters with the four dyes used in this study showing absorption maxima at *ca.* 633 nm. Reference cuvettes (same composition but with zero radiation dose) for each formulation were used as a baseline to establish the zero value.

The measured absorbance changes at 633 nm wavelength versus the radiation absorbed doses for all of the dosimeters are displayed in figure 3. Absorption values were obtained by subtracting the relevant value of a reference cuvette for the same batch with zero radiation dose from that of the irradiated cuvettes. Consequently, the intercepts of the dose–response plots are nearly zero. In all cases a very good correlation coefficient linearity ($R^2 > 0.99$) for the dose response was observed for PRESAGE® compositions with all of the leuco dyes investigated over the applied radiation dose range.

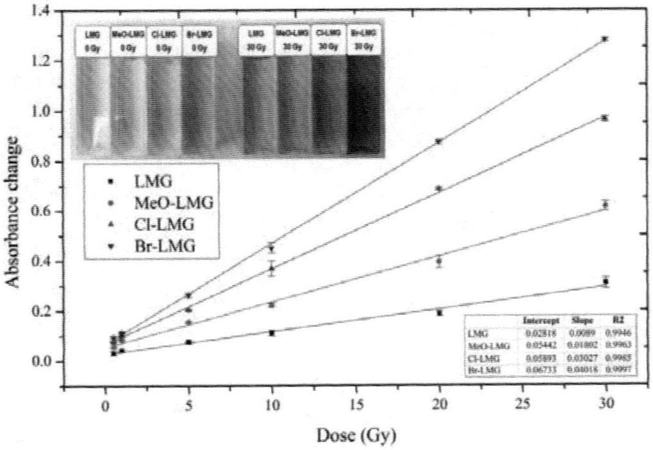

Figure 3: Recorded absorbance changes as a function of absorbed radiation dose for the PRESAGE® dosimeter with LMG dye and its derivatives. Correlation coefficient parameters of each fitted line are shown in the bottom right inset. Error bars represent the standard deviation in the measurement and are smaller than most points. Photograph of the dosimeter cuvettes with LMG dye and its derivatives pre- and post- exposure to 30 Gy are shown in the top left inset.

It was found that the change in absorbance varied significantly between LMG and its derivatives (figure 3), with the dose–response curves for PRESAGE® with Br-LMG showing the highest sensitivity followed by those with Cl-LMG and MeO-LMG (450 %, 340 %, and 200 % increase, respectively) relative to the commercial LMG. The significant increase observed for halogen substituted LMG derivatives could be attributed to the fact that halogens (Cl, Br) are weak deactivators that would withdraw electron density and weaken the methine C-H bond making it easier for radicals to abstract the proton. These electronic effects would influence the dose sensitivity and post-response photofading. In Addition, The 110% increase in sensitivity with Br-LMG than Cl-LMG coud be attributed the lower C–Br bond dissociation energy (72.1 kcal.mol^{-1}) compared to C–Cl (83.7 kcal.mol^{-1}) [4, 11]. Thus, less energy is required to cause oxidation of Br-LMG than Cl-LMG. The 200 % increase in sensitivity observed with the MeO-LMG derivative is attributed to its methoxy substituent (OCH_3), which is known as a moderate activator whereby the lone pair of the oxygen would be expected to stabilize the colored cation intermediate formed after radical abstraction of the proton of the methine group.

The post-response photofading of the PRESAGE® dosimeters with different leuco dyes are shown in figure 4. The results demonstrate that, the LMG dye and its synthesised derivatives were fairly stable over a period of one week with MeO-LMG showing a slight improvement in the retention of the post-response absorption value over the period studied.

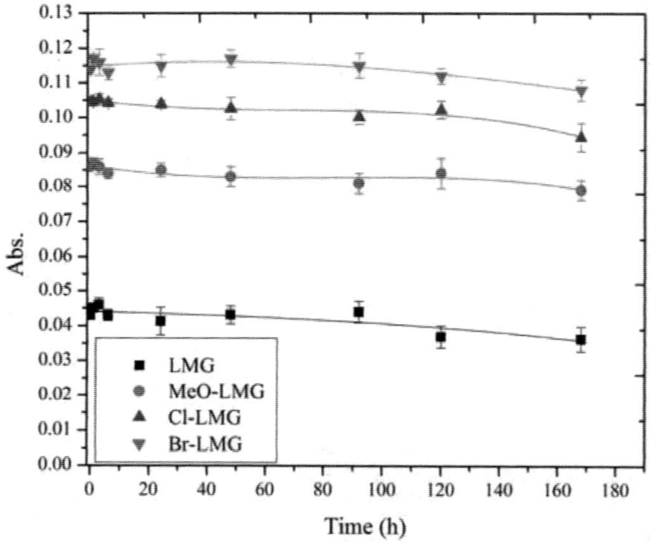

Figure 4: Variation in absorbance over time for the PRESAGE® dosimeters with the four dyes used in this study after exposure to 1 Gy. Error bars represent the standard deviation values in the measurement.

4. Conclusion

In this work, newly synthesised LMG derivatives were investigated for potential use in the composition of the commercial PRESAGE® dosimeter. In general, all new LMG derivatives showed significant improvement over the commercial version. In addition, they also had no influence on the typical characteristics of the PRESAGE® dosimeter. Therefore, where higher sensitivity is required, it is recommended that the commercial LMG used in the composition of the PRESAGE® dosimeter be replace by any of the newly synthesised derivatives, especially Br-LMG.

5. References

[1] Clift C *et al* 2010 *Phys. Med. Biol.* **55** 1279-93
[2] Abdul Rahman AT *et al* 2011 *Phys. Med. Biol.* **56** 4177-99
[3] Baldock C *et al* 2010 Polymer gel dosimetry *Phys. Med. Biol.* **55** R1-63
[4] Alqathami M *et al* 2012 *Radiat. Phys. Chem.* **81** 867-73
[5] Adamovics J and Maryanski M J 2006 *Radiat. Prot. Dosim.* **120** 107-12
[6] Guo P *et al* 2006 *Med. Phys.* **33** 1338-45
[7] Brown S *et al* 2008 *Appl. Radiat. Isot.* **66** 1970-4
[8] Gorjiara T *et al* 2011 *Med. Phys.* **38** 2265-74
[9] Alqathami M *et al* 2012 *Radiat. Phys. Chem.* (in press)
[10] Rintoul L *et al* 2003 *Appl. Spectrosc.* **57** 51-7
[11] Blanksby S J and Ellison G B 2003 *Acc. Chem. Res.* **36** 255-63

Journal of Physics: Conference Series **444** (2013) 012035 doi:10.1088/1742-6596/444/1/012035

Measurements for dose distribution with a photo-stimulated luminescence dosimeter sheet

C Kurokawa[1], A Urushiyama[2]
[1]Graduate School of Medicine, Juntendo University, Tokyo, Japan
[2]Department of Chemistry, College of Science, St. Paul's (Rikkyo) University, Tokyo, Japan

E-mail: ckuroka@juntendo.ac.jp

Abstract. Dose distributions for photon beam were measured using a Photo-Stimulated Luminescence Dosimeter (PSLD) sheet, which has 18 x 24 cm^2 dimension and 0.2 mm thickness. Its density and effective atomic number are 1.0 g/cm^3 and 7.6, respectively. The read out was performed by blue LED lights for 20 seconds, which was much shorter than the readout time for TLD. The percent depth dose and dose profile were obtained as smooth curve after the calibration and it reproduced the measurements with ionization chamber, except for the tail region in the dose profile. We demonstrated the measurement of the prostate VMAT dose distribution also. The result reproduces the calculation by treatment planning system (TPS) qualitatively. It is shown that the PSLD sheet has the potential to measure the dose distribution.

1. Introduction

Accurate determination of the absorbed dose at the surface and in the buildup region requires careful considerations in the selection of detectors. It is known that the size of the detector along the beam direction should be as small as possible in order to minimize the effect of dosimeter volume. Recently, such a future is considered to be necessary for the QA of electron intensity modulated radiation therapy (IMRT) also since the surface dose should be measured with high accuracy. The report of AAPM TG-106 recommends the extrapolation chambers for the surface dose measurement [1], however, its availability is limited and the usage is time consuming. Instead, there have been measurements of the surface dose by various dosimeters, such as diode, film, TLD, gel, etc. [2-6]. The commercial TLD materials, such as rods, chips, powder, disks and so on, are usually used only for point measurements. Thus we have developed a TLD sheet for the verification of the spatial dose distribution [7-13] and showed that the sheet could be used for measurements of 3D dose distributions [14]. Although the TLD sheets have many attractive characters, it takes few minutes for heating the TLD crystal. On the other hand, we have investigated another type of dosimeter, a Photo-Stimulated Luminescence Dosimeter (PSLD) sheet [15]. A significant advantage of the PSLD sheet over the TLD sheet is that the PSLD sheet requires less time for light stimulation and it is easy for handling.

The purpose of this study is to demonstrate the PSLD sheet for dose distribution measurements of photon beam.

2. Method and Materials

Dose distributions of 10 x 10 cm^2 field size for 6 MV photon were measured using the PSLD sheet. In practice, the PSLD sheet can be formed in any size, and in this study we used 18 x 24 cm^2 dimension and 0.2 mm thickness as shown in Figure 1. The PSLD sheet is softer than film and it is tough flexible. The characters of PSLD sheet are summarized in Table 1 comparing with that of TLD sheets. We inserted these PSLD sheets in a tough water phantom (Kyoto Kagaku, Japan) in the same manner as film. The PSLD sheet does not have high uniformity across the sheet. Therefore, before the measurements of dose distributions, the irradiation of 200 cGy with 40 x 40 cm^2 field size was carried out to calibrate the response over 2D area of each PSLD sheet. Each sheet was characterized by an individual sensitivity value and we applied the calibration coefficients to each PSLD sheet.

A light stimulation system includes computer controlled blue LED lights and CCD camera. The read out was performed by blue LED lights (100W) for 20 seconds and generated light was detected with a CCD camera which has 1024 x 1024 pixels (FLIML 1001E-2, Finger Lakes Instrument).

For the data comparison, a small thimble ionization chamber (PTW Semiflex ionization chamber) was also used.

Figure 1. PSLD sheet with the dimension of 18 x 24 cm^2 and thickness of 0.2 mm thickness is developed. This sheet can be used in the same manner as film for planar dose measurements.

Table 1. Characteristics of TLD and PSLD sheets.

	TLD	**PSLD**
Physical Density	1.0 g/cm^3	1.0 g/cm^3
Effective atomic number	7.4	7.6
Dose range	Up to 1000 Gy	1 cGy to 10 Gy
Readout	Heat 180 °C for 240 sec.	Blue LED light for 20 sec.

3. Results

3.1. PDD and Profile

The present depth dose (PDD) and dose profile were obtained as smooth curve after the response calibration as shown in Figures 2 and 3. Obtained PDD well reproduced the results with ionization chamber measurement. For dose profile, the discrepancies were observed in the tail region and PSLD shows lower response comparing to the ionization chamber. In the tail region, the component of the lower energy and the low dose contribute. Thus it is necessary to study carefully the energy response and response to low dose of the PSLD sheet. The penumbra measured with PSLD is sharper than that with ionization chamber. It is because the PSLD has higher resolution compared to the ionization chamber. The measured filed sizes with the PSLD and ionization chamber showed agreement with 1%.

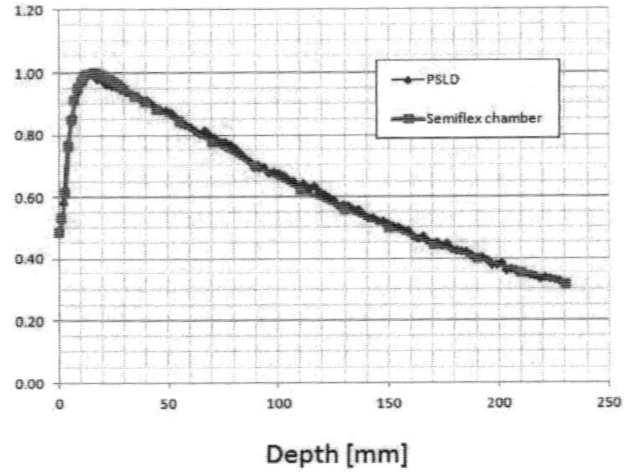

Figure 2. Percept depth dose (PDD) for 6MV photon beam measured by the PSLD sheet and PTW Semiflex ionization chamber.

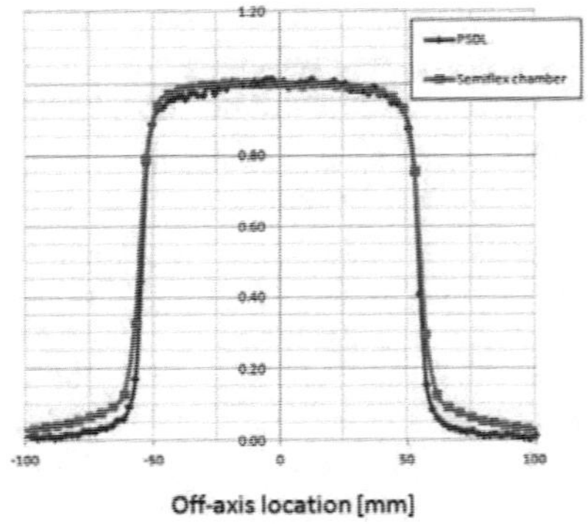

Figure 3. Dose profile for 6MV photon beam measured by the PSLD sheet and PTW Semiflex ionization chamber.

3.2. VMAT dose distribution

The comparison of dose distribution was carried out between the dose calculated by Pinnacle[3] treatment planning system (Philips Medical Systems, Madison, WI) and the dose measured by PSLD sheets in the phantom. The prostate plan of volumetric modulated arc therapy (VMAT) was used for the comparison as shown in Figures 4. Although the measurement of PSLD reproduces qualitatively the calculated result of Pinnacle[3] TPS, it was found that there were a couple of problems for the quantitative comparison. Our PSLD sheet has some non-uniformity over a 2D area since the PSLD powder were coated by hand for each piece, which was not performed by manufacture. The fluctuation of the PSLD coating influenced the result of the dose distribution comparison even though the response calibration over the 2D plane was applied before the measurement. We thus need to improve the uniformity of the PSLD sheet or the method of data processing for the application to the dose distribution measurement. It is expected that some manufacturer can improve the uniformity of the PSLD sheet.

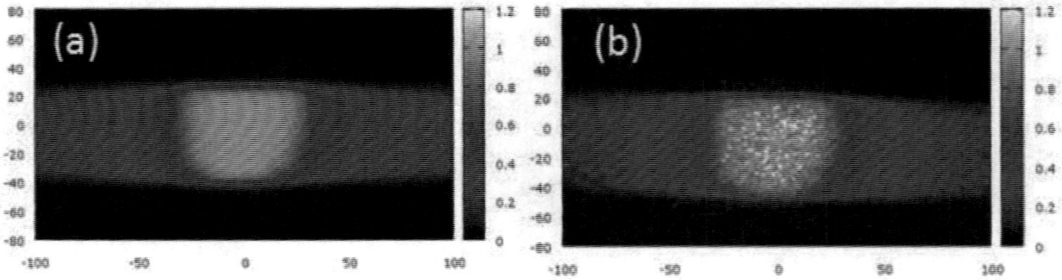

Figure 4. 2D dose distributions of prostate VMAT plan (a) calculated in Pinnacle[3] Treatment planning system, and (b) measured with PSLD sheet. Dose normalization is performed at isocenter.

4. Conclusion

We have developed the PSLD sheet for the measurement of dose distributions. The PSLD has significant advantage of the readout time over the TLD. Through the measurements, it has been shown that the PSLD sheet has the potential to measure the 3D dose distribution including low dose at surface. The measured results with PSLD qualitatively reproduced the calculated dose distribution.

However, the uniformity variation of the PSLD sheet over 2D area influenced the PSLD readout and it caused some discrepancies between the measurements and TPS results. We thus need to improve the PSLD itself in order to use it for the measurement with high accuracy, such as VMAT and IMRT QA. After some improvements, it can be considered that the PSLD sheet are used for a surface dose measurement of cone-beam CT, dose distribution of electron IMRT, and the dose measurements where it is difficult for the film to access.

5. References

[1] Das I J *et al* 2008 *Med. Phys* **35** 4186
[2] Heukelom S *et al* 1991 *Phys. Med. Biol.* **36** 47
[3] Butson M J *et al* 1996 *Med. Phys* **23** 655
[4] Butson M J *et al* 1996 *Phys. Med. Biol.* **41** 1073
[5] Mather M L *et al* 2002 *Phys. Med. Biol.* **47** 4397-409
[6] Hurley C *et al* 2006 *Nucl. Instrum. Methods Phys. Res. A* **565** 801-11
[7] JP Patent No. 4431701
[8] JP Patent No. 4457219
[9] JP Publication No. 2012-52070
[10] JP Publication No. 2012-52071
[11] JP Publication No. 2012-52072
[12] JP Patent Application No. 2011-077599
[13] US Publication No. US2011-0253899A1
[14] C. Kurokawa *et al* 2011 *Med. Phys* **38** 3517
[15] JP Patent Application No. 2011-223026

Exploring the dose response of radiochromic dosimeters

PS Skyt[1,2], I Wahlstedt[2], ES Yates[1], L P Muren[1], JBB Petersen[1] and P Balling[2]
[1]Dept of Medical Physics, Aarhus University /Aarhus University Hospital, Denmark
[2]Dept of Physics and Astronomy, Aarhus University, Denmark

E-mail: skyt@phys.au.dk

Abstract. The aim of this study was to explore the dose response of a newly developed radiochromic hydrogel dosimeter based on leuco malachite green dye in a gelatine matrix. The original dosimeter composition was first investigated in terms of dose response and dose-rate dependence. In addition, the initiating compounds producing chlorine radicals were substituted with compounds producing fluorine radicals, oxygen-centered radicals, carbon-centered radicals and bromine radicals. Also the surfactant was substituted by other compounds of different molecular size and charge. The original composition gave a dose response of $3.5 \cdot 10^{-3}$ $Gy^{-1}cm^{-1}$ at 6 Gy/min with a dose rate dependence giving a 27 % increase when decreasing the dose rate to 1 Gy/min. None of the substituted initiating components contributed to an increase in dose response while only one surfactant increased the dose response slightly.

1. Introduction

Radiochromic dosimeters are promising tools for three-dimensional dosimetry [1] since they i) are insensitive to oxygen, ii) offer non-scattering dose response, and iii) are well suited for optical readout [2]. In recent years, new dosimeter compositions have been proposed based on leuco malachite green dye suspended in a gelatine gel [3, 4]. However, the dosimeters suffer from low dose response, dose-rate dependence, and a relatively high auto-oxidation. To assess these issues we have in this study investigated the effect on the dose response when substituting chemical components of the dosimeter. Two components were substituted, the initiator in order to change its reactivity, as well as the surfactant to change the micelle structures.

2. Materials and Methods

2.1. Dosimeter fabrication

The original composition used in this study is based on that proposed in [4]. It consists of 6 % (w/w) gelatine that forms the matrix of the volume dosimeter. The active component is 0.37 mM leuco malachite green (LMG) dissolved in 80 mM trichloromethane ($CHCl_3$) while 5 mM trichloroacetic acid (TCA) was added as an initiator. To dissolve the LMG and $CHCl_3$ in the gelatine solution 50 mM sodium dodecyl sulfate (SDS) was added as a surfactant. Since SDS consists of a hydrophilic head region and a hydrophobic tail these molecules forms so-called micelles, making it possible to dissolve the non-polar LMG and $CHCl_3$ in an aqueous solution. A description of the procedure for manufacturing the dosimeter can be found in [4]. The gel was prepared in standard PMMA cuvettes (1x1x4.5 cm) and placed in a refrigerator until the next day.

2.2. *Substituting initiator*

The initiator, TCA, was substituted to investigate the effect of changing the reactivity of both the compound and the radicals produced. TCA was substituted with triflouroacetic acid (TFA) to yield fluorine radicals, 2,4-pentadione peroxide (abbreviated 2,4-peroxide) and hydrogen peroxide (H_2O_2), to yield oxygen-centered radicals, 2,2'-azobis(2-methylpropionamide) dihydrochloride (abbreviated 2,2-azo) and 4,4'-azobis(4-cyanovaleric acid) (abbreviated 4,4-azo) to yield carbon-centered radicals, and HBr and CBr_4 to yield bromine radicals. NaOH was added to very acidic gel formulations since such acidic solution gave a thick opaque gel with two phases. The change in pH was not observed to affect the dose response for the original version.

In addition, a batch without TCA as well as batches with both TCA and $CHCl_3$ substituted with CCl_4 or dimethylformamide (DMF) were made to investigate the initiating effect of TCA and $CHCl_3$.

2.3. *Substituting the surfactant*

Since LMG is expected to be inside the micelles due to its non-polarity, the shape of the micelles might be of influence for the sensitivity of the dosimeter. Therefore the surfactant SDS was substituted in order to change the micelle size. It was hypothesized that larger heads or shorter tails of the surfactant would lead to smaller micelles increasing the surface-to-volume ratio. Three surfactants were investigated: Sodium dodecylbenzenesulfate (SDBS), sodium octyl sulfate (SOS), and dodecyltrimethylammonium bromide (DTAB). SDBS contains a larger head region since unlike SDS a benzene ring is present in the head region while SOS has a shorter tail than SDS. DTAB was similar to SDS but with a cationic head contrary to the anionic SDS.

2.4. *Irradiation*

Irradiation of the dosimeters was performed with x-rays from a linear accelerator set at 6 MV. The dosimeters were placed in a source-to-surface distance of 94.5 cm behind 5 cm of solid water. Additionally 5 cm solid water was placed behind the dosimeters to ensure backscatter. The dose rate was 6 Gy/min and dose sequences up to 80 Gy were given.

2.5. *Read-out and data analysis*

The optical densities of the dosimeters were measured with a spectrophotometer (Helios Alpha, Thermo Spectronic) at 633 nm both a few hours before irradiation and about 16 hours after irradiation. The two measurements were subtracted in order to obtain the optical response, i.e. the optical density change caused by the irradiation. The dose response was then obtained by plotting the optical response as a function of dose and fitting to a linear equation. The dosimeters were assessed in terms of dose response as well as transparency.

3. Results

The mean dose response of 9 batches of the original composition was found to be $(3.5\pm0.1)\cdot10^{-3}$ $Gy^{-1}cm^{-1}$ at a dose rate of 6 Gy/min that increased with 27% when the dose rate was decreased to 1 Gy/min. When TCA was omitted the dose response was $(2.78\pm0.05)\cdot10^{-3}$ $Gy^{-1}cm^{-1}$ and when both TCA and $CHCl_3$ were substituted by DMF or CCl_4 dose responses of $(1.7\pm0.2)\cdot10^{-3}$ $Gy^{-1}cm^{-1}$ and $(1.72\pm0.07)\cdot10^{-3}$ $Gy^{-1}cm^{-1}$ were obtained. The dose response of all other dosimeter formulations where the initiator TCA was substituted is summarized in table 1.

When adding TFA at the same concentration as that used for TCA in the original formulation, a 12% lower dose response was observed. When doubling the concentration the gel turned opaque, but by adjusting the pH by adding NaOH the transparency was increased. The dose response, however, decreased with increasing concentration. Similar tendencies were observed for both HBr and 4,4-azo with a decreased dose response of 3% and 18%, respectively. CBr_4, 4,4-azo, and both peroxides gave lower responses even at the original TCA concentration.

The highest dose response (15% increase compared to the original formulation) and the only increase compared to the original formulation was obtained with SDBS which, however, resulted in opaque gels. Of the three surfactants only DTAB resulted in transparent gels but gave very low dose response. The results for dosimeter formulations with substitutes for the surfactant SDS are summarized in table 2.

Initiators	Concentration	Notes	Dose response (10^{-3} $Gy^{-1}cm^{-1}$)
TFA	5 mM		3.09 ± 0.06
	10 mM	Hazy	2.91 ± 0.05
	10 mM	pH adjusted	2.2 ± 0.1
2,4-peroxide	5 mM		1.86 ± 0.05
	10 mM		2 ± 1
H_2O_2	5 mM	Hazy	2.4 ± 0.3
	10 mM	Hazy	1.7 ± 0.1
2,2-azo	5 mM	Opaque	2.6 ± 0.8
4,4-azo	5 mM		2.86 ± 0.02
	10 mM	Opaque	0.4 ± 0.6
CBr_4	5 mM		2.4 ± 0.2
	10 mM	pH adjusted	No response
HBr	7 mM		3.39 ± 0.04
	30 mM	pH adjusted	2.9 ± 0.2

Table 1: The original initiator, TCA, was substituted with the compounds and concentrations shown in this table. In addition, notes on their appearance as well as the dose response are shown. 4,4-azo was dissolved in DMF since it could not be dissolved in $CHCl_3$.

Surfactants	Concentration	Notes	Dose response (10^{-3} $Gy^{-1}cm^{-1}$)
DTAB	40 mM		0.42 ± 0.02
	80 mM		0.3 ± 0.1
SOS	80 mM	Opaque	2.07 ± 0.04
SDBS	25 mM	Opaque	3.2 ± 0.1
	50 mM	Opaque	3.91 ± 0.05
	100 mM	Opaque	4.1 ± 0.2

Table 2: The original surfactant, SDS, was substituted with the compounds and concentrations shown in this table.

4. Discussion

In this study we have explored the dose response of a radiochromic hydrogel dosimeter using a range of different compositions including a range of initiators as well as surfactants. Initially, using the same composition as Vandecasteele et al. [4] resulted in a mean dose response 20 % lower than that reported in [4]. However, this difference is probably partly due to a difference in the dose rate used. In addition, only half the dose response was observed with CCl_4 as initiator compared to the same study.

In [3] the surfactant Triton X-100 was used instead of SDS which resulted in a dose response similar to Vandecasteele et al. [4]. In addition, in both [4] and [5] a dosimeter based on leuco crystal violet dye instead of LMG was investigated, resulting in higher dose responses. They are, however, still lower than that of the commercially available LMG based polyurethane dosimeter Presage™ where a dose response of $(2.2\pm0.3)\cdot10^{-2}$ $Gy^{-1}cm^{-1}$ has previously been reported for photon irradiation [6]. A similar dose response for Presage™ was obtained in [7] and this study showed, in addition, a high sensitivity to irradiation temperature. The mentioned dose responses from [6] and [7] as well as all measurements in this study were all performed at room temperature.

When modifying the original composition by omitting the initiator TCA a 21% lower response than the original formulation was obtained. However, the fact that a response is observed even without TCA as initiator indicates that a considerable part of the LMG reaction is not caused by TCA but by $CHCl_3$ and other components. Substituting both TCA and $CHCl_3$ with DMF or CCl_4 decreased the dose response to half that of the original version. However, adding DMF or CCl_4 to the original composition (data not shown) did not considerably change the dose response and it therefore seems that they do not contribute. A considerable part of the dose response therefore has to be ascribed to other components than TCA and $CHCl_3$.

When substituting the initiator TCA with other compounds the general trend was that the dose response decreased when the initiator concentration was increased. At the same concentration as TCA all initiators gave lower dose response, however, considerably closer to the original response than with higher concentrations. It therefore seems that the chemical reaction scheme does not depend on the specific initiator used. The initiator is therefore probably not the limiting factor for the chemical reactions causing the dose sensitivity.

Of the three surfactants used, only DTAB produced transparent gels but the dosimeters gave very low dose response. Only the surfactant SDBS gave a slightly higher dose response than the original version, but at all concentrations the gels were opaque. Large structures therefore must be present in the gel maybe due to large micelles or micelle aggregates. Since opaque dosimeters cannot be used for optical read-out this was not investigated further.

5. Conclusion

The measurements in this study indicate that while a part of the dose response originates from the initiator TCA and $CHCl_3$, a considerable part of the dose response origins from other components. In addition, it was not possible to increase the dose response by substituting the two compounds. When substituting the surfactant a small increase was observed and it cannot be concluded whether the micelle structures are connected with the limitation in dose response. However, due to the dose rate dependence, it seems that the dose rate and the reaction rates giving the dose response are on similar time scales.

6. Acknowledgement

Anastasia Bochenkova, Grethe Vestergaard Jensen and Jan Skov Pedersen, Aarhus University are gratefully acknowledged for their contribution with knowledge regarding radical and micelle chemistry. This work was supported by CIRRO - The Lundbeck Foundation Center for Interventional Research in Radiation Oncology and The Danish Council for Strategic Research.

7. References

[1] Baldock C *et al* 2010 *Phys. Med. Biol.* **55** R1-63
[2] Bosi S *et al* 2007 *Phys. Med. Biol.* **52** 2893-903
[3] Jordan K and Avvakumov N 2009 *Phys. Med. Biol.* **54** 6773
[4] Vandecasteele J *et al* 2011 *Phys. Med. Biol.* **56** 627
[5] Babic S *et al* 2009 *Phys. Med. Biol.* **54** 6791
[6] Yates E S *et al* 2011 *Acta. Oncol.* **50** 829
[7] Skyt P S *et al* 2011 *Med. Phys.* **38** 2806

Photometric calibration of a radiochromic gel dosimeter

JB Davies[1,2], SG Bosi[1,3,4] and C Baldock[1,5]
[1]Institute of Medical Physics, School of Physics, University of Sydney, NSW 2006, Australia.
[2]Australian Nuclear Science and Technology Organisation, Lucas Heights, NSW 2234, Australia.
[3]School of Science and Technology, University of New England, Armidale, NSW 2351, Australia.
[4]Radiation Oncology Prince of Wales Hospital, High Street, Randwick, NSW 2031, Australia.
[5]Faculty of Science, Macquarie University, NSW 2109, Australia.

E-mail: jbd@ansto.gov.au

Abstract. Measurement traceability of dosimetry systems to standards of measurement underpins reliable radiation treatments. A method to determine the product of the molar linear absorption coefficient and the radiation chemical yield is presented, which provides a means of photometric calibration of chemical dosimetry systems in order that the absorbed dose be determined directly from photometric measurements.

1. Introduction

Polymer gel dosimeters are manufactured from radiation sensitive chemicals, which upon irradiation polymerize as a function of the absorbed radiation dose [1]. These gel dosimeters which record the radiation dose distribution in three-dimensions (3D) have specific advantages when compared to one-dimensional dosimeters and two-dimensional dosimeters [2, 3]. These 3D dosimeters are radiologically soft-tissue equivalent [4] with properties that may be modified depending on the application. The 3D radiation dose distribution in polymer gel dosimeters may be imaged using magnetic resonance imaging (MRI) [5, 6, 7], optical-computerized tomography (optical-CT) [8, 9], x-ray CT [10-12], ultrasound [13-17] or vibrational spectroscopy [18, 19].

To be meaningful as a dosimetry system, the determination of absorbed dose should be traceable to standards of measurement through an unbroken chain of comparisons from a primary standard to the end user measurement. Calibration of dosimetry systems to standards of measurement underpins reliable radiation treatments [20]. The quantity of interest in radiation therapy is absorbed dose to tissue, usually with reference to the measurement of absorbed dose to water [21].

Gel dosimeters are essentially chemical dosimeters, in which chemical changes are induced by the interaction of ionising radiation in the material. In some gel dosimeters (and the case examined here), this radiation-induced chemical change is accompanied by a change in the optical absorbance, allowing dose to be read photometrically. A general technique for the calibration of photometrically-readable chemical dosimetry systems, including gel dosimeters, is presented. The aim of this is to determine a constant of proportionality that can be used to calculate the absorbed dose to gel directly from photometric measurements.

2. Theory

In any chemical dosimetry system, the number of moles of a particular chemical species changed, destroyed or created, Δn, per unit of energy absorbed, ΔE, is the radiation chemical yield, G:

$$G = \frac{\Delta n}{\Delta E} \tag{1}$$

In a given material of density ϱ, the G value may be expressed in terms of the change in the concentration of the analyte of interest, Δc, and a proportional change in the absorbed dose, ΔD:

$$G = \frac{\Delta c}{\Delta D \rho} \tag{2}$$

One technique to determine the concentration is photometric measurement of the change in optical density, or absorbance, ΔA, defined by the Beer-Lambert Law:

$$\Delta A = \varepsilon_m \, d \, \Delta c \tag{3}$$

where ε_m is the molar linear absorption coefficient for the molecular species of interest at a given wavelength of monochromatic light and d is the optical pathlength through the sample. Combining equations 2 and 3, absorbed dose to the material, in this case gel, ΔD_g, becomes

$$\Delta D_g = \frac{\Delta A}{\rho \, d \, \varepsilon_m G} \tag{4}$$

In practice, ΔD_g is converted from the traceable measurement of absorbed dose to water, ΔD_w,

$$\Delta D_w = \frac{\Delta A}{\rho \, d \, \varepsilon_m G} \left(\frac{\mu_{en}}{\rho} \right)_g^w \tag{5}$$

where $(\mu_{en}/\rho)_g^w$ is the ratio of the mass-energy absorption coefficients for water and gel, assuming charged particle equilibrium conditions exist.

Many chemical dosimeters have dependencies on either the temperature during measurement, t, or the temperature during irradiation, t', or both. For this reason, the molar linear absorption coefficient is denoted as $(\varepsilon_m)_t$ and the radiation chemical yield is denoted as $G_{t'}$. In the standard Fricke solution [22], the product $(\varepsilon_m)_t G_{t'}$ was recommended [23] to be used rather than individually determined $(\varepsilon_m)_t$ and $G_{t'}$ values, due to a correlation in their respective uncertainties. Grouping these quantities as though they were one quantity avoids the difficulty in considering correlated uncertainties. Grouping of these quantities also helps avoid another complication - in some gel dosimeters, there may be multiple species responsible for the radiation-induced change observed in the optical absorbance measurement. Therefore, in such cases, the product $(\varepsilon_m)_t G_{t'}$ is, strictly speaking, an effective value, and is the sum of this product for all species responsible for optical absorption at the measured wavelength:

$$(\varepsilon_m)_t G_{t'} = \sum_i (\varepsilon_m)_{t,i} G_{t',i} \tag{6}$$

A reference measurement temperature, t_r, and a reference irradiation temperature, t_r', are arbitrarily chosen to define the molar linear absorption coefficient, $(\varepsilon_m)_{t_r}$, and the radiation chemical yield, $G_{t_r'}$, respectively [24]. At these reference conditions, the product $(\varepsilon_m)_{t_r} G_{t'_r}$ may be determined, to which all other values of $(\varepsilon_m)_t G_{t'}$ are compared. Equation 4 may then be expanded to give

$$\Delta D_g = \frac{\Delta A}{\rho d (\varepsilon_m)_{t_r} G_{t'_r} \left[1 + k_1 (t - t_r)\right]\left[1 + k_2 (t' - t'_r)\right]} \tag{7}$$

where k_1 is the measurement temperature coefficient and k_2 is the irradiation temperature coefficient. Thus, once the product $(\varepsilon_m)_{t_r} G_{t'_r}$ has been evaluated, the absorbed dose to gel may be calculated directly from photometric measurements.

$u_c(y)$ is the combined standard uncertainty in the quantity of interest, y, and is calculated following the Guide to the expression of uncertainty in measurement [25]:

$$u_c^2(y) = \sum_{i=1}^{N} \left(\frac{\partial y}{\partial x_i} \right)^2 u^2(x_i) \tag{8}$$

The partial derivative is referred to as the sensitivity coefficient, also denoted c_i, and is effectively a weighting factor of each independent influence quantity x_i. The relative uncertainty in the influence quantity is then $c_i u(x_i)/y$.

3. Discussion

Measurement of each influence quantity should be traceable to standards of measurement. For the purposes of dosimeter calibration, absorbance readings are usually obtained for gels prepared in 1 cm cuvettes. The repeatability of multiple absorbance readings, expressed as the standard deviation, indicates only the precision in the measurement and does not account for a possible systematic departure in the measured absorbance from the 'true' or traceable measurement of absorbance. Calibration of the spectrophotometer would provide information on the magnitude of this uncertainty.

The traceable measurement of the absorbed dose standard will have an uncertainty that must be included in the combined uncertainty. Reproducibility of the dose response for multiple batches of the same gel formulation accounts for random variability in the gel preparation. If a dose rate or energy dependence exists, the combined uncertainty should be evaluated under the changed conditions.

This approach only considers the dosimetric calibration of the gel as a radiation dosimeter. It does not consider other influences affecting the dosimeter uncertainty related to variations in the scanning of the dose distribution in 3D discussed elsewhere in the literature [26-28].

4. Conclusion

A technique is presented to determine the product $(\varepsilon_m)_{t_r} G_{t'_r}$, for any chemical dosimeter, including gel dosimeters, measured photometrically. Calibration of this constant of proportionality allows the absorbed dose [29] to gel to be calculated directly from photometric measurements.

5. References

[1] Baldock C *et al* 1998 *Phys. Med. Biol.* **43** 695-702
[2] Baldock C *et al* 2010 *Phys. Med. Biol.* **55** R1-63
[3] Murry P and Baldock C 2000 *Australas. Phys. Eng. Sci. Med.* **23** 44-51
[4] Keall P and Baldock C 1999 *Australas. Phys. Eng. Sci. Med.* **22** 85-91
[5] Lepage M *et al* 2001 *Phys. Med. Biol.* **46** 2827-39
[6] Gustavsson H *et al* 2004 *Phys. Med. Biol.* **49** 227-41
[7] Lepage M *et al* 2002 *Phys. Med. Biol.* **47** 1881-90
[8] Bosi S *et al* 2007 *Phys. Med. Biol.* **52** 2893-903
[9] Bosi S G *et al* 2009 *Phys. Med. Biol.* **54** 275-83

[10] Trapp J V *et al* 2002 *Phys. Med. Biol.* **47** 4247-58
[11] Brindha S *et al* 2004 *Phys. Med. Biol.* **49** N353-61
[12] Hill B *et al* 2005 *Br. J. Radiol.* **78** 623-30
[13] Mather M L *et al* 2002 *Phys. Med. Biol.* **47** 1449-58
[14] Mather M L *et al* 2002 *Phys. Med. Biol.* **47** 4397-409
[15] Mather M L and Baldock C 2003 *Med. Phys.* **30** 2140-8
[16] Mather M L *et al Ultrasonics* **41** 551-9
[17] Mather M L *et al* 2003 *Phys. Med. Biol.* **48** N269-75
[18] Baldock C *et al* 1998 *Phys. Med. Biol.* **43** 3617-27
[19] Rintoul L *et al* 2003 *Appl. Spectrosc.* **57** 51-7
[20] Willowson K *et al 2008 Phys. Med. Biol.* **53** 3099-112
[21] International Atomic Energy Agency (IAEA) 2000 *Absorbed Dose Determination in External Beam Radiotherapy: An International Code of Practice for Dosimetry Based on Standards of Absorbed Dose to Water Technical Reports Series* **398**
[22] Fricke H and Hart EJ 1966 *Chemical Dosimetry* in *Radiation Dosimetry* 2nd ed. Vol. II. FH Attix and WC Roesch (ed.) (Academic Press, New York)
[23] International Commission on Radiation Units and Measurements 1984 *ICRU Report* **35**
[24] Davies J B and Baldock C 2010 *Rad. Phys. Chem.* **79** 660-2
[25] International Organization for Standardization 2008 Uncertainty of measurement - Part 3: Guide to the expression of uncertainty in measurement *ISO/IEC Guide* **98-3**
[26] Baldock C *et al* 1999 *Phys. Med. Biol.* **44** N243-6
[27] Baldock C *et al* 2001 *Phys. Med. Biol.* **46** 449-60
[28] De Deene Y 2006 *J. Phys.: Conf. Ser.* **56** 72-85
[29] Hill R et al 2005 *Phys. Med. Biol.* **50** N331-44

Diffusion properties of a radiochromic hydrogel dosimeter

PS Skyt[1,2], TS Kinnari[2], I Wahlstedt[2], ES Yates[1], LP Muren[1], JBB Petersen[1] and P Balling[2]
[1]Dept of Medical Physics, Aarhus University /Aarhus University Hospital, Denmark
[2]Dept of Physics and Astronomy, Aarhus University, Denmark

E-mail: skyt@phys.au.dk

Abstract. The aim of this study was to investigate the diffusion properties of a radiochromic hydrogel dosimeter based on leuco malachite green dye in a gelatine matrix. One half of each dosimeter was irradiated while the other half was left un-irradiated creating dose gradients over which diffusion could be investigated. Read-out of the optical response was performed with a high-resolution optical scanner. The dosimeters were found to exhibit a low diffusion rate but a high auto-oxidation level leading to a fading of the contrast in the dose response with time.

1. Introduction
The high conformity of modern radiotherapy techniques has motivated research into dosimeters that offer full three-dimensional dose measurements. Dosimeters capable of such measurements include both polymer and radiochromic dosimeters [1, 2]. Radiochromic hydrogel dosimeters are promising tools within three-dimensional dosimetry since i) they are easy to make, ii) they are insensitive to oxygen, and iii) they possess good properties for optical read-out due to a non-scattering dose response. However, the accuracy of such dosimeters can potentially be deteriorated due to diffusion of the optical response. Diffusion in radiochromic hydrogel dosimeters has previously been reported in the literature for various hydrogel dosimeters [3-7], however to our knowledge only for a single dose and temperature. In order to obtain a broader description of the diffusion of the optical response, we have in this study investigated the diffusion properties of a radiochromic hydrogel dosimeter based on leuco malachite green dye in a gelatine matrix for a range of doses as well as a range of post-irradiation storage temperatures.

2. Materials and Methods

2.1. Dosimeter fabrication
The dosimeter used in this study was a radiochromic hydrogel dosimeter as proposed in Ref. 3. The chemical composition consists of the active component leuco malachite green dye (0.37 mM) which was dissolved in 80 mM trichloromethane while 5 mM trichloroacetic acid was added as an initiator. This solution together with the surfactant sodium dodecyl sulphate (50 mM) was mixed with a 6 % (w/w) aqueous gelatine solution to form a three-dimensional dosimeter. The surfactant made it possible to dissolve the non-polar leuco malachite green dye and the trichloromethane in the aqueous gelatine solution due to formation of so-called micelles. The gel was prepared in standard PMMA cuvettes (1x1x4.5 cm) and placed in a refrigerator after fabrication.

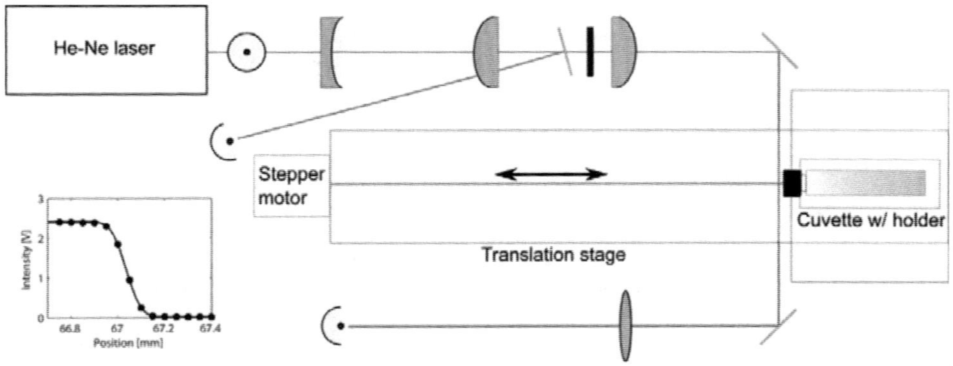

Figure 1: Illustration of the high resolution optical scanner. The dosimeter is translated through a thin laser beam from a helium-neon laser while the transmitted intensity is measured with a detector. The graph in the bottom left shows a knife-edge measurement (i.e. the integrated beam profile) at the position of the dosimeter giving a beam waist of 0.102±0.005 mm.

2.2. *Irradiation and temperature*

Irradiation of the dosimeters was performed with a dose rate of 14.6 Gy/min using x-rays from a linear accelerator set at 6 MV. The dosimeters were placed on the gantry center axis with one jaw covering half the dosimeter, creating a dose gradient in the dosimeter. The upper half of the dosimeters were then irradiated to doses of 20 Gy, 40 Gy, 60 Gy or 80 Gy while the bottom half was left un-irradiated. Irradiation was performed at room temperature while the dosimeters were stored in a refrigerator before irradiation and at a series of temperatures ranging from 5.4 °C to 29.8 °C after irradiation. The storage temperatures after irradiation were stabilized with a precision of ±0.5 °C.

2.3. *Read-out of the optical response*

In order to perform high resolution measurements of the optical response in three dimensional dosimeters, a high resolution optical scanner was built. The scanner was designed to measure the optical density of a volume dosimeter along the axis of a standard cuvette or similar dosimeter container. A schematic illustration of the scanner is shown in figure 1. The laser beam from a helium-neon laser (632.8 nm) is shaped vertically with the use of two cylindrical lenses (-5.8 mm and 150 mm focal length). Together with an iris this creates a vertical beam diameter of 8 mm. Horizontally, a cylindrical lens with 150 mm focal length focuses the beam to 0.102±0.005 mm beam waist, i.e. a FWHM of 0.120±0.006 mm, positioned in the middle of the dosimeter. This was a compromise between a narrow beam waist and limited beam divergence (i.e. a sufficiently large Rayleigh length). When a cuvette is scanned along its axis, the vertically extended, horizontally narrow beam will result in a high resolution along the measurement direction while integrating the signal perpendicular to the cuvette axis. The latter reduces uncertainties due to impurities in the dosimeter as well as on the cuvette. In addition, a spherical lens (f = 58 mm focal length) is positioned behind the cuvette at a $2 \cdot f$ distance from both the dosimeter and the detector in order to image the middle of the dosimeter onto the detector to maintain a high spatial resolution of the scanning system.

The dosimeter was placed on a motion stage motorized with a stepper motor with a minimum step length well below the resolution limit of the optical system. The laser intensity was measured with two detectors, with a detector placed after the dosimeter measuring the transmitted light, and a reference detector, in order to compensate for possible instabilities in laser power. Both the data acquisition (Data Translation DT9834) and the motorized stage were computer controlled, and in this study the dosimeters were translated 0.2 mm between each step. At each position, a data sequence was recorded and averaged.

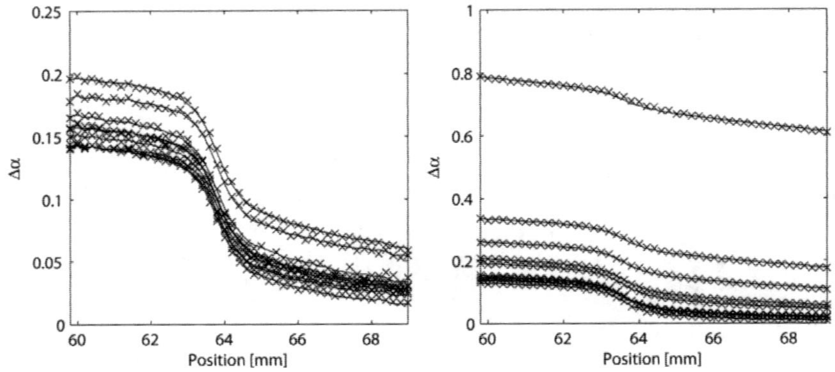

Figure 2: Dose gradients from 40 Gy (left side of the gradient) to 0 Gy in dosimeters stored after irradiation at 5.4 °C (left graph) and 20.8 °C (right graph). The lowest curves in the two graphs were measured 2 hours and 1 hour after irradiation, respectively. The following curves from bottom to top were measured 3, 5, 7, 19, 25, 45, 68, and 143 hours after irradiation. The solid lines are fits to Eq. 1.

The dosimeters were measured with the optical scanner 1-2 hours before irradiation as well as several times after irradiation to quantify the change in dose gradient with time due to diffusion of the optical response.

2.4. Data analysis
The change in optical density due to the irradiation, i.e. the optical response, was found by subtracting the optical densities of the pre-irradiation measurement from the post-irradiation measurements. The diffusion properties of the dosimeters were then analysed by fitting the optical response, or change in attenuation coefficient, to

$$\Delta\alpha(x) = AO + s\int \Delta\alpha_{irr}(x') \cdot \frac{1}{\sqrt{2\pi}\sigma}\exp\left(\frac{-(x'-x)^2}{2\sigma^2}\right)dx', \qquad (1)$$

where AO is a second order polynomial expression describing auto oxidation of the leuco dye and s is a scaling factor. The integral describes a convolution of the first post-irradiation measurement, $\Delta\alpha_{irr}$, and a normal distribution describing the change from $\Delta\alpha_{irr}$, i.e. changes in optical response due to diffusion.

3. Results
Figure 2 shows the optical response as a function of position in two dosimeters for a range of post-irradiation measurements including fits to Eq. 1. The left graph shows a dosimeter stored at 5.4 °C after irradiation while the right graph shows a dosimeter stored at 20.8 °C. Both dosimeters have received 40 Gy with a gradient down to 0 Gy. The lowest curve in each graph was used as $\Delta\alpha_{irr}$ when fitting all other curves for each dosimeter. These curves were measured 2.1 hours and 0.9 hours after irradiation, respectively. The following curves shows post-irradiation measurements with the top curve approximately 143 hours after irradiation.

As clearly visible on the graphs, the observed diffusion is low. Values for the broadening of the optical response, σ, as determined from the fits was found to be constant in time for both temperatures with mean values of 0.08±0.01 mm and 0.09±0.01 mm at 5.4 °C and 20.8 °C, respectively. In addition, no considerable change in σ was observed in the remaining temperature range of 11.3 °C, 14.9 °C, 25.0 °C and 29.8 °C or in the remaining dose range of 20 Gy to 80 Gy. The overall mean across all measurements was found to be 0.073±0.003 mm.

A pronounced auto-oxidation of the optical response was observed within the time range of the measurements (figure 2). The auto-oxidation not only resulted in an offset in optical response but also a curvature through the dosimeter, described by a second-order polynomial expression (*AO* in Eq. 1).

4. Discussion

In this study we have investigated the diffusion properties of the dose response in a radiochromic hydrogel dosimeter using a high resolution optical scanner. Since the dosimeter used in this study contains the non-polar leuco malachite green dye in a hydrogel, the dye is initially expected to be captured within micelles formed by the surfactant. In the expected reaction upon irradiation, the leuco dye is transformed to a coloured dye that is charged. The coloured dye might therefore acquire sufficient polar characteristics that it may diffuse out of the micelles. This, however, seems not to be a pronounced effect since a very low diffusion of the optical response was observed in this study. The change in dose gradients with time due to diffusion was observed to be lower than the resolution of the scanner. The constant value for the broadening of the initial gradient, σ, which was obtained for all fits to Eq. 1 independent of time after irradiation, post-irradiation temperature or dose, is therefore attributed to a measure for the reproducibility of the optical scanner.

Diffusion of the optical response has previously been reported for radiochromic hydrogel dosimeters, however with the use of another surfactant than sodium dodecyl sulfate [3, 4]. In addition diffusion in dosimeters with the same chemical composition as in this study has been reported [5]. In that study a diffusion rate was observed when taking into account offsets due to auto oxidation, and the remaining change with time was then attributed to diffusion. This is contrary to our study, where the only visible effect within the time range of the experiments of 143 hours was due to auto-oxidation as well as a relative fading of the signal caused by irradiation. In addition, the auto-oxidation in our study varied through the dosimeters and had to be described by a second-order polynomial expression. Similar auto-oxidation behaviour was also observed in a set of un-irradiated dosimeters, however, with different offsets between the optical response measured in the irradiated dosimeters and the un-irradiated dosimeters.

5. Conclusion

In terms of diffusion of the optical response, the radiochromic hydrogel dosimeter investigated here was observed to be stable with a diffusion lower than the resolution of the optical scanner. However, the dosimeter showed a high auto-oxidation in addition to a relative fading of the optical response with time.

6. Acknowledgement

This work was supported by CIRRO - The Lundbeck Foundation Center for Interventional Research in Radiation Oncology and The Danish Council for Strategic Research.

7. References

[1] Doran S J 2009 *Appl. Radiat. Isotopes* **67** 393
[2] Baldock C *et al* 2010 *Phys. Med. Biol.* **55** R1-63
[3] Jordan K and Avvakumov N 2009 *Phys. Med. Biol.* **54** 6773
[4] Babic S *et al* 2009 *Phys. Med. Biol.* **54** 6791
[5] Vandecasteele J *et al* 2011 *Phys. Med. Biol.* **56** 627
[6] Harris P J *et al* 1996 *Phys. Med. Biol.* **41** 1745-53
[7] Baldock C *et al* 2001 *Australas. Phys. Eng. Sci. Med.* **24** 19-30

Journal of Physics: Conference Series **444** (2013) 012039 doi:10.1088/1742-6596/444/1/012039

Non-diffusing photochromic gel for optical computed tomography phantoms

K Jordan
London Regional Cancer Program (London Health Sciences Centre), Department of Medical Biophysics, Western University, London, ON, Canada

E-mail: kevin.jordan@lhsc.on.ca

Abstract. This study examines photochromic response in radiation sensitive hydrogels. Genipin, crosslinked, gelatin gel can support high resolution images because the chromophores do not diffuse. A low power, 633 nm He-Ne laser was used to write lines into the gels by a photobleaching reaction. Optical cone-beam computed tomography (CBCT) scans mapped the high resolution images in 3D with 0.25 mm voxel resolution. A straight line was written into a deformed gel and then readout in its relaxed, initial shape. The curved, photo-bleached line demonstrated deformable 3D dosimetry is possible with this system to the balloon edge. High resolution, photochromic images provide key information for characterizing optical CT scanners and 3D dosimeters. Many, ionizing radiation, dosimeter materials demonstrate either a photochromic or photothermal response, allowing this approach to be widely used in quantitative 3D scanning.

1. Introduction

Optical computed tomography (CT) with visible light can achieve spatial resolutions from 1 micron to greater than 1 mm depending on the application. In order to characterize these systems test images of similar spatial dimensions are required. Images that have similar 3D refractive indices and similar absorption and scatter characteristics are simpler to interpret. In the case of PRESAGE®, high resolution test images have been written with synchrotron beams and UV photo-lithography equipment [1]. In this initial report, a red light, low power laser is used to write high resolution images into a hydrogel that is sensitive to both MeV (radiochromic) and eV (photochromic) photons. Genipin hydrogel was chosen because of chromophore is crosslinked to the gelatin polymer chains, eliminating diffusion [2-4]. Genipin gel radiation response has been characterized and from keV to MeV energies and very similar to Fricke solutions and water [5-7]. It is anticipated that writing photochromic images will be a simple, inexpensive method for preparing calibration images that can be produced in any facility with an optical CT scanner. While not the focus of this study, these results apply to other forms of microscopy such as optical projection tomography.

2. Methods

Genipin gel was prepared with: 3% gelatin by mass, genipin 0.1% mass of gelatin, and 200 mM H_2SO_4. The genipin and gelatin were dissolved in distilled water and stirred at 45 C for 18 hours and acid was added to form the stock gel. A stock gel sample was diluted with blank gel (200 mM H_2SO_4, 3% gelatin) to achieve acceptable levels of transmission at 594 nm for optical CT scans. A commercial

"5 inch clear" latex balloon and a rigid PETE plastic jar from Modus Medical Devices were filled with genipin gel.

The jar sample was scanned with a modified Vista10 optical cone-beam computed tomography (CBCT) scanner (590 nm light, 512 projections per 360 degrees, camera lens aperture of F4 and 8 minute scan time). Reconstruction with Vista software was performed at 0.25 mm^3 voxel size with Ramp filter, generating a 512^3 array of attenuation coefficients. While the larger, gel filled–balloon was scanned with an in-house prototype laser scanning CBCT scanner (594 nm laser light, 512 projections per 360 degrees, 200 slices, 30 minutes scan time). Propylene glycol aqueous solutions were prepared for refractive index matching. Reconstruction was performed with in-house software implementing the FDK algorithm with 0.34 mm^3 voxel size and Ramp filter.

Following pre-illumination (reference) optical CT scans the gel samples were illuminated with a, He-Ne laser (633 nm, 13 mW, 0.7 mm diameter beam) for 30 minutes per position. The gels were positioned in a cubic water phantom to minimize refraction effects. The cylindrical jar was first placed upright in the corner of the water phantom and illuminated with the laser beam near the half radius position and then rotated 90 degrees for second beam to generate orthogonal coplanar rays for illustration purposes. Finally the jar was placed horizontally and a third beam illumination formed the last mutually orthogonal ray. The gel-filled balloon was deformed by another water-filled PETE jar and the laser beam passed within 2 mm in the inner surface of the deformation, see figure 1.

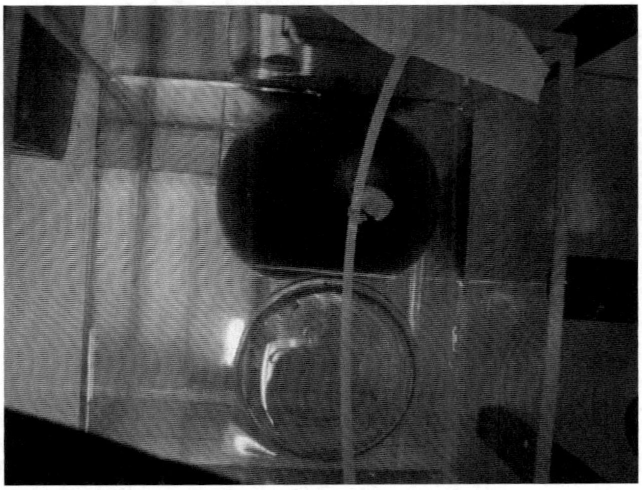

Figure 1: Photograph of deformed genipin gel balloon in a water phantom during laser photobleaching illumination, viewed from above.

The balloon relaxed for 30 minutes before acquiring the post illumination optical CT scan. Segments from several transverse slices of the 3D reconstruction were stitched together to record the path of the laser beam through the deformed balloon. A centering base ring was made for the balloon and fixed to the CT rotation platform. The liquid level was adjusted so that several centimeters of the balloon neck was above the liquid to provide sufficient friction between the gel and the rotation platform. A black, ink-spot fiducial was placed on the balloon. The laser beam was directed onto the fiducial and the position recorded. After irradiation the balloon was again placed in the scanner and adjusted until the fiducial aligned with the laser beam. Note, this genipin gel balloon was previously irradiated with crossed beams and results were reported in a separate submission related to optical CT scanning of latex balloons.

3. Results and Discussion

A transverse slice from the reconstruction of the 3 orthogonal laser beam illumination is shown in figure 2. Note that the intersecting laser beams do not exactly lie in this plane as seen by the varying attenuation changes. Also, the out of plane beam is not obvious and an arrow points to its location. This ray is offset by ~2 mm from the intersection of the first two beams. If required, much smaller features can be written by focusing the laser beam.

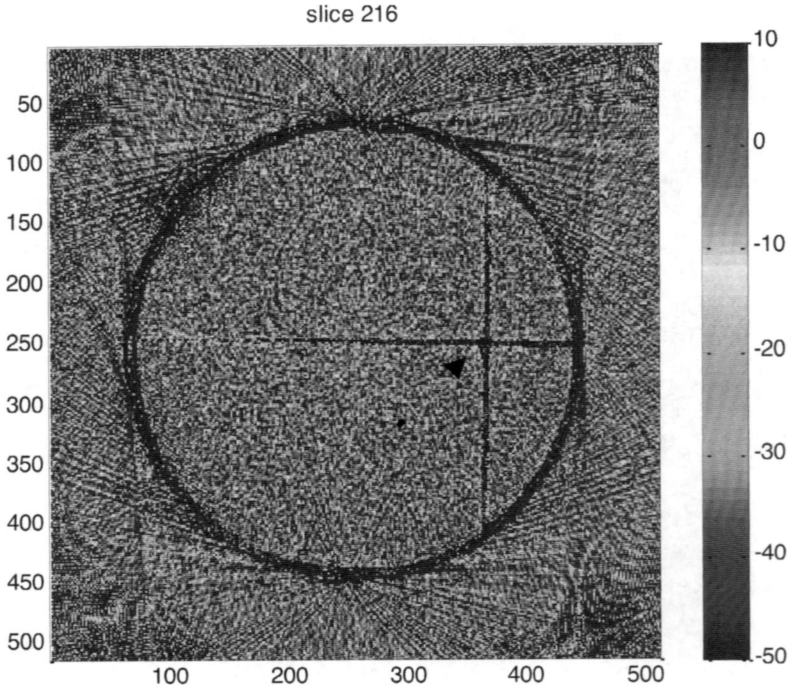

Figure 2: Transverse slice containing the photochromic image of the intersecting laser beams. Note the arrow pointing to third mutually orthogonal beam.

In figure 3, a composite slice from the 3D reconstruction shows the deformation of the path the laser beam. The photo-bleached path contains the voxels that were illuminated by the laser beam with the gel deformed within the balloon. The ability to probe near the surface of the dosimeter is especially important for deformation experiments since the surface regions experience the greatest deformations. Also present within the slice is an artifact image from a previous dose distribution due to intersecting MV x-ray beams. The dosimeter has been irradiated for another study presented at this conference. The position of dosimeter between the reference and data scans was slightly different allowing the edges of the distribution to be evident in the ratioed images. Examples of other photochromic materials are: Radgel, leuco crystal violet and leuco malachite green micelle gels with Laponite® and Presage®. Since many of the radiosensitive dosimeter materials are also photochromic and thermochromic, this is a general technique for preparing test patterns. While, polymerization gels were not part of this study, it is anticipated that lasers could initiate polymer image formation in these materials as well. Materials with diffusion of reactants or photoproducts can also be studied by monitoring growth or fading of written images. Photochromic or photothermal responses may be alternatives to radiosensitivity calibration procedures providing further simplification of 3D dosimetry.

Journal of Physics: Conference Series **444** (2013) 012039 doi:10.1088/1742-6596/444/1/012039

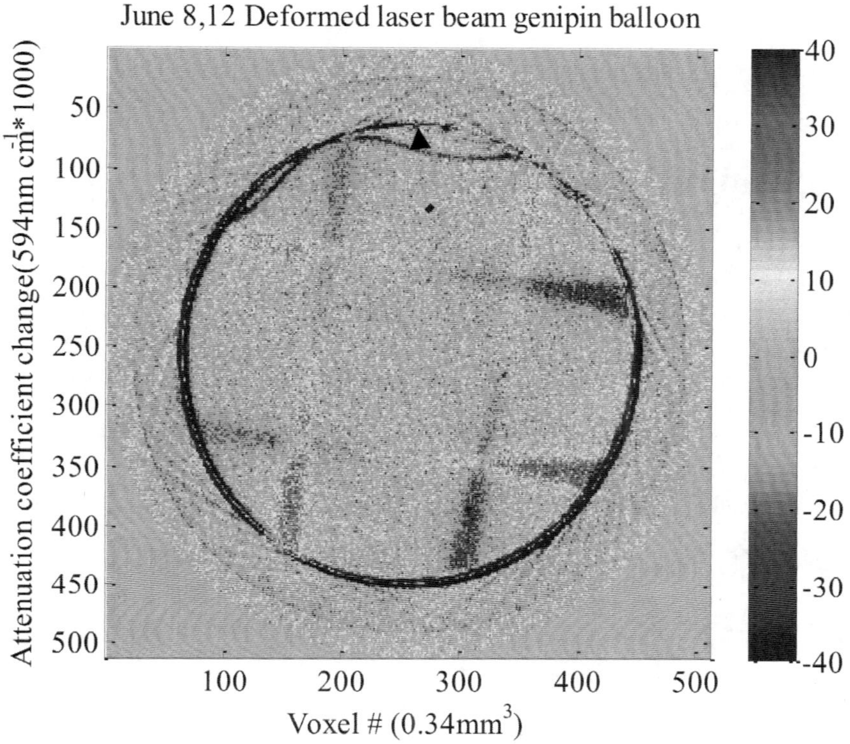

Figure 3: Composite slice of the photobleached line following relaxation of the deformed gel-filled balloon. Arrow highlights path.

4. Conclusion

Non-diffusing photochromic gels are versatile materials for writing high resolution 3D images. One application of this process is development of test patterns to characterize optical CT performance. Optically probing deformation properties of 3D dosimeters is another application.

5. References

[1] Abdul Rahman AT *et al* 2011 *Phys. Med. Biol.* **56** 4177-99
[2] Jordan K 2009 *J. Phys.:Conf. Ser.* **164** 012029
[3] Baldock C *et al* 2001 *Australas. Phys. Eng. Sci. Med.* **24** 19-30
[4] Harris P J *et al* 1996 *Phys. Med. Biol.* **41** 1745-53
[5] Davies J B *et al* 2012 *Radiat. Phys. Chem.* **81** 1263-5
[6] Gorjiara T *et al* 2011 *Phys. Med. Biol.* **56** 4685-99
[7] Bosi S G *et al* 2009 *Phys. Med. Biol.* **54** 275-83

Verification of a non-diffusing gel dosimeter

JB Davies[1,2], SG Bosi[1,3,4] and C Baldock[1,5]
[1]Institute of Medical Physics, School of Physics, University of Sydney, NSW 2006, Australia.
[2]Australian Nuclear Science and Technology Organisation, Lucas Heights, NSW 2234, Australia.
[3]School of Science and Technology, University of New England, Armidale, NSW 2351, Australia.
[4]Radiation Oncology Prince of Wales Hospital, High Street, Randwick, NSW 2031, Australia.
[5]Faculty of Science, Macquarie University, NSW 2109, Australia

E-mail: jbd@ansto.gov.au

Abstract. An evaluation of the diffusion coefficient of a genipin-gelatin gel dosimeter was carried out by fitting an inverse square root function to image profile data. A comparison was made with a Fricke-gelatin-xylenol orange (FGX) gel dosimeter, in which the ions are known to diffuse. The diffusion coefficient for this FGX gel, consisting of 0.5 mM ferrous ammonium sulphate, 50 mM sulphuric acid, 0.15 mM xylenol orange and 3 % by weight gelatin was 0.70 ± 0.05 mm^2 h^{-1} at 5 Gy. The genipin-gelatin gel consisted of 50 μM genipin, 4 % by weight gelatin and 100 mM sulphuric acid. The fitted parameter that is proportional to the diffusion coefficient did not significantly change over time, demonstrating that this genipin-gelatin gel is a non-diffusing dosimeter.

1. Introduction

Polymer gel dosimeters are manufactured from radiation sensitive chemicals, which upon irradiation polymerize as a function of the absorbed radiation dose [1]. These gel dosimeters which record the radiation dose distribution in three-dimensions (3D) have specific advantages when compared to one-dimensional dosimeters and two-dimensional dosimeters [2]. These 3D dosimeters are radiologically soft-tissue equivalent [3] with properties that may be modified depending on the application. The 3D radiation dose distribution in polymer gel dosimeters may be imaged using magnetic resonance imaging (MRI) [4, 5], optical-computerized tomography (optical-CT) [6, 7], x-ray CT [8, 9], ultrasound [10-14] or vibrational spectroscopy [15, 16].

Gelatin-based gel dosimeters are made from radiation-sensitive materials that change as a function of absorbed dose, stabilised in a gelatin matrix. The resulting dose distribution within the gel volume needs to be spatially stable for a time sufficient to allow measurement of the dose distribution in three-dimensions (3D). A limiting feature of any gel dosimeter infused with radiation-sensitive species is the diffusion of those species in regions of dose gradients within the gel. This effect will eventually lead to a loss of 3D dosimetric information in the material.

In Fricke-infused gels, the radiation-sensitive ferrous ion (Fe^{2+}) is oxidised upon irradiation to a ferric ion (Fe^{3+}), and these ions will diffuse. Addition of the chelating agent, xylenol orange (XO), which binds to Fe^{3+} and forms a larger molecule, can reduce the diffusion coefficient [17-19]. It has been reported that a genipin-gelatin gel does not diffuse post-irradiation [20]. Diffusion in a new genipin-gelatin gel dosimeter formulation [21] was quantitatively evaluated and compared with the diffusion coefficient determined in a Fricke-gelatin-xylenol orange (FGX) gel dosimeter, in which the ions are known to diffuse across dose gradients.

2. Methods

Genipin gel was prepared according to a new formulation [21] made from 50 μM genipin (078-03021, Wako Pure Chemical Industries), 4 % by weight gelatin from porcine skin (300 bloom, G2500, Sigma-Aldrich) and 100 mM sulphuric acid (102761C, BDH) in the final gel. FGX gel was prepared according to a previous formulation [22] and the final gel consisted of 0.5 mM ferrous ammonium sulphate (09719, Fluka), 50 mM sulphuric acid (102761C, BDH), 0.15 mM xylenol orange (XO) (398187, Sigma-Aldrich) and 3 % by weight gelatin from porcine skin (300 bloom, G2500, Sigma-Aldrich). All gels were pipetted into 1 cm UV-grade methacrylate cuvettes (Z188018, Sigma-Aldrich) and capped with disposable plastic caps (9020, Starna).

All gels were brought to room temperature prior to recording baseline transmission images using a modified flatbed optical scanner (Canon, MX850). Image processing settings were disabled in the scanner software to obtain as closely as possible, the raw image data in tagged image file format (tiff). The cuvette was placed in the centre of the scanner platen where the standard deviation in the noise of a background image was 0.3 %. Profile slices for each sample were averaged from up to 10 consecutive images to improve the signal to noise ratio.

In order to create a sharp dose boundary, cuvettes were attached to the depleted uranium (DU) trimmer bars of a jaw collimator on an Eldorado-6 Atomic Energy of Canada (AECL) cobalt-60 therapy unit so that approximately half of the gel was shielded by the DU and the remainder exposed to the radiation beam at a dose rate of approximately 0.5 Gy min^{-1}. The gels were exposed to the source for a dose sufficient to clearly define a dose boundary. Three genipin-gelatin cuvettes were irradiated for 90 minutes and one FGX cuvette was irradiated for 10 minutes.

Image profiles were taken at various times after irradiation. The temperature during all measurements was 22 ± 1 °C. The average baseline image profile was subtracted from all post-irradiation profiles to obtain the raw data as relative pixel values. As we postulated that diffusion would not be observed in this genipin-gelatin gel dosimeter [20], a relatively simple technique was chosen to evaluate the diffusion coefficient. Data were fitted to an inverse square root function (equation 1). Terms used are analogous to the equation defined previously [19].

$$dA = dA_0 + \frac{1}{2}\left(dA_1 - dA_0\right)\left\{1 + \frac{x}{\sqrt{x^2 + n}}\right\} \tag{1}$$

where dA is the measured value at the distance, x, from the inflection point (the inflection point being at $x = 0$). dA_0 is the minimum value of dA and dA_1 the maximum value of dA. n is a curvature parameter that varies inversely with the slope at the inflection point. dA_0, dA_1 and n are all fitted parameters. The slope of n versus time, multiplied by 0.212, is equal to the diffusion coefficient [23].

3. Results and discussion

Image profile data were fitted to the inverse square root function given in equation 1. For the FGX gel, a change in the slope at the inflection point over time can be seen in the fitted curves (figure 1), as a result of ion diffusion. Plots in figure 1 are also indicative of the lower dose sensitivity of the genipin-gelatin gel as compared to the FGX gel.

The curvature parameter, n, increased linearly with time (figure 2a), and the diffusion coefficient was 0.70 ± 0.05 mm^2 h^{-1} at 5 Gy. In the genipin-gelatin gel, n did not significantly change over the 45 h following irradiation (figure 2b). This confirms that there is no significant diffusion occurring in these genipin gels.

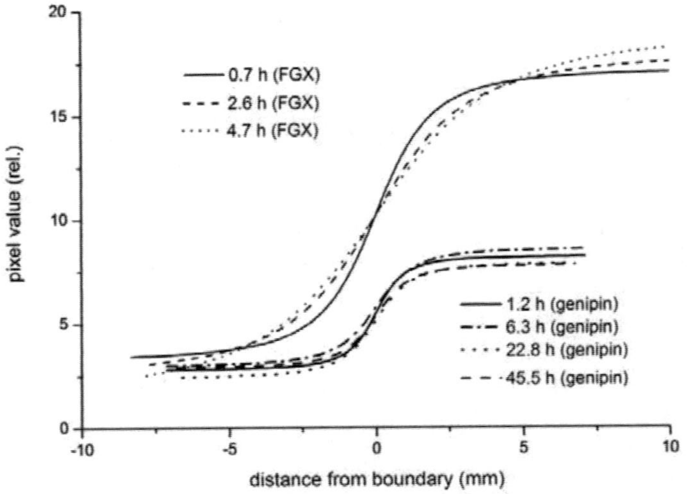

Figure 1. Fitted curves of image profiles of FGX gel (upper curves) and genipin-gelatin gel (lower curves) obtained at various times.

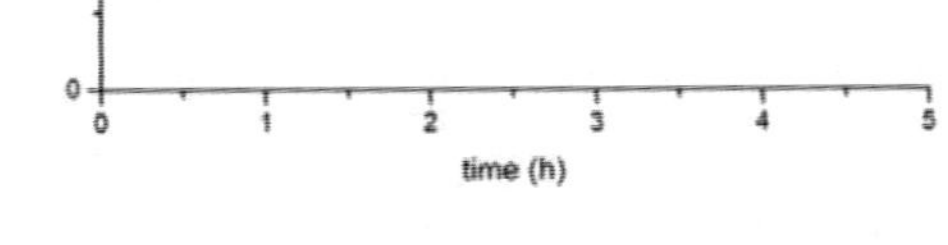

Figure 2a. The diffusion coefficient in this FGX gel was calculated from the slope of the linear fit of n versus time and was 0.70 ± 0.05 mm^2 h^{-1} at 5 Gy.

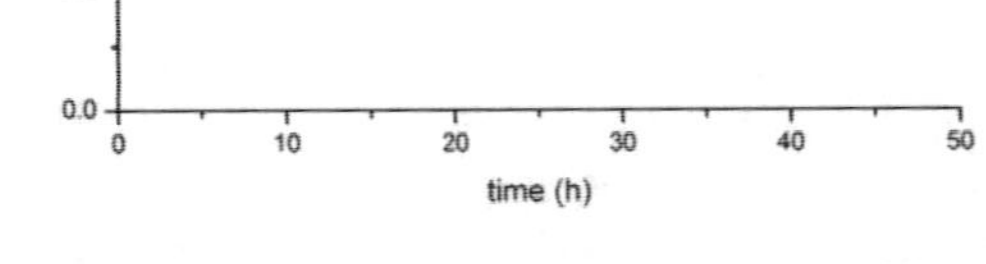

Figure 2b. The curvature parameter, n, did not significantly change over time for the genipin-gelatin gel dosimeter. Plotted are the average values (error bars indicate one standard deviation) of measurements of three samples from the same batch of gel.

4. Conclusion

The effect of diffusion on a genipin-gelatin gel dosimeter was evaluated and compared with the known diffusion in an FGX gel dosimeter. Genipin forms blue pigments as a result of cross-linking some of the amino acids that make up gelatin. The genipin-gelatin gel therefore differs from other gel dosimeters infused with radiation-sensitive species (such as Fricke-based gels) in that it will not diffuse in regions of dose gradients. A clear advantage of this genipin-gelatin gel over other types of hydrogel dosimeter systems is that this gel is non-diffusing. Other influences that have a more significant impact on this genipin-gelatin gel dosimeter in evaluating its usefulness for 3D dosimetry include the dosimetric sensitivity, measurement uncertainty, stability and sensitivity to light and heat [24, 25].

5. References

[1] Baldock C *et al* 1998 *Phys. Med. Biol.* **43** 695-702
[2] Baldock C *et al* 2010 *Phys. Med. Biol.* **55** R1-63
[3] Keall P and Baldock C 1999 *Australas. Phys. Eng. Sci. Med.* **22** 85-91
[4] Gustavsson H *et al* 2004 *Phys. Med. Biol.* **49** 227-41
[5] Lepage M *et al* 2002 *Phys. Med. Biol.* **47** 1881-90
[6] Bosi S *et al* 2007 *Phys. Med. Biol.* **52** 2893-903
[7] Bosi S G *et al* 2009 *Phys. Med. Biol.* **54** 275-83
[8] Trapp J V *et al* 2002 *Phys. Med. Biol.* **47** 4247-58
[9] Hill B *et al* 2005 *Br. J. Radiol.* **78** 623-30
[10] Mather M L *et al* 2002 *Phys. Med. Biol.* **47** 1449-58
[11] Mather M L *et al* 2002 *Phys. Med. Biol.* **47** 4397-409
[12] Mather M L and Baldock C 2003 *Med. Phys.* **30** 2140-8
[13] Mather M L *et al Ultrasonics* **41** 551-9
[14] Mather M L *et al* 2003 *Phys. Med. Biol.* **48** N269-75
[15] Baldock C *et al* 1998 *Phys. Med. Biol.* **43** 3617-27
[16] Rintoul L *et al* 2003 *Appl. Spectrosc.* **57** 51-7
[17] Rae W I D *et al* 1996 *Med. Phys.* **23** 15-23
[18] Pedersen T V *et al* 1997 *Phys. Med. Biol.* **42** 1575-85
[19] Kron T *et al* 1997 *Magn. Reson. Imaging* **15** 211-21
[20] Jordan K 2008 *Med. Phys.* **35** 3413
[21] Davies J B *et al* 2012 *Radiat. Phys. Chem.* **81** 1263-5
[22] Davies J B and Baldock C 2008 *Radiat. Phys. Chem.* **77** 690-6
[23] Jordan K and Avvakumov N 2009 *Phys. Med. Biol.* **54** 6773-89
[24] Willowson K *et al 2008 Phys. Med. Biol.* **53** 3099-112
[25] Hill R et al 2005 *Phys. Med. Biol.* **50** N331-44

Preliminary evaluation of diacetylene-based 3D micelle gel dosimeters

A Nasr[1], T Olding[2], LJ Schreiner[2,3] and KB McAuley[1]
[1]Chemical Engineering Dept., Queen's University, Kingston, Canada, K7L 3N6
[2]Cancer Centre of Southeastern Ontario, Kingston, Canada, K7L 5P9
[3]Depts. of Oncology and Physics, Queen's University, Kingston, Canada, K7L 3N6

E-mail: kim.mcauley@chee.queensu.ca

Abstract. Emulsified diacetylenes as reporter molecules in micelle gel dosimeters were evaluated in the current article. It was observed that gels containing PCDA emulsified in deionized water using SDS changed from colourless to blue upon irradiation. Unfortunately, recipes that led to turbid gels resulted in a colour change, but transparent gels did not change colour. The colour change may be due to the oligomerization of precipitated solid PCA crystals, rather than emulsified PCDA.

1. Introduction

New radiochromic micelle gel dosimeters [1] for optical readout were recently developed by Jordan and coworkers [2-4]. These dosimeters consist of colourless leuco-dyes emulsified in a hydrogel matrix using a surfactant. The leuco-dye molecules react with free radicals, generated by water radiolysis, and change from colourless to deeply coloured as the radiation dose increases.

Micelles are self-assembled aggregates of surfactant molecules that have both hydrophilic and hydrophobic parts. Above the critical micelle concentration (CMC), surfactant molecules orient themselves so that their hydrophobic parts are at the centres of the micelles and their hydrophilic parts are in contact with water. The main purpose of using micelles in radiochromic micelle gel dosimeters is to emulsify water-insoluble reporter molecules to distribute them throughout the gel volume [2]. Micelles are significantly larger than individual reporter molecules. As a result, the micelles have low diffusivity within the gel matrix, which improved the spatial stability of dose information, compared with micelle-free optical dosimeters such as Fricke gel dosimeters [3]. A key benefit of dosimeters that display changes in optical properties is that they can be read out using optical CT rather than MRI, which is expensive and unavailable in many locations [4-6]. Optical CT scanners are low in cost, and could be readily accessible in most clinical environments [7, 8].

Micelle gels (mainly composed of water) should be more tissue equivalent than competing plastic (e.g., PRESAGE) dosimeters [4]. Moreover, the simplicity of fabricating water-based dosimeters, as opposed to polyurethane-based PRESAGE dosimeters, gives considerable advantages to micelle gels for use in anthropomorphic phantoms [4]. Current micelle gel dosimeters could benefit from further improvements as they are light-sensitive and temperature-sensitive during irradiation and tend to fade over time [2]. They also have a relatively low dose sensitivity and may have significant dose-rate dependence [3].

In an attempt to develop improved micelle gel dosimeters, the use of emulsified diacetylenes, especially PCDA (see Fig. 1), as reporter molecules is investigated in this paper. PCDA molecules are the reporter molecules used in radiochromic GafChromic® films which are widely used for 2D dosimetry [9, 10]. Diacetylenes are organic molecules that contain two conjugated acetylene groups [10]. Upon irradiation, diacetylene molecules can react to give polymers that have conjugated double and triple bonds, which result in formation of colour, with intensity that is proportional to the absorbed dose [10, 11]. Diacetylenes other than PCDA have also been used as reporter molecules in self-developing films [12]. Figure 1 shows the three diacetylenes considered in this study, along with their melting points. Note that these diacetylenes are solids at room temperature.

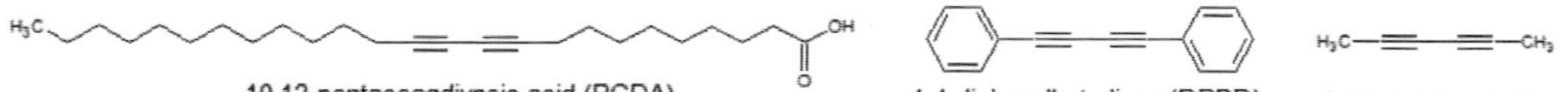

Figure 1: Chemical structure of some diacetylenes with their Molecular weight and melting point.

2. Materials and methods

2.1. Preliminary experiments

Preliminary experiments focused on emulsifying PCDA (GFS chemicals, Ohio, USA) using three different common surfactants at different concentrations. The three surfactants (Fisher Scientific, Ottawa Canada) are the anionic surfactant Sodium Dodecyl Sulfate (SDS), the nonionic surfactant Triton x-100, and the cationic surfactant Cetyl Trimethyl Ammonium Bromide (CTAB). All were dissolved in deionized water. A variety of concentrations for both the PCDA and the surfactants were tested as shown in table 1. Bovine gelatin (300 Bloom bovine bone, Eastman Gelatin Corp., MA, USA)[13] was added to make the gels. All emulsions and gels were manufactured by first preparing stock solutions of surfactants then adding the monomer, and stirring at 58 °C for 15 minutes. Mixtures were then heated to slightly above the melting point of the monomer and stirred for 15 minutes until the emulsion was clear. The mixtures were cooled to room temperature then gelatin was added and left to swell for 30 minutes. Mixtures were then stirred at 45 °C until they became clear. All samples were then refrigerated at 4 °C. Samples of the prepared emulsions and gels were irradiated 48 h after manufacturing to 100 Gy (Varian Clinac 6EX, 400 MU/min, 10x10 cm^2 field size). This large dose was used so that any oxygen inhibition would be unimportant.

Table 1: Transparent solutions and gels manufactured for preliminary testing.

Vial #	PCDA Conc.	Surfactant Type	Surfactant Conc.	Gelatin Added	Irradiated
1	3 mM	SDS	51 mM	0 wt%	Y
2	3 mM	SDS	51 mM	0 wt%	N
3	3 mM	SDS	51 mM	5 wt%	Y
4	3 mM	SDS	51 mM	5 wt%	N
5	16 mM	SDS	327 mM	0 wt%	Y
6	16 mM	SDS	327 mM	0 wt%	N
7	16 mM	SDS	327 mM	5 wt%	Y
8	16 mM	SDS	327 mM	5 wt%	N
9	8 mM	Tx-100	60 mM	0 wt%	Y
10	8 mM	Tx-100	60 mM	0 wt%	N
11	8 mM	Tx-100	60 mM	5 wt%	Y
12	8 mM	Tx-100	60 mM	5 wt%	N
13	5 mM	CTAB	54 mM	0 wt%	Y
14	5 mM	CTAB	54 mM	0 wt%	N
15	5 mM	CTAB	54 mM	5 wt%	Y
16	5 mM	CTAB	54 mM	5 wt%	N

2.2. Experiments comparing PCDA and other diacetylenes
The three diacetylenes in figure 1 were emulsified in SDS using the procedure above. Gels of PCDA, DPBD (Aldrich) and HD (Aldrich) were prepared using overall reporter concentrations of 3 mM in emulsions containing 51 mM SDS. Bovine gelatin was added at 5 wt%. Samples of the prepared gels were irradiated 24 h after manufacturing to 0, 5, 10, 20 and 40 Gy. Spectra of irradiated samples containing PCDA were measured using a SpectroVis Plus spectrophotometer (Vernier Software & Technology, OR, USA). In addition, PCDA gels from different batches were replicated (manufactured and irradiated) 10 times using 3 mM PCDA and 10 Gy.

2.3. Gel ageing and phase behaviour experiments
PCDA gels with concentrations 3 and 9 mM were prepared using 5 wt% gelatin and 51 mM SDS. Gel samples were irradiated to 10 Gy at the following times after manufacturing: 0.5, 1.5, 2.5, 3.5 and 7.5d.

3. Results and Discussion
Figure 2 shows a photograph of non-irradiated and irradiated vials (two hours after irradiation) from the experiments in table 1. Vials containing CTAB did not show any noticeable colour change. SDS and Triton x-100 vials, with and without gelatin, turned blue in response to radiation. Irradiated vials containing SDS showed deeper blue colour compared to those containing Triton x-100. Note that the samples that showed colour change are cloudy. Based on these preliminary experiments, the next set of experiments focused on emulsions containing various diacetylenes using SDS as the emulsifier.

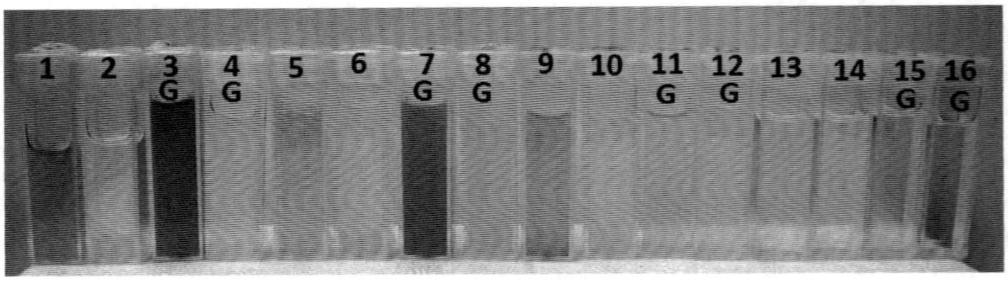

Figure 2: Irradiated and non-irradiated vials containing emulsions and gels from table1 irradiated to 100 Gy. The symbol G indicates vials containing gelatin.

During the second set of experiments, vials containing DPBD and HD did not change colour in response to irradiation. Vials containing PCDA (Fig. 3) showed a smooth response to radiation, with no apparent threshold that would indicate oxygen inhibition (Fig. 3c) [14, 15]. This type of response occurred in three of the 10 replicated experiments. In seven trials, no colour change was observed, indicating poor reproducibility. Note that all vials that exhibited a colour change were turbid before irradiation. Vials that remained clear during the 24 h period between manufacturing and irradiation did not change colour.

Gel aging experiments were designed to further investigate whether the colour change was related to turbidity (i.e, oligomerization of precipitated PCDA). The goal was to track gel clarity over time and to check whether colour change occurs only in precipitated PCDA gels, or whether it could also occur in PCDA micelles (transparent gels). The two PCDA gels (3 mM and 9 mM) remained clear during the first 24 h after manufacturing, and the 9 mM PCDA gel showed some cloudiness after 1.5 d. The 3 mM PCDA gel started to become cloudy about 2.5 days after manufacturing. Turbidity increased with time for both PCDA concentrations. Figure 4 shows the absorbance (at 670 nm) for vials of PCDA gels irradiated to 10 Gy at different times. Note that gels that were transparent showed no colour change. Turbid gels turned blue during irradiation and the colour intensity was higher for gels that were more turbid. Also, the colour intensity was higher for the 9 mM gel samples than for the 3 mM samples, presumably due to a larger amount of precipitated PCDA in the 9 mM gels.

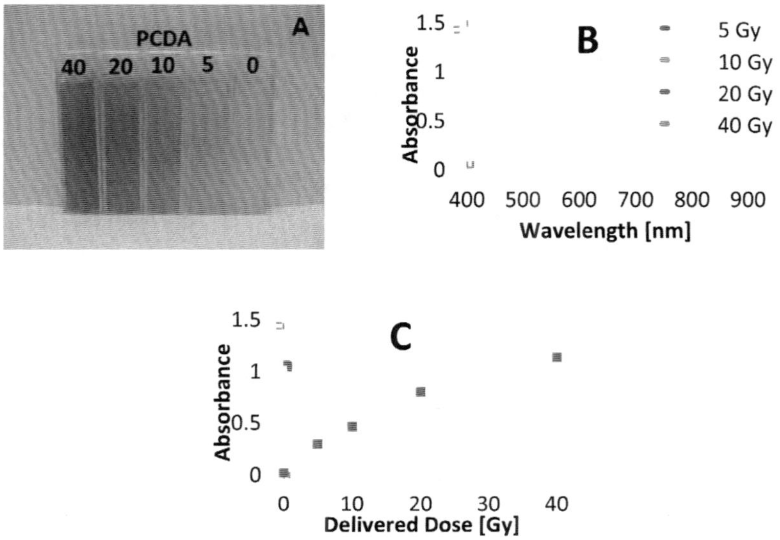

Figure 3: Response of PCDA (3 mM) gel with SDS (51 mM) to 0, 5, 10, 20, and 40 Gy doses, irradiated 24 h after manufacturing. The figure shows a) a photograph of the vials immediately after irradiation, b) measured spectra for these vials and c) absorbance at 680 nm for the irradiated vials from two experiments. Of the 10 replicate experiments, only three showed a colour change in response to irradiation.

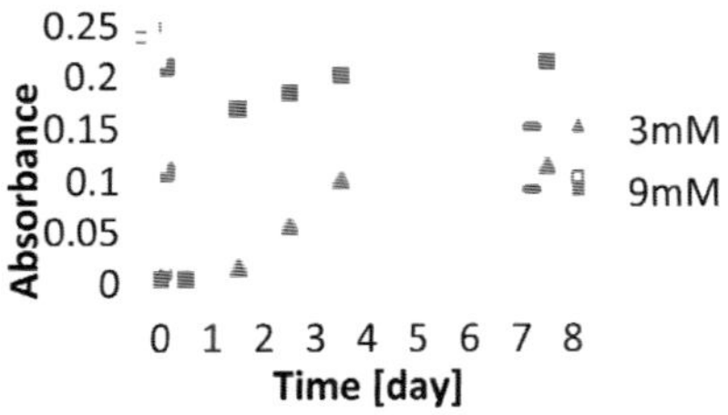

Figure 4: Absorbance at 670 nm of PCDA gels (3 mM and 9 mM) emulsified in SDS at 51 mM irradiate to 10 Gy. The absorbance of the 9 mM PCDA gel is higher than that for the 3 mM gel indicating more PCDA precipitation.

4. Conclusions

The use of emulsified diacetylenes as reporter molecules in micelle gel dosimeters was evaluated. Many gels containing PCDA changed from colourless to blue upon irradiation. For the emulsifier concentrations tested, the colour was deeper when PCDA gels were prepared using SDS than using Triton x-100. Gels produced using CTAB remained clear and did not show any colour change due to radiation. Gels containing SDS and the other diacetylenes (DPBD and HD) did not show any colour change upon irradiation. The response of gels containing PCDA and SDS to irradiation was smooth with no apparent dose threshold that would indicate oxygen inhibition.

Throughout the experiments, only turbid gels changed colour in response to irradiation. When samples of transparent gels were aged, resulting in PCDA precipitation over a period of days, the cloudy PCDA gels changed colour in response to irradiation. These results indicate that the colour change is due to oligomerization of solid PCDA particles, rather than oligomerization of PCDA in micelles. Since PCDA crystallization is not very reproducible and turbid gels are undesirable for

Journal of Physics: Conference Series **444** (2013) 012041 doi:10.1088/1742-6596/444/1/012041

optical CT readout, the use of emulsified diacetylenes for three-dimensional gel dosimetry does not seem to be promising.

5. References

[1] Baldock C *et al* 2010 try *Phys. Med. Biol.* **55** R1-63
[2] Babic S *et al* 2009 *Phys. Med. Biol.* **54** 6791
[3] Jordan J and Avvakumov N 2009 *Phys. Med. Biol.* **54** 6773
[4] Vandecasteele J *et al* 2011 *Phys. Med. Biol.* **56** 627
[5] Bosi S *et al* 2007 *Phys. Med. Biol.* **52** 2893
[6] Bosi S G *Phys. Med. Biol.* 2009 *Phys. Med. Biol.* **54** 275
[7] Doran S 2009 *J. Phys.: Conf. Ser.* **164** 012020
[8] Olding T *et al* 2011 *Med. Phys.* **36** 3
[9] Rad A *et al* 1998 *Med. Phys.* **25** 2093
[10] Rink A *et al* 2008 *Med. Phys.* **35** 4545
[11] Ogawa K 1989 *J. Phys. Chem.* **93** 5305
[12] Watanabe Y *et al* 2006 *Rad. Prot. Dosimetry* **120** 121
[13] Olding T *et al* 2010 *J. Phys.: Conf. Ser.* **250** 012028
[14] Lepage M *et al* C 2002 *Phys. Med. Biol.* **47** 1881-1890
[15] Rintoul L *et al* 2003 *Appl. Spectrosc.* **57** 51-7

3D Dose reconstruction: Banding artefacts in cine mode EPID images during VMAT delivery

HC Woodruff[1] and PB Greer[1,2]
[1]School of Medicine and Public Health, University of Newcastle, Newcastle NSW 2308, Australia
[2]Calvary Mater Newcastle Hospital, Newcastle NSW 2298, Australia

Email: peter.greer@newcastle.edu.au

Abstract. Cine (continuous) mode images obtained during VMAT delivery are heavily degraded by banding artefacts. We have developed a method to reconstruct the pulse sequence (and hence dose deposited) from open field images. For clinical VMAT fields we have devised a frame averaging strategy that greatly improves image quality and dosimetric information for three-dimensional dose reconstruction.

1. Introduction

Volumetric modulated arc therapy (VMAT) is capable of delivering very precise 3D dose distributions by combining multi-leaf collimator (MLC) beam shaping, dose rate variation and continuous gantry rotation[1]. The advantages of better target coverage and faster delivery are being explored by a growing number of centres. The complexity of radiation dose delivery has increased and will continue to increase as these systems are further developed and researched.

Electronic portal imaging devices (EPIDs) have been extensively studied as a tool for intensity modulated radiotherapy (IMRT) QA [2-5] and are currently being used clinically for patient-specific IMRT verification [6-8]. The EPID therefore potentially represents a powerful tool for dose monitoring for IMRT and IMAT. Three-dimensional dose reconstruction from EPID images acquired during VMAT would be an ideal approach [9] McCurdy *et al.* demonstrated the feasibility of using cine (continuous) mode imaging by comparing real-time EPID response and real-time ion-chamber data for selected points in the IMRT deliveries [3]. However, during VMAT delivery the EPID readout is synchronized with the pulsed radiation generated by the linear accelerator. To maintain the dose delivered between control points (which correspond to sampled gantry angles), individual or series of pulses need to be "dropped". Due to the serial, line-by-line read-out of EPID pixels, some rows of pixels will receive more pulses than others, leading to horizontal stripes or 'bands' as shown in figure 1. These bands considerably degrade image quality as well as encoding dose pulse information [10].

We have investigated the characteristics of two types of amorphous silicon portal imaging devices (aS500 and aS1000, Varian Medical Systems) when used in cine-mode and during VMAT delivery. Pulse sequence information was extracted from open field images, allowing us to recover the exact dose delivered to each frame assuming a constant dose per pulse. To reduce banding artefacts a frame averaging strategy was devised and tested which makes use of the cyclical banding phase relations in consecutive frames.

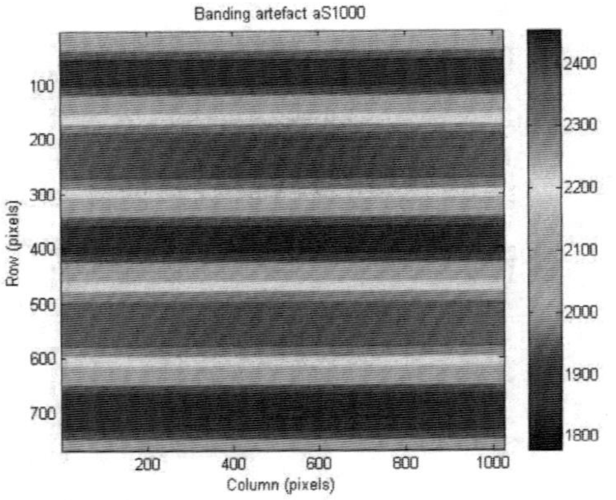

Figure 1. Varian PortalVision aS1000 cine-mode image of an open field delivered in RapidArc mode showing banding artefacts.

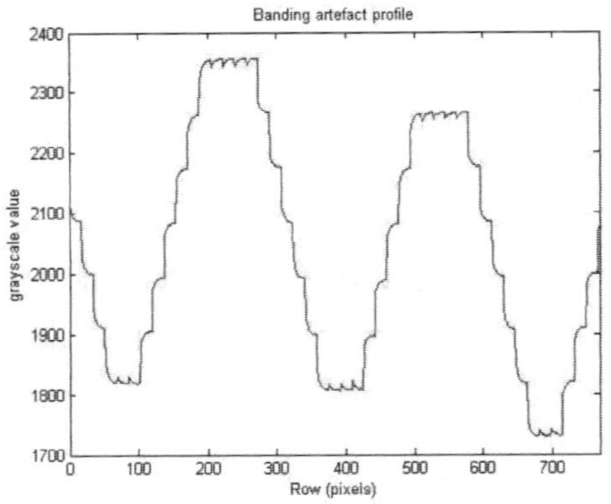

Figure 2. Banding artefact profile of frame in figure 1.

2. Methods and materials

2.1. Linear accelerator, EPID and acquisition mode

The EPIDs used in this study are the commercially available aS500 and aS1000 mounted on a Varian Clinac iX (Varian Medical Systems, Palo Alto, CA). The detectors have an area of 40x30 cm^2 with a matrix of 512x386 pixels (aS500) and 1024x768 pixels (aS1000), each square pixel having a side length of 0.0784 cm and 0.0392 cm, respectively. All data was acquired in continuous (cine) acquisition mode controlled by the IAS3 (Image Acquisition System version 7.3.15) AM-Maintenance software module (Varian Medical Systems, Palo Alto, CA). The VMAT fields were acquired with 600 monitor units per minute (MU/min) and the EPID positioned at 150 cm source-to-detector-distance (SDD). All images were automatically dark-field and flood-field corrected by the IAS3.

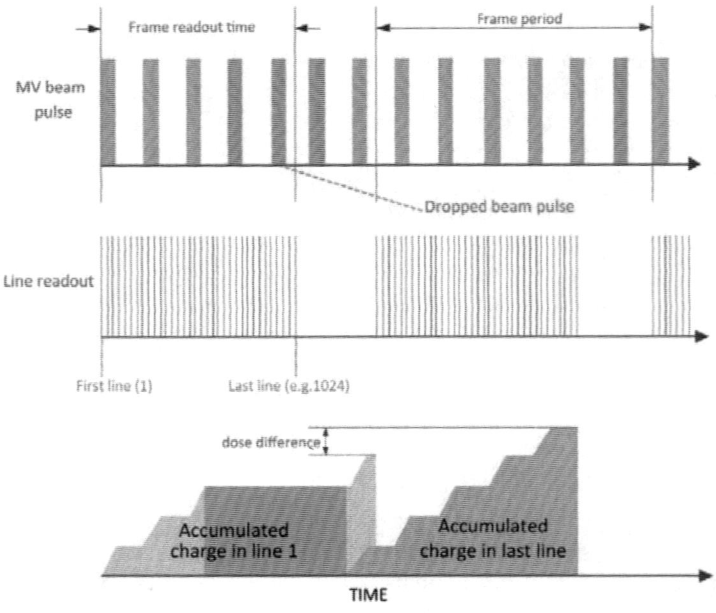

Figure 3. Conceptual timing diagram showing concepts of frame readout time, frame period and line readout. As different lines count different pulses, a dose gradient appears across the detector. The pulse information encoded in line-to-line variation of grayscale (dose) intensity can be used to derive pulse sequence through a deconvolution.

2.2. *Banding artefacts and deconvolution*

Figure 3 shows a conceptual timing diagram illustrating the principles underlying the formation of grayscale (dose) gradients across a single frame acquired during VMAT delivery. After removing pixel sensitivity differences through flood-field correction and disregarding effects of backscatter, the detector can be regarded as having a homogeneous response function. Figure 4 is a graphic representation of how the grayscale value at a given row $G(row)$ can be understood as the convolution of the detector response function D and the pulse sequence P.

$$G(row) = D * P. \qquad \text{Equation 1}$$

For a known beam shape and detector response function, it is possible to reconstruct the beam pulse sequence through deconvolution of the image and the detector response function. In order to minimise noise and computing time, the pixel values of each row can be integrated to deliver a profile of the banding artefacts (see figure 2).

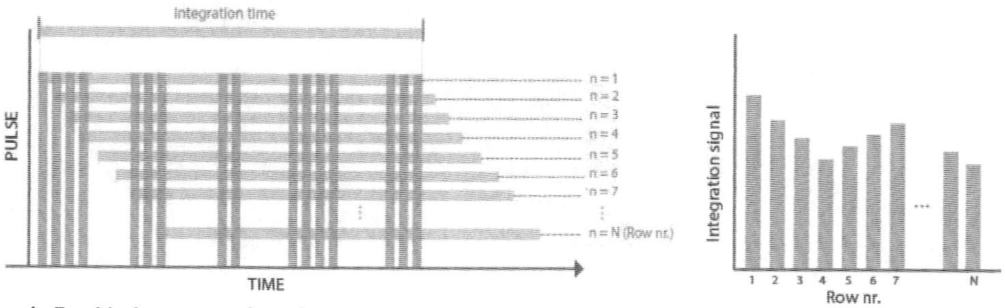

Figure 4. Graphical representation of the convolution between the homogeneous response function of the detector and the beam pulses of the linear accelerator.

3. Results

We were able to reconstruct beam pulse sequences (figure 5) from open field images using the deconvolution outlined above. Assuming all pulses deliver the same dose, this can be used to calculate the exact dose delivered to individual frames. In cases where beam shaping devices (jaws, multi-leaf collimators) obscure entire rows of the EPID, this method becomes increasingly inexact with decreasing field size.

However, we found it is possible to adjust the settings in the IAS3 to achieve a 120° phase shift between the approximately sinusoidal banding patterns in sequential frames. This facilitates a 3 frame averaging strategy which virtually eliminates the artefacts as the dose differences cancel each other out (see figure 6). Although there are many combinations that can achieve useful phase shifts, the number of frames averaged should be kept small to allow for real-time dosimetric monitoring.

4. Discussion and Conclusion

We analysed the banding artefacts observed during cine-mode imaging of VMAT delivery and were able to reconstruct the beam pulse sequence from the band structure using a simple and fast deconvolution. The total dose delivered to each EPID frame equals the total number of pulses delivered multiplied by the dose contained in each pulse.

As a method to improve image quality and dosimetric properties when using the IAS 3 (Varian Medical Systems, Palo Alto, CA), we propose machine settings and a 3 frame averaging strategy which removes the banding artefacts by making use of an inherent 120° phase shift in the sinusoidal bands.

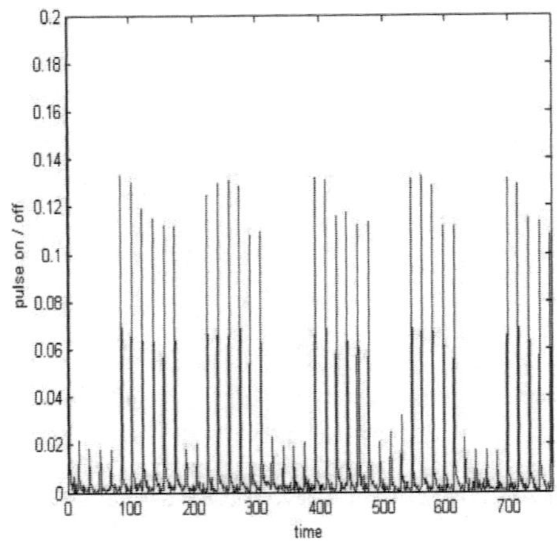

Figure 5. Beam pulse sequence reconstructed from a single open field aS1000 image. The height fluctuations of the pulse peaks are due to the discrete deconvolution process.

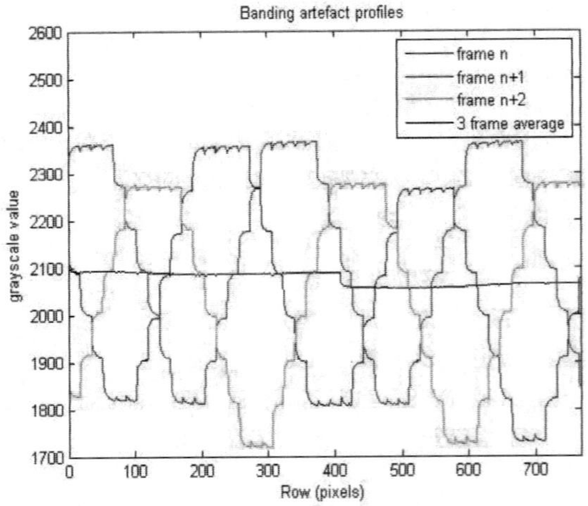

Figure 6. Profiles of three consecutive frames applying the settings outlined in table 1.

Table 1: IAS3 settings to achieve 120° phase shift of banding artefacts between successive frames.

EPID	Dose rate [MU/min]	Frame rate [fps]	Lines / pulse	Sync freq. [Hz]	Spare time [μs]
aS1000	600	7.499	17	359.97	117
aS500	600	7.499	15	359.97	126

5. References

[1] Schreibmann E *et al* 2009 *Med. Phys.* **36** 4530-5
[2] Mohammad M and Eva B 2006 *Phys. Med. Biol.* **51** 2971
[3] McCurdy B M C and Greer P B 2009 *Med. Phys.* **36** 3028-39
[4] Vieira S C *et al* 2002 *Med. Phys.* **29** 2034-40
[5] James H V *et al* 2000 *Phys. Med. Biol.* **45** 495
[6] McDermott L N *et al* 2007 *Int. J. Radiat. Oncol. Biol. Phys.* **67** 1568-77
[7] van Elmpt W *et al* 2008 *Radiother. Oncol.* **88** 289-309
[8] Mans A *et al* 2010 *Med. Phys.* **37** 2638-44
[9] Ansbacher W *et al* 2010 *J. Phys.: Conf. Ser.* **250** 012022
[10] Wertz H *et al* 2010 *Phys. Med. Biol.* **55** 4203

Clinical validation of an in-house EPID dosimetry system for IMRT QA at the Prince of Wales Hospital

M Tyler[1,2], P Vial[3,4], P Metcalfe[2] and S Downes[1]
[1]Department of Medical Physics, Prince of Wales Hospital, Randwick, NSW, Australia
[2]Centre for Medical Radiation Physics, University of Wollongong, NSW, Australia
[3]Department of Medical Physics, Liverpool and Macarthur Cancer Therapy Centres, NSW, Australia
[4]Institute of Medical Physics, School of Physics, University of Sydney, NSW, Australia

E-mail: madelaine.tyler@sesiahs.health.nsw.gov.au

Abstract. In this study a simple method using standard flood-field corrected Electronic Portal Imaging Device (EPID) images for routine Intensity Modulated Radiation Therapy (IMRT) Quality Assurance (QA) was investigated. The EPID QA system was designed and tested on a Siemens Oncor Impression linear accelerator with an OptiVue 1000ST EPID panel (Siemens Medical Solutions USA, Inc, USA) and an Elekta Axesse linear accelerator with an iViewGT EPID (Elekta AB, Sweden) for 6 and 10 MV IMRT fields with Step-and-Shoot and dynamic-MLC delivery. Two different planning systems were used for patient IMRT field generation for comparison with the measured EPID fluences. All measured IMRT plans had >95% agreement to the planning fluences (using 3 cGy / 3 mm Gamma Criteria) and were comparable to the pass-rates calculated using a 2-D diode array dosimeter.

1. Introduction

Intensity Modulated Radiation Therapy (IMRT) is used routinely for prostate, head and neck, and abdominal treatment sites at the Prince of Wales Hospital (POWH). Per-patient pre-treatment QA on treatment fields is recommended in ESTRO Booklet 9 [1], ICRU Report 83 [2] and the ASTRO practice guideline for IMRT [3]. All IMRT treatments at the POWH have QA performed with the MapCHECK (Sun Nuclear Corporation, Fl, USA) two-dimensional diode array dosimeter for a per-field comparison with treatment planning system (TPS) planar dose fluences. The comparison method used is Gamma Analysis with 3 cGy / 3 mm criteria, with greater than 95% of the dose points being in the criteria. Ongoing increases in IMRT patient workload puts pressure on physics resources and has led to investigations for alternatives to the standard diode array based method.

Amorphous silicon (a-Si) Electronic Portal Imaging Devices (EPIDs) have been implemented for two-dimensional dosimetry in many departments globally. EPIDs have many characteristics that make them well suited for IMRT dosimetry [4-8] and the reduced equipment set-up time compared to other dosimeters has the potential to improve efficiencies. Many approaches have been implemented for portal dosimetry as described in a review paper by Van Elmpt [9]. Despite numerous research papers on the topic, and several commercial products being available, portal dosimetry is not yet widespread across all vendor equipment.

Lee [10] implemented a simplified model for Varian EPID dosimetry that calculated a reference depth in water (d_{ref}) at which the EPID response matched conventional dosimeters. The EPID dose response was matched to dose in water at d_{ref} based on field size factors and profile shape. This method eliminates the need for complex scatter corrections and EPID modeling and makes use of an existing TPS to predict EPID response, providing a simple QA tool for IMRT available to any department. This method has not yet been implemented on non-Varian equipment. In this work we implement a similar model across a variety of planning and delivery platforms available at POWH.

2. Materials and Methods

2.1. EPID image acquisition

EPID images were acquired using non-transit methods (no object in the path of the beam), with the panel set at an SSD of 145 cm for the Siemens OptiVue1000ST and 160 cm for the Elekta iViewGT. Based on preliminary measurements a 1 cm solid-water plate was required on the EPID panel for 10 MV measurements to allow sufficient build-up for electronic equilibrium in the detector. No build-up was required for 6 MV measurements.

Images were exported as frame-averaged images in DICOM format for the Siemens OptiVue 1000ST, and lossless JPEG format for the Elekta iViewGT. Image analysis software was used for conversion of frame-averaged images to integrated images and for analysis of measured EPID fields.

2.2. EPID Calibration

Methods for determining the reference depth in water (d_{ref}) were adapted from the work by Lee [11] for Siemens and Elekta EPID images. One difference in the method we used to that of Lee *et al.* is that we did not implement a profile correction. In our method the FF corrected images were used as acquired. Details of determining d_{ref} will be reported elsewhere.

Pixel values for a measured field must be converted to absorbed dose using a measured pixel-to-dose calibration factor ($CF_{EPID:Dose}(x,y)$) for EPID dosimetry. The calibration factor was calculated for 6 and 10 MV based on measurements with the EPID and the absorbed dose in water at d_{ref} for a 10 x 10 cm^2 field. The $CF_{EPID:Dose}(x,y)$ was calculated by dividing the absorbed dose at d_{ref} ($Dose_{dref}$) by the average EPID pixel value (in a 0.4 x 0.4 cm region) at the beam central axis (Pixel (x,y)).

$$CF_{EPID:Dose}(x,y) = \frac{Dose_{d_{ref}}(cGy)}{Pixel(x,y)} \quad (1)$$

EPID calibration factors were subsequently applied to all measured images for conversion of pixel values to dose (cGy).

2.3. IMRT Fluence map generation

Dose fluences were generated for each IMRT patient field using XiO v4.64 and Monaco v3.1 planning systems (Elekta AB, Stockholm, Sweden). XiO uses a superposition algorithm for dose computation whereas Monaco dose calculation is based on a Monte Carlo algorithm.

QA fields were created by transferring the patient plan to a virtual water phantom with dimensions 40 x 40 x 40 cm on XiO, and onto a CT scan of a solid water phantom with dimensions of 30 x 30 x 30 cm for Monaco (due to software restrictions, Monaco cannot calculate dose on a virtual phantom).

Gantry, Collimator and Couch angles were set to zero degrees (0°) planar dose corresponding to d_{ref} for EPID QA beams and at a depth of 3.0 cm for MapCHECK QA beams. Beams were exported at these depths for EPID and MapCHECK plans respectively.

2.4. IMRT Analysis

IMRT field-by-field comparisons were performed using Gamma Analysis with clinical criteria of 3 cGy / 3 mm, and a 10% low dose threshold. A minimum of 95% of points passing the criteria was a requirement for plan approval. The IMRT suite in RIT v5.3 (Radiological Imaging Technology, Colorado Springs, USA) was used for Elekta iViewGT Gamma Analysis. OmniPro I'mRT (IBA Dosimetry GmbH, Schwarzenbruck, Germany) is a pre-treatment verification system for IMRT and rotational therapies and was used for Siemens OptiVue 1000ST Gamma analysis. Different programs were used for IMRT comparisons due to compatibility issues with the different vendor image formats.

3. Clinical Results

IMRT QA fields for each patient were measured with the EPID and the MapCHECK diode array and compared to TPS-generated fluence maps at calculation depths stated in Section 2.4. The percentage pass rate presented in this paper is an average Gamma Analysis result for all treatment fields in an individual patient plan.

EPID vendor-type, beam energy, and the percentage of pixels passing the Gamma criteria of 3 cGy / 3mm with EPID and MapCHECK systems is presented in table 2 for 6 and 10 MV IMRT fields. Note that 10 MV fields were measured with the OptiVue 1000ST only as the iViewGT cannot currently acquire 10 MV images. The delivery method for all listed patients was Step-and-Shoot, with the exception of Patient 9 having a dynamic MLC delivery.

Table 2: Gamma Analysis results for EPID and MapCHECK IMRT QA.

				Percentage pixels Passing Gamma Analysis Per Plan (%)	
Patient	Treatment Site	Machine	Energy (MV)	EPID	MapCHECK
1	Hypopharynx	Siemens	6	99.3	98.4
2	Head & Neck	Siemens	6	98.5	100.0
3	Base of Tongue	Siemens	6	98.6	99.3
4	Head & Neck	Siemens	6	99.4	98.5
5	Tonsil	Siemens	6	99.4	99.2
6	Nasopharynx	Siemens	6	97.5	99.1
7	SCC	Siemens	6	97.4	98.8
8	Head & Neck	Elekta	6	97.8	99.0
9	Head & Neck	Elekta	6	97.2	99.6
10	Sacrum	Elekta	6	95.5	98.2
11	Prostate Bed	Siemens	10	96.5	97.2
12	Anus	Siemens	10	97.6	99.5
13	Anus	Siemens	10	96.6	96.9

Gamma Analysis results for IMRT QA fields measured with the EPID were compared to Gamma analysis results for fields measured with the MapCHECK. Pass rates for the EPID were comparable to the MapCHECK using clinical criteria of 3 cGy / 3mm. All EPID results for the total treatment were within a maximum of ±2.4% from the MapCHECK gamma analysis results. The average percentage of pixels in a treatment plan complying with gamma criteria was over 95% in all cases tested.

2 out of 86 EPID and 3 out of 86 MapCHECK measured beams failed the 95% pass rate. One of these beams did not pass either measurement method.

4. Discussion and Conclusion

A simple EPID dosimetry procedure was implemented and tested for clinical IMRT fields using a variety of common equipment and techniques, including both Step and Shoot and dynamic-MLC delivery techniques, Siemens and Elekta linear accelerators with a-Si EPIDs, Xio and Monaco treatment planning systems, and 6 MV and 10 MV energies.

Using a Gamma Criteria of 3 cGy / 3mm for IMRT field comparison, all patient plans had an average Gamma pass rate above 95%. EPID results agreed with the MapCHECK diode array with an average difference of 0.7% and a maximum difference of 2.4% in gamma pass rates.

Results from this study indicate that the EPID IMRT QA procedure can be used as an accurate and reliable tool for per-patient pre-treatment QA across different platforms and delivery techniques. Further assessment is required to determine the procedure efficiency compared to the current standard method, and to implement the system across multiple-departments with a still wider range of vendor equipment.

5. References

[1] Alber M *et al* 2008 *ESTRO Booklet 9: Guidelines for the Verification of IMRT* (Brussels, Belgium : ESTRO Publications)
[2] Gregoire V *et al* 2010 *Journal of the ICRU* **10** 1473-6691
[3] Hartford A C *et al* 2011 *Int. J. Rad. Onc. Biol. Phys.* **73** 9-14
[4] Budgell G J *et al* 2005 *Med. Phys.* **32** 3267-78
[5] Deshpande S *et al* 2011 *Radiat. Meas.* **46** 1916-19
[6] Winkler P *et al* 2005 *Med. Phys.* **10** 3095-105
[7] Nijsten S M J J G *et al* 2007 *Med. Phys.* **34** 3872-84
[8] Siebers J V *et al* 2004 *Med. Phys.* **31** 2135-46
[9] Van Elmpt W *et al* 2008 *Radiother. Oncol.* **88** 289-309
[10] Lee C *et al* 2009 *Med. Phys.* **36** 984-92

Journal of Physics: Conference Series **444** (2013) 012044 doi:10.1088/1742-6596/444/1/012044

3-D dose verification by cone-beam optical CT scanning of PRESAGE dosimeter

C-S Wuu[1], SJ Hoogcarspel[1], K Deh[1], W-Y Hsu[1] and J Adamovics[2]
[1]Department of Radiation Oncology, Columbia University, New York, NY, USA
[2]Department of Chemistry and Biology, Rider University, Lawrencsville, NJ. USA

E-mail: csw6@columbia.edu

Abstract. There is a clear need for an accurate and practical dosimeter that is able to verify 3-D dose distributions from complex radiation treatments. The purpose of this study is to evaluate the dosimetric performance of PRESAGE radiochromic plastic dosimeter in conjunction with a cone-beam optical CT scanner, Vista™, for 3-D dosimetry. The cone-beam optical CT scanner presented in this study can perform optical readout in less than 30 min, which makes same day dose verification for treatment possible. For dosimetric accuracy, a complete 3-D dose verification procedures were performed, including dose response calibration with a 12 MeV electron beam, a 6 MV photon square field irradiation, as well as a 5-field IMRT irradiation. This study shows that 3-D dose verification with fast optical CT scan of PRESAGE dosimeter is feasible. First-generation optical CT scanner can be used as a gold standard for optical readout of dosimeters.

1. Introduction

Complex radiation treatment modalities such as intensity-modulated radiation therapy (IMRT) and stereotactic radiosurgery offer the benefit of dose conformity to the target volume and dose sparing for critical organs. Currently, verification of complex dose distribution is predominantly done using dosimeters that only sample part of the three-dimensional dose information [1-3]. This can be partially attributed to the fact that all the reliable and fast QA tools are 2-D dosimeters. This raises the need for an accurate and practical dosimeter that is able to verify 3D dose distribution within 1-2 hours, or allow for same-day treatment. A first-generation optical CT scanner, in conjunction with gels or radiochromic plastic dosimeters, has been successfully used for 3-D dose verification of complex treatments. However, true 3-D scan with a high spatial resolution and a large field-of-view takes many hours to acquire the data for dose reconstruction, which makes same-day dose verification for treatment impossible.

This study evaluates the dosimetric performance of a commercial fast cone-beam optical CT scanner, Vista™ (Modus Medical Devices Inc.), in conjunction with a radiochromic plastic dosimeter, PRESAGE® [4, 5] for 3-D photon dosimetry.

2. Materials and Methods

Radiochromic plastic dosimeters, PRESAGE®, were used for dose measurements in this study. Three sets of phantoms (11.4 cm diameter × 10 cm length) were irradiated. One was irradiated with a 12MeV electron beam for dose response calibration; one was irradiated with a 6MV, 4 x 4 cm square field, and one was irradiated with a 5-field IMRT plan. The PDD matching method using electron depth dose curve from ion chamber measurement as a benchmark has been demonstrated to be a convenient and relatively accurate way for dose response calibration [6]. The dosimeters were maintained at a temperature of 7°C before and after irradiation. The dosimeters were set to reacclimatize to room temperature of about 22°C, before irradiation and optical-CT readout.

Optical readout was performed using a Vista™ cone beam optical CT scanner (Modus Medical Devices Inc.). The method has been described in a few articles [7, 8]. Additionally, Olding *et al* investigated its basic properties [9]. The scanner utilizes a LED diffuse light panel and a lens based CCD camera to capture a series of 2D optical projections through the region of interest while the object is being rotated 360°. Initial investigations indicated that cone-beam optical computed tomography of polymer gels is complicated by scattered stray light perturbation [10]. Radiochromic dosimeters such as PRESAGE could be more suitable for readout by a cone-beam optical CT scanner since the optical density change is due to light absorption, not scattering.

Dosimeters are suspended in a liquid-filled aquarium for refractive index matching. A mounting post, glued on the dosimeter, can be locked onto the rotation table of the scanner in only one manner, resulting in an identical position of the dosimeter for each scan. This allows for accurate registration of the pre-irradiation and post-irradiation scans.

A Feldkamp filtered back projection was used to reconstruct the 3D dose map, with a spatial resolution of 1 mm. Each dosimeter was scanned before and after irradiation in order to eliminate artifacts caused by light scattering and dosimeter impurities.

The dose distributions obtained from PRESAGE/Vista scanner were compared with the calculated dose distributions from ECLIPSE treatment planning system (Varian Medical Systems Inc, Palo Alto, CA). One dosimeter was scanned with a CT simulator and the image was transported to the treatment planning system. A five-field IMRT plan was then used to generate a hybrid phantom plan. Both measured and calculated dose distributions were exported to an in-house dose comparison plugin written for ImageJ. (National Institute of Health, Bethesda, MD, USA)

3. Results and Discussion

The comparison of optical densities at depths from the PRESAGE measurement and the ECLIPSE depth dose curve, for a 12 MeV electron field, is shown in Fig. 1a. The dose response curve as determined from electron PDD matching is presented in Fig. 1b. The PDD curves, measured and calculated, agree within 2% and 1mm. The "depth-dose matching" technique has the advantage of acquiring many more data points in modeling the dose response.

a b

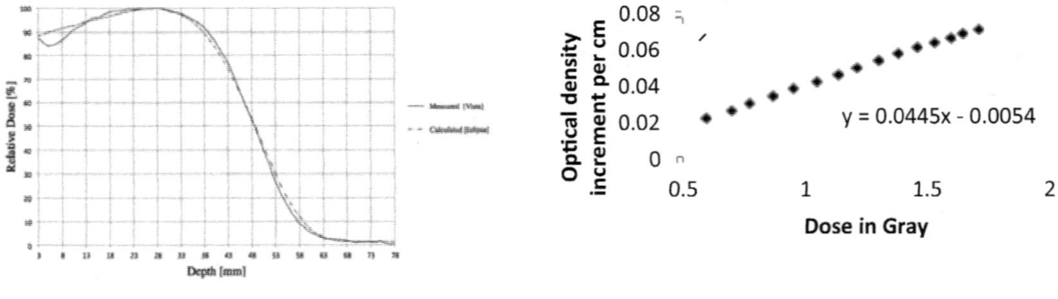

Figure 1: (a) Central axis depth dose comparison between PRESAGE measurement and Eclipse treatment planning calculation for a 12 MeV, 6x6 cm^2 electron field; (b) Dose response curves of the PRESAGE dosimeter obtained from the electron PDD matching method.

Figure 2 shows a good agreement in isodose lines between PRESAGE measurement and planning calculation, for a 6 MV, 4x4 cm square field. Dose comparison for a single photon square field provides a good baseline of accuracy in dose measurement, since the dose distribution in the planning system is obtained from commissioning beam data.

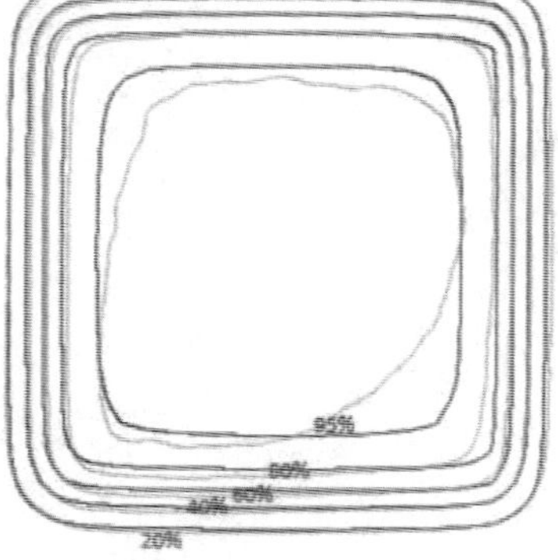

Figure 2: Comparison of the isodose lines from the PRESAGE measurement (red) and Eclipse planning system (blue lines)

Figure 3 shows dose distribution comparison, for a 5-field IMRT plan, at the central axial plane between measured and calculated dose distributions, with maximum dose normalized to 100%. The agreement is very good from 70 to 100% isodose lines. However, the measured isodose lines are consistently a few percent higher than the calculated isodose line in the region of 50 to 70%. The reconstruction protocol for the cone-beam CT scanner involves subtracting the data set from pre-irradiation reference scan with the data set from post-irradiation scan before it is fed into the back projection algorithm. Considering the lesion used in this study is only 2.5 cc and the dose gradient in such a small area is significant, the reference scan is crucial for correct dose reconstruction.

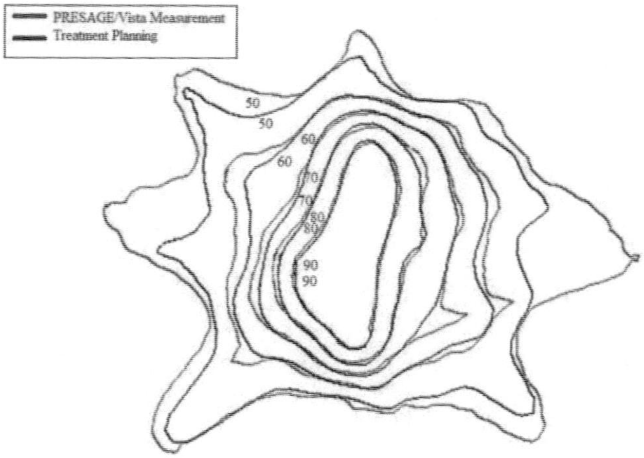

Figure 3: Comparison of the isodose lines obtained from the PRESAGE measurement with Vista cone-beam optical CT scans (Red) and the Eclipse planning calculation (green lines) at the central axial slice.

Figure 4 shows two grayscale images of the calculated (a) and measured (b) dose distributions of the IMRT plan at central slice. Streaking artifacts are observed in the measured dose distribution (yellow arrows). These artifacts can result from debris in the optical matching tank or from scratches/impurities in the dosimeter.

The pre-irradiation reference scan method is utilized to eliminate most artifact resources, including scattering, scratches on the optical-matching tank and any inhomogeneities contained within the PRESAGE material.

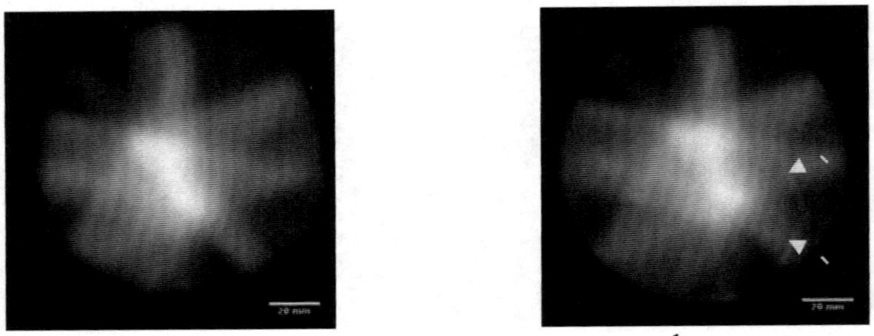

a b

Figure 4: Calculated (a) and measured (b) dose image of the 5-field IMRT plan at the central slice. Streaking artifacts can be observed in the measured dose distribution

The interest for developing an optical-CT system that utilizes a broad light beam instead of a small laser beam is driven by the demand for a faster optical read out modality. The cone-beam optical scanner presented in this study had a scan time of less than 30 min, which makes same day dose verification for treatment possible. However, all 'broad beam' scanners that are currently described in the literature utilize an LED light panel to illuminate the dosimeter during image acquisition. This light source has a distinct polychromatic light emission spectrum. The absorption spectrum of the PRESAGE material is non-uniform. As a result, spectral artifacts arising from preferential absorption can be introduced [11]. In this study, our results showed that the cone-beam optical CT scanner can produce reasonable reconstructed dose, but slightly inferior to the first-generation optical-CT scanner. This result is not surprising since it has been suggested that results obtained with first-generation optical scanner can be used as the 'gold standard' when it comes to evaluating the results obtained with broad-beam optical-CT scanners.

4. Acknowledgement

The authors would like to thank John Miller and Jen Dietrich at Modus Medical Devices Inc. for their assistance in phantom scans.

5. References

[1] Baldock C *et al Phys. Med. Biol.* **55** R1-63
[2] Hurley C *et al* (MAGAT) *Appl. Radiat. Isot.* **63** 443-56
[3] Gustavsson H *et al* C 2004 *Phys. Med. Biol.* **49** 227-41
[4] Brown S *et al* 2008 *Appl. Radiat. Isot.* **66** 1970-4
[5] Gorjiara T *et al* C 2011 *Med. Phys.* **38** 2265-74
[6] Xu Y *et al* 2010 *Med. Phys.* **37** 861-8
[7] Jordan K and Battista J 2006 *J. Phys.: Conf. Ser.* **56** 217-20
[8] Wolodzko J G *et al* 1999 *Med. Phys.* **26** 2508-13
[9] Olding T *et al* 2010 *Phys. Med. Biol.* **55** 2819-40
[10] Doran S J *et al* 2001 *Phys. Med. Biol.* **46** 3191-213
[11] Thomas A *et al* 2011 *Phys. Med. Biol.* **56** 3403-16

Impact of patient rotational errors on target and critical structure dose in IMRT: A 3D simulation study

S Arumugam[1], A Xing[1], P Vial[1,2], A Scotti[1], R Stirton[1], G Goozee[1,3] and Lois Holloway[1,2,4]

[1]Liverpool and Macarthur Cancer Therapy Centres and Ingham Institute, Liverpool Hospital, Sydney, New South Wales, Australia

[2]Institute of Medical Physics, School of Physics, University of Sydney, Sydney, New South Wales, Australia

[3]University of New South Wales, Sydney, Australia

[4]Centre for Medical Radiation Physics, University of Wollongong, Wollongong, New South Wales, Australia

E-mail: Sankar.Arumugam@sswahs.nsw.gov.au

Abstract. The impact of 3D rotational errors in patient positioning on dose delivered target volumes and critical structures in IMRT was studied. Patient rotational errors ranging from -3^0 to $+3^0$ was introduced to IMRT treatment plans of pelvis, head and neck and brain treatment sites and the impact of rotational error on DVH metrics was assessed. The magnitude of impact of rotational error on the error in dose delivered to the target volume and critical structures depends on the location of the structures from plan isocentre. In studied plans, a maximum percentage difference of up to -9.8(1s=13.4) % in D95 to PTV was observed for head and neck treatments. Similarly, in Brain treatments a maximum difference of up to 24.0(1s=33.0) % in maximum dose of Optic chiasm was observed. The results suggest that failure to correct patient's rotational error results in under-dosage to target volumes and over-dosage to the critical structures in some specific treatment scenarios.

1. Introduction

During the past two decades radiotherapy planning, delivery and verification methods have evolved rapidly[1]. While a wide variety of treatment planning, delivery, and image guidance methods including Intensity Modulated Radiation Therapy (IMRT) have been implemented in modern radiotherapy, all these techniques are aiming to achieve a common goal of delivering highly conformal dose to tumor volumes whilst sparing normal critical structures[2].

Highly conformal dose distributions require accurate positioning of patients. Of utmost importance is the agreement between the patient anatomy at the time of treatment to the planning CT. Inaccurate patient set-up may result in geographical miss of target volumes (TVs) and possible overdose to critical structures[3]. Modern linear accelerators (LAs) come equipped with image guidance systems that help to ensure the accuracy of patient position at the time of treatment[4]. While the majority of image guidance systems can determine the positional offsets in three translational directions (left-right, anterior-posterior and superior-inferior) only, some systems also calculate the offsets in rotational directions (pitch, roll and yaw). Conventional patient support systems (PSS) that come with LAs allow positional corrections of patients along three translational and one rotational (yaw) direction. Recently robotic couch systems have been introduced for patient positioning in all six degrees of freedom with sub mm/degree accuracy[5]. While the accuracy of these robotic couch systems has been reported, there is little evidence of the clinical significance of rotational corrections. The procurement and implementation of new technologies such as robotic couch systems should be justified by assessing clinical benefit versus cost as well as the impact on clinical processes and patient safety. In this work we investigate the impact of rigid body rotational errors in patient position on IMRT TV and critical structure dosimetric endpoints for various treatment sites.

2. Materials and Methods

A set of IMRT treatment plans that include three different treatment sites and various levels of complexity was considered for this study. The Pinnacle treatment planning system, V9.0, (Philips, WI, USA) was used to generate step and shoot IMRT plans for Elekta Synergy (Elekta Ltd, Crawly, UK) and Siemens Oncor (Siemens Ltd, Germany) LAs. Table 1 shows the details of treatment sites and number of patient plans considered in each treatment site.

Table1: Details of treatment sites investigated to study the impact of rotational errors on planned dose.

Treatment site	Treatment case	Target volume dose levels	CTV to PTV margin (mm)	Number of patient plans studied
Pelvis	Intact prostate	Single	7	3
	Prostate Bed	Two	10	3
	Prostate with pelvic nodes	Two	7	3
Head and Neck	Base of tongue with bilateral neck nodes	Three	6	2
	SCC lower lip with bilateral neck nodes	Three	6	1
Brain	Brain stem Glioma	Single	5	1
	Left temporal Glioblastoma	Single	5	1

In all studied IMRT plans the isocentre was placed at the geometric centre of the total target volume.The rotational errors in patient position were simulated by rotating the RT structures about the plan isocentre. The planning CT was imported into Pinnacle as primary and secondary data sets. The structures were rotated by copying the structures from the primary data to rotated secondary data. The secondary data was then reset to achieve structures with rotation. Symmetrical rotational errors ranging from -3^0 to $+3^0$, in steps of 1^0, along each axis were introduced to all structures of the studied plans. The Clinical Target Volume (CTV) to Planning Target Volume (PTV) margins were fixed at those used clinically in our department for each site (see table 1). The impact of rotational errors on dose to PTVs was studied by comparing D95, V95, and mean dose to PTV. Similarly, D98, V98 and mean dose to CTV were compared. For plans where multiple dose levels existed, the DVH metric of target volumes with their respective high dose volumes subtracted were analysed. Mean and maximum dose to critical structures was also assessed.

3. Results and Discussion

Figure 1 shows the mean percentage difference (standard deviation) in DVH metrics for TVs in pelvis, head and neck and brain IMRT plans with rotational errors. In general, the percentage difference in DVH metrics for PTVs increases as the rotational error increases. In pelvic treatments, the mean dose to TVs differed by a maximum of -0.4(0.3) % for low dose PTV with -3^0 rotational error. A maximum difference of -2.0(1.0) % and -3.6(2.3) % in D95 and V95 was observed for the high dose PTV with - 3^0 error (figure1a). For CTVs a maximum difference of -0.9(0.7) % in D98 was observed for the low dose CTV.

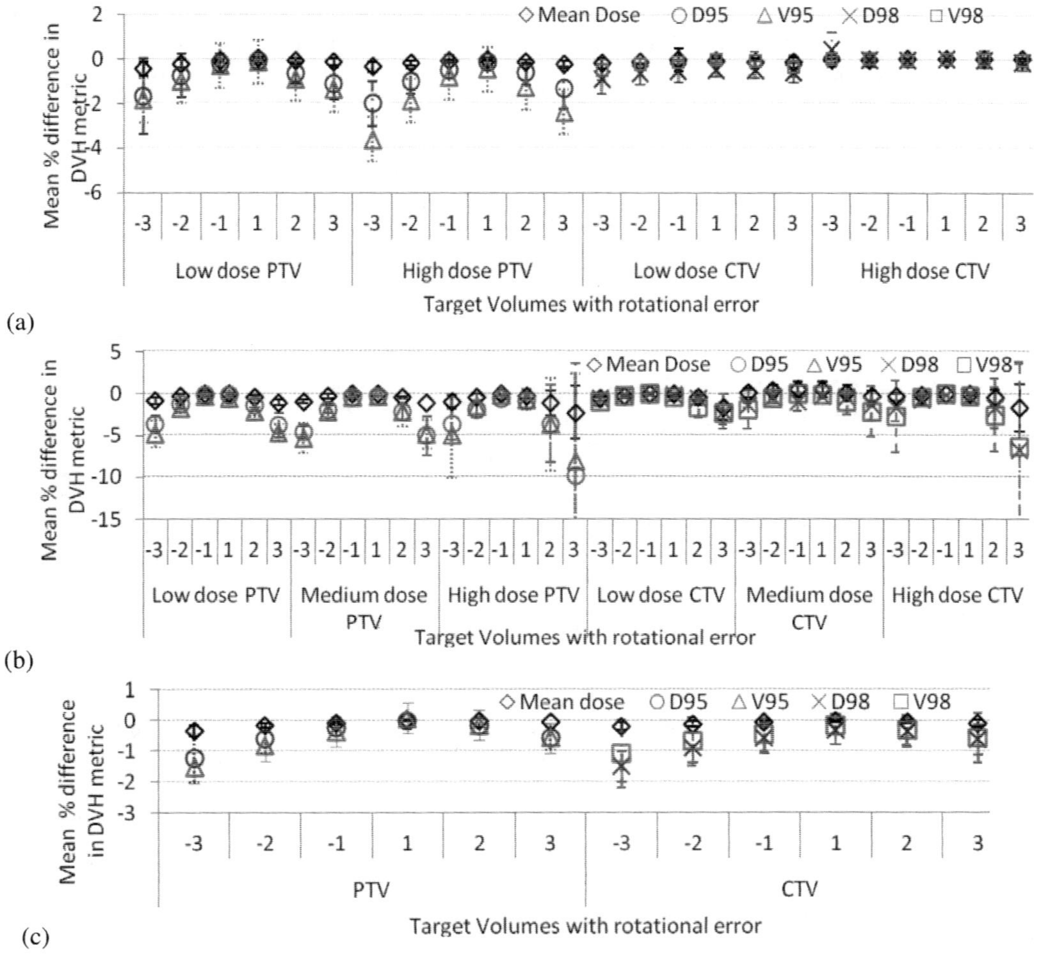

Figure1: Mean percentage difference in DVH metrics (with 1σ) for rotational error introduced target volumes for (a) Pelvis, (b) Head and Neck and (c) Brain IMRT treatments.

Figure 1b shows the difference in DVH metrics for TVs in head and neck treatments. The percentage difference in mean dose to PTVs and CTVs varied from 0.0(0.2) % to -2.3(3.2) % and from 0.5(0.9) % to-1.6(2.8) % respectively. Increased mean doses were observed in the medium dose CTVs, suggesting overlapping in high dose regions with rotational errors. The difference in D95 and V95 for PTVs ranged from -0.3(0.0) % to -9.8(13.4) % and -0.3(0.1) to -8.1(10.5) % respectively. Larger differences in D95 and V95 for high dose PTVs was observed. This large difference was mainly contributed from the SCC lower lip head and neck case (table 1), due to the fact that the geometric centre of high dose PTV was at 90 mm superior and 52 mm anterior from the plan isocentre. The introduction of rotational errors to this structure about isocentre resulted in relatively high 3D dislocation of PTV from its original position and larger differences in DVH metrics. For the same reason the DVH metrics of the high dose CTV also showed major differences (figure 1b).

Figure 1c shows the mean percentage difference in DVH metrics for brain IMRT treatments. Mean dose showed very small change due to rotational error in all target volumes. A maximum difference of - 1.5(1.3) % and -1.1(1.1) % in D98 and V98 was observed for CTVs.

The mean percentage difference in maximum and mean dose to some of the important critical structures in the studied treatment sites is shown in figure 2. In general, the maximum and mean dose to critical structures varied from planned dose based on the direction of rotational error introduced to the structures. Particularly larger differences, up to 24.3(33.0) %, in maximum dose were observed for optic chiasm in Brain treatments (figure 2).

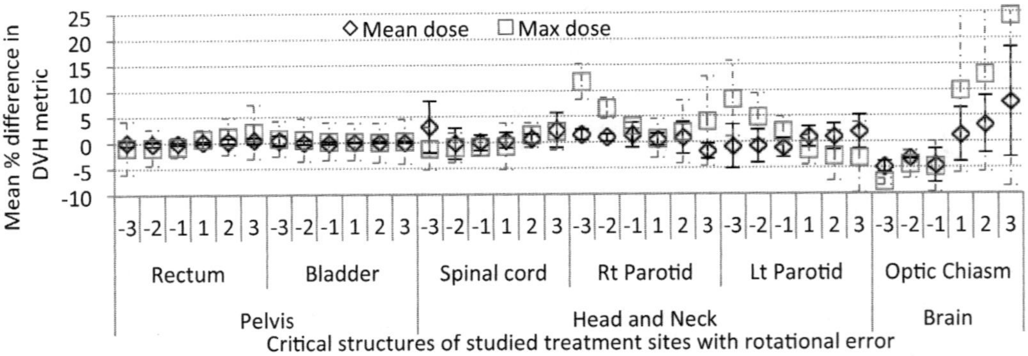

Figure2: Mean percentage difference in DVH metrics (with 1σ) for critical structures due to introduced rotational error for Pelvis, Head and Neck and Brain IMRT treatments.

The introduction of rotational errors resulted in under-dosage to PTVs in all treatment sites. In general the magnitude of dose difference depends on factors such as the dose conformity offered by particular IMRT delivery technique, the deformation in patient anatomy, the magnitude of residual translational error and complexity in target shape etc. In this work we studied the impact of rigid body rotational errors on the dose delivered to structures with the assumption that the residual translational error in patient set-up is zero. The magnitude of impact also depends on the position of geometric centre of the structure with respect to the plan isocentre. PTV margins sufficiently covered the prescription dose to CTVs whose centre was within the vicinity of plan isocentre. However plans with CTVs not centred at the plan isocentre were more sensitive to rotation errors. The dose received by critical structures with respect to rotation errors was also very sensitive to its location (figure2). In pelvis treatments, rotational errors less than 2^0 resulted in clinically less important differences in DVH metrics for target volumes and critical structures (figure 1 a and 2) . However in brain treatments rotational error as little as 1^0 had a significant clinical impact on the DVH metrics for critical structures (figure 2). These results suggest the importance of rotational corrections in patient positioning for these special scenarios.

4. Conclusion

The impact of rotational error on the dose delivered to TVs and critical structures depends on the location of the structures from plan isocentre. This study demonstrates that failure to correct patient's rotational error may result in clinically unacceptable underdosage to target volumes and overdosage to the critical structures in some specific treatment scenarios. These findings will inform the implementation of robotic couch corrections in our clinic.

5. References

[1] Van Dyk J 1999 *The modern technology of radiation oncology* (*Medical Physics Publishing)*
[2] Muzik J *et al* 2008 *Med. Phys.* **35** 1580
[3] Dawson L A and Sharpe MB 2006 *Lancet. Oncol.* **7** 848-58
[4] Xing L *et al* 2006 *Med. Dosim.* **31** 91-112
[5] Guckenberger M *et al* 2007 *Strahlenther Onkol.* **183** 307-13

Image guidance during breast radiotherapy: a phantom dosimetry and radiation-induced second cancer risk study

A Quinn[1,2], L Holloway[1,2,3] and P Metcalfe[1,2]
[1]Centre for Medical Radiation Physics, University of Wollongong, Wollongong, Australia
[2]Liverpool and Macarthur Cancer Therapy Centres, Australia
[3]Institute of Medical Physics, School of Physics, University of Sydney, Sydney, Australia

E-mail: aq908@uowmail.edu.au

Abstract. Imaging procedures utilised for patient position verification during breast radiotherapy can add a considerable dose to organs surrounding the target volume on top of therapeutic scatter dose. This study investigated the dose from a breast kilovoltage cone-beam CT (kV-CBCT), a breast megavoltage fan-beam CT (MV-FBCT), and a TomoDirect™ breast treatment. Thermoluminescent dosimeters placed within a female anthropomorphic phantom were utilised to measure the dose to various organs and tissues. The contralateral breast, lungs and heart received 0.40 cGy, 0.45 cGy and 0.40 cGy from the kV-CBCT and 1.74 cGy, 1.39 cGy and 1.73 cGy from the MV-FBCT. In comparison to treatment alone, daily imaging would increase the contralateral breast, contralateral lung and heart dose by a relative 12%, 24% and 13% for the kV-CBCT, and 52%, 101% and 58% for the MV-FBCT. The impact of the imaging dose relative to the treatment dose was assessed with linear and linear-quadratic radiation-induced secondary cancer risk models for the contralateral breast. The additional imaging dose and risk estimates presented in this study should be taken into account when considering an image modality and frequency for patient position verification protocols in breast radiotherapy.

1. Introduction

Three-dimensional image guidance during breast radiotherapy is becoming more common and has been found to be more effective in comparison to two-dimensional image verification based on bony anatomy information [1]. However, these additional images will increase the dose to organs and tissues adjacent the treated breast, which would not receive a large dose from treatment alone. This may increase the likelihood of long-term side effects such as secondary cancers occurring [2]. An increased long-term risk of secondary cancer in the contralateral breast has been observed for breast radiotherapy patients who receive a dose of 1 Gy or more to the contralateral breast [2].

This study investigated the dose to various organs and tissues from a single kV-CBCT and a single MV-FBCT acquired for breast patient position verification as well as a TomoDirect™ breast tomotherapy treatment. The risk of developing a radiation-induced secondary contralateral breast cancer was estimated with a general linear quadratic model [3], a linear quadratic model incorporating fractionation [4], and a excess relative risk model [5].

2. Materials and methods

2.1. Dosimetry, phantom, treatment and imaging

The dose from a kilovoltage cone-beam CT (kV-CBCT), megavoltage fan-beam CT (MV-FBCT), and a 2-field TomoDirect™ breast treatment were evaluated for a female anthropomorphic phantom (CIRS Inc., Norfolk VA, USA).

kV-CBCT imaging was performed on an Elekta Synergy XVI system (Elekta; Stockholm, Sweden) with the phantom setup on a breast board. The image protocol involved a 270° arc rotation (clockwise 270° to 180°), with 120 kVp, 20 mA and 20 ms per frame, field-of-view size S20, no filter (F0), and with approximately 350 frames collected per scan.

The breast treatment and MV-FBCT scan were performed on a Tomotherapy unit (Tomotherapy Inc., Madison WI, USA). The MV-FBCT was acquired with 3.5 MV photons, pitch of 8.0 (normal), jaw width 1 mm, and field length 15.6 cm. The TomoDirect™ treatment consisted of 2.5 cm jaw width, modulation factor 3.0, pitch 0.287, and beam angles 117° and 308°.

Lithium-fluoride thermoluminescent dosimeters (TLD-100H; Harshaw, Germany), size 3.2 x 3.2 x 0.4 mm^3 were placed throughout the phantom at locations representing organs and tissues. A Harshaw 5500 reader was utilised with a read-out cycle which included a preheat of 100°C for 10 seconds (s) and a linear ramp of 20°Cs^{-1} to 235°C for 20 s. The chips were annealed at 240 ± 2°C for 15 minutes and cooled rapidly on a steel block. For the kV-CBCT measurements the TLDs were calibrated with an orthovoltage machine (Pantak Therapax 300; Elimpex-Medizintechnik, Austria), with beam energy 135 kVp. TLDs were calibrated on a Varian Clinac linear accelerator (Varian Medical Systems Inc., Palo Alto CA, USA) for the MV-FBCT and TomoDirect™ measurements. Individual chip calibration factors were determined and only TLD chips with a reproducibility of ± 3% were used. Dose measurements were repeated twice with two TLD chips at each measurement location.

2.2. Contralateral breast secondary cancer risk

The risk of developing a radiation-induced cancer in the contralateral breast was estimated for treatment with nil imaging and treatment plus daily imaging (25 scans) for both image modalities. Three risk models were utilised; the UNSCEAR general linear quadratic model [3], the Dasu *et al* linear quadratic model incorporating the influence of fractionation [4], and the Preston *et al* excess relative risk (ERR) model [5]. The UNSCEAR [3] general linear quadratic model:

$$Effect = (\alpha_1 D + \beta_1 D^2) \times e^{-(\alpha_2 D + \beta_2 D^2)} \quad (1)$$

where D is the mean contralateral breast dose (Gy) from total treatment. The first term of the equation accounts for radiation-induced DNA mutations, α_1 and β_1 values of 0.02 Gy^{-1} and 0.005 Gy^{-2} were utilised in this study, and the second term represents cell survival, α_2 and β_2 values of 0.25 Gy^{-1} and 0.06 Gy^{-2} were utilised in this study.

The Dasu *et al* [4] linear quadratic model with a fractionation parameter:

$$Effect = \left(\alpha_1 D + \frac{\beta_2 D^2}{n}\right) \times e^{-\left(\alpha_2 D + \frac{\beta_2 D^2}{n}\right)} \quad (2)$$

where n in the number of fractions (25 for this study). To enable comparison with the ERR model below, the risk values calculated above were multiplied with a factor B [6] which depends upon demographic and cohort factors, such that $ERR = B \times Effect$.

The Preston *et al* ERR model [5] was developed from work which examined the breast incidence associated with imaging and low therapeutic doses:

$$ERR = D \times \beta \times \left(\frac{age}{50}\right)^{\gamma} \quad (3)$$

where β is the ERR/Gy and γ is an exponent of attained age. The ERR model parameters were obtained from cohorts receiving similar dose values to those investigated in this study, this included the New York acute post-partum mastitis cohort and the Massachusetts tuberculosis fluoroscopy cohort and its extension cohort. From this, parameters β and γ were determined to 0.68 and -1 respectively [5].

3. Results

The dose to organs from a single breast TomoDirect™ fraction, single kV-CBCT, and a single MV-FBCT acquired for patient position verification in breast radiotherapy are displayed in table 1. The cumulative dose and secondary cancer risks to the contralateral breast from treatment with nil imaging, treatment plus daily kV-CBCT and treatment plus daily MV-FBCT are outlined in table 2.

Table 1: Organ doses from a single breast TomoDirect™ fraction, a single breast kV-CBCT, and a single breast MV-FBCT.

Organ	Dose (cGy)		
	TomoDirect™	kV-CBCT	MV-FBCT
Ipsilateral breast	199.91 ± 8.32	0.51 ± 0.03	1.74 ± 0.11
Contralateral breast	3.32 ± 0.29	0.40 ± 0.01	1.74 ± 0.11
Ipsilateral lung	23.59 ± 2.92	0.45 ± 0.02	1.34 ± 0.15
Contralateral lung	1.38 ± 0.07	0.33 ± 0.01	1.39 ± 0.14
Heart	2.98 ± 0.14	0.40 ± 0.01	1.73 ± 0.08

Table 2: The cumulative dose and excess relative risk to the contralateral breast from treatment with nil imaging, treatment plus daily kV-CBCT imaging and treatment plus daily MV-FBCT imaging.

	Treatment with nil imaging	Treatment plus daily kV-CBCT	Treatment plus daily MV-FBCT
Dose (Gy)	0.83 ± 0.07	0.99 ± 0.11	1.27 ± 0.08
Excess relative risk			
UNSCEAR [3]	0.53	0.62	0.75
Dasu *et al* [4]	0.46	0.53	0.63
Preston *et al* [5][a]	0.47	0.56	0.72

[a]estimated for attained age 60-years, for women treated at age 50-years. A 10-year latency period for radiation-induced secondary malignancy is within the time period, 5-14 –years where Clarke et al [7] found an excess of contralateral breast cancers after breast radiotherapy.

4. Discussion

For the breast patient position verification image protocols investigated in this study, the kV-CBCT was found to deliver a lower dose in comparison to the MV-FBCT. If utilised for daily image guidance the patient dose would be increased by up to a maximum 12.8 cGy for the kV-CBCT and up to 43.5 cGy for the MV-FBCT. Patient position verification does not increase the dose to the target volume greatly, daily imaging increases the dose to the treated breast by a relative 0.3% and 0.9% for the kV-CBCT and MV-FBCT respectively. However for organs surrounding the treated breast, treatment with daily imaging in comparison to treatment with nil imaging, increases in the contralateral breast, contralateral lung and heart dose by a relative 12%, 24% and 13% for the kV-CBCT and 52%, 101% and 58% for the MV-FBCT. For this reason, imaging dose needs to be quantified in order to assess potential long-term risks, such as secondary cancer.

The significance of the imaging dose relative to the treatment dose was assessed for the contralateral breast with three secondary cancer risk models. The ERR estimates were similar for all three models, agreeing to within 9%. This is attributed to the low dose received by the contralateral breast for the treatment and imaging protocols. For low doses the second term in the linear quadratic models [3, 4] becomes negligible and thus the risk estimates linear. The dose-risk relationship for low doses (less than 1 Gy) is controversial [8] however a linear-no-threshold model is generally recommended for low doses [3, 9, 10]. MV-FBCT daily imaging was determined to increase the ERR by up to a relative 53% in comparison to treatment with nil imaging.

5. Conclusion

The MV-FBCT delivered a greater dose in comparison to the kV-CBCT for breast radiotherapy image guidance. Daily imaging was found to deliver a dose of 10.0 cGy and 12.8 cGy to the contralateral breast and ipsilateral breast for the kV-CBCT, and 43.5cGy and 43.5 cGy for the MV-FBCT. In comparison to treatment alone, the treatment plus daily imaging would increase the contralateral breast and ipsilateral breast dose by a relative 12% and 0.3% for the kV-CBCT, and 52% and 0.9% for the MV-FBCT. Thus, daily imaging increased the estimated risk of developing a radiation induced contralateral breast cancer relatively by up to 53% (MV-FBCT with the Preston *et al* ERR model). The dose and risk estimates presented in this study may help clinicians make informed decisions on the image modality type and image frequency required for patient position verification in breast radiotherapy.

6. References

[1] Topolnjak R *et al* 2010 *Int. J Radiat. Oncol. Biol. Phys.* **78** 1235
[2] Stovall M *et al* 2008 *Int. J Radiat. Oncol. Biol. Phys.* **72** 1021
[3] UNSCEAR 2006 *Report to the General Assembly, with scientific annexes* New York: United Nations
[4] Dasu A *et al* 2005 *Acta. Oncol.* **44** 339
[5] Preston D L *et al* 2002 *Radiat. Res.* **158** 220
[6] Sachs R K and Brenner DJ 2005 *Nat. Acad. Sci. Proc.* **102** 13040
[7] Clarke M *et al 2005 Lancet.* **366** 2087
[8] Tubiana M *et al* 2006 *Radiat. Environ. Biophys.* **44** 245
[9] BEIR, *Health Risks from Exposure to Low Levels of Ionizing Radiation: BEIR VII Phase 2* (The National Academies Press, Washington, DC, 2006).
[10] ICRP Publication 103 2007 *Ann. ICRP* **37**

Experimental verification of the planned dose perturbation algorithm in an anthropomorphic phantom

V Feygelman[1], D Opp[1], G Zhang[1], C Stevens[1] and B Nelms[2]
[1]Department of Radiation Oncology, Moffitt Cancer Center, Tampa, Florida 33612, USA
[2]Canis Lupus LLC, Merrimac, Wisconsin 53561, USA

E-mail: vladimir.feygelman@moffitt.org

Abstract. 3DVH software is capable of generating a volumetric *patient* VMAT dose by applying a volumetric perturbation algorithm based on comparing measurement-guided dose reconstruction and TPS calculated dose to a cylindrical *phantom*. The primary purpose of this paper is to validate this dose reconstruction method on an anthropomorphic heterogeneous thoracic phantom by direct comparison to independent measurements. Reconstructed "patient" doses compare well with the independent dose profile measurements in the unit-density target inside a thoracic phantom lung. The largest differences are observed in lung and are associated with the highly modulated plan with narrow (few mm) MLC openings. Such a plan is instructive as a stress test of the algorithm but is not likely to be clinically encountered in lung. This residual disagreement underscores the fact that 3DVH is not designed to correct the errors related to the TPS dose calculations in the low-density media.

1. Introduction

Historically, the most frequently employed method for practical dosimetric QA in highly conformal radiation therapy, such as IMRT and VMAT, has been comparison of the calculated and measured dose in a phantom. Ideally, the dose should be measured with high resolution throughout the entire phantom volume. But even then it is not trivial to correlate the detected dose-errors in a phantom to the clinically relevant dose deviations in the patient. A software package called 3DVH (Sun Nuclear Corp., USA) attempts to bridge this gap by employing a planned dose perturbation algorithm to estimate the dose deliverable to the *patient* from the *phantom* measurement. A full density absolute 3D dose in a phantom can be reconstructed from the relatively sparse detector data if the shape of the dose profiles is known *a priori*. In 3DVH software, this is achieved by an independent calculation by a convolution/superposition method of the relative dose distribution, which is then converted to absolute dose based on the measured point doses. The ratio of the actual to TPS dose for any voxel in the *patient* is about the same as the ratio of the reconstructed and TPS dose in the *phantom*, for the corresponding voxel in relation to the isocenter. This was successfully demonstrated in homogeneous phantoms representing a patient for IMRT [1] and VMAT [2]. While this approach is fairly intuitive for a homogeneous "patient", it is less obvious when a substantial heterogeneity is present. Hence we set out to specifically validate the 3DVH method in a heterogeneous anthropomorphic thoracic phantom with an SBRT target suspended in the lung material.

Journal of Physics: Conference Series **444** (2013) 012047 doi:10.1088/1742-6596/444/1/012047

2. Methods

2.1. The phantom

The novel detector insert was designed in this work for the commercial IMRT Thorax Phantom (CIRS Inc., USA). The design pursued a dual goal of accomplishing measurements necessary for this work but also of creating a phantom useful for periodic end-to-end testing of the SBRT delivery chain. The phantom is made of unit-density tissue equivalent plastic with 0.21 g/cm^3 lung inserts (Figure 1). The right lung can accommodate an 8.5 cm diameter cylindrical lung density insert.

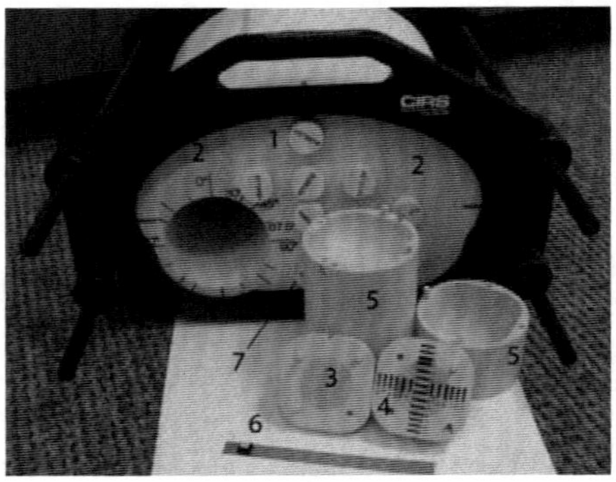

Figure 1. Thoracic phantom. 1: Plastic Water elliptical body. 2: "Lungs" (~0.21 g/cc). 3 and 4: Lung material dosimetric 2-piece insert with the Plastic Water target and nanoDot slots. 5: A two-piece lung cylinder accommodating the dosimetric insert in different orthogonal orientations. 6: A nanoDot. 7: Bone-density insert.

One version of the insert contains a 4 cm diameter spherical Plastic Water target with a cavity for an ion chamber in the center. The other version (5) is overall geometrically equivalent but has a removable two-piece (3 and 4 in Figure 1) target insert designed to accommodate orthogonal linear arrays of optically stimulated luminescent dosimeters (OSLD) known as nanoDots (Landauer Inc., USA). The dosimeters are 5 mm apart in the target and 4.1 mm in the lung.

2.2. OSLD calibration

The OSLDs were calibrated individually. The average angular correction was determined by exposing the nanoDots to an open 360° arc irradiation in the Thoracic phantom. The angular corrections are separate for the nanodots parallel to the transverse plane ("edge on") and two other cardinal planes ("standard orientation").

2.3. Treatment planning

We used 5 VMAT plans for measurements in lung. The first two were optimized to deliver 200 cGy to a 2 and 7 cm diameter targets centered inside the Plastic Water sphere. The next three plans were not created anew but rather copied from our previous work on the homogeneous phantom [2]. They were modeled after some of TG-119 plans [3], including the most challenging C-shape one.

2.4. Dose reconstruction and comparison

3D dose matrix in the patient (Thorax Phantom) was reconstructed from the measurements in the ARCCHECK (Sun Nuclear) diode array phantom using 3DVH software v. 2.2. For the ion chamber point dose measurements, the Thorax Phantom was squared on the room lasers and aligned on the lung target by CBCT. The doses in the middle of the spherical target were extracted from both the initial TPS calculation and 3DVH dose reconstruction on the Thorax Phantom.

By using the dosimetric insert in three different orientations for each plan, two nanoDot profiles in each of the cardinal directions (SI, AP, LR) were acquired and averaged. The individual nanoDot doses were compared to the TPS and 3DVH by composite % dose-difference (local)/DTA analysis. The data were further subdivided for the measurement points inside the unit density sphere and in the lung. In addition, the nanoDot doses in the center of the target were compared to the ion chamber measurements.

3. Results and discussion

3.1. OSLD angular dependence
The average angular corrections for the "edge-on" and standard irradiation geometries were 1.020±0.008 (1SD) and 1.008±0.004, respectively. The first value is 1.5% lower than reported by Kerns et al [5] and almost splits the difference between the values obtained in that work and by Jursinic [6]. The second value has no direct counterpart in the cited work but our numerical integration of the figure 5 curve in [5] yields the value of 1.022, again 1.4% different from our result. The values may not be really different as the 1SD error bars overlap.

3.2. 3D dose comparison – TPS vs. 3DVH
The 3DVH reconstructed patient dose is generally close to the TPS calculations. This means that in these cases, the TPS dose does not need to be perturbed much. The lowest γ-analysis passing rate among the 5 plans ranged from 98.3% for the most lenient 3%(global)/3mm thresholds to 90.9% for the most stringent 2%(local)/2mm combination. The lowest passing rates were always associated with the C-shape plan, with disagreement primarily localized in the low dose central area corresponding to the avoidance structure.

3.3. Point dose comparison – ion chamber
The ion chamber measurements for the most part agree well with both 3DVH and the TPS (Table 1), with the notable exception of the C-shape plan, where the isocenter is in the low dose (avoidance) region. It must be mentioned that all percentage errors here are normalized locally. If normalized globally as in the TG-119 report [3], the % error values for the C-shape would be about ¼ of that presented in Table 1. For this plan, 3DVH substantially reduced the difference from the measured value compared to the TPS.

3.4. Dose profiles in lung - nanodots
The ratio of the average nanoDot dose from the two innermost detector positions to the ion chamber was 0.999±0.015 (1SD, range 0.976-1.013), thus validation our overall OSLD correction methodology.

The composite analysis passing rates for all 5 VMAT plans in the lung are given in Table 1, while dose profiles in the SI and LR directions are presented in for the 4 cases in Figure 2. The AP profiles are included in the numerical analysis but omitted from the figures for clarity. Overall, for the 4 plans that do not include the C-shape, the results of the 3%/3mm composite analyses are similar between the TPS and 3DVH. In the unit density target, 3DVH provides very good agreement (100% of the points) while the TPS agrees with the measured profiles for 93% of the points in the worst case scenario. The lower passing rates in the lung are associated with the plans where there is no sharp dose gradient (7 cm target and HN). In those plans, the TPS errors in the heterogeneous media are not hidden in the distance to agreement analysis. However the difference between 3DVH and measurement in the low gradient region in lung never exceeded 3.9%. The last case in the lung – the C-shape– shows a persistent area of disagreement between the TPS and 3DVH in the low dose area (central avoidance structure). In the water-equivalent target 3DVH reconstruction agrees well with the measured dose, unlike the TPS. However the disagreement between the measured dose and 3DVH in lung remains substantial (up to 14% normalized locally), although smaller than with the TPS (up to 28%).

4. Conclusions
3DVH measurement-guided VMAT dose reconstruction compared favorably with the independent measurements in the unit density mass inside the lung volume of the thoracic phantom. Composite

analysis with 3% (local)/3mm thresholds yielded 100% agreement for 4 VMAT plans, and 96% passing rate for the last, exceedingly complex one. Dosimetric agreement in lung is slightly worse, reflecting the fact that by design 3DVH does not attempt to correct the errors related to the TPS dose calculations in the low-density media. The largest errors are associated with the highly modulated plan, which is useful as a stress test of the algorithm. These errors are primarily localized in the low-dose region corresponding to a central avoidance structure where it has lung density. Such a combination is rather unlikely in a patient

Table 1: Percent difference in point doses on the Thorax Phantom: measurement-guided dose reconstruction (3DVH) and TPS vs. the ion chamber (IC), and composite analysis passing rates overall (All), in the unit density central sphere (Water), and Lung. Local % dose-error normalization is used.

	3DV. – IC (%)	3DVH Composite analysis pass rate 3%/3mm (%)			TPS – IC (%)	TPS Composite analysis pass rate 3%/3mm (%)		
		All	Water	Lung		All	Water	Lung
2 cm	0.0	100	100	100	0.8	98	96	94
7 cm	-0.2	93	100	83	-0.1	98	100	94
H&N	1.0	93	100	83	-1.8	88	93	78
Multi-Target	-1.9	100	100	100	-1.6	98	100	94
C-shape	4.9	85	96	72	-12.5	59	48	72
Bone	1.5				2.6			

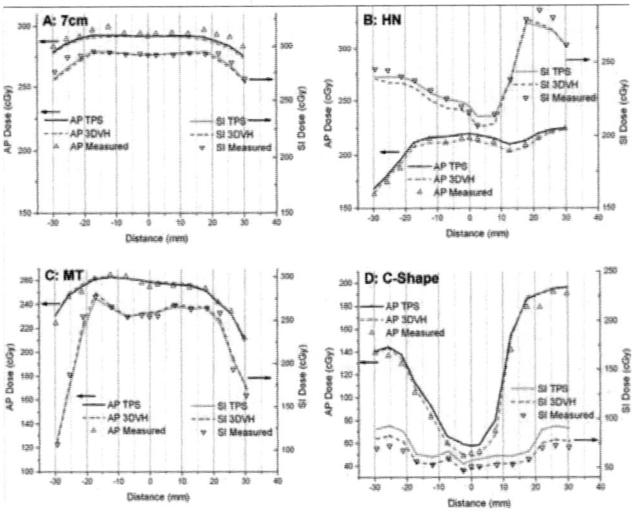

Figure 2: AP (left axes) and SI (right axes) dose profiles for 7-cm target (A), Mock Head and Neck (B), Multi-Target (C), and C-shape (D) VMAT plans. Pinnacle calculations (TPS), 3DVH measurement-guided reconstruction (3DVH) and independent measurements (OSLD + ion chamber in the center) are compared. The left and right scales are slightly different to separate the profiles for better visualization. Otherwise AP and SI profiles would intersect at zero distance from the center.

5. Acknowledgement
This work was supported in part by a grant from Sun Nuclear Corp.

6. References
[1] Olch A J 2012 *Med. Phys.* **39** 81-6
[2] Nelms B E *et al* 2012 *Med. Phys.* **39** 4228-38
[3] Ezzell G A *et al* 2009 *Med. Phys.* **36** 5359-73
[4] Kozelka J *et al* 2011 *Med. Phys.* **38** 5021-32
[5] Kerns J R *et al* 2011 *Med. Phys.* **38** 3955
[6] Jursinic P A 2007 *Med. Phys.* **34** 4594

Journal of Physics: Conference Series **444** (2013) 012048 doi:10.1088/1742-6596/444/1/012048

IMRT patient-specific QA using the Delta^4 dosimetry system and evaluation based on ICRU 83 recommendations

J Nilsson, A Karlsson Hauer and A Bäck

Department of Therapeutic Radiation physics, Sahlgrenska University Hospital, Göteborg, Sweden

E-mail: julia.c.nilsson@vgregion.se, anna.hauer@vgregion.se, anna.back@vgregion.se

Abstract. Patient-specific IMRT QA is dependent on the dosimetry system and the evaluation procedure. The ICRU report 83 provides recommendations of tolerated deviations between measured and calculated absorbed dose distributions for QA of IMRT treatment plans. The result of doing IMRT patient-specific QA with the Delta^4 dosimetry system and using the ICRU recommendations for evaluation is studied. To be able to investigate the QA procedure the original IMRT treatment plans were modified in the treatment planning system to create calculated dose distributions with dosimetric deviations from the original treatment plans. The modified dose distributions were compared to the dose distributions from the Delta^4 measurements of the original treatment plans and the differences were evaluated with criteria and tolerance levels according to the recommendations from ICRU. The evaluation for all 28 modified dose distributions have gamma passing rates higher than the tolerance level recommended from ICRU and will therefore pass the patient-specific QA. More than half of the evaluations have a gamma passing rate of 100 %. Evaluation of the differences between the modified and the original calculated dose distributions revealed in several cases large unacceptable dose differences in the PTV volumes and the organs at risk, for example an increase in the near-maximum dose $D_{2\%}$ to the spinal cord of 5.5 Gy. This study indicates that patient-specific QA with the Delta^4 dosimetry system and the ICRU recommendations for evaluation can not be used to distinguish differences between planned and measured dose of dosimetrical relevance.

1. Introduction

In order to ensure the accuracy of the delivered absorbed dose distribution during intensity modulated radiation therapy (IMRT) treatment, a patient-specific quality assurance (QA) is often performed. The QA is most often based on measurements of delivered dose to find possible deviations between the planned and the delivered dose distribution. The method used for patient-specific IMRT QA measurements depends on the equipment available at the clinic but the result of the QA is also dependent on the evaluation procedure, for example the choice of acceptance criteria and tolerance levels. The QA could be designed to have different intentions, either to reveal major errors, such as no multileaf collimator (MLC) in the beam or use of the wrong plan, but it could also be designed to find smaller deviations between planned and delivered dose, due to for example inaccuracies in the absorbed dose calculation procedure, to keep the differences between planned and delivered dose within a specified uncertainty. It is important to understand what the combination of dosimetry system and evaluation procedure implies in terms of what kind of deviations and/or errors that will be detected.

In the International Commission on Radiation Units and Measurements (ICRU) report 83[1], recommendations of tolerated deviations between measured and calculated absorbed dose distributions for QA of IMRT treatment plans are suggested. In the recommendations, the tolerance levels are specified separately for high and low absorbed dose gradient regions. Low gradient regions are defined to have a relative absorbed dose change less than 20 %/cm. High gradient regions are defined to have a relative change in absorbed dose of more than 20 %/cm. The recommendation by the ICRU of tolerated dose difference between measured and calculated IMRT dose distributions is ±3.5 % in low gradient regions. Assuming that the dose difference frequency histogram has a normal distribution centered on a dose difference of 0 % with a standard deviation of 3.5 %, it means that 85 % of the data in the low gradient region should have a dose difference within ±5 %. In high gradient regions distance-to-agreement (DTA) is the suggested choice of accuracy criteria. Accuracy in DTA of 3.5 mm is recommended by ICRU 83, meaning that 85 % of the data in the high gradient region should be within 5 mm DTA.

The objective is to evaluate the result of doing IMRT patient-specific QA with the Delta^4 dosimetry system from ScandiDos based on ICRU report 83 recommendations for evaluation.

2. Material and method

Three head and neck IMRT treatment plans, using dynamic MLC with sliding window, that were used for treatment of three different patients at Sahlgrenska University Hospital were selected for this study. The treatment plans were generated in the Eclipse treatment planning system (Eclipse version 10.0, Varian Medical Systems). Patient 1 and 2 were treated for malignancies in the nasopharynx, and patient 3 was treated for secondary and unspecified malignancies in the lymph nodes. All patients were planned for simultaneous treatment of two target volumes PTV-T (including the gross tumour volume) and PTV-N (including subclinical disease). The treatment plan for patient 1 was designed to also give a simultaneous boost dose to a smaller volume within PTV-T. To be able to investigate the QA procedure these original plans were modified by varying the MLC leaf transmission factor, dosimetric leaf gap, normalization and the isocenter position in the treatment planning system. The intention with the modifications was not to introduce actual possible errors likely to occur during treatment planning but to introduce different kinds of deviations between the dose distributions to be able to study whether unacceptable dose deviations would be detected by the QA procedure. A total of 28 modified dose distributions were created and studied.

The absorbed dose distributions of the three original IMRT treatment plans were measured with the Delta^4 dosimetry system. The Delta^4 phantom consists of two orthogonal planes with a total of 1069 diodes in a cylindrical PMMA phantom. The spacing between the diodes is 0.5 cm in the central area of 6 cm x 6 cm in each plane and 1 cm outside this area. The measured dose distributions were normalized by a mean value of the relative dose deviation in all measurement points between the measured and calculated original dose distributions of the treatment plans. The measurements constitute the basis for the evaluation study and were compared to the modified calculated dose distributions to investigate the ability of the evaluation method to distinguish dosimetric deviations.

The recommended evaluation criteria from the ICRU 83 are specified separately for low and high absorbed dose gradient regions and evaluation using the dose difference and DTA is recommended for the low and high gradient regions, respectively. This is not strictly the same as using the gamma evaluation (Low et al. 1998)[2] with the same criteria, but according to the ICRU; "setting the gamma criterion to the same values would result in a more conservative test". The Delta^4 software is provided with a tool to separate dose distributions in high gradient regions to evaluate with a DTA criterion. It is not possible to evaluate low gradient regions separately in the Delta^4 software. Therefore the recommendation from the ICRU must be transformed to the gamma evaluation with the criteria 5 % / 5 mm and passing rate 85 % if the Delta^4 software is to be used. The Delta^4 software also provides an interpolation algorithm that estimates the dose to regions where no detectors are located. However, the gamma evaluated passing rate is only given for the measurement points in the two detector planes wherefore the analysis is limited to the measurement points. The evaluation was carried out both with all the diodes included, but also when excluding diodes that receive less than 20 % of the absorbed dose in the isocenter. For comparison, the dose distributions were also evaluated strictly according to ICRU by separating the dose distributions in high and low dose gradient regions. Data of relative dose difference and DTA for the measurement points were exported from the Delta[4] software and the high and low dose gradient regions were separated and evaluated separately outside the Delta[4] software.

Evaluation of the QA results is done by analyzing the dose differences between the modified and the original calculated dose distributions in structures of interest. The investigated dose values are the near-minimum ($D_{98\,\%}$), mean and near-maximum ($D_{2\,\%}$) dose to the PTV-T and PTV-N volumes, near-maximum dose ($D_{2\,\%}$) to the spinal cord and the brainstem (brainstem were only evaluated for patient 1 and 2) and the mean dose to the parotid glands. The dose values are obtained from dose volume histograms generated by the treatment planning system. This analysis was done with the assumption that it is reasonable to desire less than 5 % absorbed dose difference in these dose values, and larger dose differences are considered as relevant for the treatment of the patient, and should be detected by the QA procedure.

3. Results and discussion

When using the gamma evaluation in the Delta[4] software and the criteria 5 % / 5 mm, the evaluation of all 28 modified dose distributions have a gamma passing rate higher than 85 % and 18 of the evaluations have a gamma passing rate of 100 %. Only one of the evaluations has a gamma passing rate below 90 %. This means that all of the investigated dose distributions passed the gamma evaluation according to the recommendations of the ICRU report 83 with the use of the Delta[4] dosimetry system.

The analysis of the dose differences in structures of interest show, on the other hand, large unacceptable dose differences in some cases. Four of the evaluated modified dose distributions have more than a 5 % increase of the $D_{2\,\%}$ for the spinal cord compared to the original plan. A 5 % increase in $D_{2\,\%}$ for the spinal cord corresponds to a mean absorbed dose of 2.2 Gy for the whole treatment, for the three plans studied. The maximum dose increase in $D_{2\,\%}$ found for the spinal cord was 12.5 %, which corresponds to an absorbed dose of 5.5 Gy for the whole treatment. Seven of the modified dose distributions have an increase of more than 5 % in the mean dose to one or both of the parotid glands. Nine of the evaluated modified dose distributions have more than a 5 % decrease of the minimum dose to one or both of the PTV volumes studied. Overall it was found that 15 of the 28 IMRT modified dose distributions have large dosimetric differences >5 % that are not detected in the QA procedure based on the gamma evaluation and the Delta[4].

When the modified dose distributions were evaluated separately for the high and low absorbed dose gradient regions the result is somewhat different. In this case the evaluation for 6 of the 15 modified dose distributions with large dosimetric differences were found to have passing rates lower than 85 %. For all 6 cases this was found for the evaluation in the low gradient region with the criterion of relative dose differences < ±5 %. The passing rate in the high gradient region using criterion of DTA < 5 mm was higher than 99 % in all cases. In the 6 cases with passing rates lower than 85 %, between 24 - 48 % of the points failing in the low gradient region will pass the gamma evaluation due to the included DTA criteria. This indicates that for the cases evaluated in this study, the gamma evaluation in the Delta[4] software with the criteria 5 % / 5 mm is not more conservative than evaluation with separate criteria for low and high gradient regions but rather the contrary. There are, however, several dose distributions with large dosimetric deviations that will not be detected with either evaluation method.

Examples of the relative dose differences in the PTV volumes and the organs at risk together with the passing rates for the different evaluation methods are shown in table 1 for 5 of the worst cases that had large dose differences between the original and the modified dose distributions. For one case, an increased dose of 12.5 % to the spinal cord will pass both the gamma evaluation and the evaluation separated in high and low gradient regions. Another case have a decreased absorbed dose in all investigated values, $D_{2\,\%}$, mean and $D_{98\,\%}$, to both PTV volumes but pass both evaluations with high passing rates. Two of the examples shown in table 1 did not pass the evaluation when the high and low absorbed dose gradient regions were separated.

Table 1. Examples of the relative dose difference analysis in structures of interest and the passing rates of the QA based on gamma evaluation (see column labeled "Gamma") and the evaluation separated in the high and low gradient regions (see column labeled "ICRU") for five of the evaluated modified dose distributions. The relative dose difference [%] between the modified and the original dose distributions where the modified plans have more than a ±5 % dose difference in the PTV volumes and the organs at risk are marked with a grey shade. Passing rates below 85 % are marked with bold numbers.

Patient	Relative dose difference in PTV [%]						Relative dose difference in organs at risk [%]				Passing rate [%]		
	PTV-T			PTV-N			Spinal cord	Left parotid	Right parotid	Brainstem	Gamma	ICRU	
	$D_{2\,\%}$	Mean	$D_{98\,\%}$	$D_{2\,\%}$	Mean	$D_{98\,\%}$	$D_{2\,\%}$	Mean	Mean	$D_{2\,\%}$	Total[a]	Low[b]	High[c]
1	6.8	7.0	6.6	6.6	7.6	7.2	12.5	16.8	16.3	8.2	100	98.6	100
1	0	-0.7	-5.6	-0.3	-0.9	-6.0	2.5	5.7	-12.7	3.0	100	99.3	100
2	-5.7	-6.1	-6.3	-5.9	-6.2	-6.2	-7.3	-9.2	-7.1	-6.5	94.5	**78.2**	100
3	5.6	5.0	5.4	5.9	5.8	5.4	11.6	19.0	9.6		100	98.7	100
3	-6.0	-6.3	-6.3	-5.8	-5.6	-5.9	-6.2	-11.6	-9.0		92.0	**69.5**	99.0

[a]Gamma evaluation when all the diodes are included.
[b]Low gradient region, dose change less than 20 %/cm, evaluated with dose difference criterion of ±5 %.
[c] High gradient region, dose change more than 20 %/cm, evaluated with distance to agreement criterion of ±5 mm.

4. Conclusion

Patient-specific QA with the Delta[4] dosimetry system, using the evaluation limited to the diode measurement points, and the ICRU 83 recommendations for evaluation can not be used with the intention to distinguish differences between planned and measured dose of dosimetrical relevance for the treatment of the patient.

References

[1] International Commission on Radiation Units and Measurements, ICRU Report 83 2010 *Prescribing, Recording, and Reporting Photon-Beam Intensity-Modulated Radiation Therapy (IMRT)*
[2] Low D A *et al* 1998 *Med. Phys.* **25** 656-6

The effect of motion on IMRT – looking at interplay with 3D measurements

A Thomas[1], H Yan[1], M Oldham[1], T Juang[1], J Adamovics[2] and FF Yin[1]
[1]Duke University Medical Center, Durham, NC, USA
[2]Rider University, Lawerenceville, NJ, USA

E-mail: mark.oldham@duke.edu

Abstract. Six base of skull IMRT treatment plans were delivered to 3D dosimeters within the RPC Head and Neck Phantom for QA verification. Isotropic 2mm 3D data was obtained using the DLOS-PRESAGE system and compared to an Eclipse (Varian) treatment plan. Normalized Dose Distribution pass rates were obtained for a number of criteria. High quality 3D dosimetry data was observed from the DLOS system, illustrated here through colormaps, isodose lines, profiles, and NDD 3D maps. Excellent agreement with the planned dose distributions was also observed with NDD analysis revealing > 90% NDD pass rates [3%, 2mm], noise < 0.5%. This paper focuses on a detailed exploration of the quality and use of 3D dosimetry data obtained with the DLOS-PRESAGE system.

1. Introduction

The effect of interplay has been well documented and written into TG reports with warnings of its negative consequences[1-6]. However, the effect has only been observed in 2D or a series of 2D devices, and has never been measured with a truly 3D device. Research on a 3D device may lead to different conclusions as to the outcome and possible consequences of motion management with leaf motion present. This work shows our initial investigations of interplay as documented by a 3D dosimeter building on the work of Brady *et al* [7].

2. Methods

Isotropic 2mm 3D dosimetry data was measured using the Presage/DLOS (Duke Large Field-of-view Optical-CT Scanner) [8]. Measured data was compared to eclipse calculation (AAA algorithm) in the CERR software[9]. Comparison methods included comparing isodose lines and dose-volume-histograms (DVH).

2.1. Dosimeter/Phantom

A 5cm diameter, 5cm long PRESAGE cylinder was placed within a CIRS dynamic thorax phantom[7], see Fig 1. The phantom was set to oscillate with a period of 4s and an amplitude of ±5mm. The CIRS phantom-dosimeter combination was scanned on a GE Lightspeed X-ray-CT scanner with a 1.25mm slice thickness. Contours were made on the CT for a GTV/CTV and then expanded in the direction of motion creating an ITV/PTV as shown in Fig 1. From these contours a 5-field, 12 Gy IMRT and a 1-arc, 12 Gy RapidArc treatment plans were created for the phantom-dosimeter combination. Calculation was performed with the Eclipse planning system, using the Analytic-Anisotropy-Algorithm (AAA), with 2.5mm^3 isotropic resolution, with the heterogeneity correction applied.

The delivery of the plans occurred on a Novalis Tx. The delivery of each plan (IMRT & RapidArc) occurred twice; once with the motion phantom in a static position acting as a control, and once with the phantom in motion using the parameters stated above.

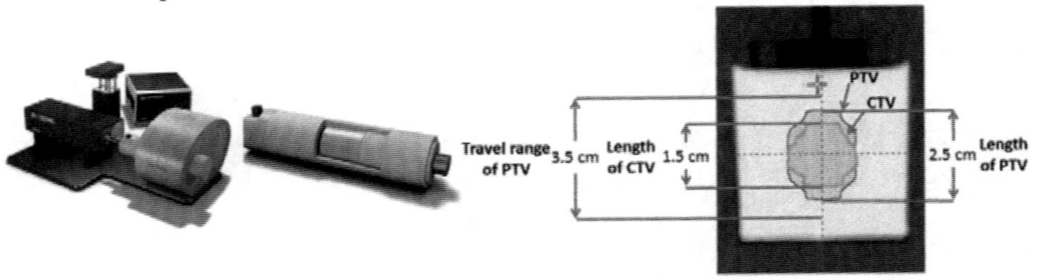

Figure 1: CIRS motion phantom (left) and dosimeter insert (middle) with CT of dosimeter and GTV/CTV and ITV/PTV volumes contoured (right).

2.2. *3D Data Acquisition*

3D dose measurements were performed with the cylindrical PRESAGE® dosimeters which were compatible with the CIRS dynamic phantom. PRESAGE® (Heuris Pharma, Skillman, NJ) is a polyurethane based dosimetry material which changes color when exposed to ionizing radiation, causing an optical density change (ΔOD) for visible light. The radiation induced ΔOD is proportional to the locally absorbed dose, and can be imaged using optical-CT[10-12]. Recent comprehensive evaluations have shown PRESAGE® to be an excellent material for accurate 3D dosimetry [13, 14].

To obtain the delivered 3D dose distribution, the radiation-induced ΔOD throughout the dosimeter was imaged using the DLOS system). The ΔOD is determined by reconstructing corrected projection images utilizing parallel beam filtered back-projection as described in [8]. 3D data are acquired through the accumulation of 360 projection images over 360° to meet Nyquist sampling criteria for $2mm^3$ isotropic reconstructions. Each projection image is captured 20 times and averaged to increase the SNR at each angle. The projection images are dark noise and flood field corrected to reduce the effects of particulates on glass surfaces, non-uniformity of CCD pixel response, dark current and readout noise.

2.3. *3D IMRT QA Data Analysis*

The 3D ΔOD distribution was converted to dose by normalizing the IMRT static delivery to the treatment plan and using that normalization value to scale the remaining dosimeters. Upon normalization of the data DVHs and isodose overlays of the plan, static and motion deliveries were created to quantify the interplay effect in 3D.

Figure 2 shows results of the DVH plots according to the contours used in creating the treatment plans. Inspection of the figure shows the least coverage for the ITV/PTV volumes with the motion delivery. Of interest, however, the GTV/CTV coverage is on par with the stationary delivery meaning the expanded volumes performed their purpose even with interplay. This was expected to break down due to interplay, but given the relatively low modulation of the beams and small volume irradiated the effect was not as dramatic as seen previously by other authors[1, 3, 6].

3. Results

3.1. DVH Analysis

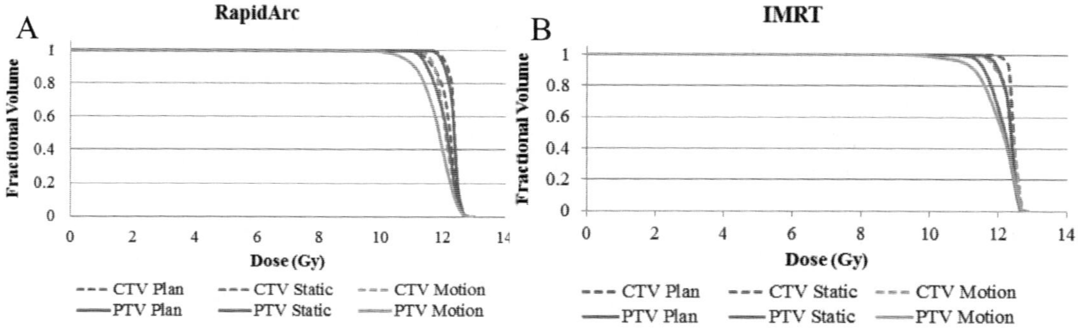

Figure 2: DVHs of (A) The RapidArc plan and (B) The IMRT plan. In both instances dashed lines indicate GTV/CTV coverage and solid lines represent ITV/PTV coverage.

3.2. Isodose Map

Figure 3 shows the RapidArc and IMRT axial and coronal isodose overlays of the planned delivery with the static and motion deliveries.

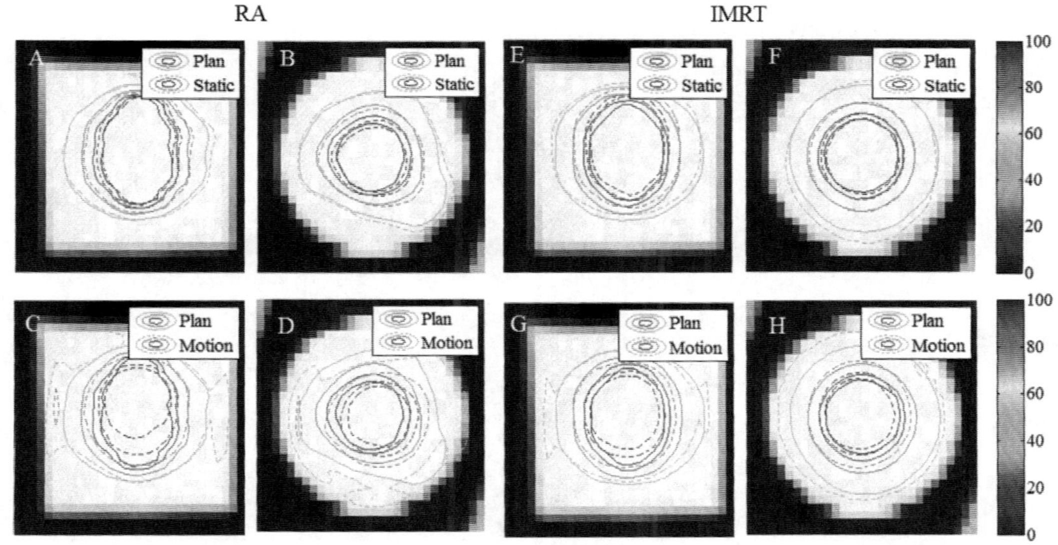

Figure 3: Static (Top Row) and Motion (Bottom Row) delivery overlaying the planned Rapid Arc and IMRT deliveries. The 1st column is the coronal view and the 2nd is the axial view. The colorbar indicates dose delivered in % of maximum planned dose. In all plots above the solid lines equal the planned doses and the dashed lines equal the measured doses.

Inspection of figure 3 shows the results of interplay in the coronal and axial planes. In both the RapidArc and IMRT cases, the Static delivery (dashed lines – top row) matches the Planned delivery (solid lines – bottom row) to within a few millimeters, however the motion cases, significant deviation in the form of shifting and stretching are seen – leading to under-dosing of the PTV as seen in the

DVHs. The plans with motion also appear to be noisier, but it is not due to the readout. Given the motion during delivery this is simply a redistribution of dose creating a more chaotic distribution. The noise in all dosimeters was < 0.5% of the high dose regions.

4. Conclusions

The data acquired for this work was both promising and initially surprising. 3D data documenting interplay with motion during delivery has been acquired with high quality and 2mm isotropic resolution. It was surprising in the sense that when expanding the GTV/CTV for motion, the drop in coverage of the volume was not significantly reduced as seen by a stationary delivery of the same volume. Perhaps this was simply due to the small volume. This work shows that further investigation into the effects of interplay can be done in 3D and perhaps allow for better understanding of motion effects with different frequencies (ie. Leaf speed, Breathing period, Gantry rotation, etc) and lead to updated recommendations on techniques of motion management.

5. Acknowledgments

This work was supported by NIH R01CA100835

6. References

[1] Bortfeld T *et al* 2004 *Semin. Radiat. Oncol.* **14** 41-51
[2] Chui C S *et al* 2003 *Med. Phys.* **30** 1736-46
[3] Jiang S *et al* 2003 *Phys. Med. Biol.* **48** 1773-84
[4] Keall P J *et al* 2006 *Med. Phys.* **33** 3874-900
[5] Kissick M W and Mackie T R 2009 *Med. Phys.* **36** 5721-2
[6] Yu C X *et al* 1998 *Phys. Med. Biol.* **43** 91-104
[7] Brady S L *et al* 2010 *Phys. Med. Biol.* **55** 2187-201
[8] Thomas A *et al* 2011 *Med. Phys.* **38** 4846-57
[9] Deasy J O *et al* 2003 *Med. Phys.* **30** 979-85
[10] Oldham M 2006 *J. Phys.: Conf. Ser.* **56** 58-71
[11] Krstajic N and Doran S J 2007 *Phys. Med. Biol.* **52** N257-63
[12] Babic S *et al* 2009 *Phys. Med. Biol.* **54** 2463-81
[13] Sakhalkar H S *et al* 2009 *Med. Phys.* **36** 71-82
[14] Sakhalkar H *et al* 2009 *Med. Phys.* **36** 3371-7

Comprehensive quality assurance for base of skull IMRT

A Thomas[1], J O'Daniel[1], J Adamovics[2], G Ibbott[3] and M Oldham[1]
[1]Duke University Medical Center, Durham, NC, USA
[2]Rider University, Lawerenceville, NJ, USA
[3]University of Texas - MD Anderson, Houston, TX, USA

E-mail: mark.oldham@duke.edu

Abstract. Six base of skull IMRT treatment plans were delivered to Presage dosimeters within the RPC Head and Neck Phantom for quality assurance (QA) verification. Isotropic 2mm 3D data were acquired by optical-CT scanning with the DLOS system (Duke Large Optical-CT Scanner) and compared to the Eclipse (Varian) treatment plan. Normalized Dose Distribution (NDD) pass rates were obtained for a number of criteria. High quality 3D dosimetry data was observed from the DLOS system, illustrated here through colormaps, isodose lines, and profiles. Excellent agreement with the planned dose distributions was also observed with NDD analysis revealing > 90% pass rates (with criteria 3%, 2mm), and noise < 0.5%. The results comprehensively confirm the high accuracy of base-of-skull IMRT treatment in our clinic.

1. Introduction

The accuracy of base-of-skull (BOS) IMRT treatment was evaluated in six patients treated at Duke between 2008 and 2011. A condensed summary analysis was presented in Oldham et al [1], which focused on evaluating the clinical significance of 3D verification QA data after transform back to the patients CT data set [2]. The current work compliments this analysis by focusing on a direct dosimetry comparison, measured to calculated, in the RPC head phantom geometry. Particular attention is paid to investigating the quality of the 3D dosimetry data acquired with the DLOS system [3], and on a detailed comparison of the 3D measured and calculated (Eclipse) distributions.

2. Methods

Six patients that received base of skull IMRT treatment were randomly selected. The fractionation scheme and target volumes varied, with PTVs ranging from $4.4cm^3$ up to $46cm^3$, and dose per fraction ranging from 1.8 Gy to 12.5 Gy. All data was anonymized prior to analysis, and all patients were treated with a Novalis Tx system commissioned for radiosurgery treatment at Duke. The prospective verification measurements described here were made on the same machine for consistency.

2.1. 3D IMRT QA Data Acquisition

3D dose measurements were performed with cylindrical PRESAGE® dosimeters (10cm diameter, 12cm long) which were compatible with the RPC head-and-neck IMRT credentialing phantom [4]. PRESAGE® (Heuris Pharma, Skillman, NJ) is a polyurethane based dosimetry material which changes color when exposed to ionizing radiation, causing an optical density change (ΔOD) for visible light [5, 6]. The radiation induced ΔOD is proportional to the locally absorbed dose, and can be imaged using optical-CT [7-9]. Recent comprehensive evaluations have shown PRESAGE® to be an excellent material for accurate 3D dosimetry [5, 10].

The RPC phantom-dosimeter combination was scanned on a GE Lightspeed X-ray-CT scanner with a 1.25mm slice thickness. A verification treatment plan was created for each patient, by re-calculating the patient's original treatment plan onto the RPC phantom CT data. The isocenter of the verification plan was positioned centrally in the PRESAGE® dosimetery insert to ensure a 3D measurement of the whole distribution would be feasible. Re-calculation was performed with the Eclipse planning system, using the Analytic-Anisotropy-Algorithm (AAA), at 2.5mm^3 isotropic resolution, with the heterogeneity correction applied. Each verification plan was delivered to the RPC phantom containing the dosimeter. Each plan (except plan HN5 which had high dose/fx) was delivered 4 times to the phantom in order to get close to ~8Gy, the optimal dose for the dynamic range of the DLOS system for small treatment volumes.

To obtain the delivered 3D dose distribution, the radiation-induced ΔOD throughout the dosimeter was imaged using the DLOS system (Duke Large Field-of-view Optical-CT Scanner). The ΔOD is determined by reconstructing corrected projection images utilizing parallel beam filtered back-projection. A corrected projection consists of a post-irradiation projection that has been divided by the corresponding pre-irradiation projection to obtain the radiation-induced change. Details of this method are given in [3]. 3D data are acquired through the accumulation of 360 projection images over 360° to meet Nyquist sampling criteria for 2mm^3 isotropic reconstructions. Each projection image is captured 20 times and averaged to increase the SNR at each angle. The projection images are dark noise and flood field corrected to reduce the effects of particulates on glass surfaces, non-uniformity of CCD pixel response, dark current and readout noise. Each pre- and post-projection image is corrected for stray light following the method described in [11].

2.2. *3D IMRT QA Data Analysis*

The methods described above yielded isotropic 2mm relative 3D dosimetry data throughout the Presage dosimeter volume for all patients. The dose response of Presage is known to be linear [10], and the 3D relative ΔOD distribution was converted to dose by scaling to a point of known absolute dose within the head-phantom. This point of known dose was determined by a second delivery to the head-phantom, this time incorporating a Presage dosimeter of identical dimensions but containing a central channel for a micro-ion-chamber (CC01 Scandatronix/Wellhofer). The location of the active volume of the chamber was in the middle of the uniform high-dose treatment region. The measured and planned absolute dose distributions were then compared using the CERR tool (Deasy et al. [12]). Comparison methods included line-profiles, isodose lines, and NDD plots (criteria 3%,3mm and 3%, 2mm, with 5% threshold) [13]. The outer 5mm rind of the dosimeter was not included in the analysis due to loss of superficial data from edge artifacts.

3. Results

3.1. *Overall Pass Rates*

The pass rates according to the NDD algorithm are recorded in Table 1. Pass rates for [3%, 3mm] [3%, 2mm] and [5%, 3mm] are given. Even with the stringent [3%, 2mm] criteria all plans tested recorded pass rates over 90%. Noise for each scan is also given and was obtained as the standard deviation in a high dose region divided by the mean in that same high dose region.

3.2. *Specific Cases*

Figures of each patient are not shown due to space limitations, however, the cases with the highest and lowest NDD pass rates are shown as well as a case that seemed more typical. Each set of figures contains a sagittal view of the measured dose wash accompanied by a vertical and horizontal line profile to give the reader an appreciation of the low noise and close agreement as indicated by the NDD pass rates.

Table 1: NDD results for all six patient cases analysed with a 5% dose threshold.

Patient	1	2	3	4	5	6
NDD Pass Rate 3%, 3mm	98.5%	98.1%	99.7%	96.5%	98.7%	95.7%
NDD Pass Rate 3%, 2mm	93.9%	97.2%	98.9%	93.1%	96.0%	90.1%
NDD Pass Rate 5%, 3mm	99.1%	99.6%	99.9%	96.9%	99.9%	97.6%
Noise	0.35%	0.64%	0.30%	0.32%	0.58%	0.57%

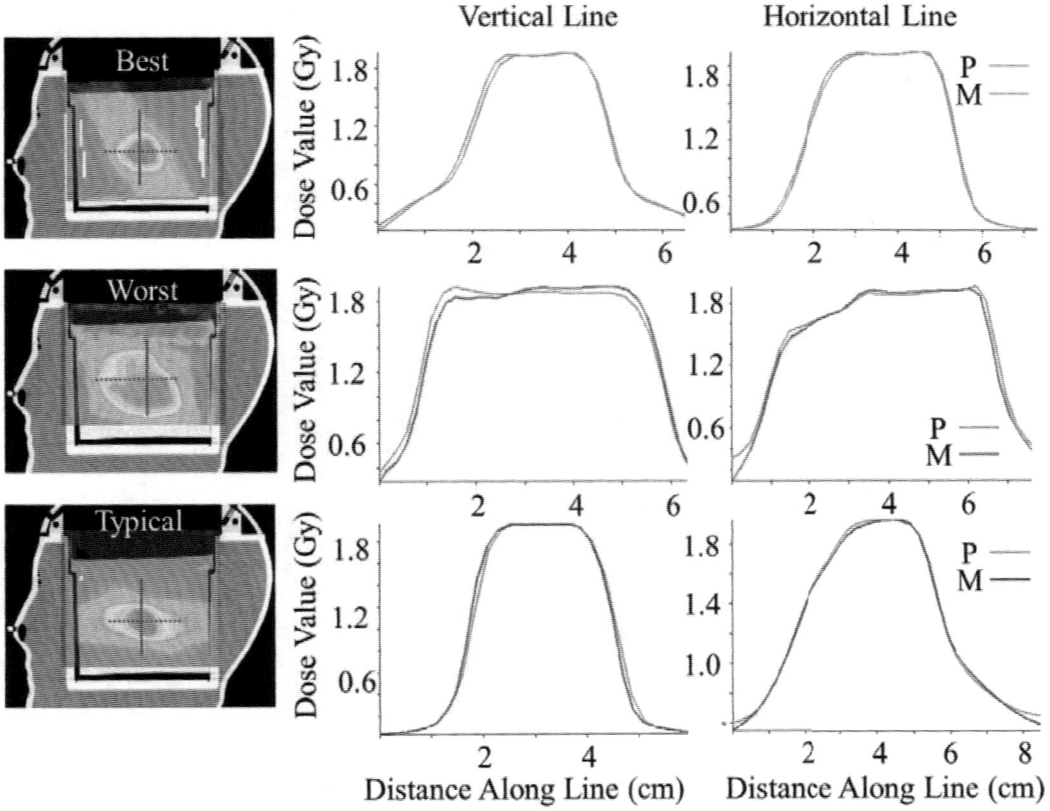

Figure 1: 3 cases – Best (1st row), Worst (2nd row) and Typical (3rd row) as indicated by NDD pass rates. Sagittal dose planes showing the PRESAGE measurements overlayed on the CT of the RPC phantom (1st column). Vertical line profiles (2nd column) and horizontal line profiles (3rd column). Legend has plan (P) and measurement (M) showing the nearly identical agreement of the treatment planning system calculation and the 3D measurement.

Best Case – patient case #3. The delivery consisted of 5 fields and 4 couch angles with 1.8 Gy delivered to the target 4 times for proper contrast in the PRESAGE. Low noise from the data acquisition can be seen from both the line profiles and the dose plane displayed.

Worst Case – patient case #4. This plan gave the lowest overall passing rate of 93.7%. Visual inspection line profiles show disparities within this treatment verification plan relative to the other two displayed. The other plans verified in this study lead us to believe the differences seen in the measured distribution, while slight, are real and most likely due to using the same CT scan for each patient case. Each dosimeter top is not molded identically and thus any beams initiating superiorly may not be modeled correctly in the planned dose.

Typical Case – patient case #1. The delivery consisted of 9 fields and 3 couch angles with a 1.8 Gy delivery to the isocenter 4 times. The low noise in the data can be seen qualitatively from the reconstructions and isodose lines.

4. Conclusions

This work is the first application of high resolution 3D dosimetry to verify the accuracy of base-of-skull IMRT delivery. The high NDD pass rates (Table 1) comparing the planned and measured distributions illustrates the high accuracy of this technique in our clinic. 3D data acquisition time with the DLOS was under 20 minutes (including set-up time) and produced low noise data (< 0.5%) throughout the dosimeters to within just a few millimetres from the edge. The convenience, practicality and cost of the system is approaching clinical feasibility for routine use.

5. Acknowledgments

This work was supported by NIH R01CA100835

6. References

[1] Oldham M *et al* 2012 *Int. J Radiat. Oncol. Biol. Phys.* (in Press)
[2] Baldock C *et al* 2010 *Phys. Med. Biol.* **55** R1-63
[3] Thomas A *et al* 2011 *Med. Phys.* **38** 4846-57
[4] Sakhalkar H *et al* 2009 *Med. Phys.* **36** 3371-7
[5] Brown S *et al* 2008 *Appl. Radiat. Isot.* **66** 1970-4
[6] Gorjiara T *et al* 2011 *Med. Phys.* **38** 2265-74
[7] Oldham M 2006 *J. Phys.: Conf. Ser.* **56** 58-71
[8] Krstajic N and Doran S J 2007 *Phys. Med. Biol.* **52** N257-63
[9] Babic S *et al* 2009 *Phys. Med. Biol.* **54** 2463-81
[10] Sakhalkar H S *et al* 2009 *Med. Phys.* **36** 71-82
[11] Thomas A 2011 *Phys. Med. Biol.* **56** 4433-51
[12] Deasy J O *et al* 2003 *Med. Phys.* **30** 979-85
[13] Jiang S B *et al* 2006 *Phys. Med. Biol.* **51** 759-76

Validation of an IMRT beam model using a secondary treatment planning system as a 3D dosimeter

A Xing[1], S Arumugam[1], L Holloway[1,2,3] and G Goozee[1]
[1]Liverpool and Macarthur cancer therapy centres and Ingham Institute, Liverpool Hospital, New South Wales, Sydney, Australia
[2]Institute of Medical Physics, School of Physics, University of Sydney, Sydney, New South Wales, Australia
[3]Centre for Medical Radiation Physics, University of Wollongong, Wollongong, New South Wales, Australia

E-mail: aitang.xing@sswahs.nsw.gov.au

Abstract. A treatment planning system IMRT beam model is usually validated using phantom-based measurement, however this will not detect errors related to patient anatomy and inhomogeneity. In this study a secondary treatment system (CMS XIO) was used as a 3D dosimeter to verify an IMRT beam model recently commissioned in a Philips Pinnacle treatment planning system. Data sets from three head-neck and two prostate patients previously treated were utilised. The IMRT plans for these patients were planned in Pinnacle and transferred to XIO. The dose at each voxel in the patient volume was calculated in both XIO and Pinnacle. The 2D dose gamma maps for three orthogonal planes passing through the isocenter were calculated with a criteria of 3%/3mm. The mean gamma pass rate for all patients was 96.86% with maximum and minimum values of 99.6% and 95%. One coronal dose plane at 5.5 cm depth in the phantom was also measured and compared with dose calculated by the Pinnacle IMRT beam model using same gamma criteria. The measured mean gamma pass rate for this coronal plane dose was 96.7% with maximum and minimum of 98.41% and 95.3%. This was comparable with the gamma map pass rates for the three orthogonal dose planes calculated by XIO for the patient data. A secondary treatment planning system was shown to provide a supplementary verification tool based on calculation-based 3D dosimetry using patient anatomy.

1. Introduction

Intensity modulated radiotherapy (IMRT) has been used as a routine treatment technique in many centres since the first IMRT treatment was delivered in 1996 [1]. It is critical that the treatment planning system (TPS) IMRT beam model is fully commissioned before it is released for clinical use. AAPM TG 119 includes a set of test cases for this purpose [2]. These tests were recommended to be performed using a square water equivalent phantom with artificial targets and organs at risk (OAR) drawn on CT images. These tests, however, cannot detect errors related to patient anatomy as these structures are unable to represent real patient anatomy. For example, test No.12 involves simulating a prostate treatment. The prostate clinical target volume (CTV) and rectum are represented by roughly ellipsoidal volumes with posterior concavity and a cylinder with diameter of 1.5 cm, respectively. In reality, the CTVs and OARs are complicated and vary from patient to patient, especially for head-neck cases.

As a component of commissioning tests, it is desirable to verify the IMRT beam model within patient anatomy. However, it is very difficult to directly measure the dose inside patients. This can be achieved using an independent calculation-based method which has been previously verified. One study showed that the beam model in a TPS could be verified using a full Monte Carlo calculation instead of phantom-based measurement [3]. Recently in our department, an IMRT beam model was commissioned for the Philips Pinnacle TPS. The purpose of this study was to evaluate the suitability of using a secondary commercial TPS to verify an IMRT beam model considering true patient anatomy and inhomogeneity as a supplemental test to phantom-based measurement verification.

2. Materials and Methods

To verify an IMRT beam model for an Elekta Synergy linear accelerator, considering true patient anatomy and inhomogeneity a process was developed and analysed for transferring patient data and generated treatment plans including all beam sizes, shapes and weights from the Pinnacle TPS which was being commissioned, to a CMS XIO TPS which had been utilized in our department for IMRT treatment planning for a number of years.

2.1. CMS XIO as a 3D dose calculator

CMS XIO is a three-dimensional (3D) TPS for external beam radiotherapy able to calculate dose to each voxel in a patient volume. A superposition algorithm is utilized in this system, in which the total energy fluence of the primary beam in the patient volume is first calculated and then convolved with a kernel that takes into account the transport of secondary photons and electrons within the given patient anatomy. The accuracy of this algorithm meets the current requirements for radiotherapy and has been verified by measurements and full Monte Carlo calculations [4, 5]. In our department, XIO has been used for IMRT planning and treatment since 2008.

Using this XiO TPS 310 IMRT patients have been planned and treated on Elekta linear accelerators (Linac) within our department. For each patient, an ion chamber point dose measurement was undertaken using the CIRS phantom. The plane dose was also measured at 5.4 cm depth in a water equivalent square phantom using a chamber array (I'mRT Matrix). The gamma map comparing TPS and measurement was analyzed using 3%/3mm criteria. The maximum and minimum difference to date between measured and XIO calculated point dose is 2.8 % and 0.3 % with a mean value of 1.35 %. The mean gamma pass rate is 96.7 %. The maximum and minimum pass rates are 95.3 % and 99.2 %, respectively. These patient specific quality assurance results with phantom provided an extensive verification of XIO against the IMRT delivery system. In this investigation, XIO was used as an independent 3D dose calculator for verifying the Elekta IMRT beam model commissioned in Pinnacle.

2.2. Calculation of Pinnacle IMRT plans in XIO

To verify the IMRT beam model in Pinnacle using XIO, five typical datasets from previous patients were utilised. Three of them were head-neck and two were prostate cases. The patient CT data set and associated radiotherapy (RT) structures were first imported into Pinnacle. An IMRT plan was generated for each patient using the Pinnacle Elekta IMRT beam model following current clinical protocols. Each plan was reviewed and considered acceptable by the treating radiation oncologist. The same patient CT data and RT structures were then transferred from Pinnacle to XIO using DICOM, thus ensuring plan objects such as interest points, weight point and isocenter have the same coordinates for Pinnacle and XIO.

XIO does not have a function or interface that allows direct importing of a Pinnacle IMRT plan. To overcome this in-house software was developed. The Pinnacle IMRT plan was first exported as a RTP file. The in house software then converted the RTP file for the Pinnacle IMRT plan into a tel.1 as required by XIO, enabling patient datasets to be set-up and viewed in XIO with the exact plan details e in Pinnacle. No additional optimization or plan adjustment was undertaken in XIO. The 3D dose distribution was calculated in XIO using the same calculation settings as those in Pinnacle. The calculation grid used was 2 mm and the calculation volume was chosen to be large enough to cover the entire patient volume. The 3D dose distributions calculated in XIO were exported in DICOM format and compared with Pinnacle calculated dose distributions.

A 2D gamma map for three orthogonal planes passing through the isocenter was calculated for each patient dataset comparing Pinnacle and XiO calculated dose distributions. A gamma map analysis criteria of 3%/3 mm was used as recommended by AAPM TG 119 for IMRT beam model commissioning and used by most radiotherapy centres for measurement-based patient-specific QA [2]. Dose for one coronal plane at 5.4 cm depth in a water equivalent square phantom using the I'mRT Matrix was also measured for each patient Pinnacle IMRT plan. The measured dose was compared with the same dose plane calculated in the phantom by Pinnacle using the Pinnacle IMRT model.

3. Results and Discussion

An example 3D dose cube calculated by XIO and Pinnacle using the commissioned IMRT beam model for one head-neck patient is shown in figure 1. The 2D gamma map for one coronal plane passing through the isocenter was also calculated. The gamma analysis passed for all pixels except for a very small area near the patient surface, which was also indicated by a 2D dose profile comparison in the same dose plane. The failure of the gamma analysis near the patient surface is mainly due to the difference in the electron contamination model used by XIO and Pinnacle.

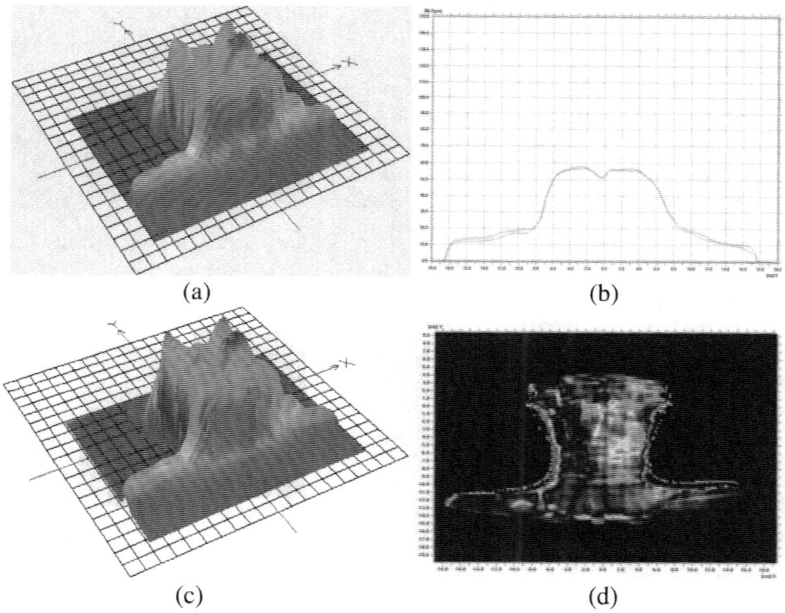

Figure1. Comparison of a coronal dose plane for a 3D dose distribution in patient anatomy for one head-neck patient: (a) The dose to patient calculated by XIO (c) The dose in patient calculated using the Elekta IMRT beam model by Pinnacle (b) A dose profile passing through isocenter in the coronal plane (d) The gamma map for the coronal plane passing through isocenter.

Table 1 shows the gamma pass rate for the three orthogonal planes passing through the isocenter for all patient datasets. The first three patients were head-neck and the last two prostate cases. The average gamma pass rate for all beams and all patients was 96.86%. The maximum and minimum gamma pass rates were 99.6% and 94.3%, respectively. Table 2 presents the gamma pass rate for one coronal plane dose measured at 5.4 cm depth in a water equivalent phantom against the same plane dose calculated by Pinnacle. The gamma map was calculated for each IMRT beam for all patients. The average gamma pass rate for this measured coronal plane dose for five patients was 96.7% with minimum and maximum gamma pass rates of 95.3% and 98.41%, respectively. The verification results using XIO as a 3D dosimeter for verifying the IMRT beam model in Pinnacle were comparable to results measured using the phantom.

Table 1. Gamma pass rates for three orthogonal planes passing through isocenter comparing Pinnacle and XiO calculated plans

	Gamma pass rate (%)		
Patient #	Transverse plane	Sagittal plane	Coronal plane
1	96.5	94.9	94.3
2	95.0	95.7	96.1
3	99.6	98.6	96.0
4	98.3	97.0	97.6
5	97.4	97.6	98.3

Table 2. Measured gamma pass rate for one coronal plane at 5.4 cm depth using a square solid water phantom and compared with Pinnacle calculated plans

	Gamma pass rate for each Pinnacle IMRT beam (%)							
	beam #							
Patient #	1	2	3	4	5	6	7	Mean pass rate (%)
1	95.9	97.6	95.8	96.2	97.4	98.4	96.6	97
2	96.92	96.84	95.51	97.31	95.3			96.4
3	96.04	95.93	95.83	97.07	96.95	97.87	96.74	96.5
4	97.1	96.75	96.21	97.04	96.21	98.41	97.76	97.1
5	95.53	97.02	96.1	95.96	98.23			97

The advantage of using a secondary commercial TPS for verifying an IMRT beam model is that it is a calculation-based 3D dosimetry. The dose at each voxel in the patient volume can be obtained. Realistic complicated inhomogeneities in real patient anatomy, often regions where limitations in treatment planning systems are seen can be analysed.

4. Conclusion

A process for using a secondary treatment planning system as a 3D dosimeter to verify an IMRT beam model has been developed and verified. This provides an alternative method for verifying an IMRT beam model based on true patient anatomy. The methodology presented here can be applied to other commercial treatment systems.

5. References

[1] Carol M *et al* 1996 *Stereotact Funct. Neurosurg.* **66** 310
[2] Ezzell G A *et al* 2009 *Med. Phys.* **36** 5359
[3] Wieslander E and Knoos T 2008 *J Phys.: Conf. Ser.* **102** 012027
[4] Fotina I *et al* 2009 *Radiother. Oncol.* **93** 645
[5] Francescon P *et al* 2003 *Med. Phys.* **30** 144

A multi-configurational cylindrical phantom based evaluation of patient-specific IMRT QA tools

T Olding[1], T Halsall[1], L J Schreiner[1,2] and A Kerr[1,2]
[1]Cancer Centre of South Eastern Ontario at Kingston General Hospital, 25 King Street West, Kingston, ON, Canada, K7L5P9
[2]Department of Physics, Queen's University, Kingston, Ontario, Canada, K7L3N6

E-Mail: Tim.Olding@krcc.on.ca

Abstract. A custom in-house built multi-purpose phantom has been designed and built to investigate the integrity of the 2D Matrixx ion chamber (Scanditronix-Welhoffer, Bartlett, TN) and 3D electronic portal image device (EPID) techniques employed for patient specific IMRT delivery QA at our centre. Single ion chamber, EBT3 film and FXG gel dose measurements from the common phantom system were found to be consistent with the Matrixx and EPID measurements except in the limit of highly modulated plan deliveries.

1. Introduction

Intensity modulated radiation therapy (IMRT) has become a standard treatment approach for a number of cancer sites. Quality assurance (QA) of IMRT dose is considerably diverse, ranging from simple independent point dose calculations, to complex measurements using different 2D and 3D dosimeters. These QA tools/tasks may be utilized together or on their own, depending on the IMRT technique complexity and level of past clinical experience for that technique. At our centre, we use a commercial 2D matrix ion chamber array and an EPID based dose prediction protocol[1] for patient specific IMRT delivery QA. Gel dosimetry has also been used when we introduce and validate new IMRT techniques. While these different approaches complement each other, they often differ slightly in their response, and it would be useful to better understand the factors that give rise to these differences. The purpose of this work is to implement a common phantom system to compare 2D and 3D dose data from different IMRT QA measurement tools. The system was made as simple as possible to remove possible variations that occurred because of phantom characteristics.

2. Materials and Methods

2.1. IMRT QA Tool Comparison

The IMRT QA tools under consideration in this study were: a) Single ion chamber (Capintec PR-05P 0.07 cc/Capintec electrometer); b) 2D Matrixx ion chamber; c) Gafchromic EBT2/3 film (ISP, Wayne, NJ); d) EPID based fluence measurement and dose prediction algorithm[1]; and e) Fricke-xylenol orange-gelatin (FXG) based gel dosimeter with optical CT readout[2]. The EPID based dosimetry system (EPIDOSE) utilizes the approach developed by Ansbacher *et al.*[1], where fluence is measured using an MV imager, with nothing in the beam. A dose algorithm then is applied to the Dicom fluence data, back-projecting 3D dose to a virtual cylindrical phantom of 20.4 cm diameter.

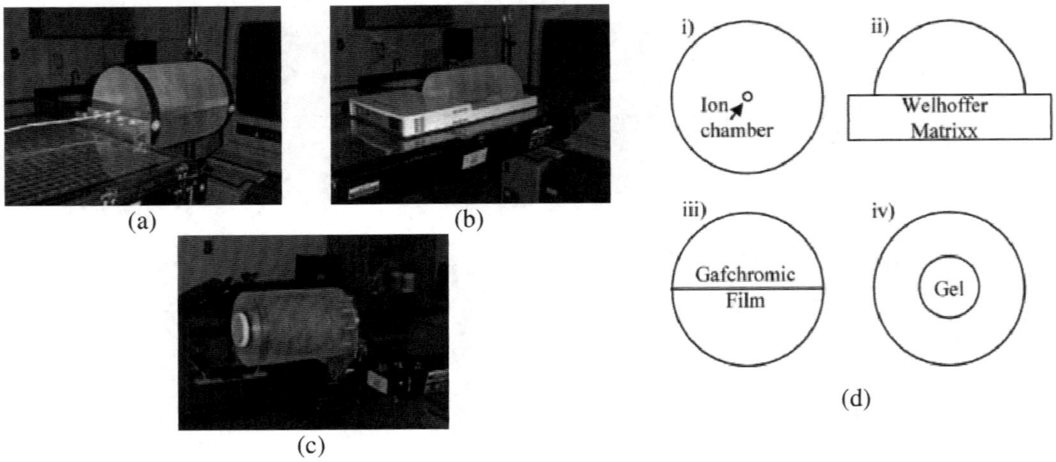

Figure 1: (a) Photographs of the phantom on the treatment unit with ion chamber and film, (b) 2D Matrixx, and (c) 3D FXG gel. (d) Cross-sections of the different configurations. The EPIDOSE configuration is not shown as it is a virtual phantom approach.

We have constructed an equivalent 20.4 cm diameter, 30 cm long density acrylic phantom for comparison of the various test tools to the EPIDOSE measurement geometry. Several versions of the cylindrical phantom have been constructed, all with same overall size, but with different inner configurations to accommodate the various dosimeters. Photographs of selected set up geometries on a Varian linac (Varian Medical Systems, Palo Alto, CA), and schematic cross sections of the different configurations are shown in Fig 1. The CT number-to-electron density calibration was adjusted slightly to model more precisely the known electron density of the acrylic phantom material. All plans were calculated in Eclipse (Varian Medical Systems) and delivered on a Clinac 2100iX linear accelerator to one or more phantom configurations. The intention was to begin with simple fields and simple IMRT patterns to provide baseline data for comparison, then move to studying more clinically relevant deliveries

Each dose measurement was compared against the corresponding Eclipse dose using commercial software designed for that dosimeter. Matrixx and film measurements were analyzed using Welhoffer QA software (OmniPro v 1.6.0002-2008-03-12). EPIDOSE dose was compared using MATLAB (Mathworks, Newark, NJ) based software provided by Ansbacher. The reconstructed gel dose data from the Vista^{TM} optical CT scanner (Modus Medical Devices Inc, London, ON), was converted to Dicom format, and compared to Eclipse within the CERR environment (Computational Environment for Radiotherapy Research, Washington State University, St.Louis, MO).

3. Results

3.1. Absolute Dosimetry

Table 1 shows the results of absolute point dose comparisons between dosimeter measurements and calculated dose from Eclipse. The measurements were obtained from plan deliveries on a single linac with appropriate corrections (temperature, pressure, readings normalized to a 10 x 10 cm^2 field reading under reference conditions, etc.). Good agreement is observed between the dosimeters except for the highly modulated sliding window plans, where EPIDOSE reads low and Matrixx reads high.

3.2. 2D/3D Comparisons

2D and 3D comparison data are shown in Fig.2 for a test prostate delivery. The film data were acquired using EBT2 film, and the disagreement recorded in the film fiducial mark regions was excluded from the analysis. Figure 3 shows the dose data from the delivery of another prostate plan to the Matrixx, with an EBT3 film inserted beneath the phantom material placed on the Matrixx.

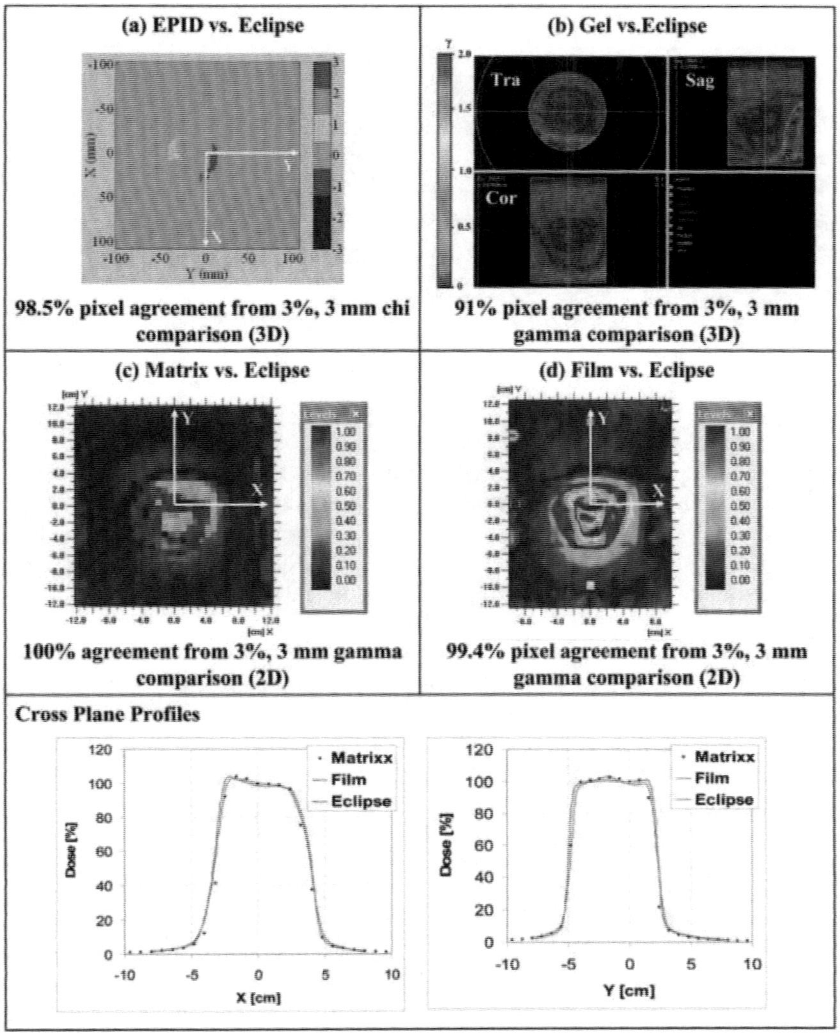

Figure 2: Dose comparisons for a prostate IMRT plan using different dosimeter tools. Dose evaluation maps for EPIDOSE, Matrixx and EBT2 film are for the coronal plane.

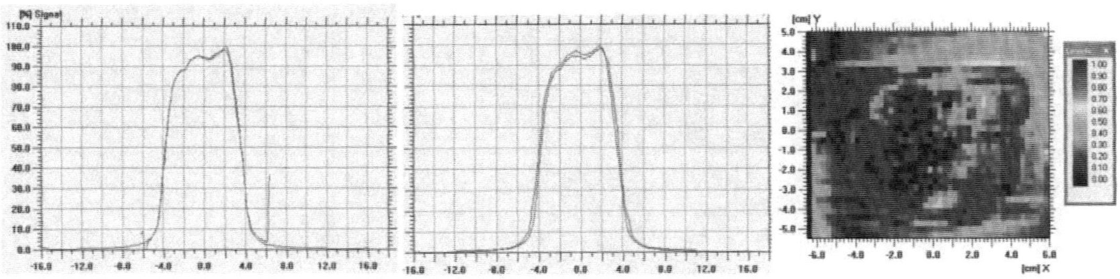

Figure 3: (left) EBT3 film (green) and Eclipse (red) dose profiles of a test prostate delivery, (centre) Matrixx (red) and Eclipse (green) dose profiles of the same delivery (on the Matrixx ion chamber measurement plane, which is 4 mm below the film plane), and (right) a 3%,3 mm 2D gamma comparison showing 100% agreement between film and Eclipse planar dose data.

Table 1: Initial absolute dose comparisons against Eclipse for a fixed point at/near isocentre

Plan	Matrixx	EPID	Chamber	Film	Gel
10 x 10 cm^2 open	-0.7	1.2	-0.8	-0.9	-
2 mm fixed width sliding window	5.8	-5.1	1.7	-1.7	-
5 mm fixed width sliding window	3.7	-3.0	0.04	-	-
20 mm fixed width sliding window	0.6	-0.3	-0.7	2.6	-
10 x 10 cm^2 open (7 field)	n/a	1.0	-1.2	-	-
6 x 6 cm^2 open (7 field)	n/a	0.9	1.8	-	1.0
5 mm fixed width sliding window (7 field)	n/a	-3.4	0.5	0.2	-
5 mm fixed width sliding window (7 field)	n/a	-0.1	-0.5	-	-
Prostate Plan at nominal gantry angles (7 fields)	n/a	-0.8	-1.6	-2.4	-0.5
Prostate Plan with gantry fixed at zero (7 fields)	-0.2	-0.6	-1.7	-1.9	-

4. Discussion & Conclusions

The data reported in table 1 give good indication of the absolute measurement capabilities of film and gel dosimetry. The greatest differences between measurement and plan dose (at a point set at, or near, isocentre, as dictated by the different measurement configurations) were observed in the Matrixx and EPID measurements of narrow sliding window deliveries. In determining the source of the Matrixx differences, film and ion chamber measurements (table 1, Fig.3) strongly suggest that the result is, in fact, a Matrixx over-response and not an under-calculation of Eclipse dose. This is consistent with the literature, which reports several factors contributing to an over-response of the Matrixx to highly modulated deliveries[4,5]. For the EPID sliding window measurements, it is possible that the observed under-response could be due to variable response of the imager to scatter dose and/or the limitations of the EPIDOSE reconstruction model (which has fitted parameters based on open field data)[1].

Some variations were observed in the EBT2 film data (see profiles in Fig.2), which after careful evaluation of scanner and film performance, have been mainly attributed to non-uniform film response. EBT3 film was investigated as an alternative, and found to be more reliable in its dosimetry. The disagreement in the low-to mid dose posterior region of the gel dosimeter for the prostate delivery (Fig.2) has been identified as primarily due to a resolvable issue in the gel calibration. Another set of gel and film measurements are forthcoming.

Overall, the Eclipse and measurement data from the different dosimeters are in good agreement for simple open field and relatively low modulation prostate IMRT deliveries. The value of the common cylindrical phantom system has also been demonstrated. Next steps include a similar cross-comparison for more highly modulated complex head and neck IMRT deliveries.

5. Acknowledgements

Research funding for this work has been provided by the Canadian Institutes of Health Research (CIHR) and the Cancer Centre of Southeastern Ontario.

6. References

[1] Ansbacher W 2006 *Med. Phys.* **33** 3369-82
[2] Olding T *et al* 2010 *Phys. Med. Biol.* **56** 1259-79
[3] Bakai A *et al* 2003 *Phys. Med. Biol.* **48** 3543-53
[4] Han S K Ng *et al* 2010 *Med. Phys.* **37** 3704-14
[5] Gago-Arias A *et al* 2012 *Phys. Med. Biol.* **57** 2005–20

Journal of Physics: Conference Series **444** (2013) 012053 doi:10.1088/1742-6596/444/1/012053

Optical laser scanning of a leucodye micelle gel: preliminary results of a 3D dose verification of an IMRT treatment for a brain tumor

J Vandecasteele[1] and Y De Deene[1,2]
[1]Department for Radiation Oncology and Experimental Cancer Research, Ghent University, De Pintelaan 185, 9000 Gent, Belgium
[2]Institute of Medical Physics, School of Physics, University of Sydney, Sydney NSW, Australia

E-mail: Jan.Vandecasteele@UGent.be

Abstract. In the present study an in-house developed leucodye micelle gel was used in combination with an in-house developed optical laser scanner for the 3D dose verification of an IMRT treatment of a pituitary adenoma. In an initial prospective study, a gel measured depth dose distribution of a square 6 MV photon beam was compared with an ion chamber measurement. In a second experiment, the gel and scanner were used to verify a clinical dose distribution on a recently installed linear accelerator. The calibration procedure is identified as the major source of dose deviations.

1. Introduction

To make a quantitative comparison between MRI and optical read-out techniques, a head phantom (figure 1) was constructed with a cylindrical cavity (10 cm in diameter) for the placement of a 3D gel dosimeter. In this exploratory study the optical read-out technique using an in-house developed micelle gel [1] was evaluated. Sources of error in the dose maps dose are identified and will be used in a final optimization.

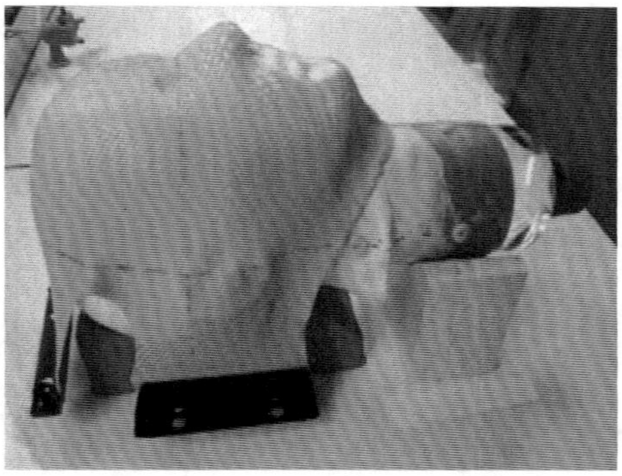

Figure 1: An in-house built head phantom with cylindrical cavity for the insertion of a 3D dosimeter.

2. Materials and methods

2.1. Gel fabrication

A leucodye micelle gel was fabricated consisting of 6% gelatin, 80 mM chloroform ($CHCl_3$), 50 mM sodium dodecyl sulphate (SDS), 5 mM trichloroacetic acid (CCl_3COOH) and 0.37 mM leucomalachite green (LMG) all dissolved in deionised water. Gelatin is dissolved in 60% (w/w) of the total water volume at room temperature and is left to swell for 10 minutes. Thereafter the gelatin-water solution is heated to 45°C. The remaining 32% of total water volume is used to dissolve SDS, CCl_3COOH, $CHCl_3$ and LMG. The solution is stirred for 30 minutes until all LMG is dissolved. After cooling down the gelatin-water solution to approximately 40°C, the two solutions are added together and stirred for 30 minutes in a dark room. Finally, the gel is poured into cylindrical Teflon recipients with a diameter of 9.6 cm and a height of 20 cm.

2.2. Depth dose measurement

A micelle gel dosimeter was irradiated with a 4 cm x 4 cm 6 MV photon beam to a total dose of 43.7 Gy at the dose maximum (SSD 90 cm) using a clinical linear accelerator (Elekta Synergy). The dose distribution in the micelle gel dosimeter was read-out using an in house built optical CT laser scanner: the OPTOSCAN (figure 2, [2]).

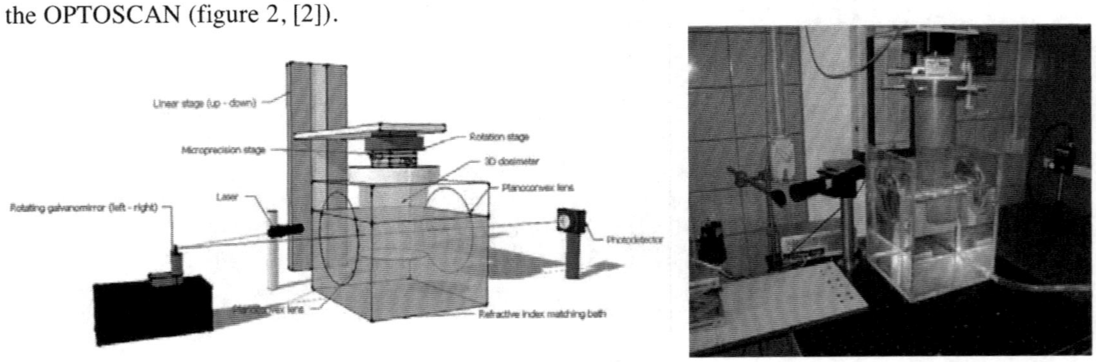

(a) **(b)**

Figure 2: Schematic overview of the in-house built optical CT laser scanner (a) and a photograph (b).

180 projections of 120 mm were recorded per slice with a phantom rotation increment of 2°. 600 spatial increments were acquired per projection yielding a profile resolution of 0.2 mm. All sinograms were averaged to a resolution of 1 mm. Twenty-three slices were acquired over a total range of 115 mm (each separated by 5 mm). The total scan time amounted to 90 minutes. The refractive index matching solution consisted of approximately 9% propylene glycol in water. In-house developed Matlab code was used to control the optical scanner and construct optical density images. All optical density images were calibrated to dose images via a linear dose response relationship determined using the optical density values of the region of dose maximum and zero dose. A first scan of the phantom was acquired 19 hours post-irradiation. A second scan was acquired 21 hours post-irradiation. The first 12 hours post irradiation the phantom was stored in a refrigerator at 4°C. Afterwards, the gel was kept at 23°C.

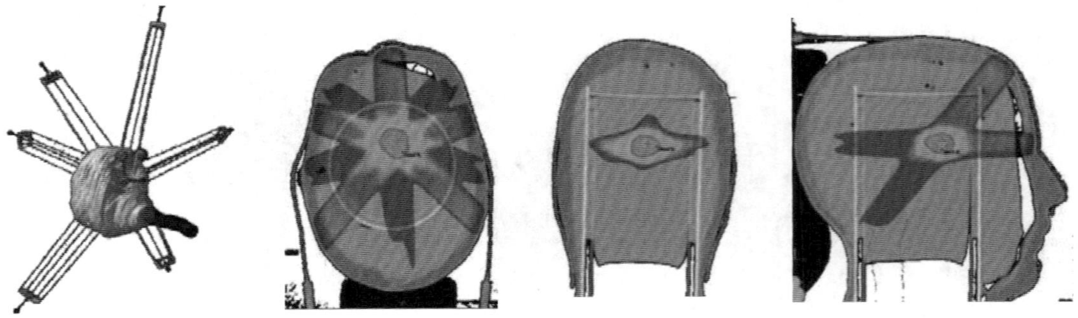

(a) (b) (c) (d)

Figure 3: An overview of the important anatomy relative to the 6 beams used for the IMRT treatment in (a). In b, c and d a transverse, coronal and sagittal slice through the irradiation isocentre are shown.

2.3. *Clinical experiment*

A 3D dose verification of an IMRT treatment for a brain tumour was performed with a leucodye micelle gel and the OPTOSCAN. A head phantom was casted from epoxy resin and filled with 6% gelatin gel. Inside this head phantom, a cylindrical cavity was made for the 3D dosimeter phantom. A helical CT scan was acquired of the head phantom filled with a 3D micelle gel dosimeter. Using these CT images an IMRT plan for a pituitary adenoma was planned in Eclipse (Varian) in collaboration with the radiation oncologist (figure 3). A CTV, PTV, and critical organs such as cerebrum, optic chiasm, eye lenses, optical nerves and brainstem were identified (figure 3a). A sliding window IMRT treatment was optimized for six beam angles of which 5 angles were coplanar. A total dose of 15.8 Gy was planned at the isocentre calculated by the AAA_8908 algorithm (figure 3b, 3c and 3d).

The irradiation was performed on a Varian Clinac 2300IX (figure 4). Prior to the treatment delivery, a cone beam CT was acquired to match the phantom position to the treatment planning CT. The phantom was irradiated with 3 fractions resulting in a dose of 47.4 Gy to the isocentre.

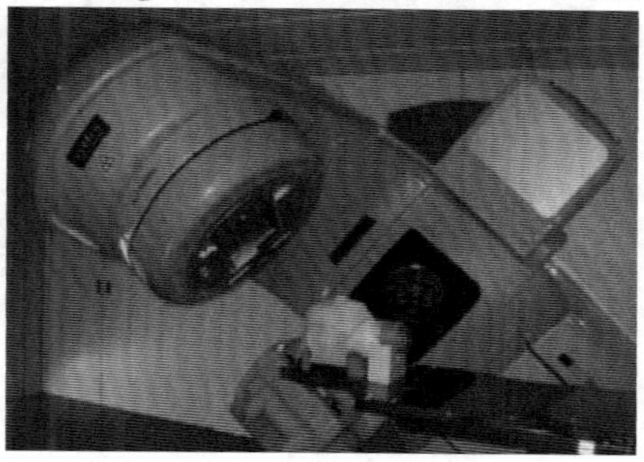

Figure 4: Irradiation treatment of the gel dosimeter head phantom

A stack of 30 slices (2 mm slice distance) of the irradiated micelle gel dosimeter was optically acquired using the same imaging parameters as in the first experiment. The total scan time amounted to 119 minutes. The optical density maps were calibrated to dose maps relative to the isocentre dose predicted by the TPS. An ion chamber measurement is planned in the near future to independently verify the delivered dose at the isocentre.

3. Results and discussion

3.1. *Depth dose measurement*

Both gel measurements show excellent agreement with the ion chamber recorded dose values (figure 5). Remarkably no clear signs of diffusion of the leucodye molecules can be observed post-irradiation (figure 5a). A minor change has been made in the fabrication protocol. Instead of mixing the leucodye with pure chloroform both products are separately added to the water - SDS - CCl_3COOH solution. A precise measurement of the diffusion coefficient will be performed in the near future.

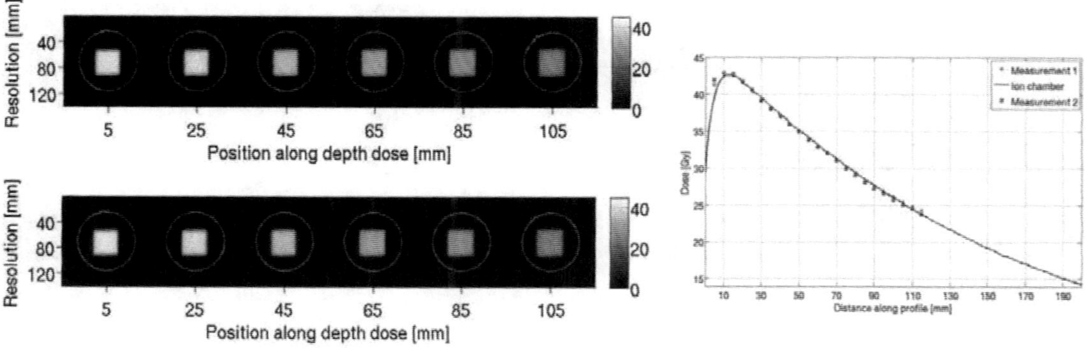

(a) (b)

Figure 5: In (a) 6 out of 23 measured dose maps are shown of the fist measurement 19 minutes post-irradiation (top) and the second measurement 21 hours post-irradiation (bottom). The mean of the measured dose in a circular ROI in the centre of the square field with radius 0.7 cm is plotted against an ion chamber dose measurement in (b).

3.2 Clinical experiments

An edge artefact caused by a small mismatch of the refractive index matching solution results in a large dose deviation in an annular region of 0.5 cm from the edge of the phantom. This region was omitted from our results. In figure 6a and 6b, the TPS calculated and gel measured dose distribution near the isocentre can be seen respectively. In figure 6c a percentage difference map is shown were maximum dose underestimations in the order of 6% are found. In figure 6d a 3D gamma evaluation using the 3%/3mm criterion is shown displaying the failing pixels.

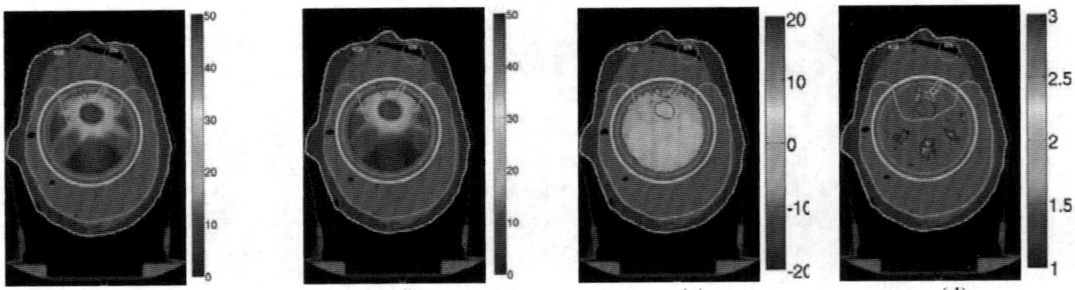

(a) (b) (c) (d)

Figure 6: The dose distribution of the TPS (a), the gel measurement (b), the difference map between the TPS and gel (expressed in percent) (c) and a gamma map (3%/3mm, d) are shown as an overlay figure on the CT images at the level of the isocentre. Anatomical structures are shown: cerebrum, eyes, eye lenses, optical nerves, optic chiasm, PTV and CTV.

Additionally, in a coronal and sagittal direction a good qualitative agreement was found (figure 7). However, on average 5% of the pixels in the scanned volume (300ml) deviate by more than 7% in absolute dose from the TPS. In the total scanned volume 9% of the pixels fail the 3%/3mm criteria, while only 2.8% of the pixels fail the 5%/3mm criteria. The majority of these pixels are located in the low dose regions.

These errors are presumed to be attributed to the linear calibration process. Furthermore, temperature during readout and post-irradiation stability need to be further investigated.

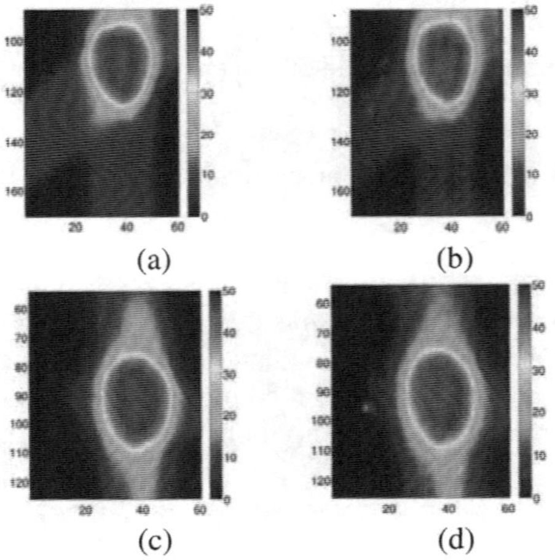

(a) (b) (c) (d)

Figure 7: A comparison between a TPS calculated (a,c) and gel measured (b,d) slice along sagittal (a,b) and coronal (c,d) direction through the isocentre.

4. Conclusions

The accuracy of our system amounts to approximately 7%. However, the origins of dose deviations are being investigated. Accurate measurements within 5% in a total 3D volume are expected to be feasible.

5. Acknowledgements

This research is supported by the Institute for the Promotion of Innovation though Science and Technology in Flanders (IWT -Vlaanderen).

6. References

[1] Vandecasteele J *et al* 2011 *Phys. Med. Biol.* **56** 627

[2] Vandecasteele J and De Deene Y 2009 *J. Phys.: Conf. Ser.* **164** 012024

Journal of Physics: Conference Series **444** (2013) 012054 doi:10.1088/1742-6596/444/1/012054

High resolution dosimetry in monoenergetic proton beam therapy on a normoxic polymer gel: the importance of high spatial resolution for reduced Bragg-Peak-quenching.

A Berg[1,2], M Wieland[1,2,3], J Naumann[4] and O Jaekel[4]
[1]Center for Med. Phys. & Biomed. Eng., Med. Univ.of Vienna, Vienna, Austria.
[2]High-Field MR-Centre of Excellence, Medical University of Vienna, Austria
[3]Institut für Radioonkologie, Sozialmedizinisches Zentrum Ost - Donauspital Wien, Austria
[4]Heidelberg Ion Therapy Center (HIT), Germany

E-Mail: andreas.berg@meduniwien.ac.at

Abstract. Proton ion beam therapy demands for high resolution dosimetry due to the high dose gradients present in lateral confinement and final Bragg-peak. In polymer gels the reduction of the linear dose response in the area of the Bragg-peak is reported (Bragg-peak quenching), which is assumed to be mainly due to the high linear energy transfer (LET). We here investigate the impact of the spatial resolution in T2-mapping for accurate Magnetic Resonance Imaging (MRI)-based polymer gel dosimetry in the Bragg-peak for monoenergetic ion beams. We implemented MR-protocols for T2-mapping at microscopic resolution on a High-Field 7T human MR-scanner using an insert gradient system and sensitive rf-coils. The best results are obtained for an optimzed polymer gel based on THPC with an optimized MR-protocol for reduced measurement time and sufficient SNR at 0,547 mm pixel size. The dose in the fine Bragg-peak could be measured correctly for a monoenergetic proton beam as confirmed by Monte Carlo dose simulations. Such high spatial resolutions at minimum are necessary for an accurate measurement of the dose in the sharp Bragg-peak for monoenergetic ion beams. We demonstrate that at higher pixel size the dose levels may be underestimated due to spatial averaging in MRI-based polymer gel dosimetry.

1. Introduction

Hadron beam irradiation concepts aim on the fine dose placement at the tumour site leaving nearby healthy tissue at minimum dose. The main advantages of ion beams are represented by the possibility for precise lateral confinement and dose enhancement at penetration depth ("Bragg-peak", see fig. 1). Ionization-chambers represent the standard for the verification of treatment planning systems for proton therapy [1]. Also 2D-arrays of pin-point ionization chambers, film dosimetry and polymer gel dosimetry are used for the 3D-dosimetry of complex fields [2-4]. With regard to 3D-dose verification for patient treatment high spatial resolution is possible but critical [5-7]. The MR-Parameter T2 has shown to be most sensitive to the polymerization process in polymer gels subsequent to irradiation [8-10]. However the application of MR-based polymer gel dosimetry (MRPD) on ion beams is restricted by mainly three physical principles: 1.) the spatial resolution in T2-MRI [5] 2.) the signal-to-noise ratio (SNR) in MRI and consequently dose 3.) edge enhancement effects due to monomer diffusion [11] and 4.) a quenching of the dose response in the Bragg-peak-regime due to high LET [12].

In this contribution we investigate: 1.) the point spread function (PSF) in T2-micro-imaging on a human high-field (7T) MR-scanner equipped with a micro-imaging insert, 2.) the achievable T2-noise within reasonable MR-scanning time (11min), 3.) the influence and importance of the spatial resolution in MRI on the detected dose level in the Bragg-peak. Moreover we present a quantitative comparison between the depth dose profile obtained by a fast multi-slice T2-protocol with the result of Monte-Carlo dose Simulations (MCS) for a "normoxic" polymer gel [13, 14] modification featuring a reduced linear energy transfer (LET)-quenching. A 3-dimensional (3D) dose data-set is obtained for a mono-energetic proton beam with a voxel-size, 3 orders of magnitude (10^3) smaller than those available for 2D-arrangements of pin-point ionization-chambers used for the verification of treatment plans in intensity modulated radiotherapy with heavy ions [1].

2. Materials and methods

2.1 MRI

The high spatial resolution in parallel with acceptable SNR was achieved on a 7T human MR-scanner, equipped with a small sized (i.d. 90mm) strong gradient system (G=750 mT/m) and a linear TR-birdcage coil (i.d. 72 mm) [15].

2.1.1 High-resolution protocol

For investigating the spatial resolution in quantitative T2-maping the point spread-function around a Boron wire (o.d. 100 μm) in a polymer gel irradiated with neutrons (T2=74ms) was evaluated using a high-resolution multi-slice-multi-echo sequence. The MR-measurement parameters for this high-resolution protocol are listed in the following: FOV:$(30\text{mm})^2$, Mtx: 320^2, Voxel-size: 94x94x1000 μm^3, TE=10ms, 20 echoes.

2.1.2 3D-MRI-protocol

For investigating the 3D-dose response of several types of polymer gels (8 flacons in parallel) a protocol with reduced but still high spatial resolution was established with acceptable SNR (TM=11 min); FOV $(50\text{mm})^2$, Mtx: 128x128x23, Vs: 547x547x1000 μm^3.

2.2 Polymer gels

Several sets of "normoxic" polymer gels with different additives for sensitivity and dose range modifications are maufactured. The presented data belongs to a variant using THPC [16] as oxygen scavenger, methacrylic acid (6% w/w) as monomer, gelatin (14%w/w) and water.

2.3 Irradiation

Heidelberger Ion-beam-Therapy-center (HIT); Monoenergetic (E=59,8MeV) proton beam for human therapy; beam diameter: 6cm; D = 0-4Gy (plateau). The Monte-Carlo-simulations (MCS) are performed at fine raster (VS:500x500x500 μm^3) to resolve the fine Bragg-peak area.

3. Results

3.1 The full-width-half-maximum (FWHM) for the PSF in high resolution T2-maps (ps: 94x94μm^2) was calculated to be FWHM=214μm, assumed to be sufficiently high for the fine Bragg-peak.

3.2 The MR-protocol, optimized for measurement time and scan volume offered low T2-noise (ROI analysis: about 1-3%, see. also Fig. 1 for dose noise). The results concerning depth dose distribtion and Bragg-peak enhancement ratio are sensitive to the polymer gel composition in specific oxygen scavenging. The measurement data for several different types of polymer gels exhibits also aging effects, when the normoxic polymer gels are stored at room temperature.

Journal of Physics: Conference Series **444** (2013) 012054 doi:10.1088/1742-6596/444/1/012054

3.3 Best results are obtained for a THPC-scavenging polymer gel composition. The medium slice taken from the 3D-data-dose set (fig. 1b) evidences the possibility for 3D-high-resolution and low-noise dose-imaging for this polymer gel type. Using the optimized MR-protocol the depth dose distribution was calculated from calibrated data between 0-4 Gy using linear inter- and extrapolation for the proton beam. It is compared to the results of MCS (fig.1). The depth is corrected for the density of the polymer gel. MCS data is shifted in depth such that the Bragg-peak coincides with the dose maximum from experiment.

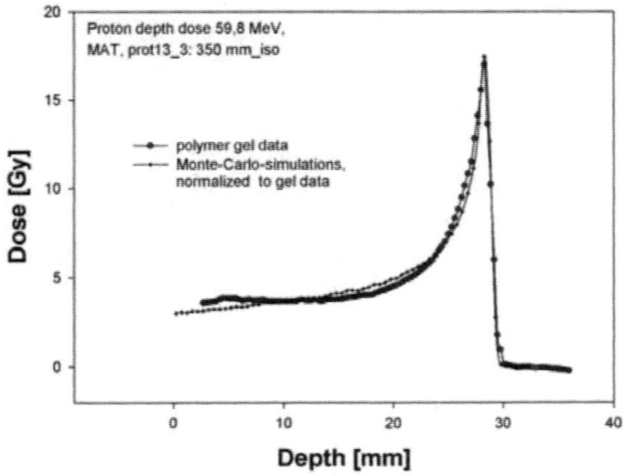

Figure 1a: Depth dose profile as obtained from polymer gel data (fig. 1.b). For comparison the results of Monte-Carlo-simulations normalized to the plateau-level at 10 mm depth are indicated.

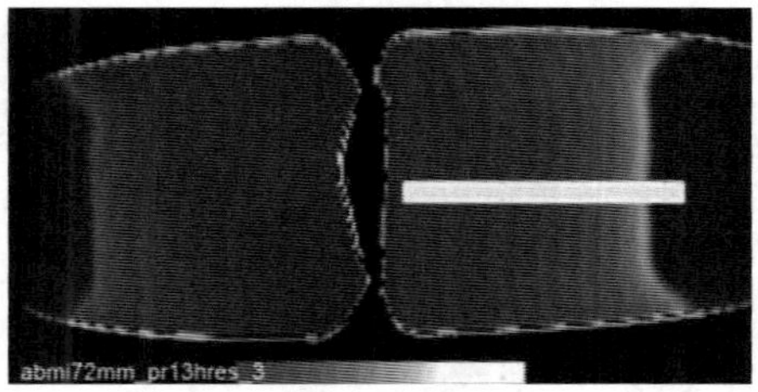

Figure 1b: Dose image calculated from the medium slice of a 3D-T2-data-set after calibration. **Right:** polymer gel irradiated at a dose level of 4Gy (plateau). The path of the profile for fig. 1 and 2 is indicatedTh. e proton beam has entered the sample from left. The bright line at right indicates the Bragg-peak dose enhanced penetration depth. **Left:** dose image of a sample irradiated at D = 2 Gy with opposite direction.

The depth dose distribution for an irradiation at 4 Gy is characterized by a slightly more flat distribution in the plateau regime with comparison to MCS. The reason is not yet clear. It might be related to a slight divergence of the proton beam, which is not included in MCS. The difference in Bragg peak is therefore sensitive to the normalization area of the MCS dose level.

The detected dose in the Bragg-peak (max.dose) is sensitive to the spatial resolution of MRI (pixel size). The result of voxel volume avaraging connected to reduced MRI-measurement matrix size and subsequent increase in pixel size is plotted in fig. 2.

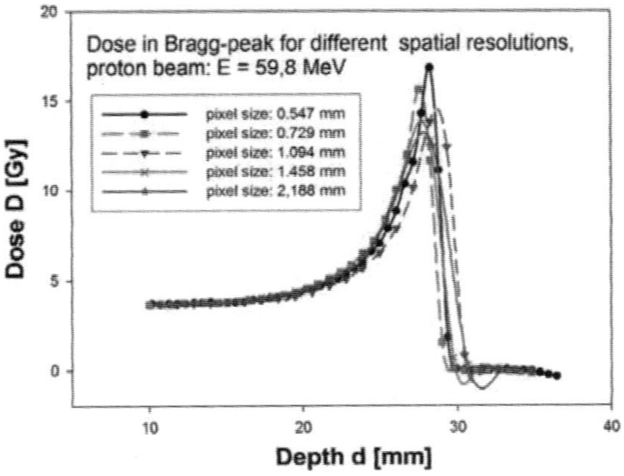

Figure 2: Depth dose profiles (lines are spline interpolated) for different spatial resolutions (pixel size). The detected dose level in the Bragg peak is strongly underestimated when the pixel size is increased from ps=547 μm (black line) up to 2188 μm (green line).

Reducing MR-scan resolution from ps: 547μm down to 1094 μm reduces the measured dose level by about 18% relative to the more accurate measurement (fig. 2). Edge enhancement in the polymer gel [11] was minimized by gel composition but could not be quantified.

4. Discussion/Conclusion

High spatial resolution in dosimtric imaging is critical for the accurate determination of the dose level in regions of high dose gradient as the Bragg-peak area. Using additional hardware components on a high-field human 7T-scanner T2-imaging can be implemented even at microscopic pixel size. Quantitative multislice-T2-micro-imaging at reduced spatial resolution and tolerable measurment time can be performed and applied to high-resolution 3D-polymer gel dosimetry for proton beam therapy, which poses high demands on spatial precision and signal/noise. The dose response of the specific polymer gel recipe is very sensitive to the oxygen scavenger concentration. A high resolution and pixel size of about 500 μm at minimum is necessary for an accurate measurement of the dose in the Bragg-peak regime. Otherwise the dose levels are underestimated in the Bragg-peak dose enhanced region due to spatial averaging of the dose in the voxel. The absolute accuracy in the Bragg-peak may be also be influenced by edge enhancement [11] in the polymer gel, by LET-quenching [12] and the temporal stability of the polymer gel dose response.

5. References

[1] ICRU 1998 Report No. **59** Bethesda, MD
[2] Karger C P *et al* 1999 *Med. Phys.* **26** 2125-32
[3] Baldock C *et al* 2010 *Phys. Med. Biol.* **55** R1-63
[4] Gustavsson H *et al* 2004 *Phys. Med. Biol.* **49** 227-41
[5] Berg A *et al* 2005 *Biomedizinische Technik/Biomedical Engineering* **50** S1 118-9
[6] Berg A *et al* 2004 *Phys. Med. Biol.* **49** 4087-108
[7] Zeidan O A *et al* 2010 *Med. Phys.* **37** 2145-52
[8] De Deene Y and Baldock C 2002 *Phys. Med. Biol.* **47** 3117-41
[9] Lepage M *et al* 2002 *Phys. Med. Biol.* **47** 1881-90
[1] Hurley C *et al* 2006 *Nucl. Instrum. Methods Phys. Res. A* **565** 801-11
[11] Vergote K *et al* 2004 *Phys. Med. Biol.* **49** 4507-22
[12] Gustavsson H *et al* 2004 *Phys. Med. Biol.* **49** 3847-55
[13] De Deene Y *et al* 2002 *Phys. Med. Biol.* **47** 3441-63
[14] De Deene Y *et al* 2002 *Phys. Med. Biol.* **47** 2459-70
[15] Berg A *et al* 2010 *Proc. ISMRM 2012* (Stockholm, Sweden, May 2010) 1048
[16] Bayreder C *et al* 2006 *Med. Phys.* **33** 2506-18

Dosimetric assessment of the PRESAGE dosimeter for a proton pencil beam

C-S Wuu[1], Y Xu[2], X Qian[1], J Adamovics[3], E Cascio[4] and H-M Lu[4]
[1]Department of Radiation Oncology, Columbia University, New York, NY, USA
[2]Department of Radiation Oncology, University of Pittsburgh Medical Center, Pittsburgh, PA, USA
[3]Department of Chemistry and Biology, Rider University, Lawrenceville, NJ, USA
[4]Department of Radiation Oncology, Massachusetts General Hospital, Harvard Medical School, Boston, MA, USA

E-mail: csw6@columbia.edu

Abstract. The objective of this study is to assess the feasibility of using PRESAGE dosimeters for proton pencil beam dosimetry. Two different formulations of phantom materials were tested for their suitability in characterizing a single proton pencil beam. The dosimetric response of PRESAGE was found to be linear up to 4Gy. First-generation optical CT scanner, OCTOPUS™ was used to implement dose distributions for proton pencil beams since it provides most accurate readout. Percentage depth dose curves and beam profiles for two proton energy, 110 MeV, and 93 MeV, were used to evaluate the dosimetric performance of two PRESAGE phantom formulas. The findings from this study show that the dosimetric properties of the phantom materials match with basic physics of proton beams.

1. Introduction

A proton beam has the advantage in its characteristic distribution of dose with depth. The depth dose peaks to highest value near the end of its range followed by a rapid falloff to zero. These physical characteristics of proton beams offer higher dose conformity to the tumor and lower dose to the surrounding normal tissues than a photon beam can achieve. There are two beam delivery systems in terms of beam spreading to its desired field cross section: 1) passive beam spreading by thin sheets of high-atomic materials; or 2) pencil beam scanning by magnets. Proton pencil beam scanning can provide precise and efficient treatment delivery for both conventional and intensity modulated proton therapy (IMPT) techniques. Detailed three-dimensional dosimetric information for a proton beam is crucial for accurate treatment planning design and treatment delivery due to the sharp dose falloff at the end of the beam range and laterally at the field edges [1-3]. It is more challenging to measure the 3-D dose distribution of a pencil beam (1 - 1.5 cm diameter) because of its small field size as well as the complexity of dose changes in various directions.

Both polymer gels [4] and PRESAGE [5, 6] radiochromic plastic dosimeter, in conjunction with optical CT scanner [7, 8], have been employed to verify 3-D dose distribution for complex treatments with photon beams, such as IMRT and stereotactic radiosurgery [9, 10]. These studies have shown promising results in providing a 3-D dosimetry tools for photon external beam treatments. There are some recent studies on proton dosimetry using polymer gels [11] and PRESAGE dosimeters [12], with optical CT scanners as readout systems. However, there has not been any report on dosimetry of a single proton pencil beam. This study is the first investigation in characterizing the dosimetry of a single proton pencil beam using PRESAGE radiochromic dosimeters and an optical CT scanner.

2. Materials and Methods

In this study, cylindrical PRESAGE® dosimeters and an optical CT scanner, OCTOPUS™ (MGS Research Inc., Madison, CT) were used to implement dose distribution measurements for proton pencil beams at the Northeast Proton Therapy Center (NPTC), Massachusetts General Hospital (MGH). Each cylindrical PRESAGE phantom of 11.4 cm diameter and 10 cm height, modified to optimal dose-response characteristics, was used for proton pencil beam measurements. Two different dosimeter formulas were used for measurements. One formula has a physical density of 1.09 gm/cm^3 and an average atomic number of 8.9, and the other formula has a physical density of 1.06 gm/cm^3 and an average atomic number of 7.5. One objective of this study is to assess which PRESAGE formulation can be clinically used as a proton phantom. Two single proton pencil beams, with energies of 110 MeV and 93 MeV, were used for studies. These two proton energies were chosen because their ranges, 9.5 cm and 6.5 cm, respectively, are within the dosimeter length dimension. The dosimeters were maintained at a temperature of 7°C before and after irradiation. The dosimeters were set to reacclimatize to room temperature of about 22°C before irradiation and optical-CT readout.

In order to exam the optical density response of PRESAGE dosimeter versus proton dose, the optical density dose response was studied by irradiating the dosimeters with a 4 x 4 cm square field, uniform scattered clinical beam, to various known doses (up to 400 cGy). Four cylindrical phantoms were used for proton pencil beam dosimetry study. Each phantom was irradiated with three horizontal proton pencil beams (FWHM 1.1-1.4 cm) from one end, since the size of the phantom can easily accommodate three pencil beams. For each energy, a wide range of doses from 50 cGy up to 150 cGy at the entrance were delivered among 6 beams. This arrangement allows us to find the optimal dose range for best dose image reconstruction. The dosimeter sample was scanned with 1mm pixel resolution using the optical CT scanner, OCTOPUS™. Details of the operating principles, 3-D dose reconstruction, and instrument architecture of the optical scanner were described previously [13, 14]. For comparative dosimetry, Gafchromic films, placed at various depths in a solid water phantom, were irradiated with a proton pencil beam using the same beam parameters.

2. Results and Discussion

As shown in figure1, the optical density of the irradiated PRESAGE versus dose was found to be linear from 0 to 2 Gy. Therefore, the optical density distribution in the irradiated phantoms represents the dose distribution.

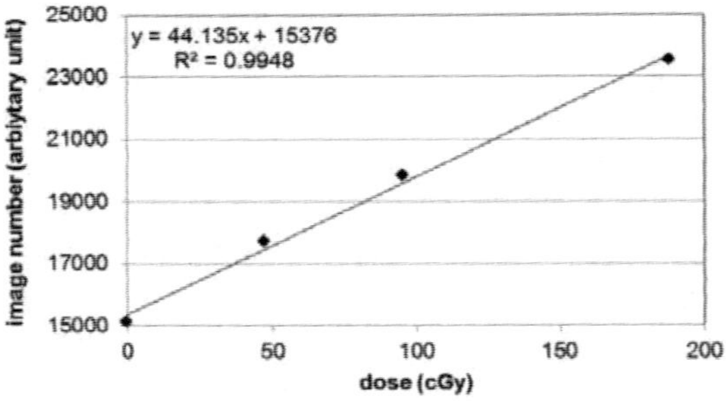

Figure1: Dose response curve for PRESAGE, calibrated with 4x4 cm uniform scattered beams at various doses.

Figure 2: PRESAGE phantom irradiated with proton pencil beams. The Bragg peak with a sharp dose falloff can be observed.

Journal of Physics: Conference Series **444** (2013) 012055 doi:10.1088/1742-6596/444/1/012055

Figure 2 shows a clear image of Bragg peak for a 93 MeV proton pencil beam. The phantom formula used in this irradiation has a physical density of 1.06 gm/cm^3, and an average atomic number of 7.5. Note that there is a beam spread near the end of the proton range.

In figure 3, the integral dose at each depth was obtained over the beam cross section. The phantom depth was scaled by its physical density relative to water (1.09), and yields an equivalent depth in water. The integral doses for PRESAGE, at the Bragg peak and its upstream region, are 10-25% lower compared to the results obtained from film measurements in solid water. The differences can be largely attributed to the mass stopping power ratios between PRESAGE material and water. Mass stopping power for protons is greater in low Z materials than in high Z materials. The phantom material used for the study in figure 3 has an average atomic number of 8.9, and a physical density of 1.09 gm/cm^3.

Figure 4 presents the integral percentage depth dose measured from PRESAGE for a 93 MeV proton pencil beam. The 93 MeV proton pencil beam was achieved by passage of 110MeV proton pencil beam through Lucite slabs with density thickness of 2.125 gm/cm^2. As expected, the distal dose falloff for 93 MeV is much less sharper than that for 110 MeV. The phantom material used for the study in figure 4 has an average atomic number of 7.5, and a physical density of 1.06 gm/cm^3. Future study will be focused on the development of a new formulation that is optimal for proton pencil beam dosimetry.

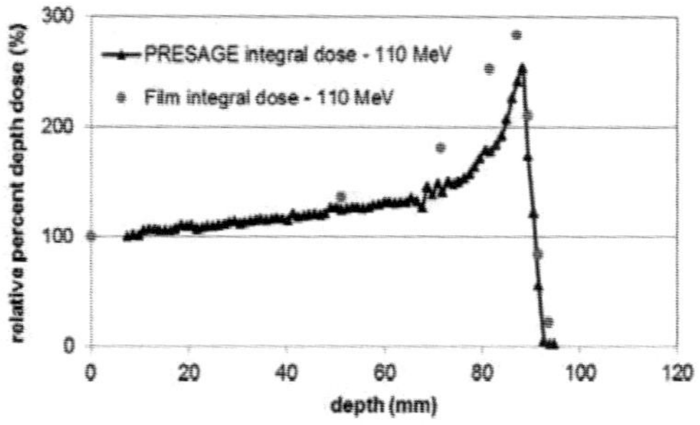

Figure 3: Percentage depth dose measured with PRESAGE dosimeter and Gafchromic film for a 110 MeV pencil beam.

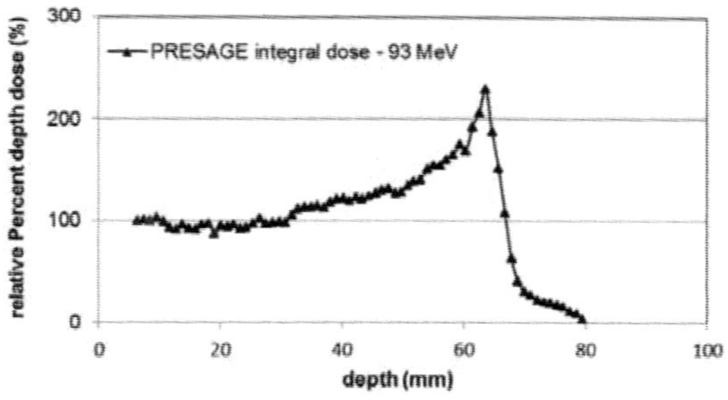

Figure 4: Percentage depth dose measured with dosimeter for a 93 MeV pencil beam.

The relative dose profiles at different depths for a 110 MeV pencil beam was plotted in figure 5. The lateral spread increases with the depth since scattering in a given material increases as the proton slows down. The angular spread is approximately a Gaussian distribution. Figure 6 shows 2-D dose distribution in coronal direction. The FWHM is between 1.3-1.4 mm. The lateral dose falloff compared to the results from film measurement is under investigation.

3. Conclusion

We have evaluated the dosimetric performance of PRESAGE dosimeters for proton pencil beam dosimetry, at 110 MeV and 93 MeV. The findings from this study suggest that it is feasible to use PRESAGE dosimeter for proton pencil beam study. Future study will be focused on comparison with proton pencil beam data, as well as evaluation of dosimeter formulation.

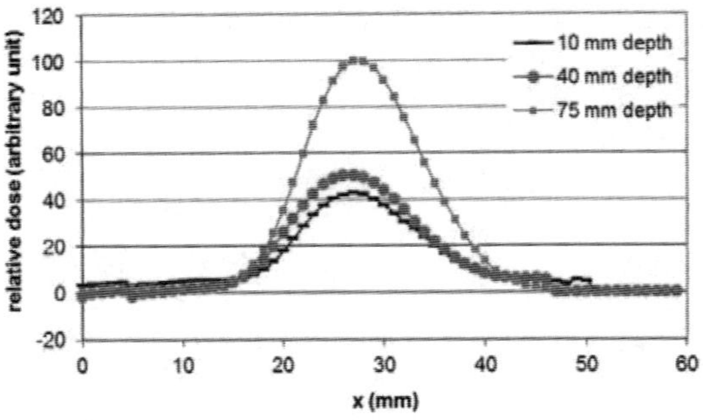

Figure 5: Dose profiles at different depths for a 110 MeV proton pencil beam, obtained from PRESAGE measurements.

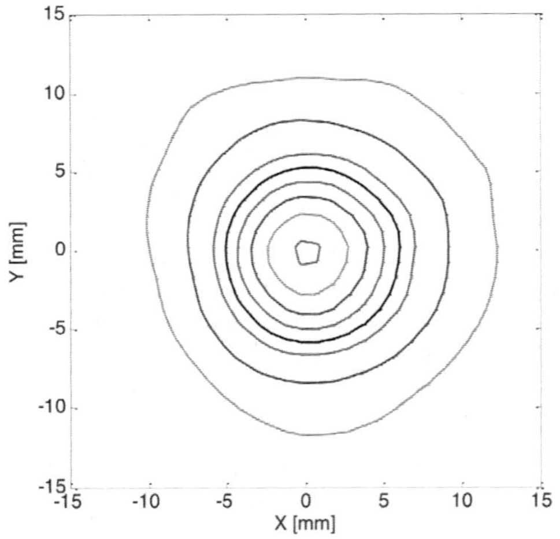

Figure 6: Coronal isodose lines for a 110 MeV proton pencil beam at 7.5 cm depth, from PRESAGE measurements (99%, 90%, 80%, 70%, 60%, 50%, 30%, 10%)

4. References

[1] Miller D W 1995 *Med. Phys.* **22** 1943-54

[2] Gotschalk B and Pedroni E 2008 In: Delaney TF, Kooy HM, eds. Proton and Charged Particle Radiotherapy Philadelphia: Lippincott Williams & Wilkins 33-49

[3] Kooy H M *et al* 2008 In: Delaney TF, Kooy HM, eds. Protons and Charged Particle Radiotherapy. Philadelphia: Lippincott Williams & Wilkins 70-107

[4] Baldock C *et al* 2010 *Phys. Med. Biol.* **55** R1-63

[5] Brown S *et al* C 2008 *Appl. Radiat. Isot.* **66** 1970-74

[6] Gorjiara T *et al* 2011 *Med. Phys.* **38** 2265-74

[7] Bosi S *et al* 2007 *Phys. Med. Biol.* **52** 2893–903

[8] Bosi S G *et al* 2009 *Phys. Med. Biol.* **54** 275-83

[9] Wuu CS and Xu Y 2006 *Med. Phys.* **33** 1412-1419

[10] Oldham M *et al* 2008 *Med. Phys.* **35** 2072-80

[11] Zeidan O A *et al* 2010 *Med. Phys.* **37** 2145-52

[12] Zhao L *et al* 2010 *J. Phys.: Conf. Ser.* **250** 1

[13] Xu Y *et al* 2003 *Med. Phys.* **30** 2257-63

[14] Xu Y *et al* 2004 *Med. Phys.* **31** 3024-33

Feasibility study of proton-based quality assurance of proton range compensator

S Park[1], C Jeong[1], BJ Min[1], J Kwak[2], J Lee[3], S Cho[3], D Shin[1], YK Lim[1], SY Park[4] and SB Lee[1]
[1]Proton Therapy Center, National Cancer Center, Republic of Korea
[2]Department of Radiation Oncology, Asan Medical Center, Republic of Korea
[3]Department of Nuclear and Quantum Engineering, KAIST, Republic of Korea
[4]Proton Therapy, McLaren Cancer Institute – Flint, USA

E-mail: seyjoon@ncc.re.kr

Abstract. All patient specific range compensators (RCs) are customized for achieving distal dose conformity of target volume in passively scattered proton therapy. Compensators are milled precisely using a computerized machine. In proton therapy, precision of the compensator is critical and quality assurance (QA) is required to protect normal tissues and organs from radiation damage. This study aims to evaluate the precision of proton-based quality assurance of range compensator. First, the geometry information of two compensators was extracted from the DICOM Radiotherapy (RT) plan. Next, RCs were irradiated on the EBT film individually by proton beam which is modulated to have a photon-like percent depth dose (PDD). Step phantoms were also irradiated on the EBT film to generate calibration curve which indicates relationship between optical density of irradiated film and perpendicular depth of compensator. Comparisons were made using the mean absolute difference (MAD) between coordinate information from DICOM RT and converted depth information from the EBT film. MAD over the whole region was 1.7, and 2.0 mm. However, MAD over the relatively flat regions on each compensator selected for comparison was within 1 mm. These results shows that proton-based quality assurance of range compensator is feasible and it is expected to achieve MAD over the whole region less than 1 mm with further correction about scattering effect of proton imaging.

1. Introduction

Proton therapy offers some distinct advantages compared to photon therapy. This is because the proton beam's enhanced dose deposition region, known as the Bragg peak, may be exploited to irradiate tumor while the rapid fall-off in dose just beyond the Bragg peak may be exploited to increase sparing of normal tissue from damage (Smith 2006). As in conventional radiotherapy, treatment fields are shaped to a desired target profile using custom milled blocks generally made of brass. In addition to conventional blocks, a patient-specific device, known as a range compensator, is required in proton therapy to modify the beam range according to each patient's anatomy and tumor shape. Compensators are field modifiers placed between the radiation source and the patient's skin to modulate the beam range. Using a compensator, the distal end of a high-dose region in the dose distribution can be made to closely conform to the shape of the target. Therefore, it is critically important to ensure that compensators are manufactured correctly as planned. Recently, Li *et al* (2010) and Yoon *et al* (2008) proposed a CT-based compensator QA tool. In this study, we present a novel method for the compensator QA, providing for the first time the possibility of using proton beam. Proton beam was modulated to have a photon-like PDD so proton energy can be deposited linearly according to depth. Comparisons were made between coordinate information of compensator extracted from DICOM RT and proton imaging of compensator on the EBT film. We evaluate the agreement using mean absolute difference (MAD).

2. Materials and methods

2.1. Proton therapy system and beam current modulation (BCM)

At the Korea National Cancer Center, an Ion Beam Applications (IBA) PROTEUS 235 proton therapy machine has been operating since 2007. Our proton system features a 230 MeV proton cyclotron, a beam line, two gantry rooms, and a fixed room. Our cyclotron is 220 tons in weight, 210 cm in height, and 434 cm in diameter.

The spread-out Bragg peaks, which are typical of double scattering and uniform scanning treatment modes, are formed by integration of individual Bragg peaks with different ranges. The relative heights of these pristine Bragg peaks can be adjusted by controlling the incident proton beam current as a function of small wheel angle. A database containing the BCM is recorded for each hospital configuration; the values are determined by empirical measurement conducted by the manufacturer when the equipment is installed.

In this study we modified BCM to make photon-like proton beam. The individual Bragg curves were multiplied by the new weighting factors and the beam current intensity was modulated in synchronization with range modulation wheel rotation, providing a nearly linear slope of depth dose distribution (see figure 1). This depth dose distribution similar to that of x-ray attenuation was measured by ion chamber in the water phantom.

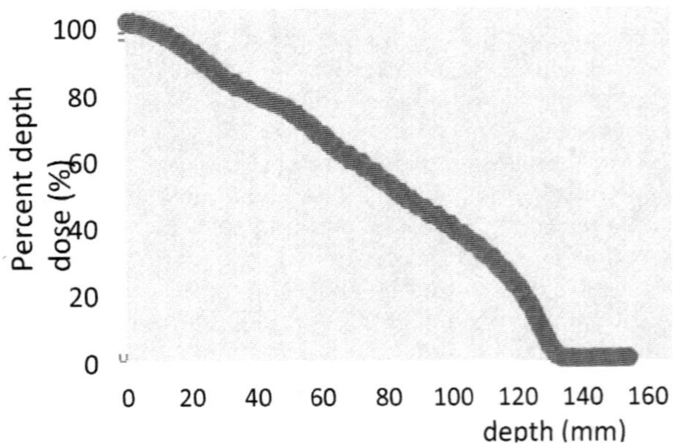

Figure 1: Percent depth dose of a photon-like modulated proton beam

2.2. Proton imaging of range compensator

Two proton range compensators were randomly chosen from patients who were treated with proton radiotherapy at our institution. The compensator matrix is produced within the calculation process in TPS (Eclipse; Varian Medical Systems, Salt Lake City, UT) and is then converted into a drilling pattern based on drill bit definitions found in a special milling machine parameter. The proton compensators were made of methyl methacrylate and manufactured using a milling machine (V-CNC500; Cincinnati Lamb, Birmingham, England) with a drill of radius 2.5 mm.

Geometry information of each range compensator comes directly from the DICOM RT plan and was imported to MATLAB software (Mathworks, Natick, MA) (see figure 2). Maximum depth of each compensator was 35.6 and 50.2 mm. Depth information in the z-axis were to be compared with optical density (OD) information of the gafchromic EBT film (International Specialty Products, Wayne, NJ). In order to make this comparison possible, step phantoms ranging from 4 mm to 50 mm in 2 mm increments were irradiated by photon-like proton beam on the EBT film to generate OD-depth calibration curve as a lookup table.

After converting OD to given depth, six flat region was chosen and the mean absolute difference (MAD) was calculated to quantify the similarity of chosen area between DICOM and film images:

$$MAD = \frac{1}{mn}\sum_{p=1}^{m}\sum_{q=1}^{n}\left| I_{DICOM}\left[p,q\right] - I_{EBT}\left[p,q\right]\right| \qquad (1)$$

The m and n is maximum pixel number of x-axis and y-axis. I_{DICOM} and I_{EBT} are the pqth depth for the DICOM and EBT.

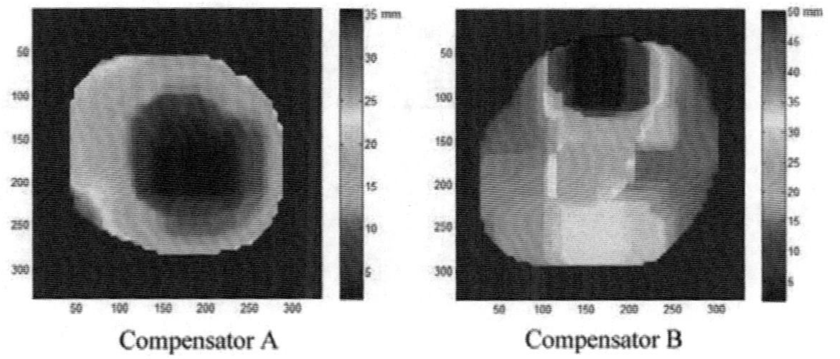

Figure 2: Two range compensators extracted from RT DICOM.

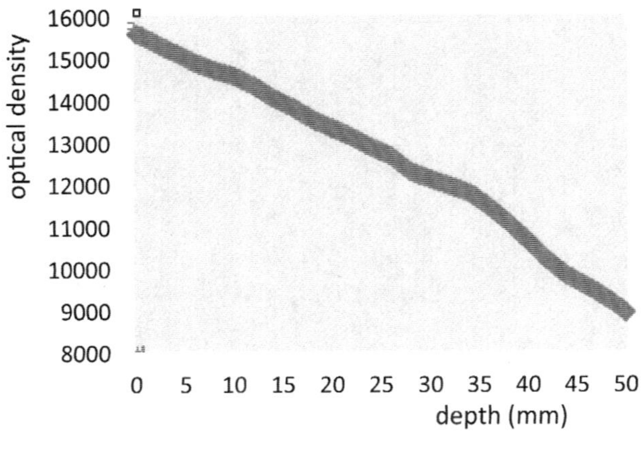

Figure 3: OD-depth calibration curve obtained from step phantoms

3. Results

Converted film images and chosen flat areas for comparison were shown in figure 4. Calculated MADs over chosen areas were listed in table 1.

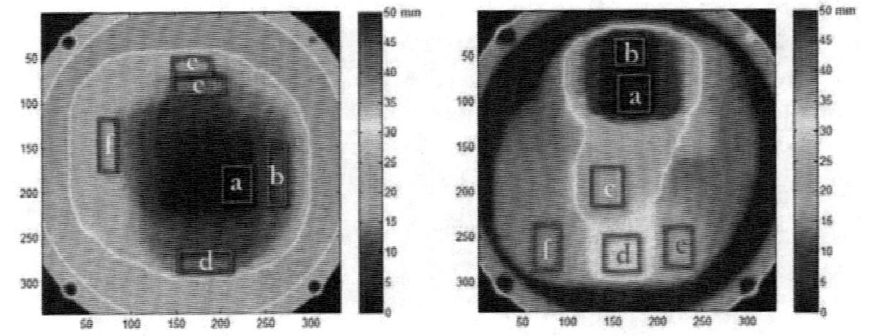

Compensator A Compensator B

Figure 4: Two range compensators image converted from OD to depth

Table 1: Depth over the selected area and calculated MAD.

	Compensator A						Compensator B					
Region	a	b	c	d	e	F	a	b	c	d	e	f
MAD (mm)	0.7	1.0	0.2	0.6	0.1	0.8	0.3	0.1	0.8	0.9	0.3	0.7

4. Discussion

The purpose of the project was to conduct a feasibility study of alternate method for QA of proton range compensator. For evaluating the precision of manufactured compensators we developed QA method using photon-like proton beam. Our comparison using MAD showed good agreement within 1 mm. The MAD over the high gradient area was over 1 mm, but we in this study focused on evaluation of precision of compensator depth over the flat region with low gradient. This proton-based QA method is expected to provide more reasonable precision with further correction of scattering effect on proton imaging.

5. Reference

[1] Li H *et al* 2010 *Phys. Med. Biol.* **55** 6759-81
[2] Smith A R 2006 Review: proton therapy *Phys. Med. Biol.* **51** 491-504
[3] Yoon M *et al* 2008 *Med. Phys.* **35** 3511-7

Application of a PAGAT/$MgCl_2$ gel for dose measurements in a 150 MeV proton beam

T Tominaga[1], S Hayashi[1], S Usui[1], H Kawamura[2] and K Katahira[3]
[1]Hiroshima International University, Department of Clinical Radiology, Higashi-Hiroshima, Hiroshima, Japan
[2]Ibaraki Prefectural University of Health Sciences, Inashiki, Ibaraki, Japan
[3]Hyogo Ion Beam Medical Center, Tatsuno, Hyogo, Japan

E-mail: t-tomi@hs.hirokoku-u.ac.jp

Abstract. The purpose of this study is to evaluate the dose response of polyacrylamide-based gel (PAGAT) when irradiated with clinical proton beams. Recently inorganic salt additive in gel has been reported to improve dose sensitivity substantially. We attempted to add $MgCl_2$ (0.5M) to regular PAGAT gel in order to compensate its lower radiation sensitivity. The spin-spin relaxation rates (R_2) as dose readout was calculated from MR imaging after irradiation with 150MeV proton beam. The dose sensitivity was discussed from the slope at dose-R_2 response curve. As the result, the sensitivity of the gel with $MgCl_2$ is approximately 3 times higher than that of regular PAGAT gel without spoiling dose response stability under the various irradiation conditions such as dose rate and dose integration.

1. Introduction

Recent studies show that polymer gel detectors are the most promising three-dimensional dosimetric tools for external photon beam radiotherapy [1]. However, in dosimetric verification of clinical proton beams, strong suppression of Bragg peak due to LET has been reported in polymer gel measurements [2].

We have investigated the radiological characteristics of gel in an irradiation to proton beam, and reported the method to estimate dose using multi-Dose-R_2 relations [3] in consideration of beam location in the gel detector from incident point of beam [4]. In previous studies, a methacrylic-acid-based gel (MAG) has been used, which has advantage in radiation sensitivity, but disadvantage in dose response stability under the various irradiation conditions such as dose rate, dose integration and the thermal condition. While it has been reported that polyacrylamide-based gel (PAG) [5] had produced low-contrast images due to the low radiation sensitivity, but had an advantage in dose response stability [6].

Recently Hayashi et al. [7] have investigated the effect of inorganic salts on the dose sensitivity of MAGAT gel [8], and reported substantial increase in the radiological response of the gel with containing inorganic salt, especially $MgCl_2$ improved dose sensitivity approximately 3 times. Their results indicated that inorganic salts as additive acted as an accelerator for radiation-induced free-radical polymerization in gel. They have reported that inorganic salts are also effective in PAGAT gels [9] too [10].

In this study, we attempt $MgCl_2$ to add to PAGAT gel in order to compensate its lower radiation sensitivity without spoiling dose response stability in the irradiation of 150 MeV proton beam, and examine its improved response on the suppression of Bragg peak.

2. Materials and methods

We prepared two type of polyacrylamide-based gel for this study. One is a regular PAGAT gel consisting of water, gelatin, acrylamide as a monomer, *N,N'*-methylene-bis-acrylamide as a cross-linker and tetrakis-hydroxymethyl-phosphonium chloride (THPC) as an oxygen scavenger [11, 12]. The other has basically same composition as a regular PAGAT gel, but with $MgCl_2$ (0.5M) additive. We will refer to this gel from here on as PAGAT/$MgCl_2$ gel. The compounded gelatin solution was divided into 12 cm test tubes for the measurements of spread-out depth dose in the gel. All samples were irradiated in a fixed horizontal 150 MeV proton beam at the facility of the Hyogo Ion Beam Medical Center. In irradiations, a test-tube-sample was set between two tough water phantoms (Kyoto Kagaku Co., WE-3015) and was placed along the beam line including 3 cm SOBP. First of all doses of 2, 4, 8, and 16 Gy were delivered to compare dose response of the above two type of PAGAT gels. At this time, the dose rate of 4 Gy/min typically applied in treatments was fixed in SOBP irradiations. Secondary the dose rate was varied to 1 Gy/min to examine the dose rate dependence on the response. And finally the dose integration property of the gel detector in the multiple field irradiations is examined by irradiating 3 Gy after 1 minute post-irradiation of 3 Gy to be compared with that of 6 Gy single exposure. MR analysis [13] was performed at 1.5T TOSHIBA Excelart Vantage scanner using Quadrature Torso coil 3 days after irradiation. MR images of samples were recorded with a double spin-echo sequence at echo times TE_1=30 ms, TE_2=250 ms with a repetition time TR=4 sec. Maps of the R_2 were calculated from the signal intensity at TE_1 and TE_2. Relative R_2 resolution was estimated to 5% in the R_2 range less than eight.

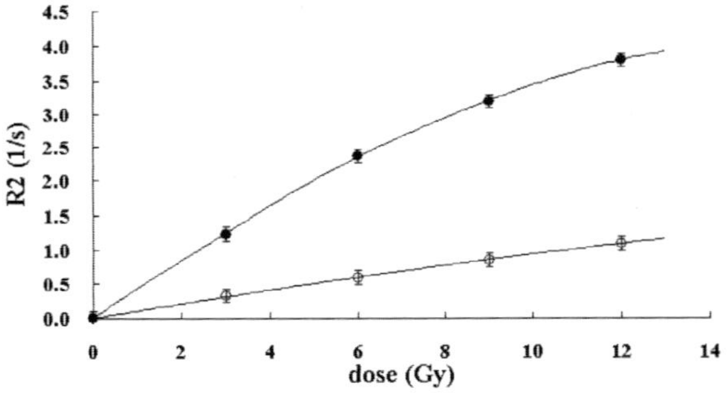

Figure 1: Dose-R_2 responses

3. Results and Discussions

3.1. Dose-R_2 sensitivity

The dose responses of two types gel applied to proton beams are shown in Figure1. The data points were obtained by averaging R_2 value at the central region of SOBP irradiated with different dose. The open circles show the R_2 response in the regular PAGAT gel, and the solid circles show that in the PAGAT/$MgCl_2$ gel in an irradiation. These were fitted by polynomial relation, but Dose-R_2 response can be regarded as linear at dose less than 10 Gy. The R_2 value in both gels at 0 Gy has no difference, but the average gradient at dose response curve for PAGAT/$MgCl_2$ is about 3 times larger than that of regular PAGAT, and the gradient is getting lose above 10 Gy closing to R_2 maximum.

3.2. Dose profiles

Figure 2 shows depth dose profiles in PAGAT/$MgCl_2$ gel at post proton irradiations with various amount of dose. A total of 12, 9, 6, 3 Gy were delivered to the SOBP of proton beam.

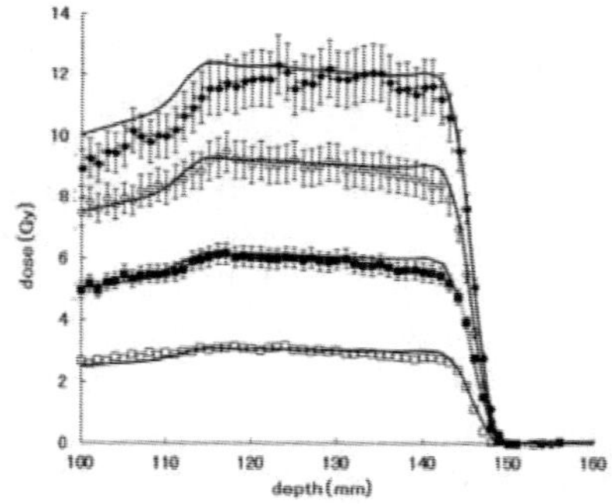

Figure 2: Depth dose profile in PAGAT/$MgCl_2$ gel.
Solid circles: 12 Gy
Open circles: 9 Gy
Solid squares: 6 Gy
Open squares: 3 Gy

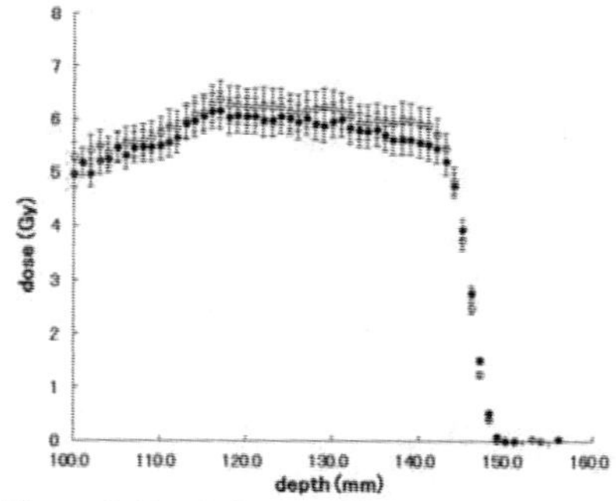

Figure 3: Depth dose profile in PAGAT/$MgCl_2$ at dose rate 4 Gy/min and 1Gy/min.

Each error bar was calculated from the relative R_2 resolution and the gradient at dose response curve. The solid lines correspond to dose estimated from ion chamber scans in water.

Though there still small suppression, less than 5%, are seen at the distal end of SOBP, the measured gel profiles show good overall agreement at depth with ion chamber profile except the profile for a beam of 12 Gy, where gel profile is lower than ion chamber profile from the entrance portion to the end.

3.3. Dose-rate dependence

A comparison of depth dose profiles for a beam of 6 Gy at dose rate 4 Gy/min (open circles) and that at 1 Gy/min (solid circles) in PAGAT/$MgCl_2$ gel is presented in Fig. 3. The dose profile at 1 Gy/min gives overall 2-3% higher dose than that at 4 Gy/min. This result corresponds to the dose-rate dependence of the dose sensitivity in a proton beam which has been previously reported in MAG type gel, but it is very small in PAGAT/$MgCl_2$ gel compared to 25% deference in MAG type gel [4].

3.4. Dose-integration dependence

A comparison of depth dose profile of total of 6 Gy by single exposure and by double exposure (3 Gy + 3 Gy) is presented in Fig. 4. The profile in the double exposed PAGAT/$MgCl_2$ gel shows good over all agreement at depth with that in the single exposed gel.

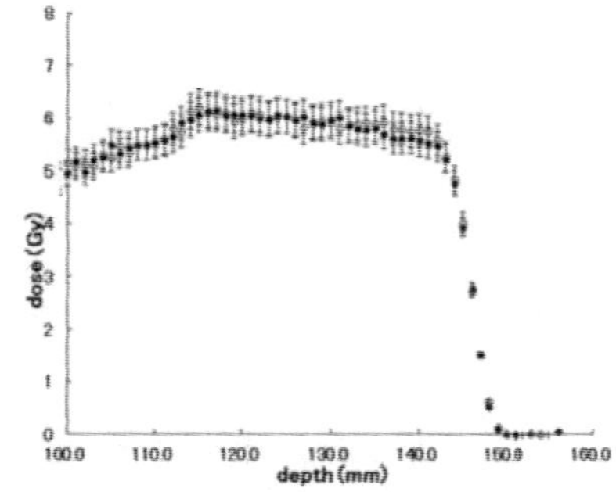

Figure 4: Depth dose profile in PAGAT/$MgCl_2$ exposed proton beam by 6 Gy single exposure (solid circle) and 3+3 Gy double exposure (open circle).

4. Conclusion

In this work, we concerned with the R_2 dose response of PAGAT/$MgCl_2$ gel in the irradiation of proton beam. So far PAG type gel has been said not to be suitable for dose measurement for proton beam due to its low dose sensitivity [14]. However it became the most promising three-dimensional dosimetric tools for proton beam radiotherapy as the results of sensitivity and dose resolution improved by $MgCl_2$ additive. And for the SOBP profile in PAGAT/$MgCl_2$ gel, the suppression at the distal end is a quite small such as less than 5%. Recently the BANG3-Pro2 gel (MGS Research Inc), which was formulated specifically for proton beam dosimetry, showed no dose quenching in the Bragg peak region in the measurement by optical CT readout system [15]. But it is not sure that this gel has not observed dose quenching even in the R_2 dose response using MRI, which is more convenient readout in the clinical use. At the further direction of our study we will pursue PAGAT/$MgCl_2$ gel's possibility to be used as QA tool for 3D dose verification in proton therapy by means of comparison the gel dose map to treatment planning system calculations.

5. References

[1] Baldock C *et al* 2010 *Phys. Med. Biol.* **55** R1-63
[2] Gustavsson H *et al* 2004 *Phys. Med. Biol.* **49** 3847-55
[3] Murry P and Baldock C 2000 *Australas. Phys. Eng. Sci. Med.* **23** 44-51
[4] Yoshioka M *et al* 2011 *Radiat. Meas.* **46** 64-71
[5] Baldock C *et al* 1998 *Phys. Med. Biol.* **43** 695-702
[6] De Deene Y *et al* 2006 *Phys. Med. Biol.* **51** 653-73
[7] Hayashi S *et al* 2012 *Radiat. Phys. Chem.* **81** 884-88
[8] Hurley C *et al* 2005 *Appl. Radiat. Isot.* **63** 443-56
[9] Venning A J *et al* 2005 *Phys. Med. Biol.* **50** 3875-88
[10] Hayashi S *et al* 2012 (IC3Dose 2012)
[11] Jirasek A and Duzenli C 2002 *Med. Phys.* **29** 569–77
[12] De Deene Y *et al* 2002 *Phys. Med. Biol.* **47** 2459-70
[13] De Deene Y *et al* 2002 *Phys. Med. Biol.* **47** 3441-63
[14] Lepage M *et al* C 2002 *Phys. Med. Biol.* **47** 1881-90
[15] Zeidan O *Phys. Med. Biol.* 2010 *Med. Phys.* **37** 2145-52

Journal of Physics: Conference Series **444** (2013) 012058 doi:10.1088/1742-6596/444/1/012058

Preliminary characterization of PRESAGE® for 3D dosimetry of 62 MeV proton beam

T Gorjiara[1], A Kacperek[2], Z Kuncic[1], C Baldock[1,4] and S Doran[3]
[1]Institute of Medical Physics, School of Physics, University of Sydney, NSW 2006, Australia
[2]Douglas Cyclotron, Clatterbridge Centre for Oncology, Wirral, UK
[3]CRUK Clinical Magnetic Resonance Research Group, Institute of Cancer Research, Surrey, UK
[4]Faculty of Science, Macquarie University, NSW 2109, Australia

E-mail: simon.doran.icr.ac.uk

Abstract. PRESAGE® has previously shown potential for 3D dosimetry of heavy particles. A new formulation has specifically developed for dosimetry of protons/heavy ions. This work provides a preliminary characterization the new formulation of PRESAGE® by measuring optical absorbance and dose response after irradiating by a 62 MeV proton beam for a dose range of 0.5 – 20 Gy. Results show linear dose response and the evolution over time of the optical density of the 3D dosimeter.

1. Introduction

Polymer gel dosimeters are manufactured from radiation sensitive chemicals, which upon irradiation polymerize as a function of the absorbed radiation dose [1]. These gel dosimeters which record the radiation dose distribution in three-dimensions (3D) have specific advantages when compared to one-dimensional dosimeters and two-dimensional dosimeters [2]. These 3D dosimeters are radiologically soft-tissue equivalent [3] with properties that may be modified depending on the application. The 3D radiation dose distribution in polymer gel dosimeters may be imaged using magnetic resonance imaging (MRI) [4, 5], optical-computerized tomography (optical-CT) [6, 7], x-ray CT [8, 9], ultrasound [10, 11, 12] or vibrational spectroscopy [13, 14].
Using proton beams for radiotherapy is an advanced method for cancer treatment. Because of the high dose gradients at the distal end of the delivered distribution, 3D verification is of great importance in proton dosimetry, to ensure correct coverage of tumor volume and sparing of healthy tissues. Ionization chambers, which are the recommended dosimeters for proton beams [15, 16], are not ideal for measuring two dimensional (2D) or 3D dose distributions. Therefore, treatment with proton beam can benefit from having an accurate 3D dosimeter with reproducible measurements with comparable with an ionization chamber, and having similar dimensions to the target volume [2, 17].

PRESAGE® is a polyurethane-based radiochromic 3D dosimeter with a linear dose response [2, 18-21], which has already shown considerable promise for proton/heavy ion dosimetry [22-26]. However, similar to 3D polymer gel dosimeters, dose response saturation (lower dose measurement) has been observed at the Bragg peak of protons/heavy ions due to high LET (Linear Energy Transfer)

[2, 23-27]. The saturation effect has been reported less pronounced in the preliminary studies on a new formulation of PRESAGE®, developed for heavy particle dosimetry [28, 29].

This work is a preliminary study on the characterization of the new formulation of PRESAGE® for a 62 MeV proton beam, commonly used to treat different ocular melanomas. Previous formulations have shown good temporal stability when stored correctly, but the manufacturer recommends minimizing exposure to visible light and it has been demonstrated that ultraviolet radiation is particularly efficient in causing a colour change [30]. It has also previously been shown that there is a marked temperature-related effect [31] One of the aims of this study was thus to investigate the impact of different storage conditions on the optical response of the dosimeter.

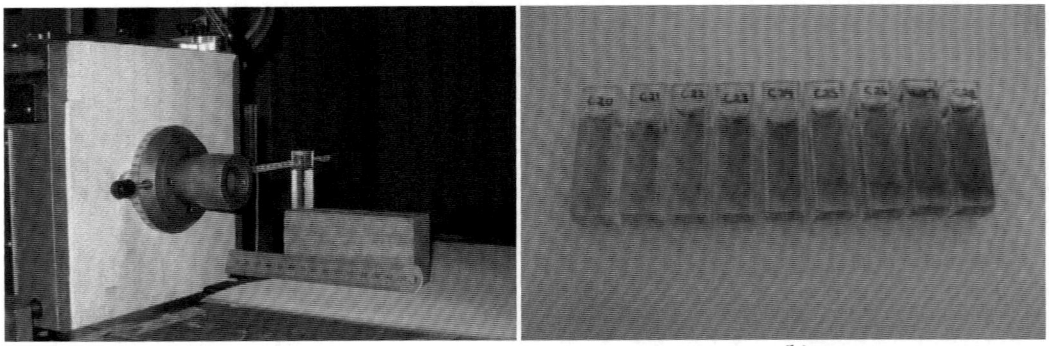

(a) (b)

Figure 1: (a) Placement of PRESAGE® cuvettes in front of an unmodulated 62 MeV proton beam and (b) irradiated PRESAGE® cuvettes with a dose range of 0.5 Gy – 8 Gy.

2. Method and Materials

Eight cuvettes filled with the new formulation of PRESAGE® were irradiated by an unmodulated 62 MeV proton beam from the Douglas Cyclotron at the Clatterbridge Center for Oncology, UK. The duration of irradiation was adjusted in a way to deliver 0.5, 1, 2, 6, 4, 8, 10, 15 and 20 Gy dose to the nine PRESAGE® cuvettes. The set-up of the experiment and irradiated cuvettes are presented in figure 1. To investigate stability and provide a calibration curve, the optical absorbance of samples was measured 12 hours and 108 hours after irradiation using a CAMSPEC M350 double beam spectrophotometer. Three sets of four PRESAGE® cuvettes were also irradiated by a 6 MV x-ray beam with 2, 4, 10 and 20 Gy dose to evaluate how storage condition (i.e., temperature and light) effects the optical of PRESAGE® samples. After irradiation, one of these sets stored in the refrigerator and the second set stored in the dark at room temperature. The third set of cuvettes was stored at room temperature, but experienced periods of both light and darkness. Initially, the samples were exposed to light from a fluorescent tube and darkened very rapidly, to the extent that it became impossible to read them correctly with our spectrophotometer. They were then placed in the dark and, finally, placed in a sunlit position. The optical absorbance of these samples was measured over a period of time (up to 760 hours post-irradiation) using the spectrophotometer.

3. Results and discussion

Figure 2(a) shows the optical absorbance of irradiated PRESAGE® cuvettes versus the wavelength (nm). The maximum absorbance was observed for the 633 nm wavelength. The plotted calibration curves at this wavelength versus dose for the spectrophotometer readings 12 hours and 108 hours after irradiation are presented in figure 2(b). Linear response with dose increase and good stability with time are evident.

Figure 3(a) shows the optical absorbance of the PRESAGE® cuvettes at 633 nm wavelength versus dose in the different storage conditions 80 hour post-irradiation. As anticipated, optical absorbance of the PRESAGE® dosimeter changes significantly with light due to sensitivity of PRESAGE® dosimeter to ultraviolet (UV) radiation. The optical density of PRESAGE® stored at dark in room temperature seems slightly lower than the one kept in the refrigerator. In figure 3(b) the optical absorbance of the 20 Gy irradiated PRESAGE® cuvettes is presented over a period of post-irradiation time in the different storage conditions. Changes over the time are observed in optical absorbance of PRESAGE® for all the storage conditions, but are most marked for the sample which was stored in the light. Here, an initial rapid increase is followed by significant fading while the dosimeter was in the dark and another large rise when the sample was returned to the light.

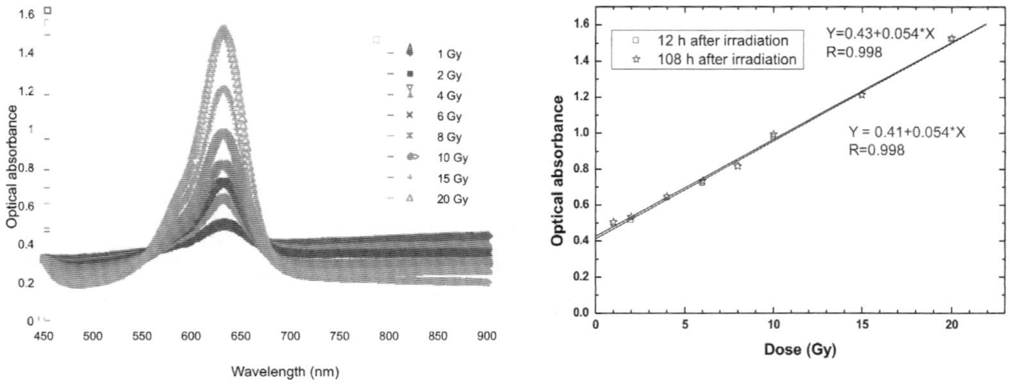

(a) (b)

Figure 2: The optical absorbance of nine irradiated PRESAGE® cuvettes by an unmodulated 62 MeV proton beam (a) 12 hours after irradiation and (b) at 633 nm 12 hours and 108 hours after irradiation by.

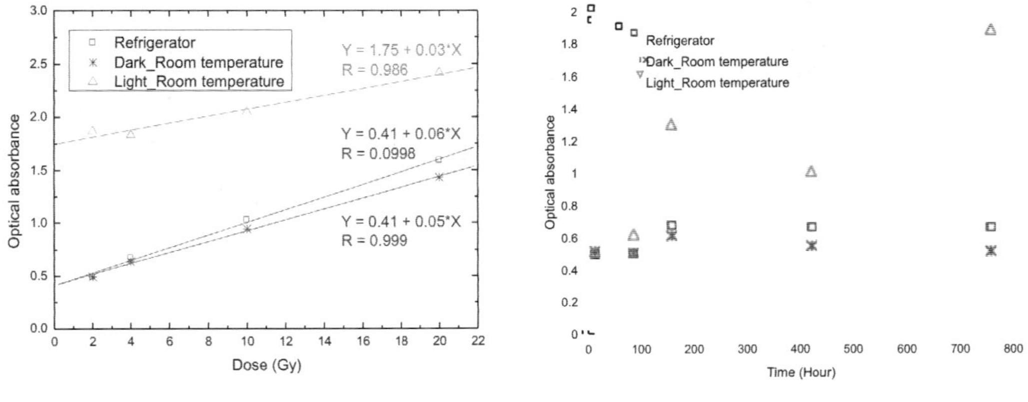

(a) (b)

Figure 3: The optical absorbance of irradiated PRESAGE® cuvettes (a) versus dose, at 633 nm, 80 hours after the irradiation and (b) for 20 Gy dose, at 670 nm, over a period of post-irradiation time in the fridge, in the dark at room temperature and in the light at room temperature.

4. Conclusion

The new formulation of PRESAGE®, developed for proton dosimetry, shows linear response with dose increase and good dose stability with time, providing that it is stored appropriately. Post-irradiation storage of PRESAGE® in the refrigerator seems the best way of preserving the dose data in the dosimeter. Storage in the light significantly degrades dose data of the PRESAGE® 3D dosimeter. The effect has been shown to be reversible, but it is highly unlikely that the originally stored dose information could be recovered in this way.

5. References

[1] Baldock C *et al* 1998 *Phys. Med. Biol.* **43** 695-702
[2] Baldock C *et al* 2010 *Phys. Med. Biol.* **55** R1–63
[3] Keall P and Baldock C 1999 *Australas. Phys. Eng. Sci. Med.* **22** 85-91
[4] Gustavsson H *et al* 2004 *Phys. Med. Biol.* **49** 227-41
[5] Lepage M *et al* 2002 *Phys. Med. Biol.* **47** 1881-90
[6] Bosi S *et al* 2007 *Physics in Medicine and Biology* **52** 2893–2903
[7] Bosi S G *et al* 2009 *Phys. Med. Biol.* **54** 275-83
[8] Trapp J V *et al* 2002 *Phys. Med. Biol.* **47** 4247-58
[9] Hill B *et al* 2005 *Br. J. Radiol.* **78** 623-30
[10] Mather M L and Baldock C 2003 *Med. Phys.* **30** 2140-2148
[11] Mather M L *et al* 2002 *Phys. Med. Biol.* **47** 1449-58
[12] Mather M L *et al* 2003 *Ultrasonics* **41** 551-559
[13] Baldock C *et al* 1998 *Phys. Med. Biol.* **43** 3617-27
[14] Rintoul L *et al* 2003 *Appl. Spectrosc.* **57** 51-7
[15] Andreo P *et al* 1995 IAEA TRS-381. The use of plane parallel ionization chambers in high energy electron and photon beams. An international code of practice for dosimetry.
[16] Millwater C J *et al* 1998 *Br. J. Radiol.* **71** 661-8
[17] Zeidan O A *et al* 2010 *Med. Phys.* **37** 2145-52
[18] Sakhalkar H S *et al* 2009 *Med. Phys.* **36** 3371-7
[19] Sakhalkar H S *et al* 2009 *Med. Phys.* **36** 71-82
[20] Doran S J 2009 *Appl. Radiat. Isot.* **67** 393-8
[21] Guo P Y *et al* 2006 *Med. Phys.* **33** 1338-45
[22] Rikner G and Grusell E 1987 *Phys. Med. Biol.* **32** 1109-17
[23] Feygelman V *et al* 2010 *Med. Dosim.* **35** 324-9
[24] Feygelman V *et al* 2009 *J. Appl. Clin. Med. Phys.* **10** 64-78
[25] Al-Nowais S *et al* 2009 *Appl. Radiat. Isot.* **67** 415-8
[26] Rosenfeld A B 2002 *Radiat. Prot. Dosimetry.* **101** 393-8
[27] Gustavsson H *et al* 2004 *Phys. Med. Biol.* **49** 3847-55
[28] Zhao L *et al* 2011 *Med. Phys.* **38** 3506
[29] Devic S 2011 *Physica Medica* **27** 122-34
[30] Abdul Rahman A T *et al* 2011 *Phys. Med. Biol.* **56** 4177-99
[31] Skyt P S *et al* 2011 *Med. Phys.* **38** 2806-11

Journal of Physics: Conference Series **444** (2013) 012059 doi:10.1088/1742-6596/444/1/012059

Investigation of photon and proton overlapping fields in PRESAGE® dosimeters

M Carroll[1,2], G Ibbott[1], R Grant[1], M Gillin[1] and J Adamovics[3]
[1]Department of Radiation Physics, University of Texas MD Anderson Cancer Center, 1515 Holcombe Blvd., Houston, TX 77030 USA
[2]University of Texas at Houston Graduate School of Biomedical Sciences, Houston, TX 77030 USA
[3]Heuris Pharma, LLC, 412 Sunset Rd., Skillman, NJ, 08558 USA

E-mail: gibbott@mdanderson.org

Abstract. To evaluate the effects of overlapping dose volumes for varying field arrangements in PRESAGE®, several sequential beam irradiations were delivered each to formulations intended for, and irradiated with, proton beams as well as photon beams. The dosimeters were irradiated within timespans consistent of overlapping fields in clinical treatment plans. Dose profiles taken along the beam direction indicated slight over-responses in higher dose regions relative to similar irradiations given in a single fraction. These results will aid future measurements of overlapping field treatment plans delivered to PRESAGE® for treatment verification of proton and photon 3D dosimetry.

1. Background

In previous studies, PRESAGE® dosimeters [1, 2] have been evaluated in open beams [3, 4]. With the introduction of stereotactic radiosurgery (SRS), intensity modulated radiotherapy (IMRT), and other advanced treatments, these dosimeters have been tested for three-dimensional verification [5, 6]. Additional studies have been conducted to determine temperature, dose rate, and dose sensitivity [7, 8]. In order to fully assess their quality for clinical verification purposes of radiotherapy treatments with several beam segments, this study was implemented to evaluate the dose integration [9] of overlapping dose volumes. Similar studies have been performed on polymer gel dosimeters [10]. Two formulations of PRESAGE® were used: one intended for, and irradiated with, proton beams and the other photon beams.

2. Methods

Cylindrical PRESAGE® dosimeters approximately 9 cm in height and 6 cm in diameter were used in this study. Photon irradiations were performed on Varian 2100 series linacs using 6 MV open beams at the University of Texas at M. D. Anderson Cancer Center in Houston, Texas. Field sizes ranged from 3x3 cm^2 to 5x5cm^2. Each dosimeter was irradiated using table-top conditions to a consistent dose at d_{max}. The proton dosimeter irradiations were performed at the M. D. Anderson Proton Therapy Center (PTC). These were irradiated using a fixed gantry, 4x4 cm^2 field size, and 200 MeV beam in a water tank at approximately 15.8 cm surface-to-dosimeter distance. A spread-out Bragg peak (SOBP) of 3 cm was delivered approximately 6 cm into the dosimeter in order to fully capture all relevant dose regions. All irradiations were delivered perpendicular to the flat top surface rather than the curved sides to reduce dose profile inhomogeneities.

For each treatment modality (photon, proton), two overlapping field setups were performed. These included a stationary dosimeter irradiated over six fractions and a dosimeter shifted laterally to the field to deliver a dose plateau in two fractions. All subsequent fractions were given within ten minutes and never less than one minute apart to simulate a lengthy clinical treatment delivery. Two dosimeters were irradiated for each setup. The dosimeters were paired, with one dosimeter given total dose by a single fraction while the other followed one of the overlapping field setups. The dosimeters were analyzed using the Duke Medium field-of-view Optical CT Scanner (DMOS) 24 hours after irradiation and exported to the Computational Environment for Radiotherapy Research (CERR) software platform where the doses were compared between paired dosimeters. Dose profiles were taken parallel to the beam path for all setups. Additionally, cross dose profiles of the of the dose plateau studies were taken.

3. Results

Paired dosimeter agreement was determined by the dose variations along the central axis to the overlapped irradiated fields of the fractionated and dose plateau field setups. The profile was taken laterally to the fields and at the center of the dose volume created by overlapping two and four field box irradiations. Because of inconsistencies in dosimeter sizes between pairs and data lost at the edges of the dosimeters, spatial agreement between dose profiles was corrected for by matching distal ends of those irradiated with protons and d_{max} of those irradiated with photons.

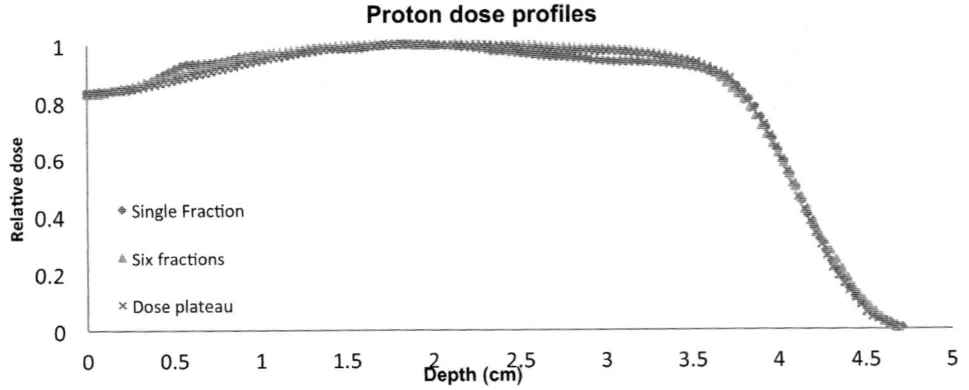

Figure 1: Relative dose profiles of the proton formulation of PRESAGE® showing agreement between single and multiple fraction irradiations from the proximal through the distal regions.

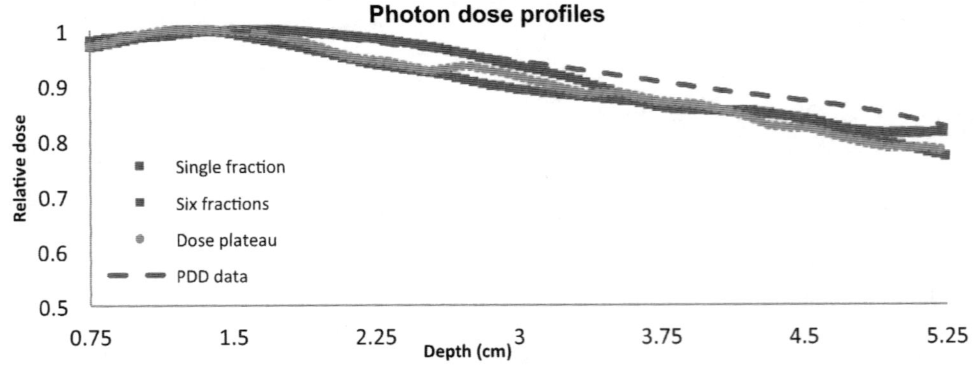

Figure 2: Relative dose profiles of photon formulation of PRESAGE® showing the agreement between single fractions, multiple fractions, and PDD data in the dose peak region.

Dose profile comparisons showed relative dose agreement between paired dosimeters within 5% along the SOBP region of the proton formulation and shown in figure 3. Both of the overlapping field irradiations showed an under-response in the proximal region and an over-response in the distal side of the SOBP relative to the single fraction irradiation with no PTC ion chamber data for this setup to compare. Dose agreement between the photon dosimeter treated with overlapping fields also over-responded by as much as 11% relative to the single fraction irradiation. Additionally, cross-beam dose plateau profiles further characterized the change in recorded dose between the overlapping volume and the integration of the sequential fields as illustrated in figure 3 and table 1. These showed a discrepancy between added and recorded absorbed dose.

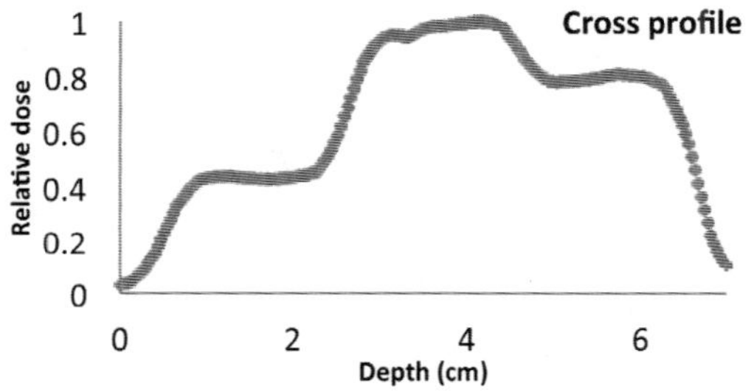

Figure 3: Cross profile of overlapping fields (100 and 200 cGy) treated with photons.

	Dose 9 (cGy)	Average relative dose	Stdv
Photon	100.00	0.43	0.01
	200.00	0.79	0.01
	Plateau (300)	0.98	0.02
Proton	100.00	0.40	0.00
	200.00	0.72	0.00
	Plateau (300)	0.96	0.02

Table 1: Averaged relative doses taken over each dose plateau along the cross profile of overlapping fields.

4. Discussion and Conclusions

The proton formulation of PRESAGE® showed good dose agreement between single and overlapping field irradiations. Dose volumes treated with sequential beams, primarily in the distal end of the SOBP, actually resulted in dose profiles more consistent with expected results. The photon formulation had slightly less agreement, while the sequential field irradiations again showed a closer agreement with PDD data. Repetition of the dose plateau irradiations to remove irregularities seen in the dose profile as well as comparison with treatment planning system data will allow further verification.

5. References

[1] Brown S *et al* 2008 *Appl. Radiat. Isot.* **66** 1970-4
[2] Gorjiara T *et al* 2011 *Med. Phys.* **38** 2265-74
[3] Adamovics J *et al* 2006 *J. Phys.: Conf. Ser.* **56** 172-5
[4] Adamovics J *et al* 2006 *J. Phys.: Conf. Ser.* **56** 176-8
[5] Oldham M *et al* 2008 *Med. Phys.* **35** 2072-80
[6] Baldock C *et al* 2010 *Phys. Med. Biol.* **55** R1-63
[7] Skyt P S *et al* 2011 *Med. Phys.* **38** 2806-11
[8] Pierquet M *et al* 2010 *J. Phys. Conf Ser* **250** 1-4
[9] Hurley C *et al* 2005 *Appl. Radiat. Isot.* **63** 443-56
[10] Karlsson A *et al* 2007 *Phys. Med. Biol.* **52** 4697-706

Laser CT evaluation on normoxic PAGAT gel dosimeter

DS Kumar[1,2], EJJ Samuel[1] and Y Watanabe[2]
[1]Photonics Division, School of Advanced Sciences, VIT University, Vellore-632014, Tamil Nadu, India
[2]Department of Therapeutic Radiology, University of Minnesota, Minneapolis, (Twin cities), MN-55455, USA

E-mail: kalaimeenasen@gmail.com

Abstract. Optical computed tomography has been shown to be a potentially useful imaging tool for the radiation therapy physicists. In radiation therapy, researchers have used optical CT for the readout of 3D dosimeters. The purpose of this paper is to describe the initial evaluation of a newly fabricated laser CT scanner for 3D gel dosimetry which works using the first generation principle. A normoxic PAGAT (Polyacrylamide Gelatin and Tetrakis) gel is used as a dosimeter for this analysis. When a laser passes through the gel phantom, absorption and scattering of photon take place. The optical attenuation coefficient of the laser can be obtained by measuring its intensity after passing through the gel by a sensor. The scanner motion is controlled by a computer program written in Microsoft Visual C++. Reconstruction and data analysis on the irradiated gel phantom is performed by suitable algorithm using Matlab software.

1. Introduction

Modern radiation delivery techniques such as stereotactic radiotherapy, three-dimensional (3D) conformal radiotherapy and intensity modulated radiation therapy aim at precise delivery of radiation dose. These techniques are based on increasing complexity of radiation delivery methods and therefore necessitate verification of computer calculated dose distribution by an accurate dosimetric method. Gel dosimeters are capable of recording 3D dose distributions [1, 2]. Previously, magnetic resonance imaging (MRI) was the only available method for reading the dose distribution recorded in polymer gels [3, 4]. However, due to the cost and problem of availability, optical scanning of gel has been used by several researchers [5-9] in preference to MRI, x-ray CT [10-13], ultrasound [14-16] and vibrational spectroscopy [17, 18] readout. We present some initial investigations using normoxic PAGAT [19, 20] gel dosimeter with an in-house built laser CT scanner.

2. Materials and Method

2.1. Gel Preparation

The normoxic PAGAT gel described elsewhere [19] was prepared under normal atmospheric conditions. The gel preparation was carried out as follows: 5% gelatin was added to the magnetic stirred glass flask containing 89% of triple distilled water and allowed to swell for 10–15 minutes. Then the gelatin solution was heated to the temperature of 50 ^{0}C for complete dissolution of the gelatin. After obtaining the clear gelatin solution 3.5% BIS and 3.5% Acrylamide were added and stirred continuously until the mixture was dissolved homogeneously. The antioxidant 10 mM THP was added to the solution under stirring just before transferring the gel into polyethylene terephthalate cylindrical containers of size 10 cm diameter and 13 cm long. All the cylindrical containers containing the PAGAT gel were sealed properly to avoid the further diffusion of oxygen and were kept in a refrigerator at 5°C for one full night for gelling [21]. All chemicals were from Sigma Aldrich.

2.2. Gel Irradiation

To evaluate the dose response of polymer gel to gamma radiations, the polymer gel dosimeters were irradiated after complete gelation [19]. The polymer gel dosimeters were irradiated to 12 Gy with a square field size of 5 x 5 cm^2 using Co-60 gamma rays. The source-to-surface distance (SSD) was 100 cm. One dosimeter was kept unirradiated for background subtraction. The irradiated gel dosimeters were left in a refrigerator for one day before imaging to allow the complete post irradiation polymerization [22].

2.3. Laser CT scanning system

The Laser CT scanner was constructed on the principle of a first generation X-ray CT is shown in Figure 1. The construction part of this scanner was briefly explained and a single slice scanning has been reported previously using this scanner [23]. We extended the work to scan multiple slices on the PAGAT gel phantom for different depths. Figure 3 represents the multi slice scanning process.

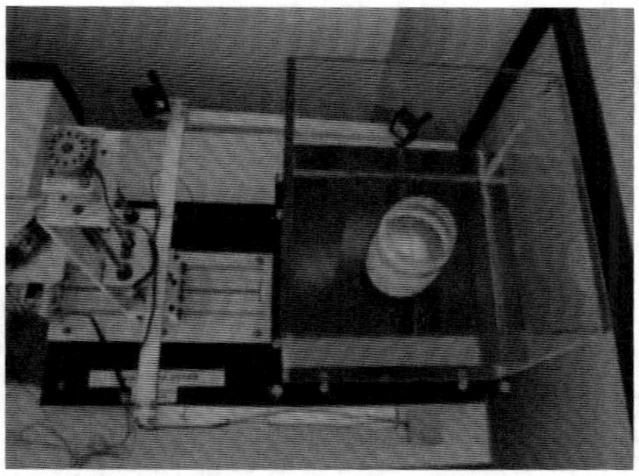

Figure 1: In-house fabricated Laser CT scanner.

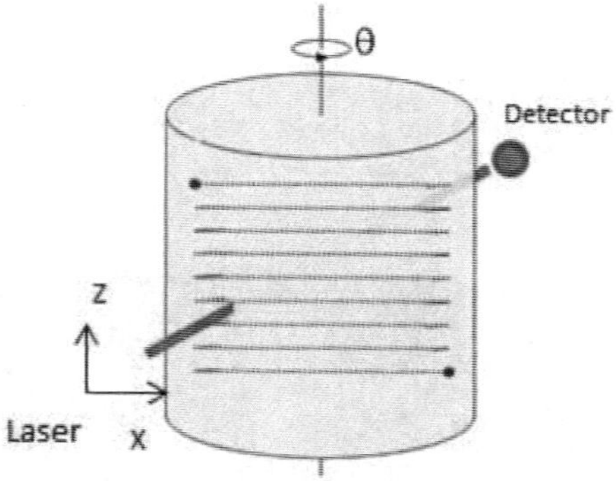

Figure 2: Multi-slice scanning process.

2.4. Scanner Control and Image Reconstruction

The software controls the scanner, acquires and stores the data and performs the necessary calculations for data procession and transformation [24]. The image reconstruction software converts the scanner data into 3-dimensional image data. The purpose of this software is to apply an Inverse Radon Transformation to the scanner raw data. Matlab readily provides the Filtered Back-Projection algorithm, an efficient algorithm for the Inverse Radon Transformation.

3. Results and Discussion

Figure 3 & figure 4 represents the top view of pre & post-irradiated gel phantoms. The pre-irradiation image was subtracted from the post-irradiation scan and the subtracted image were then analysed. The reconstruction for a single slice was undertaken using the Filtered Back Projection algorithm implemented in Matlab. Dose profiling the reconstructed slices and subtraction of post irradiated images from pre irradiated images was performed with in-house code developed in Matlab. Figure 5 represents the image of the pre irradiated gel slice. Figure 6, 7 and 8 represents the 2D reconstructed image slices of the post-irradiated gel phantom at 4, 6 and 8 cm depths. A line is drawn on the pre- and post-irradiated gel slice to get their corresponding dose profile. Figure 9 represents the dose profile for the pre-irradiated gel slice with distance across the gel dosimeter (mm) versus the optical density (cm⁻

^{1}). Figure 9, 10, 11 and 12 represents their corresponding dose profile with distance across the gel dosimeter (mm) versus the optical density (cm^{-1}). From the distance versus optical density characteristics, it can be clearly observed that, as the distance across the gel phantom (mm) increases the optical density (cm^{-1}) also increases. The distance across the gel phantom has been plotted in the x-axis and the optical density has been plotted on the y-axis. The time taken for scanning and reconstructing single slice was ~ 10 minutes.

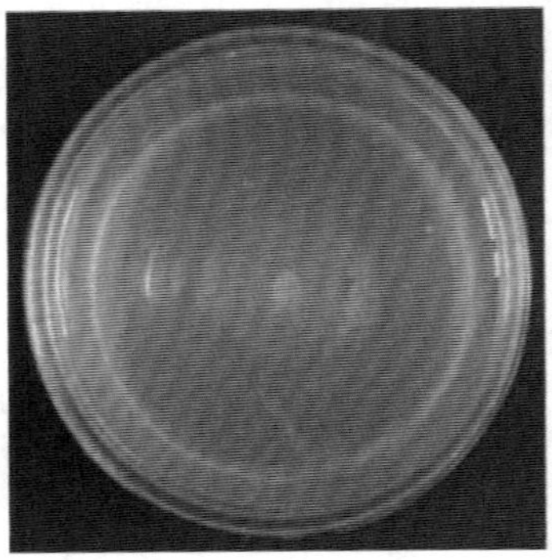

Figure 3: Top view of Pre irradiated Gel Phantom.

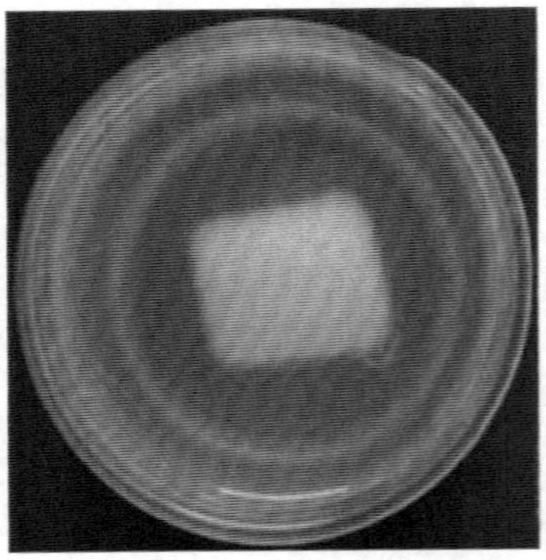

Figure 4: Top view Post irradiated Gel Phantom.

Figure 5: Reconstructed slice of the Pre-irradiated gel

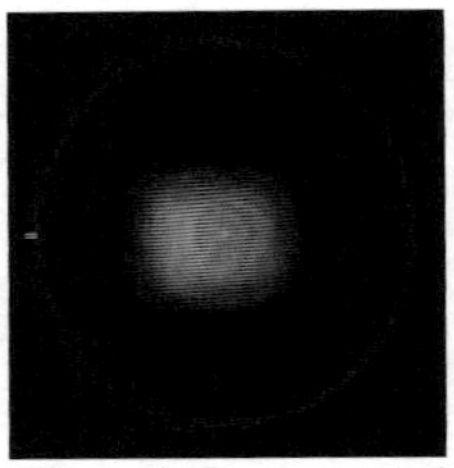

Figure 6: Reconstructed slice of the Post-irradiated gel at 4 cm depth

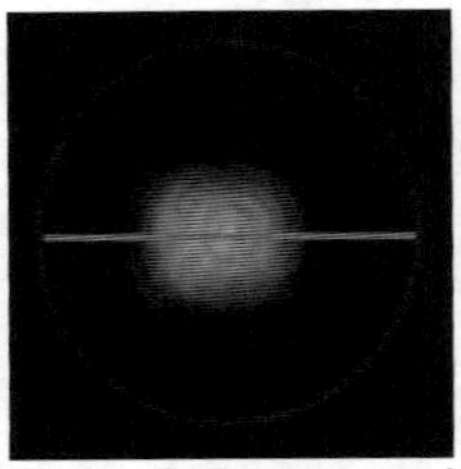

Figure 7: Reconstructed slice of the Post-irradiated gel at 6 cm depth

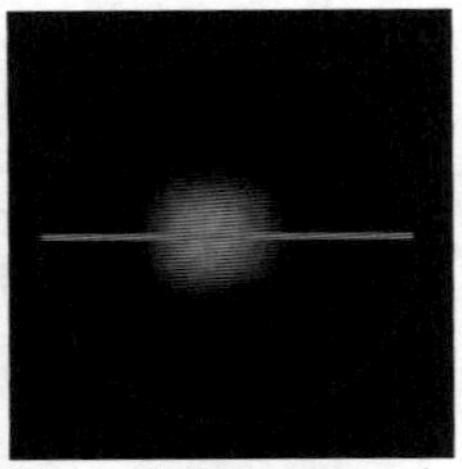

Figure 8: Reconstructed slice of the Post-irradiated gel at 8 cm depth

4. Conclusions

A laser CT scanner for reading the absorbed dose in a tissue equivalent gel was constructed for the total material cost of about $3,000 (US) and its initial performance for reading an irradiated gel dosimeter was investigated. The reconstructed image for the irradiated slice and its data analysis was performed with in-house code developed in Matlab. From the results, we conclude that the scanner is suitable for reading normoxic gels. However, the scanning time is relatively long and this has to be reduced in the future.

5. Acknowledgements

The authors are grateful to the management of VIT University for their continuous support and encouragement. The authors also acknowledge the guidance provided by Prof. B. Paul Ravindran, Prof. S.L. Keshava, Dr. R.M. Nehru and Dr. P. Sellakumar in performing this work.

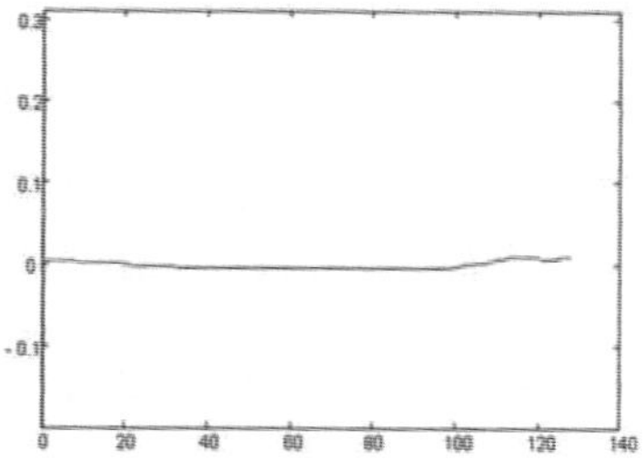

Figure 9: Dose Profile for Pre- irradiated gel slice (Distance across the gel phantom vs Optical Density).

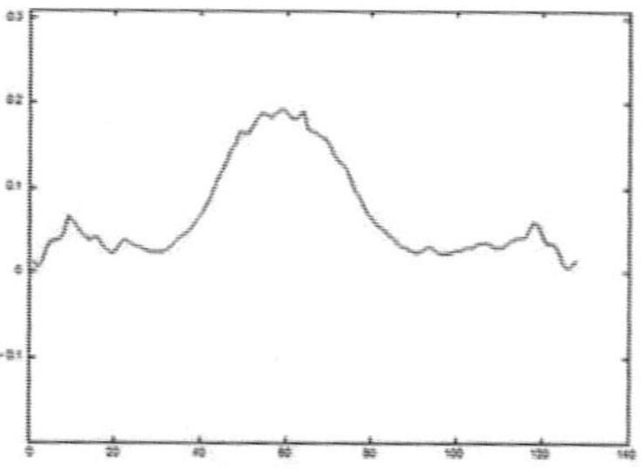

Figure 10: Dose Profile for Post- irradiated gel slice at 4 cm depth (Distance across the gel phantom vs Optical Density).

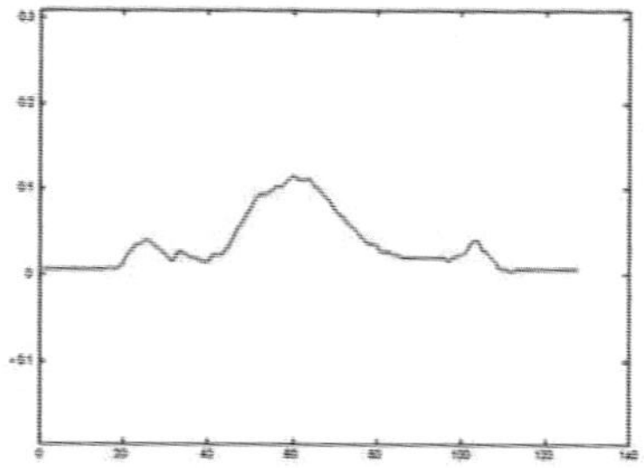

Figure 11: Dose Profile for Post-irradiated gel slice at 6 cm depth (Distance across the gel phantom vs Optical Density).

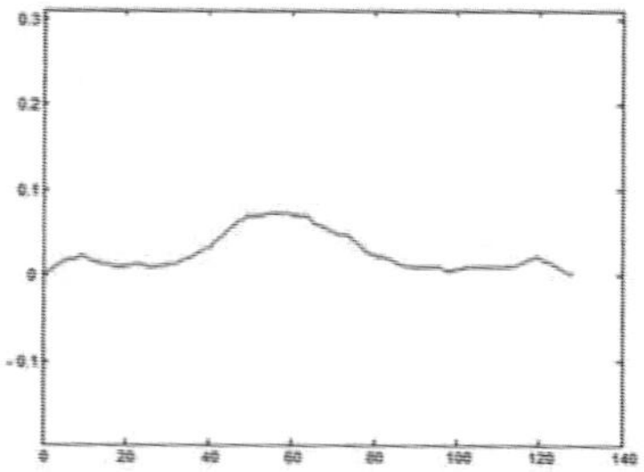

Figure 12: Dose Profile for Post-irradiated gel slice at 8 cm depth (Distance across the gel phantom vs Optical Density).

6. References

[1] Baldock C *et al* 1998 *Phys. Med. Biol.* **43** 695-702
[2] Baldock C *et al* 2010 *Phys. Med. Biol.* **55** R1–63
[3] Gustavsson H *et al* 2004 *Phys. Med. Biol.* **49** 227-41
[4] Lepage M *et al* 2002 *Phys. Med. Biol.* **47** 1881-90
[5] Oldham M *et al* 2001 *Med. Phys.* **28** 1436-45
[6] Oldham M *et al* 2003 *Med. Phys.* **30** 623-34
[7] Oldham M and Kim L 2004 *Med. Phys.* **31** 1093-104
[8] Xu Y *et al* 2004 *Med. Phys.* **31** 3024-33
[9] Bosi S G *et al* 2009 *Phys. Med. Biol.* **54** 275-83
[10] Trapp J V *et al* 2002 *Phys. Med. Biol.***47** 4247-58
[11] Hill B *et al* 2005 *Br. J. Radiol.* **78** 623-30
[12] Brindha S *et al* 2004 *Phys. Med. Biol.* **49** N353-61
[13] Brindha S *et al* 2006 *J. Med. Phys.* **31** 72-7
[14] Mather M L and Baldock C 2003 *Med. Phys.* **30** 2140-8
[15] Mather M L *et al* 2002 *Phys. Med. Biol.***47** 1449-58
[16] Mather M L *et al* 2003 *Ultrasonics* **41** 551-9
[17] Baldock C *et al* 1998 *Phys. Med. Biol.* **43** 3617-27
[18] Rintoul L *et al* 2003 *Appl. Spectrosc.* **57** 51-7
[19] Venning A J *et al* 2005 *Phys. Med. Biol.* **50** 3875-88
[20] Brown S *et al* 2008 *Appl. Radiat. Isot.* **66** 1970-4
[21] Michael G J *et al* 2000 *Phys. Med. Biol.* **45** N133-8
[22] Lepage M *et al* 2001 *Phys. Med. Biol.* **46** 2827-2839
[23] Kumar D S and Samuel E J J 2010 *J. Phys.: Conf. Ser.* **250** 012040
[24] Murry P and Baldock C 2000 *Australas. Phys. Eng. Sci. Med.* **23** 44-51

Journal of Physics: Conference Series **444** (2013) 012061 doi:10.1088/1742-6596/444/1/012061

Array of square waveguides for scintillation dosimetry in external radiotherapy

P Naseri[1], DR McKenzie[1], P Liu[1], S Fleming[1,2] and N Suchowerska[1,3]
[1]School of Physics, University of Sydney, NSW, 2006, Australia
[2]Institute of Photonics and Optical Science, School of Physics, University of Sydney, NSW, 2006, Australia
[3]Department of Radiation Oncology, Royal Prince Alfred Hospital, Sydney, NSW, 2050, Australia

E-mail: p.naseri@physics.usyd.edu.au

Abstract. An array of air core scintillation dosimeters (of round or square cross section) is an efficient solution for managing the problem of Cerenkov background light in megavoltage radiation. This array generates a high-resolution dose map in a way that satisfies ICRU dosimetric accuracy recommendations without the need for correction factors. Efficient scintillation signal transportation is vital to sensitivity of the dosimeter. The attenuation of the light irradiance as a function of waveguide length in PMMA and silver hollow square and round waveguides is studied experimentally and theoretically. In practice, the silvered square waveguide has the least attenuation while the PMMA square waveguide performs almost as well as commercially sourced silvered tubes. The attenuation of the commercially sourced tubes is increased by the rough internal silver surfaces.

1. Introduction

Modern external beam treatment methods such as intensity modulated radiation therapy employ dynamic radiation fields with high dose gradients. Water equivalent dosimeters do not have any beam perturbation effect and an array of them can be used to provide multi dimensional dose mapping without the need for correction factors. Polymer gels are excellent water equivalent dosimeters for high-resolution dose measurements in three-dimension [1]. However, they do not give real-time measurements and may have complications in the interpretation of the readout [2-4].

Table 1: A summary of characteristics for a selection of dosimeters that have high spatial resolution and give real time measurements.

	Water Equivalence	**Energy Independence**	**Angular Independence**	**Dose rate Independence**
MOSFET	×	×	×	×
Diode	×	×	√	√
Diamond	√	√	√	×
Scintillation	√	√	√	√

Array dosimeters that give real time measurements have been designed with resolutions between 5 mm to 20 mm, depending on the technology they use and the beam orientation [5-8]. Table 1 compares dosimeters that have high resolution and give real time measurement in terms of their desirable properties for modern radiotherapy and compares metal oxide semiconductor field effect transistor (MOSFET) dosimeters [9-10], diodes [11], diamond [12-13] and scintillation dosimeters [6, 14]. In a megavoltage radiation beam, scintillation dosimeters comprised of a plastic scintillator and a silvered air core waveguide avoid the generation of Cerenkov background in the radiation field [15]. However, the incorporation of silver and air in the construction of the dosimeter raise questions concerning its water equivalence and its suitability for array dosimetry. Moreover, air core waveguides suffer from signal attenuation specially with decreasing their cross section. The efficiency of light transmission in hollow waveguides while maintaining a high resolution determines the sensitivity of the dosimeter and is therefore central to air core dosimeter performance.

In this paper, the water equivalence of a round air core scintillation dosimeter as a single dosimeter and in array arrangement is investigated using Monte Carlo (MC) simulations. The light attenuation as a function of distance in a round air core waveguide is compared with that for the square waveguide using theory and experiments. A novel design of a dosimeter array with integrated square waveguides is that uses square waveguides is tested under a therapeutic radiation field.

2. Methods and Materials

Figure 1 shows a schematic diagram of a 1 mm diameter scintillation air core dosimeter. The EGSnrc MC simulation package was used to calculate the dose deviation from water when the dosimeter was used alone or in an array arrangement under a 6 MV beam. Figure 2 shows diagrams of two types (I and II) of integrated array dosimeters using square cross section air core waveguides. The square array waveguides were made from uncoated black PMMA surfaces. The light guide module was built by embedding PMMA sheets into two other sheets of PMMA. The BC 400 Saint-Gobain scintillators were sections of clear polyvinyl toluene doped with anthracene which were fitted into the end of the light guide.

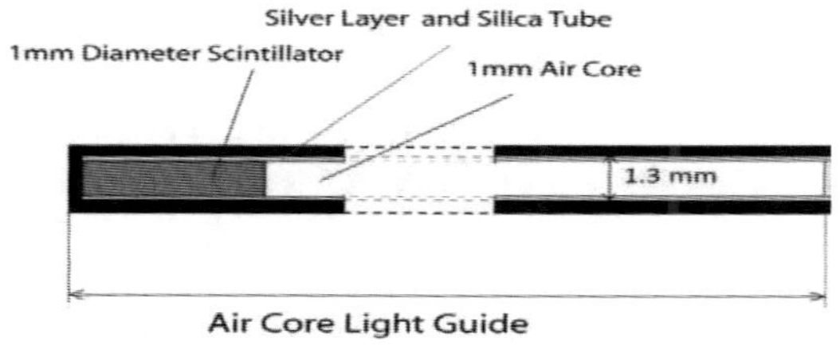

Figure 1: Schematic diagram of a scintillation dosimeter consisting of an 18 mm long silvered tube.

The performance of 3 types of the air core dosimeters (Figures 1 and 2) was tested at the Royal Prince Alfred Hospital (NSW, Australia) under 6 MV photon beams produced by a Varian linear accelerator. Measurement procedure for array prototype I dosimeter is described here. Similar procedures were conducted for other dosimeters. 1 mm diameter PMMA optical fibres held at the exit aperture of the waveguides conducted the scintillation signal to an array of H3164-10 Hamamatsu photomultiplier (PMT) array located outside the treatment bunker and its signals were read by an SIB032D Vertilon array electrometer.

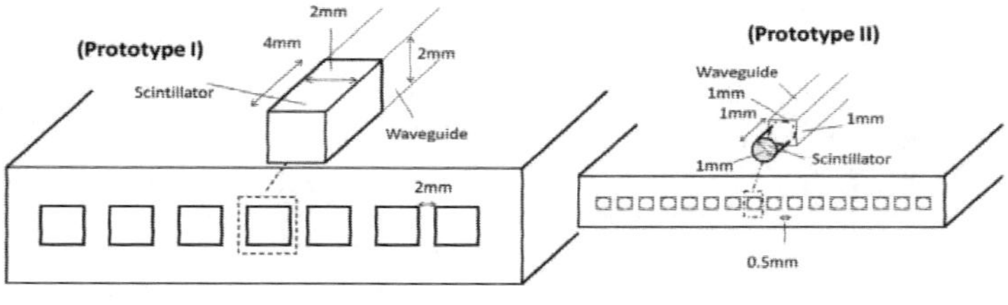

Figure 2: schematic diagrams of two prototypes of integrated array dosimeters consisting of 18 mm square air core waveguides. A magnified image of an array element is shown above the array.

Light attenuation as a function of air core waveguide length was measured by a method of insertion into the air core waveguides using 1 mm diameter optical fibre, a Jarrell-Ash monochromator set at wavelength of 455 nm, R372 Hamamatsu PMT and 617 Keithley electrometer. The waveguides included commercially sourced 1 mm diameter Polymicro silver tube, silver coated 1 mm square waveguides made from optical flats and glass slides, and the 2 mm square PMMA waveguide that was an element of the array prototype I waveguide (figure 2).

A non-imaging geometric optics method based on the formation of images in a square kaleidoscope [16] and a ray tracing method in a circular tube [17] was used to calculate attenuation as a function of waveguide length.

3. Results and Discussions

Table 2 shows that the deviation from dose to water in the 1 mm silvered air core dosimeter is less than 2.5%. The dosimeter elements can be arranged in an array configuration such that the deviation from water remains less than 2.5%. This satisfies the IRCU recommendation on accuracy in dosimetry, without the need for corrections to be applied [18].

Table 2: MC simulation results for dose deviation in the sensitive volume of parallel 1 mm diameter silvered air core dosimeters with beam orientation of normal, N, and parallel, P, to the array plane.

Number of Dosimeters	Beam Orientation	Spacing (mm)	Deviation from water (%)
1	N, P	-	0.5
2	N, P	1.3	1.5
4	P	2.6	1
14	N	2	2

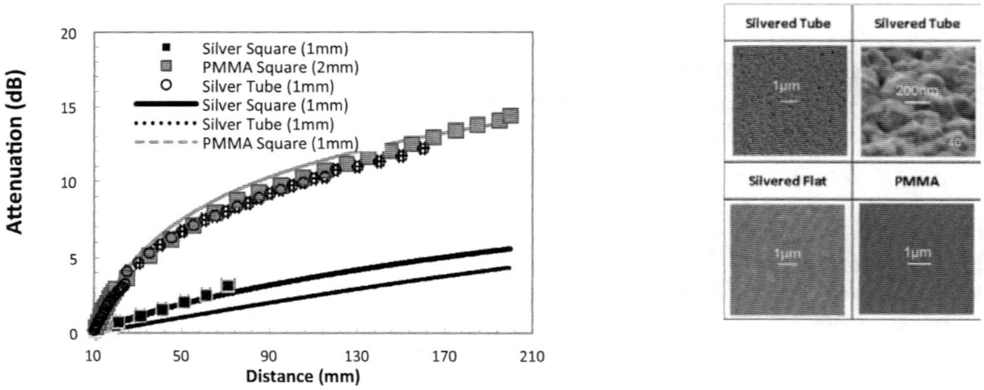

Figure 3: Attenuation as a function of waveguide distance for air core waveguides experimentally (in symbols) and theoretically (in lines), (on the left). The numbers in the brackets show the inside meter of the silver tube or the side length of the square waveguides. SEM images of the surfaces of the commercially sourced silvered tube, the silver coated glass and the PMMA are shown on the right.

Figure 3 shows that in theory, attenuation in 1 mm diameter silvered waveguide is the lowest, followed by 1 mm wide silvered square and the 1 mm wide square PMMA waveguides. However, in practice, the square silvered waveguides have better performance due to their superior reflecting surfaces.

Table 3: A comparison of the signal to noise ratio for an air core dosimeter with 1 mm diameter silvered waveguide (figure 1) with that for an air core dosimeter with 2 mm wide square PMMA waveguide (prototype I). The depth, field size and dose rate are 60 mm, 8×8 cm^2, respectively.

	Signal/ Noise	Sensitive volume (mm^3)
Round silver dosimeter	7.5 ± 0.9	3.1
Square PMMA dosimeter	3.6± 0.25	16

Table 3 shows that the uncoated 2 mm width square PMMA waveguide has a signal to noise ratio that is comparable to the 1 mm diameter commercially sourced silvered waveguide. Figure 4 shows the uniform dose distribution across the square array (prototype I) for large field sizes and the penumbra region of the fields. This figure shows that the square array dosimeter prototype I is able to quantify the field distribution with a single rapid measurement. Figure 5 shows that the array prototype II is able to provide a depth dose curve with a resolution of 1 mm that agrees with ionisation chamber measurements within 0.5%.

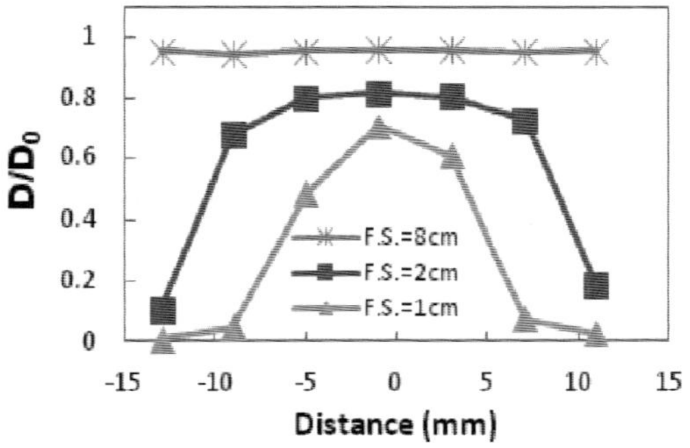

Figure 4: Relative dose as a function of distance from the beam centre at depth 53 mm for array prototype I. The dose in the scintillators, D, is normalized to their values at the 10 cm ×10 cm field size, D_0.

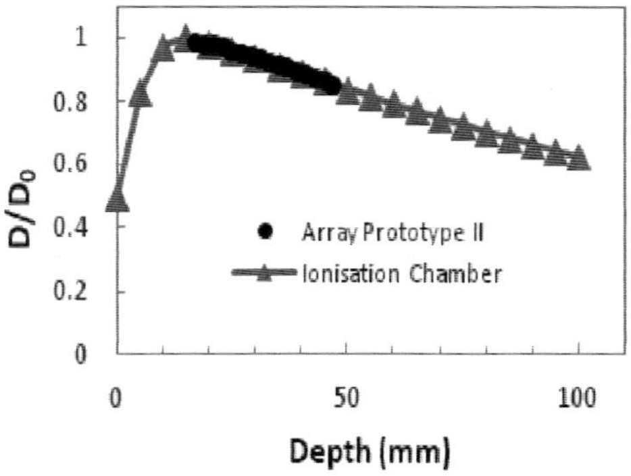

Figure 5: Percentage depth dose in Perspex measured using the array. Measurements using an ionisation chamber dosimeter are added for comparison. Measurements were made with a field size of 40× 40 mm.

4. Conclusions

It is shown that the array of 1 mm diameter air core dosimeters is water equivalent and provides a resolution of 2.6 mm which outperforms the resolution in currently commercially available array technologies. A novel integrated scintillation dosimeter using PMMA square air core waveguides provides a dose map in a large range of field sizes, presenting an efficient solution for managing the problem of Cerenkov background light in scintillation dosimetry for megavoltage beams. It is predicted that silver coating the PMMA square waveguides will decrease the signal attenuation by about 10 dB, facilitating dose mapping for even larger field sizes.

5. References

[1] Baldock C *et al* 2010 *Phys. Med. Biol.* **55** R1-63
[2] Bosi S G *et al* 2009 *Appl. Optics.* **48** 2427-34
[3] Bosi S *et al* 2007 *Phys. Med. Biol.* **52** 2893-903
[4] Bosi S G *et al* 2009 *Phys. Med. Biol.* **54** 275-83
[5] Bloemen-van Gurp E J 2009 *Int. J. Radiat. Oncol.* **73** 314-21

[6] Archambault L 2007 *Med. Phys.* **34** 1583-92
[7] Cilla S *et al* 2008 *Nucl. Instrum. Meth. B* **266** 2643-50
[8] Zhou X 2009 *Med. Phys.* **36** 523-9
[9] Lavallee M-C *et al* 2006 *Med. Phys.* **33** 3683-9
[10] Zilio V O 2006 *Med. Phys.* **33** 1532-9
[11] Colussi V C *et al* 2001 *J. Appl. Clin. Med. Phys.* **2** 210-8
[12] Hoban P W *et al* 1994 *Phys. Med. Biol.* **39** 1219-29
[13] Planskoy B 1980 *Phys. Med. Biol.* **25** 519-32
[14] Clift M A 2000 *Phys. Med. Biol.* **45** 1165-82
[15] Lambert J *et al* 2010 *Phys. Med. Biol.* **55** 1115-26
[16] Cheng YK and Chern JL 2006 *J. Opt. Soc. Am. A* **23** 427-34
[17] Miranda C P 1996 *J Heat Trans-T Asme* **118** 495-7
[18] Dyk JV 1999 *The modern technology of radiation oncology: A compendium for medical physicist and radiation oncologists* (Madison, Wisconsin: Medical Physics Publishing)

Journal of Physics: Conference Series **444** (2013) 012062 doi:10.1088/1742-6596/444/1/012062

Laser cone beam computed tomography scanner geometry for large volume 3D dosimetry

KJ Jordan[1,2], D Turnbull[1] and JJ Batista[1,2]

[1]London Regional Cancer Program, London Health Sciences Centre
[2]Department of Medical Biophysics, Western University, London, ON, Canada

E-mail: kevin.jordan@lhsc.on.ca

Abstract. A new scanner geometry for fast optical cone-beam computed tomography is reported. The system consists of a low power laser beam, raster scanned, under computer control, through a transparent object in a refractive index matching aquarium. The transmitted beam is scattered from a diffuser screen and detected by a photomultiplier tube. Modest stray light is present in the projection images since only a single ray is present in the object during measurement and there is no imaging optics to introduce further stray light in the form of glare. A scan time of 30 minutes was required for 512 projections with a field of view of 12 x 18 cm. Initial performance from scanning a 15 cm diameter jar with black solutions is presented. Averaged reconstruction coefficients are within 2% along the height of the jar and within the central 85% of diameter, due to the index mismatch of the jar. Agreement with spectrometer measurements was better than 0.5% for a minimum transmission of 4% and within 4% for a dark, 0.1% transmission sample. This geometry's advantages include high dynamic range and low cost of scaling to larger (>15 cm) fields of view.

1. Introduction

There are several distinct geometries of optical CT scanners that have been reported for 3D dosimetry applications [1]. These geometries range from broad beam – 2D detector array to *single ray – single detector* systems [2, 3]. Scan speed is an important factor for practical implementation of 3D dosimetry especially in larger volumes. For example, two hours from beginning of an irradiation to completion of data analysis is considered a reasonable goal for 3D gel dosimetry [4]. Fast scanning may be essential for certain applications such as small field dosimetry with strong dose gradients that drive diffusion [5]. Fast scanning is a convenient method to obtain accurate 3D doses compared with conventional dosimeters that require hours or days to map the irradiation field.

While there are many optical CT geometries that provide accurate scanning for small transparent dosimeters, there are few reports on scanning larger dosimeters. For the purposes of this report, large dosimeters are considered to have diameters between 10 and 30 cm. Accuracy is to be maintained within 2% of true attenuation coefficient at the required spatial resolution (typically 1 mm isotropic voxel size). A recent example of performance in a transparent, radiochromic plastic dosimeter has been reported [6]. As the size of the dosimeter increases, the amount of optical scatter and zero dose absorption increase. This places greater importance on stray light correction and dynamic range of scanners. Large polymerization dosimeters that have an increase in optical scatter with dose are particularly challenging for optical CT readout. Efficient phantom scatter rejection has been achieved using telecentric optics and scanned laser beams. However, as the field of view (FOV) increases the cost of imaging quality, optical components becomes prohibitive for instruments intended for radiotherapy clinics. Fresnel lenses are generally not considered adequate for imaging purposes. But the spatial resolution required for 3D megavoltage dosimetry is lower, approximately 1 mm. At this resolution, Fresnel lenses may be a cost effective approach [7]. There is now a commercial optical CT geometry that includes Fresnel lenses from MGS Research Inc.

In this report, a new optical CT geometry is reported that provides fast and accurate scanning of large dosimeters with less expensive optics. It is based on raster scanning of a single, narrow laser beam in a cone-beam (CBCT) geometry through a large dosimeter, immersed in a refractive index matching aquarium. The transmitted beam is scattered by a thin white film and detected by a single photomultiplier tube avoiding stray light introduced by camera lenses.

2. Methods

A photograph of the laser CBCT scanner is shown in Figure 1. The prototype scanner includes: yellow He-Ne laser (attenuated to ~10 microwatts), spectral filter (10 nm bandpass, central wavelength = 594 nm), 1 mm diameter spatial filter, 2D scanning galvo mirror system (Thorlabs), aquarium (FOV = 18 cm), sample jar (15 cm diameter PETE jar, Modus Medical Devices Inc.), diffusive screen detector (white Mylar film), secondary diffuser of white plastic sheet and a photomultiplier tube (PMT). The current from the PMT was amplified with an in-house electronic circuit and digitized with a 12 bit PC based data acquisition card. A Matlab program was developed for cone beam reconstruction using a graphics processing unit (GPU, Nividia) and CUDA routines. Aqueous carbon black solutions were prepared similar to previous reports, Jordan [8] using Triton X100 (2 mM) and hydrogen peroxide as an anti-microbial agent (1 mM). Absorption coefficients were measured independently with a visible light absorption spectrometer (Hitachi-Perkin Elmer Model 204).

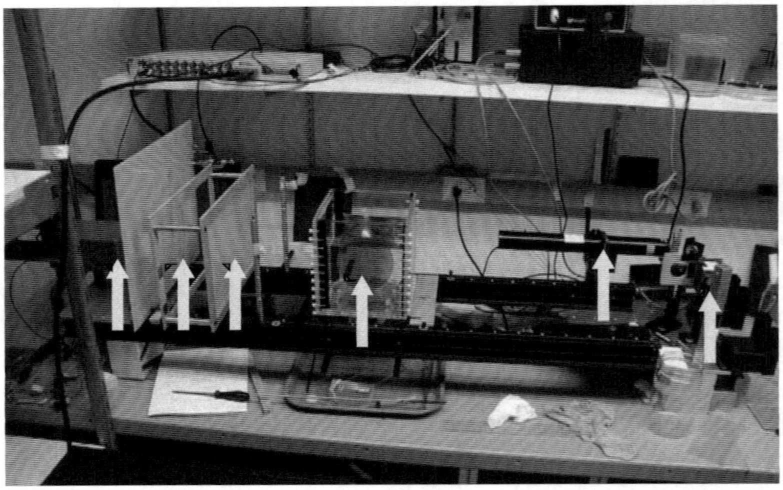

Figure 1: Photograph of laser CBCT scanner. Components highlighted by arrows from left to right: photomultiplier tube, secondary diffuser, primary diffuser, aquarium with gel-filled balloon, laser, galvo mirror scanner.

3. Results and discussion

Scan time for acquisition of 512 projections, 200 slices, 512 projection samples per slice over a FOV of 17 cm x 12 cm is 30 minutes. Full 3D reconstruction including reading of input data is <20 seconds for a 512x512x200 array.

3.1. Spatial resolution

Figure 2, is a transmission image through a plastic ruler, demonstrating submillimeter spatial resolution for the projection images. The full width half maximum of the laser beam was 0.6 mm in the rotation axis reference plane.

Figure 2: Transmission image of plastic ruler in scanner rotation axis plane.

3.2. Uniformity and accuracy

Figure 3, presents a central slice from two carbon black solutions representing a dark object (0.1% transmission) and an intermediate opacity object with transmission of 4%. Figure 4, is a plot of the corresponding horizontal profiles, showing uniformity within 85% of radius. This uniformity varies by only 2% from top to bottom of the jar. Reconstruction ring artifacts are due to reflections within the aquarium system and correction methods including antireflective windows and flat black surfaces are being considered. A large cone of stray light, corresponding to 2% of the laser beam intensity, illuminated the aquarium. The source of this stray light is scatter from the galvo mirrors. Mirror replacement is expected to minimize this stray light component. A stray light measurement was performed by placing a beam block at the entrance of the aquarium and sampling the signal in the shadow. This value was subtracted for each pixel in the projection image, approximating the stray light as a constant. Mean attenuation coefficients for the solutions measured with spectrometer and laser CBCT, were 0.220/ 0.220 and 0.453/ 0.473 (cm^{-1}) respectively. An error in the stray light estimate for the dark solution may be the cause of the CT attenuation being greater than the spectrometer measurement.

4. Conclusion

The combination of raster sweeping a single laser beam and a single, large-area detector provides a quantitative, tomographic optical CBCT scanner geometry for fast 3D measurements of large dosimeters. A scan time of 30 minutes was demonstrated for a FOV of 12 x18 cm. Uniformity to within 2% of the mean value was recorded over the useful 12 cm jar height and the central 85% radially. The use of vessels with lower refractive indexes would allow quantitative scanning of a larger fraction of the dosimeter volume. The single-ray single-detector geometry provides excellent stray light rejection for large FOV's. Stay light rejection is important for quantitative, high dynamic range optical CT scanning of large dosimeters that can represent patient volumes exposed in radiotherapy.

5. Acknowledgements

This work was funded by: London Regional Cancer Program small grants, CIHR grant # 200403MOP, D Turnbull was partially supported by Ontario R&D Challenge Fund (ORDCF-OCAIRO Project)

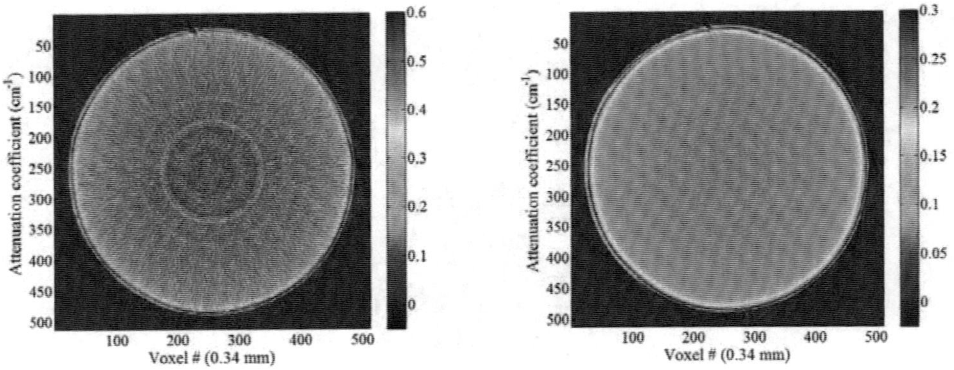

Figure 3: Central transverse slices for reconstructions with carbon black solutions in 15 cm jars with 0.1% (left) and 4% (right) transmission along central ray. Note that the two rings in darker, left image have been attributed to stray reflections within the aquarium.

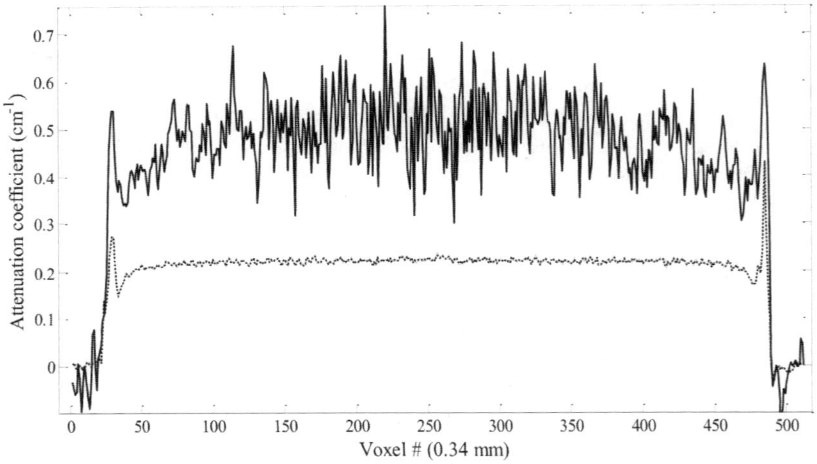

Figure 4: Horizontal, central profiles for reconstructed slices of phantoms of Figure 3. Note the uniform response is limited to the central 85% of the diameter due to the refractive mismatch of the contained solution and jar wall.

6. References

[1] Baldock C *et al* 2010 Polymer gel dosimetry *Phys. Med. Biol.* **55** R1-63
[2] Doran S 2009 *Appl. Radiat. Isotop.* **67** 393-8
[3] Jordan K 2005 *J. of Med. Phys.* **30** 15-20
[4] Oldham M *et al* 2003 *Med. Phys.* **30** 623-34
[5[Babic S *et al* 2009 *Phys. Med. Biol.* **54** 2463-81
[6] Thomas A *et al* 2011 *Med. Phys.* **38** 4846-57
[7] Conklin J *et al* 2006 *J. Phys: Conf Ser* **56** 030
[8] Jordan K and Battista J 2009 *J. Phys: Conf. Ser.* **164** 012045

Effect of refraction on dose reconstruction in optical-CT gel dosimetry

L Florescu[1], G Ambartsoumian[2] and C-S Wuu[1]
[1]Department of Radiation Oncology, Columbia University, New York, NY 10032, USA
[2]Department of Mathematics, University of Texas at Arlington, Arlington, TX 76019, USA

E-mail: csw6@columbia.edu

Abstract. We address the problem of dose reconstruction based on limited experimentally accessible data due to the effect of refraction in optical-CT gel dosimetry. The refractive index mismatch between the components of the optical-CT scanner result in light scattering and ultimately in the inability to capture parts of the projection datasets. We determine the maximum loss of data and the corresponding refractive index mismatch for which accurate dose reconstruction in the central part of the phantom is still possible. Also, a mathematical formalism that indicates how exact reconstructions can be obtained using a priori knowledge of the optical attenuation coefficient of the gel is presented. This study establishes rigorous design principles for accurate 3D dose reconstruction.

1. Introduction

Advances in radiation therapy and in particular new radiotherapy procedures such as intensity modulated radiation therapy or high dose rate brachytherapy call for the development of new techniques for accurate 3D dose reconstruction [1]. In this context, optical-CT gel dosimetry has emerged as a promising candidate for 3D dose verification. Gel dosimeters utilize sensitive chemicals (gels), which, upon irradiation, polymerize as a function of absorbed dose and therefore have the capability to uniquely record dose distribution in three dimensions. In optical-CT gel dosimetry, dose distribution is obtained from the CT 3D reconstruction of the optical attenuation coefficient of a gel phantom [2,3].

In an optical-CT scanner for gel dosimetry [1], a laser beam illuminates a cylindrical gel phantom and a photo-detector detects the transmitted light on the opposite side of the phantom. The light source and photo-detector are mechanically coupled and translated synchronously laterally with respect to the sample. After each lateral scan, the gel phantom is rotated. The scan-rotate process is repeated until the phantom has been rotated through 180°. Multiple optical projections acquired this way enable the reconstruction of the optical attenuation coefficient of the gel in a 2D transverse slice. Image reconstruction in different transverse slices is obtained by moving the gel phantom vertically, and 3D reconstruction is performed slice-by-slice. Absorbed dose is then reconstructed from the attenuation coefficient assuming a linear dependence between the two.

To avoid artifacts resulting from the light refraction at the air-gel interface, the gel phantom is placed in a square tank containing a liquid having the same refractive index as the gel. However, in practice, the refractive indexes of the gel and liquid do not match perfectly, resulting in scattering of the laser beam at the gel-liquid interface and therefore in spatial and angular deflections of the beam passing through the gel phantom. This ultimately results in loss of data. For large enough deflections, corresponding to beams incident at positions close to the edge of the phantom, the detector fails to capture the transmitted light. Therefore, it is of interest to investigate how the inability to scan the lateral sides of the phantom influences the reconstruction of the optical attenuation coefficient (and then dose reconstruction) in the central portion of the phantom that is usually irradiated.

The purpose of this study is to investigate the effect of incomplete data on dose reconstruction, and to identify the optimal design for an accurate gel dosimeter. Numerical experiments are carried out for a series of model attenuation coefficients of the gel phantom and various degrees of data availability corresponding to situations where only projections through certain central sections of the sample are available. A mathematical study for dose reconstruction based on truncated data is also performed in the framework of the so-called interior problem of Radon transform inversion.

2. Dose reconstruction

Since in optical-CT gel dosimetry dose is determined from the optical attenuation coefficient of the gel phantom, the problem of dose reconstruction becomes the problem of reconstruction of the optical attenuation coefficient. Optical CT tomography is based on inverting the Radon transform $R\mu(t,\theta)$ (line integral) of the total attenuation coefficient $\mu(x,y)$,

$$\Phi(t,\theta) = \mathcal{R}\mu(t,\theta) = \int_{-\infty}^{\infty} \mu(t\cos\theta - s\sin\theta, t\sin\theta \quad s\cos\theta)ds, \tag{1}$$

to obtain the attenuation coefficient in a given transverse slice from measured data $\Phi(t,\theta)$. Here t and θ are the position of the incident laser beam with respect to the central axis and the rotation angle, respectively. The problem of image reconstruction in the case of limited data due to the refractive index mismatch consists on recovering $\mu(x,y)$ from its Radon transform data $\Phi(t,\theta)$ known for all possible angles θ but limited range of distances from the origin t.

First we address the problem of image reconstruction based on limited data numerically. For this purpose, we consider a model $\mu_{\text{model}}(x,y)$ for the attenuation coefficient of the gel. The measured data is calculated as the Radon transform (1) of the model attenuation coefficient $\mu_{\text{model}}(x,y)$. We assume that there are available only optical projections for a central cylindrical volume of radius $L \le R$, where R is the radius of the gel cylinder. This is equivalent to setting $\Phi(t,\theta) = 0$ for $|t| > L$. Finally, using data calculated this way, the attenuation coefficient is obtained by inverting (1) numerically using a filtered back projection algorithm with the Shepp-Logan filter.

In our study, the model attenuation coefficient is decomposed as $\mu_{\text{model}}(x,y) = \mu_0 + \delta\ (x,y)$, where μ_0 is the optical attenuation coefficient of the non-irradiated gel, and $\delta\mu(x,y)$ corresponds to the inhomogeneity produced by irradiation. We assume that the irradiated volume is a cylinder of radius $r < R$ having the same central axis as the gel phantom, and model $\delta\mu(x,y)$ as

$$\delta\mu(x,y) = \begin{cases} \mu_0, & (x^2+y^2)^{1/2} \le r \\ 0, & \text{otherwise} \end{cases} . \tag{2}$$

Reconstructions of the total attenuation coefficient $\mu(x,y)$ are shown in figure 1 for $L/R = 100\%$, 85% and 75%. We note that image quality for the sides of the phantom depends on data availability. However, the image in the central part of the sample is little influenced by the loss of data. This can be also seen in figure 2, where we present cross sections through the reconstructed images. Also, figure 2 shows that $\mu(x,y)$ can be reconstructed with a relative error of less than 3% in the central part of the sample even when only about 82.5% of the projections are available, and this result does not depend on the size of the irradiated area (the graphs for different values r of overlap).

Journal of Physics: Conference Series **444** (2013) 012063 doi:10.1088/1742-6596/444/1/012063

Using geometrical optics considerations, we calculate the spatial deflection δy of the laser beam at the detector position as a result of the refractive index mismatch $\delta n = n_g - n_l$, where n_g and n_l are the refractive indices of gel and liquid, respectively. This deflection is presented in figure 3 for $n_g = 1.3556$. Assuming a detector spatial width of 0.5 cm, we can estimate from figure 3 that more than 82.5% of data can be collected if the refractive index mismatch is less than 1.5%.

Figure 1: Going from left to right, the figures show the model attenuation coefficient $\mu_{\text{model}}(x, y)$ and the reconstructed attenuation coefficient $\mu(x, y)$ for $L/R = 1$, 0.85, and 0.75, for a sample where $r = R/2$.

Next, we address the problem of image reconstruction in the limited data case mathematically. It can be shown that the difference between any two solutions of this "interior problem" varies very little in the interior of the imaging domain [4]. For this purpose, assume that $\mu_1(x, y)$ and $\mu_2(x, y)$ are the true image function and the image function reconstructed from truncated data, respectively. If $u = \mu_1 - \mu_2$, then by the standard inversion formula of Radon transform one can obtain [4]

$$u(x, y) = \frac{1}{4\pi^2} \int_0^{2\pi} \int_{|t|>L} \frac{\mathcal{R}u(t, \theta)}{(t - x\cos\theta - y\sin\theta)^2} \, dt \, d\theta \tag{3}$$

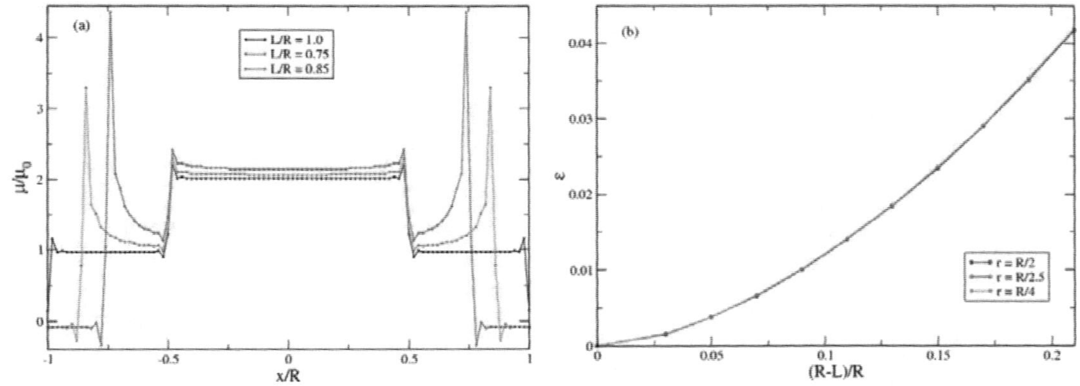

Figure 2: a) The reconstructed attenuation coefficients $\mu(x, y)$ presented in Fig. 1 shown as cross sections along a line drawn through the center of the phantom in the directions parallel to the x-axis. b) The relative error, $\varepsilon = |\mu - \mu_{\text{model}}| / \mu_{\text{model}}$ for the central part of the sample for various sizes of the irradiated area.

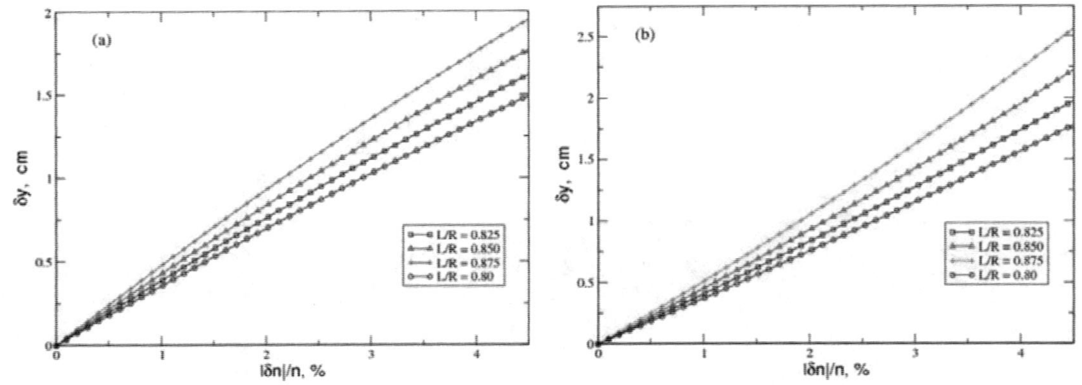

Figure 3: Spatial deflection δy of the laser beam caused by the refractive index mismatch $\delta n = n_g - n_l$ for (a) $\delta n > 0$ and (b) $\delta n < 0$, for the case where $n_g = 1.3556$, the tank side is 22 cm, and the detector is placed 2 cm away from the tank.

Using Cauchy-Schwarz inequality it follows that if $|(x,y)| \le r < L$ then $|u(x,y) - u(0,0)| \le C(L,r)\|\mathcal{R}u\|_{L_2}$, where $\|\mathcal{R}u\|_{L_2}$ is the L_2 norm of the function $\mathcal{R}u$. The value of $C(L,r)$ is small and depends on how much data is lost and how close to the center we want the estimate. For example, if $L = 0.8R$, and $r = 0.2R$ then $C(L,r) \approx 0.016$ [4]. On the other hand, using geometric considerations one can show [4] that $\|\mathcal{R}u\|_{L_2} \le 2\sqrt{\pi}(R^2 - L^2)^{1/4}\|\mu_1\|_{L_2}$. The norm $\|\mu_1\|_{L_2}$ of the image function is a bounded quantity that can be estimated depending on the properties of the material. Since L is close to R, and $C(L,r)$ is small, it follows that $u(x,y)$ is almost a constant, $u(x,y) \approx u(0,0) = c$ in the interior of the cylinder. Moreover, since certain parts of the image area are not irradiated, one may know ahead of time the values of $\mu_1(x_0, y_0)$ in that area. The values $\mu_2(x_0, y_0)$ will be reconstructed from truncated data. Hence one can compute the constant error $u(x_0, y_0) = \mu_1(x_0, y_0) - \mu_2(x_0, y_0) = c$. Finally, adding the computed constant c to the reconstructed image μ_2 we get $\mu(x, y) = \mu_2(x, y) + c$, which is a good approximation of the original image in the interior of the cylinder. The quantitative analysis of the relation between L and r to obtain the desired accuracy can be done numerically.

3. Conclusion

It was demonstrated numerically that the optical attenuation coefficient of the polymer gel and dose in an optical CT gel dosimeter can be reconstructed accurately in the central part of the dosimeter even if only incomplete experimental data are available. In particular, dose reconstruction with uncertainty less than 3% is possible even if only a central area of radius 82.5% of the radius of the gel cylinder is scanned. For usual gel dosimeters, this fraction of available data corresponds to a refractive index mismatch between the gel and liquid of 1.5%. It was also shown that a priori knowledge of the attenuation coefficient at a point in the sample enables an almost exact reconstruction of the attenuation coefficient at all points. This study provides a guideline for optimal design of optical-CT gel dosimeters.

4. References

[1] Gore J C *et al* 1996 *Phys. Med. Biol.* **41** 2695
[2] Bosi S *et al* 2007 *Phys. Med. Biol.* **52** 2893-903
[3] Bosi S G *et al* 2009 *Phys. Med. Biol.* **54** 275-83
[4] Natterer F 1986 *The Mathematics of Computerized Tomography* (New York: Wiley)

Characterization of a cone beam optical scanner

PB Ravindran[1] and HM Thomas[2]
[1]Department of Radiation Oncology, Christian Medical College, Vellore, India and
[2] Photonics Division, School of Advanced Sciences, Vellore Institute of Technology, Vellore, India

E-mail: paul@cmcvellore.ac.in

Abstract. The use of radiochromic FX gel for mapping 3D dose distribution is hampered by the diffusion of gel and the slow scanning techniques. The development of fast optical cone beam scanning has improved the chances of using radiochromic gel as a feasible dosimeter for radiotherapy applications. In this work an optical cone beam scanner has been developed in-house and its performance characteristics have been studied. The reconstructed image of the optical scanner was analyzed by studying the resolution, signal-to-noise ratio and contrast to noise ratio (CNR). The resolution of the optical cone beam CT scanner was studied by scanning a catheter of 1 mm outer diameter and the scanner was able to detect the catheter. The geometrical accuracy of the reconstruction was studied by placing catheters in spiral geometry in the gel phantom and measuring the distances. It has been observed that the in-house Optical Cone beam scanner is suitable for scanning radiochromic gels for radiotherapy applications.

1. Introduction

Over the half decade, external beam radiation therapy has improved due to i) the advance of high-energy linear accelerator with computer controlled multileaf collimators (MLCs) and on-line imaging, ii) the development of improved imaging to localize tumors for treatment planning, and iii) significant advancement in the treatment planning system with more robust dose calculation algorithms and the implementation of the inverse planning modules. One of the challenges with newer radiation delivery techniques that aim at precise delivery of radiation dose is to develop an accurate and efficient dosimetry system capable of providing full three-dimensional dose distribution. It has been accepted that patient specific quality assurance of radiation dose delivered is to be performed for intensity modulated radiotherapy techniques. Ion Chamber, TLD and film have been routinely used for these measurements. However, ion chamber can only provide point dose and although film can provide two dimensional dose distributions; has its limitations for absorbed dose measurements. Several authors have investigated the use of Fricke gel dosimeters. However, due to issues of diffusion [1, 2], alternative polymer gel dosimeters were developed.

Polymer gel dosimeters are manufactured from radiation sensitive chemicals, which upon irradiation polymerize as a function of the absorbed radiation dose [3]. These gel dosimeters which record the radiation dose distribution in three-dimensions (3D) have specific advantages when compared to one-dimensional dosimeters and two-dimensional dosimeters [4]. These 3D dosimeters are radiologically soft-tissue equivalent [5, 6] with properties that may be modified depending on the application. The 3D radiation dose distribution in polymer gel dosimeters may be imaged using magnetic resonance imaging (MRI) [7, 8], optical-computerized tomography (optical-CT) [9-11], x-ray CT [12, 13], ultrasound [14-16] or vibrational spectroscopy [17,18].

Due to the practical problems in availability and cost of MRI, optical scanning of gel has been used by several researchers as a more accurate and practical alternative to MRI.

Also, the use of radiochromic gel with optical scanning has issues such diffusion and hence optical scanning has not been predominantly used for obtaining dose distribution. However, volumetric scanning with optical cone beam CT has considerably reduced the scanning time and also provides full volumetric information of the dose matrix embedded in the gel. An optical cone beam CT (CBCT) scanner has been developed in this department and it is necessary to characterize the scanner before use [19, 20]. In this paper we discuss the construction and performance characteristics of an in-house optical cone beam CT scanner for use in radiotherapy dosimetry.

2. Methods and Materials

2.1. Development of Optical Cone Beam Scanner

2.1.1. Optical Cone Beam Scanner specifications

An optical cone beam CT scanner consists of a uniform light source, an aquarium with a turntable and a CCD camera (figure 1). LED array containing 192 LEDs of wavelength 540 nm is placed in a circular geometry and modified as a flat light source by placing a diffuser at 20 cm distance in front of the LEDs. The light source illuminating the container was collimated to 10 x 15 cm to reduce the scattered light reaching the camera. The turntable is attached to a four-pole stepper motor with reduction gear that reduces the step angle from 1.8 degrees to one degree. The stepper motor is interfaced to a computer through a simple driver circuit that uses UNL 2004 IC and is connected to the parallel port of the PC. The PointGray® CCD camera is also controlled by this PC. The gain, brightness, shutter timing and image acquisition are controlled by the software.

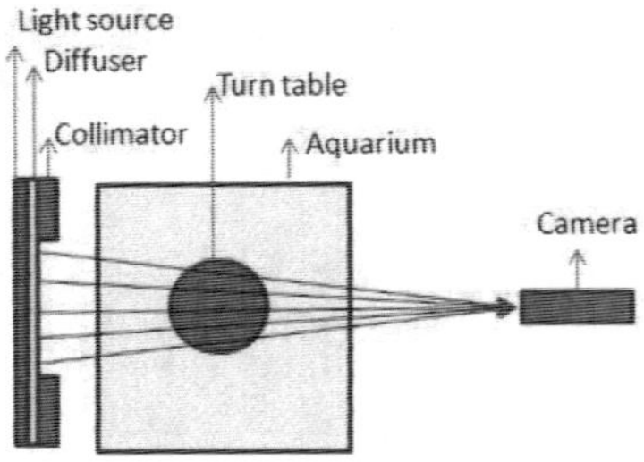

Figure 1: Schematic representation of the Optical Scanner.

2.1.2. Gel Phantom

In this work we used FX gel prepared by adding 4% Gelatin to 50 mM H_2SO_4, 0.5 mM ferrous ammonium sulphate and 0.05mM xylenol orange at 37°C in a water bath. The gel phantom was refrigerated at 5°C for about 4 hours for gelation and brought to room temperature before irradiation. The gel was prepared in a thin walled polystyrene cylinder of diameter 10 cm and height 15 cm. However, in this work on characterization of the scanner, for certain experiments, liquid that simulate the optical attenuation properties of gel has been used.

2.1.3. Image acquisition Software

The software for driving the stepper motor and for acquiring the image from the camera has been developed with visual basic. The user is required to select between the two scanning modes available, i.e half rotation scan or full rotation scan. Half rotation scan acquires images for 194° (i.e 180° plus the cone angle) and full cone beam acquires images for 360°. In order to obtain the optical attenuation coefficients, the gel phantom is scanned twice, once before irradiation of the gel (pre-scan) and then after the irradiation (post scan). The software has a provision to either select the pre-scan or the post-scan. The images are acquired at every degree of rotation and stored in bitmap format in the respective folder created for pre-scan or post scan. The software also has provision to adjust the gain and the shutter values of the camera for each scan.

2.1.4. CT reconstruction

The image reconstruction codes have been developed using MATLAB®. Two routines have been developed in Matlab, one to stack the acquired images in to a 3D dataset as pre-scan and post-scan image sets and the other to perform the reconstruction. Each image acquired is of size 640 x 480 pixels and for a half-rotation scan there will be 194 such images. This routine stacks all the 2D image matrices together to form a 3D matrix of size 194 x 640 x 480 for a half cone beam acquisition and similarly a matrix of 360 x 640 x 480 for full cone beam acquisition. This is performed for both the pre-scan and post-scan images and stored in separate user defined files. The second routine reads these files as input and determines the attenuation coefficient for each pixel and reconstructs the selected plane using the 'ifanbeam' function implemented in Matlab. A bespoke FDK based 3D reconstruction software was also developed for the optical cone beam scanner and the details of which are mentioned elsewhere [21].

2.1.5. Characterization of the Optical cone beam CT scanner

Uniformity, resolution, contrast and sensitivity of the optical CBCT scanner were verified using methods similar to those suggested by Oldham *et al* [22]. Uniformity was checked by scanning a liquid with a color solution having attenuation coefficient similar to gel irradiated at 1 Gy. The uniformity of the 2D reconstruction of the attenuation coefficient (μ) was verified by measuring the mean μ in the region of interest (ROI) of 10 x 10 pixels.

To quantify the spatial resolution and geometrical distortion of the scanner, a phantom similar to the spiral needle phantom [22] was fabricated using capillaries of 1.5 mm diameter. The modulation transfer function (MTF) was used to quantify the spatial-frequency response characteristics of the optical-CT. The contrast resolution at various points in the gel phantom was verified using a finger phantom [22]. Three different liquids with varying attenuation were used. The SNR and CNR for the fingers of varying attenuation coefficients were calculated using the formula given below.

$$SNR=20_{log}\,(A_{mean}/\,\sigma_B)\ in\ dB$$

Where, A is the region of interest in the image and σB is the standard deviation of points in the background region.

$$CNR=A_{mean}\text{-}B_{mean}/\sigma_B$$

Where, A is the region of interest in the image and B is the background region in the image.

3. Results

The uniformity was within ±5% in the central region except for one point where it was 8% and varied to about 20% at points close to the wall of the gel container. The contrast resolution observed with the finger phantom was found to be acceptable. The measurement of spatial resolution showed that a capillary of size 1.5mm diameter could be well resolved with the Opt. CBCT scanner. However, the artefacts near the wall were seen to affect the measurement.

The magnified image of the 0.9 mm copper wire phantom is shown in Fig. 2(a). The full-width at half-maximum of the reconstructed wire was 1.06 mm. the radon transform of the images were obtained using the radon function of MatLAB. The MTF was computed by fast Fourier transform of the Radon transform of the images, and the results are shown in Fig 2(c). The system has sub millimetre resolution as the MTF was greater than 10% out to approximately 0.9 mm-1 on the average.

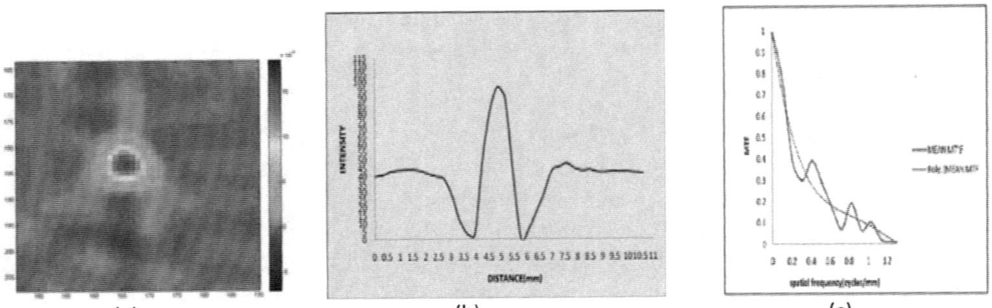

(a) (b) (c)

Figure 2: (a) Magnified reconstructed image of thin copper wire of thickness 0.9mm (b) FWHM profile of copper wire (c) MTF plot of copper wire

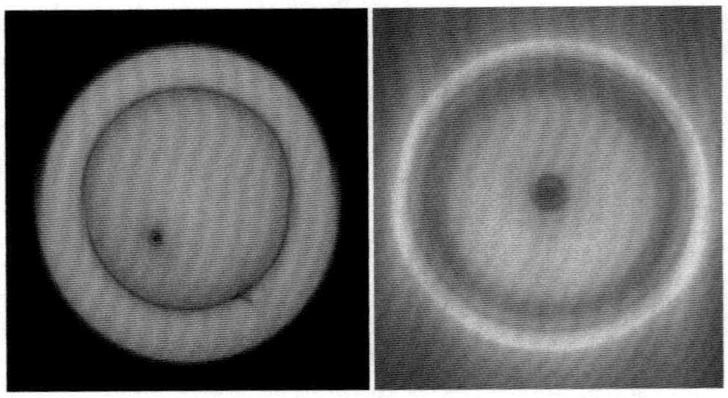

Figure 3: a) reconstructed image of the capillary b) reconstructed image of the finger phantom using FDK based reconstruction software

4. Conclusion

An optical CBCT scanner has been constructed for reading the dose distribution embedded in the gel matrix. Its performance characteristics like uniformity, spatial and contrast resolutions and sensitivity were studied and results were found to be promising. Software based on FDK algorithm for volumetric reconstruction has been developed and ensured acceptable results.

5. References

[1] Harris P J *et al* 1996 *Phys. Med. Biol.* **41** 1745-53
[2] Baldock C *et al* 2001 *Australas. Phys. Eng. Sci. Med.* **24** 19-30
[3] Baldock C *et al* 1998 *Phys. Med. Biol.* **43** 695-702
[4] Baldock C *et al* 2010 *Phys. Med. Biol.* **55** R1-63
[5] Keall P and Baldock C 1999 *Australas. Phys. Eng. Sci. Med.* **22** 85-91
[6] Brown S *et al* 2008 *Appl. Radiat. Isot.* **66** 1970-74
[7] Gustavsson H *et al* 2004 *Phys. Med. Biol.* **49** 227-41
[8] Lepage M *et al* 2002 *Phys. Med. Biol.* **47** 1881-90
[9] Bosi S *et al* 2007 *Phys. Med. Biol.* **52** 2893-903
[10] Bosi S G *et al* 2009 *Phys. Med. Biol.* **54** 275-83
[11] Bosi S G *et al* 2009 *Appl. Opt.* **48** 2427-34
[12] Trapp J V *et al* 2002 *Phys. Med. Biol.* **47** 4247-4258
[13] Hill B *et al* 2005 *Br. J. Radiol.* **78** 623-30
[14] Mather M L and Baldock C 2003 *Med. Phys.* **30** 2140-2148
[15] Mather M L *et al* 2002 *Phys. Med. Biol.* **47** 1449-58
[16] Mather M L *et al* 2003 *Ultrasonics* **41** 551-9
[17] Baldock C *et al* 1998 *Phys. Med. Biol.* **43** 3617-27

[18] Rintoul L *et al* 2003 *Appl. Spectrosc.* **57** 51-7
[19] Olding T *et al* 2010 *Phys. Med. Biol.* **55** 2819–40
[20] Olding T and Schreiner LJ 2011 *Phys. Med. Biol.* **56** 1259–79
[21] Thomas H M *et al* 2007 IEEE Nuclear Science Symposium Conference Record **6** c1
[22] Oldham M and Kim L 2004 *Med. Phys.* **31** 1093-1104

How effective can optical-CT 3D dosimetry be without refractive fluid matching?

L Rankine and Oldham
Department of Radiation Oncology,
Duke University Medical Center, Durham, USA

E-mail: leith.rankine@duke.edu

Abstract. Achieving accurate optical CT 3D dosimetry without the use of viscous refractive index (RI) matching fluids would greatly increase convenience. Software has been developed to simulate optical CT 3D dosimetry for a range of scanning configurations including parallel-beam, point and converging light sources. For each configuration the efficacy of 3 refractive media were investigated: air, water, and a fluid closely matched to Presage (RI = 1.00, 1.33 and 1.49 respectively). The results revealed that the useable radius of the dosimeter (i.e. where data was within 2% of truth) reduced to 68% for water-matching, and 31% for dry-scanning in air. Point source incident ray geometry produced slightly more favourable results, although variation between the three geometries was relatively small. The required detector size however, increased by a factor six for dry-scanning, introducing cost penalties. For applications where dose information is not required in the periphery, some dry and low-viscous matching configurations may be feasible.

1. Introduction

Immersion of dosimeters in a refractively matched fluid is a ubiquitous technique to minimize refraction artifacts in optical-CT 3D dosimetry [1-5]. Radiochromic dosimeters [6] like PRESAGE® [7] have high refractive index (RI) of ~1.51. Transparent solutions with this magnitude RI tend to be viscous, and require regular cleaning as they accumulate dust and particles from dosimeters. Obtaining an accurate match can be labor intensive, and this match may need to be changed depending on the RI consistency of different dosimeters. Although refractive matching with fluids is generally considered highly effective, it does introduce a substantial inconvenience factor [8]. Early work reported the feasibility of dry-scanning for small volumes with a laser microbeam CT [5, 9]. Recently several works have begun to explore the feasibility of dry-scanning for larger volumes [10-12]. The present work builds on these efforts, and presents simulations of various optical-CT configurations and refractive matching scenarios.

2. Method

The ScanSim software [11] was used to simulate optical CT imaging of a PRESAGE® dosimeter typically used in clinical RPC head phantoms [13]. Fixed dosimeter parameters were radius, $r_0 = 50$ mm, RI = 1.501, sensitivity = 0.023 OD/cm/Gy, and background un-irradiated Presage optical density (OD) = 0.003 /cm. The simulated *true* dose distribution was a uniform 5 Gy to the entire dosimeter.

Attenuation was calculated using integrals on rays that were spaced 0.2 mm apart on incidence, and all rays were assumed detectable, regardless of refracted angle. Filtered back-projection (FBP) was used to reconstruct the *measured* dose distribution from the simulation's pre- and post-scan intensity sinograms. The reconstructed image gives values in optical density (OD), which are then converted back to dose by normalizing to the true dose at a known point, in this case D_{max} at the center of the dosimeter.

The controlled variables of the simulations were RI of the media surrounding the PRESAGE® dosimeter and the incident ray geometry. A total of five ray geometries were investigated to simulate point, parallel, and converging ray optical-CT scanning configurations; both point and converging sources were examined at focal lengths, f, of ±200 mm and ±500 mm. For each geometrical configuration, simulations were run with four different RI values of the matching fluid surrounding the PRESAGE® dosimeter: 1.501, 1.49, 1.33 and 1.0. These values represented ideal, minor-mismatch, water and air media respectively.

To evaluate the effectiveness of each scanning and RI configuration, we defined the term useable radius of the PRESAGE® dosimeter, r_u, for which the optical-CT measured dose is within 2% of the true dose. The usable fraction of the radius of the dosimeter is thus r_u/r_0.

3. Results

A total of 20 combinations of geometry and RI were simulated and the usable fraction r_u/r_0 and required detector size $\Delta\theta$ are given in table 1. Figure 1 shows illustrative simulations of parallel, converging, and point incident ray geometries. Reconstructed images corresponding to each scanning configuration are shown, together with line profiles comparing measured (simulated) and true dose.

Table 1: Data from simulations of a 5 Gy uniform dose on r_0 = 50 mm PRESAGE® dosimeter. For each setup, incident ray geometry is described by focal length, f, and the useable fraction r_u/r_0 and required detector size $\Delta\theta$ are recorded. The RI column represents the refractive index of the media surrounding the dosimeter during optical-CT scanning.

	Point Source				**Parallel Source**		**Converging Source**			
RI	f = 200 mm		f = 500 mm		$f = \pm\infty$		f = -500 mm		f = -200 mm	
	r_u/r_0	$\Delta\theta$	r_u/r_0	$\Delta\theta$	r_u/r_0	$\Delta\theta$	r_u/r_0	$\Delta\theta$	r_u/r_0	$\Delta\theta$
1.501	99%	28°	99%	12°	99%	0°	99%	12°	93%	28°
1.49	99%	22°	98%	10°	97%	14°	97%	24°	91%	36°
1.33	68%	78°	68%	92°	66%	90°	66%	102°	62%	98°
1.00	31%	160°	31%	174°	31%	174°	31%	178°	29%	176°

4. Discussion

As shown in figures 1(a) and 1(b), pronounced variation between measured and true dose distributions is seen near the dosimeter edges when there is an RI mismatch between the dosimeter and surrounding fluid. These artifacts are caused by undersampling due to refraction; rays are diverted towards the dosimeter center, leaving the simulation with no attenuation line integral data near the dosimeter edges. Larger RI mismatch results in a larger section of the dosimeter being incompletely sampled. In addition to edge effects, this also affects any attempt to make an absolute dose measurement, due to the nature of FBP. To overcome this, our reconstructed dose distributions were normalized to the true dose D_{max} (in practice, a point of known dose would be needed in the dosimeter).

The simulation results in table 1 demonstrate that an inner fraction of the dosimeter remains useable for all scanning configurations. Optical CT scanning in a water medium (RI = 1.33) may recover a usable fraction of up to 68% with point source geometry, or 66% with parallel ray geometry. Optical CT of a dosimeter in a dry medium (air, RI = 1.00) was shown to recover a useable fraction of up to 31%. As expected, the detector size $\Delta\theta$ is dependent on incident ray geometry. In some cases, generally when RI of surrounding media is much less than RI of the dosimeter, point-source geometry is advantageous by requiring a smaller detector.

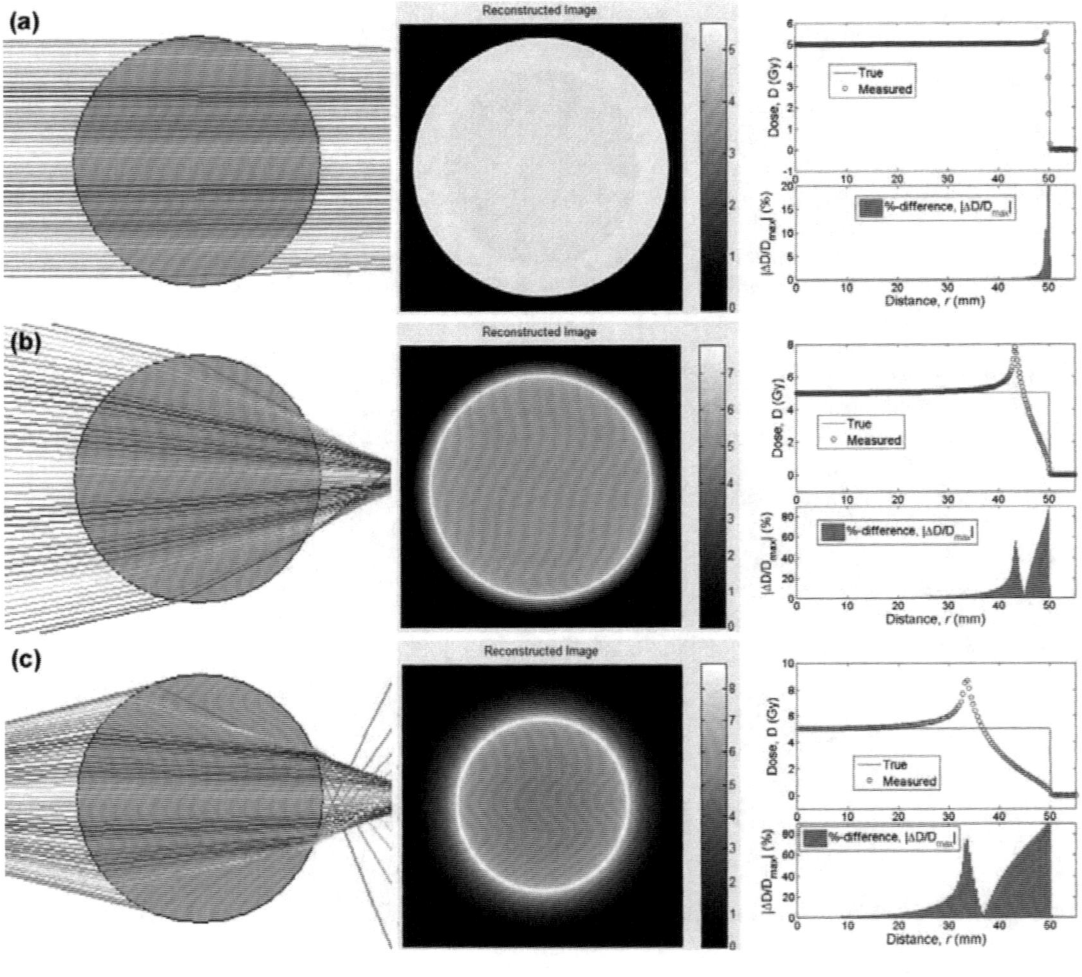

Figure 1: **(a)** Parallel telecentric geometry with surrounding media of RI = 1.49; **(b)** Converging light ray geometry (f = -200 mm), surrounding RI = 1.33; and **(c)** Point source geometry (f = 200 mm), surrounding RI = 1.00. [Left column] Ray diagram showing incident light geometry and refracted path through PRESAGE® (RI = 1.501) dosimeter. [Center column] Measured (simulated) dose distribution (Gy). [Right column] Radial line profile through the reconstruction. Plot includes measured dose (blue) vs. true dose (green), with difference $|\Delta D/D_{max}|$ shown beneath to visualize reconstruction artifacts and quantify useable radius fraction, r_u/r_0.

5. Conclusion

We have simulated optical-CT of three scanning geometry configurations: parallel, point and converging. For each geometry, we reconstructed the measured dose using four different media surrounding the dosimeter with refractive indices of 1.501, 1.49, 1.33 and 1.00. Our results have shown that optical CT of a uniformly irradiated dosimeter, using a water medium (RI = 1.33), may recover a usable fraction r_u/r_0 of up to 68%. This may be suitable for some clinical applications of 3D dosimetry where accurate dosimetry is not required near the periphery. Scanning the dosimeter in a dry medium (air, RI = 1.00) was shown to recover a useable fraction of only up to 31%. A significant concern for dry and low-viscous matching fluid configurations is the required detector size $\Delta\theta$ increases substantially (more than a factor 6 in some cases – table 1). This would likely add a cost penalty.

6. References

[1] Gore J C *et al* 1996 *Phys. Med. Biol.* **41** 2695-2704
[2] Doran S J *et al* 2001 *Phys. Med. Biol.* **46** 3191-213
[3] Conklin J *et al* 2006 *J. Phys.: Conf. Ser.* **56** 211
[4] Wuu C *et al* 2004 *J. Phys.: Conf. Ser.* **3** 297
[5] Maryanski M J and Ranade M K 2001 *Proc. SPIE* **4320** 764
[6] Baldock C *et al* 2010 *Phys. Med. Biol.* **55** R1-63
[7] Adamovics J and Maryanski M J 2004 *J. Phys.: Conf. Ser.* **3** 172
[8] Bosi S *et al* 2007 *Phys. Med. Biol.* **52** 2893-2903
[9] Maryanski M J, *et al* 1998 *Med. Phys.* **25** A107
[10] Papadakis A E *et al* 2010 *IEEE Trans. Med. Imaging* **29** 1204-12
[11] Oldham M 2010 *J. Phys.: Conf. Ser.* **250** 012064
[12] Doran S J and Yatigammana D N B 2012 *Phys. Med. Biol.* **57** 665
[13] Oldham M *et al* 2012 *Int. J. Radiat. Oncol. Biol. Phys.* (in press)

Journal of Physics: Conference Series **444** (2013) 012066 doi:10.1088/1742-6596/444/1/012066

Recent developments with a prototype fan-beam optical CT scanner

WG Campbell[1], A Jirasek[1] and D Wells[2]
[1]Dept. of Physics and Astronomy, University of Victoria, Victoria, BC, Canada
[2]BC Cancer Agency, Vancouver Island Centre, Victoria, BC, Canada

E-mail: warren.g.campbell@gmail.com

Abstract. The latest design of a prototype fan-beam optical computed tomography scanner is presented. A new beam creation system consists of a 635 nm laser diode module with variable, DC voltage-controlled beam intensity. A change in scanner alignment allows for the elimination of ring artefacts caused by data corruption that is spaced symmetrically across the detector array. These artefacts, as well as a pair of streaking artefacts caused by flask seams, are removed in sinogram space. A flask registration technique has been developed that allows for accurate, reproducible dosimeter placement. Protocol investigations with gel dosimeters have indicated the importance of: i) proper cooling techniques during gel manufacture, and ii) scanning the dosimeter while it is at room temperature. Latest reconstructions of a normoxic polymer gel dosimeter are presented as an indicator of current system performance.

1. Introduction

A variety of near tissue-equivalent materials have been developed that increase in opacity when exposed to radiation. Relationships can be established between their optical density (OD) and absorbed dose to allow for their use in radiation therapy dosimetry investigations. Notable examples currently under evaluation for 3D dosimetry [1] include polymer gels, dye-based polyurethane (PRESAGE™), and dye-based gels [2-6]. For readout of optical density, many optical computed tomography (CT) systems have been proposed. The first system used a translating pencil-beam analogous to first generation x-ray CT scanners [7]. More recent designs have opted for broad-beam geometries that are capable of faster scans by simultaneously illuminating the entire dosimeter [6, 7]. However, these systems find themselves restricted to using absorption-based dosimeters due to errors associated with imaging scatter-based dosimeters [10-12].

This work gives an update to a previously presented prototype fan-beam optical CT scanner [13, 14]. Its design intends to strike a balance between speed and versatility. Its beam simultaneously illuminates an entire slice of the dosimeter and its multi-hole collimator ensures accurate readout of highly attenuating dye-based and scatter-based samples [13]. Changes to its design are described. Two new artefact removal techniques are performed in sinogram space. A flask registration technique has been devised to finely reproduce the placement of dosimeters. Tests with gel dosimeters indicate the importance of proper cooling techniques during gel manufacture and scanning dosimeters while they are at room temperature. Finally, latest results with a normoxic polymer gel dosimeter are presented.

Journal of Physics: Conference Series **444** (2013) 012066 doi:10.1088/1742-6596/444/1/012066

2. Materials & Methods

Two main changes have been made to the prototype since it was last presented [13]. First, the green HeNe laser, pair of linear polarizers, and line generating lens have all been replaced by a red laser diode module (LDM) with 60° line-creating head (λ = 635 nm, 3 mW, Edmund Optics; Boulder, CO, USA); the beam intensity of the LDM is modulated by a DC voltage input. The second change in design was a slight shift in alignment. The axis of rotation of the dosimeter mount was moved slightly off-center (10.4 mm) to allow for a new artefact removal technique to be implemented.

The system has suffered from ring artefacts caused by data corruption that affects specific detector elements. The detector array is an approximate arc consisting of 5 linear photodiode arrays. Data was found to be unreliable in the first few detector elements of each array, resulting in unreliable data that is located roughly symmetrically across the 5 arrays. Shifting the scanner's axis of rotation off-center allows for the collection of an asymmetrical 360° sinogram that consists of two 180° sinograms containing redundant information. However, due to the shift in alignment, the locations of unreliable data no longer overlap when the second sinogram is flipped. This allows the two sinograms to be combined into a complete sinogram of reliable data. The values of the first 8 pixels in each 64-element linear array are ignored. Currently, to avoid residual ringing caused by their removal, values for these pixels are replaced using linear interpolation within each projection of both sinograms. Then, the two 180° sinograms are averaged, ray-by-ray.

Another artefact the system has been afflicted by is caused by two seams that run down the sides of the flasks being used. These seams are suspected to cause refractive errors that result in multiple streaking artefacts. The removal of these artefacts is also performed in sinogram space. Upon re-binning fan data into a parallel geometry, the locations of these seams follow easy-to-trace sinogram curves. These curves are traced and their values are replaced using interpolation of nearby pixels.

Early tests indicated that comparing "before" and "after" scans (i.e. pre-irradiation and post-irradiation scans, respectively) ray-by-ray lowered image noise considerably, and that even slight positional mismatches of these scans would reintroduce this noise. In order to finely reproduce the placement of dosimeters in the scanner, a flask registration technique was devised. To start, the central ray of the fan-beam is roughly aligned with a fiducial mark on the outside of the flask. Then, the flask is shifted a known amount so that the beam is targeting a blank region of the flask. This position is considered the "zero" position. Here, a light profile is acquired and saved, and the "before" scan can proceed. When the dosimeter is ready for its "after" scan, the fiducial mark is used again for rough alignment and the flask is shifted by the known amount to be roughly in the vicinity of the zero position. Next, a survey of many light profiles with positions slightly shifted vertically and angularly is acquired. Each light profile in the survey is then divided ray-by-ray by the light profile that was saved at the zero position during the "before" scan. Thus, a perfect alignment would give a homogenous profile with a value of 1 (noise aside), and any mismatches in position would introduce noise to the profile. Ergo, the zero position is located by locating the least noisy profile in the survey.

Tests were performed on normoxic polymer gel dosimeters housed in cylindrical flasks (1 L, 95 mm diameter). Gels consisted of (w/w%): 91.8% deionized water, 4.0% gelatin, 2.0% acrylamide, 2.0% bisacrylamide, and 0.2% (9 mM) tetrakis (hydroxymethyl) phosphonium chloride (all chemicals from Sigma-Aldrich; Oakville, ON, Canada). Two different protocols have been used for fabricating and imaging these dosimeters. Protocol #1 used the original green HeNe laser setup [13]. In this case, the flask was placed into a cold bath immediately after pouring the mixture into the flask. Six hours later, the dosimeter was removed from the fridge and scanned with the optical CT. Scanning occurs in a room temperature matching solution and takes ~45 minutes. The dosimeter was then irradiated, returned to the cold bath, and allowed to react overnight. Roughly 24 hours after irradiation, the dosimeter was removed from the cold bath and its "after" scan was acquired. Protocol #2 used the red LDM setup. In this case, after pouring the mixture into the flask, it was allowed to cool overnight in a room temperature bath. Roughly 24 hours later, the dosimeter was scanned in the room temperature matching solution. It was then irradiated, returned to the room temperature bath, and allowed to react overnight. Roughly 24 hours after irradiation, its "after" scan was acquired.

Simple 3x3 cm^2 square fields were used for irradiations (Clinac 21EX linear accelerator; Varian Medical Systems; Palo Alto, CA, USA). In protocol #1, 6 MV photons were used to create single-beam and cross-beam patterns (doses to isocenter: 3.7 Gy and 4.4 Gy, respectively). In protocol #2, 18 MV photons were used to also create single-beam and cross-beam patterns (doses to isocenter: 8.7 Gy and 10.0 Gy, respectively). Switching from green light to red light resulted in decreased beam attenuation by the polymers in the gel. Therefore, dose levels in protocol #2 were increased to achieve absorbance profiles of roughly equal magnitudes between the two protocols.

3. Results & Discussion

The effectiveness of both artefact removal techniques is shown in Figure 1. Images are of a synthetic polymer (Duramax B-1000; Rohm and Haas; Philadelphia, PA, USA) uniformly dispersed in water. Reconstructions are from the same data with and without the artefact removal techniques being implemented. As is apparent, the techniques effectively eliminate both artefacts. While removal of the streaking artefacts is more qualitatively evident in the images, profiles through the center of the flask show that ring errors with magnitudes of up to ~40% the uniform value are suppressed.

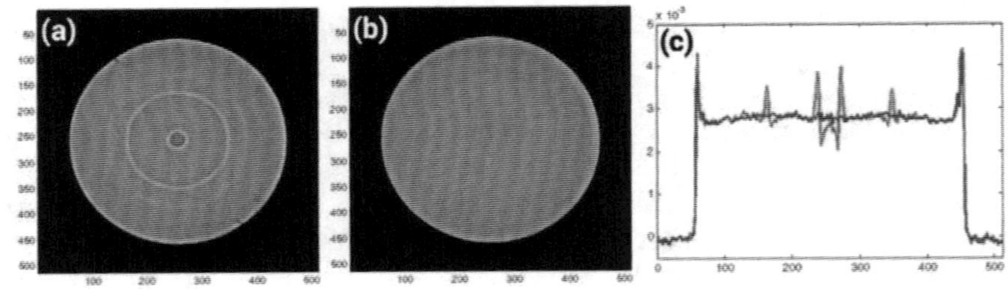

Figure 1: Images of a uniform scatterer (a) without and (b) with artefact removal implemented; (c) central profiles through (a) in red, and (b) in blue.

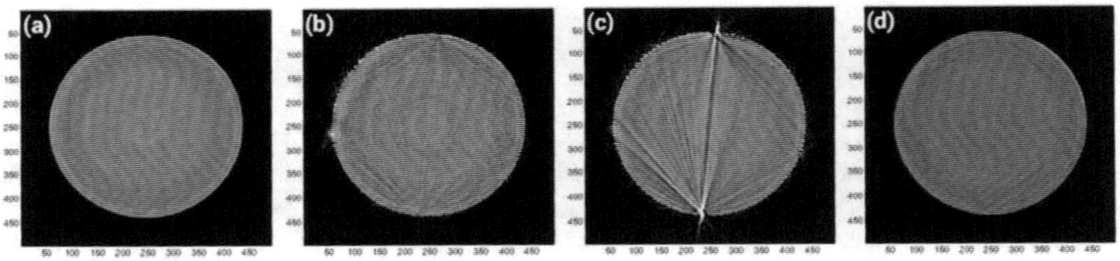

Figure 2: Images illustrating the performance of the flask registration technique: (a) flask left unmoved between I_0 and I scans, (b) data deliberately mismatched 1 mm vertically, (c) data deliberately mismatched 1° angularly, and (d) flask fully removed and registered using the technique.

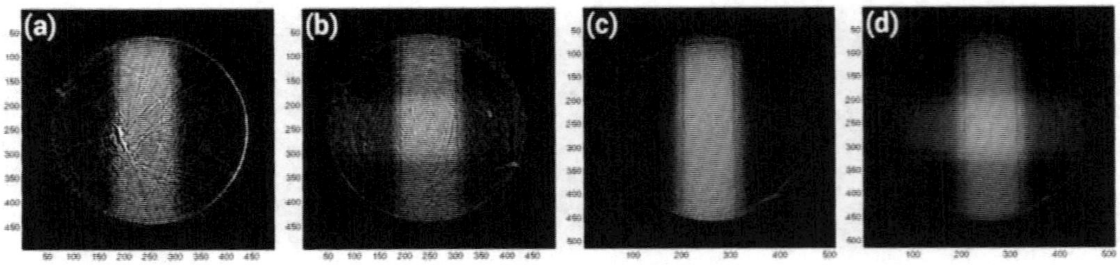

Figure 3: Reconstructions of irradiated dosimeters manufactured and scanned using protocol #1 (a & b) and protocol #2 (c & d).

Figure 2 illustrates the effectiveness of the flask registration technique. Figure 2a shows an image that was reconstructed by acquiring both scan data and reference data without disturbing the placement of the flask. Figures 2b and 2c show images where positional shifts were deliberately introduced. Figure 2d shows a reconstruction where the flask was fully removed from the scanner between acquisitions of scan data and reference data and the registration technique was implemented. The technique effectively reproduces the positioning of the flask well enough to avoid excess noise. Although this technique could overlook positional errors caused by axial tilt that might occur when the dosimeter is inserted into its mount, such errors have not yet shown themselves to be significant.

The importance of temperature concerns are demonstrated in Figure 3. Images in Figure 3a and 3b are from protocol #1 and images in Figure 3c and 3d are from protocol #2. Images obtained using protocol #1 suffer from increased schlieren errors caused by refractive index inhomogeneities in the dosimeter. Two factors are suspected to cause the errors seen in protocol #1. First, it is believed that, by placing the flask containing a warm mixture (~43 °C) into a cold bath (~5 °C), convection currents were caused in the dosimeter as it set. Second, after placing the cold dosimeter in a room temperature matching bath to be scanned, the dosimeter begins warming. Hence, the gel may be gradually morphing while the flask is being registered and the dosimeter is being scanned. Both of these issues are avoided by giving the gel more time to set and by scanning it while it is at room temperature.

Images of the dosimeter manufactured under protocol #2 are compared with dose data provided by a commercial treatment planning system (TPS) (Eclipse™; Varian Medical Systems) in Figure 4. A linear relationship between OD and dose was assumed, so a single conversion factor was used to convert reconstructions to dose distributions. Methods of filtering and binning are now being considered to reduce noise that remains in images. For example, images in Figures 4b and 4d are the same data from Figures 3c and 3d, only binned into larger pixels (from ~0.25×0.25 mm^2 to ~2×2 mm^2). Preliminary tests indicate that best results would be obtained by filtering in sinogram space.

It is worth noting the streaking artefacts that occur along the sides of each beam in Figures 3c and 3d. These artefacts are believed to be caused by radiation-induced refractive index changes, which are known to occur in polymer gel dosimeters [2]. This theory is the subject of a separate submission.

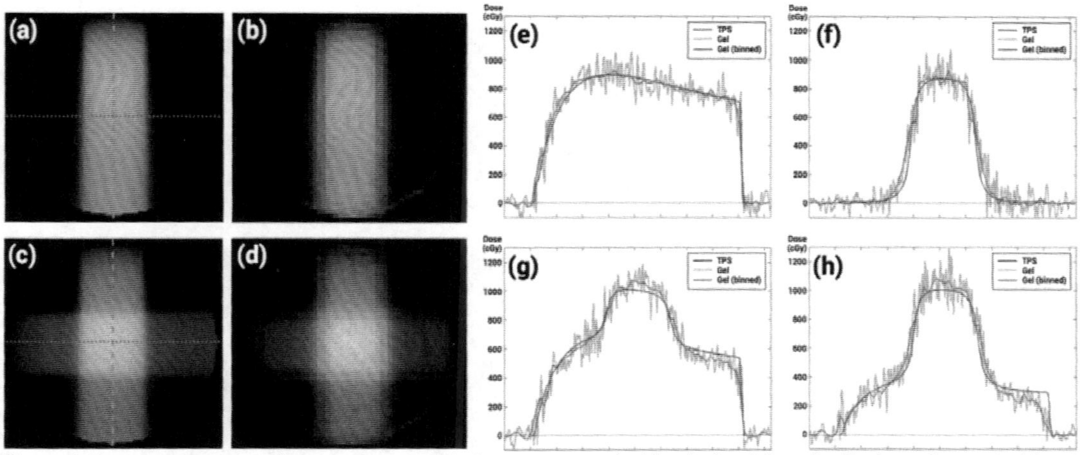

Figure 4: Dose reconstructions from a polymer gel dosimeter compared with calculated dose distributions. Images in (a) and (c) are from TPS, (b) and (d) were reconstructed from the dosimeter. Also indicated in (a) and (c) are the positions of the profiles compared in (e-h).

4. Conclusions

An update to a prototype fan-beam optical CT scanner has been provided. Artefact removal techniques that remove ring and streaking artefacts in sinogram space were described and their results presented. A flask registration technique was introduced and its performance was shown. Scans acquired using two different dosimetry protocols have indicated the need to be mindful of temperature during gel manufacture and scanning. Most recent results indicate that data filtering and binning will be necessary to reduce remaining image noise. Finally, a streaking artefact was shown that is believed to be caused by radiation-induced RI changes.

5. Acknowledgements

Work on this scanner was started by previous University of Victoria (UVic) student David Rudko and continues with the support of Dave Yardley and Steve Gray at the BCCA and Nick Braam at UVic. The presenting author would also like to acknowledge funding support from the Canadian Institutes of Health Research, the Natural Sciences and Engineering Research Council of Canada, the National Aboriginal Achievement Foundation, and the University of Victoria.

6. References

[1] Baldock C *et al* 2010 *Phys. Med. Biol.* **55** R1–63
[2] Maryañski M J *et al* 1996 *Phys. Med. Biol.* **41** 2705
[3] Adamovics J and Maryañski M J 2004 *J. Phys.: Conf. Series* **3** 172
[4] Jordan K and Avvakumov N 2009 *Phys. Med. Biol.* **54** 6673
[5] Brown S *et al* 2008 *Appl. Radiat. Isot.* **66** 1970-4
[6] Gorjiara T *et al* 2011 *Med. Phys.* **38** 2265-74
[7] Gore J C *et al* 1996 *Phys. Med. Biol.* **41** 2695
[8] DeJean P *et al* 2006 *J. Phys.: Conf. Series* **56** 179
[9] Thomas A *et al* 2011 *Med. Phys.* **38** 4846
[10] Olding T *et al* 2010 *Phys. Med. Biol.* **55** 2819
[11] Bosi S G *et al* 2009 *Phys. Med. Biol.* **54** 275-83
[12] Bosi S G *et al* 2009 *Appl. Opt.* **48** 2427-34
[13] Campbell W G *et al* 2010 *J. Phys.: Conf. Series* **250** 012024
[14] Jirasek A *et al* 2009 *J. Phys.: Conf. Series* **164** 012025

Journal of Physics: Conference Series **444** (2013) 012067 doi:10.1088/1742-6596/444/1/012067

A study on the reproducibility and spatial uniformity of N-isopropylacrylamide polymer gel dosimetry using a commercial 10X fast optical-computed tomography scanner

YJ Chang[1,2], JQ Lin[1], BT Hsieh[3], CH Chen[2]
[1]Institute of Biomedical Engineering and Material Science, Central Taiwan University of Science and Technology, Taichung, Taiwan (R.O.C.)
[2]Department of Management Information Systems, Central Taiwan University of Science and Technology, Taichung, Taiwan (R.O.C.)
[3]Department of Medical Imaging and Radiological Science, Central Taiwan University of Science and Technology, Taichung, Taiwan (R.O.C.)

E-mail: ronchang@ctust.edu.tw

Abstract. This study investigated the reproducibility and spatial uniformity of N-isopropylacrylamide (NIPAM) polymer gel as well as the reproducibility of a NIPAM polymer gel dosimeter. A commercial 10X fast optical computed tomography scanner (OCTOPUS-10X, MGS Research, Inc., Madison, CT, USA) was used as the readout tool of the NIPAM polymer gel dosimeter. A cylindrical NIPAM gel phantom measuring 10 cm (diameter) by 10 cm (height) by 3 mm (thickness) was irradiated by the four-field box treatment with a field size of 3 cm × 3 cm. The dose profiles were found to be consistent at the depths of 2.0 cm to 5.0 cm for two independent gel phantom batches, and the average uncertainty was less than 2%. The gamma pass rates were calculated to be between 94% and 95% at depths of 40 mm for two independent gel phantom batches using 4% dose difference and 4 mm distance-to-agreement criterion. The NIPAM polymer gel dosimeter was highly reproducible and spatially uniform. The results highlighted the potential of the NIPAM polymer gel dosimeter in radiotherapy.

1. Introduction

The development of three-dimensional (3D) polymer gel dosimetry and fast dose readout tools has slowly progressed in recent decades [1, 2]. Complex chemical reactions and interactions among gel compositions occur during the radiation-induced polymerization process [3-5]. Thus, the dose-response characteristics of polymer gels vary [7]. The temperature program of the gel-preparation process also affects the gel sensitivity because of the exothermal polymerization reaction of the polymer gel [8, 9]. The adoption of appropriate readout tools with the corresponding calibration gel dosimetry method is also important in obtaining accurate 3D dose distribution [11-14]. According to a previous study [12], the sources of uncertainty are as follows: (1) digitalization and electronic noise in the data acquisition system, (2) mechanical and optical source variation, (3) gel uniformity, and (4) reconstruction algorithm. In the present study, N-isopropylacrylamide (NIPAM) polymer gel was used with a commercial 10X fast optical-computed tomography (CT) scanner to evaluate the uncertainties in polymer gel dosimetry.

2. Materials and Methods

2.1. NIPAM gel fabrication

The gel was prepared inside a fume hood under normal atmospheric conditions according to the process described by Senden and De Deene [15, 16]. The NIPAM polymer gel used in the study was composed of 5% gelatin, 5% NIPAM (97% pure; Sigma–Aldrich), 3% N,N'-methylene bisacrylamide, and 5 mM Tetrakis (hydroxymethyl) phosphonium chloride. After completing the gel preparation, the gel was poured into cylindrical acrylic phantoms with the following dimensions: 10 cm (diameter), 10 cm (height), and 3 mm (wall thickness). The gel phantom was wrapped in aluminum foil to prevent photopolymerization, and then cooled to 7 °C a water bath placed inside a refrigerator for 5 h until complete solidification was achieved. Two independent gel phantom batches were prepared for the replicated experiments.

2.2. Treatment planning system and irradiation

A simple field (4 cm × 4 cm) irradiation treatment plan was generated using the Eclipse planning system (Varian Corporation, Palo Alto, CA, USA). The gel phantom was fixed on a table with crossmarks for alignment. The gel phantom was irradiated using a Varian Clinac IX linear accelerator (Varian Corporation, Palo Alto, CA, USA). The treatment field was 4 cm × 4 cm square fields with gantries aligned at 270°. The prescribed dose was 5 Gy. The dose rate was 300 cGy (559 MU), the focal point setting was at the center of the gel phantom, the source surface distance was 96 cm, and the depth was 4 cm.

2.3. Dose readout tool and data analysis

In this study, a commercial 10X fast optical-CT scanner (OCTOPUS-10X, MGS Research, Inc., Madison, CT, USA) was used as the readout tool of the NIPAM polymer gel dosimeter. The special scanning system comprises collimating Fresnel lens and an oscillating mirror located at the focus of the Fresnel lens. One gel phantom scan only takes about 30 min. A single-beam laser (780 nm wavelength, 30 mW) was used to scan the gel phantom. The laser beams scan the gel that is mounted on a turntable inside the scanning tank, which is filled with a refractive-index matching liquid. The refractive index of NIPAM gel was measured using an ATAGO refractometer (model PAL-RI). Refractive index matching ensures that the gel-transmitted rays propagate in a straight direction. After scanning, the image reconstruction program "reconQexp.m" written in MATLAB (The MathWorks, Natick, MA, USA) was used to reconstruct the projection data using the filtered back-projection technique. The final data presentation of the reconstructed image and the dose calibration were performed using the image-processing program ImageJ (version 1.43) developed by the National Institute of Health. Quantitative evaluation of dose distributions was performed using gamma analysis [17, 18]. The measured NIPAM gel dose distributions were compared with the calculated treatment planning (TPS) dose distribution using 4% dose difference and 4 mm distance-to-agreement (DTA) comparison criteria.

3. Results and Discussion

Figure 1 shows the results of gel containers filled with un-irradiated NIPAM gel. Figure 1(a) indicates the reconstructed image of a transverse slice at 40 cm depth. Figure 1(b) indicates the line profiles at various depths (2.0, 3.0, 4.0, and 5.0 cm) along the line shown in Fig. 1(a). The gel uniformity can be evaluated using the uncertainty among these profiles at various depths [12]. The average uncertainties are 1.77% and 1.91% in the central 5.5 cm region for gel phantom batches 1 and 2, respectively. However, the maximum uncertainties are located at the edge of the same region, and are 3.72% and 3.13% for gel phantom batches 1 and 2, respectively. The spatial uniformity of NIPAM gel is found to be similar with BANG 3 polymer gel [12]. The large uncertainty at the edge of the central region of the gel may be caused by the scanner [12]. Due to the adoption of a smaller phantom diameter (10 cm), the refractive effect at the phantom wall is more serious. Thus, a larger phantom diameter should be adopted.

Figure 2 shows the results of irradiated gel phantom 48 h post-irradiation. Figure 2(a) indicates the reconstructed image of a transverse slice at 4.0 cm depth. Figure 2(b) indicates the line profiles of gel phantom batch 1 for different scanning results along the line shown in Fig. 2(a). The deviations of two scans for batches 1 and 2 at various depths are listed in Table 1. The maximum mean deviation is from 3.44% to 3.79% in the center 5.5 cm region, and the uniformities of the gel measured in different scans and different batches are found to be consistent in various depths.

Figure 3 compares the results of the measured NIPAM gel dose distributions and the TPS. The gamma pass rates are between 94% and 95% at 40 mm depth for the two gel phantom batches using 4% dose difference and 4 mm DTA criteria. The gamma map in Fig. 3 presents the rejected regions at the central region and edge of the 4 cm × 4 cm square field, where high-dose gradients are present. This erroneous phenomenon is probably a result of the edge enhancement of NIPAM gel, which causes the disagreement at the high-dose-gradient region. However, another reason may be the repositioning error between the un-irradiated and irradiated gel scans, which needs further verification.

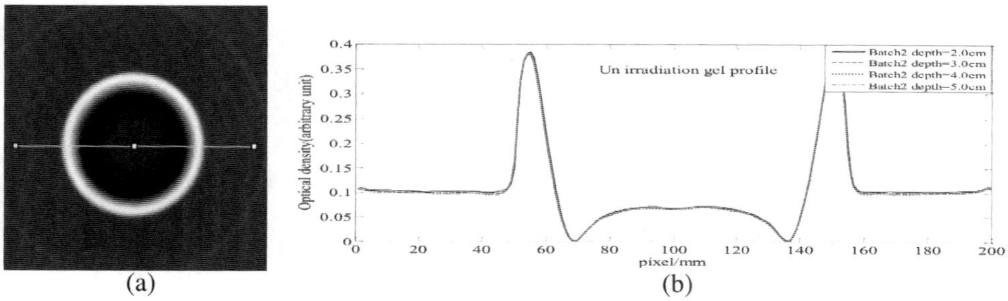

(a) (b)

Figure 1. Scanning results of gel containers filled with un-irradiated NIPAM gel: (a) reconstructed image of a transverse slice (40 cm depth), (b) line profile at various depths along the line shown in (a).

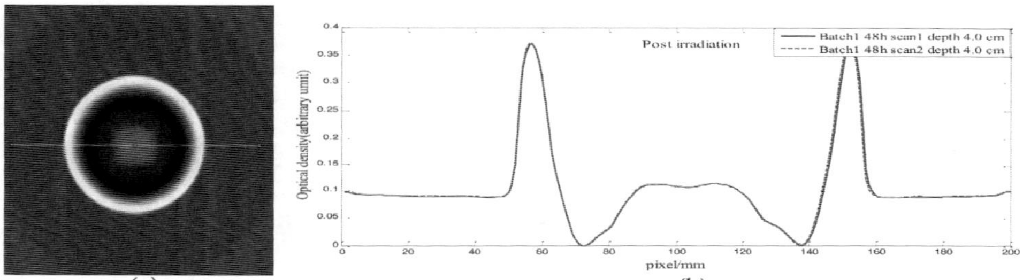

(a) (b)

Figure 2. Scanning results of gel containers filled with irradiated NIPAM gel: (a) reconstructed image of a transverse slice (40 cm depth), (b) line profile at various depths along the line shown in (a).

Table 1. Measurement deviation of two scans at various depths for different batches of irradiated NIPAM polymer gel.

Batch No.	Batch 1		Batch 2	
Depth (cm)	3	4	3	4
Deviation	3.54%	3.44%	3.45%	3.79%

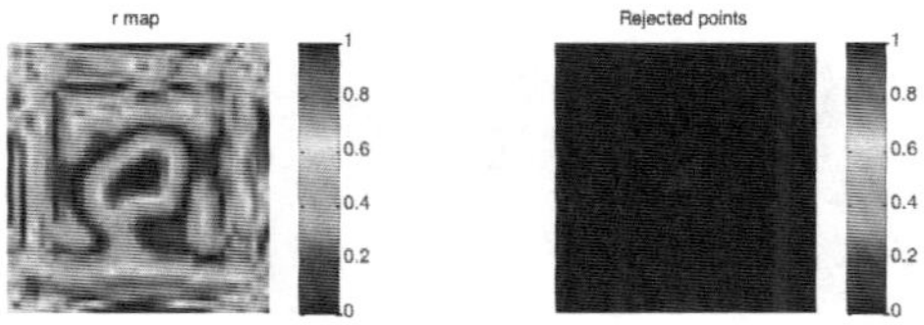

Figure 3. Representative 3D gamma comparison map of calculated treatment planning and the measured NIPAM gel dose using 4% dose difference and 4 mm distance-to-agreement criteria.

4. Conclusion

The present study investigated the reproducibility and spatial uniformity of a NIPAM polymer gel dosimeter. The spatial uniformity of NIPAM polymer gel was similar with the commercial BANG 3 polymer gel. Highly reproducible NIPAM polymer gel was obtained. The gamma index calculation showed that the gamma pass rates were between 94% and 95%. The prepared NIPAM polymer gel dosimeter has great potential in radiotherapy.

Journal of Physics: Conference Series **444** (2013) 012067 doi:10.1088/1742-6596/444/1/012067

5. References

[1] Baldock C *et al* 2010 *Phys. Med. Biol.* 55 R1-63
[2] Trapp J V *et al* 2002 *Phys. Med. Biol.* **47** 4247-58
[3] Lepage M *et al* 2001 *Phys. Med. Biol.* **46** 2827-39
[4] Brindha S *et al* 2004 *Phys. Med. Biol.* **49** N353-61
[5] Gustavsson H *et al* 2004 *Phys. Med. Biol.* **49** 227-41
[6] Hill B *et al* 2005 *Br. J. Radiol.* **78** 623-30
[7] Chang Y J *et al* 2011 *Nucl. Instrum. Methods Phys. Res. A* **652** 783-5
[8] Salomons G J *et al Phys. Med. Biol.* **47** 1435-48
[9] De Deene Y *et al* 2007 *Phys. Med. Biol.* **52** 2719-28
[10] Rintoul L *et al* 2003 *Appl. Spectrosc.* **57** 51-7
[11] Xu Y 2003 *Med. Phys.* **30** 2257-63
[12] Xu Y 2004 *Med. Phys.* **31** 3024-30
[13] Xu Y 2010 *Med. Phys.* **37** 861-8
[14] Murry P and Baldock C 2000 *Australas. Phys. Eng. Sci. Med.* **23** 44-51
[15] De Deene Y *et al* 2002 *Phys. Med. Biol.* **47** 3441-63
[16] Senden R J 2006 *Phys. Med. Biol.* **51** 3301-14
[17] Low D A 1998 *Med. Phys.* **25** 656-61
[18] Low D A 2003 *Med. Phys.* **30** 2455-64

Dosimetric characteristics of N-vinylpyrrolidone based polymer gels: utilization depending on dose range

M-V Papoutsaki[1], E Pappas[2], AE Papadakis[3], C Varveris[4], J Damilakis[1] and TG Maris[1]

[1]Department of Medical Physics, University of Crete, Medical School, Staurakia 71110, Crete, Greece

[2]Department of Medical Radiological Technologists, Technological Educational Institute of Athens, Egaleo, 12210, Athens, Greece

[3]Department of Medical Physics, University Hospital of Heraklion, Staurakia, 711110, Crete, Greece

[4]Department of Radiation Therapy, University of Crete, Medical School, Staurakia, 711110, Crete, Greece

E-mail: vasp3479@yahoo.gr

Abstract. In this study, two types of normoxic N-vinylpyrrolidone based polymer gel, VIPET^{1v} and VIPET^{2v}, were presented. VIPET^{2v} had the double amount of monomer compared to the VIPET^{1v}. The influence of this difference in the dosimetric characteristics of the polymer gel dosimeters was investigated. The doubling in N-vinylpyrrolidone concentration increased the dose sensitivity in the linear dose range region and improved the dose resolution at 95% confidence level for lower doses (D<10Gy). Unfortunately, this increase in concentration reduced the dose range response. VIPET^{1v} was sensitive in the dose range of 2 to 60 Gy, while VIPET^{2v} was sensitive in the dose range of 1 to 30 Gy. Moreover, both gels showed a linear R2-dose response for one month post-irradiation time period. During this time, variations in dose sensitivity and offset values of both dosimeters were observed. Finally, the dose resolution temporal stability was studied and was eventually improved by doubling the N-vinylpyrrolidone concentration.

1. Introduction

Nowadays, several investigators have introduced polymer gel dosimeters with different chemical compositions [1, 2]. Each dosimeter should accomplish specific dosimetric characteristics in order to be used safely in clinical practice of radiotherapy [3, 4]. Recently, there were efforts on the improvement of the dose sensitivity and dose resolution [5, 6] of the polymer gels dosimeters by: (a) the addition of cosolvents or (b) the increase of the total monomer amount of the chemical composition of the dosimeter [7-9].

Pappas et al [10] introduced the anoxic N-vinylpyrrolidone based polymer gel dosimeter (VIPAR) and its dosimetric characteristics. A few years later, a normoxic N-vinylpyrrolidone based polymer gel (VIPET) was presented by Papadakis et al [11]. The aim of this study is to explore the dosimetric characteristics and the usefulness in clinical practice of the two types of normoxic N-vinylpyrrolidone based polymer gels, VIPET^{1v} and VIPET^{2v}. VIPET^{2v} has the double amount of monomer compared to the VIPET^{1v}. In the current study, the influence of the monomer concentration in the normoxic N-vinylpyrrolidone based polymer gels is presented.

2. Materials and methods

2.1. Gel preparation
Two types of normoxic N-vinylpyrrolidone based polymer gels were produced. The first one VIPET^{1v} was introduced by Papadakis et al and the second one, VIPET^{2v} was based on the similar formula of VIPAR [10] and VIPET [11] polymer gels. In both gels, the gelatin concentration was 5% w/w, the N,N'-methylenebisacrylamide concentration was 4% w/w according to the initial formula of VIPAR and THPC was 5mM according to the VIPET's formula. The only difference between the two types was the monomer concentration. The N-vinylpyrrolidone was 4% w/w for VIPET^{1v} (8%T, 50%C) and 8% w/w for VIPET^{2v} (12%T, 33%C).

Both manufacture procedures were performed under normal atmospheric conditions inside a laminar flow hood. Firstly, the gelatin was added to the double distilled water and left to dissolve, following by heating to 50^{0}C using a hot magnetic stirrer. Then, the bis was added. After the bis dissolved, the mixture was cooled down to approximately 32^{0}C and the monomer was added. When the solution became transparent, THPC was added. Each solution was filled in two 100mL cylindrical vials. The four vials were stored at 20^{0}C ambient temperature for 24h in a cool and dark place to solidify before irradiation.

2.2. Gel irradiation
The gel phantoms were irradiated in a Primus Clinical Accelerator (Primus, LINAC, Siemens, Germany) using a 6MV photon beam, inside an in-house build solid water structure. Each phantom was irradiated separately in six different areas perpendicular to their length. The field size used to irradiate each area was 2x4 cm^2, the source sample distance was 100cm and the dose rate was 3Gy/min. The first vial of each type was irradiated along the central axis of the beam delivering doses at a range of 0.5 to 8 Gy and the second vial of each type was received doses from 10 to 60 Gy. Similar dose measurements were performed separately using a water phantom and an ion chamber and were considered as a dosimetric reference.

2.3. MRI measurements
All polymer gels were repeatedly scanned utilizing a 1.5T whole body MR Imager (Vision/ Sonata, Siemens, Germany) one day, one week, two weeks and one month post-irradiation. A 2D multi-slice-multi echo (32 echoes) spin echo Phase Alternating Phase Shift train sequence (PHAPS) was utilized for the acquisition of the parametric T2 maps. The sequence parameters were: TR=9000 ms, TE=40-1280ms in symmetric TE intervals of 40 ms (echo train), FA=180^0, slice thickness=4mm, FOV= 250x156 mm^2, matrix size= 256x256, NEX=1. MR images were transferred and analyzed to a PACS workstation (EvoRad, Athens, Greece) for the specification of dose response, dose resolution of both types of polymer gels.

3. Results

3.1. Dose response and dose resolution
Figure 1 represents the dose response of VIPET^{1v} and VIPET^{2v} at one day post-irradiation in a dose range of 0.5 to 60 Gy. VIPET^{1v} polymer gel exhibits a linear dose response in the dose range of 2-60 Gy. It is observed that VIPET^{1v} is not sensitive for doses lower than 2 Gy. On the other hand, VIPET^{2v} polymer gel exhibits a linear dose response in the dose range of 1-30 Gy. The VIPET^{2v} gel saturates for doses greater than 30 Gy.

In the common linear dose-response region, the dose-R2 data were fitted according to the linear equation: R2 (D) =αD+R_0. The calibration procedure for VIPET^{1v} yielded a dose sensitivity: α=

(0.059 ±0.001) $Gy^{-1}s^{-1}$ and an offset: R_0= (1.385 ±0.018) s^{-1}, while the calibration procedure for VIPET^{2v} yielded a dose sensitivity: α= (0.063±0,003) $Gy^{-1}s^{-1}$ and an offset: R_0= (1.148±0.049) s^{-1}.

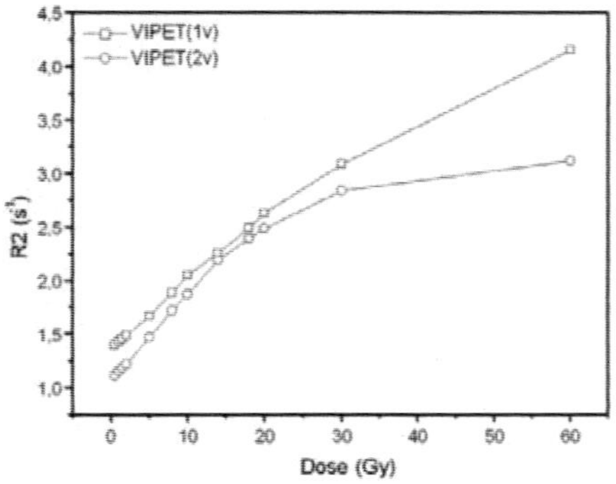

Figure 1: Dose response of VIPET^{1v} and VIPET^{2v} at one day post-irradiation.

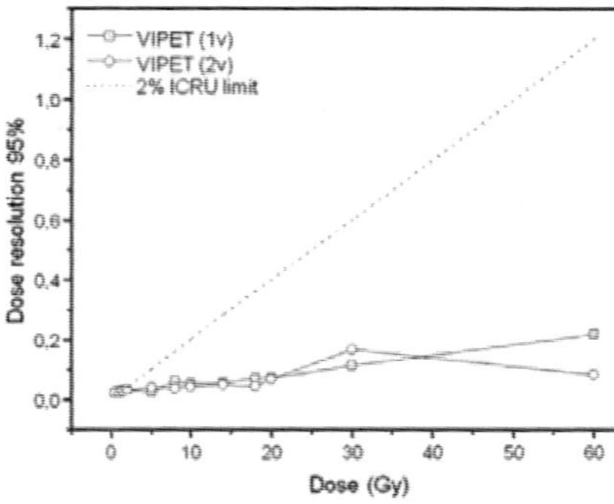

Figure2: Dose resolution at 95% confidence level of VIPET^{1v} and VIPET^{2v} and the 2% ICRU limit at one day post-irradiation.

Figure 2 shows the dose resolution at 95% confidence level at one day post-irradiation for both polymer gels including the 2% ICRU limit (dashed line). VIPET^{1v} can fulfill the 2% limit for doses above 2 Gy. While, VIPET^{2v} can accomplish the 2% limit for doses below 2 Gy.

3.2. Temporal stability in a time period of one month

Table 1 presents the dose sensitivity and offset values at different time intervals post-irradiation for the time period of one month for both polymer gels. It is obvious that the sensitivity values of VIPET^{2v} remain larger than the values of VIPET^{1v} in a month. For both types of polymer gels, the sensitivity reaches a maximum after one week and then ranges in lower values. On the other hand, for both gels the offset value is constantly increased with the post-irradiation time interval.

Table 1: Dose sensitivity and offset values of VIPET^{1v} and VIPET^{2v} gels in dose range (0-30 Gy) for a time period of a month.

Time interval	Dose sensitivity for dose range (0-30Gy)		Offset (R_0) for dose range (0-30Gy)	
	VIPET^{1v} ($s^{-1}Gy^{-1}$)	VIPET^{2v} ($s^{-1}Gy^{-1}$)	VIPET^{1v} (s^{-1})	VIPET^{2v} (s^{-1})
1 day	0.059 ± 0.001	0.063 ± 0.003	1.395 ± 0.024	1.148 ± 0.049
1 week	0.060 ± 0.001	0.064 ± 0.003	1.477 ± 0.009	1.565 ± 0.047
2 weeks	0.059 ± 0.001	0.058 ± 0.003	1.566 ± 0.014	1.695 ± 0.046
1 month	0.060 ± 0.001	0.062 ± 0.003	1.636 ± 0.008	2.103 ± 0.0045

Figures 3 and 4 show the dose resolution at 95% confidence level for VIPET^{1v} and VIPET^{2v} polymer gels for a time period of one month. The dose resolution of VIPET^{1v} for doses below 20 Gy does not change. For the same gel for doses above 20 Gy the dose resolution ranges from 0.07 to 0.4 Gy. Whereas, the dose resolution of VIPET^{2v} remains constantly stable for a period of one month.

4. Discussion

It has been shown that the doubling of the N-vinylpyrrolidone concentration increased the dose sensitivity in the linear dose range region and improved the dose resolution at 95% confidence level for lower doses (D<10 Gy). Unfortunately, this increase in concentration resulted in a dose range response restriction. It was also observed that the R2-dose response for VIPET^{1v} and VIPET^{2v} polymer gels was maintained linear for a time period of one month post-irradiation. It was suggested that the changes in dose sensitivity were mainly attributed on the post-irradiation polymerization reactions in the N-vinylpyrrolidone based polymer gels [11-13]. The increase in monomer concentration precipitated the variations in offset values. Finally, the dose resolution temporal stability was improved by the doubling of N-vinylpyrrolidone concentration.

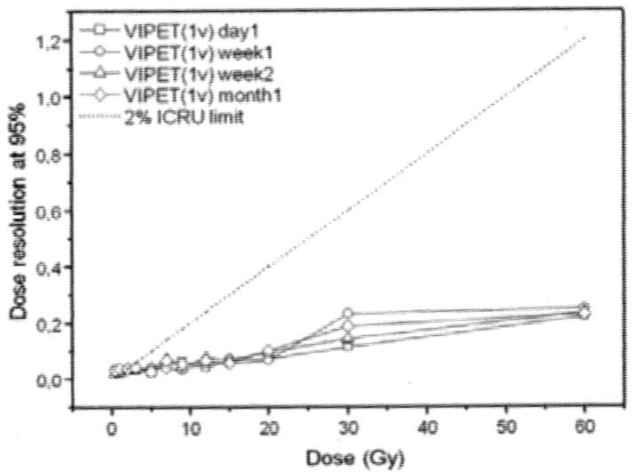

Figure 3: Dose resolution at 95% confidence level of VIPET[1v] and the 2% ICRU limit for a time period of one month post-irradiation.

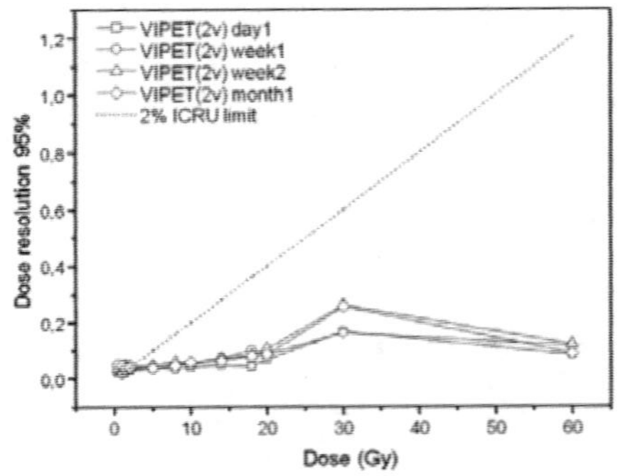

Figure 4: Dose resolution at 95% confidence level of VIPET[2v] and the 2% ICRU limit for a time period of one month post-irradiation.

5. Conclusion

N-vinylpyrrolidone based polymer gels constitute reliable dosimeters in 3D dose verification in radiotherapy. The increase in monomer concentration alters the dosimetric characteristics in terms of dose range, dose sensitivity and dose resolution. As a final statement, it was concluded that the selection of the monomer concentration should be associated with the desired dosimetric characteristics of the polymer gel.

6. References

[1] Baldock C *et al* 2010 *Phys. Med. Biol.* **55** R1-63
[2] Gustavsson H *et al* 2004 *Phys. Med. Biol.* 49 227-241
[3] Hill B *et* al 2005 *Br. J. Radiol.* **78** 623-30
[4] Rintoul L *et al* 2003 *Appl. Spectrosc.* **57** 51-7
[5] Baldock C *et al* 2001 *Phys. Med. Biol.* **46** 449-60
[6] Lepage M *et al* 2001 *Phys. Med. Biol.* **46** 2665-80
[7] Koeva V I *et al* 2008 *Macromol. Sym.* **261** 157-66
[8] Jirasek A *et al* 2010 *Phys. Med. Biol.* **55** 5269-81
[9] Chain J N M *et al* 2011 *Phys. Med. Biol.* **56** 2091-102
[10] Pappas E *et al* 1999 *Phys. Med. Biol.* **44** 2677-84
[11] Papadakis A E *et al* 2007 *Phys. Med. Biol.* **52** 5069-83
[12] DeDeene Y *et al* 2002 *Phys. Med. Biol.* **47** 2459-70
[13] Lepage M *et al* 2001 *Phys. Med. Biol.* **46** 2827-39

Journal of Physics: Conference Series **444** (2013) 012069 doi:10.1088/1742-6596/444/1/012069

A novel optical calorimetry dosimetry approach applied to an HDR Brachytherapy source

A Cavan[1,2] and J Meyer[1,3]
[1]University of Canterbury, Department of Physics & Astronomy, Private Bag 4800, Christchurch 8140, New Zealand, Phone: +64 3 364 2987-7588, Fax: +64 3 364 2469
[2]Canterbury District Health Board, Christchurch 8140, New Zealand
[3]University of Washington, Department of Radiation Oncology, 1959 NE Pacific Street, Box 356043, Seattle, WA 98195, USA, Phone: +1 206-598-4236, Fax: +1 206-598-6218

E-mail: alicia.cavan@cdhb.health.nz

Abstract. The technique of Digital Holographic Interferometry (DHI) is applied to the measurement of radiation absorbed dose distribution in water. An optical interferometer has been developed that captures the small variations in the refractive index of water due to the radiation induced temperature increase ΔT. The absorbed dose D is then determined with high temporal and spatial resolution using the calorimetric relation $D=c\Delta T$ (where c is the specific heat capacity of water). The method is capable of time resolving 3D spatial calorimetry. As a proof-of-principle of the approach, a prototype DHI dosimeter was applied to the measurement of absorbed dose from a High Dose Rate (HDR) Brachytherapy source. Initial results are in agreement with modelled doses from the Brachyvision treatment planning system, demonstrating the viability of the system for high dose rate applications. Future work will focus on applying corrections for heat diffusion and geometric effects. The method has potential to contribute to the dosimetry of diverse high dose rate applications which require high spatial resolution such as microbeam radiotherapy (MRT) or small field proton beam dosimetry but may potentially also be useful for interface dosimetry.

1. Introduction

Interferometry is a widely applied investigative technique in many fields of science and industry where it is typically utilized to measure small displacements and refractive index changes. As the refractive index of water is temperature dependent, interferometry can be used to measure very small temperature changes in a transparent water cell. The concept of our method, namely Digital Holographic Interferometry (DHI), is illustrated in figure 1. A laser beam is split into two arms, with one arm traversing the water cell and the other serving as a reference beam. Both arms are recombined on the camera plane to produce an interferogram, also referred to as a hologram. A temperature rise in the cell due to irradiation is encoded in subsequent interferograms as a change in the phase difference of the interfering light waves. From this information it is possible to determine the spatial distribution of the change in refractive index Δn, from which the temperature change and ultimately the absorbed dose to water can be inferred. The output of DHI is digitally reconstructed and unwrapped 2D phase maps from which a time-series of two dimensional maps of the absorbed dose at each time instance can be obtained.

The idea to use holographic interferometry for radiation dosimetry dates back to 1971 and was demonstrated by Hussman [1] and Miller *et al.* [2] using holographic photo paper as the recording medium. Nicolau *et al.* [3] and more recently Ackerly *et al.* [4] also worked on related optical calorimetry. The current work uses a different optical set-up and modern digital image recording and reconstruction by means of the Fresnel transformation which enables the acquisition of images in video frequency rather than at a single time instance.

The ultimate aim of this work is to develop a digital interferometer capable of accurately measuring absolute 4D deposition of radiation dose in water. The current work describes our initial prototype which we applied to dosimetry of a conveniently available high dose rate (HDR) brachytherapy source for first proof-of-principle results. Results are presented for dosimetry in two spatial dimensions plus a times-series.

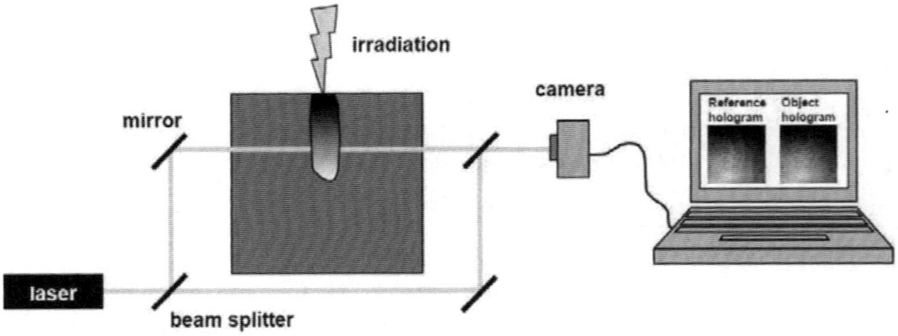

Figure 1: Illustration of the principles of DHI as applied to radiation dosimetry.

2. Methods and Materials

A Lensless Fourier Transform Digital Holography configuration was selected to simplify both the experimental set-up and the reconstruction algorithm [5]. The main difference to previous approaches is that for convenience the reference beam is converted into a spherical wave. A standard lab He-Ne laser light source was used with a PixeLINK™ 1.3 Megapixel CMOS Monochrome camera as the holographic recording medium, with 6.7μm pixel resolution. Numerical reconstruction and analysis was carried out in in-house developed MATLAB (The Mathworks, Natick, MA, USA) code. Figure 2 shows the optical components of the detector.

Figure 2: Set-up with the red line marking the laser beam.

The temperature distribution ΔT in the water cell after irradiation was determined by reconstructing the phase difference encoded in the holograms, unwrapping of the phase information [6] and by making use of the relationship between optical pathlength/refractive index and temperature.

Journal of Physics: Conference Series **444** (2013) 012069 doi:10.1088/1742-6596/444/1/012069

Dose D was then calculated by the calorimetric relation: $D= c\Delta T$, where c is the specific heat capacity of water. As the sensitivity of the approach scales with the magnitude of the temperature change a high dose rate (HDR) Ir-192 brachytherapy source with 480.2 GBq activity was selected for a proof-of-principle measurement. 1 Gy absorbed dose corresponds to a rise in temperature of ~0.0003°C and a phase change of ~0.002 radians or ~0.1°. The source was housed within a GammaMed Plus afterloader and the total irradiation time was 420 seconds. Images were taken at a frequency of 1.4 Hz, with an exposure time of 15 ms. The initial absolute temperature was obtained by the use of a digital thermometer. Measured results were compared to the dose distribution as modelled by the Brachyvision treatment planning system.

3. Results

Figure 3a) shows the recorded hologram as captured by the CMOS sensor, with the tip of the HDR source applicator visible at the bottom. Figure 3b) shows the hologram after reconstruction, phase unwrapping and conversion to a 2D temperature map. The location of the profile in the Brachyvision model is shown in c).

The relative dose distribution along a profile measured by DHI and that calculated in Brachyvision is compared in figure 4a. To show the time resolving capability of this approach several instances from the time-series obtained with DHI are shown in figure 4b, revealing the increasing dose and the subsequent flattening out of the distribution post-irradiation due to heat diffusion.

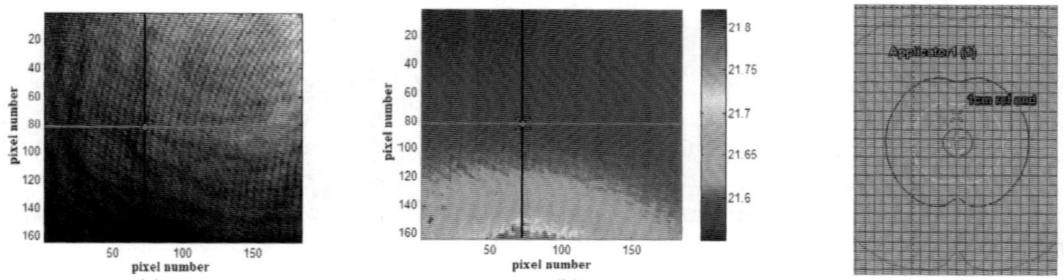

(a) (b) (c)

Figure 3: a) Hologram, b) temperature distribution, c) measurement location in Brachyvision

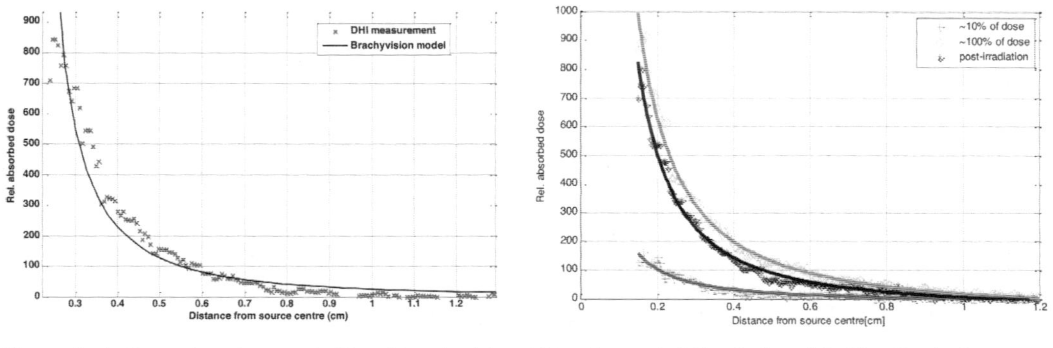

Figure 4: a) Comparison between DHI and Brachyvision (along the vertical line indicated in Fig. 3b). b) Dose as a function of time. Absorbed doses are relative and normalised to the most distant measurement point.

4. Discussion

Research into many new therapeutic applications of radiation, each with its unique dosimetric needs, has presented us with dosimetry challenges that have not been faced before. Spatial calorimetry has a possibility to contribute to the dosimetry toolbox in fields such as microbeam radiation therapy (MRT), narrow field proton beam dosimetry, high dose rate brachytherapy (HDR), or interface dosimetry, amongst others, due to its potential to non-invasively resolve small temperature gradients in water to sub-millimetre accuracy.

The application of digital holographic interferometry for radiation dosimetry, a novel optical calorimetry method, has been presented and initial results with an HDR brachytherapy source show that it is indeed possible to use DHI for radiation dosimetry. The results obtained and shown in figures 3 and 4 represent only the relative dose distribution as no reliable absolute calibration had been carried out. Nevertheless the results are encouraging as the relative dose distribution is in agreement with the model from the treatment planning system despite some inherent noise in the signal. The results presented were essentially the raw data as no noise filtering was applied to the captured holograms nor in the Fourier domain during the reconstruction. However, a 3x3 median filter was applied to the phase map. It is also anticipated that with higher quality optical components as well as a higher spec camera the signal-to-noise ratio can be improved. Figure 4 shows the time resolving capabilities of the approach and indicates the magnitude of the heat diffusion. It is indeed an area of further work to model the well-defined heat dissipation as a function of time.

A current limitation of the approach is that the dose determined at each point represents a line integral perpendicular to the direction of radiation incidence and thus underestimates the dose slightly. If the radiation source has rotational symmetry such as for narrow pencil beams or at a distance from a Brachytherapy source, the Abel Integral [7] can be used to approximate the corrected 3D dose distribution. For non-symmetric dose distributions full volumetric resolution may be achieved through the use of multiple angles and tomographic reconstruction.

5. Conclusion

A prototype digital holographic interferometer has been successfully applied to measure radiation dose distributions from an HDR source, thus establishing the potential for use of the technique as a radiation dosimeter for HDR applications [8]. A time-series of two dimensional maps proportional to absorbed dose were acquired and relative dose profiles taken from these were in agreement with modeled results from Brachyvision. The method is calorimetric, directly measuring the spatial distribution of absorbed dose to water at high resolution, independently of beam type, energy, and fundamentally the dose rate (apart from heat diffusion), without perturbing the radiation beam. The next step is an absolute calibration of the system and the application of the DHI set-up to the measurement of radiation fields where the above properties may contribute towards overcoming current dosimetric difficulties.

6. References

[1] Hussmann EK 1971 *Appl. Opt.* **10** 182-6
[2] Miller A *et al* 1974 *Radiat. Res.* **59** 225
[3] Nicolau S *et al* 1999 in *Interferometry 99 Applications* Jüptner W P O and Patorski K Eds *SPIE* Poland 393-9
[4] Ackerly T *et al* 2011 *J. Instrum.* **6** P03003
[5] Hossain M M *et al* 2006 *Opt. Eng.* **45** 106203
[6] Herraez M A *et al* 2002 *Appl. Opt.* **41** 7437-44
[7] Bracewell R 1965 *The Fourier Transform and its Applications* New York McGraw-Hill
[8] Hurley C *et al* 2006 *Nucl. Instrum. Methods Phys. Res. A* **565** 801-11

How accurate is image guided radiation therapy (IGRT) delivered with a micro-irradiator?

M Oldham[1,2], J Newton[1,2], L Rankine[1,2], J Adamovics[3], D Kirsch[1,2] and S Das[1,2]
[1]Department of Radiation Oncology, [2]Duke University Medical Center, Durham, USA
[3]Rider University, NJ, USA

E-mail: mark.oldham@duke.edu

Abstract. There is significant interest in delivering precisely targeted small-volume radiation treatments, in the pre-clinical setting, to study dose-volume relationships with tumor control and normal tissue damage. In this work we investigate the IGRT targeting accuracy of the XRad225Cx system from Precision x-Ray using high resolution 3D dosimetry techniques. Initial results revealed a significant targeting error of about 2.4mm. This error was reduced to within 0.5mm after the IGRT hardware and software had been recalibrated. The facility for 3D dosimetry was essential to gain a comprehensive understanding of the targeting error in 3D.

1. Introduction

We recently commissioned the treatment beams from a small field biological irradiator, the XRad225Cx from Precision x-Ray, for pre-clinical applications in our department [1, 2]. The XRad225 can produce a variety of small circular or square 225kVp photon beams ranging in size from 1mm to 40mm in maximum dimension. Initial commissioning focused on characterizing the radiation beams in terms of percent-depth-dose (PDD), two-dimensional (2D) profiles at various depths, and output factors. In the present work we extend this effort to evaluate the accuracy and capability of the cone-beam-CT (CBCT) image-guided radiation therapy (IGRT) capability of the system.

2. Methods

The accuracy of IGRT treatments was evaluated on Presage 3D dosimeters (5cm diameter cylinders of 5cm length), each containing 5 physical target points dispersed within the dosimeter. The target points were created by drilling channels (1.5mm diameter) of varying depth into the Presage dosimeter, parallel to the cylinder axis, but spatially offset. The end of each drill channel tapered to a sharp point, corresponding to a target point. The Presage cylinder was taped horizontally onto the XRad225 treatment stage to simulate a rat treatment. CBCT IGRT was then performed to position the tip of each drill-channel (the target point) at the isocenter. Each target point was treated with a 7 field equi-angular coplanar treatment delivered with the small 2.5mm cone. Each beam delivered about 1Gy to the isocenter, leading to a total isocentric dose of about 7Gy. Immediately after irradiation of all targets, the Presage dosimeter was imaged in the DLOS optical-CT scanner [3]. The DLOS system produced isotropic 0.5mm data throughout the Presage cylinder. Post-irradiation scans were corrected by pre-irradiation scans acquired the day before treatment. In both scans the drill channels in the Presage dosimeter were filled with refractive matching fluid, to minimize refraction artifacts around the channels. Scanning involved acquiring 720 projections over 360 degrees, which took about 20minutes.

The 0.5mm isotropic 3D images returned from the DLOS scanner report the change in optical-density (OD) throughout the Presage dosimeter, which is well established to be proportional to absorbed dose [4]. The OD distribution corresponds to the relative dose distribution throughout the dosimeter. The accuracy of the IGRT treatments in 3D was then determined by studying positional discrepancies between the target points and the intersection point of the corresponding treatment beams. The experiment was repeated on two more Presage dosimeters using 1mm and 5mm cones for delivery. All other dosimeter parameters, prescribed dose and irradiation methods were kept constant. This provided the grounds for a statistical analysis on IGRT targeting accuracy

3. Results

Single projection images through an arbitrary view through the Presage dosimeter before and after irradiation is shown in figure 1. The four target points are visible in figure 1a, as the tapered ends of the vertical drill-channels. Figure 1b shows the same projection after irradiation, where now the co-planar beam tracks and intersection point of each treatment are visible.

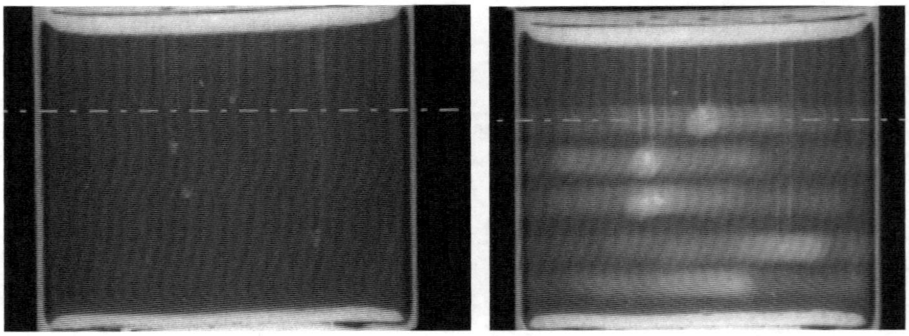

Figure 1: Illustrative projection images through (a) an un-irradiated, and (b) an irradiated Presage dosimeter.

Axial views through the results obtained on the initial run of this experiment are shown in figure 2. A misalignment is apparent because the center lines of each beam do not intersect at the same point. For our analysis, we defined *cone misalignment* as the radius of the smallest circle drawn at the isocenter which intersects all beams. *Targeting accuracy* was defined as the distance from mechanical isocenter to the tip of drilled holes. This misalignment was consistent for all 4 target points.

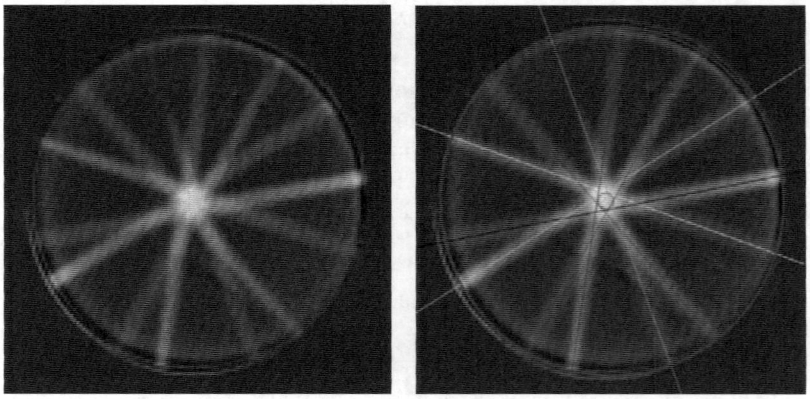

Figure 2: (a) an axial view through the Presage dosimeter at a plane through one of the 4 target points. (b) the same image but with central beam line annotations to highlight the cone misalignment.

A statistical analysis averaging the discrepancies over all four target points found an average cone misalignment of 2.4±0.6mm and targeting error of 2.1±0.6mm (see Table 1). This was substantially

Journal of Physics: Conference Series **444** (2013) 012070 doi:10.1088/1742-6596/444/1/012070

greater than the nominal advertised system accuracy of 0.5mm. The system and online targeting software was re-calibrated. The experiment was repeated on three more Presage dosimeters also containing the 4 target drill channels. Delivery of the 7 beam coplanar irradiation was performed using cone sizes of 2.5mm, 1mm, and 5mm. Results from this second experiment are shown in figure 3 and 4.

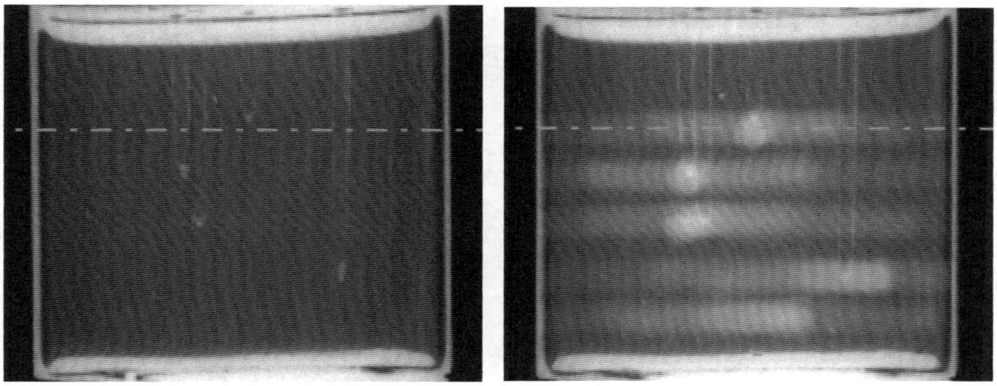

Figure 3: Illustrative projection images, after recalibration of the IGRT system, through (a) an un-irradiated, and (b) an irradiated Presage dosimeter.

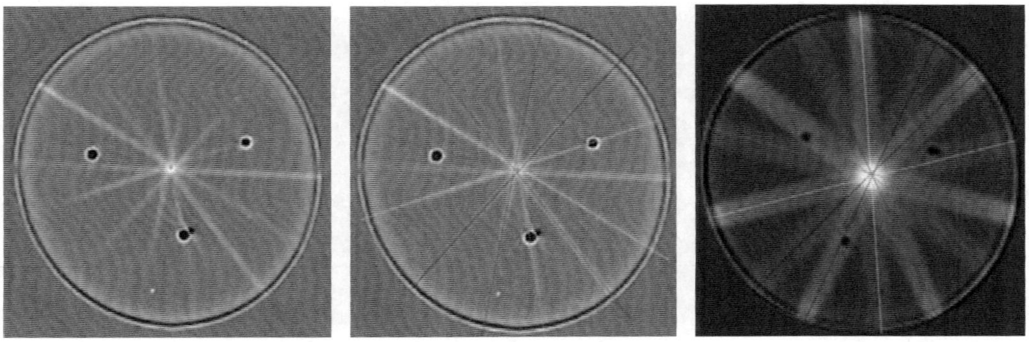

Figure 4: (a) axial view through the Presage dosimeter at a plane through one of the 4 target points treated with a 1mm cone. (b) the same image but with central beam line annotations to highlight much improved alignment compared with Figure 2. (c) similar image through another target point treated with a 5mm cone.

Statistical analysis averaging the discrepancies over all four target points after the IGRT system had been re-calibrated was performed for each of the three dosimeters and recorded in Table 1. The cone misalignment and targeting accuracy of these three dosimeters provided the means for further statistical analysis. Results indicate that after recalibration the cone misalignment was $\langle r \rangle$ = 0.5mm ($\sigma_{\langle r \rangle}$ = 0.3mm) and targeting accuracy was $\langle dr \rangle$ = 0.5mm ($\sigma_{\langle dr \rangle}$ = 0.4mm), which is in line with the nominal advertised system accuracy of 0.5mm.

4. Conclusions

The DLOS/Presage 3D dosimetry system enabled rigorous evaluation of the IGRT targeting accuracy of the XRad225 system. The ability to image with isotropic 0.5mm resolution (0.3mm for cone size 1mm) was vital for understanding the true accuracy of the system, especially for very small fields like the 1mm diameter beams shown in figure 4.

Table 1: Analysis of the cone misalignment and targeting accuracy for the four dosimeters. The xy-plane was defined as the plane traversed by the radiation beams (e.g. figure. 4). The z-axis was the longitudinal direction, normal to the axial xy-plane. Each entry shows the average over 4 target points and one standard deviation in parentheses

Cone Size [mm]	Recon. px size [mm]	Cone Misalignment [mm] Average (1.s.d.)			Targeting Accuracy [mm] Average (1.s.d.)			
		xy-plane	*z**	*r*	*dx*	*dy*	*dz**	*dr*
2.5	0.5	1.3 (0.1)	2.1 (0.6)	**2.4** (0.6)	0.05 (0.1)	0.08 (0.04)	2.1 (0.6)	**2.1** (0.6)
2.5	0.5	0.18 (0.04)	0.4 (0.1)	**0.5** (0.1)	0.07 (0.2)	-0.09 (0.2)	0.4 (0.1)	**0.5** (0.3)

*In the *z*-direction, *Cone Misalignment* and *Targeting Accuracy* were obtained from the same measurement.

5. Acknowledgements:

This work was supported by NIH R01CA100835, S103034539, NCI 10RR027610, NIAID K02AI093866, and NASA NSCOR NNX11AC60G.

6. References

[1] Newton J *et al* 2011 *Med. Phys.* **38** 6754-62
[2] Chow J C *et al* 2010 *Med. Phys.* **37** 5322-9
[3] Thomas A *et al* 2011 *Med. Phys.* **38** 4846-57
[4] Sakhalkar H S *et al* 2009 *Med. Phys.* **36** 71-82

Potential dosimetric benefit of dose-warping based 4D planning compared to conventional 3D planning in liver stereotactic body radiotherapy (SBRT)

UJ Yeo[1], ML Taylor[1,2,3], T Kron[1,2], D Pham[2], S Siva[4] and RD Franich[1]
[1]School of Applied Sciences and Health Innovations Research Institute, RMIT University, Melbourne, VIC, Australia
[2]Physical Sciences, Peter MacCallum Cancer Centre, Melbourne, VIC, Australia
[3]William Buckland Radiotherapy Centre, The Alfred Hospital, Melbourne, VIC, Australia
4Sir Peter MacCallum Department of Oncology, University of Melbourne VIC, Australia

E-mail: unjin.yeo@rmit.edu.au

Abstract. Respiratory motion induces dosimetric uncertainties for thoracic and abdominal cancer radiotherapy (RT) due to deforming and moving anatomy. This study investigates the extent of dosimetric differences between conventional 3D treatment planning and path-integrated 4D treatment planning in liver stereotactic body radiotherapy (SBRT). Respiratory-correlated 4DCT image sets with 10 phases were acquired for patients with liver tumours. Path-integrated 4D dose accumulation was performed using dose-warping techniques based on deformable image registration. Dose-volume histogram analysis demonstrated that the 3D planning approach overestimated doses to targets by up to 24% and underestimated dose to normal liver by ~4.5%, compared to the 4D planning methodology. Therefore, 4D planning has the potential to quantify such issues of under- and/or over-dosage and improve treatment accuracy.

1. Introduction

Stereotactic body radiotherapy (SBRT) for treatment of liver lesions, in principle, allows high dose conformity to the tumour with minimal dose to the normal tissue [1]. However, this is subject to the uncertainties introduced by delivering inherently inhomogeneous dose distributions to targets that deform and move due to patient respiration. Consequently, this could lead to reduced dose conformity and the potential for reduced tumour control. One approach for calculation of cumulative doses in moving and deforming targets (for inter- and intra-effects) is a 'dose-warping' technique [2-5]. In this study, we address the potential benefit of dose-warping based 4D treatment planning over conventional 3D for liver SBRT.

2. Materials and Methods

Two patients were chosen from a representative cohort of liver patients, representing opposite characteristics in terms of gross tumour volume (GTV) size and motion: a large GTV (54.5 cm^3) with small motion (7.5 mm) for patient A and a small GTV (6.7 cm^3) with large motion (16.6 mm) for patient B. Respiratory-correlated 4DCT image sets with 10 phases (0 – 90% of respiratory period) were acquired. The internal target volume (ITV) was defined as the sum of tumour positions at exhale (50%) and inhale (0%) respiration phases, and a 5mm margin was added for generation of the planning target volume (PTV). The 3D conventional treatment planning with nine beam geometries for each of six fractions (normal clinical practice) was conducted using the 50% phase CT set as a reference image; it was then copied over to the other nine phases and recalculated. The path-integrated 4D dose accumulation was performed using a dose-warping technique based on an *optical flow* method [6, 7] of deformable image registration (DIR) which we have previously established to be the most accurate of the algorithms studied [5]. For all registrations, the 50% phase was the 'target' image, used as a reference to which the 'source' images of the other respiration phases were morphed. Doses of each source image were transformed using deformation vector fields (DVFs) calculated from each registration. These warped doses were equally weighted to estimate the path-integrated 4D-cumulative dose distribution. Target volume coverage as well as doses to other surrounding organs were evaluated with the 4D planning methodology and compared with those from the conventional 3D planning approach.

3. Results and Discussion

The results of DIR were assessed first by visual inspection of difference maps, to affirm that the registration was acceptable. An example of DIR from extreme inhale phase (source) to extreme exhale phase (target) is shown in figure 1. In the first row, the upper (red, short dash) and lower (green, long dash) dashed lines show the alignment of the liver and the kidneys, respectively, before and after DIR. In the second row, the difference map after DIR shows the difference between the target and the calculated image, illustrating the high performance of the optical flow method.

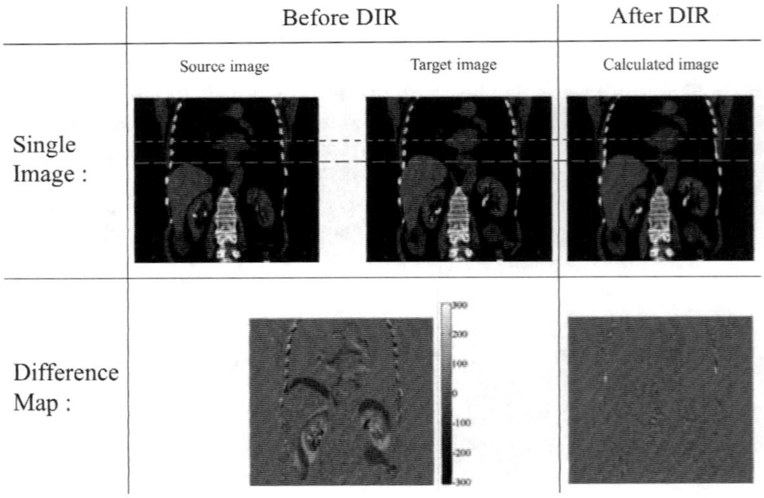

Figure 1: Example results of deformable image registration using the optical flow method are shown in the coronal view. The scale in the difference map is in HU.

Figure 2(a) and 2(b) depict dose-volume histograms (DVHs) of targets and normal liver for patients A and B, respectively. According to the dose prescriptions of the 3D-static plans, the D_{mean} of the PTV was 47.4 and 52.7 Gy for patient A and patient B, respectively. D_{mean} calculated from the 4D-cumulative plan was 2.2 and 3.8 % lower for patient A and patient B, respectively. The ratio of D_2 (2% near-maximum dose) / D_{98} (98% near-minimum dose), i.e. homogeneity indices (HI), were calculated from DVHs. The HI difference between 3D and 4D plans was less than 1 % for GTV and ITV for both patients. In contrast, HI_{3D} for PTV decreased by 14.0% and 23.5% compared to HI_{4D} for patient A and patient B. These results illustrate that the 3D-static plan overestimated dose by up to ~24%, particularly at the high-dose region around the PTV margin, compared to the 4D-cumulative plan which represents a more accurate approximation of the doses actually delivered. These discrepancies were mainly introduced by circumstances in which the presence of dose gradient was most prominent in PTV margin (but out of the ITV). Another dominant factor yielding dosimetric differences between the two planning schemes is the fact that healthy liver tissue adjacent to the PTV could move in and out of the treatment beam field over the breathing cycle, resulting in undesired dose to healthy liver while reducing dose conformity to the PTV. As evidence of the latter point, it is worth to note that the smaller volume with larger motion yielded a greater discrepancy between 3D and 4D plan than the larger volume with smaller motion.

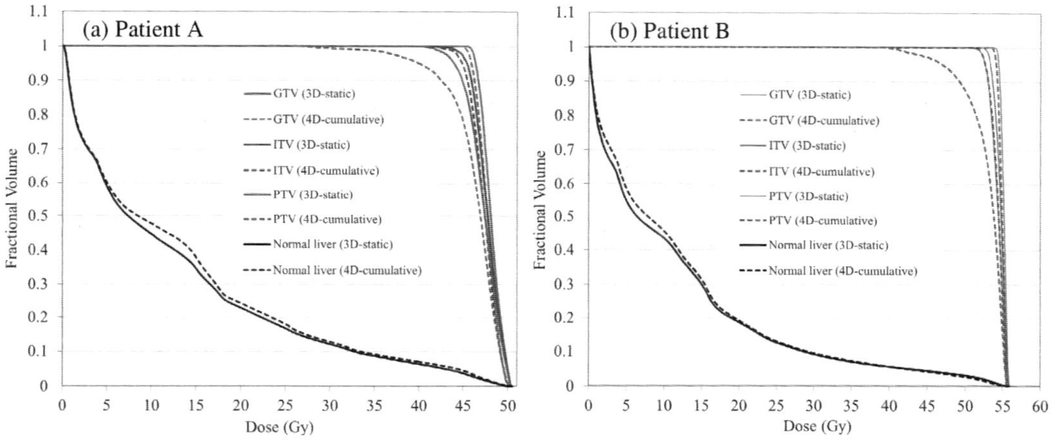

Figure 2: Dose-volume histograms (DVHs) for targets and normal liver: **(a)** patient A (large GTV with small motion) and **(b)** patient B (small GTV with large motion). Straight and dashed lines illustrate the results calculated from 3D-static and 4D-cumulative planning, respectively (red: GTV, blue: ITV, green PTV, and black: normal liver).

For normal liver, there is an observable trend such that the mean doses (D_{mean}) calculated from 3D-static planning were reduced by 4.3% and 4.7% compared to 4D-cumulative planning for patient A and B, respectively. This illustrates that 3D-static planning underestimates the delivered doses to healthy tissues. Doses to other organs are low and consequently differences between 3D and 4D methods are not as troubling in an absolute sense (see figure 3).

4. Conclusions

In this study, it is demonstrated that the typical 3D-planning approach appears to significantly underestimate dose to the PTV, while overestimating dose the surrounding healthy liver. The general consequence can thus be undesired reduction of dose conformity and tumour control. 4D-cumulative planning has, therefore, the potential to quantify such issues of under- and/or over-dosage and to benefit patient outcomes in radiotherapy treatment.

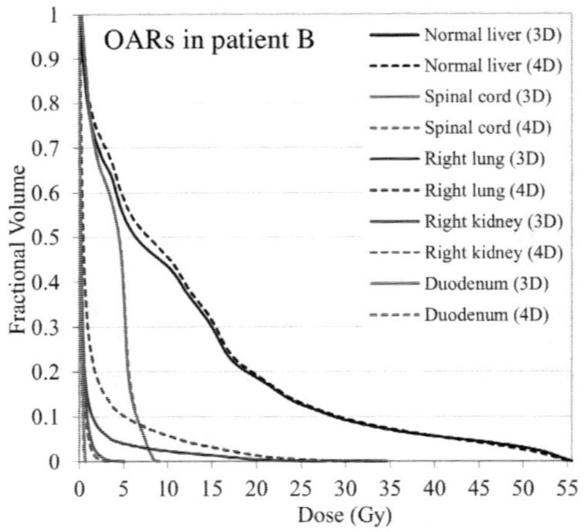

OARs	D_{mean} (3D / 4D)	D_2 (3D / 4D)	D_{98} (3D / 4D)
Normal liver	11.6 / 12.1	52.6 / 51.6	55.5 / 55.3
Spinal Cord	3.9 / 3.9	8.0 / 8.0	8.9 / 8.9
R Lung	1.0 / 2.0	11.3 / 18.0	27.7 / 34.7
R kidney	0.4 / 0.4	2.0 / 1.5	5.1 / 3.7
Duodenum	0.2 / 0.2	0.6 / 0.5	0.7 / 0.7

Figure 3: An example of dose-volume histograms (DVHs) for critical organs in patient B. Straight and dashed lines illustrate the results calculated from 3D-static and 4D-cumulative planning, respectively. The table shows the mean dose (D_{mean}), the 2% near-minimum dose (D_2), and the 98% near-maximum dose (D_{98}) in each organ for both planning schemes.

5. Acknowledgements

This work is supported by an RMIT University Research & Innovation Emerging Researcher Industry Award (Dr R Franich, Dr Taylor).

6. References

[1] Dawson L A *et al* 2006 *Acta. Oncol.* **45** 856-64
[2] Brock K *et al* 2003 *Med. Phys.* **30** 290-5
[3] Keall P 2004 *Semin. Radiat. Oncol.* **14** 81-90
[4] Rosu M *et al* 2007 *Med. Phys.* **34** 233-45
[5] Yeo U J *et al* 2012 *Med. Phys.* (in press)
[6] Horn B K P and Schunck B G 1981 *Artif. Intel.* **17** 185-203
[7] Huang T C *et al* 2010 *Radiat. Oncol.* **5** 45

7th International Conference on 3D Radiation Dosimetry (IC3DDose) IOP Publishing
Journal of Physics: Conference Series **444** (2013) 012072 doi:10.1088/1742-6596/444/1/012072

Investigating the reproducibility of a complex multifocal radiosurgery treatment

M Niebanck[1], T Juang[1], J Newton[1], J Adamovics[2], Z Wang[1] and M Oldham[1]
[1]Duke University Medical Center, Durham, NC, USA
[2]Rider University, Lawerenceville, NJ, USA

E-mail: michael.niebanck@duke.edu

Abstract. Stereotactic radiosurgery has become a widely used technique to treat solid tumors and secondary metastases of the brain. Multiple targets can be simultaneously treated with a single isocenter in order to reduce the set-up time to improve patient comfort and workflow. In this study, a 5-arc multifocal RapidArc treatment was delivered to multiple PRESAGE® dosimeters in order to explore the repeatability of the treatment. The three delivery measurements agreed well with each other, with less than 3% standard deviation of dose in the target. The deliveries also agreed well with the treatment plan, with gamma passing rates greater than 90% (5% dose-difference, and 2 mm distance-to-agreement criteria). The optical-CT PRESAGE® system provided a reproducible measurement for treatment verification, provided measurements were made immediately following treatment.

1. Introduction

The use of high single dose fractions with steep gradients requires more stringent tolerances than standard hyperfractionated schemes, which in turn requires additional set up time. Recently stereotactic techniques have emerged which enable treatment of multiple intracranial targets with a single isocenter, thereby greatly reducing treatment time [1]. Verification of multifocal, RapidArc plans is a challenge due to the complexity of the dose distributions. Providing true 3D, fast, and high resolution 3D dosimetry could be an ideal solution for SRS treatment verification, and prior works reflect this interest [2-5]. In this work we deliver the same complex treatment plan to 3 identical PRESAGE® dosimeters in order to determine the repeatability of the complex treatment delivery.

2. Methods and Materials

A single isocenter, multifocal stereotactic radiosurgery plan was delivered to three PRESAGE® dosimeters. The dosimeters were from the same formulated batch and were irradiated and imaged in sequence. The treatment plan was a 5-arc RapidArc plan with a single isocenter, and 14 Gy prescribed to 99% of the volume of the central target. In addition to the central target, four peripheral targets were treated off isocenter with doses of roughly 20 Gy. The treatment was planned using iPlan (localization and isocenter placement) and Eclipse (arc optimization) and delivered on a Novalis Tx machine to three PRESAGE® dosimeters. The dose distributions were measured using an in-house optical-CT system, the Duke Large-Field Optical Scanner (DLOS) [6], both immediately after treatment and two weeks after treatment.

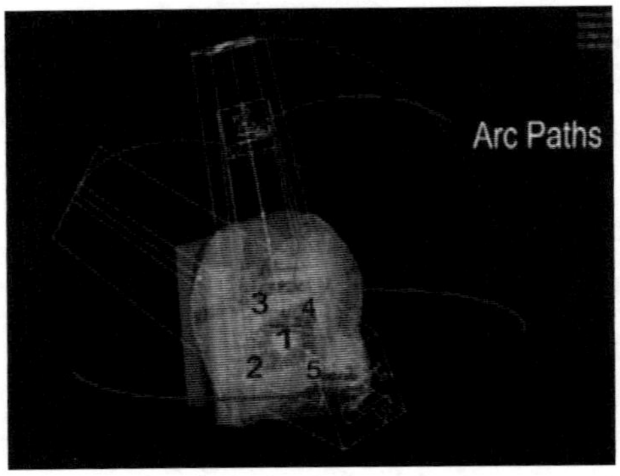

Figure 1: 5-arc RapidArc radiosurgery plan treating 5 targets displayed in Eclipse.

The batch of PRESAGE® used for this study was specially formulated to have a significantly lower sensitivity compared to normal batches in order to allow a higher prescription dose without overexposure. While the formulation had a lower sensitivity, other properties, such as effective Z value and tissue equivalency were intended to remain unchanged from previous batches [7]. The sensitivity was measured prior to setting an appropriate prescription dose for the plan by irradiating cuvettes containing the formulation with known doses.

The DLOS system uses a LED light source, a CCD camera, and two telecentric lenses which provide a central region of parallel light where the dosimeter can be imaged free of object magnification effects. The dosimeter is immersed in a fluid bath with a matched refractive index to minimize bending of the light at the dosimeter-fluid interface. The dosimeter is placed on a motorized rotating stage in the bath, allowing projection images to be captured from any angle [6]. The system has a spatial resolution of 175 μm (typically downsized to save disk space) and acquires 360 projections in 5-10 minutes.

The projections were reconstructed using a MATLAB-based GUI with 2 mm resolution and a 15x15 median filter to reduce noise. The reconstructed data cubes were exported to CERR (Computational Environment for Radiotherapy Research) along with the treatment plan from Eclipse [8]. Using CERR, the optical density distribution was registered to the planned dose distribution and scaled using an absolute dose point from an ion chamber measurement. After proper registration and scaling, the average and standard deviation of the three distributions was calculated to examine consistency. Each immediate measurement was tested against its two weak measurement counterpart using 3D gamma analysis (5% dose difference/2 mm distance-to-agreement) to examine stability of the measurement over time. Gamma analysis with the same criteria was also used to compare the Eclipse planned dose to the measured distributions.

3. Results and Discussion

3.1 Cuvette Sensitivity

The sensitivity of the PRESAGE® formulation used in this experiment was found to be 0.0029 ΔOD/(Gy□cm). The linear relationship between OD change and dose allows scaling of the entire OD change distribution to a single absolute dose point to determine the dose distribution.

3.2 Consistency Between Dosimeters

Target dose distributions varied between each of the three deliveries very little, with less than 3% standard deviation in voxels located in the high dose regions. Figure 2 shows the range of isodose contour deviations between the three deliveries, with tight agreement in high dose regions. Figure 1 also shows excellent agreement between the planned dose distribution and the average measured distribution of the three dosimeters.

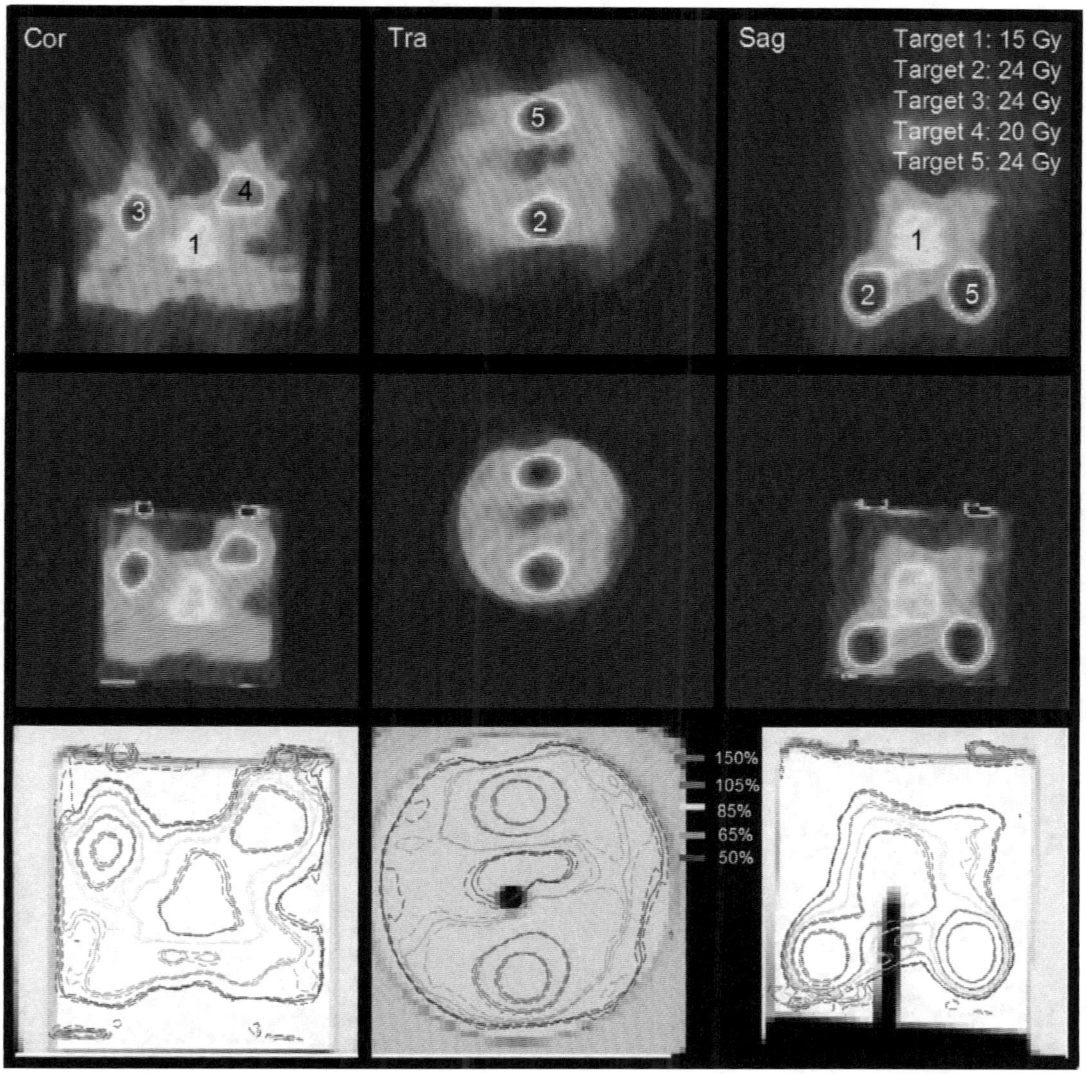

Figure 2: Top row: planned dose distributions from Eclipse to the five targets; Middle row: average of the three measured dose distributions; Bottom row: isodose lines for the average measured distribution (solid) as well as +/- 1 standard deviation (dashed lines).

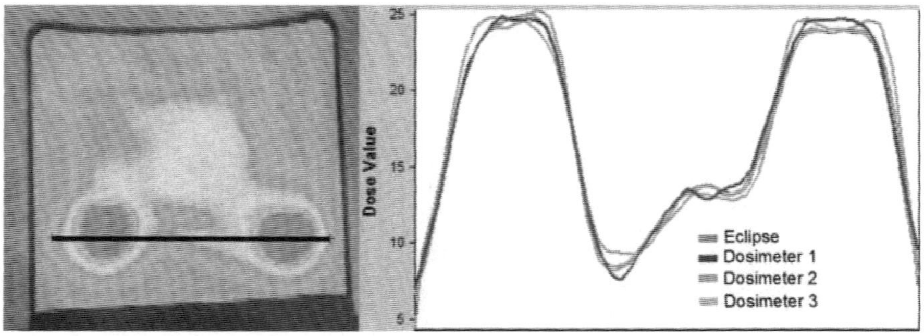

Figure 3: Line profile over two targets illustrates consistent delivery and low noise.

3.3 Consistency Over Time

The three initial measurements agreed well with the Eclipse planned dose with gamma passing rates of 92.4%, 93.0%, and 91.7% with 5% DD/2 mm DTA criteria. Closer inspection revealed the dosimeters were not stable with time. Dose measurement values in the peripheral targets changed by as much as 12% over the two week period. Gamma analysis (5%/2mm) comparing the initial measurements to measurements two weeks later yielded passing rates of 46.4%, 62.9%, and 55.7% for the three dosimeters.

4. Conclusions

The three deliveries of the radiosurgery treatment were found to be very consistent (less than 3% dose deviation in targets) with each other, demonstrating the accuracy of the treatment machine and the reproducibility of the measurement technique of PRESAGE® with optical-CT. While the three treatment deliveries agreed well with each other, as well as with the treatment plan, significant discrepancies (less than 63% gamma pass rate) were found when measuring each dosimeter at different times, limiting the window after treatment to acquire accurate measurements. Reducing the sensitivity of these dosimeters appears to have also reduced post-irradiation stability. Finding a PRESAGE® formulation with a stable optical density or at least a dose-independent change over time would solve this issue.

5. Acknowledgements

This study was supported by the NIH R01CA100835 grant.

6. References

[1] Wang Z *et al* 2012 *Med. Phys.* **39** 3910
[2] Wang Z *et al* 2010 *J. Phys.: Conf. Ser.* **250** 012058
[3] Clift C *et al* 2010 *Phys. Med. Biol.* **55** 1279-93
[4] Pourfallah T A *et al* 2009 *Med. Phys.* **36** 3002-12
[5] Watanabe Y *et al* 2005 *J. Appl. Clin. Med. Phys.* **6** 133-42
[6] Thomas A *et al* 2011 *Med. Phys.* **38** 4846-57
[7] Guo P Y 2006 *Med. Phys.* **33** 1338-45
[8] Deasy J 2003 *Med. Phys.* **30** 979-85

Stereotactic body radiation therapy delivery validation

T Olding[1], L Garcia[1], K Alexander[2], LJ Schreiner[1,2] and C Joshi[1,2]
[1]Cancer Centre of South Eastern Ontario at Kingston General Hospital, 25 King Street West, Kingston, ON, Canada, K7L5P9
[2]Department of Physics, Queen's University, Kingston, Ontario, Canada, K7L3N6

E-Mail: Tim.Olding@krcc.on.ca

Abstract. This work describes the use of a motion phantom and 1D, 2D, and 3D ion chamber, EBT3 film, electronic portal imaging device (EPID) and FXG gel measurements for dosimetric validation of a stereotactic ablative radiation therapy (SBRT) technique in our clinic. Results show good agreement between the measurements and calculated treatment plan dose.

1. Introduction

Stereotactic body radiation therapy (SBRT) is a high precision image-guided radiation therapy (IGRT) technique in which hypo-fractionated ablative doses of radiation are delivered to small, accurately delineated tumors [1]. Dose distributions for SBRT are designed such that the ablative dose is confined to the tumor and immediate vicinity, with the tumor motion characterized and compensated for. SBRT is typically offered as a treatment option for early stage (Stage I, T1/2, N0, < 5 cm), medically inoperable, non small cell lung cancer. Typical multi-fraction dose regimes include 48-60 Gy in 3-5 fractions. With such a high dose being delivered in only a few fractions, it is imperative that high geometric and dosimetric precision is maintained at each step of the treatment process. End to end validation of dose delivery during the commissioning process is, therefore, vital to a successful SBRT program. The purpose of this work is to validate a newly implemented SBRT lung treatment delivery using different dosimetric modalities. We included gel dosimetry in this work, encouraged by the success of previous gated radiation delivery validations using 3D dosimeters [2, 3].

2. Materials and Methods

2.1. SBRT Delivery Validation

The dosimetry tools used for SBRT delivery validation in this study were: a) point single ion chamber (Capintec PR-05P 0.07 cc/Capintec electrometer), b) 2D Gafchromic EBT3 film (ISP, Wayne, NJ), c) 3D EPID based fluence measurement and dose prediction algorithm [4], and d) 3.5D Fricke-xylenol orange-gelatin (FXG) based gel dosimeter with optical CT readout [5]. Static and dynamic 4D computed tomography (4DCT) scans were acquired of a Quasar motion phantom (Modus Medical Devices Inc, London, ON, figure 1a) with a cedar insert containing a 3 cm diameter acrylic sphere 'tumour' and a hole for insertion of the ion chamber. A GE Lightspeed RT16 scanner (GE Healthcare, Maple Grove, MN) was used for CT data acquisition, with the addition of a Varian RPM motion tracking system for the dynamic scan. A similar cedar insert was used for the EBT3 film (figure 1b). The scans were then repeated with a custom gel insert in the Quasar phantom (figure 1c left). The rubber ball "tumour" in the gel was used to establish a target for contouring in the CT data sets, the inside of the jar was turned to unit electron density in the CT scan for dose calculation and the gel insert was replaced with a FXG gel during the SBRT deliveries (Fig.1c centre and right). Figure 1d left shows the breathing cycle used for the ion/chamber and film 4DCT scans and SBRT deliveries (15 breaths per minute (BPM), 20 mm amplitude peak-to-peak). The sinusoidal respiratory cycle used for the gel measurements (figure 1d right) had the same rate of 15 BPM, but with reduced 10 mm amplitude due to the limitations of motion imposed by the acrylic gel holder insert. A screen capture of the 9 field ion chamber plan (with 2 non-coplanar fields) calculated in Eclipse™ (Varian Medical Systems, Palo Alto, CA) is shown in figure 1e.

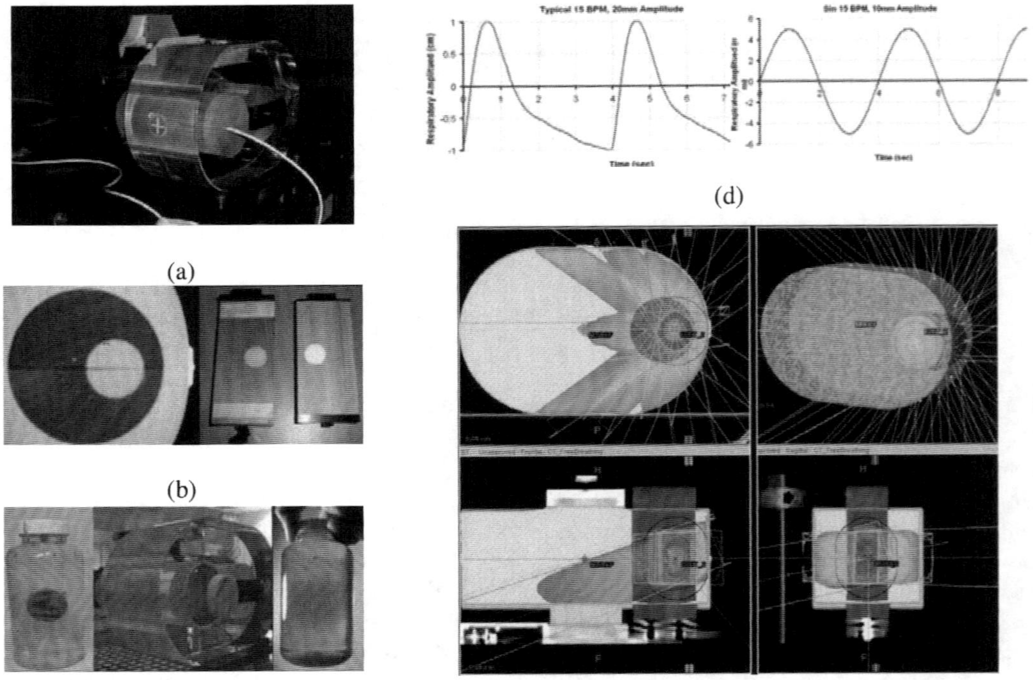

(a) (b) (c) (d) (e)

Figure 1: (a) Quasar motion phantom with ion chamber insert, (b) CT cross section and photo of cedar film insert, (c) CT gel insert and FXG gel dosimeter setup, (d) Breathing patterns for the dynamic 4DCT scans and SBRT deliveries (e) Eclipse screen capture of ion chamber plan.

gel was used to establish a target for contouring in the CT data sets, the inside of the jar was turned to unit electron density in the CT scan for dose calculation and the gel insert was replaced with a FXG gel during the SBRT deliveries (figure 1c centre and right). Figure 1d left shows the breathing cycle used for the ion/chamber and film 4DCT scans and SBRT deliveries (15 breaths per minute (BPM), 20 mm amplitude peak-to-peak). The sinusoidal respiratory cycle used for the gel measurements (Fig.1d right) had the same rate of 15 BPM, but with reduced 10 mm amplitude due to the limitations of motion imposed by the acrylic gel holder insert. A screen capture of the 9 field ion chamber plan (with 2 non-coplanar fields) calculated in Eclipse™ (Varian Medical Systems, Palo Alto, CA) is shown in figure 1e.

All treatment plans were calculated in Eclipse and delivered on a Varian Trilogy 2100iX linear accelerator (Varian Medical Systems), with Advanced OBI™ imaging cone beam CT (CBCT) phantom setup verification. Each dose measurement was compared against calculated Eclipse treatment planning system (TPS) dose using commercial software designed for that QA tool. EBT3 film measurements were analyzed using Welhoffer QA software (OmniPro v 1.6.0002-2008-03-12). EPIDOSE 3D dose was compared with Eclipse using MATLAB (Mathworks, Newark, NJ) based software provided by Ansbacher. The reconstructed, calibrated gel dose data[5] from the Vista™ optical CT scanner (Modus Medical Devices Inc., London, ON) was converted to Dicom format, and compared to Eclipse in the CERR environment (Computational Environment for Radiotherapy Research, Washington State University, St. Louis, MO).

3. Results

3.1. Dosimeter-Plan Dose Comparison

Dosimetric results for the different dosimeters used in this work are summarized in table 1. The dynamic film delivery indicated the greatest disagreement between the calculated Eclipse plan and

Table 1: Summary: measured vs. Eclipse doses. Note that the delivered dose was modified to account for the dynamic nature of the delivery and the calibration range of the dosimeter

	Chamber		**Film**		**EPID**	**Gel**	
Plan	Static	Dynamic	Static	Dynamic	Static	Static	Dynamic
Dose (Gy)	12	12	2	2	12	2	2
Fields	2POP[#]	9NCP*	9NCP	9NCP	9CP[+]	9NCP	9NCP
Dose Rate (Mu/min)	600	600	100	100	600	100	100
1D Point Dose Difference (%, measured vs. Eclipse)	-1.3	-2.3	-1.0	-1.5	-0.5	-0.8	-0.5
% Gamma ≤1 (3%/3mm)	-	-	100	82.4	-	98.5	99.9
% Chi ≤1 (2%/2mm)	-	-	-	-	98.5	-	-

[#] POP = parallel opposed pair in the anterior-posterior direction, * NCP = non-coplanar, [+] CP = coplanar

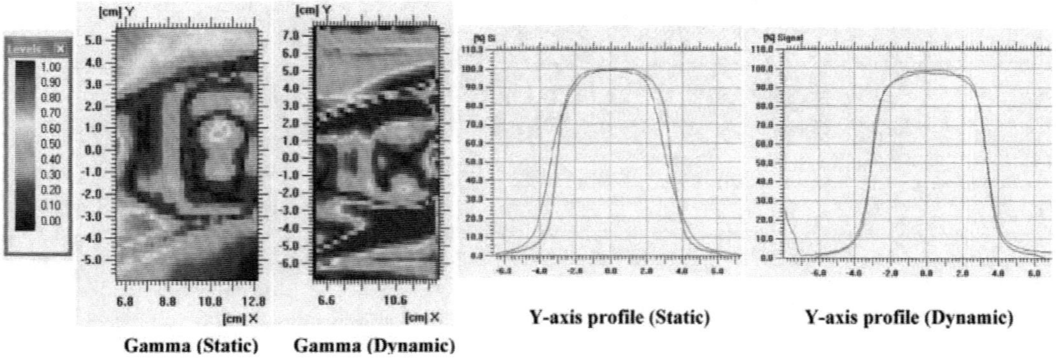

Figure 2: 3%, 3 mm gamma maps and y-axis dose profiles between the plan (red) and measurement (green) for the static and dynamic film deliveries.

measured 2D dose distribution, as shown in figure 2. Most of the disagreement is near the edges of the high dose region in the superior-inferior direction, which is the axis of phantom motion. All dosimeters are consistent in reporting a lower point dose than the TPS at or near the plan isocentre.

Figure 3 shows axial and coronal views in CERR of the Eclipse plan and measured gel dose distributions for the 2 Gy, 9 field non-coplanar dynamic delivery. While the 3%, 3mm gamma comparison indicates full agreement between the plan and measured dose, lower doses are observed in the posterior region of the dosimeter near the edge of the high dose region.

4. Discussion and Conclusions

With reference to table 1, the fact that the static point dose measurements were less than that calculated by the TPS can be attributed to a few possible factors: 1) attenuation of the photon beam through the carbon fibre couch top is not accounted for in current Eclipse dose calculations (which also could explain the decreased dose in the posterior region of the gel dosimeter); 2) TPS limitations related to dose calculations in dual electronic disequilibrium conditions (i.e. small field size + small target surrounded by inhomogeneous media; and 3) the difference in scatter conditions in Eclipse versus the measurement due to blurring of moving GTV (i.e. ITV) on 4DCT images.

Delivery of a full 12 Gy treatment fraction to both the EBT3 film and FXG gel was initially contemplated, but after an initial trial run on the EBT3 film, it was decided that the film calibration in that range was not sensitive enough to faithfully record the dose range. It was therefore decided to deliver a lower dose (2 Gy) over the same treatment time by lowering the dose rate from 600 MU/min to 100 MU/min to preserve the number of breathing cycles during the treatment. It was verified that EBT3 film and FXG gel are insensitive to dose rate in this range; these results are not presented here. The significant disagreement near the edges of the high dose region in the dynamic delivery film measurement are well-explained by the 'smearing out' of measured dose in the gradient regions due to motion during delivery. This effect is not as strongly observed in the gel dynamic delivery due to the smaller motion amplitude and the more forgiving 3D (versus film 2D) gel gamma comparison analysis.

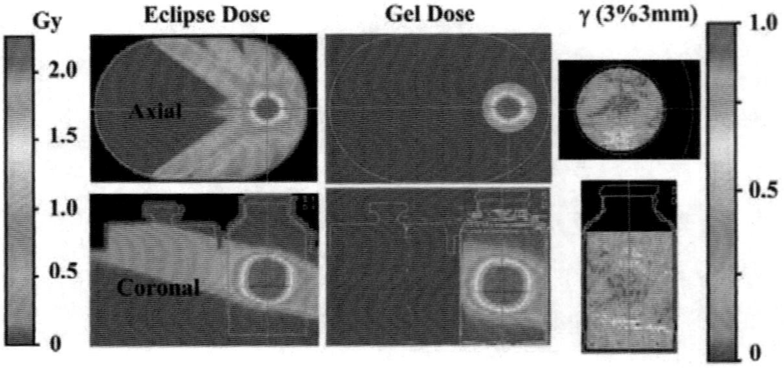

(a)

Figure 3: 3%, 3mm 3D gamma comparison (right) of the agreement between calculated plan dose (left) and measured FXG gel dose (centre) for the 2 Gy SBRT dynamic delivery.

One of the practical challenges in this work was registration of the dose data: the gel holder design had to accommodate placement of registration fiducials on the gel jar, and the film registration had to be similarly thought through. This is an important process step that is perhaps not emphasized enough in the reporting of 2D/3D dose comparisons.

There were initial problems encountered with gel when the smaller jar size for the SBRT dosimeter needed to fit in the Quasar phantom cavity, was calibrated against doses from a calibration electron beam given to a larger sized jar[5]. This was overcome by using a smaller field size calibration electron beam delivered to the same sized smaller jar as that used for the SBRT delivery. This result highlights a known limitation of optical CT based FXG gel dosimetry: the confounding effect on the measurement if the calibration and measurement jars are not the same size.

Overall, this work indicates good consistency between the data obtained from the various dosimetric methods (1D, 2D and 3D settings) used to validate the SBRT-lung dose delivery.

5. Acknowledgements
Research funding for this work has been provided by the Canadian Institutes of Health Research (CIHR) and the Cancer Centre of Southeastern Ontario.

6. References
[1] Timmerman *et al* 2010 *JAMA* **303** 1070-76
[2] Mansson S *et al* 2006 *J. Phys.: Conf. Ser.* **56** 300-3
[3] Brady S *et al* 2011 *Phys. Med. Biol.* **55** 2187-201
[4] Ansbacher W 2006 *Med. Phys.* **33** 3369-82
[5] Olding T *et al* 2010 *Phys. Med. Biol.* **56** 1259-79

Journal of Physics: Conference Series **444** (2013) 012074 doi:10.1088/1742-6596/444/1/012074

Ultra-high resolution optical CT dosimetry for the visualisation of synchrotron microbeam therapy doses

SJ Doran[1,2], AT Abdul Rahman[2,3], E Bräuer-Krisch[4], T Brochard[4] and J Adamovics[5]

[1]CRUK and EPSRC Cancer Imaging Centre, Institute of Cancer Research, Sutton, UK
[2]Department of Physics, University of Surrey, Guildford, GU2 7RB, UK
[3]Universiti Teknologi MARA Malaysia (UiTM), Campus of Negeri Sembilan, 72000 Kuala Pilah, Malaysia
[4]European Synchrotron Radiation Facility, Grenoble, France
[5]Department of Chemistry and Biology, Rider University, Lawrenceville, NJ, USA

E-mail: Simon.Doran@icr.ac.uk

Abstract. Optical CT is a method that can potentially provide both accurate dosimetry at high spatial resolution and 3-D visualisation over a large field-of-view in a single dataset. The major factors limiting spatial resolution in previous studies are analysed here and it is shown that improvements in equipment specification can overcome many of these. The need for ultra-high spatial resolution in the verification of microbeam radiation therapy verification is demonstrated and example images of a PRESAGE® sample are presented.

1. Introduction

It is well established that normal tissues can tolerate high doses of radiation over small volumes. Therefore, radiation effects in tissue are dependent not only on dose but also on the volume exposed [1]. X-ray microbeam radiation therapy (MRT) [2] is an advanced form of treatment that exploits these effects to deliver lethal doses to a tumour, while minimizing damage to the surrounding healthy tissues, by precisely controlling the size of irradiated regions. MRT using x-ray microbeams generated by a synchrotron is being pioneered at the European Synchrotron Radiation Facility (e.g., [3, 4]) and Brookhaven National Laboratory (e.g., [5]).

Many different techniques have been investigated to obtain accurate dosimetry of this challenging situation [6]. Optical CT is particularly exciting, because of the possibility of obtaining data with microscopic resolution, over the entire field-of-view of a macroscopic object. This would allow the entire microbeam treatment to be visualised with accurate dosimetry at each 3-D location. We have previously described the construction of an optical CT microscope [7]; we have developed a protocol for measuring the modulation transfer function of an optical CT scanner [8]; and we have presented preliminary results of the application of this protocol to our optical CT microscopy system, together with examples of the measurement of a synchrotron microbeam treatment applied to a PRESAGE® dosimeter [9].

The spatial resolution in our previous work was limited not by the imaging method *per se*, but rather by the camera and computing hardware available at the time that work was performed. Here, we show how advances in equipment performance — driven to a great extent by the explosive recent growth of digital imaging and image sharing in the consumer sector — have the potential to make a significant difference to the capability of high-resolution 3-D dosimetry.

The key bottlenecks in the optical CT imaging process can be summarised as follows:

- Unlike the case of MRI, in which a small sub-volume can easily be excited and imaged within a larger object, it is not straightforward (although some techniques do exist) to obtain zoomed images of small regions using CT. This means that to obtain high spatial resolution images of extended objects, one needs to acquire projection images with large matrix sizes. Until relatively recently, scientific cameras capable of recording such projection images were prohibitively expensive.
- Transferring the data from the camera to the acquisition computer has, to date, been time-consuming for large frame sizes. When combined with the Nyquist requirement on the number of projections that must be acquired for correct image reconstruction, this led to prohibitively long acquisition times with our previous equipment.
- Ultra-high resolution datasets are very large (typically, tens of GB). The first stage of CT reconstruction is to create sinograms for the projection data. From a resource perspective, the most significant step here is a *transposition* of the data and this step is hugely inefficient if the data cannot all fit in memory at the same time.
- Large datasets require a significantly increased computational power for performing the back-projection, compared with the datasets acquired for standard clinical verification tasks.
- Visualisation and display of the results of imaging with such large 3-D matrix sizes has been, and to a certain extent still is, difficult.

2. Methods

As in our previous study [9], PRESAGE® was provided by Heuris Pharma (Skillman, NJ) in the form of cylinders of diameters 22 mm and 9.7 mm. These were machined locally to give a uniform height of 60 mm. Experiment 1 was designed to demonstrate the need for high spatial resolution. A 22 mm sample was irradiated end-on with a set of dose patterns consisting of six slits for each of a number of different doses in the range 20 – 80 Gy. This pattern was repeated with four different slit widths from 52 μm to 416 μm. Experiment 2 aimed to verify the so called "cross-firing" configuration of MRT [3], in which the irradiation was delivered from four directions by orienting the sample at 0°, 45°, 90° and 135° with respect to the beam. The dose delivered was 140 Gy and the irradiation arranged to create a uniform cubic region (1mm × 1 mm × 1 mm) of dose deposited by the overlap of all four microbeam parallel arrays, each of which contained beams with a nominal FWHM 50 μm and separated by 400 μm.

Data were read out using the University of Surrey optical CT microscopy system, previously described in [7, 9]. For Experiment 1, the original Orca II BT 1024-G (Hamamatsu, Japan) was used with acquisition matrix size 512 × 512 pixels. However, for Experiment 2, a new camera, the pco.Edge (PCO AG, Kelheim, Germany) was tested. This has the capability of exceptionally rapid acquisition and output of images (up to 100 fps). For the initial trial, the system was operated in asynchronous mode, with the rotation stage in continuous motion and frames acquired every 100 ms. Two 3-D images were acquired: one with a large field-of-view to capture the full MRT treatment, and the second with a higher spatial resolution over a limited axial range. The first raw dataset consisted of 4000 projection images, each of size 960 × 1440, with isotropic pixels of size (24 × 24) μm^2 over a field of view of (23 × 34) mm^2, while for the second we acquired 5000 projections of size 2560 × 586 with non-isotropic pixels of size (4.4 × 22) μm^2 over a field of view of (11.3 × 10.3) mm^2, limited by the installed memory on the acquisition computer. Processing was performed using IDL on a 2010 Mac Pro with 24 GB of RAM and movies created using the medical imaging application OsiriX.

3. Results and discussion

Figure 1 displays a summary of the results obtained from Experiment 1. Notice how, for the two wider

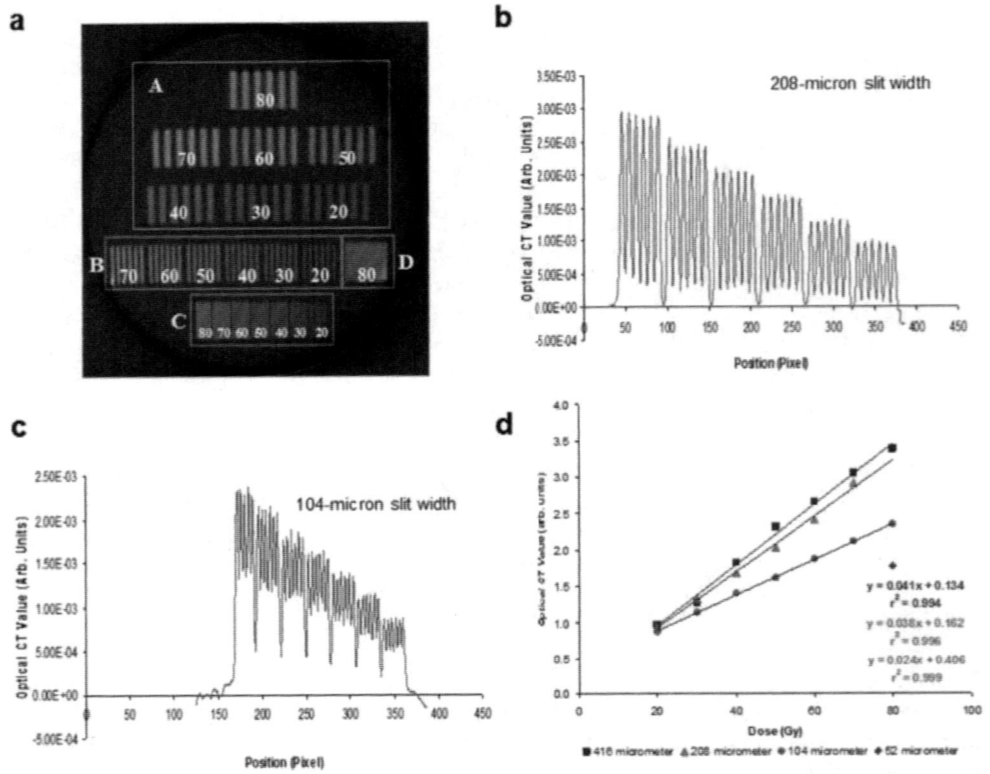

Figure 1: (a) Reconstructed image acquired with Hamamatsu camera from original projection images of dimension 512 × 512 pixels; (b) horizontal profile across slit pattern B (208 μm); (c) horizontal profile across slit pattern C (104 μm); (d) dose-response, as obtained from the four different slit sizes.

slit patterns (416 and 208 μm), the gradients of figure 1(d) are very similar, whereas for the 108 μm slit pattern, the gradient (and hence the apparent dose-response of the PRESAGE®) is around 40% lower. This is a straightforward manifestation of the effect of the modulation transfer function, but illustrates the fact that significant changes in pixel value occur even for structures that are several times the nominal spatial resolution (in this case around 20 μm). Whilst this type of effect is commonly tolerated in qualitative diagnostic imaging, resulting as it does merely in a reduction in image contrast, it leads to problems in the quantitative imaging of radiation dose and suggests that ultra-high resolution methods will be needed if MRT verification is to be successful.

Figure 2 provides a graphic illustration of the quality of data that can be obtained by imaging this type of sample at higher spatial resolution. The individual microbeams are clearly resolved and it is easy to see how this type of image, together with the full 3-D reconstruction - see accompanying multimedia files - could be used for diagnosing problems with the complex equipment required to implement MRT.

At the time of writing, our ability to handle the very large image datasets generated is still limited by the computing platform attached to our imaging camera. However, it is now possible to purchase desktop workstations with installed RAM of up to 256 GB and parallel-compute (GPU) cards that will speed up processing significantly. At the same time, the introduction of true-3D displays is on the horizon and we expect to be able to perform advanced visualisations in the near future.

4. Conclusion

Optical CT shows significant potential for ultra-high resolution dosimetry. It can already be used to

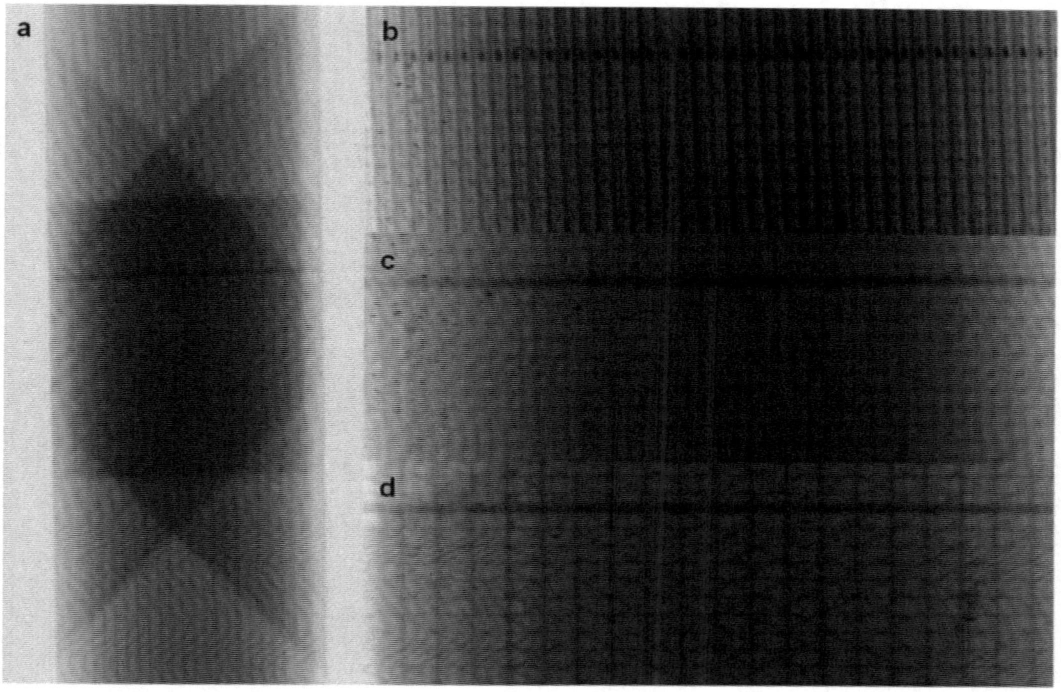

Figure 2: (a) Single projection from large field-of-view dataset from Experiment 2, showing mainly successful MRT treatment, but with a prominent horizontal band requiring further investigation; (b–d) example projections from the smaller field-of-view ultra-high resolution dataset, showing very clear resolution of the microbeams. Note the particular utility of view (b), which demonstrates that the dark horizontal band seen in (a), (c) and (d) is due to a plane of extra-wide microbeams.

visualise the results of sophisticated microbeam radiation therapy treatments in 3D and future work will focus on the degree to which we can achieve truly quantitative dosimetry measurements.

5. References

[1] Brauer-Krisch E *et al* 2010 *Mutat. Res.* **704** 160-6
[2] Slatkin D N *et al 1992 Med. Phys.* **19** 1395-400
[3] Brauer-Krisch E *et al 2005 Phys. Med. Biol.* **50** 3103-11
[4] Serduc R *et al* 2010 *PLoS One* **5** (2) e9028
[5] Dilmanian F A *et al 2002 Neuro. Oncol.* **4** 26-38
[6] Brauer-Krisch E *et al* 2010 *AIP Conference Proceedings* **1266** 89-97
[7] Doran S J *et al* 2010 *Phys. Med. Biol.* **55** 1531-47
[8] Abdul Rahman A T *et al* 2011 *Phys. Med. Biol.* **56** 4177-99
[9] Rahman A T A *et al* 2010 *J. Phys.: Conf. Ser.* **250** 012083

Scintillators for 3D and 4D dosimetry: current status and future potential for clinical translation

N Suchaworska[1], PZY Liu[1], A Ralston[1], P Naseri[1], P Abolfathi[2], J Lee[2], K Warrener[2] and DR McKenzie[2]
[1]Royal Prince Alfred Hospital, Camperdown, NSW, Australia
[2]School of Physics, University of Sydney, NSW, Australia

E-mail: natalka@email.cs.nsw.gov.au

Abstract. New radiotherapy treatment techniques make unprecedented demands on dosimetry. Water equivalence, radiation hardness, high temporal and spatial resolution in 4D are no longer desirable but essential for safe implementation of several techniques. Plastic scintillation dosimetryhas the capabilities to meet all of the above criteria. Despite several attempts, a system has not yet been commercialised for widespread availability in the clinic. Successful translation of good science to the benefits of clinical practice demands that the primary focus be shifted from the technological innovation to the patient needs. We will present the clinically beneficial features of scintillation dosimetry and some remaining challenges. Then we will discuss the barriers to widespread availability of scintillation dosimeters.

1. The Problem: unmet needs in clinical dosimetry

New treatment approaches have placed additional demands on clinical dosimeters. As an example, small field and dynamic composite treatments are now mainstream in contemporary radiotherapy practice, however, the dosimetry of these fields is often difficult. It is argued that this is due to the lack of lateral electron equilibrium for fields below 30 mm [1] and that the measurements of non-water equivalent dosimeters is affected by the detector density [2]. Insufficient care with dosimetry of such fields could substantially affect the quality of radiotherapy treatments (Figure 1).

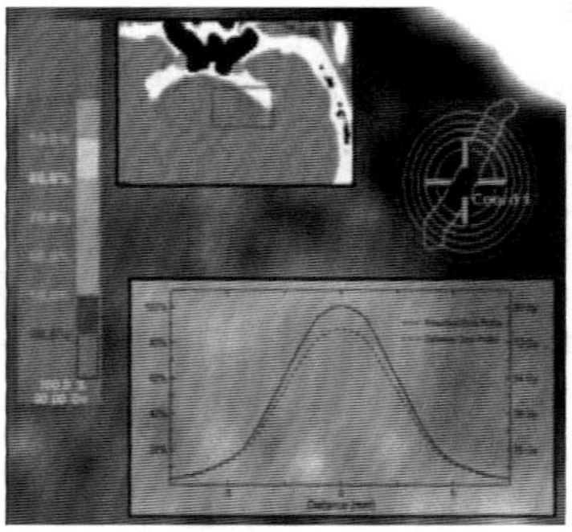

Figure 1: Radiation dose distribution calculated in iPlan for a trigeminal neuralgia, treated with a 4mm cone on the Varian Novalis linear accelerator. The dose profile prescribed to is based on measurements using a PTW 60012 diode, while the delivered dose (dotted line) is determined from scintillator detector measurements. The difference in the measured dose on the central axis is 13%.

Although correction factors for non-water equivalent dosimeters used in small radiation fields have been published [3, 4], Scott *et al* [2] argues against their use, suggesting a better approach would be to manufacture small-field relative detectors with small active volumes and a density similar to water. Scintillation dosimetry offers a solution with plastic scintillators, which have a water equivalent response in megavoltage beams.

2. A Solution: scintillation dosimetry

The concept of scintillation dosimetry involves capturing the light from a scintillator with an optic fibre and measuring its magnitude with a photodetector. The measured light signal is proportional to the rate of deposition of ionizing radiation energy in the volume of the scintillator. While conceptually simple, the choice of scintillator, optic fibre and photodetector can affect the performance of the dosimeter and each component needs to be optimised for the dosimeter's specific application [5, 6].

Historically, the clinical uptake of scintillation detectors in external beam radiotherapy has been complicated by the unwanted generation of Cerenkov radiation at megavoltage energies. In the megavoltage beam of a linear accelerator, Cerenkov light background is generated in the optic fibre used to transport the scintillation signal to the photodetector. Several methods have been proposed to account for the Cerenkov signal. These methods include: the use of a parallel background fibre to estimate the Cerenkov radiation in its partner [7]; utilizing spectral differences between scintillation and Cerenkov light [8, 9]; separation using temporal differences in the prompt generation of Cerenkov light and the delayed generation of scintillation light [10, 11]. A recent successful approach is the use of an air core dosimeter, Figure 2 [12, 13].

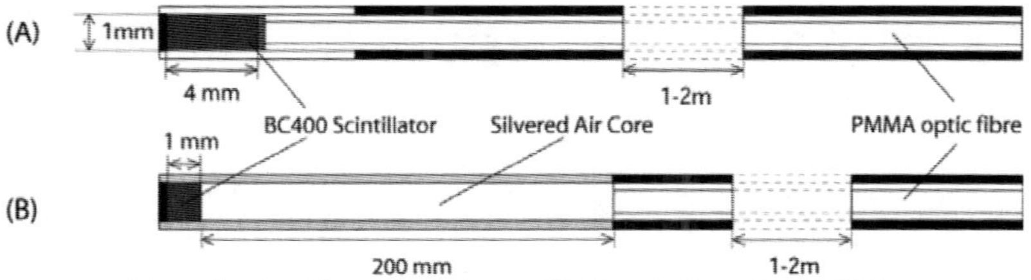

Figure 2: (A) a fibre optic dosimeter (FOD) and (B) an air-core FOD.

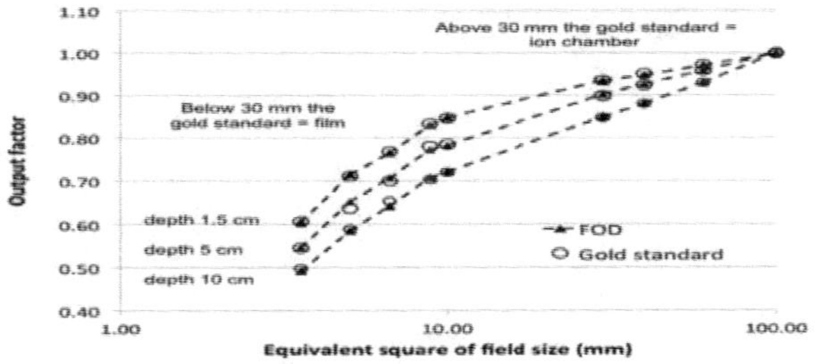

Figure 3: Relative output factors in water (S_p), measured with an air core scintillation dosimeter, ion chamber and radiochromic film.

These methods have all been shown capable of obtaining an accurate measurement of dose in megavoltage beams, promoting scintillation dosimeters as an effective point dosimeter in external beam therapy. For example, in the measurement of relative output factors in water (S_p), scintillation dosimeters have excellent agreement with ion chambers for field sizes 30 mm – 100 mm, and with radiochromic film down to 4 mm, Figure 3 [4].

Journal of Physics: Conference Series **444** (2013) 012075 doi:10.1088/1742-6596/444/1/012075

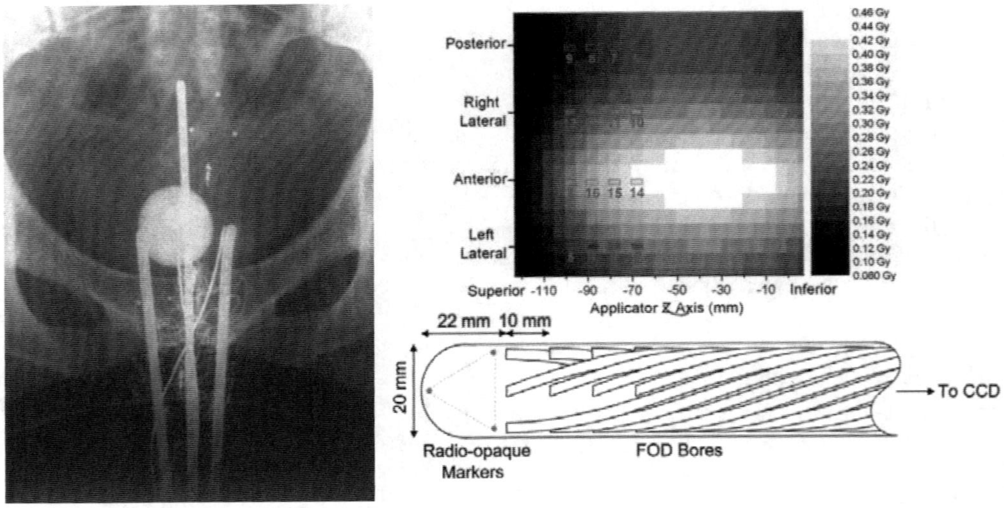

Figure 4: FOD array applicator, used in a clinical trial to map the rectal wall dose [14]

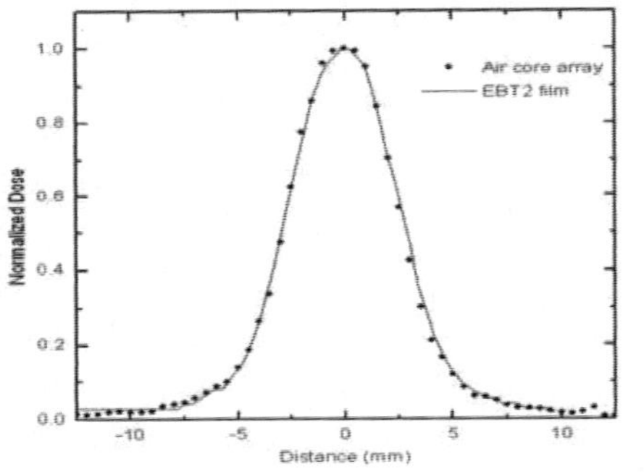

Figure 5: Dose profile of a 5 mm MLC field measured with two dosimeters

Plastic scintillators are water equivalent, angle and dose rate independent. Furthermore their small size minimizes volume averaging and as a result, scintillation dosimeters can be seen as the ideal component to build high spatial and temporal resolution dosimeter arrays. We have carried out a successful clinical trial of an array of 16 FODs to measure, in real time, the dose to the rectal wall during a brachytherapy treatment (Figure 4), providing conclusive evidence of the clinical benefits.

Figure 5 shows a dose profile for an SRS field on a Varian Novalis linear accelerator, measured with an air-core FOD array, validated with radiochromic film. This array provides high speed data processing and progress towards a 4D scintillation dosimetry system is well underway.

3. Challenges to clinical implementation of scintillation dosimetry

For clinical availability, dosimeters need to be commercialised. To date the only fully commercialised dosimeter based on plastic scintillators has been the "Daily Constancy Tool" (Gammex model 444), in which the sensing element was coupled to a silicon photodiode via a short fibre [15]. Due to inadequate clinical performance the scintillator was replaced with a silicon detector. A 2D scintillating dosimeter DosiMap [9] using spectral discrimination was specifically designed for IMRT QA. DosiMap made strong progress towards clinical translation, however it did not become available to the clinical community. Recently, Standard Imaging have promised a single point scintillation dosimeter, using photodiodes for readout in the linac bunker. Lack of commercialisation success emphasises the importance of fully evaluating the performance of the materials and technology in a routine clinical environment before going to market with a device. Any proposed device must be subjected to a trial under clinical rather than laboratory conditions. Reproducibility of the system day to day, sterilisation capability and temperature dependence (for in-vivo), hospital cleaner hardness, immediate response (no warm up), radiation hardness, intuitive interface and ergonomic handling and set up of the device are all important along with dosimetric accuracy, which has been the main focus to date.

4. Conclusion

Radiation dosimeters must serve the radiotherapy treatment process and not vice versa. As the field of scintillation dosimetry matures with several groups introducing innovative and exciting applications, it is timely for clinical physicists to engage in developing the applications which best serve currently inadequately met clinical dosimetry needs.

5. References

[1] McKerracher C and Thwaites D I 1999 *Phys. Med. Biol.* **44** 2143-60
[2] Scott A J *et al* 2012 *Phys. Med. Biol.* **57** 4461-76
[3] Francescon P *et al* 2008 *Med. Phys.* **35** (2) 504-13
[4] Ralston A *et al* 2012 *Phys. Med. Biol.* **57** 2587-602
[5] Archambault L *et al* 2005 *Med. Phys.* **32** 2271-8
[6] Liu P Z *et al* 2012 *Phys. Med. Biol.* **57** 3667-73
[7] Beddar A S *et al* 1992 *Phys. Med. Biol.* **37** 1901-13
[8] Fontbonne J *et al* 2002 *IEEE Trans. Nucl. Sci.* **5** 2223-7
[9] Frelin A M *et al* 2008 *Med. Phys.* **35** 1651-62
[10] Clift M A 2002 *Phys. Med. Biol.* **47** 1421-33
[11] Andersen C E *et al* 2006 *Radiat. Prot. Dosimetry* **120** 7-13
[12] Lambert J *et al* 2008 *Phys. Med. Biol.* **53** 3071-80
[13] Konnoff D C *et al* 2011 *Radiat. Meas.* **46** 76-87
[14] Cartwright L E *et al* 2010 *Med. Phys.* **37** 2247-55
[15] Beddar A S 1994 *Phys. Med. Biol.* **39** 253-63

Journal of Physics: Conference Series **444** (2013) 012076 doi:10.1088/1742-6596/444/1/012076

Optical computed tomography liquid calibration phantom

K Jordan
London Regional Cancer Program (London Health Sciences Centre), Department of Medical Biophysics, Western University, London, ON, Canada

E-mail: kevin.jordan@lhsc.on.ca

Abstract. Fluorinated ethylene propylene tubing is investigated as a method of preparing a contrast-resolution phantom for quantitative characterization of optical CT scanners and hydrogel dosimeters. Two sizes of tubing were examined: 6 and 13 mm inner diameter with 0.75 and 0.5 mm wall thicknesses, respectively. Water solutions of carbon black, nanoparticles in micelles provided continuously adjustable absorption contrast. Cross-sectional slices from two phantoms scanned with two different optical CT scanners are presented. Reconstructions from these simple phantoms can be used to identify scanner artefacts and improve instrument design. These phantoms represent a more reproducible approach than casting "gel fingers" into gel phantoms for system characterization. The thinner walled tubes have fewer optical artefacts.

1. Introduction

Optical computed tomography (CT) for 3D radiation dosimetry has many features in common with x-ray CT. In fact, a dedicated optical CT scanner, DeskCAT Modus Medical Devices Inc. is under development for the purpose of teaching the principles of x-ray CT. Clinical x-ray CT scanners provide a market for the commercial development of dedicated phantoms for standardized performance testing. For example, the Catphan® phantom from The Phantom Laboratory, is commonly used to test contrast, linearity, uniformity and spatial resolution. This company also provides several liquid-filled phantoms for nuclear medicine and magnetic resonance imaging QA. Optical CT is a much smaller market with limited commercial phantoms for QA purposes. There are several types of 3D dosimeters under investigation. These include transparent radiochromic hydrogels and plastics and polymerization hydrogels. The performance of these dosimeters is partly limited by the optical properties of the vessel containing the material. In summary, the scanner, dosimeter and dosimeter vessel all contribute to the overall performance of various 3D dosimetry systems. In x-ray CT, scanners and reconstruction software are commonly compared using a standard QA phantom, such as the Catphan® phantom. Analogous phantoms will assist in deployment of quantitative optical CT.

Contrast "finger phantoms" prepared by pouring coloured gel into tube shaped holes in gelatin phantoms for optical CT studies have be reported. Food colouring dyes were first reported by Oldham [1] and Kristajic [2]. However the dye diffused from the fingers limiting there value for comparative studies. A non-diffusing colloidal gel finger phantom based on contrast from scatter was characterized by Bosi [3, 4]. Carbon black (CB), nanoparticle, non-diffusing gel finger phantoms were then reported by Jordan [5]. With hydrogen peroxide antimicrobial treatments, this latter phantom remained stable for over two years. However, preparation of gel finger phantoms is time consuming and each phantom is likely to have unique optical characteristics. Clear fluorinated ethylene propylene (FEP) thin-walled tubes has been used as cylinder walls for gels in optical CT because of the near refractive index match to several hydrogel formulations [6]. In this report initial results from aqueous "finger" phantoms for characterizing 3D dosimetry systems are presented. The work focuses on water-equivalent, hydrogel dosimeters with refractive indices near 1.34 but could be adapted to higher refractive indices for other materials. The tubes can be filled with coloured solutions to provide a range of contrasts for scanner system evaluation.

Rahman et al [7] have reported high resolution test phantoms based on irradiation of the radiochromic plastic Presage® by synchrotron beams. The paper provides a thorough discussion of the developments and needs related to phantoms for QA of optical CT – 3D dosimetry systems.

2. Methods

Two samples of FEP tubing were investigated: 1) tubing, VWR #63014-695 OD:ID 7.9:6.4 mm 2) thin-walled heat shrink tubing, Newark SKU# 92N6000, OD:ID 13.5:12.5 mm. The smaller diameter tubing was straightened by sliding over 5 mm diameter glass rod and annealing at 100 C. Top and base rings were machined from polymethymethacrylate to hold the tubes rigid. In the first case, this 4 tube phantom was placed in a PETE jar for scanning with a modified Vista10 optical cone-beam (CBCT) scanner, Modus Medical Devices Inc. The tubes, jar and aquarium were filled with an 8.3% by mass, propylene glycol in water solution. The sample was scanned with 590 nm light, 512 projections per 360 degrees, camera lens aperture of F4 and 8 minutes scan time. Next, the reference solution was replaced with refractive index matched carbon black solutions in 3 of the 4 tubes and the CT scan repeated. Reconstruction was performed with 0.25 mm^3 voxel size and Ramp filter, 512^3 array. A photograph of the phantom is shown in Figure 1. A second fan beam scan set was collected by inserting a pair of horizontal slots to vertically collimate the beam.

The second case consisted of 4 thin-walled tubes pressed into holes drilled in a plastic plate. Thermal annealing to straighten theses tube was not performed. The phantom was mounted in an in-house prototype laser scanning cone-beam CT scanner (details of scanner presented in separate conference submission). The tubes and aquarium were filled with water and scanned with 594 nm laser light, 512 projections per 360 degrees, 30 minutes scan time. Next, the water was replaced with carbon black solutions in 2 of the 4 tubes and the CT scan repeated. Reconstruction was performed with 0.34 mm^3 voxel size and Ramp filter, 512x512x200 array. A photograph of the phantom in the aquarium is shown in Figure 2.

Figure 1: Photograph of 6 mm diameter FEP tube phantom with 3 tubes filled with carbon black solutions and 4th with refractive index matching liquid.

Figure 2: Photograph of 13 mm diameter FEP tube phantom in water filled aquarium: 2 tubes filled with carbon black solutions and 2 tubes with water (see tops of tubes above water level).

3. Results and Discussion

The middle slice of the 3D reconstruction for the smaller diameter tube phantom is presented in Figure 3. Several artefacts are present in the reconstruction examples include streaking between the tubes, gradients within tubes (note the higher attenuation of lightest CB solution on side facing darkest CB solution) and ghost images of CB tubes. The ghost images are likely the result of small differences in the refractive indices of the 3 CB solutions and position of the phantom for the reference and data scans. The fourth tube reconstructed as a single tube and contained the same solution for both scans. Independent comparison of attenuation coefficients with an absorption spectrometer were not performed at the time of CT scans. The next day black films were observed on the beakers storing the samples indicating instability of these CB solutions. Attenuation coefficients for the fan versus cone beam geometry scans were approximately 10% greater, consistent with other stray light studies. The reconstructed and nominal diameters of the reference tube agreed within one voxel or 0.25 mm of each other.

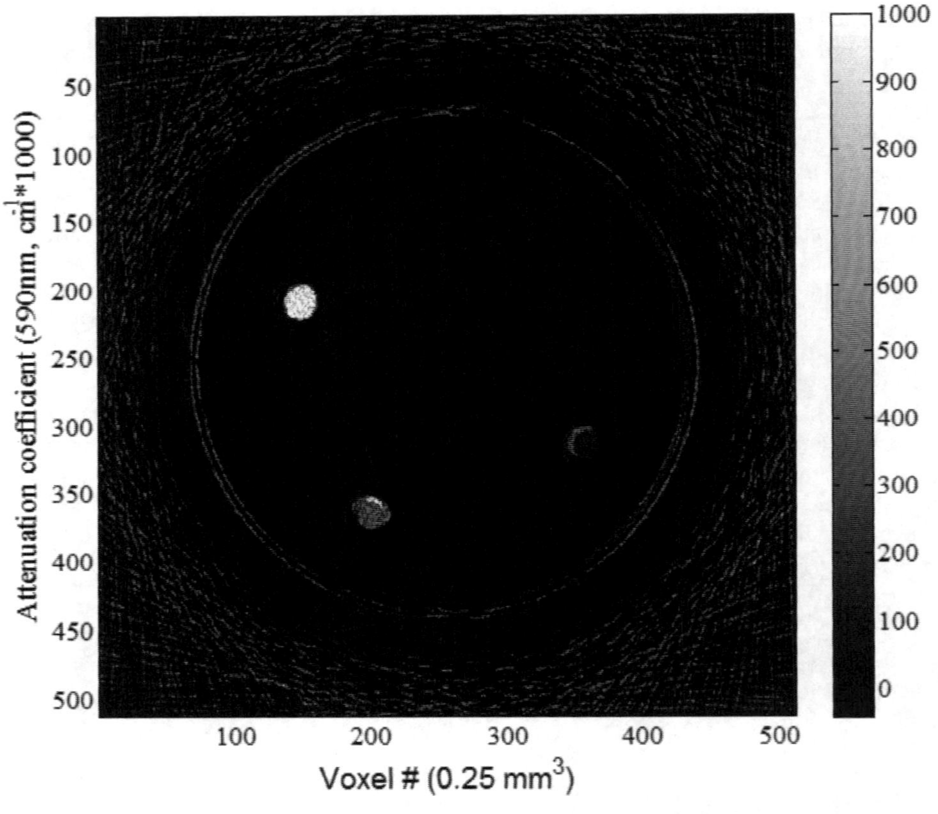

Figure 3: Middle slice from 6 mm tube phantom with carbon black solutions, scanned with modified Vista10 CBCT. Note artifacts in tube uniformity due to position changes between reference and data scans. Phantom mounted in PETE jar, walls and seams are evident.

In Figure 4, the middle slice of the 3D reconstruction for the 13mm diameter tube phantom is shown. The refractive index mismatch is larger with water, but the perturbations are smaller due to the thinner tube walls. This phantom highlights a simpler system with no vessel to introduce additional artifacts, such as external reflections near grazing incidence. Adjusting window and level allows small ring artifacts to be observed as well as an increasing background with an inverse radial dependence, indicating some artifacts are due to reflections within the aquarium itself. Mean attenuations in the tubes averaged over $2mm^3$ volume at mid height were 0.090, 0.210 and 0.425 compared to spectrometer readings in 100mm long cuvettes of 0.098, 0.220 and 0.452 cm^{-1}. Results are around 5% low indicating scatter within these tube samples is not entirely eliminated in the laser cone beam CT data set.

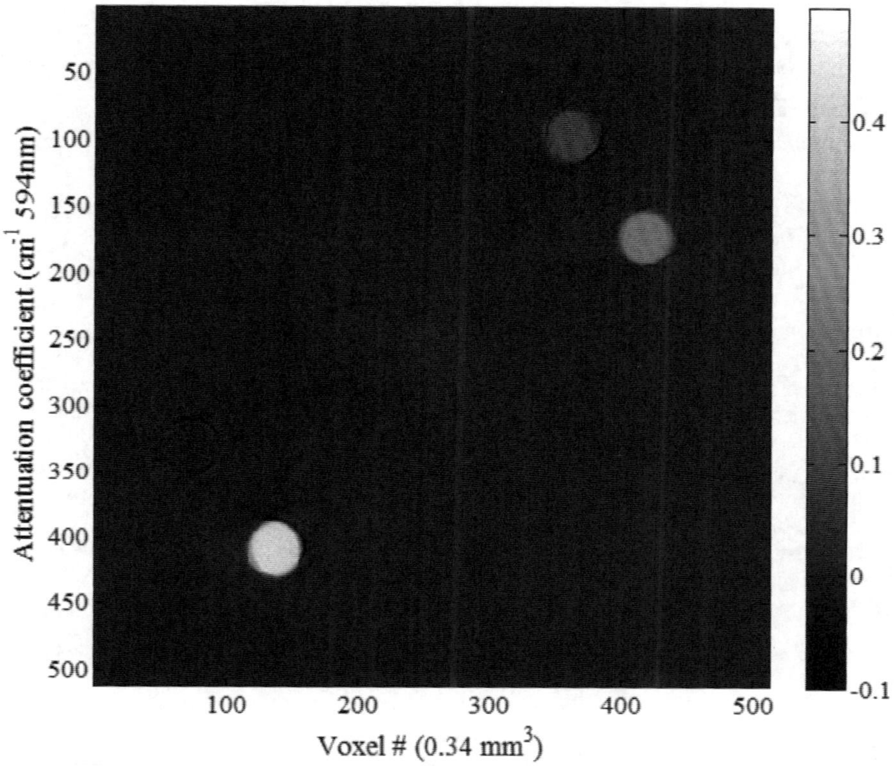

Figure 4: Middle slice from 13 mm tube phantom with carbon black solutions, scanned with laser CBCT. Note ring artefacts due to reflections within aquarium. Increased attenuation near rotation axis also related to reflections.

4. Conclusion

FEP tubing has several favourable properties for preparation of optical CT QA phantoms. Such as clarity, refractive index, chemical inertness and plasticity. Aqueous solutions are inexpensive to prepare and biodegradable formulations of different colours and degree of scatter can be prepared. Black solutions are of particular interest because of the low sensitivity to wavelength and bandwidth of various optical CT light sources. These phantoms are useful since refractive index, scatter and absorption coefficients can be continuously tuned for specific tests.

5. References

[1] Oldham M *et al* 2003 *Med. Phys.* **30** 623-34
[2] Krstajic N and Doran SJ 2006 *Phys. Med. Biol.* **51** 2055-75
[3] Bosi S *et al* 2007 *Phys. Med. Biol.* **52** 2893-903
[4] Bosi S G *et al* 2009 *Phys. Med. Biol.* **54** 275-83
[5] Jordan K and Battista J 2008 *J. Phys.: Conference Series* **164** 012046
[6] Jordan K and Battista J 2006 *J. Phys.: Conference Series* **56** 214-6
[7] Rahman A T *et al* 2012 *Phys. Med. Biol.* **56** 4177-99

Revealing the impact of radiation-induced refractive index changes in polymer gel dosimeters

WG Campbell[1], A Jirasek[1] and D Wells[2]
[1]Dept. of Physics and Astronomy, University of Victoria, Victoria, BC, Canada
[2]BC Cancer Agency, Vancouver Island Centre, Victoria, BC, Canada

E-mail: warren.g.campbell@gmail.com

Abstract. This study examines effects of radiation-induced refractive index (RI) changes in polymer gel dosimeters. A prototype fan-beam optical computed tomography scanner was used to image a normoxic polymer gel dosimeter that was irradiated with two simple irradiation patterns—one single beam, one cross beam. A combed fan-beam was used for rayline tracing. Scans revealed that notable rayline errors occur when a steep, side-to-side dose gradient (i.e. RI gradient) is encountered. When the gradient occurs in the plane of the detector system, distinctive streaks in images are observed. When the gradient occurs perpendicular to the plane of the detector system, much more severe image errors are observed.

1. Introduction

The optical properties of gelatin dosimeters (both absorption-based and scatter-based) continue to be researched for 3D radiation dosimetry [1, 2]. Due to their pliability, they could potentially fill a multifaceted role as a deformable 3D dosimeter [2]. Such dosimeters would be clinically attractive, as a number of tumour sites change position and shape on inter-fraction and intra-fraction timescales (e.g. prostate and lung, respectively). If effective methods to accommodate tumour motion and tumour deformation are to be prudently developed, a deformable 3D dosimeter would be an appealing tool to verify these methods.

It has long been known that polymer gel dosimeters can exhibit slight changes in refractive index (RI) when irradiated [3]. The potential for geometric distortion in optical computed tomography (CT) caused by radiation-induced RI changes has been tested previously [4]. However, those tests used the positions of completely opaque needles to evaluate geometric distortion. The work here will show that, in certain cases, radiation-induced RI changes can lead to readout errors caused by rayline bending that only occurs from certain projection angles; these would have little effect on the positions of needles in a needle-phantom. This work will contend that rayline errors occurring over a short range of projection angles can result in distinctive streaking artefacts near steep, side-to-side dose gradients. In other cases, radiation-induced RI changes can cause errors that are far more severe, particularly for the fan-beam optical CT geometry.

2. Material and Methods

A normoxic polymer gel dosimeter (cylindrical, 1 L, 95 mm diameter) was irradiated with two simple dose-distributions (see figure 1a, 1b). The gel (w/w%: 91.8% deionized water, 4.0% gelatin, 2.0% acrylamide, 2.0% bisacrylamide, 0.2% (9mM) tetrakis (hydroxymethyl) phosphonium chloride (all chemicals from Sigma-Aldrich; Oakville, ON, Canada)) was mixed and allowed to set in a room

Journal of Physics: Conference Series **444** (2013) 012077 doi:10.1088/1742-6596/444/1/012077

temperature bath for 24 hours prior to a pre-irradiation scan with a prototype fan-beam optical CT scanner [5]. Then, the dosimeter was immediately irradiated and given 24 hours to polymerize before a post-irradiation scan was acquired. Pre-irradiation and post-irradiation scans were compared ray-by-ray to calculate transmission data. Two artefact removal techniques were performed in sinogram space to remove ring artefacts that are distinct to the prototype and streaking artefacts attributable to a pair of seams that run down the sides of the flask housing the gel. These artefact removal techniques are described in more detail in an accompanying submission.

A simple line-pair pattern was printed on a piece of transparency plastic (1.5 lp/mm, see figure 1c) to allow rayline tracing to be easily performed. To do so, one light sinogram was acquired with the plain fan-beam. Then, the line-pattern ("comb") was taped in placed on the inside of the entry window to create a combed fan-beam. Next, a combed light sinogram was acquired. A rayline sinogram was then calculated by dividing the plain sinogram by the combed sinogram.

Figure 1: Square photon beams (18 MV, 3×3 cm^2, Clinac 21EX linear accelerator; Varian Medical Systems; Palo Alto, CA, USA) were used for simple (a) single-beam, and (b) cross-beam irradiation patterns with isocenter doses of 8.7 Gy and 10.0 Gy, respectively. Distributions shown were calculated by treatment planning software (Eclipse™ ; Varian Medical Systems). The comb pattern used for rayline tracing is shown in (c) with a ruler for scale.

3. Results and Discussion

Figure 2 displays scanner reconstructions of slices located in the middle of the single-beam and cross-beam irradiated regions of the dosimeter. It is contended that the streaking artefacts that follow along the sides of each beam (approximately -30% in magnitude) are caused by radiation-induced RI errors. In order to initially assess this interpretation, rayline tracing was used for each irradiated slice, as well as an unirradiated slice. Light sinograms and rayline sinograms for each slice are shown in figure 3.

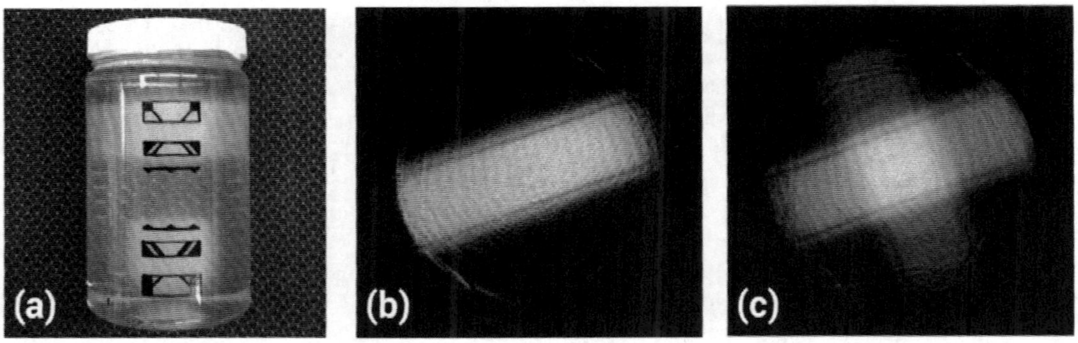

Figure 2: The dosimeter is shown in (a). Opacity reconstructions for slices in the middle of the single-beam and cross-beam irradiated regions are shown in (b) and (c), respectively.

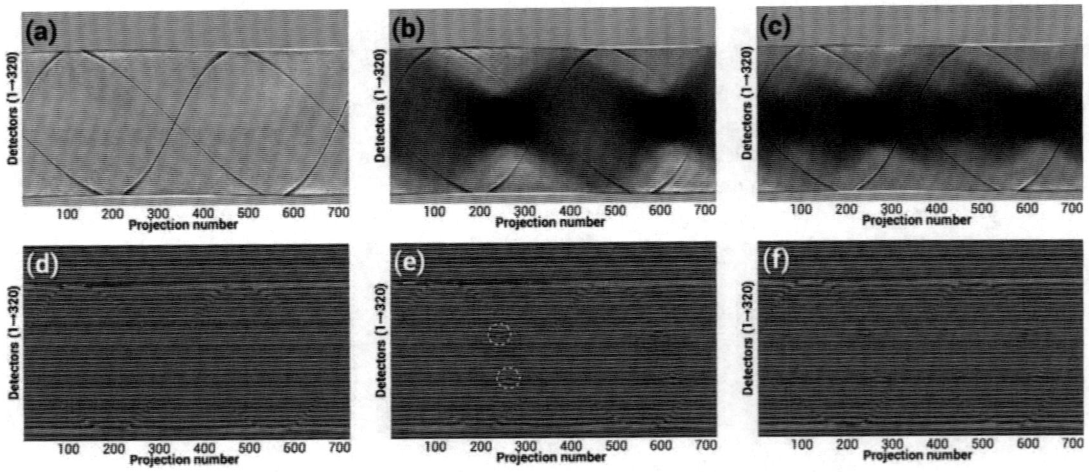

Figure 3: Top row: light sinograms of (a) unirradiated slice, (b) single-beam slice, and (c) cross-beam slice. Bottom row: (d), (e) and (f) are rayline sinograms of (a), (b) and (c), respectively. Two dashed circles in (e) indicate the locations of the first set of bent raylines near steep RI gradients.

Rayline tracing indicates two different errors in these slices. First, tracing in all three slices elucidates the errors that are caused by the pair of seams that run vertically down the sides of the flask. These seams bend rays off of their initial trajectory, which causes errors that must be removed in order to avoid streaks in reconstructions. Second, raylines around projections 250 and 600 in figure 3e show ray bending that corresponds to steep dose gradients, as can be seen in figure 3b. These errors occur each time the scanner faces down the irradiated beam. Similar errors occur four times in the sinogram in figure 3f near projections 75, 260, 430, and 615. Again, the locations of these errors correspond to steep dose gradients in projections in figure 3c.

To further investigate the ramifications of radiation-induced RI changes, two additional slices of the dosimeter were scanned. Figures 4a and 4b display reconstructions of slices located in the penumbrae of the single-beam and cross-beam irradiated regions of the dosimeter. The locations of these slices, as well as those from figures 2b and 2c, are indicated in figure 4c. As is evident, severe errors result. Here, rather than dose gradients occurring side-to-side in the plane of the detector system, dose gradients occur vertically, perpendicular to the detector system. In these cases, rays are redirected out of the plane and away from the detector system. This results in an overestimation of opacity. It should be noted that identical window and level values have been used for all reconstructions. Despite these slices being in the penumbral region of the irradiation patterns, maximum optical density values in figures 4a and 4b are more than double the maximum values in figures 2b and 2c.

Correction methods are being considered for these artefacts. Ray tracing could be used to pinpoint rayline errors that result from RI gradients that occur side-to-side in the detector plane. In these instances, one could replace values of bad raylines using interpolation of nearby pixels in sinogram space. However, RI gradients that occur perpendicularly to the detector plane cause three main problems in the fan-beam geometry: 1) rayline tracing with a combed beam cannot denote raylines that bend vertically out of the detector plane, 2) many more rays of the sinogram are affected, and 3) the magnitude of these errors are much greater. Theoretically, one could determine a value for the maximum dose gradient that would still allow for proper raylines. This could then be taken into consideration when comparing measured dose distributions with planned dose distributions.

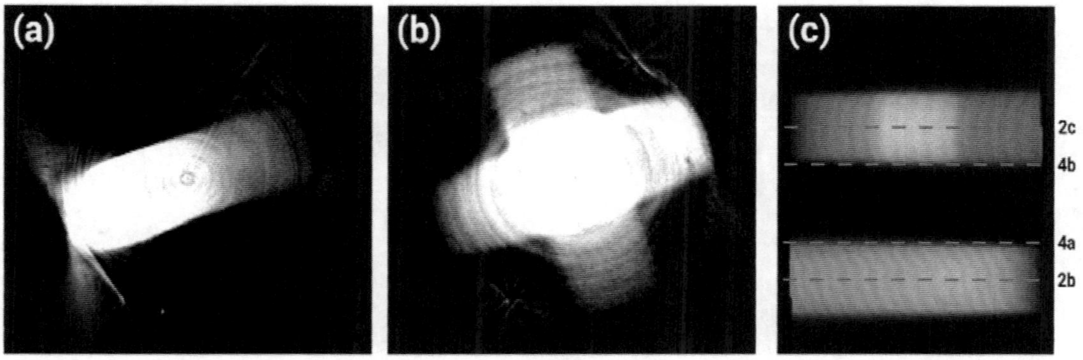

Figure 4: Opacity reconstructions for slices near the edges of the (a) single-beam, and (b) cross-beam irradiated regions. The slice locations of all four reconstructions are indicated in (c) in the sagittal plane of the dose distribution calculated by treatment planning software.

4. Conclusions

New image artefacts have been observed in a normoxic polymer gel dosimeter. It is believed that these are the first artefacts that are clearly attributable to radiation-induced refractive index changes. This attribution has been demonstrated in two ways. Rayline tracing was used to illustrate the bending of rays caused by dose gradients that occur side-to-side in the detector plane; this results in distinctive streaking artefacts. Dose gradients that occur perpendicular to the detector plane result in more problematic errors due to multiple raylines being redirected off of the detector system; this results in severe errors in reconstructions.

5. Acknowledgements

Work on this scanner was started by previous University of Victoria (UVic) student David Rudko and continues with the support of Dave Yardley and Steve Gray at the BCCA and Nick Braam at UVic. The presenting author would also like to acknowledge funding support from the Canadian Institutes of Health Research, the Natural Sciences and Engineering Research Council of Canada, the National Aboriginal Achievement Foundation, and the University of Victoria.

6. References

[1] Vandecasteele J *et al* 2011 *Phys. Med. Biol.* **56** 627-51
[2] Yeo U J *et al* 2012 *Med. Phys.* **39** 2203-13
[3] Maryañski M J *et al* 1996 *Phys. Med. Biol.* **41** 2705-17
[4] Oldham M and Kim L 2004 *Med. Phys.* **31** 1093-104
[5] Campbell W G *et al* 2010 *J. Phys.: Conf. Series* **250** 012024

Journal of Physics: Conference Series **444** (2013) 012078 doi:10.1088/1742-6596/444/1/012078

Laser beam optical CT scanner for in-air gel readout: imaging artefacts

D Ramm
Department of Medical Physics, Royal Adelaide Hospital Cancer Centre, South Australia, Australia
School of Chemistry and Physics, University of Adelaide, South Australia, Australia

E-mail: daniel.ramm@health.sa.gov.au

Abstract. Ongoing progress on development of an in-air scanning optical CT is reported, specifically dealing with the minimization of scanner imaging artefacts. Improved scratch resistance of the PMMA gel cylinder was a primary goal, so that routine cleaning would not degrade the polished surfaces. This was achieved by the addition of a hard coating to the cylinder surfaces. New artefacts were introduced and subsequently reduced by alternative processing of projection data. The outcome was a gel cylinder of much greater practicality for routine use while maintaining similar signal to noise ratios and uniformity in the image reconstruction field of view.

1. Introduction

Optical CT scanning of dosimetric gels [1] usually requires the use of a RI matching fluid bath to reduce refraction and reflection effects at the surface of the cylindrical gel container. Recently we have developed an 'in-air scanner' that avoids the use of a fluid bath while obtaining effectively parallel ray projection data for 90 % of the gel sample diameter [2]. Previously Maryanski and Ranade developed the 'dry scanner' intended mainly for brachytherapy source applications [3]. This scanner cannot accurately image optical attenuation outside a small central region of the gel dosimeter since it only acquires projection data for about 36 % of the gel diameter. Another approach using the PRESAGE [4] dosimeter has been proposed, using algebraic reconstruction to account for non-parallel light rays [5]. Limited projection data (65 % of PRESAGE diameter) was still a limiting issue for that simulation study. Our in-air scanning approach is the only one realized to demonstrate the feasibility of scanning external beam irradiated gels without RI matching fluid.

While the principle of the prototype in-air scanner's operation has been shown, further developments are required in preparation for radiotherapy clinical application. One aspect is imaging artefacts [6, 7] that degrade 3D dosimetric data and can present a significant obstacle in gaining high quality measured data and general acceptance of the dosimetry method. Preliminary results indicated some readily observable artefacts [2]. These artefacts have been under investigation, together with some others that have become apparent.

2. Methods

2.1. Optical CT scanner

The scanner is detailed in [2], with brief summary as follows. A He-Ne laser beam is fanned by a rotating mirror across the key component, a machined and polished PMMA cylinder (76 mm outer diameter (OD), 51 mm inner diameter (ID)). Inside the cylinder is the gel, and after the laser beam passes through with close to parallel ray geometry, it is refracted by the cylinder walls to a focal point at a silicon photodiode detector. With the ID/OD ratio of around 67 % and FXG inside, parallel rays extend to about 90 % of the gel diameter. Extrapolation of the missing projection ends is applied prior to back-projection.

Changes to the original scanner details include the following: (i) A revised bearing assembly for the sample rotation mechanism, giving improved rotational precision. (ii) The PMMA cylinder internal and external walls were hard coated with a scratch resistant coating (Crystalcoat MP-101, SDC Technologies). This polysiloxane based coating had a measured thickness of 3.8 ± 0.1 μm over the cylinder surfaces. (iii) Improved gel optical quality by using a syringe with filter (2 μm nominal pore size) to fill the gel containers. (iv) An option for low pass filtering of projection data and re-sampling to 0.4 mm. (v) An option for point by point division of measurement projection data by reference scan data for the entire cylinder, as opposed to the original method of a single averaged reference scan projection for normalization.

2.2. Imaging artefacts

The three most apparent artefacts identified previously were: (i) An outer ring in the extrapolation region due, in part, to mechanical imprecision of the sample rotation. (ii) Small spots in the sample due to foreign particles (dust). (iii) Low level streaking due to scratches or dust on the cylinder walls. Additionally, cylinder wall scratching through routine cleaning was a chief concern because severe streak artefacts can result. Attempts to improve these artefacts have been made by the aforementioned changes to the scanner.

Another potential source of artefacts is inter-reflection, ie. multiple reflections between optical components. One instance was reflected light from the photodiode surface back-reflected to itself by the cylinder. This was previously addressed by tilting the photodiode so that the reflection was not incident upon the cylinder. There is a more subtle effect where the beam is reflected from the exit wall of the cylinder (PMMA-air interface) back to the entry wall, and then towards the photodiode. Since air-PMMA interface reflectance is about 4 % in the worst case (for the acquired range of data) this signal can only amount to 0.2 %, and less with gel attenuation.

Optical interference effects are particularly important to avoid because signal modulation of several percent may be introduced. This has been observed at the central axis of the cylinder where cylinder inter-reflections can, at times, result in constructive and destructive interference. A significant transient waveform in the centre of projection data can occur, particularly when there is minimal attenuation in the cylinder. Interference effects can also be introduced where coatings are applied to components, such as the hard coating on the PMMA cylinder. With the current hard coating there is a mismatch in RI ($n_{coating} = 1.45$, $n_{PMMA} = 1.49$), leading to a change in transmission at an air-coating-PMMA interface of about 0.5 % for a 0.1 μm (typical) change in coating thickness ($\lambda = 594$ nm).

To investigate and illustrate the artefacts, the cylinder was filled with filtered distilled water and green dye water solutions. Undyed scans were used as the reference scan throughout, and an undyed measurement scan was used to examine low level effects (mainly interference). These effects are more evident in this case than scans with significant optical attenuation in the cylinder. Scans of water samples with uniform attenuation were made with dye solutions of moderate ($OD_{moderate} = 0.18$ cm^{-1}) and high ($OD_{high} = 0.50$ cm^{-1}) levels. Reconstructed dose images were viewed and analyzed using ImageJ (National Institutes of Health, USA).

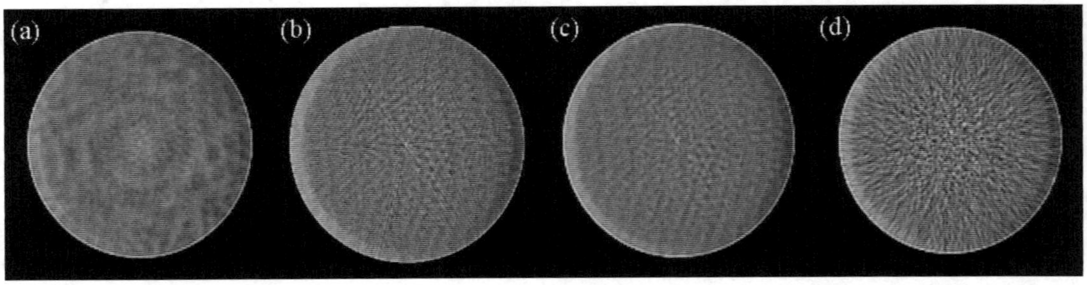

Figure 1: Reconstructed images of one slice for the green dye water filled cylinder, with viewing window widths of 20 % of the mean and centred on the mean value. (a) $OD_{moderate}$, 0.4 mm pixel size and averaged normalizing reference projection. (b) $OD_{moderate}$, 0.2 mm pixel and point by point reference scan normalization. (c) Same as 1(b) except for projection filtering and 0.4 mm pixel. (d) Same as 1(c) except for OD_{high}.

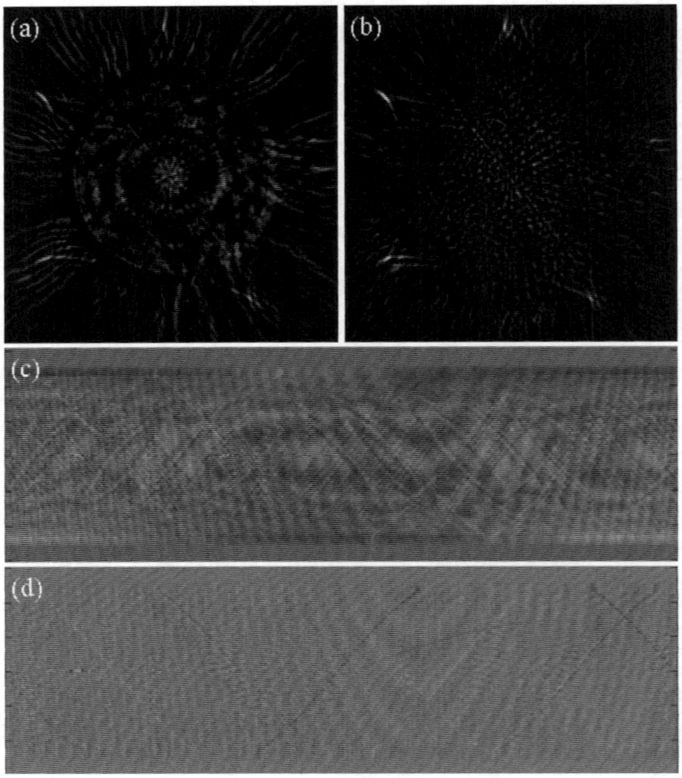

Figure 2: Undyed water scan normalized to another undyed water scan, where the ideal result would be a blank image. (a) Reconstructed slice using an averaged reference projection. Viewing window width 5 % of mean pixel value for $OD_{moderate}$. Interference artefacts appear inside and outside the internal volume of the cylinder. (b) Same as 2(a) except for point by point reference division, giving reduced interference artefacts. (c) Sinogram for 2(a) showing superimposed multiple interference patterns. (d) Sinogram for 2(b) with less interference patterns. The thin lines (dark from reference scan and light from measurement scan) are due to dust specks on the outer cylinder surface.

3. Results and discussion

A reconstructed slice of dyed water ($OD_{moderate}$) is shown in figures 1(a), 1(b) and 1(c). The same projection data was used for these three images, just differing in data processing. The most obvious non-uniformity feature of figure 1(a) was the patterned appearance that was new to this version of the scanner. The reason for this was due to interference effects from the addition of the hard coating on the cylinder. Figure 2 shows more detail of these interference patterns. Figures 2(a) and 2(b) are reconstructed slices of undyed water scans normalized to another undyed scan with averaged reference projection (2(a)) and point by point reference projections (2(b)). The sinograms are shown in figures 2(c) and 2(d) where optical interference patterns become more discernible, particularly figure 2(c). Figure 2(b) and the corresponding sinogram, figure 2(d), show that the point by point reference division largely removes the coating interference effects. Figures 1(b) and 1(c) confirm the minimization of this interference effect. The hard coating is substantially more durable than a PMMA surface, meaning the cylinder surfaces can be routinely cleaned with reduced scratching concerns.

Another optical interference effect was an interference signal on the central axis shown in the sinograms of figure 2(c) and 2(d). There is a broken line of signals of high and low relative intensities running down the central axis. These manifest in reconstructions as radial lines as shown in figure 1(b). Since these interference waveforms in the projection data are of high frequency, filtering assists in their minimization as shown by figure 1(c). This interference effect is dependent upon the degree of attenuation by the sample with the worst case occurring for undyed water as used here for all reference projection data. When gel is used, the presence of attenuation even in FXG pre-irradiation reference scans, means this effect is usually minimal.

Figure 1(d) is a high OD example with central axis OD of about 2.5 (0.3 % transmission). Noise in the projection data becomes more significant, hence a random noise appearance dominates the residual artefacts of figure 1(c). The residual artefacts mainly comprise a mottled appearance originally thought to be caused by multiple low level streaking, however further investigation is required to confirm the source(s). There is still a remaining artefact in the extrapolation region, however it is not considered reliable FOV in any case, and does not affect the full data FOV.

To put the fluctuations observed in the images of figure 1 into perspective, 1(a) had a standard deviation in the FOV of 0.8 % of the mean value, 1(b): 1.3 %, 1(c): 0.8 % and 1(d): 2.0 %.

4. Conclusions

Optical interference has been found to be a source of artefacts in the process of continued development of an in-air optical CT scanner. The addition of a hard coating to the gel cylinder provided greater scratch resistance, reducing the likelihood of streak artefact inducing scratches. However interference artefacts were also introduced. Point-by-point reference scan division helped minimize these effects. RI matched hard coating is planned for the future. A remaining structured mottle artefact in the FOV requires further investigation, however presently a signal to noise ratio of around 130 is obtained for a uniform moderate optical density.

5. Acknowledgements

Application of the cylinder hard coating by Colin Hall, University of South Australia, and support by John Schneider, Royal Adelaide Hospital, Medical Physics workshop is gratefully acknowledged.

5. References

[1] Baldock C *et al* 2010 *Phys. Med. Biol.* **55** R1-63
[2] Ramm D *et al* 2012 *Phys. Med. Biol.* **57** 3853-68
[3] Maryanski M J and Ranade M K 2001 *Proc. of SPIE Int. Soc. Optic. Eng.* **4320** 764-767
[4] Gorjiara T *et al* 2011 *Med. Phys.* **38** 2265-74
[5] Doran S J and Yatigammana D N 2012 *Phys. Med. Biol.* **57** 665-83
[6] Bosi S *et al* 2007 *Phys. Med. Biol.* **52** 2893–2903
[7] Bosi S G *et al* 2009 *Phys. Med. Biol.* **54** 275-83

Journal of Physics: Conference Series **444** (2013) 012079 doi:10.1088/1742-6596/444/1/012079

Feasibility study using MRI and two optical CT scanners for readout of polymer gel and Presage™

H Svensson[1], PS Skyt[2], S Ceberg[1,3], S Doran[4], LP Muren[2], P Balling[5], JBB Petersen[2] and SÅJ Bäck[1,3]

[1]Department of Radiation Physics, Skåne University Hospital, Malmö, Sweden
[2]Dept of Medical Physics, Aarhus University/Aarhus University Hospital, Denmark
[3]Department of Medical Radiation Physics, Lund University, Malmö, Sweden
[4]Department of Physics, University of Surrey, Guildford, Surrey, UK
[5]Dept of Physics and Astronomy, Aarhus University, Denmark

E-mail: henrik.svensson@skane.se

Abstract. The aim of this study was to compare the conventional combination of three-dimensional dosimeter (nPAG gel) and readout method (MRI) with other combinations of three-dimensional dosimeters (nPAG gel/Presage™) and readout methods (optical CT scanners). In the first experiment, the dose readout of a gel irradiated with a four field-box technique was performed with both an Octopus IQ scanner and MRI. It was seen that the MRI readout agreed slightly better to the TPS. In another experiment, a gel and a Presage™ sample were irradiated with a VMAT field and read out using MRI and a fast laser scanner, respectively. A comparison between the TPS and the volumes revealed that the MRI/gel readout had closer resemblance to the TPS than the optical CT/Presage™ readout. There are clearly potential in the evaluated optical CT scanners, but more time has to be invested in the particular scanning scenario than was possible in this study.

1. Introduction

Traditionally, three-dimensional (3D) dosimetry [1] has been based on polymer gel that is read out by magnetic resonance imaging (MRI) [2], but the introduction of a promising radiochromic plastic dosimeter have created interest in optical computed tomography (optical CT) scanners as readout modality. The advantage of optical CT scanners is the ability to scan both plastics and polymer gel dosimeters and the fact that it is dedicated to 3D dosimetry [3]. It could also be located close to the clinical environment in a radiotherapy department, thus making MRI readout obsolete. In this initial feasibility study, two optical CT scanners are tested in two different experiments and compared to the well-established combination of polymer gel and MRI.

2. Material and methods

2.1. Readout modalities

The two optical CT scanners used were the fast laser scanner from the University of Surrey [4] and an Octopus IQ scanner (MGS. Research, Inc., Madison, CT, USA) located at Aarhus University. The MRI machine used to scan the gels were a conventional 1.5 T unit (Symphony, Siemens, Germany) located in Skåne University hospital in Malmö.

2.2. Preparation, irradiation and readout

In the first experiment, a normoxic polyacrylamide gel (nPAG gel) [5] was manufactured as described elsewhere [6] and poured into an acrylic cylindrical container (Ø 15 cm, length 17 cm). The gel was X-ray CT scanned and a four-field box technique with 3x3 cm 6 MV photon beams (0°, 90°, 180° and 270°) was planned using the CT data. The prescribed dose to the four-field box was 3 Gy, delivered with a dose rate of 600 MU min^{-1} with a Varian Trilogy Tx (Varian Medical Systems Palo Alto, CA, USA). After irradiation, the gel was left to polymerize for 16 hours before the optical CT scan with the Octopus IQ scanner was initialized. Identical readout scans were performed pre- and post-irradiation with the gel placed upright on exactly the same position, using default Octopus IQ driver settings to control the motors and detector. A number of 300 projections with 145 slices, each divided on 180°, were acquired with a total scan time of approximately 48 minutes. The pre-irradiation scan was used to account for potential defects in the optical density of the gel. The refractive index liquid used was optimized for nPAG gel dosimeters and based on a mix of glycerol, water and a dye. Reconstruction was performed with a resolution of 1x1x1 mm^3 with units of optical density, which was converted to relative absorbed dose by normalization to a region in the centre of the four-field box using MATLAB (MathWorks, Natick, MA, USA). A 3x3x3 box-filter was applied to smooth the raw data.

The MRI readout of the gel was conducted 24 hours after irradiation. The image acquisition was carried out using a 32-echo multi-spin echo sequence with 25 ms echo time and 9780 ms repetition time. The voxel size was 1x1x3 mm^3 with two averages and four concatenations. The imaging time was 3 hours. In-house developed software was used to create R2 maps from the MRI data and the additional image processing was done using MATLAB. The R2 maps were converted to relative absorbed dose by background subtraction using the R2 value from an unirradiated region in the gel and normalization to a central volume in the four-field box. The dose distribution was linearly interpolated to 1x1x1 mm^3 (as was the dose images from the treatment planning system (TPS) with an original resolution of 2x2x2 mm^3). The raw data was smoothed with a 3x3x3 box-filter.

In the second experiment, a Presage™ sample (Heuris Pharma LLC, NJ, USA) [7] was procured and wrapped in two layers of 0.5 cm thick SuperFlab bolus material. This was done to achieve the same thickness as the gel container it should be compared with. The gel in question was manufactured and poured into two identical 0.5 l cylindrical glass containers (Ø 8 cm, length 12 cm). One of the bottles was left unirradiated and later used to obtain a background R2 value for the irradiated gel. The samples were X-ray CT-scanned and a clinical prostate Varian VMAT plan was used to receive clinically relevant dose modulation. However, the prostate target was reduced in size to fit into the small Presage™ sample. The target dose was prescribed to 3 Gy given with one full clockwise arc rotation and a collimator rotation of 30° using 450 MU and a varying dose rate between 200 and 600 MU/min. The plan was delivered to both dosimeters using a Varian Clinac iX linear accelerator. Due to low sensitivity to absorbed dose in the Presage™ sample, three identical dose deliveries were carried out consecutively.

The readouts of the Presage™ sample were carried out with the fast laser scanner at the University of Surrey with a total scan time of 25 minutes. Due to the setup arrangements of the scanner, the projections had to be corrected for distortion using the WARP_TRI algorithm in IDL 6.3 (ITT Visual Information Solutions, Boulder, Colorado, USA). Reconstruction of the absorbance projections was done using IDL 6.3. To reduce the optical and reconstruction artifacts, a 5x5x9 box-filter was applied to the reconstructed data. In MATLAB, linear interpolation to increase the pixel side from 0.3 mm to 1 mm was carried out and the resulted matrix were smoothed with a 3x3x3 box-filter. The TPS data was linearly interpolated from 2.5x2.5x2 mm^3 to 1x1x1 mm^3.

The scanning of the gel with MRI was carried out 24 hours post irradiation. A 32-echo multi-spin echo sequence was used with the following parameters: echo time 25 ms, repetition time 6000 ms, voxel size 1x1x3 mm^3, two averages and four concatenations. The scan time was 1 hour and 43 minutes. A smaller part of the unirradiated gel was scanned with the same parameters. Further image processing was performed as described above.

3. Results and discussion

Overall, good agreement was found between the readout of the polymer gel, irradiated with a simple 3x3 cm four-field box, carried out with MRI and the Octopus IQ scanner (figure 1). To investigate the dose difference between the two data sets and the data from the TPS, the volumes were positioned in approximately the same position, thus excluding any possible set-up variations. The relative absorbed dose differences were calculated voxel-by-voxel within a volume of interest, i.e. a volume enclosed by the 90% isodose surface. A relative dose difference histogram for Octopus and MRI versus the TPS (X versus Y means that $X_{voxel}(i)$-$Y_{voxel}(i)$ in the $VOI_{90\%}$ of Y) are presented (figure 1, right column). The mean dose deviation for Octopus versus TPS and MRI versus TPS were (-1 ± 3) % and (0 ± 3) %, i.e. a little bit better for the MRI readout. For the gel batch tested in this study the dose response was confirmed to be linear in the relevant dose range, which meant that the background subtracted R2 data and optical density could be directly translated to relative absorbed dose.

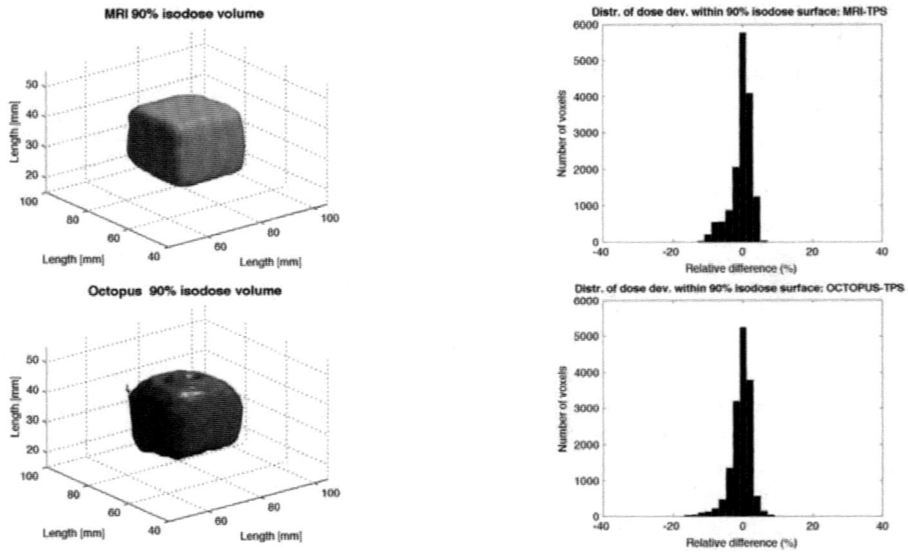

Figure 1: Isodose volumes of the four-field box irradiated on a polymer gel dosimeter from MRI (top left) and Octopus IQ (bottom left) readouts. The corresponding distributions of voxel-by-voxel dose deviations in a volume enclosed by the 90% isodose surface between the Octopus IQ readout and the TPS (top right) and between the MRI readout and the TPS (bottom right).

The MRI readout of the gel irradiated with the VMAT field agreed well to the TPS (figure 2, upper row). The 90% isodose volume read out with MRI and the one calculated by the TPS were almost identical and the relative absorbed dose differences between the two volumes calculated voxel-by-voxel within the 90% isodose surface, resulted in a mean and standard deviation of (0 ± 2) % (figure 2, top right). Due to limitations in the distortion correction, only approximately half the irradiated volume of the scanned Presage™ sample with the fast laser scanner at the University of Surrey could be corrected and are thus presented here. The two volumes (TPS and fast laser readout) presented at 90% isodose volume, share the same features (figure 2, bottom row). The distribution of dose deviations for the fast laser scanner versus the TPS resulted in a mean deviation of (-2± 10) % within the 90% isodose surface (figure 2, bottom right). The skewed appearance of the distribution suggests that the volume measured was smaller than the one from the TPS. The optomechanical complexity of the scanner requires very accurate alignment to work properly and if that was not achieved, it is possible that the beam in some spots missed the detector and gave lower signal in some positions in the projections. This could be one explanation to the divergence between the investigated volumes. Limitations on the time available to perform the experiment meant that it was not possible to optimise the refractive index of the matching fluid used, which led to the well-known edge artefact in the form of dark stripes at the edge of the projections and the useful field-of-view in the resulting reconstructions was reduced. Similarly, whilst it is recommended to do a scan of the sample both prior to irradiation and afterwards [8], this was unfortunately not possible in this study.

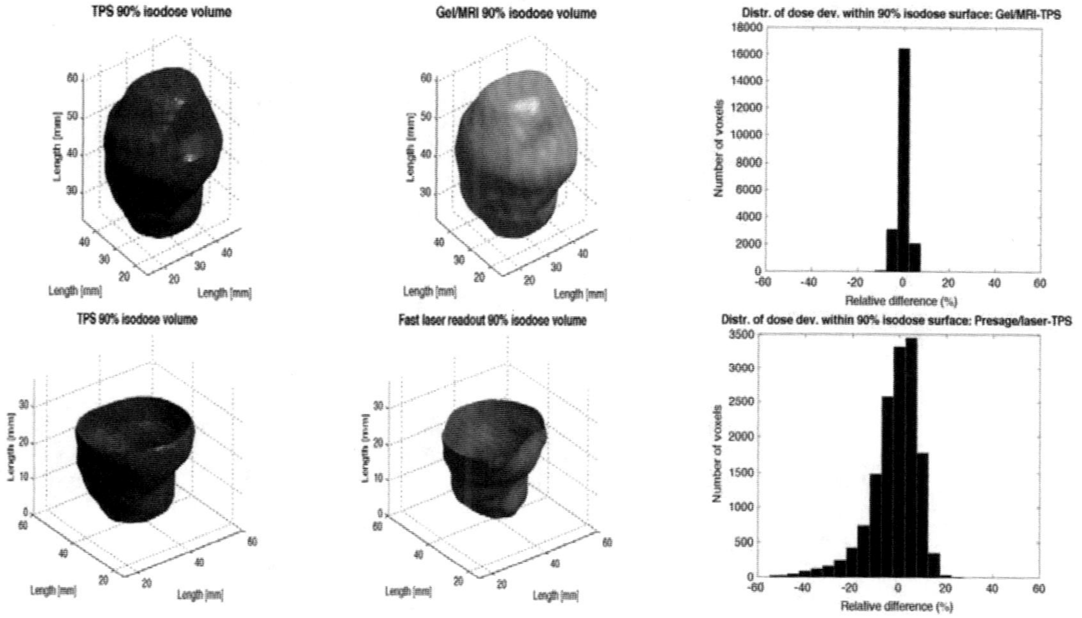

Figure 2: Isodose volumes projected into 3D views for the TPS and gel/MRI and corresponding distributions of dose deviations inside the 90% isodose volumes (upper row) and the TPS and a Presage^{TM} sample read out by the fast laser scanner from the University of Surrey (bottom row).

4. Conclusion

The results of this study showed that the conventional gel and MRI dosimeter/readout combination correlated better to the TPS than optical CT readouts. This illustrates that it is non-trivial to implement an optical CT programme and more time is needed to "work up" the optical techniques for this particular scenario, coupled with ongoing improvements to the scanners and readout methods.

5. Acknowledgements

Supported by the Swedish Cancer Society.

6. References

[1] Baldock C *et al* 2010 *Phys. Med. Biol.* 55 R1-63
[2] Gustavsson H *et al* 2004 *Phys. Med. Biol.* 49 227-41
[3] Bosi S G *et al* 2009 *Phys. Med. Biol.* 54 275-83
[4] Krstajic N and Doran S J 2007 *Phys. Med. Biol.* 52 N257-63
[5] De Deene Y *et al* 2006 *Phys. Med. Biol.* 51 653-73
[6] Karlsson A *et al* 2007 *Phys. Med. Biol.* 52 4697-706
[7] Adamovics J and Maryanski M J 2006 *Radiat. Prot. Dosimetry* 120 107-12
[8] Doran S J 2010 *J. Phys.: Conf. Ser.* 250 012086

Journal of Physics: Conference Series **444** (2013) 012080 doi:10.1088/1742-6596/444/1/012080

Preliminary investigation and application of a novel deformable PRESAGE® dosimeter

T Juang[1], J Newton[1], S Das[1], J Adamovics[2] and M Oldham[1]

[1]Duke University Medical Center, Durham, NC, USA
[2]Rider University, Lawrenceville, NJ, USA

E.mail: mark.oldham@duke.edu

Abstract. Deformable 3D dosimeters have potential applications in validating deformable dose mapping algorithms. This study evaluates a novel deformable PRESAGE® dosimeter and its application toward validating the deformable algorithm employed by VelocityAI. The deformable PRESAGE® dosimeter exhibited a linear dose response with a sensitivity of 0.0032 ΔOD/(Gy/cm). Comparison of an experimental dosimeter irradiated with an MLC pencilbeam checkerboard pattern under lateral compression up to 27% to a non-deformed control dosimeter irradiated with the same pattern verified dose tracking under deformation. CTs of the experimental dosimeter prior to and during compression were exported into VelocityAI and used to map an Eclipse dose distribution calculated on the compressed dosimeter to its original shape. A comparison between the VelocityAI dose distribution and the distribution from the dosimeter showed field displacements up to 7.3 mm and up to a 175% difference in field dimensions. These results highlight the need for validating deformable dose mapping algorithms to ensure patient safety and quality of care.

1. Introduction

Use of deformable image registration algorithms has introduced a method of calculating cumulative dose feedback for organs of interest subject to internal organ motion [1,2]. While many methods of deformable registration exist, there is currently no method of validating how well these algorithms meet the challenge of correctly mapping dose within a volume. This could be remedied, however, with a 3D dosimeter capable of capturing dose under deformation, thus providing a physical distribution for comparison [3–5]. In this study, we evaluate a novel deformable PRESAGE® dosimeter and investigate the feasibility of applying this dosimeter to a new method for validating dose tracking algorithms in deforming tissues.

2. Methods

2.1. Materials and Dose Sensitivity

Two cylindrical dosimeters (6 cm diameter, 4.75 cm long), as shown in Figure 1, were manufactured from a novel PRESAGE® (Heuris Pharma, Skillman, NJ) formulation – consisting of an elastic polyurethane matrix doped with radiochromic leuco dye – which regains its original shape following physical deformation. This deformable PRESAGE® formulation has a density of 1.02 g/cm, leuco dye content 2% (wt/wt) o-MeO-LMG, and Z_{eff} of 7.5. The Shore A hardness of the dosimeter is 10-20.

Journal of Physics: Conference Series **444** (2013) 012080 doi:10.1088/1742-6596/444/1/012080

Dose sensitivity was characterized by irradiating cuvettes containing PRESAGE® material from the same batch to known doses up to 8 Gy.

2.2. Validating Quantitative Dose Tracking Under Deformation

Two 6 cm diameter cylindrical dosimeters were irradiated with a checkerboard arrangement of 5 mm square pencil beams created by MLC fields (Figure 3A). One dosimeter served as a control and was irradiated without deformation. The second, experimental dosimeter was irradiated under non-uniform lateral compression by up to 27% (1.6 cm) to simulate a deformed organ. Compression was removed after irradiation, allowing the dosimeter to return to its original shape.

High-resolution 3D dose distributions (isotropic 1 mm resolution) were obtained by optical-CT imaging [6]. Physical dose deformation was quantified by comparing pencil beam shapes and positions in the control and deformed dosimeters in MATLAB.

2.3. Application in Validating Deformation Dose Tracking Algorithm

CT images taken of the experimental dosimeter both with and without compression (Figure 2) were imported into VelocityAI (Velocity Medical Solutions, Atlanta, GA), a deformation dose tracking software tool under evaluation in our clinic. The compressed dosimeter CT was deformed back to the shape of the uncompressed dosimeter using VelocityAI's deformable image registration capabilities. This deformation was then applied to a dose distribution calculated on the CT of the compressed dosimeter (exported from Eclipse) to yield a deformed dose distribution for the dosimeter with compression removed.

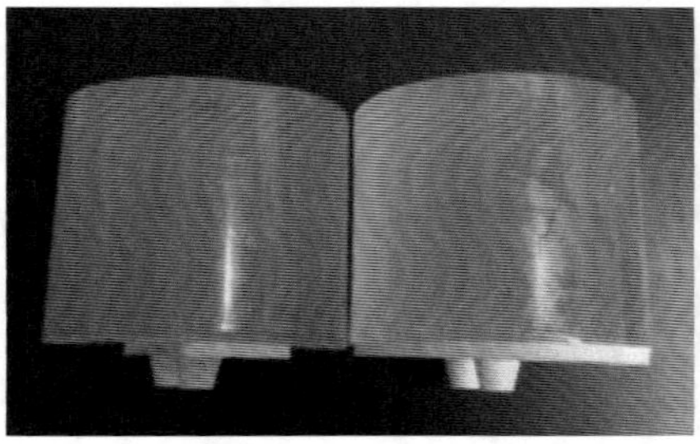

Figure 1: 6 cm diameter cylindrical deformable dosimeters with attached plates keyed for directional registration in the optical-CT scanner.

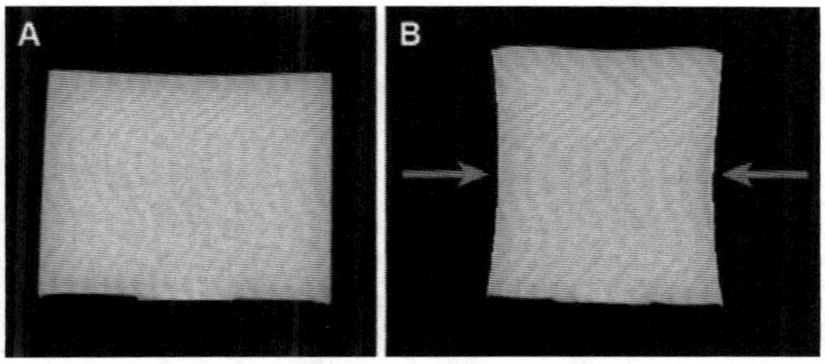

Figure 2: (A) CT of 6 cm diameter dosimeter without compression. (B) CT of the same dosimeter with compression up to 1.6 cm (27%). Arrows show direction of compression.

The difference between the dose distribution from Velocity and our physical dosimeter data was determined at an axial slice in the region of greatest compression and quantified in MATLAB using 3 metrics: (1) the displacement of the centroid of each irradiated checker, and the differences in the full width at half maximum (FWHM) for a given checker along its (2) horizontal (along the axis of compression) and (3) vertical (perpendicular to the axis of compression) profiles.

3. Results

3.1. Deformable Formulation Characterization

The dose response of the deformable PRESAGE® formulation across a range of 0 – 8 Gy demonstrated a linear response with a sensitivity of 0.0032 ΔOD/(Gy/cm).

3.2. Validation of Quantitative Dose Tracking Under Deformation

A single slice of the optical-CT scan of the control (non-deformed dosimeter) is shown in Figure 3B. The corresponding slice through the second dosimeter, which was compressed laterally by 1.6 cm during irradiation, is shown in Figure 3C. This post-compression dose tracking is highlighted with overlaid gridlines that bisect the irradiated fields. Lateral compression by 27% resulted in expansion of the dose pattern by approximately 8% - 42% along the axis of compression and reduction of the dose pattern by approximately 11% - 13% perpendicular to compression. This clearly indicates that deformed dose tracking was successfully captured in the deformed dosimeter.

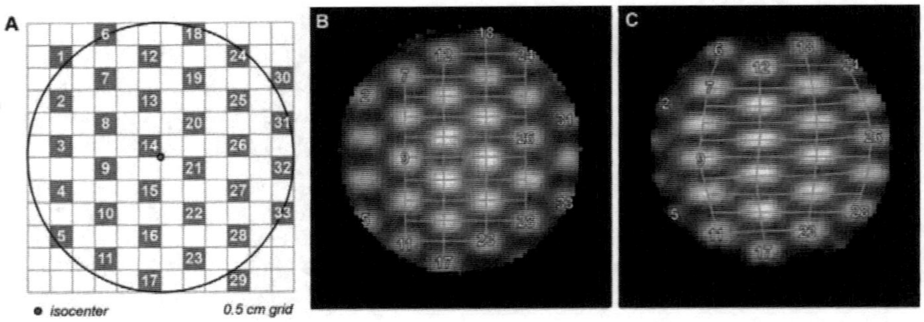

Figure 3: Geometric dose tracking in the deformed dosimeter is quantified with an overlaid grid. Edge artifacts from dosimeter optical-CT images (B-C) have been cropped. (A) Diagram illustrating the MLC checkerboard pattern. 5 mm x 5 mm radiation fields are shown in blue. (B) Axial cross-section of an uncompressed dosimeter irradiated with the pattern in (A). (C) Axial cross-section of a dosimeter laterally compressed by 27% during irradiation with the same pattern.

3.3. Comparison Between VelocityAI Deformation Algorithm and Deformable Dosimeter

Axial cross-sections of dose distributions in the region of greatest compression from the actual dosimeter and the VelocityAI deformation algorithm are shown in Figure 4A and 4B. Figure 4C displays the magnitude of displacement of each field's centroid between the two. The displacement between VelocityAI's calculated fields and the actual fields captured in the dosimeter ranged between 2 - 7.3 mm with a mean displacement of 4.5 mm.

The differences in field shapes between VelocityAI and the dosimeter were evaluated with a selection of representative fields and summarized in Table 1. Within this sample, horizontal FWHM (along the axis of compression) varied from 40% narrower to 175% wider than the actual fields. Vertical FWHM (perpendicular to the axis of compression) varied from 33.3% shorter to 50% taller than the actual fields. Magnitude of centroid displacement did not correlate with differences in field shape.

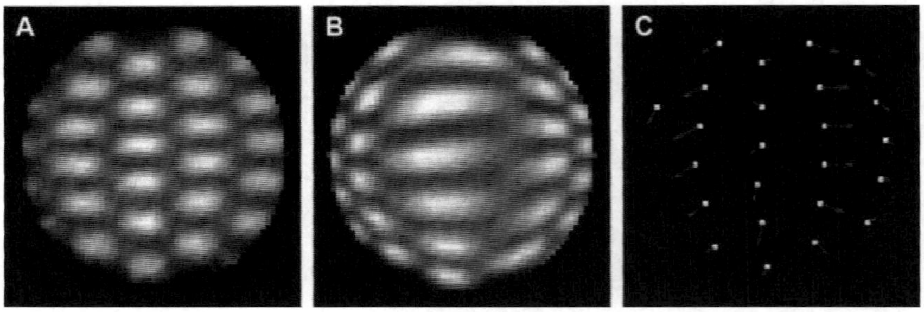

Figure 4: (A) Measured deformed dose distribution - axial cross-section of dosimeter laterally compressed by 27% during irradiation with the pattern shown in Figure 3A. (B) Corresponding axial cross-section showing the deformed dose distribution as predicted by VelocityAI deformable algorithm. (C) Quiver plot displaying magnitude of centroid displacement errors for each field in the VelocityAI distribution. White pixels mark the true location of centroids of each field in the dosimeter (i.e. positions in A).

Journal of Physics: Conference Series **444** (2013) 012080 doi:10.1088/1742-6596/444/1/012080

Table 1: Differences between the dose distributions from the Velocity AI deformable dose tracking algorithm and the physical dose tracking data from the deformable dosimeter.

Field[a]	Centroid Displacement[b] (mm)	Horizontal FWHM		Vertical FWHM	
		Difference[c] (mm)	Percent Error[d]	Difference[c] (mm)	Percent Error[d]
9	5.8	-4	-40.0%	0	0.0%
13	2.2	14	175.0%	2	50.0%
17	2.2	1	12.5%	-1	-20.0%
18	7.3	10	125.0%	-2	-33.3%
21	7.0	2	22.2%	1	33.3%

[a] Field numbers refer to enumerated irradiated MLC fields as labelled in Figure 3D.
[b] Magnitude of displacement between the centroid of a given field in the Velocity distribution versus the corresponding field in the dosimeter.
[c] $FWHM_{Velocity} - FWHM_{dosimeter}$
[d] $(FWHM_{Velocity} - FWHM_{dosimeter})/FWHM_{dosimeter}$

4. Conclusion

A novel, deformable 3D dosimeter was able to successfully track dose under deformation. These results are significant in that this deformable 3D dosimeter has potential to validate deformable dose tracking and registration algorithms. Applying this method to the deformable dose tracking algorithm in VelocityAI demonstrated significant limitations in accurately predicting deformable dose distributions within a homogenous structure. These results highlight the need for validating deformable dose mapping algorithms to ensure patient safety and quality of care.

Acknowledgements
This work was supported by NIH R01CA100835.

References

[1] Yan D *et al* 1999 *International Journal of Radiation Oncology, Biology, Physics* **44** 665-75
[2] Yan D 2010 *Seminars in Radiation Oncology* **20** 79-83
[3] Walsh A *et al* 2011 *med. phys.* **38** 3728
[4] Niu C J *et al* 2012 *med. phys.* **39** 765-76
[5] Yeo U J *et al med. phys.* **39** (4) 2203-13
[6] Thomas A *et al* 2011 *med. phys.* **38** 4846

Performance of latex balloons for optical computed tomography

K Jordan[1,2], A Walsh[1], M Peng[2] and J Battista[1,2]
[1]London Regional Cancer Program (London Health Sciences Centre)
[2]Department of Medical Biophysics, Western University, London, ON, Canada

E-mail: kevin.jordan@lhsc.on.ca

Abstract. Latex balloons filled with radiation sensitive hydrogels were evaluated as 3D dosimeters with optical computed tomography (CT) readout. Custom balloons, with less than 10 cm diameters, were made from latex sheets. Commercial, 13 cm diameter, clear balloons were investigated for larger volumes. Ferrous-xylenol orange and genipin gelatin gels selected for 1 and 30 Gy experiments, respectively. The thin stretched latex membrane allowed optical imaging to within 1 mm of the interior balloon edge. Reconstructed dose distributions demonstrated valid measurements to within 2 mm of the balloon surface. The rubber membrane provides a hybrid approach to deforming hydrogels. Uniform irradiation of a deformed gel resulted in a uniform dose being measured when scanned in the relaxed, initial balloon shape. The 13 cm diameter balloons were also effective and inexpensive vessels for hydrogels due to their high clarity, thinness and mechanical strength. Latex balloons represent an inexpensive method to obtain useful information from nearly the entire dosimeter volume.

1. Introduction

Optical computed tomography (CT) for 3D radiation dosimetry involves either radiation sensitive hydrogels or plastics. The hydrogels generally require a plastic sealed vessel to provide mechanical rigidity and chemical isolation. One important advantage of radiochromic plastics, such as Presage, is the elimination of the vessel and optical artifacts that are associated with refractive index mismatch between the gel and vessel wall. The optical artifacts include: reflections, refractions, internal wave-guiding in the wall and polarization effects. The exception is genipin hydrogel which has been optically scanned to the surface without support in a vessel [1]. FEP Teflon has a refractive index similar to hydrogel formulations and is used when dosimetry to edge of vessel is required. However, oxygen permeability is a limitation for some of the radiation chemistries under investigation. High refractive index plastics are also used for optical CT vessels and if the wall is thin enough most of the gel interior can be probed. This effect was characterized by Krstajic [2] and has important implications for optical CT design and performance. Latex balloons when stretched to near bursting become very transparent and are now being used for gel dosimetry with optical CT readout. Results with radiochromic gels and polymerization gels in latex balloons have been reported [3,4]. Because of the thin latex membrane surrounding the gel it is expected that optical sampling to the inner edge of the balloon should be possible. This feature increases the efficiency of 3D gel dosimetry by measuring dose throughout the entire dosimeter volume. The hybrid mechanical properties of the gels and balloon allow deformable dosimetry to be possible.

2. Methods

Initial tests indicated that commercial clear latex balloons were more scattering that latex sheet prepared in the laboratory without mould-release coatings. Latex sheets were prepared by pouring raw latex onto glass plates and annealing at 65 C for one day. Circles were cut from the sheet and attached to a custom fill tube. Ferrous-xylenol orange (FX) gel was injected into the tube to form balloons that ranged in diameters from 40 to 100 mm diameters. Commercial "5 inch clear" latex balloons were filled with genipin gel. Ferrous xylenol orange gels were prepared with 4% gelatin by mass, 50 mM sulfuric acid (H_2SO_4), 0.3 mM ferrous ammonium sulphate and 0.05 mM xylenol orange. Genipin gel was prepared with 3% gelatin by mass, genipin 0.1% mass of gelatin, and 200 mM H_2SO_4. The genipin and gelatin were dissolved in distilled water and stirred at 45 C for 18 hours and acid was added to form the stock gel. A stock gel sample was diluted with blank gel (200 mM H_2SO_4, 3% gelatin) to achieve acceptable level of transmission at 594 nm for optical CT scans.

The smaller FX gel-balloon was scanned by an modified Vista10 optical CBCT scanner (590 nm light, 512 projections per 360 degrees, camera lens aperture of F4 and 8 minute scan time) from Modus Medical Devices. Reconstruction with Vista software was performed at 0.25 mm^3 voxel size with Ramp filter, generating a 512^3 array of attenuation coefficients. While the larger, genipin gel – balloon was scanned with an in-house prototype laser scanning CBCT scanner (594 nm laser light, 512 projections per 360 degrees, 200 slices, 30 minutes scan time). Propylene glycol aqueous solutions were prepared for refractive index matching of the balloon phantoms. Reconstruction was performed with in-house software implementing the FDK algorithm with 0.34 mm^3 voxel size, Ramp filter and 512x512x200 output array. A photograph of the phantom in the aquarium is shown in figure 1. A centering base ring was made for the balloon and fixed to the CT rotation platform. The liquid level was adjusted so that several centimeters of the balloon neck was above the liquid to provide sufficient friction between the gel and the rotation platform. A black, ink-spot fiducial was placed on the balloon. The laser beam was directed onto the fiducial and the position recorded. After irradiation the balloon was again placed in the scanner and adjusted until the fiducial aligned with the laser beam.

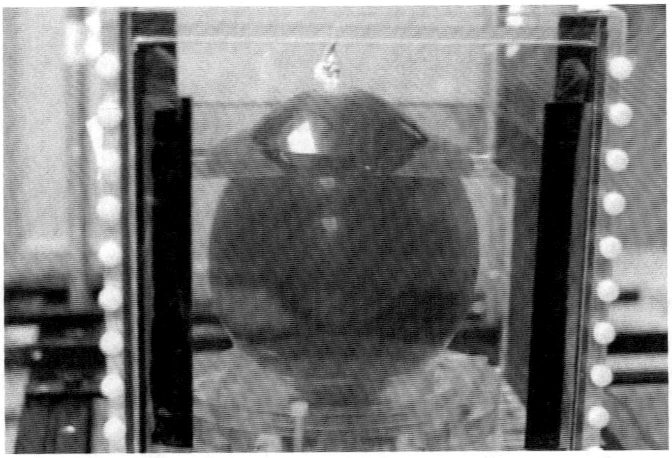

Figure 1: Photograph of latex balloon filled with genipin hydrogel positioned in refractive index matched laser CBCT scanner. Note: laser beam aligned with black ink spot fiducial above liquid surface and reflection centered on optical axis.

The gels were placed at isocentre of the radiotherapy machines in a 20x20x20 cm water phantom for irradiation. Parallel opposed beams 20x20 cm cross section delivered 1 and 30 Gy respectively to the FX and genipin balloons. The FX gel was deformed prior to irradiation to a uniform dose of 1 Gray and then allowed to relax before the data scan was obtained. A second irradiation was delivered to the genipin balloon, orthogonal, 5x5 cm, 10 MV beams of 15 Gray each at isocentre.

3. Results and Discussion

The optical performance of a small 5.5 cm diameter FX gel-filled latex balloon is shown in figure 2. The projection image was acquired with a modified Vista10 optical CBCT scanner. A transmission image through the gel and balloon was taken of a clear, plastic ruler placed at the entrance window to the aquarium. The 1 mm gradations on the ruler are clearly resolved, demonstrating the low optical scatter of this dosimeter.

Figure 2: Transmission image of plastic ruler placed between diffuse light source and a custom latex balloon filled with ferrous-xylenol orange gel mounted in a modified Vista10 CBCT scanner.

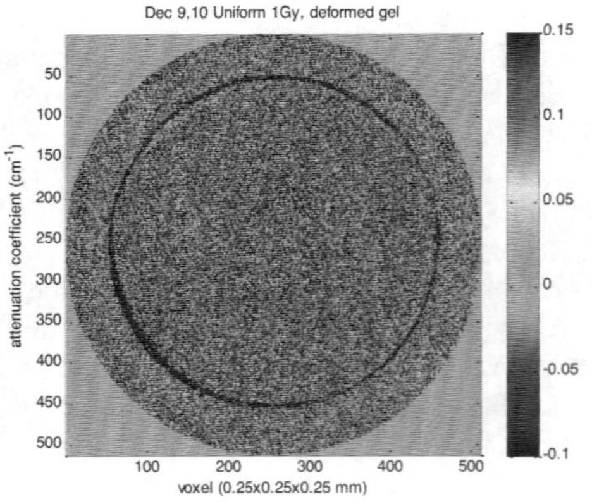

Figure 3: Central transverse slice from reconstruction of uniform 1 Gy dose, delivered to deformed balloon. Note, lower left perimeter where balloon shape had not completely relaxed.

At the edge of the balloon all gradations can be counted and no distortion of the lines is observed. This demonstrates that light passing less than 1 mm from the edge of the balloon can be imaged. In figure 3, a reconstructed slice of a uniform dose distribution is shown demonstrating that quantitative dosimetry can be obtained to within 2 mm of the balloon edge with FX gels. Note the segment of the perimeter where the gel-balloon had not returned to its original, reference scan shape.

A genipin gel-filled commercial "5 inch clear" balloon, irradiated to a uniform dose of 30 Gray, was scanned with a prototype laser CBCT scanner. In figure 4, the reconstructed attenuation indicates a lower sensitivity towards the balloon dosimeter edge. Further experiments are required to verify this

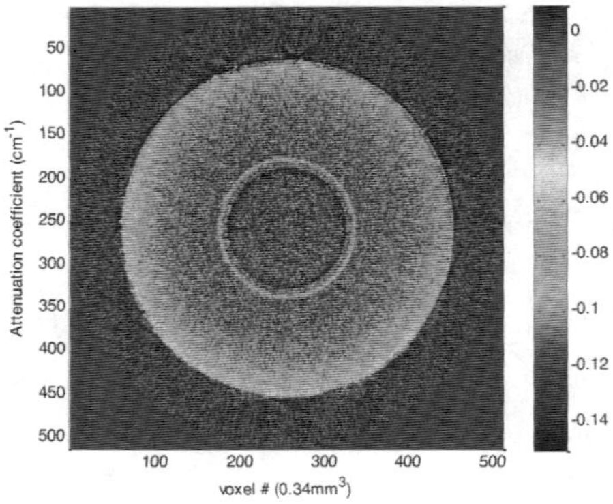

fure 4: Central transverse slice from genepin gel filled commercial latex balloon. The balloon was irradiated to a uniform 30 Gy dose. Note the central ring artifact is due to reflection, see figure 1.

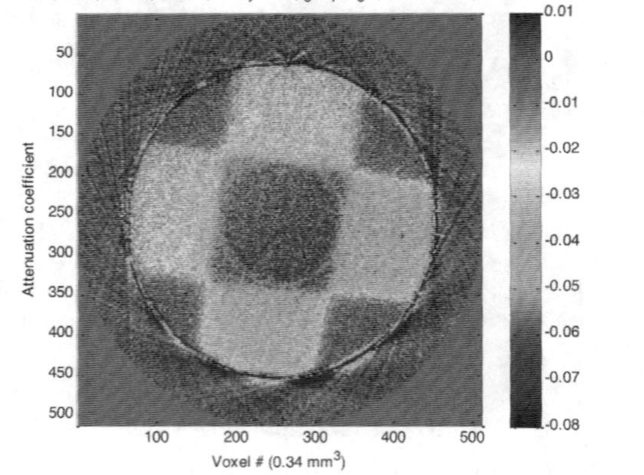

Figure 5: Central slice following irradiation of dosimeter (figure 4) with crossed 10 MV photon beams. Balloon was centered in water phantom during irradiation.

Journal of Physics: Conference Series **444** (2013) 012081 doi:10.1088/1742-6596/444/1/012081

result. Generally, non-uniform gel dosimeters can be improved by pre-irradiation with a uniform dose. Next the same balloon was with irradiated with orthogonal beams. The middle slice of the 3D reconstruction is shown in figure 5. The dose distribution is continuous to the edge of the dosimeter indicating that quantitative measurements to the edge of this dosimeter are possible. A similar conclusion with nPAG gel has been reported [4]. The dominant ring artifact is related to a reflection from the balloon.

4. Conclusion

Latex balloons provide a high efficiency vessel for optical CT scanning of hydrogels because of the thin rubber membrane. This allows optical sampling to within 2 mm of the balloon edge. Reconstructions are then useful to similar proximity of the dosimeter's perimeter. Both ferrous-xylenol orange and genipin hydrogels in latex balloons were used to demonstrate that dose distributions could be measured to within 2 mm of edge. This optical and chemical performance allows nearly the entire gel to be used as a dosimeter. Hydrogels within latex balloons can be reversibly deformed without tearing. This allows handling of the gel filled balloons without damage to the dosimeter and provides a method to perform deformable 3D dosimetry.

5. Acknowledgements

A Walsh was partially supported by funding from CIHR grant #200403MOP and M Peng from a FEDDEV industrial partnership grant.

6. References

[1] Jordan K 2009 *J. Phys.: Conf. Ser.* **164** 012029
[2] Krstajic N and Doran S J 2006 *Phys. Med. Biol.* **51** 2055-75
[3] Walsh A *et al* 2011 *Med. Phys.* **38** 3728
[4] Yeo U J *et al Med. Phys.* **39** 2203-13

A note on the interpretation of the gamma evaluation index

Crister Ceberg
Medical Radiation Physics, Department of Clinical Sciences Lund, Lund University, Lund, Sweden

E-mail: crister.ceberg@med.lu.se

Abstract. The gamma evaluation method has become the gold standard for the comparison between measured and calculated absorbed dose distributions. However, test criteria and failure rate tolerance levels have hitherto normally been based on empirical evidence, rather than rigorous statistical analysis. In this work, it is proposed that the gamma-evaluation method could be reinterpreted such that the absorbed dose difference and distance-to-agreement criteria are replaced by the standard deviations of the associated uncertainties. By comparison between absorbed dose calculations and simulated measurements for clinically realistic test cases in 1D and 2D, it is then shown that the resulting squared gamma distribution follows a chi-squared distribution with one degree of freedom. This result can be used to test the statistical significance of measured deviations, and to determine proper failure rate tolerance levels in clinical radiotherapy quality control.

1. Introduction

One of the most important quality control (QC) procedures in radiation therapy is the experimental verification of calculated treatment plans by actual measurements, and the standard method for the evaluation of such comparisons is the so-called gamma-evaluation test method [1-2]. Due to the unavoidable uncertainties of absorbed dose measurements and detector positioning, some points in the gamma evaluation will always fail the test criteria even if there is no true deviation, and a certain failure rate has to be tolerated. Although test criteria and failure rate tolerance levels have hitherto normally been based only on empirical evidence [3], proper figures should preferably be chosen based on a statistical analysis of the actual uncertainty associated with the absorbed dose measurement and the positioning of the detector. For this purpose, some authors have adopted an error propagation approach, incorporating the local absorbed dose gradient into the analysis [4-7]. In practical implementations, however, such methods are generally limited to a linear gradient approximation, which may introduce anomalies in convex or concave parts of the dose distribution.

In case the complete probability distribution of the gamma index was known, the test criteria could be determined based on probability analysis [8-9]. Under certain conditions, an approximate gamma distribution can be derived theoretically, and a three-moment approximation of the squared gamma distribution has been developed by applying general results about quadratic forms of normal random variables and the chi-squared distribution [9]. This method could potentially be implemented in conventional gamma evaluation procedures, requiring only slight modifications and few extra calculations.

Journal of Physics: Conference Series **444** (2013) 012082 doi:10.1088/1742-6596/444/1/012082

In this work, a further simplified method was presented, which could be used together with current gamma test procedures without modifications. The method was evaluated by using calculated dose distributions in 1D and 2D for clinically realistic QC situations, and by simulating measurements based on statistical uncertainties of absorbed dose measurements and detector positioning. The resulting squared gamma distribution was investigated and compared with the chi-squared distribution with one degree of freedom, and the results was used as a basis for the determination of gamma evaluation acceptance criteria.

2. Material and methods

The calculations in this work were based on clinically realistic absorbed dose distributions in 1D and 2D. In both cases, the relative absorbed dose was given as a continuous function of the spatial coordinates. For the 1D case, the equation provided by Low et al. [1] was used as a representation of a calculated relative absorbed dose profile, see figure 1a. The 2D case was based on an IMRT beam calculated by the Oncentra Master Plan treatment planning system, with a pixel resolution of 1mm and linear interpolation between the points of the dose matrix, see figure 1b.

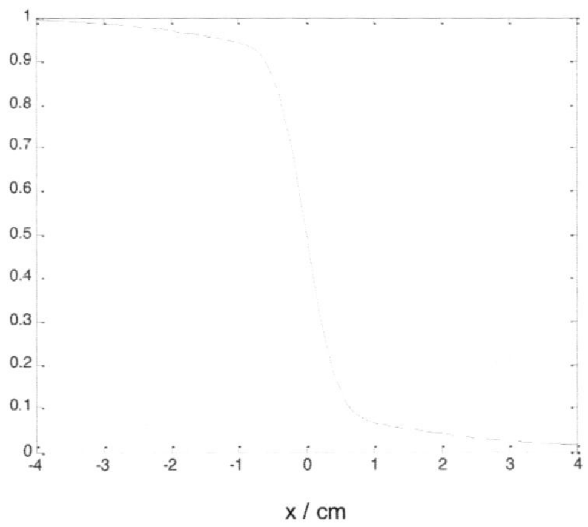

Figure 1a: The dose distribution in 1D.

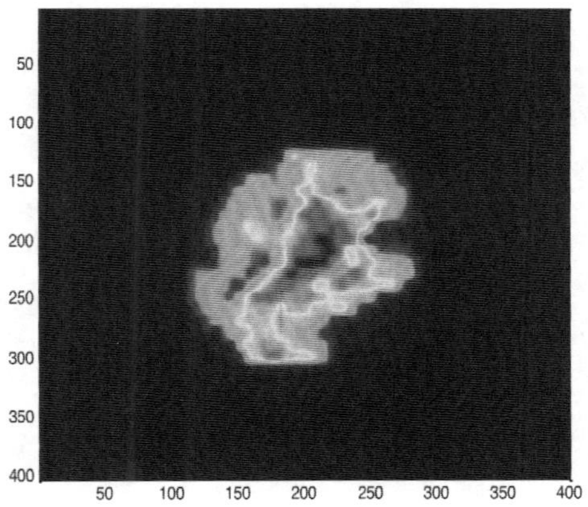

Figure 1b: The dose distribution in 2D.

Measured absorbed dose values were simulated by assuming that the calculated value at (and in the vicinity of) the measurement point represented the expectation value, and by adding a random measurement uncertainty. Thus, for each nominal measurement point, a measured value was simulated by taking the calculated dose at a randomly displaced position, and by adding a random measurement noise. Both the detector displacement and the measurement noise were drawn from normal distributions, with standard deviations σ_D=1cGy and σ_r=1mm, respectively. Then, the nearest calculation point was determined according to the gamma evaluation procedure, and the smallest gamma index was calculated. The acceptance criteria were set to $\Delta D_{crit}=\sigma_D$ for the dose difference and to $DTA_{crit}=\sigma_r$ for the distance-to-agreement. By repeating this simulation a large number of times (10^5 times in 1D and 10^4 times in 2D), a squared gamma (γ^2) distribution was obtained, which was then compared with the chi-squared distribution with one degree of freedom.

3. Results and discussion

The resulting γ^2-distributions are shown (blue circles) for the 1D case in figure 2a (at the position x=0), and for the 2D case in figure 2b (at the position marked with a cross in figure 1b). In both cases it can be seen that the γ^2-distribution follows very well the chi-squared distribution with one degree of freedom (red curve).

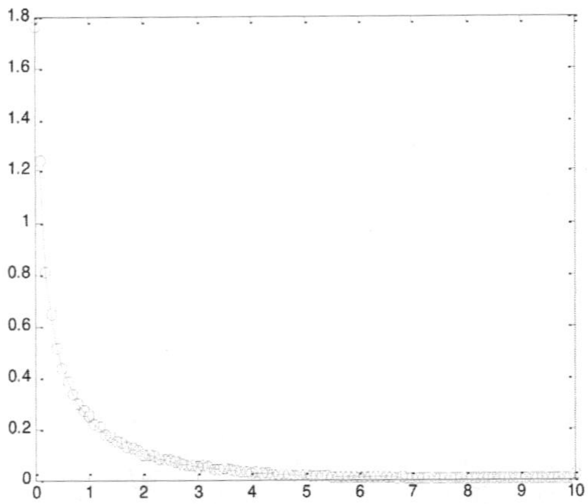

Figure 2a: Distribution of gamma values in 1D.

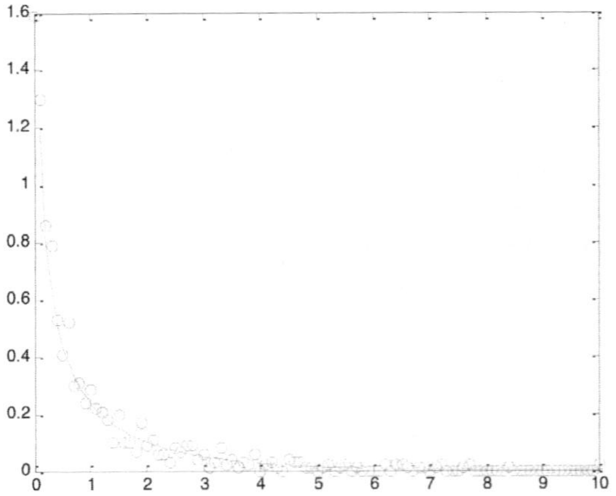

Figure 2b: Distribution of gamma values in 2D.

According to the chi-squared distribution with one degree of freedom, 5% of the values should be greater than 3.841 (given that there is no true deviation). The actual fraction of values greater than 3.841 for these particular simulations were 5.1% in the 1D case and 5.0% in the 2D case.

Importantly, these numbers remained close to 5% also at other measurement points. In figure 3, the fraction of values above 3.841 for the acquired γ^2-distribution is shown for positions from -2.0 cm to +1.0 cm along the x-axis. It can be observed that a 5% failure rate is approximately maintained (within about 0.5%) at all positions along the x-axis, even across the convex and concave parts of the penumbra region, where there is a non-linear absorbed dose gradient.

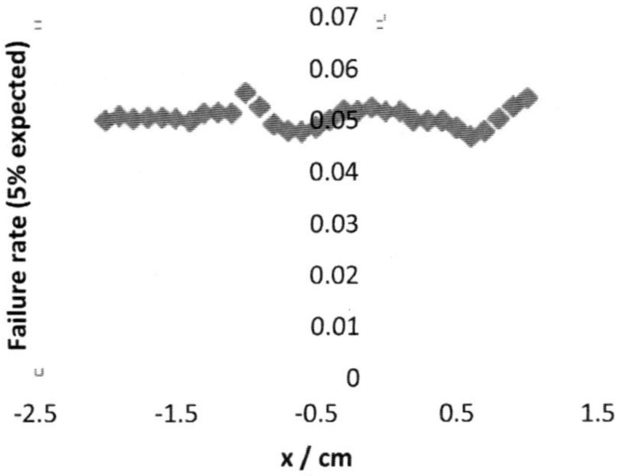

Figure 3: Failure rate for the 1D case as a function of position along the x-axis

The chi-squared distribution is intimately associated with quadratic forms of normally distributed variables. Assuming that the deviations between the observed and expected absorbed dose and detector position, relative to the test criteria, are independent standard normal random variables, the squared gamma index should follow (by definition) a chi-squared distribution with two degrees of freedom. However, since the absorbed dose and detector position are not entirely independent variables (and the relationship between them is unknown), this is no longer the case. While an approximate solution to this situation can be obtained by using more refined mathematics [9], a simpler approach using reduced degrees of freedom was tested in this work. The squared gamma distribution was investigated and found to agree well with the chi-squared distribution with only one degree of freedom. This result is very useful, as it makes it straightforward to implement a statistical interpretation of the gamma evaluation index into current gamma-evaluation procedures, without modifications.

4. Conclusions

The gamma evaluation method has become the gold standard for the evaluation of comparisons between measured and calculated absorbed dose distributions. However, in clinical practice the acceptance criteria have been based largely on experience, without any underlying statistical analysis. In this work, it is proposed that the gamma-evaluation method could be reinterpreted such that the absorbed dose difference and distance-to-agreement criteria are replaced by the standard deviations of the associated uncertainties. Then, the squared gamma index can be compared with the chi-squared distribution with one degree of freedom in order to test the statistical significance of the measured deviation. The results can be used to determine proper failure rate tolerance levels in clinical radiotherapy QC.

5. References

[1] Low D A *et al* 1998 *Med. Phys.* **25** 656-61
[2] Low D A and Dempsey J F 2003 *Med. Phys.* **30** 2455-64
[3] Basran P S andAn analysis of tolerance levels in IMRT quality assurance procedures *Med. Phys.* **35** 2300-7
[4] Bakai A *et al* 2003 *Phys. Med. Biol.* **48** 3543-53
[5] Jin H et al 2005 *Med. Phys.* **32** 1747-56
[6] Jin H et al 2008 *Med. Phys.* **35** 982-96
[7] Jin H et al 2012 *Int. J. Radiat. Oncol. Biol. Phys.* **78** 9208
[8] Stock M *et al* 2005 *et al Phys. Med. Biol.* **50** 399-411
[9] Henríquez F C and Castrillón S C 2011 *Comput. Math. Methods Med.* 861869 1-10

Journal of Physics: Conference Series **444** (2013) 012083 doi:10.1088/1742-6596/444/1/012083

Measurement of a PAGAT gel dosimeter by ultrasound computed tomography

S Khoei, JV Trapp and C M Langton

School of Chemistry, Physics, & Mechanical Engineering and Institute of Health & Biomedical Innovation, Queensland University of Technology (QUT), Brisbane, Australia

s.khoei@qut.edu.au

Abstract. In this work we used a 3D quantitative CT ultrasound imaging system to characterise polymer gel dosimeters. The system comprised of two identical 5 MHz 128 element phased-array ultrasound transducers co-axially aligned and submerged in water as a coupling agent. Rotational and translational movement of the gel dosimeter sample between the transducers were performed using a robotic arm. Ultrasound signals were generated and received using an Olympus Omniscan unit. Dose sensitivity of attenuation and time of flight ultrasonic parameters were assessed using this system.

1. Introduction

Recent years have seen the development of gel dosimetry for radiotherapy purposes. Following irradiation, the physical properties of these radiosensitive gels undergo changes that can be measured using various medical imaging equipment [1-21]. Traditionally this imaging has utilised either optical computed tomography (CT) [1, 2] magnetic resonance imaging [3-5] or x-ray CT [6, 7]. Previous studies have examined the feasibility of ultrasound based CT to study dose response in PAG dosimeters [8-11]. Here we extend that previous work on polymer gel dosimeters utilising a new ultrasound scanner design. The sensitivity response was investigated through measuring ultrasound attenuation coefficients and ultrasound time of flight in irradiated PAGAT gel [12, 22] dosimeters.

2. Method and Materials

2.1. PAGAT gel preparation and irradiation

A batch of PAGAT gel was manufactured based on the Venning *et al* [10] formulation with the THPC (Tetrakis Hyydroxymethyl Phosphonium Chloride) concentration increased to 8 mM [11] under normal atmospheric conditions. Once prepared, the gel was poured into 150 ml cylindrical shape PET (Polyethylene terephthalate) containers of 10 cm height and 4.5 cm diameter and refrigerated at 4°C for about 24 hours prior to irradiation. Each gel was then irradiated to various doses from 2 to 30 Gy using a Gammacell 200 (Co-60) source. Irradiated gels were again refrigerated at 4°C for about 24 hours before imaging.

2.2. *Imaging technique*

The equipment required to set up the ultrasound CT system was an Olympus Omniscan unit, two 5 MHz 128 element phased-array transducers, Olympus TomoView software (to define beam configuration), Motoman HP6 mounted robotic arm (with 6 degrees of freedom), microcontroller, water tank and transducer fixture.

In our ultrasound CT system the phased array transducers are mounted and aligned in parallel on a frame and submerged in a water tank (figure 1). One transducer serves as the transmitter and the other as the receiver. Scans were conducted with the transducers fixed in position while rotational and translational movement of the gel dosimeter sample between the transducers was performed using a programmed robotic arm. The lateral ultrasound scan of the sample was performed by exciting individual transducer elements from first to the last without any need to displace the transducer and subsequently a projection data set was taken. After the completion of a single scan, the sample was rotated at an angle increment of one degree and the next projection was recorded. It should be noted that the Omniscan MX equipped with the PA32128 acquisition module is capable of transmitting and acquiring with 128 individual elements and for this reason only 64 elements could be allocated for each of the two transducers; this was accomplished by using a splitter connector.

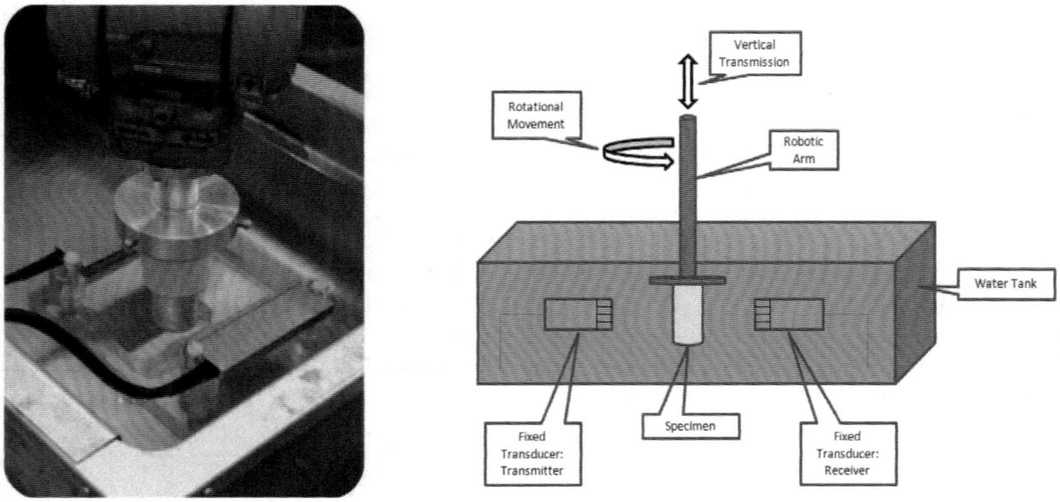

Figure 1: Ultrasound transmission CT

All signal acquisition parameters such as digitising frequency, data sample size, signal rectifications were configured through TomoView software. The rotational position of the sample and ultrasound transmission/reception were synchronised via a microcontroller.

2.3. *Data collection and image reconstruction*

Signal digitisation was performed at 25 MHz and 12 bit resolution. For 64 transmission elements acquired through 360 projections created 23040 ultrasound signals in total. 2D images of different slices through the scanned object were reconstructed using the filtered back projection technique. To determine attenuation coefficient, two sets of measurements were recorded; one taken for the sample and the other one for the reference material, chosen to be water.

3. Results and Discussion

Figure 2 shows an example of reconstructed attenuation image for one slice through one batch of gel, irradiated at 25 Gy. Ring artifacts associated with CT can be seen in the reconstructed images. The image reconstruction procedure was repeated for each batch of irradiated gel. From the reconstructed images, a consistent region of interest was selected, with average and standard deviation values calculated.

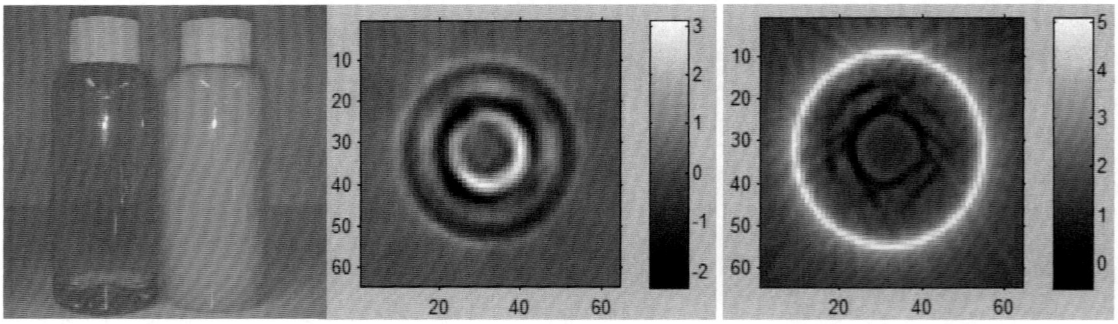

(a) (b) (c)

Figure 2: (a) non-irradiate (left) and irradiated (right) PAGAT gel at 25 Gy; (b) reconstructed attenuation map; (c) reconstructed TOF map

The variation of ultrasound attenuation as a function of dose is illustrated in Figure 3(a). Linear regression was used to determine the attenuation coefficient along with its related uncertainties in dose range of 0-30 Gy. The attenuation coefficient was calculated to be 1.44 ± 0.08 dB m-1Gy-1 and R2 (adjusted coefficient determination) of 97% (P= 6.86E-6). Even though the overall attenuation coefficient measured at various doses for PAGAT is different to that previously reported for PAG (2.9 ± 0.3 dB m-1Gy-1) and MAGIC gel (4.2± 0.3 dB m-1Gy-1), the trend is similar to those previously reported by Mather et al for attenuation-dose sensitivity with the ultrasound fundamental frequency of approximately 4MHz [8-11].

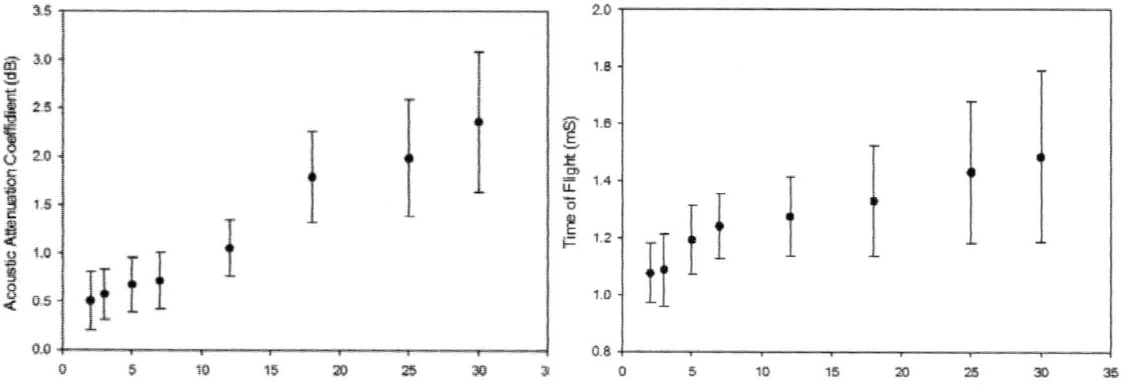

(a) (b)

Figure 3: Ultrasound attenuation and TOF measured for PAGAT gel across a variety of doses- error bars are the standard deviation of the ROI

Figure 3(b) illustrates the variation of the time of flight (TOF) as a function of absorbed dose. From TOF, inverse ultrasound speed was calculated. Linear regression was used and the dose sensitivity for inverse ultrasound speed values was determined to be $1.37 \times 10^{-4} \pm 0.10 \times 10^{-4}$ s m $^{-1}$ Gy $^{-1}$ (R^2 =95%; P=5.31E-5). This dependency of inverse acoustic speed on dose agrees with the results achieved by Mather et al for inverse acoustic speed of propagation as function of dose ($1.8 \times 10^{-4} \pm 1\times10^{-5}$ s m^{-1} Gy^{-1}) [8-11, 22].

4. Conclusion

The capability of ultrasound CT was investigated through quantitative assessment of a PAGAT gel dosimeter via attenuation and TOF measurements exhibiting sensitivity to dose comparable to those previously reported [8-11].

5. Acknowledgement

The authors would like to acknowledge the Institute of Health and Biomedical Innovation (IHBI) for their support on this project.

6. References

[1] Gore J C *et al* 1996 *Phys. Med. Biol.* **41** 2695-704
[2] Maryanski M J *et al* 1996 *Phys. Med. Biol.* **41** 2705-17
[3] Ibbott G S *et al* 1997 *Int. J. Radiat. Oncol. Biol. Phys.* **38** 1097-103
[4] Maryanski M J *et al* 1994 *Phys. Med. Biol.* **39** 1437-55
[5] De Deene Y *et al* 2006 *Phys. Med. Biol.* **51** 653-673
[6] Hilts M *et al* 2000 *Phys. Med. Biol.* **45** 2559-71
[7] Trapp J V *et al* 2001 *Phys. Med. Biol.* **46** 2939-51
[8] Mather M L *et al* 2003 *Phys. Med. Biol.* **48** N269-75
[9] Mather M L *et al* 2002 *Phys. Med. Biol.* **47** 1449-58
[10] Mather M L *et al* 2003 *Ultrasonics* **41** 551-9
[11] Mather M L *et al* 2002 *Phys. Med. Biol.* **47** 4397-409
[12] Venning A J *et al* 2005 *Phys. Med. Biol.* **50** 3875-88
[13] Murry P and Baldock C 2000 *Australas. Phys. Eng. Sci. Med.* **23** 44-51
[14] Rintoul L et al 2003 *Appl. Spectrosc.* **57** 51-7
[15] Gustavsson H *et al* 2004 *Phys. Med. Biol.* **49** 227-4
[16] Hill B *et al* 2005 *Med. Phys.* **32** 1589-97
[17] Brown S *et al* 2008 *Appl. Radiat. Isot.* **66** 1970-74
[18] Lepage M *et al* 2002 *Phys. Med. Biol.* **47** 1881-90
[19] Bosi S G *et al* 2009 *Phys. Med. Biol.* **54** 275-83
[20] Trapp J V *et al* 2004 *Phys. Med. Biol.* **49** N139-46
[21] Hill B *et al* 2005 *Med. Phys.* **32** 1589-97
[22] Khoei S et al 2010 *J. Phys.: Conf. Ser.* **250** 012019
[23] Oldham M *et al* 2003 *Med. Phys.* **30** 623-34
[24] Jirasek A. and M. Hilts 2009 *J. Phys.: Conf. Ser.* 164 012038

7th International Conference on 3D Radiation Dosimetry (IC3DDose) IOP Publishing
Journal of Physics: Conference Series **444** (2013) 012084 doi:10.1088/1742-6596/444/1/012084

Quantitative evaluation of polymer gel dosimeters by broadband ultrasound attenuation

S Khoei, JV Trapp and CM Langton

School of Chemistry, Physics and Mechanical Engineering and Institute of Health & Biomedical Innovation, Queensland University of Technology (QUT), Brisbane, Australia

E-mail: s.khoei@qut.edu.au

Abstract. Ultrasound has been examined previously as an alternative readout method for irradiated polymer gel dosimeters, with authors reporting varying dose response to ultrasound transmission measurements. In this current work we extend previous work to measure the broadband ultrasound attenuation (BUA) response of irradiated PAGAT gel dosimeters, using a novel ultrasound computed tomography system.

1. Introduction

Ultrasound has been shown to have the potential to evaluate gel dosimeters [1] due to changes in the measurement of velocity, attenuation and frequency dependent attenuation [2-6]. Here, we evaluate a measure of the linear increase in ultrasound attenuation with frequency, termed broadband ultrasound attenuation (BUA [7]), using a novel computed tomography system.

2. Method and Materials

2.1. PAGAT gel preparation and irradiation

A batch of PAGAT gel was manufactured based on the Venning *et al* [8] formulation with the THPC (Tetrakis Hyydroxymethyl Phosphonium Chloride) [9] concentration increased to 8 mM [10] under normal atmospheric conditions. Once prepared, the gel was poured into 150 ml cylindrical shape PET (Polyethylene terephthalate) containers of 10 cm height and 4.5 cm diameter and refrigerated at 4° C for about 24 hours prior to irradiation. Each gel was then located in a water tank [11] and irradiated to various doses from 2 to 30 Gy, at the depth dose of 1.5 cm, 100 cm SSD and 2x2 cm^2 field size utilising a 6 MV photon beam delivered by Linac 600 CD. The irradiated gels were again refrigerated at 4°C for about 24 hours before imaging [12].

2.2. Imaging technique and data collection

The ultrasound computed tomography (CT) system used to scan our samples comprises of two co-aligned 5 MHz, 128 element linear-array transducers; one serving as transmitter, the other as receiver. Only 64 elements of each transducer were utilised, totalling 128 elements, the maximum operational capacity of the Olympus Omniscan unit utilised. Scans were performed whereby the transducers were fixed in position whilst the sample was rotated in one degree steps using a programmed robotic arm (figure1). For each scan position an excitation pulse was applied sequentially to each transmitter transducer element (Tn) via the Omnsican unit and associated TomoView software. The propagated ultrasound signal for detected by the corresponding co-aligned receive transducer element (Rn); hence T1-R1, T2-R2, T64-R64. The total number of acquired signals per slice is 23040 resulting from 64 transmissions per projection times by 360 (number of projections).

Figure 1: Ultrasound transmission computed tomography scanner.

To determine broadband ultrasound attenuation (BUA), two sets of measurements were recorded; one taken for the gel sample and the other one for the reference material, chosen to be non-irradiated gel. For each transmitted signal, the BUA value was measured through four steps:

1. For each individual transmitted signal (time domain signal), the time-portion of signal which includes the maximum signal amplitude was extracted for both sample (irradiated gel) and reference (non-irradiated gel).
2. Fast Fourier transforms of these two extracted portions were calculated and the attenuation at each individual frequency calculated using equation 1, where A_{Ref} and A_{sam} are the signal amplitudes of the reference signal and amplitude of the sample respectively.
3. Attenuation was plotted against frequency.
4. For a defined and consistent frequency range where attenuation increased approximately linearly with frequency, BUA was calculated as the slope of ultrasound attenuation versus frequency using a least square fit.

$$\text{Attenuation}(f) = 20.\log\left(A_{Ref}(f)/A_{Sam}(f)\right)\ (\text{dB}) \qquad \text{[Eq 1]}$$

The measured BUA values corresponding to each of the 64 transducer element pairs at each projection angle, were transferred into a two dimensions matrix (64 × 360) termed a sinogram, describing BUA data as a function of distance along the projection (transducer pair signals) and projection angle. The matrix was transformed into a two dimensional BUA image using the iradon function in Matlab (Mathworks Inc).

Ultrasound frequency spectra for a) reference (non-irradiated gel) signal amplitude, b) sample (irradiated polymer gel) signal amplitude and c) attenuation are shown in figure 2. In this work, BUA values were obtained over a frequency range from 2.5 MHz to 6 MHz.

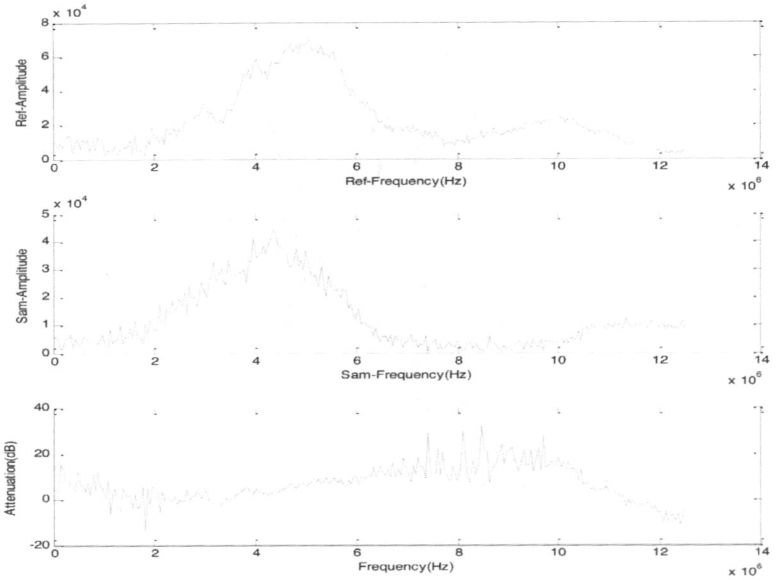

Figure 2: from top to the bottom: Ultrasound frequency spectra for a) reference (non-irradiated gel) signal amplitude, b) sample (irradiated polymer gel) signal amplitude and c) attenuation

3. Results and Discussions

Figure 3, shows an irradiated sample and reconstructed BUA map for one slice through one batch of gel, irradiated with a 2x2 cm square field at a specific dose rate of 30Gy. The image reconstruction procedure was repeated for each batch of the gel irradiated at different doses. From the reconstructed images, a consistent region of interest (located about the centre of the image) was specified, with average and standard deviation BUA values calculated.

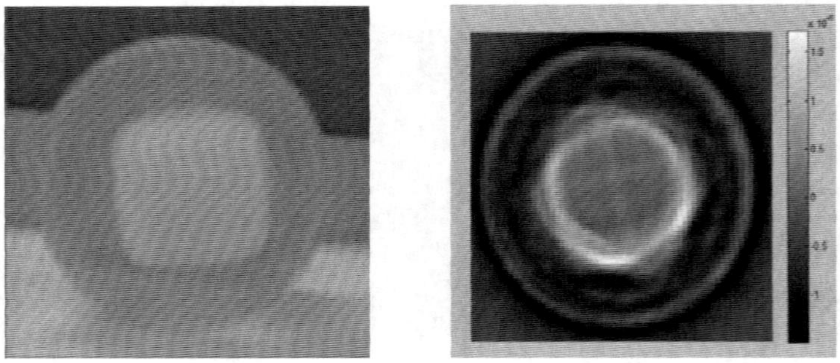

(a) (b)

Figure 3: (a) PAGAT gel sample irradiated with 2x2 cm square field at 30 Gy; (b) reconstructed BUA map

The variation of the BUA values with absorbed dose is shown in figure 4, noting that they show a dependence upon radiation dose. Further investigation is required of the dose dependency of BUA values at low doses.

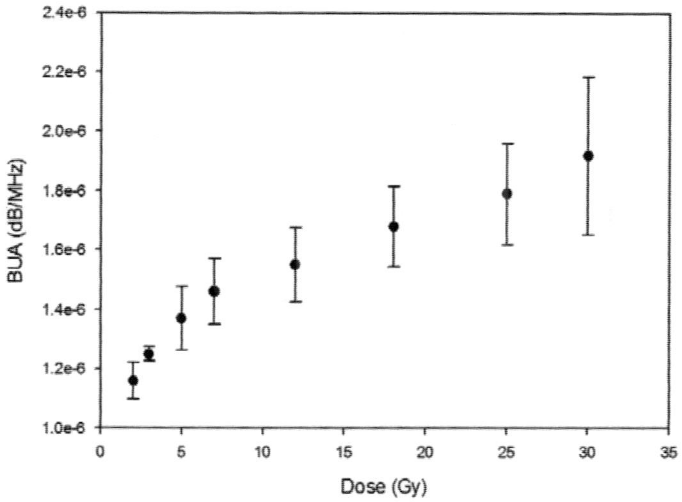

Figure 4: dose response (BUA vs Dose) for PAGAT gel scanned using USCT

4. Conclusion

BUA computed tomography has been shown to have potential for measurement of the dose response of PAGAT gel dosimeters. Further work is required to examine the non-linear response at doses below 10 Gy and to investigate other polymer gel compositions [13-15].

5. Acknowledgement

The authors would like to acknowledge the Institute of Health and Biomedical Innovation (IHBI) for their support on this project.

6. References

[1] Baldock C *et al* 2010 *Phys. Med. Biol.* **55** R1-63
[2] Mather M L *et al* 2003 *Phys. Med. Biol.* **48** N269-75
[3] Mather M L *et al* 2002 *Phys. Med. Biol.* **47** 1449-1458
[4] Mather M L *et al* 2003 *Ultrasonics* **41** 551-9
[5] Mather M L *et al* 2002 *Phys. Med. Biol.* **47** 4397-409
[6] Crescenti R A *et al* 2007 *Phys. Med. Biol.* **52** 6747-59
[7] Langton C M and Njeh C F 2008 *IEEE Trans Ultrason. Ferroelectr. Freq. Control* **55** 1546-54
[8] Venning A J *et al* 2005 *Phys. Med. Biol.* **50** 3875-88
[9] Brown S *et al* 2008 *Appl. Radiat. Isot.* **66** 1970-4
[10] Khoei S *et al* 2010 *J. Phys.: Conf. Ser.* **250** 012019
[11] Michael G J *et al* 2000 *Phys. Med. Biol.* **45** N133-8
[12] Trapp J V *et al* 2004 *Phys. Med. Biol.* **49** N139-46
[13] Gustavsson H *et al* 2004 *Phys. Med. Biol.* **49** 227-41
[14] Lepage M *et al* 2002 *Phys. Med. Biol.* **47** 1881-90
[15] Brindha S *et al* 2004 *Phys. Med. Biol.* **49** N353-61

A potential modification of the γ-evaluation: mapping dose disagreements using γ-vector fields

O Holmes[2], J Darko[1], T Olding[1] and LJ Schreiner[1,2]
[1]Cancer Centre of South Eastern Ontario at Kingston General Hospital, 25 King Street West, Kingston, ON, Canada, K7L5P9
[2]Department of Physics, Queen's University, Kingston, Ontario, Canada, K7L3N6

E-Mail: john.schreiner@krcc.on.ca

Abstract. This work describes the use of a motion phantom and 1D, 2D, and 3D ion chamber, EBT3 film, electronic portal imaging device (EPID) and FXG gel measurements for dosimetric validation of a stereotactic ablative radiation therapy (SBRT) technique in our clinic. Results show good agreement between the measurements and calculated treatment plan dose.

1. Introduction

Gamma evaluations [1] are a preferred clinical tool used in the validation of dose delivery in modern radiation therapy. The gamma comparison provides a quantitative comparison between dose-distributions, combining both dose difference and distance to agreement criteria. The gamma evaluations permit rapid analysis and visualization of the agreement between complex dose distributions.

A gamma comparison is performed between two dose maps: one distribution is the reference plan (typically from a treatment planning system) and the other is the evaluated distribution usually from a two or three-dimensional dose measuring system. The reference image is treated as the true distribution, while the evaluated image is analyzed for its agreement with the reference as follows: Every point in the reference image has a corresponding γ_r value; a measure of agreement at that location. Each possible point in the reference distribution, r_r, can be coupled with any point in the evaluated image, r_e. For all pairs a value Γ_r, defined by the vector difference between the points, exists. Tolerance criteria Δd_M and ΔD_M (e.g., 3mm and 3%) are used to normalize Γ_r along the distance and dose vector dimensions, correspondingly. The gamma index value, γ_r, is the smallest Γ_r that can be found considering the entire evaluated distribution. In the image space, $\Gamma_r^2=1$ describes an ellipsoid whose surface is defined by the tolerance criteria (figure 1). When $\gamma_r \leq 1$ the distributions agree within the stipulated tolerances. Conversely, when $\gamma_r>1$, no point in the evaluated distribution can be found within the ellipsoid and the dose distributions disagree at that location.

Although the γ comparison provides a useful measure of agreement between distributions when the index is less than one, the scalar gamma value provides little information into the clinical significance or source of disagreements of failing gamma values (i.e., when $\gamma>1$). Stock *et al* [2] previously presented the gamma angle as an indicator of the relative influence of the distance to agreement versus the dose difference on gamma. This angle represents only a portion of the available vector information. In this work, we present a modification to the gamma evaluation such that the complete gamma vector information for 2D comparisons is calculated. We will also present an initial analysis of the influence of various dose disagreements on the resulting vector field.

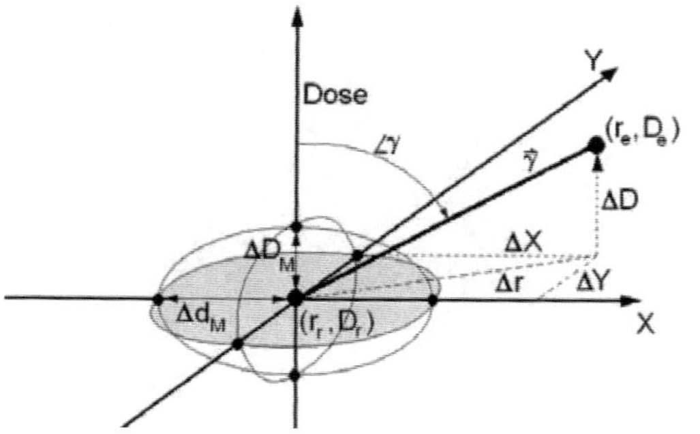

Figure 1: A graphical representation of gamma between a point on the reference distribution $r_r(r_r,D_r)$, and a point in the evaluated distribution $r_e(r_e,D_e)$. The figure illustrates the vector property of the γ evaluation.

2. Methods and Results

Gamma comparisons were performed using in-house software written in MATLAB (Mathworks, Newark, NJ). The tool calculates the magnitude of gamma, as usual, but also returns the corresponding vector information. The result for a 2D comparison is a 3D vector field comprised of two distance and one dose dimensions. Therefore, gamma vectors for 2D comparisons have three components, ΔX, ΔY and ΔD; each is normalized to the tolerance criteria (e.g., 3%3mm). The distance to agreement portion of gamma, Δr, is broken into its constituent spatial components ΔX and ΔY, while ΔD alone represents the dose difference contribution. The gamma algorithm used in this evaluation is similar to one presented previously [3]. However, in the current implementation neither interpolation nor geometric methods are employed to avoid the overestimation of gamma in regions of high dose gradient: a well known limitation that is caused by discrete, pixelized, representations of truly continuous dose distributions [3, 4].

The γ-vector evaluation tool was assayed on test images under a variety of manipulations. Each test, involved modifying the original distribution (to provide the evaluated distribution) and comparing it back to the original unmodified image (the reference). Initial tests involved the test distribution originally implemented by Low[1,4] (figures 2 and 3). Results will also be shown of tests in which the evaluated image was uniformly shifted along each of the three image dimensions one at a time (dose is the third dimension). The net effect of each misalignment was measured by calculating the mean value of each gamma component over the entire γ-vector map. Vector gamma analysis of simulated dose errors (in the form of double Gaussians doses added into a head and neck distribution at separate locations corresponding to the surrounding tissue, a target, an avoidance structure, and the target boundary area will be reported to provide more clinical examples (figure 4).

3. Results and Discussion

Our implementation of γ provides magnitude data consistent with the original test (figure 2). For tests in which the perturbation of the initial reference distribution was generated by a spatial misalignment, the spatial gamma component, Δr, clearly dominates the dose component, ΔD. The different response of the two distributions under identical manipulations indicates that exact behaviour of the resulting vector field depends on the unique properties of each distribution. Nevertheless, the response of the vector field shows the same trends regardless of the image under investigation.

In figure 4, there are two small regions where ΔD is negative which was unexpected because the disagreement was due to a known overdose. A closer examination of the reference image revealed that the unexpected negative regions were caused by variations in the local dose gradient. Therefore, a weakness in our current implementation of gamma is the sensitivity to dose gradients. A more robust implementation of gamma using geometric or interpolation methods may obviate this effect altogether.

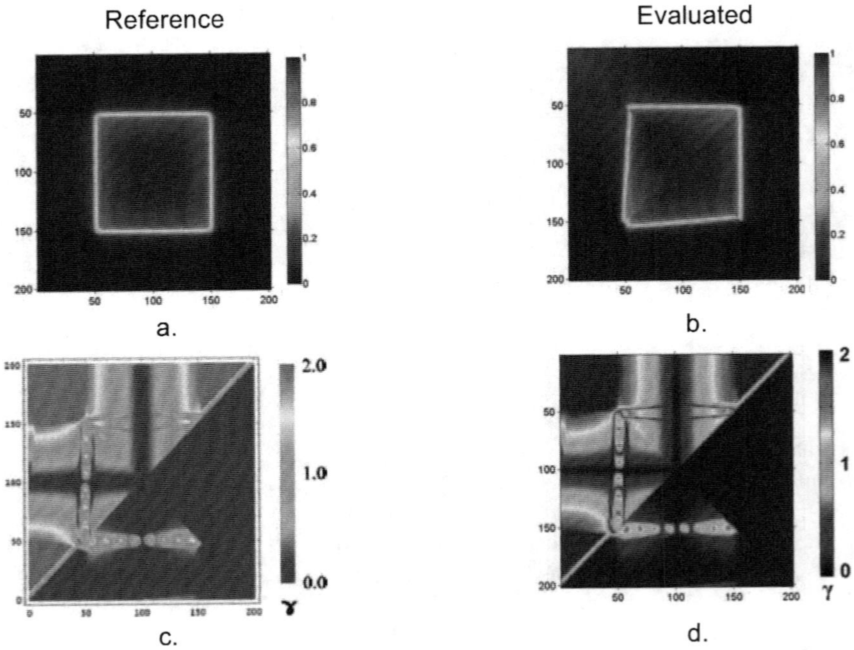

Figure 2: Gamma comparisons between test distributions used by Low and Dempsey (2003). (a) the reference distribution. (b) the evaluated distribution. (c) gamma magnitude distribution calculated with the original γ-tool (reproduced with permission[5]). (d) gamma magnitude distribution calculated with the new in-house γ-tool.

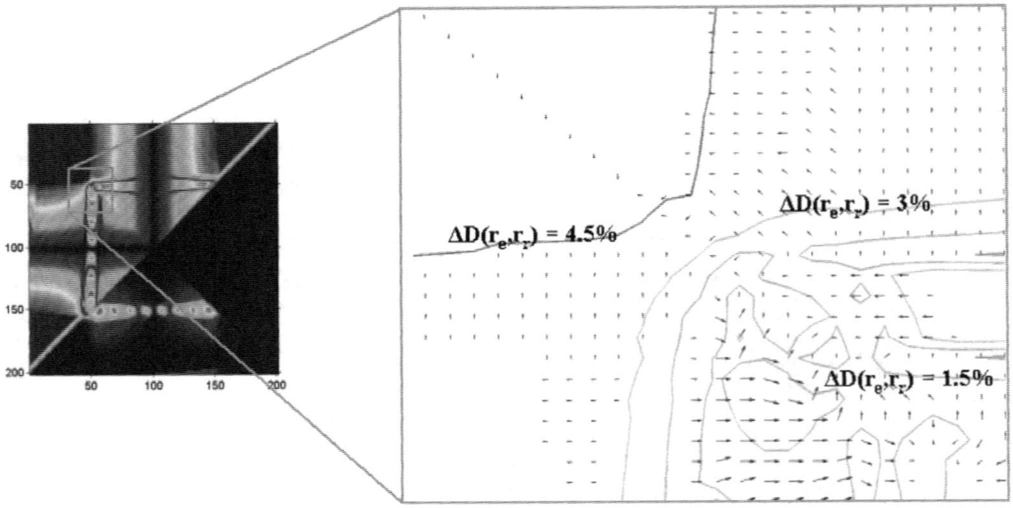

Figure 3: The γ-vector evaluation yields the same magnitude plot (Left), as well as the corresponding γ-vector information in component form (Right). The vector information shown on the right corresponds to the rectangular ROI on the top left corner of magnitude plot. The voids in the vector plot occur when the spatial components of γ are zero in magnitude..

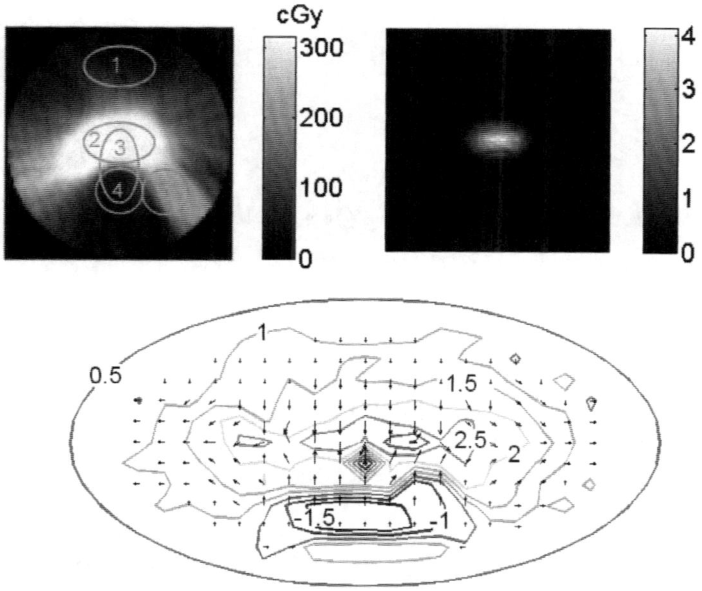

Figure 4: Top Left: A clinical head and neck distribution (simulated using in-house planning software) with five regions of interest highlighted. Top Right: When a double Gaussian 15% overdose is added to region 2 in the evaluated image a typical 2D gamma plot shows only the magnitude of the gamma vector. The enlarged gamma vector plot at the bottom corresponding to the 15% overdose in region 2 is shown above. The arrows represent the ΔX and ΔY contribution while dose component, ΔD, is indicated by the contour lines.

4. Conclusions

The gamma vector field contains information that could be useful in identifying the cause and significance of dose disagreements between plan and delivery that fail the acceptance criteria. The response of the vector field depends on properties distinctive to each distribution, such as the local dose gradient. Further analysis is underway to improve the implementation.

5. Acknowledgements

Research funding for this work has been provided by the Canadian Institutes of Health Research (CIHR) and the Ontario Consortium for Adaptive Interventions in Oncology (OCAIRO) through the Ontario Research Excellence Program. We thank Dr Low for providing the initial test distributions from his work.

6. References

[1] Low D A *et al* 1998 *Med. Phys.* **25** 656-61
[2] Stock M *et al* 2005 *Phys. Med. Biol.* **50** 399-411
[3] Wendling M *et al* 2007 *Med. Phys.* **34** 1647-54
[4] Ju T *et al* 2008 *Med. Phys.* **35** 879-87

Real-world examples of sensitivity failures of the 3%/3mm pass rate metric and published action levels when used in IMRT/VMAT system commissioning

B Nelms[1], G Jarry[2], M Chan[3], C Hampton[4], Y Watanabe[5] and V Feygelman[6]
[1]Canis Lupus LLC, Merrimac, Wisconsin 53561, USA
[2]L'Hopital Maissoneuve-Rosemont, Montreal, Quebec H1T 2M4, Canada
[3]Memorial Sloan-Kettering Cancer Center, Basking Ridge, New Jersey 07920, USA
[4]Wake Forest Baptist Medical Center, Winston-Salem, North Carolina 27157, USA
[5]University of Minnesota, Minneapolis, Minnesota 55455, USA
[6]Department of Radiation Oncology, Moffitt Cancer Center, Tampa, FL 33612, USA

E-mail: alpha@canislupusllc.com

Abstract. In IMRT/VMAT system commissioning (as with any system), quality is improved by striving for tight tolerances of stringent metrics of accuracy. For 5 cases, passing rates for 3%/3mm gamma analysis were generated following the TG119 instructions. Subsequently, more stringent/sensitive criteria combined with advanced volumetric dose analysis were applied, and in each case significant systematic errors were clearly identified despite the high 3%/3mm passing rates. In 4 of 5 cases, the error was easily remedied. These real-world examples of observed "false negatives" (insensitivities) point towards the inappropriateness of the 3%/3mm gamma passing rate metric as the basis for acceptance testing/commissioning of the IMRT/VMAT delivery chain.

1. Introduction

The AAPM TG-119 report [1] focuses on the commissioning of IMRT systems. The 3%/3mm gamma analysis passing rate metric is used as the basis for proposed action levels of 90% (per-beam) or 88-90% (composite dose) when comparing measured and calculated doses. The same 3%/3mm passing rate metric was found in a 2007 survey to be the metric most frequently used [2]. Recently, the sensitivity and specificity of this ubiquitous 3%/3mm gamma metric have been challenged [3-6]. However, these reports were focused on per-patient dose QA, which was not the topic of TG119. Although it is intuitive that a poor metric for per-patient dose QA would also have limited power in commissioning, this connection has not been directly stated and, as such, 3%/3mm continues to be a metric-of-choice in system commissioning.

Quality leaders such as Deming, Shewhart, and Taguchi stressed that driving out variation and conforming to tight tolerances improves the quality of a system more than voluminous inspection of the system's outputs [7-9]. It is vital to use sensitive metrics that identify and help to root out systematic errors and quantify, and potentially minimize, random errors in the system. An apt question is: as applied to the TPS commissioning, does the popular 3%/3mm passing rate metric meet these needs? In this work, we present real-world examples of important and clinically-relevant systematic errors that fail to be detected by the 3%/3mm passing rate-based action levels described in TG-119.

2. Methods

2.1. Case study summary and conventional 3%/3mm passing rate analyses
Passing rates for 3%/3mm gamma analysis were generated using exactly the TG-119 instructions [10] for five example cases in Table 1 (Cases 1-3: IMRT, 4: VMAT, 5: an open field). For cases 1 and 3, the MapCHECK diode array [11] (Sun Nuclear, USA) was used for per-beam planar measurements, normal to the beam CAX. For Case 2, the measurement method was EPIDose [12-13] (Sun Nuclear) with the same TG-119 criteria as MapCHECK but without "measurement uncertainty" applied. Cases 4 and 5 were measured with the ArcCHECK quasi-3D phantom [14-16] (Sun Nuclear) using the criteria specified for the 2D diode array.

Table 1: Case study summary, including plan details and 3%/3mm gamma passing rates using TG119 instructions [10]. For Cases 1-3, passing rates are the average of per-beam planar passing rates. Cases 4 and 5 are based on composite dose.

Case ID	Plan Summary	3%/3mm Passing Rate (%)	Measurement Type/ Dosimeter
Case 1	7-Field IMRT (SMLC) Head/Neck Boost 6 MV [Pinnacle TPS, Elekta 6MV 80-leaf (10 mm)]	**99.2**	Per-Beam/ 2D Diode Array
Case 2	7-Field IMRT (DMLC) Head/Neck 6 MV [MSKCC TPS, Varian 120-leaf]	**99.4**	Per-Beam/ EPID-based
Case 3	7-Field IMRT (DMLC) Head/Neck 6 MV [MSKCC TPS, Varian 120-leaf]	**99.7**	Per-Beam /2D Diode Array
Case 4	1-Arc VMAT Head/Neck 6 MV [Monaco TPS, Elekta 80-leaf (4 mm)]	**96.8**	Composite/ 3D Diode Array
Case 5	10x10 Open field 10 MV [Pinnacle TPS, Elekta 80-leaf (10 mm)]	**97.6**	Composite /3D Diode Array

2.2. Advanced diagnostics and identification of errors
In all cases, more stringent/sensitive criteria such as reducing tolerances, turning off "measurement uncertainty" padding, and using local rather than global % dose-difference were applied to probe for error patterns and regions of high local dose error. In addition, in all cases 3D patient measurement-guided dose reconstruction (MGDR) [5,17,18] was performed using 3DVH software (Sun Nuclear) to estimate volumetric dose and DVH differences, accompanied in some cases by film and/or ion chamber measurements for verification. In all cases, the errors were verified as systematic errors of noticeable impact and in 4 of 5 cases (1-3, 5) they were readily remedied.

3. Results and discussion

3.1. Case 1
Despite a high 3%/3mm average passing rate (99.2%), there was a systematic error in the TPS model of the rounded MLC leaf ends. The calculated penumbra at the MLC edges was too wide compared to measurements, resulting in TPS dose being overestimated, in particular along gradients surrounding the target. Using MGDR, the compilation of these errors were reconstructed in the 3D patient dose, highlighting the impact on patient dose and DVH, which turned out to be 3-4% lower than calculated by the TPS (e.g. D95% (Gy) of CTV was in error by 3.8%, TPS being too high). The problem was easily fixed by adjusting the leaf offset table.

3.2. Case 2
Despite a high 3%/3mm average passing rate (99.4%), there was a systematic error in the TPS: the tongue-and-groove (TnG) correction was turned off. The TPS was overestimating dose between the leaves (Figure 1a), but this was essentially hidden in the 3mm distance criterion. The impact on the

patient target volume D95% (Gy) values ranged from 2 to 4% (Figure 1b). Once the TPS TnG model was applied correctly, the differences were essentially eliminated.

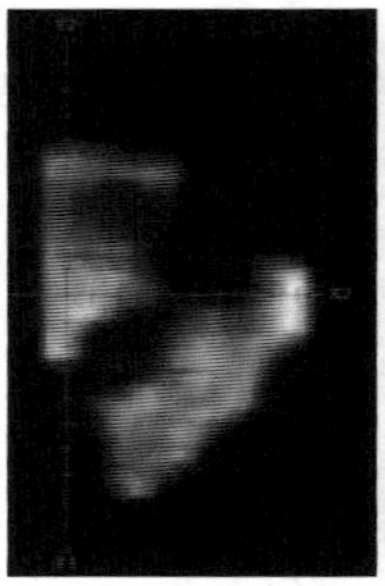

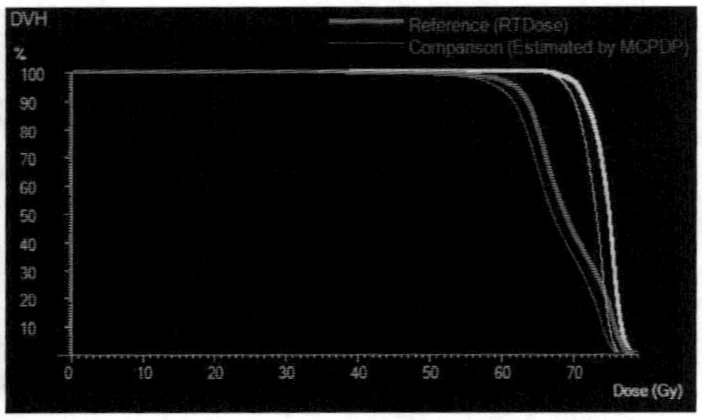

Figure 1a (Left). Measured – Calculated dose difference map, clearly indicating the TPS not correctly accounting for tongue-and-groove effects.

Figure 1b (Right). 99.4% average passing rate hides a systematic error in the TPS tongue-and-groove model which, when properly corrected, results in lower D95% (Gy) dose to the PTV70 (-2.4%) and PTV59.4 (-4.1%).

3.3. Case 3

Despite a high 3%/3mm average passing rate (99.7%), there was a systematic error in the position of the X1 jaw, which was the limiting aperture at the PTV inferior border. The actual X1 position was about 1-2 mm less than indicated, reducing the delivered dose. The MGDR showed a local difference at the inferior target border where the errors accumulated (Figure 2a). Again, the X1 readout error was clearly too small to be detected by a 3mm distance criterion in the gamma index.

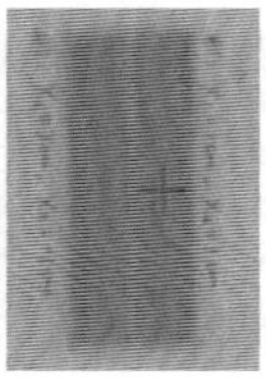

Figure 2a (Left). Measured – Calculated dose difference map using MGDR as measurement shows region of accumulation of low dose at inferior dose border due to X1 impingement.

Figure 2b (Right). Film analysis verifies an error of ~ 1 mm in the X1 jaw position.

3.4. Case 4

Despite a 3%/3mm passing rate of 96.8% (Figure 3a), there were systematic errors in the TPS dose calculation: 1) improper rendering of high gradient dose, and 2) over-estimation of dose under MLC leaves. These errors were hidden by the dose and distance methods prescribed by TG119, but clearly visible as a pattern if more stringent criteria (2% local/2mm) are used (Figure 3b) and when reconstructed in 3D using MGDR (Figure 3c).

Figure 3a: 96.8% passing rate hides a systematic error in the TPS dose calculation in high gradient regions (peaks, valleys, edges).

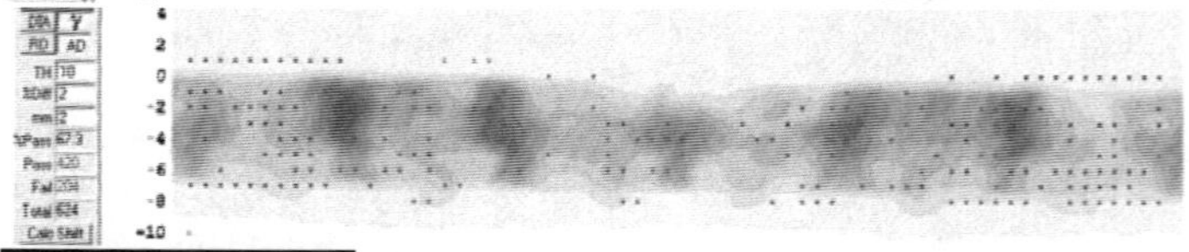

Figure 3b: The same data analyzed with more stringent 2% (local)/2mm, clearly showing the TPS dose error patterns in peaks, valleys, and edges.

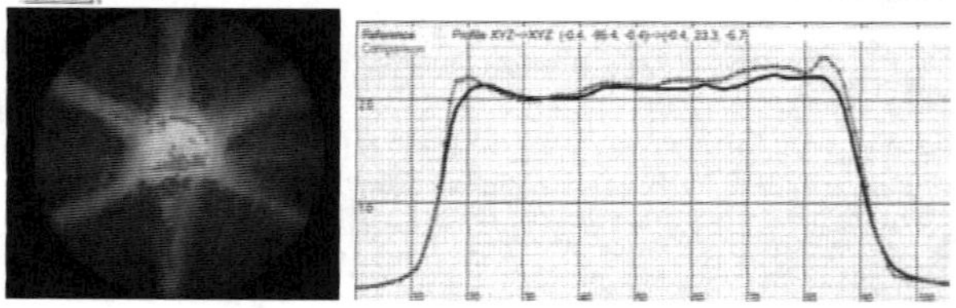

Figure 3c: MGDR reconstructed dose showing impact of TPS errors in a patient dose axial image (left) and along a sup/inf dose profile (right). Local dose errors in some small regions are in excess of 10%.

3.5. Case 5

Despite a high 3%/3mm passing rate (97.6%), there was a systematic error in how the virtual phantom was modelled in the TPS. The first sign was the 3.8% disagreement between the 0.6cc ion chamber measurement and TPS calculation at isocenter. When examining the exit diodes of the phantom (~23.7 cm deep), the TPS calculation error was almost +7%. 3D MGDR dose confirmed the overestimation of the TPS dose due to improper phantom representation in the TPS. The conventional 3% analysis failed to detect the high TPS error because the global % normalization method uses the maximum measured dose as the denominator, and therefore the exit diode dose, which is less than 0.3 of the entrance, is given an effective absolute dose error tolerance of almost 10% (not 3%). The large local error deep in the phantom was thus completely hidden. See Figures 4a and 4b for the error maps using TG-119 instructions vs. using 2% local/2mm analyses. The source of the error in this case was that instead of appropriately modelling the phantom as a uniform PMMA cylinder [16], a CT-based heterogeneity correction was applied. This resulted in an average electron-density value closer to 1.1 relative to water, instead of the required actual electron density of PMMA, 1.147. Once this was corrected, the TPS calculations became much more accurate, agreeing with both the ion chamber and exit diodes to within 1% (local error).

Figure 4a: 97.6% passing rate hides a systematic error, causing large TPS calculation errors go unnoticed.

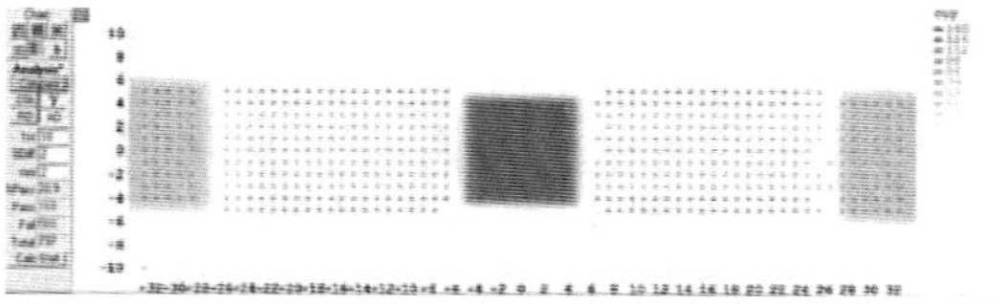

Figure 4b: The same data analyzed with more stringent 2% (local)/2mm criteria, where the results clearly show the TPS dose errors. The blue dots indicate the exit dose errors.

4. Conclusions

In IMRT/VMAT system commissioning (as with any system), quality is improved by the proven strategy of striving for tight tolerances of stringent metrics of accuracy. However, the common 3%/3mm gamma passing rate is insensitive (at any action level) when it comes to detecting systematic errors. A collection of observed "false negatives" are presented here as real-world examples of the inappropriateness of the metric if used as the basis for acceptance testing/commissioning. Overreliance on this *de-facto* current standard inhibits detection and mitigation of systematic errors that can become pervasive in the system, and thus imposes direct and indirect costs (Taguchi Loss Function).

5. References

[1] Ezzell G A *et al* 2009 *Med. Phys.* **36** 5359-73
[2] Nelms B E and Simon J A 2007 *J. Appl. Clin. Med. Phys.* **8** 76-90
[3] Kruse J J 2010 *Med. Phys.* **37** 2516-24
[4] Nelms B E *et al* 2011 *Med. Phys.* **38** 1037-44
[5] Zhen H *et al* 2011 *Med. Phys.* **38** 5477-89
[6] Carrasco P et al 2012 *Med. Phys.* (In press)
[7] Deming W E 1986 Out of the Crisis (*The MIT Press*)
[8] Aguayo R 1991 Dr. Deming: The American who taught the Japanese about quality (*Simon and Schuster*)
[9] Shewhart W A 1939 Statistical method from the viewpoint of quality control (*The Graduate School, Department of Agriculture. Washington DC*)
[10] Ezzell G A *et al* 2009 *Med. Phys.* **36** 5359-73
[11] Jursinic P A and Nelms B E 2003 *Med. Phys.* **30** 870-79
[12] Nelms B E *et al* 2010 *J. Appl. Clin. Med. Phys.* **11** 140-57
[13] Bailey D W *et al* 2012 *J. Appl. Clin. Med. Phys.* **4** 82-99
[14] Letourneau D *et al* 2009 *Med. Phys.* **36** 1813-21
[15] Feygelman V *et al* 2011 *J. Appl. Clin. Med. Phys.* **12** 146-67

[16] Kozelka J *et al* 2011 *Med. Phys.* **38** 5021-32
[17] Olch A J 2012 *Med. Phys.* **39** 81-6
[18] Nelms B E *et al* 2012 *Med. Phys.* **39** 4228-38

7th International Conference on 3D Radiation Dosimetry (IC3DDose) IOP Publishing
Journal of Physics: Conference Series **444** (2013) 012087 doi:10.1088/1742-6596/444/1/012087

A software tool for 3D dose verification and analysis

M Al Sa'd[1], J Graham[1] and GP Liney[2]
[1]Imaging Sciences, School of Cancer and Enabling Sciences, Manchester Academic Health Sciences Centre, Stopford Building - University of Manchester, Oxford Road, Manchester, M13 9PT, UK
[2]Radiation Physics Department, Queen's Centre for Oncology & Haematology - Castle Hill Hospital, Cottingham, East Yorkshire, HU16 5JQ, UK

E-mail: mohammad.alsad@manchester.ac.uk

Abstract. The main recent developments in radiotherapy have focused on improved treatment techniques in order to generate further significant improvements in patient prognosis. There is now an internationally recognised need to improve 3D verification of highly conformal radiotherapy treatments. This is because of the very high dose gradients used in modern treatment techniques, which can result in a small error in the spatial dose distribution leading to a serious complication. In order to gain the full benefits of using 3D dosimetric technologies (such as gel dosimetry), it is vital to use 3D evaluation methods and algorithms. We present in this paper a software solution that provides a comprehensive 3D dose evaluation and analysis. The software is applied to gel dosimetry, which is based on magnetic resonance imaging (MRI) as a read-out method. The software can also be used to compare any two dose distributions, such as two distributions planned using different methods of treatment planning systems, or different dose calculation algorithms.

1. Introduction

Advanced radiotherapy technologies, such as intensity-modulated radiotherapy (IMRT) and volumetric modulated arc therapy (VMAT), can provide considerable improvements to the result of radiotherapy both in terms of maximising the therapeutic effect of dose distribution on tumour, and minimising its damaging effect on surrounding healthy tissues and organs at risk (OAR). The increasing complexity of irradiation techniques has driven the development and adoption of 3D dosimetery methods, in order to optimise treatment planning and delivery systems, as well as to quality-assure their functionality. The adoption of 3D dosimetry methods has been increasing over the last decade [1, 3]. However, software applications (both freeware and commercial) that are used for dose evaluation and quality assurance (QA) purposes are primarily based on 2D evaluation methods. These 2D evaluation methods are prone to error in evaluating the accuracy of a particular dose distribution, mainly because of the mismatch that can happen in selecting the corresponding slices from the dose distribution volumes being compared.

In principle, QA based on 3D verification is assumed to provide more quality indicators for further analysis. Also, it makes it possible to define tolerance criteria in 3D in order to account for setup inaccuracies of the dosimeter phantom and/or detector. In order to derive the full benefits of using the 3D dosimeter, it is essential to use a software tool that provides analysis and evaluation results based on 3D methods and techniques. In fact, there is no software solution that provides comprehensive 3D dose evaluation and analysis. In this paper, we present a software suite that covers a wide range of 3D dose evaluation techniques. We have particularly applied the software to gel dosimetry, based on magnetic resonance imaging (MRI) as a read-out method [4]. In addition to comparing the measured and calculated dose distributions, the software can also be used to compare plans produced using different methods such as commercial treatment planning system (TPS) or Monte Carlo (MC) algorithms. The software has been evaluated using datasets of different radiotherapy plans and MRI gel dosimeter scans.

2. Materials and Methods

The software tool presented here was produced using the MATLAB® computing language and interactive environment (version R2011a), which provides convenient and flexible high-level language and advanced graphical capabilities including 3D rendering. Also, the C programming language was used along with OpenMP API in order to optimise the speed of complex computational processes. The analysis is presented in a friendly user interface, which allows manipulation of the settings of each type of analysis. The software accepts different data formats as an input for the analysis, including DICOM and Analyze 7.5. The tool was designed to meet the analysis requirements of MRI gel dosimetry, such as calculating R2 rate data (which is proportional to the absorbed dose), and applying calibration data to produce absolute dose values.

The software tool automatically detects and calculates the 3D deviation between the reference and evaluated dose distributions by using volume registration [5-7]. The user is informed about inaccuracies arising from sources of error such as misplacements of the dosimeter during radiation delivery or read-out stage. The user can choose whether to account for this deviation in the comparison calculations. Together with the 3D analysis methods, the software tool also provides some analysis in 2D so that the 3D evaluation methods can be compared to the more conventional 2D forms.

2.1. *Evaluation methods used*

The software provides both qualitative and quantitative analysis. The qualitative analysis includes various types of volume visualisation methods offered by MATLAB. The quantitative analysis includes the following: dose volume histograms (DVH), absolute dose difference, relative dose difference (either globally relative to a specific dose value or locally relative to the dose at each reference point), absolute spatial difference between each reference point and the closest point (of the same dose value) in the evaluated dataset, distance-to-agreement (DTA) test (whereby a spatial tolerance is used as a pass/fail criterion), gamma evaluation (which combines a DTA criterion with a dose difference criterion through a composite analysis) [8], gamma volume histograms [9], and gamma-angle analysis (which indicates which criterion had more influenced the calculated gamma value at each reference point) [10].

2.2. *Comparison datasets*

Three reference/evaluation 3D sample pairs were compared using the software in this paper. Sample A is a standard uniform intensity conformal treatment plan which was delivered to two MRI gel dosimeter phantoms; one was stationary during the irradiation as a reference distribution, and the other was moving to simulate human respiration whilst being irradiated at full inhalation using the respiratory gated radiotherapy technique (RGRT). Sample B is for an IMRT head and neck case, where the reference distribution was measured using MRI gel dosimetry in order to evaluate its corresponding TPS plan. Sample C is for another IMRT head and neck case with an MC calculated reference plan and an evaluated TPS plan, wherein there was no experimental uncertainty involved. All the samples share the same size of 256 mm in each direction and a voxel resolution of 1 mm, which forms cubic datasets of 256^3.

3. Results

For the entire 256^3 points and using a PC equipped with Intel i7 processor, the average calculation speed for the 3D gamma was less than 1.2 seconds. The screenshot in figure 1 shows the 3D deviation map between reference and evaluated dose distributions from sample A. For sample B an average 3D deviation of ~6mm was detected by volume registration, which may have been introduced by inaccurate positioning of the gel phantom in irradiation or read-out phases. This spatial error invalidates the entire principle of 2D evaluation, which is based on comparing the corresponding slices of the two volumes and stacking up the 2D results into a 2.5D volume. With the option to account for the spatial uncertainties selected, the proportion of points passing a 3% dose difference criterion and a 3mm DTA acceptance criterion were 88.23% and 95.71%, for the 2.5D and 3D gamma calculations, respectively. Figure 2 shows the overlapped 3D render of the planned and evaluated dose distributions from sample B. For sample C, there was no 3D deviation detected, because both datasets are for calculated plans, which did not involve experimental uncertainties. The proportion of points passing a 3% dose difference criterion and a 3mm DTA acceptance criterion were 83.44% and 98.64%, for the 2.5D and 3D gamma calculations, respectively.

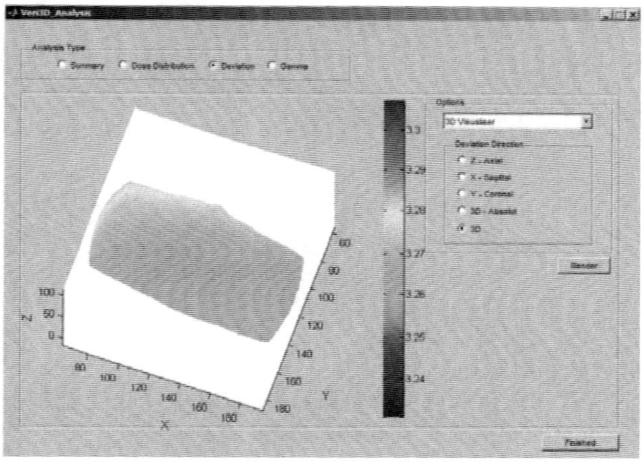

Figure 1: A screenshot showing a 3D deviation map between the two dose distributions in sample A.

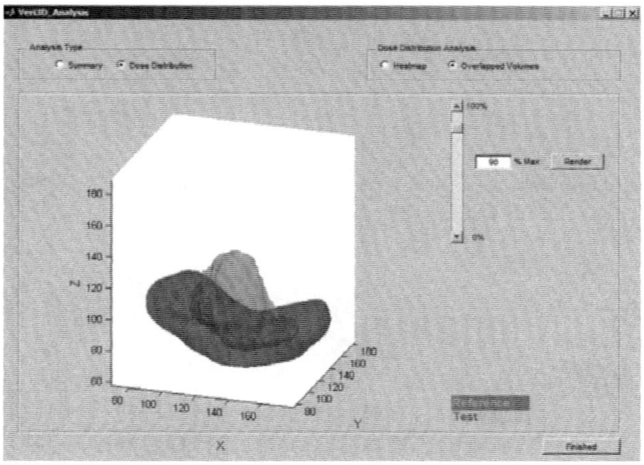

Figure 2: A screenshot for the analysis of Sample B, showing an overlay volume rendering for the reference and evaluated dose distributions at the 90% isodose surface.

4. Discussion

We present in this paper a software tool for 3D dose evaluation. In addition to 3D volume rendering for dose distributions being compared and analysis results, the software provides a catalogue of dose evaluation methods that are based on three-dimensional calculations and analysis. The settings of various analysis methods can be manipulated via a friendly graphical user interface, which allows the user to interactively examine the results of any changes in processing parameters. While the main application of the software would be to quantify the absolute accuracy of MRI gel dosimetry for planning verification, it also can be used to compare any two dose distributions. Moreover, it is planned to integrate the algorithms needed to process data obtained using other read-out techniques (such as optical CT) in future.

Without a true 3D evaluation analysis it becomes impossible to really determine and quantify the expected accuracy of gel dosimetry as a technique. It is anticipated that if this software is accepted routinely then it would become invaluable in routine QA checks. The analysis using the software to compare dose distributions which ought to be identical showed that the proportion of points passing the DTA and dose difference criteria is higher using the 3D evaluation methods than with 2.5 D analysis. This demonstrates that extending the search to points in the 3D space, rather than just in the 2D space, enhances the chance of passing the evaluation criteria. It also shows that the 3D evaluation methods account for the small movements and setup error; therefore, they produce more reliable evaluation results than the 2D evaluation methods.

5. References

[1] Baldock C *et al* 2010 *Phys. Med. Biol.* **55** R1-63
[2] Mans A *et al* 2010 3D *Radiother. Oncol.* **94** 181-7
[3] Murry P and Baldock C 2000 *Australas. Phys. Eng. Sci. Med.* **23** 44-51
[4] Liney G P *et al* 2003 *J. Radiother. Pract.* **3** 123-9
[5] Thirion J P 1998 *Med. Image Anal.* **2** 243-60
[6] Rueckert D *et al* 1999 *IEEE Trans. Med. Imaging.* **18** 712-21
[7] Kroon D J and Slump C H 2009 *IEEE International Symposium on Biomedical Imaging: From Nano to Macro* 963-6
[8] Low D A 2010 *J. Phys.: Conf. Ser.* **250** 012071
[9] Spezi E and Lewis D G 2006 *Radiother. Oncol.* **79** 224-30
[10] Stock M *et al* 2005 *Phys. Med. Biol.* **50** 399-411

Polymer gel dosimetry utilizing a 2D (SE) and a 2D (HASTE) multiple echo sequences

M-V Papoutsaki[1], E Pappas[2], AE Papadakis[3], C Varveris[4], J Damilakis[1] and TG Maris[1]
[1]Department of Medical Physics, University of Crete, Medical School, Staurakia 71110, Crete, Greece
[2]Department of Medical Radiological Technologists, Technological Educational Institute of Athens, Egaleo, 12210, Athens, Greece
[3]Department of Medical Physics, University Hospital of Heraklion, Staurakia, 711110, Crete, Greece
[4]Department of Radiation Therapy, University of Crete, Medical School, Staurakia, 711110, Crete, Greece

E-mail: vasp3479@yahoo.gr

Abstract. Two pulse sequences were used for the estimation of dosimetric characteristics of VIPET polymer gels. The first one, multi- echo spin echo (MESE) is the well-established method for T2 measurements. The other method is a new multi-echo single shot turbo spin echo pulse sequence, MEHASTE that reduces the acquisition time significantly. Both techniques showed a linear R2-dose response. With MESE sequence, the dose sensitivity was slightly enhanced as compared to MEHASTE. The linear portion of the R2-dose curve was restricted using the MEHASTE sequence. For doses above 7 Gy both methods fulfill the 2% ICRU criterion limit for dose resolution estimations (95% confidence level). Finally, for a time period of one month the temporal stability of R2-dose response was maintained stable utilizing both MESE and MEHASTE pulse sequences. MEHASTE serves as an excellent means for fast 3D polymer gel dosimetry.

1. Introduction

Precision and accuracy are crucial factors for polymer gel dosimetry applications in clinical practice. The overall polymer gel dosimetry procedure consists of: a) the preparation of a certain chemical composition polymer gel, b) the irradiation, c) the readout process and d) the mathematical analysis method used for the estimation of the final irradiation doses [1]. Nowadays, different imaging modalities applied for the readout and analysis processes are used. These include Magnetic Resonance Imaging (MRI) [2], Optical Computed Tomography (opt-CT) [3], x-ray CT [4] and Ultrasound [5]. Amongst them, MRI is an established methodology as a readout process in polymer gel dosimetry [6]. Several research groups used different pulse sequences for the acquisition of T2 measurements [1]. De Deene et al [7] was the first who introduced the multiple echo (32 echoes) spin echo (MESE) sequence. Utilizing this methodology, the uncertainty of the dose determination related to the pulse sequence was minimized. As a consequence, the last ten years MESE pulse sequence is well-established in MRI polymer gel dosimetry [8, 9]. The main drawback of this pulse sequence is the relatively long acquisition time. The aim of this work is to present a new multi-echo single shot turbo spin echo sequence (MEHASTE) for MRI polymer gel dosimetry and to compare the results with the standard (MESE) sequence utilizing normoxic N-vinylpyrrolidone based polymer gels [10].

2. Materials and methods

2.1. Polymer gel manufacture

300 mL normoxic N-vinylpyrrolidone based polymer gel, VIPET [10] was manufactured, containing 7% w/w gelatin, 4% w/w N-vinylpyrrolidone, 4% w/w N,N'-methylenebisacrylamide (8%T, 50% C) and 5 mM THPC. The polymer gel solution was prepared under normal atmospheric conditions inside a laminar flow hood.

Initially, the gelatin was added to the double distilled water and left to dissolved. Then the solution was heated to 50^0C using a hot magnetic stirrer. The temperature was stabilized and the bis was added. When the bis had fully dissolved, the solution was cooled down to approximately 32 ^{0}C and the N-vinylpyrrolidone was added. Finally, when all the constituents were completely dissolved, the THPC was added. The solution was filled in three 100 mL cylindrical vials and closed using the appropriate caps and Parafilm. The gel phantoms were stored at 20^0C ambient temperature for 24h in a cool, dark and dry place to solidify before irradiation.

2.2. Polymer gel irradiation

The VIPET dosimeters were irradiated in a Primus Clinical Accelerator (Primus, LINAC, Siemens, Germany) using a 6MV photon beam, inside an in-house build solid water structure. Each vial was irradiated at 6 different areas (perpendicular to their length) using different irradiation doses. The delivered doses covered a range from 0.5 Gy up to 60 Gy. The field size of each irradiation was 2x4 cm^2, the source sample distance was 100 cm and the dose rate was 3 Gy/min. A water phantom and an ion chamber were utilized for the accomplishment of the similar experiments and these results were considered as a dosimetric reference.

2.3. MRI measurements

All polymer gels were scanned using a 1.5T whole body MR Imager (Vision/ Sonata, Siemens, Germany). T2 measurements were produced employing two different MR pulse sequences at one day, one week, two weeks and one month post-irradiation. Firstly, a 2D non space filling multi-slice- multi echo (32 echoes) spin echo (MESE) sequence was utilized for the calculation of the parametric T2 maps. MESE sequence parameters were: TR=9000 ms, TE=40-1280ms in symmetric TE intervals of 40 ms (echo train), slice thickness=4mm, number of slices= 5, interslice interval (gap) = 4 mm, FOV= 250x156 mm^2, matrix size= 256x256, NEX=1, ETL=1, scan time=45 min. Secondly, T2 parametric maps were also calculated utilizing a 2D space filling multi-slice- multi echo (4 echoes) Half-Fourier Acquisition Single-Shot Turbo Spin Echo (MEHASTE) sequence. MEHASTE sequence parameters were: TR=2500 ms, TE=36-1230ms in asymmetric TE intervals (36, 436, 835, 1230 ms), slice thickness=2mm, number of slices= 25, interslice interval (gap) = 0 mm, FOV= 169x270 mm^2, matrix size= 256x160, NEX=16, ETL= 160, scan time=15 min. A restore RF pre-pulse was incorporated in the MEHASTE sequence, thus enabling the significant reduction of the TR interval. T2 parametric maps were calculated for both sequences utilizing a weighted linear fitting regression algorithm.

3. Results

Dose responses of VIPET gels using both sequences at one day post-irradiation are presented in figure 1. For MESE, VIPET exhibits a linear dose response for doses from 0.5 up to 50 Gy. However, for MEHASTE, VIPET exhibits a linear dose response in a slightly reduced dose region from 10 to 50 Gy.

A linear fit of the form $R2\ (D) = \alpha D + R_0$ was applied to R2-dose data at the common linear dose-response region. Dose sensitivity and offset were found to be: i) $\alpha = (0.048 \pm 0.001)\ Gy^{-1}s^{-1}$ and $R_0 =$

(1.614 ±0.050) s^{-1} for MESE and ii) α= (0.035 ±0.050) $Gy^{-1}s^{-1}$ and R_0= (1.650 ±0.037) s^{-1} for MEHASTE respectively.

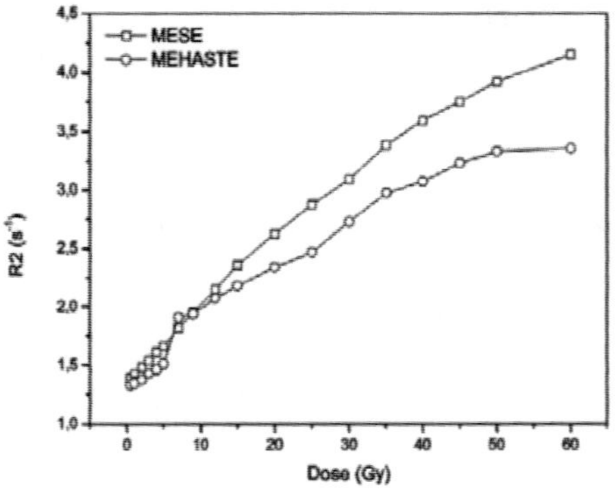

Figure1: Dose response of VIPET at one day post-irradiation using MESE and MEHASTE sequences.

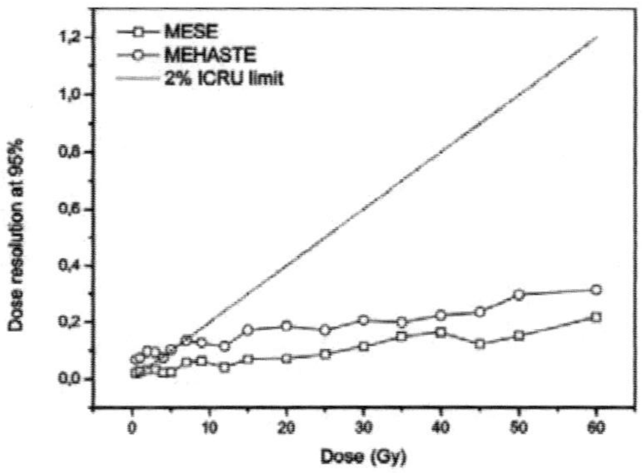

Figure2: Dose resolution at 95% confidence level of VIPET at one day post-irradiation using MESE and MEHASTE sequences, including the 2% ICRU limit.

Dose resolution at 95% confidence level at one day post-irradiation utilizing both sequences is presented in figure 2. For MEHASTE, the 2% ICRU limit can be satisfied for doses above 7 Gy, while for MESE above 2 Gy. Dose resolution for MEHASTE sequence ranged from 0.07 to 0.3 Gy and for MESE ranged from 0.02 to 0.2 Gy.

In addition, figures 3 and 4 represent the temporal stability of the R2-dose response at a time period of one month for both sequences. The linear R2-dose response of VIPET gels maintained stable for a time period of one month in both cases. Particularly, the sensitivity was slightly enhanced one week post-irradiation. Moreover, offset values were gradually increased at each time interval.

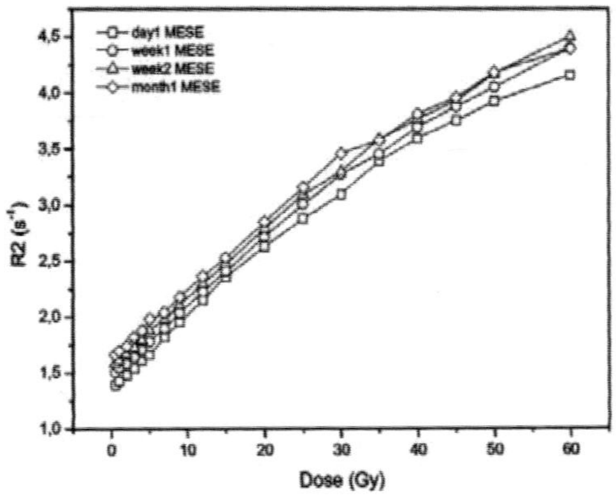

Figure 3: Temporal stability dose response of VIPET in one month post-irradiation using the MESE sequence.

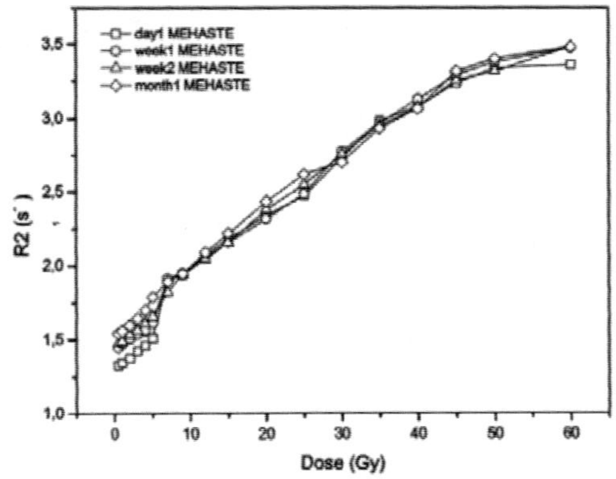

Figure 4: Temporal stability dose response of VIPET in one month post-irradiation using the MEHASTE sequence.

4. Discussion

Two MRI pulse sequences, MESE and MEHASTE were evaluated as readout processes for the estimation of the dosimetric characteristics of VIPET polymer gels. MESE is the standard sequence for MRI polymer gel dosimetry and MEHASTE is a new multi-echo single shot turbo spin echo sequence. In both sequences, a linear R2-dose response was exhibited. The sensitivity values between MESE and MEHASTE varied less than 30% in the common linear R2-dose region. MEHASTE linear R2-dose region was limited in relation to MESE linear region. For doses above 7Gy, both methods can fulfill the 2% ICRU limit for dose resolution values at 95% confidence level. The observed slight underestimation of the R2 (overestimation of the T2) values in the MEHASTE as compared to the MESE sequence is mainly due to the extent of the 180 refocusing pulses present in all turbo spin echo sequences. This effect is more prominent in longer R2 (shorter T2) values. The MEHASTE dose resolution values were 50% larger than the MESE related values. In one month period, the linear R2-dose response of the VIPET dosimeter was preserved with both MESE and MEHASTE sequences. Eventually, similar percentage differences between sensitivity and offset values per each time interval were observed using both sequences.

MEHASTE sequences are faster than MESE sequences due to their inherent sequence design (higher ETL factor). The gain in scan time is even higher due to the short TR, which in turn is a direct consequence of the incorporation of the RF restore pre-pulse in the MEHASTE sequence duty cycle. The overall scan time gain is a greater advantage for the final MEHASTE sequence design, enabling the acquisition of 25 thin (2mm) space filling (gap=0) slices in 15 min utilizing 16 excitations. Under these conditions large volume real 3D gel dosimetry is feasible in a reasonable time.

5. Conclusion

A new multi-slice multi-echo single shot turbo spin echo sequence (MEHASTE) was presented for the estimation of dosimetric characteristics of polymer gels and was compared with the standard multi-slice multi-echo spin echo (MESE) sequence. It has been shown that the dosimetric results between the two methods did not exhibit large deviations. To conclude, the overall scan time gain in combination with the superb space filling capabilities render MEHASTE sequences as an excellent means for fast 3D polymer gel dosimetry.

6. References

[1] Baldock C *et al* 2010 *Phys. Med. Biol.* **55** R1-63
[2] Murry P and Baldock C 2000 *Australas. Phys. Eng. Sci. Med.* **23** 44-51
[3] Bosi S *et al Phys. Med. Biol.* **52** 2893-903
[4] Brindha S *et al* 2004 *Phys. Med. Biol.* **49** N353-61
[5] Mather M L *et al* 2002 *Phys. Med. Biol.* **47** 4397-409
[6] De Deene Y *et al* 1998 *Radioth. Oncol.* **48** 238-291
[7] De Deene Y and Baldock C 2002 *Phys. Med. Biol.* **47** 3117-41
[8] De Deene Y *et al* 2002 *Phys. Med. Biol.* **47** 3441-63
[9] De Deene Y *et al* 2002 *Phys. Med. Biol.* **47** 2459-70
[10] Papadakis A E *et al* 2007 *Phys. Med. Biol.* **52** 5069-83

Journal of Physics: Conference Series **444** (2013) 012089 doi:10.1088/1742-6596/444/1/012089

Dose–response measurement in gel dosimeter using various imaging modalities

T Fujibuchi[1], H Kawamura[1], K Yamanashi[1], A Hiroki[2], S Yamashita[2], M Taguchi[2], Y Sato[3], K Mimura[3], H Ushiba[3] and T Okihara[3]

[1]Ibaraki Prefectural University of Health Sciences, 4669-2 Ami, Ami-machi, Inashiki-gun, Ibaraki, Japan

[2]Japan Atomic Energy Agency, 1233 Watanuki, Takasahiki, Gunnma, Japan

[3]Chiyoda Technol Corporation, 1-7-12, Yushima, Bunkyo-ku, Tokyo, Japan

E-mail: fujibuchi@ipu.ac.jp

Abstract. Measurement methods that accurately measure radiation dose distribution in a three dimensional manner in order to allow comparisons of treatment plans are needed for quality assurance. One such measurement method involves the use of a polymer gel dosimeter to measure the dose distribution in three dimensions. During irradiation, a polymerization reaction makes new chemical bonds and induces changes of the chemical structure of the gel of the gel dosimeter. In the present study, dose–response measurement of an environment-friendly material used in the gel dosimeter was performed by imaging with computed tomography (CT) and R1, R2, and fluid-attenuated inversion-recovery (FLAIR) magnetic resonance imaging (MRI) under various imaging conditions. Dose–response characteristics in the gel dosimeter used in the experiment were observed at doses of 5–20 Gy administered by X-ray CT and MRI. Although the FLAIR signal was a relative value, the dose–response values with FLAIR were excellent compared to those with R1, R2, and CT. Determination of more appropriate imaging conditions could help expand the dose–response parameters of each measurement method.

1. Introduction

Recently, radiation dose distribution has become more complicated as attempts have been made to make radiation therapy more accurate [1]. Therefore, measurement and comparison of the three dimensions of dose distribution for the purpose of treatment planning is required for quality assurance. One measurement method uses a polymer gel dosimeter to measure the three dimensions of dose distribution [1-4]. During irradiation, a polymerization reaction changes the chemical structure of the gel in the gel dosimeter. The extent of polymerization varies with dose. Development of a dose evaluation technology that utilizes X-ray computed tomography (CT) and magnetic resonance imaging (MRI) parameters is necessary for clinical use of the gel dosimeter [1, 5, 6].

In the present study, dose–response characteristics of an environment-friendly gel dosimeter composed of less toxic monomers and polysaccharide gel were evaluated by use of CT and MRI imaging.

2. Materials and Methods

2.1. The gel dosimeter

The gel dosimeter was prepared as follows. A mixture of deacylated gellan gum (0.4 g) and ultrapure water (49.6 g) was heated at about 95°C to dissolve completely the deacylated gellan gum. The aqueous solution was cooled slowly until 40°C, and then mixed with monomer solutions consisting of 2-hydroxyethyl methacrylate (2.0 g), triethylene glycol monoethyl ether monomethacrylate (2.0 g), polyethylene glycol 400 dimethacrylate (2.0 g), tetrakis hydroxymethyl phosphonium chloride (0.11 g), glucono delta-lactone (0.36 g), and ultrapure water (43.53 g). The mixed solutions were transferred into a plastic bottle before getting cold. The bottles were vacuum-packed, and then stored overnight in a refrigerator to get the gel dosimeter used in this work.

2.2. Irradiation method and conditions

The gel dosimeter was irradiated with 10-MV X-rays using EXL-15SP (Mitsubishi Electric Corporation). The gel dosimeter was arranged at the center of the 10 cm × 10-cm irradiation field in a water phantom. The source–surface distance was adjusted to 90 cm and the source–detector distance to 100 cm. Dosimeters were irradiated at 0 to 60 Gy. The gel dosimeter temperature was maintained in a refrigerator.

2.3. X-ray CT device

The ECLOS (Hitachi Medico Corporation) was used. The tube voltage was 120 kV, beam current was 300 mA, and scan time was 3.0 seconds per 1 rotation. Slice thickness was 10 mm.

2.4. MRI device

EXCELART vantage 1.5-T (Toshiba Medical Corporation) was used. A quadrature detection knee coil was used for data acquisition. The slice thickness was 5 mm. T1 and T2 values and fluid-attenuated inversion-recovery (FLAIR) images of samples were analyzed.

To measure T1 values, repetition time (TR) was set to 6000 msec and echo time (TE) was set to 10 msec. Inversion time (TI) was varied at 100, 300, 500, and 1000 msec. The reciprocal of the obtained T1 value was assumed to be R1.

To measure T2 values, TR was set to 2000 msec and TE was set to 30, 60, and 120 msec. The reciprocal of the obtained T2 value was assumed to be R2.

To measure signal intensity of FLAIR images, TR was set to 1400 msec and TE was set to 105 msec. TI was varied at 1000, 1500, 2000, 2500, and 3000 msec.

3. Results

3.1. X-ray CT device

The scanned CT image is shown in Figure 1. Figure 2 shows the radiation dose and the response characteristics. Linearity was observed from 5 to 20 Gy. The polymerization reaction became saturated as the dose approached 40 Gy, as indicated by the plateau in CT values at 40 Gy and above.

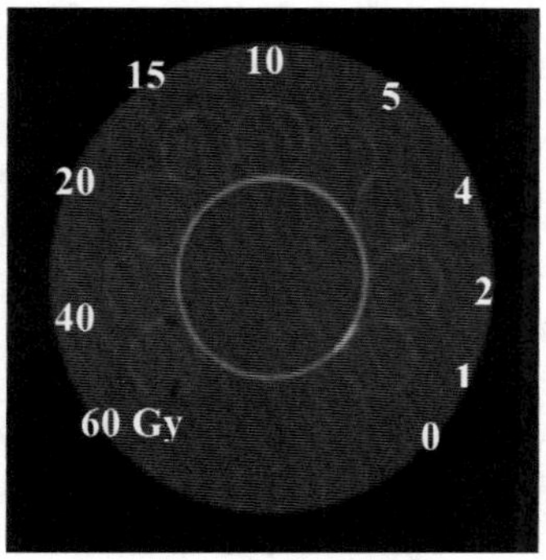

Figure 1: Computed tomography image of the gel.

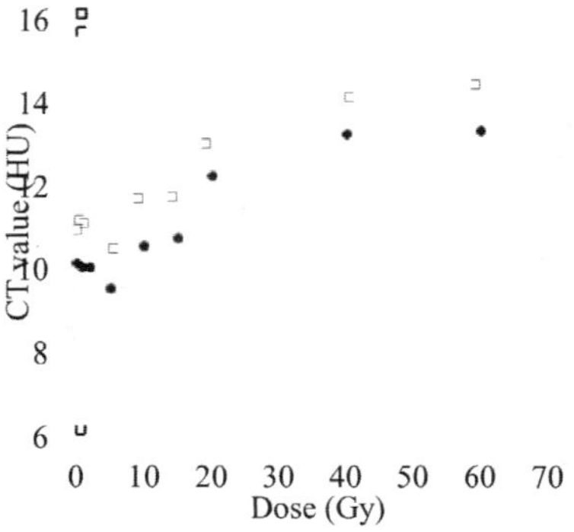

Figure 2: Dose–CT value response characteristics of the gel.

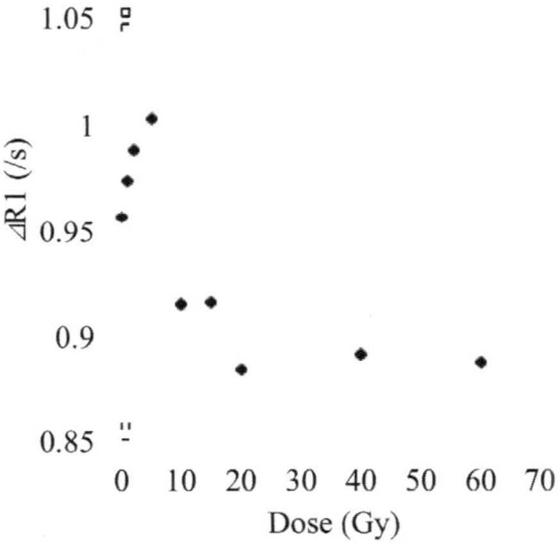

Figure 3: Dose-R1 response curve of the gel.

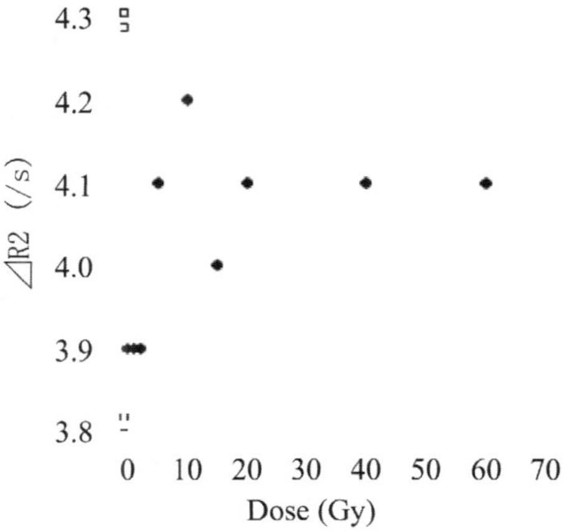

Figure 4: Dose-R2 response curve of the gel.

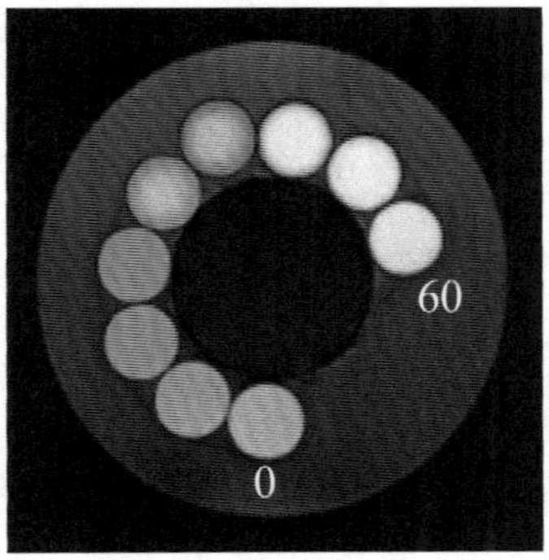

Figure 5: FLAIR image at TI = 1500 msec.

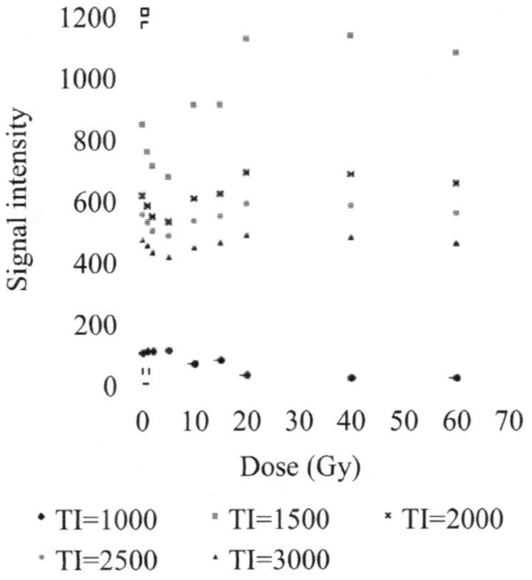

Figure 6: Dose-FLAIR signal intensity response curve of the gel.

Journal of Physics: Conference Series **444** (2013) 012089 doi:10.1088/1742-6596/444/1/012089

4. Discussion

4.1. Comparison by modality

Use of CT rather than MRI allowed the measurement time to be reduced from several minutes to just a few seconds. Compared to MRI, advantages of CT include high spatial resolution and less image distortion. On the other hand, protons were measured by MRI because changes resulting from the polymerization reaction could be more clearly described compared with CT. Although the result of FLAIR was a relative value, the dose–response values were excellent compared with R1, R2, and CT dose–response values. However, calibration is necessary for comparative assessment because FLAIR, in which evaluation by one sequence is possible, can shorten the total imaging time compared with R1 and R2 measurements.

4.2. Accuracy management of the gel dosimeter

To improve measurement precision of the gel dosimeter, accuracy control of the dosimeter is important. Accuracy management involves prevention of oxygen exposure, temperature management, measurement receptacle maintenance, shading of ultraviolet light, and accuracy of mixing, and it is necessary for experimental reproducibility. The gel dosimeter should be shaded to prevent anomalous activation of the polymerization reaction, which can be activated not only by ionizing radiation but also by ultraviolet light. However, shading coupled with exposure to oxygen weakens the polymerization reaction. This effect is gradually diminished by the deoxygenator contained in the gel dosimeter. Therefore, when the gel dosimeter is prepared, it is necessary to irradiate it ahead of time.

5. Conclusion

Dose–response characteristics in the gel dosimeter used in the experiment were observed at doses of 5–20 Gy administered by X-ray CT and MRI. Identification of more appropriate imaging conditions to expand the dose–response characteristics of each measurement method would be highly desirable. Because environmental and technical factors can greatly affect reproducibility, accuracy management of the gel dosimeter is also important.

6. References

[1] Baldock C *et al* 2010 *Phys. Med. Biol.* **55** R1-63
[2] Lepage M *et al* 2002 *Phys. Med. Biol.* **47** 1881-90
[3] Zhus X *et al* 2010 *Med. Phys.* **37** 183-8
[4] Gear J I *et al* 2006 *Phys. Med. Biol.* **51** 3503-16
[5] Hill B *et al* 2005 *Br. J. Radiol.* **78** 623-30
[6] Murry P and Baldock C 2000 *Australas. Phys. Eng. Sci. Med.* **23** 44-51

Monte Carlo water-equivalence study of two PRESAGE® formulations for proton beam dosimetry

T Gorjiara[1], Z Kuncic[1], J Adamovics[2] and C Baldock[1,3]
[1]Institute of Medical Physics, School of Physics, University of Sydney, NSW 2006, Australia
[2]Department of Chemistry and Biology, Rider University, Lawrenceville, NJ 08648, USA
[3]Faculty of Science, Macquarie University, NSW 2109, Australia

E-mail: clive.baldock@mq.edu.au

Abstract. PRESAGE® is a radiochromic solid dosimeter which shows promising potential for 3D proton beam dosimetry. Since an idea dosimeter should be water-equivalent, total depth dose distributions in two PRESAGE® formulations irradiated by a 62 MeV proton beam were compared with that in water using GEANT4 Monte Carlo simulations. The dose delivered by secondary particles was also calculated. Our results show that after water-equivalent depth scaling, PRESAGE® can be considered water equivalent for dosimetry of a 62 MeV clinical proton beam.

1. Introduction

Polymer gel dosimeters are manufactured from radiation sensitive chemicals, which upon irradiation polymerize as a function of the absorbed radiation dose [1]. These gel dosimeters which record the radiation dose distribution in three-dimensions (3D) have specific advantages when compared to one-dimensional dosimeters and two-dimensional dosimeters [2]. These 3D dosimeters are radiologically soft-tissue equivalent [3] with properties that may be modified depending on the application. The 3D radiation dose distribution in polymer gel dosimeters may be imaged using magnetic resonance imaging (MRI) [4, 5], optical-computerized tomography (optical-CT) [6, 7], x-ray CT [8, 9], ultrasound [10, 11, 12] or vibrational spectroscopy [13, 14].

PRESAGE® is a radiochromic polyurethane based dosimeter containing leuco dyes and free radical initiators. Ionizing radiation darkens PRESAGE® and changes its optical density; hence, a quantitative evaluation of the optical absorbance change of PRESAGE® can provide a measure of 3D absorbed dose [15-18]. PRESAGE® has presented potential for dosimetry of proton therapy [18-22]. Proton therapy is an external radiation treatment which delivers vey localized high dose at the distal end of the beam (Bragg peak). Precise 3D dosimetry of proton beams is important to ensure of treating tumor with minimizing damage to the healthy surrounding tissues.

To be used for clinical dosimetry, ideally a dosimeter should be water equivalent. To be considered water equivalent for proton dosimetry, it should have a similar electronic mass stopping power and secondary particle production to water [23, 24].

The water equivalence of different PRESAGE® formulations has previously been investigated for x-ray radiotherapy [25-27]. In this study, total depth dose and dose delivered by secondary particles in two different formulations of PRESAGE® by a 62 MeV proton beam are compared with those of water using Monte Carlo modeling.

2. Materials and Methods

Table 1 shows the chemical formula and fractional weight of two PRESAGE® formulations, one with the lower halogen content (formulation A), and water. To investigate water equivalency of PRESAGE® formulations, a model was developed based on the hadrontherapy advanced example in GEANT4 toolkit (a Monte Carlo code) [28] to calculate depth-dose profiles of a validated 62 MeV proton beam in two PRESAGE® formulations and water. Dosimeters and water were modeled as large cubic volumes (10×10×10 cm^3). Delivered dose was calculated along the central axis and laterally in a 4×4×4 cm^3 cube which was divided into voxels of 0.2×2×2 mm^3. The dose delivered by primary and secondary protons, secondary electrons and secondary neutrons was also calculated by setting the appropriate tracking ID number in the GEANT4 code. 2D cross-sectional total dose distributions were also calculated.

Water equivalent depth was calculated using the following formula based on IAEA report 398 and ICRU report 49 recommendations [24, 29]: Water equivalent depth = $z_m \cdot c_m$, where z_m is the material areal density (density-thickness) and c_m is a depth scaling factor defined as the ratio of the continuous slowing down approximation (CSDA) range of protons in water to the CSDA range of protons in a given material. CSDA ranges were estimated using our Monte Carlo simulated results.

Table 1: Chemical formula and fractional weight (w_k) for materials of interest.

Material	Formula	w_H	w_C	w_N	w_O	w_S	w_{Cl}	w_{Br}
PRESAGE® formulation A	$C_{1758}H_{3000}N_{121}O_{442}S_4Cl_{30}Br_1$	0.08850	0.6178	0.04960	0.2069	0.003800	0.03112	0.002300
PRESAGE® formulation B	$C_{481}H_{842}N_{30}O_{129}Cl_9Br$	0.08920	0.6074	0.04460	0.2172	-	0.03340	0.008400
Water	H_2O	0.1119	-	-	0.8881		-	-

3. Results

Figure 1 shows the calculated percentage dose versus depth for two PRESAGE® formulations and water. Protons deliver their dose more rapidly in the PRESAGE® formulations than in water and this can be attributed to the different chemical compositions which lead to the differences in stopping power. Dose delivered in PRESAGE® formulations A and B and water reaches its Bragg peak at 27.3 mm, 25.5 mm and 29.1 mm, respectively. Practical range (where dose reaches to the 10% of the Bragg Peak) of the 62 MeV proton beam is at 26.8 mm, 28.6 mm and 30.5 mm depth from the surface for PRESAGE® formulation A and B and water, respectively. As a result of these discrepancies, a water equivalent depth calculation is required to compare the depth dose of PRESAGE® formulations with water.

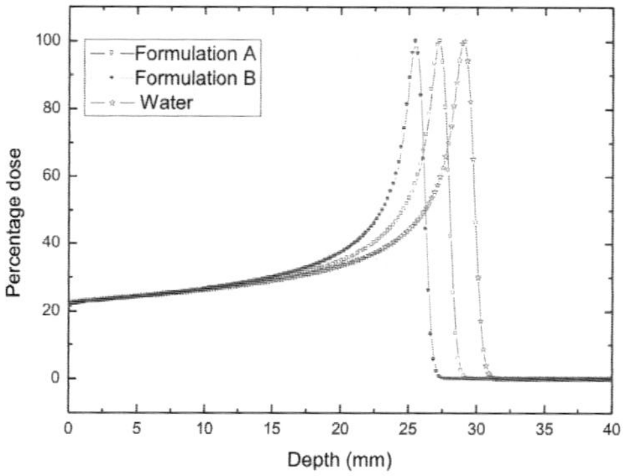

Figure1: Monte Carlo calculated total percentage dose versus depth for the two PRESAGE® formulations and water by a 62 MeV proton beam

The percentage contributions of dose delivered by secondary protons and electrons to the total dose are presented in figure 2. As expected, more than 95% of the total dose is delivered by the primary particles. Secondary protons and secondary electrons deliver less dose in PRESAGE® formulations compare to water over the entire range of the beam. Contributions of secondary protons and electrons in the total dose for formulation B is significantly lower than formulation A and water. Dose delivered by secondary neutrons is negligible in all the materials of interest for the 62 MeV proton beam.

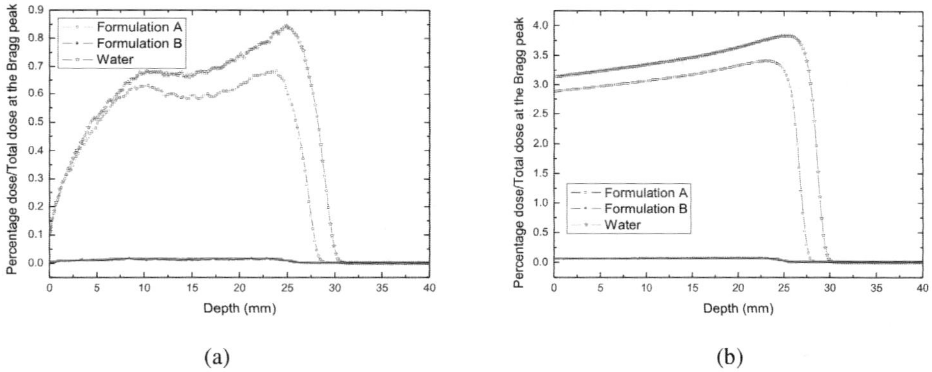

Figure 2: Monte Carlo calculated percentage dose normalized to the total dose at the Bragg peak versus depth for (a) secondary protons and (b) secondary electrons for two PRESAGE® formulations and water.

Figure 3 shows the Monte Carlo calculated percentage total dose for PRESAGE® formulations and water for a 62 MeV proton beam versus water equivalent depth. Also shown is the cross-sectional total dose distributions calculated at 29.5 mm water equivalent depth normalized to dose at the Bragg peak. Before the Bragg peak, dose delivered in PRESAGE® formulations is less than water by $\approx 2\%$ due to having less dose delivered by secondary particles. This trend is consistent with the calculated 2D cross sectional dose result. After the Bragg peak, less than 2% discrepancies from water is observable which can be attributed to the differences in the dose delivered by primary protons.

4. Conclusion

These results indicate that both PRESAGE® formulations investigated can be considered water equivalent for dosimetry of a 62 MeV proton beam if a water equivalent scaling is applied.

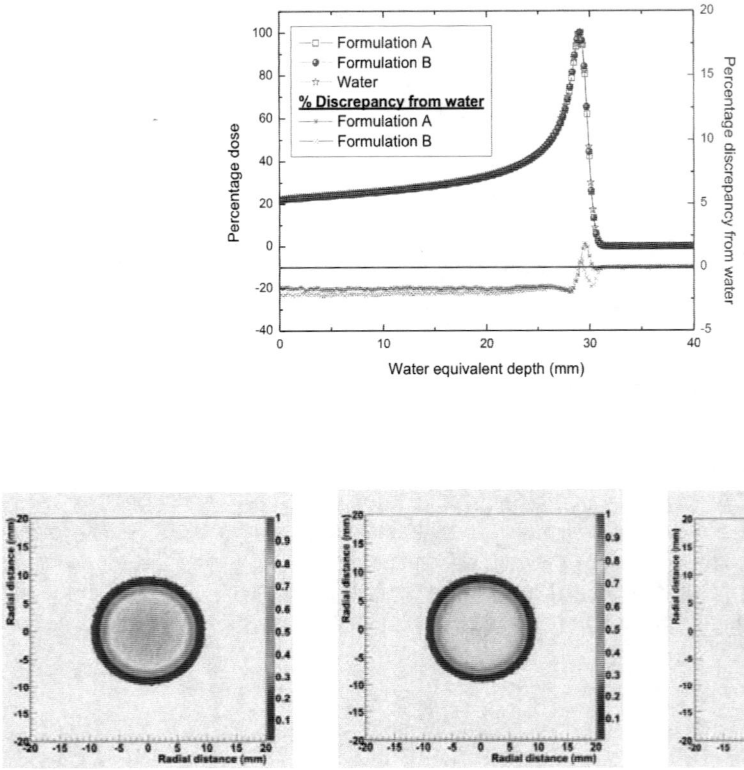

Figure 3: Top panel: Monte Carlo calculated total percentage dose versus water-equivalent depth for two PRESAGE® formulations and water. The discrepancy from water is plotted below and the scale is shown on the right axis. Bottom panel: Monte Carlo calculated cross-sectional total dose distributions normalized to dose at the Bragg peak calculated at 29.5 mm water equivalent depth for water, PRESAGE® formulation A and formulation B (left, middle and right, respectively).

5. References

[1] Baldock C *et al* 1998 *Phys. Med. Biol.* **43** 695-702
[2] Baldock C *et al* 2010 *Phys. Med. Biol.* **55** R1-63
[3] Keall P and Baldock C 1999 *Australas. Phys. Eng. Sci. Med.* **22** 85-91
[4] Gustavsson H *et al* 2004 *Phys. Med. Biol.* **49** 227-41
[5] Lepage M *et al* 2002 *Phys. Med. Biol.* **47** 1881-90
[6] Bosi S *et al* 2007 *Phys. Med. Biol.* **52** 2893-903
[7] Bosi S G *et al* 2009 *Phys. Med. Biol.* **54** 275-83
[8] Trapp J V *et al* 2002 *Phys. Med. Biol.* **47** 4247-58
[9] Hill B *et al* 2005 *Br. J. Radiol.* **78** 623-30
[10] Mather M L and Baldock C 2003 *Med. Phys.* **30** 2140-8
[11] Mather M L *et al* 2002 *Phys. Med. Biol.* **47** 1449-58
[12] Mather M L *et al* 2003 *Ultrasonics* **41** 551-9
[13] Baldock C *et al* 1998 *Phys. Med. Biol.* **43** 3617-27
[14] Rintoul L *et al* 2003 *Appl. Spectrosc.* **57** 51-57
[15] Adamovics J and Maryanski M J 2004 *J. Phys.: Conf. Ser.* **3** 172-5
[16] Adamovics J and Maryanski M J 2006 *Radiat. Prot. Dosimetry.* **120** 107-12
[17] Adamovics J *et al* 2006 *J. Phys.: Conf. Ser.* **56** 172-5

[18] Yates E S *et al* 2011 *Acta. Oncologica* **50** 829-34
[19] Zhao L *et al* 2010 *J. Phys.: Conf. Ser.* **250** 012035
[20] Doran S *et al* 2006 *J. Phys.: Conf. Ser.* **56** 231-4
[21] Al-Nowais S *et al* 2009 *Appl. Radiat. Isot.* **67** 415-8
[22] Heard M *et al* 2006 *J. Phys.: Conf. Ser.* **56** 228-230
[23] Verhey L *et al* 1998 in *ICRU report No. 59*
[24] Berger M J *et al* 1993 in *ICRU report No. 49*
[25] Gorjiara T *et al* 2010 *J. Phys.: Conf. Ser.* **250** 012053
[26] Gorjiara T *et al* 2011 *Med. Phys.* **38** 2265-74
[27] Gorjiara T *et al* 2010 *J. Phys.: Conf. Ser.* **250** 012093
[28] Agostinelli S *et al* 2003 *Nucl. Instrum. Methods Phys. Res. A* **506** 250-303
[29] Andreo P *et al* 1995 IAEA TRS-381 The use of plane parallel ionization chambers in high energy electron and photon beams. An international code of practice for dosimetry *Stockholm*

Journal of Physics: Conference Series **444** (2013) 012091 doi:10.1088/1742-6596/444/1/012091

Reconstructing 3D x-ray CT images of polymer gel dosimeters using the zero-scan method

MB Kakakhel[1,3], T Kairn[1,2], J Kenny[2] and JV Trapp[1]
[1]School of Chemistry, Physics & Mechanical Engineering, Queensland University of Technology, GPO Box 2434, Brisbane, Qld 4001, Australia
[2]Premion Cancer Care, The Wesley Medical Centre, Suite 1, 40 Chasely St, Auchenflower, Qld 4066, Australia
[3]Department of Physics and Applied Mathematics, DPAM, Pakistan Institute of Engineering and Applied Sciences, PO Nilore, Islamabad 45650, Pakistan

E-mail: j.trapp@qut.edu.au

Abstract. In this study x-ray CT has been used to produce a 3D image of an irradiated PAGAT gel sample, with noise-reduction achieved using the 'zero-scan' method. The gel was repeatedly CT scanned and a linear fit to the varying Hounsfield unit of each pixel in the 3D volume was evaluated across the repeated scans, allowing a zero-scan extrapolation of the image to be obtained. To minimise heating of the CT scanner's x-ray tube, this study used a large slice thickness (1 cm), to provide image slices across the irradiated region of the gel, and a relatively small number of CT scans (63), to extrapolate the zero-scan image. The resulting set of transverse images shows reduced noise compared to images from the initial CT scan of the gel, without being degraded by the additional radiation dose delivered to the gel during the repeated scanning. The full, 3D image of the gel has a low spatial resolution in the longitudinal direction, due to the selected scan parameters. Nonetheless, important features of the dose distribution are apparent in the 3D x-ray CT scan of the gel. The results of this study demonstrate that the zero-scan extrapolation method can be applied to the reconstruction of multiple x-ray CT slices, to provide useful 2D and 3D images of irradiated dosimetry gels.

1. Introduction

Three dimensional readout of dosimetry gels irradiated with radiotherapy beams has been achieved using a range of imaging modalities [1]. Magnetic resonance imaging (MRI) [2, 3] and optical CT imaging [4-6] are a well-established techniques for providing low-noise, high-resolution 3D images of irradiated gels, but MRI and optical CT systems are not often available in radiation oncology departments. By contrast, obtaining good quality 3D images of irradiated gels using x-ray CT [7, 8] and ultrasound [9] imaging systems remains challenging, although advantageous given the frequent availability of these systems in radiation oncology departments.

The usefulness of x-ray CT for dosimetry gel readout has been limited due to the small change in gel density with irradiation [8, 9], which results in a weak image signal that is easily overwhelmed by noise. Noise in x-ray CT images can be reduced by averaging over repeated scans, but this technique results in a loss of dosimetric fidelity as the gel is further irradiated by each successive CT scan [10].

Recently this issue has been addressed by the introduction of the 'zero-scan' method [11], summarised below, which has been shown to substantially reduce the noise in a single slice (2D) x-ray

CT image of a PAGAT gel sample [11]. The aim of the current study is to further examine the zero-scan method and to establish its utility for producing noise-reduced x-ray CT images of dosimetry gels in three dimensions. In this work we extend the zero-scan method to a multi-slice dataset.

2. Methods and Materials

The zero-scan method involves the repeated x-ray CT scanning of an irradiated gel sample. As the gel absorbs and reacts to the compounding x-ray scanning dose, the Hounsfield unit of each pixel in the resulting images slowly increases [11]. A linear fit for Hounsfield unit versus scan number is then evaluated across the repeated scans and, for each pixel, this linear fit is extrapolated back to the zero, producing the 'zero-scan' image of the gel.

For this study, a PAGAT gel sample was prepared according to the recipe reported by Venning *et al.* [12], with an increased Tetrakis (hydroxymethyl) phosphonium chloride (THPC) concentration as recommended by Khoei *et al.* [13], and set in a 10 cm high, 7 cm wide, cylindrical polyethylene terephthalate (PET) container. The gel was stored for 18 h at 4 °C before irradiation.

The gel sample was irradiated using a Varian iX linear accelerator (Varian Medical Systems, Palo Alto, USA), producing a nominal 6 MV photon beam at a dose rate of 600 MU/min dose rate. The treatment delivered to the gel consisted of three dynamically wedged fields [14], with a maximum dose of 6 Gy. The gel was then returned to storage at 4 °C.

The x-ray CT of the gel was carried out 6 h after irradiation. The gel was scanned 63 times using a GE Lightspeed RT 4 CT scanner (GE Healthcare, Waukesha, USA) with an x-ray tube load of 230 mA with a 1 s rotation, a nominal tube energy of 120 kVp, and 4 × 1 cm slices. A total of 12 contiguous 1 cm images scanned with a 25 cm Scan field of view and reconstructed on a 512 × 512 pixel image matrix.

Before performing the zero-scan extrapolation, each image of the irradiated gel was first corrected using the image subtraction protocol developed by Trapp *et al* [8]. The CT data was imported into Matlab (version 7.8.0.342, The MathWorks, Natick, MA) for further processing. The linear 12 × 512 × 512 linear fits required by the zero-scan method were obtained by applying a linear fit using the inbuilt *Polyfit* function to each pixel in the 3D data set [11].

3. Results

Transverse images of the six central slices in the irradiated region of the gel (slices 4-9) are shown in figure 1, with and without the use of the zero-scan method. The zero-scan images show a clearer signal with less noise than the raw images. The locations, orientations and overlaps between the radiation beams can be clearly identified in all of the zero-scan images, while this information cannot be easily obtained from the raw images.

Figures 2(a) to (f) show slices through the zero-scan 3D image of the gel, in the coronal, transverse and sagittal planes. The low resolution of the coronal and sagittal images arises from the 1.0 cm CT slice thickness used to obtain the data. Despite the low resolution of these images, it is nonetheless possible to observe the different intensities of the irradiated regions in the gel in figures 2(a) to (f): Region A was irradiated with one beam. Region B was irradiated with an opposed pair of beams. Region A+B was irradiated with all three beams.

It is evident from the grey scale values shown in figures 2(a) to (f), where regions of low dose appear darker than regions of high dose, that region A received a lower dose than region B, which received a lower dose than region A+B. It is also apparent, although less obvious, from an examination of figures 2(a) to (f), that in the coronal and sagittal images, the direction of dynamic wedge motion direction is up the page, with the thin edge of the wedge at the bottom of the page and the thick end of the wedge at the top of the page. In the transverse image, the wedge motion direction is into the page.

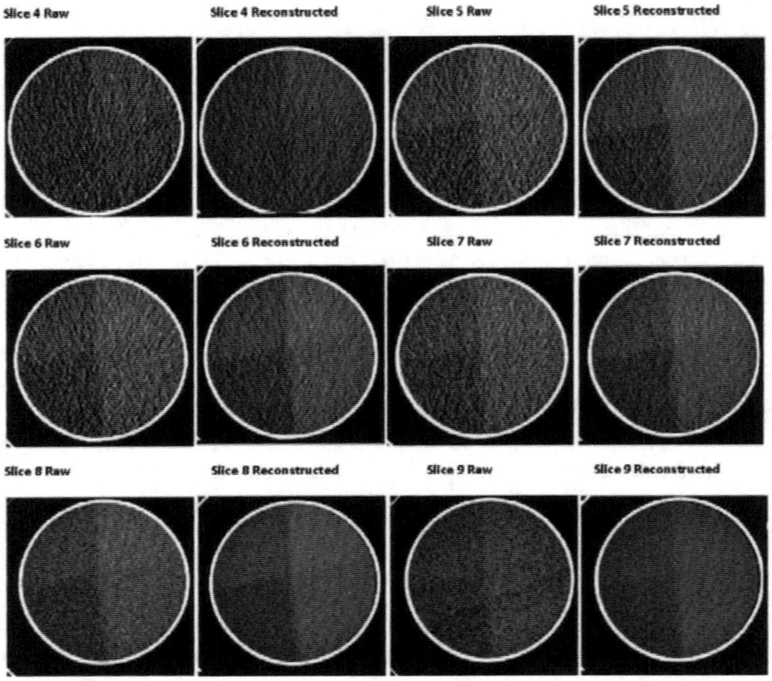

Figure 1. Image slices reconstructed with (right hand image in each pair) and without (left hand image in each pair) using the zero-scan method.

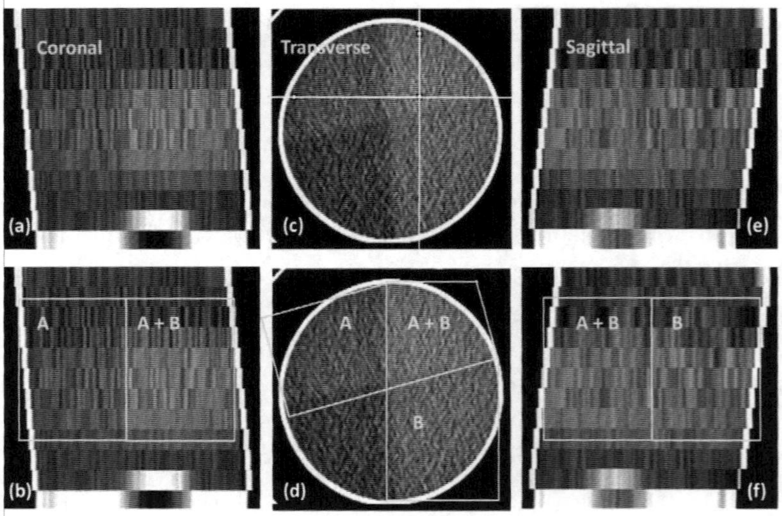

Figure 2. Coronal (a) and (b), transverse (c) and (d) and sagittal (e) and (f) views, through the zero-scan 3D image of the gel. White lines in (c) indicate the positions of the coronal (horizontal line) and sagittal (vertical line) planes shown in (a) and (e), respectively. Images (b), (d) and (f) are duplicates of (a), (c) and (e). Regions A, B and A + B are described in the text.

4. Discussion and Conclusion

The transverse plane images shown in figure 1 confirm previous observations [11], indicating that the zero-scan method can produce x-ray CT images with reduced noise compared to images from the initial CT scan of the gel. While the noise in the zero-scan images is reduced, the image intensity is unchanged, indicating that the dosimetric accuracy of the zero-scan images has not been degraded by the additional radiation dose delivered to the gel during the repeated scanning.

The full, 3D image of the gel has a low spatial resolution in the longitudinal direction, due to the selected scan parameters. Nonetheless, important features of the dose distribution, including the locations of the beams and the orientations of the dynamic wedges are apparent in the 3D x-ray CT scan of the gel.

Higher-resolution images of dosimetric gels could be potentially be obtained by using a smaller slice thickness and the noise in these images could be further reduced by performing additional CT scans and obtaining the zero-scan images from a fit to a larger data set, as shown for Kakakhel *et al*'s single-slice scan [11].

The results of this study demonstrate that the zero-scan extrapolation method can be applied to the reconstruction of multiple x-ray CT slices, to provide useful 2D and 3D images of irradiated dosimetry gels. Such detailed dosimetric information could be used to verify the accuracy and deliverability for more-complex plans such as stereotactic [4], intensity-modulated or dynamically wedged radiotherapy treatments [14].

5. Acknowledgments

Queensland Cancer Physics Collaborative is acknowledged for their support.

6. References

[1] Baldock C *et al* 2010 *Phys. Med. Biol.* **55** R1–63
[2] Gore J C and Kang Y S 1984 *Phys. Med. Biol.* **29** 1189-97
[3] Maryanski M J 1994 *Phys. Med. Biol.* **39** 1437-55
[4] Gore J C *et al* 1996 *Phys. Med. Biol.* **41** 2695-704
[5] Kairn T *et al* 2012 *Phys. Med. Biol.* **57** 3359
[6] Bosi S *et al* C 2007 *Phys. Med. Biol.* **52** 2893–2903
[7] Hilts M *et al* 2000 *Phys. Med. Biol.* **45** 2559-71
[8] Trapp J V *et al* 2001 *Phys. Med. Biol.* **46** 2939-51
[9] Mather M L *et al* 2002 *Phys. Med. Biol.* **47** 1449-58
[10] Hindmarsh J *et al* 2010 *J. Phys.: Conf. Ser.* **250** 012073
[11] Kakakhel M B *et al* 2011 *Med. Phys.* **38** 5130-5
[12] Venning A J et al 2005 *Phys. Med. Biol.* **50** 3875-88
[13] Khoei S , *et al* 2010 *J. Phys. Conf. Ser.* **250** 012019
[14] Kakakhel M B *et al* 2012 *Physica Medica*

Characterization of the essential dosimetric properties of cosolvent-free polymer gel dosimeters: Recent progress in x-ray CT based normoxic polymer gel dosimetry

H Johnston[1], M Hilts[1,2], J Carrick[1] and A Jirasek[1]
[1]Department of Physics and Astronomy, University of Victoria, Victoria, BC V8W 3P6, Canada
[2]Medical Physics, BC Cancer Agency - Vancouver Island Centre, Victoria, BC V6R 2B6, Canada

E-mail: hjohnston3@bccancer.bc.ca

Abstract. This work evaluates the temporal stability, spatial stability, batch reproducibility and dose rate dependence of a new cosolvent-free polymer gel dosimeter optimized for use with x-ray computed tomography readout. Temporal and spatial stability investigations reveal the new gel formulation should be imaged between 15-36 hours after irradiation. Intra- and inter-batch reproducibility were found to be excellent over the entire range of doses examined. A dose rate dependence was found for gels irradiated with machine dose rates between 100-600MU/min. An intensity-modulated radiation therapy treatment validation is also presented to illustrate an example clinical application using the new gel formulation.

1. Introduction

Until recently, x-ray computed tomography (CT) polymer gel dosimetry (PGD) has been hindered by low signal-to-noise CT images, largely resulting from gel formulations with low radiation sensitivity. Work to improve sensitivity has focused on identifying new dose-sensitive monomers [1, 2] and finding methods to increase the concentration of N, N'-methylenebisacrylamide (BIS) in the gel system [3-6]. A significant improvement was made using isopropanol as a cosolvent to increase the amount of BIS in the gel formulation [5]. It was further found that comparable sensitivity could be obtained by increasing BIS concentration using N-isopropylacrylamide (NIPAM) [6]. This new, cosolvent-free formulation, consisting of 15% NIPAM, 4.5% BIS, 5% gelatin, 75.5% deionized water and 5mM tetrakis hydroxymethyl phosphonium chloride (THPC), provides superior radiological tissue equivalence over the isopropanol recipe [7] and is the formulation of choice for this work. The aim of this study is to characterize the essential dosimetric properties of the new cosolvent-free polymer gel formulation. Temporal and spatial stability are investigated to evaluate the reliability of gel response. Intra-batch reproducibility is examined locally, within a single region of gel, and globally, throughout a gel volume. Inter-batch reproducibility is also examined using 3 independently fabricated gel batches. The dependence of gel response on machine dose rate is evaluated for irradiations delivered between 100-600MU/min. An example clinical application of the new gel formulation is also presented using a test intensity-modulated radiation therapy (IMRT) treatment validation.

Journal of Physics: Conference Series **444** (2013) 012092 doi:10.1088/1742-6596/444/1/012092

2. Materials and Methods

2.1. Gel fabrication

All polymer gels were manufactured as described previously [6]. For the temporal stability study, the final gel solution was poured into 20mL scintillation vials (Wheaton Scientific, Millville, NJ) and sealed in acrylic phantoms [8]. For the remaining studies, completed gel solutions were transferred to 1L polyethylene terephthalate jars (Modus Medical Devices, London, ON). Blank gels for background subtraction were manufactured for the temporal and spatial stability studies as above using 5% gelatin, 0.5% NIPAM and 0.5% BIS. For the remaining studies, backgrounds were taken from central locations in each 1L gel that received less than 3% scattered dose from neighboring irradiations.

2.2. Gel irradiation

All gels were irradiated with 6MV photons using a Varian Clinac linear accelerator (Varian Medical Systems, Palo Alto, CA) at a machine dose rate of 400MU/min unless otherwise noted. Treatment planning for each experiment was performed using the Eclipse treatment planning system (Varian Medical Systems, Palo Alto, CA).

For the temporal stability study, gel vials were irradiated to a uniform dose of 10Gy as described previously [8]. For the remaining studies, a custom-built head and neck phantom, optimized for low CT image noise and reproducible localization [9], was used to position gels at CT simulation and radiation delivery.

The 1L gel used to investigate spatial stability was irradiated to 10Gy over half its volume using two parallel-opposed 20x8cm^2 wedged fields.

For the batch reproducibility and dose rate studies, a range of doses up to 28Gy were delivered using three intersecting 3x3cm^2 fields. The distribution was delivered to one end of a 1L gel to examine local intra-batch reproducibility and to both ends of a second 1L gel to examine global intra-batch reproducibility. Inter-batch reproducibility was examined by delivering the 3-field distribution to three independently fabricated gels at identical locations in each cylinder. For the dose rate study, the distribution was delivered to both ends of a 1L gel at 100 and 600 MU/min, respectively.

For the IMRT treatment validation, both calibration and test irradiations were delivered to the same 1L gel. The calibration irradiation consisted of the 3-field distribution described above. The IMRT treatment plan was designed to treat a concave target wrapped around an avoidance structure and consisted of 5 beams spread over anterior gantry angles, with a prescription dose of 22Gy.

2.3. Gel imaging

Imaging for all experiments was performed using a GE HiSpeed FX/i CT scanner (GE Medical Systems, Milwaukee, WI) operating at 120kVp and 200mAs [9].

Gel vials used to investigate temporal stability were imaged one at a time between 3-45 hours post-irradiation using a purpose-built phantom described previously [8]. Sixteen images were acquired through the centre of each vial using 10mm slice thickness.

The head and neck phantom used for CT simulation and radiation delivery was used to position gels at CT readout for all remaining investigations. For the spatial stability study, the 1L gel was imaged at 4 locations along the length of the cylinder between 15-47 hours after irradiation. Scans acquired at different times were separated by 3.0cm to avoid CT dose between neighboring acquisitions [10]. Thirty-two images were acquired for each scan using 5mm slice thickness.

For the batch reproducibility, dose rate and IMRT validation studies, 25 images were acquired using 3mm slice thickness. The local intra-batch gel was scanned at 3 locations within the irradiated region. For the remaining studies, images were acquired through the centre of each distribution.

2.4. Data processing

Data processing was completed using Matlab (The Mathworks, Natick, MA). Image averaging and background subtraction were performed for all data sets followed by image filtering using an adaptive mean filter (3x3, n=1) [11] and remnant artefact removal (parameters optimized for each study) [12].

Processed images for the temporal stability study were analyzed by extracting the mean and standard deviation of CT numbers (N_{CT}) within a circular region-of-interest for the irradiated gel vial and an unirradiated blank gel. The radiation-induced change in N_{CT} above an unirradiated gel (ΔN_{CT}) was then computed for each vial and plotted as a function of time between irradiation and imaging.

For the spatial stability study, 7 adjacent ΔN_{CT} profiles through the diameter of the 1L cylinder were extracted for each scan. The mean and standard deviation of ΔN_{CT} were then computed over 7 pixels in a direction perpendicular to the profiles and plotted as a function of distance in the image.

Processed images for the batch reproducibility and dose rate gels were used to generate dose-response curves using a new pixel-by-pixel calibration technique outlined previously [13]. A dose-response curve was also generated for the IMRT distribution and used to convert ΔN_{CT} to dose for the test irradiation image. Measured and planned IMRT doses were then compared using an isodose plot, 2D gamma map [14] and dose profiles.

3. Results and Discussion

3.1. Gel Characterization

Figure 1a illustrates the time-course of post-irradiation polymerization for the cosolvent-free gel. An initial increase in ΔN_{CT} is observed between 0-15 hours after irradiation. Beyond 15 hours, the polymerization reaction stabilizes. It is therefore recommended that at least 15 hours is allowed between irradiation and imaging for this gel formulation.

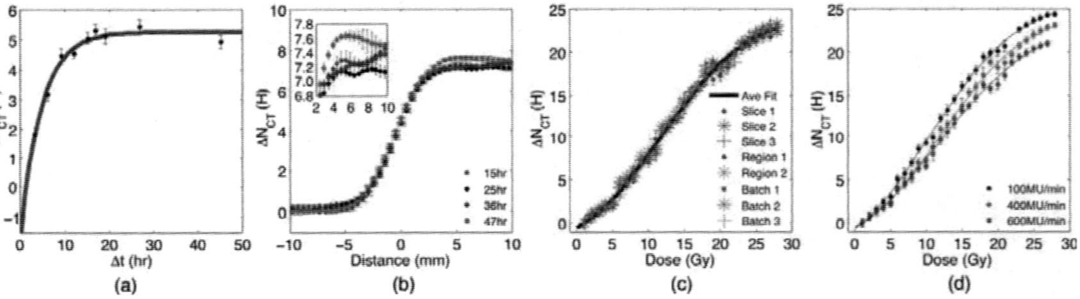

Figure 1: (a) Time-course of post-irradiation polymerization for gel vials irradiated to 10Gy. (b) Profiles of ΔN_{CT} extracted along the diameter of a 1L gel irradiated to 10Gy over half its volume. (c) Dose-response curves for the intra- and inter-batch reproducibility studies. (d) Dose-response curves for gels irradiated using machine dose rates between 100 - 600MU/min.

Profiles of ΔN_{CT} extracted along the diameter of the 1L gel for images acquired between 15-47 hours post-irradiation are illustrated in figure 1b. Profiles agree within experimental uncertainty for scans acquired 15-36 hours after irradiation, indicating spatial stability is excellent within this time frame. Beyond 36 hours, a small overshoot in response is observed. Further investigation is required to determine the exact mechanism causing the observed overshoot. Nevertheless, an upper limit on the time between irradiation and imaging of 36 hours is recommended for the new gel recipe.

Figure 1c shows dose-response curves for the local intra-batch (slices 1-3), global intra-batch (regions 1-2) and inter-batch (batches 1-3) reproducibility studies plotted with the average fit function as a reference. Excellent dose-response reproducibility is achieved over the entire dose range studied. The intra-batch reproducibility achievable within a single region of gel and throughout a 1L volume prove calibration and test irradiations can be delivered to the same gel dosimeter. Similarly, the inter-batch reproducibility shown indicates the new gel is insensitive to small variations in the PGD process.

Dose-response curves for gels irradiated using machine dose rates between 100-600MU/min are illustrated in figure 1d. The shape of the dose-response curve changes as a function of dose rate, indicating a machine dose rate dependence for the new gel formulation. As dosimetric error could occur in a clinical application if machine dose rate was not consistent between calibration and test irradiations, caution must be exercised when using the new gel for clinical applications. The

Journal of Physics: Conference Series **444** (2013) 012092 doi:10.1088/1742-6596/444/1/012092

implication for variations in dose rate beyond machine dose rate (i.e. variations in deposited dose rate, mean dose rate over treatment delivery, intensity modulation, etc.) are the subject of current study.

3.2. Clinical application

Figure 2a compares gel measured doses to the corresponding planned isodose levels for the IMRT treatment validation. It is readily seen that agreement between gel and planned doses is excellent. Similarly, the gamma distribution shown in figure 2b gives a nearly complete pass rate (99.3%). Excellent agreement between planned and measured doses is again observed in the row and column profiles through the middle of the distribution, shown in figures 2c and 2d, respectively. By all parameters, the test treatment validation shown here illustrates a remarkable improvement over previous CT PGD clinical applications [15-16].

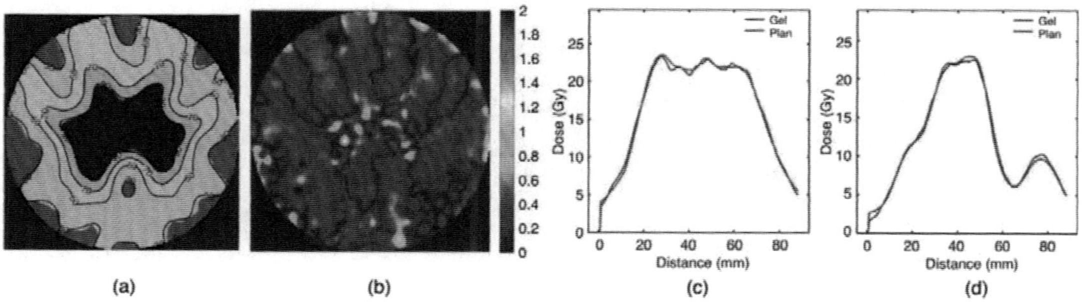

Figure 2: Comparison of planned and measured doses for the IMRT treatment validation using (a) an isodose plot, (b) 2D gamma map and (c) row and (d) column profiles.

4. Conclusions

The results of this work illustrate that the new cosolvent-free polymer gel formulation has promising dosimetric characteristics for CT PGD. Temporal and spatial stability investigations reveal the response of the gel is reliable between 15-36 hours post-irradiation and excellent batch reproducibility is achieved for doses between 0-28 Gy. The implication of the observed machine dose rate dependence remains to be determined and is the focus of current study.

5. References

[1] Senden R J *et al* 2006 *Phys. Med. Biol.* **51** 3301-14
[2] Koeva V I *et al* 2008 *Macromol. Symp.* **261** 157-66
[3] Koeva V I *et al* 2009 *Phys. Med. Biol.* **54** 2779-90
[4] Jirasek A *et al* 2009 *Phys. Med. Biol.* **54** 907-18
[5] Jirasek A *et al* 2010 *Phys. Med. Biol.* **55** 5269-81
[6] Chain J N M *et al* 2011 *Phys. Med. Biol.* **56** 2091-102
[7] Gorjiara T *et al Phys. Med. Biol.* (In press)
[8] Hilts M *et al* 2004 *Phys. Med. Biol.* **49** 2477-90
[9] Hilts M *et al* 2011 *Med. Phys.* **38** 3507
[10] Hill B *et al Med. Phys.* **32** 1589-97
[11] Hilts M and Jirasek A 2008 *Med. Phys.* **35** 12
[12] Jirasek A *et al* 2012 *Phys. Med. Biol.* **57** 3137-53
[13] Johnston H *et al* 2012 *Phys. Med. Biol.* **57** 3155-75
[14] Low D *et al* 1998 *Med. Phys.* **25** 656–61
[15] Audet C *et al* 2002 *J. App. Clin. Med. Phys.* **3** 110-8
[16] Hilts M *et al* 2000 *Radiother. Oncol.* **56** S80

Journal of Physics: Conference Series **444** (2013) 012093 doi:10.1088/1742-6596/444/1/012093

Influence of preparation and calibration method of PAGAT dosimeter on TSE MR readout results

K Vávrů[1], P Dvořák[2], J Tintěra[3] and V Spěváček[1]
[1]Czech Technical University in Prague, Faculty of Nuclear Sciences and Physical Engineering, Břehová 7, 115 19 Prague 1, Czech Republic
[2]The London Clinic, 22 Devonshire Place, London, W1G 6JA, United Kingdom
[3]Institute for Clinical and Experimental Medicine, Vídeňská 1958/9, 140 21 Prague 4, Czech Republic

E-mail: katerina.vavru@fjfi.cvut.cz

Abstract. In this study PAGAT dosimeter evaluation by TSE sequence was tested. PAGAT dosimeter preparation procedure was modified to increase the dosimeter sensitivity. Because THPC reacts with gelatin, adding THPC to monomer solution prior to dissolved gelatine helps exploit THPC as an antioxidant. Turbo spin echo sequence enables to evaluate gel dosimeter with 3D equidistant resolution in a reasonable scanning time. Glass walls of the phantom cause problems both by computing inaccuracies and MR imaging artefacts. The inner dosimeter volume is not affected by these inaccuracies and should be used for radiotherapy plan verification.

1. Introduction

Polyacrylamide gel [1] containing tetrakis (hydroxymethyl) phosphonium chloride as an antioxidant [2, 3] is widely used but still should be more studied [4]. We show that only chemicals order modification can increase the dosimeter sensitivity that was evaluated by turbo spin echo magnetic resonance measurement.

2. Materials and methods

2.1. *Gel preparation*

A bulk of 3000 ml of PAGAT gel was prepared using 5% gelatin (from porcine skin, Type A, Sigma-Aldrich, USA), 3% (w/w) acrylamide (AA) (electrophoresis grade, >99%, Sigma-Aldrich, India), 3% (w/w) N,N'-methylenebisacrylamide (BIS) (4K ultrapure, AppliChem, Darmstadt, Germany), and 4,5 mM tetrakis (hydroxymethyl) phosphonium chloride (THPC) (80% solution in water, Sigma-Aldrich, Switzerland).

To prepare PAGAT dosimeter [5], distilled water was heated to 47°C using water bath with temperature control. Two separated Erlenmeyer flasks were used; 150 g gelatin was allowed to dissolve in 1500 ml and 45 g BIS in 1000 ml of hot distilled water. When gelatin completely dissolved, 45 g BIS was added, and after BIS dissolving 90 g AA and appropriate amount of water was added as well. When all components completely dissolved, the temperature was lowered to 37°C and both solutions were bubbled with nitrogen gas for about 10 min. Subsequently, during rapid stirring of the BIS solution 2,5 ml 80% aqueous solution of THPC was injected, the solution was transfused into the first flask and bubbled again with nitrogen gas for about 20 min. Four purpose-made glass flasks (700 ml, 9.5 cm in diameter and 9.5 cm the cylinder part long) were filled with the gel in a nitrogen environment to avoid oxygen contamination and placed in a refrigerator (8°C) to solidify over the night. Whole gel dosimeters preparation took about 5 h.

2.2. Irradiation
Gel dosimeters were irradiated 2 days after preparation. Before irradiation, gels were allowed to equilibrate to a room temperature. Each glass flask was placed in a purpose-built holder and irradiated by a Varian linac with a Millenium 120 MLC in The London Clinic, United Kingdom.

A highly complex VMAT H&N plan with spine sparing was used. Although not applied clinically, the dose was delivered in a gated mode with approx. 50% duty cycle. 6 MV photon beam was used and the dose was proportionally adapted to dynamic dose range of the gel. One flask was irradiated by four 2x2 cm fields and appropriate depth dose curves were used for gel dosimeter calibration. For comparison purposes, EBT2 Gafchromic films (ISP, Wayne, NJ, USA) were irradiated by identical calibration fields. The films were held vertically inside the water filled flask by a narrow slit in the rubber cap.

2.3. Magnetic resonance imaging
MR readout was performed on a 1,5 T whole body scanner Magnetom Avanto (Siemens, Germany) in the Institute for Clinical and Experimental Medicine, Prague, Czech Republic. Each glass flask was placed in a purpose-built holder and scanned using a head coil. 3D Turbo-Spin-Echo (TSE) sequence with TR = 4000 ms, TF = 90, 1 acquisition and effective echo times TE = 31, 102, 204, 307, 409, 511, 613, 707 was used. Field of view was 150x192 mm and the whole flask was scanned with voxel size (1x1x1)mm within 1,5 h (11 min per echo).

For comparison purposes, a few slices were also measured with multiple spin-echo (CPMG) sequence with following parameters: TR = 4000 ms, 32 equidistant echos with TE = 30, 60, 90, ... 960, 1 acquisition, pixel size 1x1 mm, slice thickness 3 mm, FOV = (156x256)mm, 12 min per slice.

2.4. Data processing
The R2 relaxation rate of each pixel was derived by fitting the time course of the pixel value in the consecutive base images to a mono-exponential decay using a linear fit of log values. For data processing Matlab (The MathWorks, Inc., USA) and Microsoft Office Excel (Microsoft Corporation, USA) were used.

Films were digitized 3 times with an EPSON Perfection V700 Photo scanner (Seiko Epson Corporation, Japan) in the transmission mode. Results were averaged and the calibration curve was derived from max values of each film compared with Eclipse AAA plan data.

3. Results and discussion

3.1. PAGAT preparation
We modified preparation method of PAGAT dosimeter. Our experiments showed that over 95% THPC added in dosimeter was exhausted due to its reaction with gelatin. Therefore, a system of separated reservoirs and THPC adding first in solely monomer solution prior to dissolved gelatin was used. Moreover, to reduce oxygen contamination of dosimeter we bubbled both solutions with nitrogen gas prior and after THPC adding and final glass flasks filling was performed in nitrogen environment as well.

Dose response study of PAGAT dosimeter prepared in an anoxic environment made by Sedaghat et al. [4] showed decreasing sensitivity of the gel dosimeter with increasing THPC concentration. We point out to the fact that only chemicals order modification increases the dosimeter sensitivity again. For 4.5 mM THPC concentration the slope change was more than 30%. Our preparation method is only partially anoxic, so THPC acts as the rest of dissolved oxygen cleaning agent. It is presumable that lower THPC concentration would be sufficient to scavenge all dissolved oxygen in this case. On the other hand, an excess of THPC reduce the post-irradiation polymerization, which leads to flatter profiles next to edges [6] and which is proposed to stabilize crosslink density of the polymer [7].

3.2. TSE vs CPMG sequence

Turbo spin echo sequence as the fast imaging protocol was proposed to offer great promise in polymer gel dosimetry [8], but is still rarely used [9, 10]. We test this protocol because of the time effectiveness as well as the 3D approach. A sample slice of the glass flask scanned in air acquired by CPMG and TSE sequence is shown in Figure 1. Although the TSE data near the glass interface are affected by ring artifact, the middle part response is stable and comparable to the CPMG data.

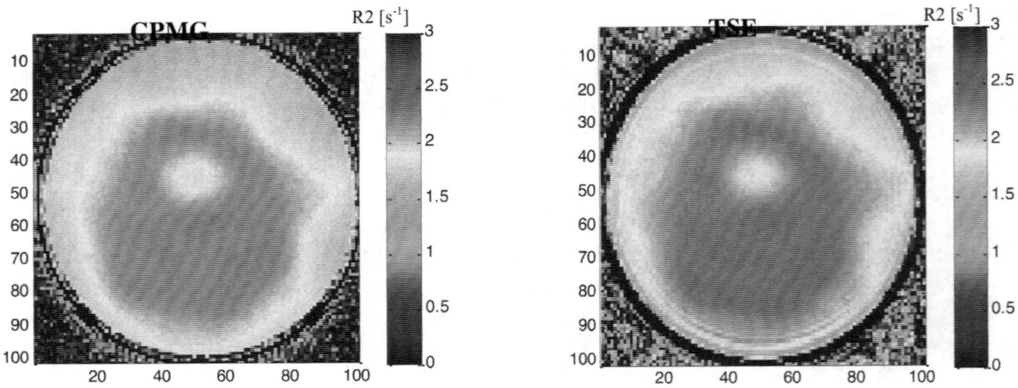

Figure 1: Comparison of the same slice of the PAGAT gel acquired by CPMG and TSE sequence.

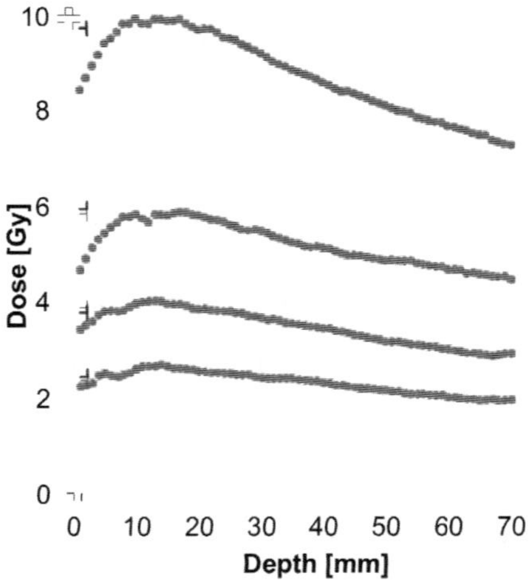

Figure 2: Depth dose curves in glass flask calculated by Eclipse AAA (—) and measured by EBT2 Gafchromic film (---) and PAGAT gel dosimeter (….)

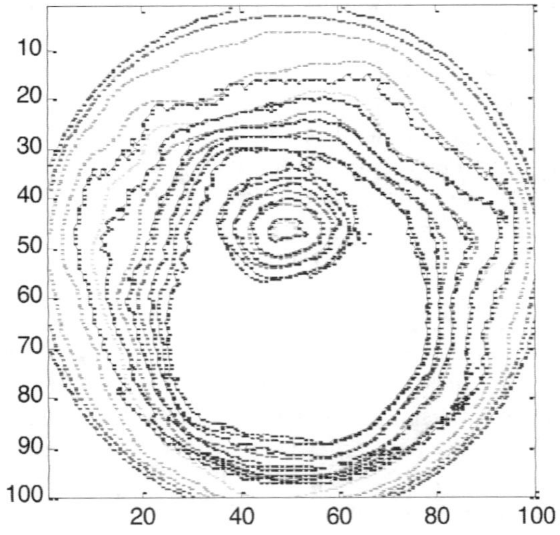

Figure 3: Isodose comparison of VMAT H&N plan, Eclipse AAA (---) and PAGAT gel dosimeter measured by TSE sequence (—)

3.3. Gel calibration and plan verification

To eliminate volume effects, PAGAT gel dosimeter calibration was performed by depth dose curve measurement of four fields irradiated in one flask. Although it is not recommended [7], the use of maximum dose data for calibration was more accurate comparing film measurements (Figure 2) than calibration curve created by plan and gel data from the depth behind maximum [8]. In our case, the area before dose maximum does not correspond with planned data. It is probably due to incorrect interpretation of the glass walls by the planning system.

The same glass flask with PAGAT gel dosimeter was irradiated by complex VMAT H&N plan with spine sparing. Resulting comparison of one slice is depicted in Figure 3.

4. Conclusion

Regarding PAGAT dosimeter properties we would like to point out to possible preparation procedure modification which slightly increases the dosimeter sensitivity. Because THPC reacts with gelatine, adding THPC to monomer solution prior to dissolved gelatin helps exploit THPC as an antioxidant.

Turbo spin echo sequence enables to evaluate gel dosimeter with 3D equidistant resolution in a reasonable scanning time. Using glass flask for TSE gel measurements does not allow to compare data near the glass walls with plan because of: 1) wrong plan computing next to the glass wall and 2) the TSE ring artefact. The inner dosimeter volume is not affected by these inaccuracies and should be used for radiotherapy plan verification.

5. Acknowledgements

This work was supported by the Grant Agency of the Czech Technical University in Prague, grant No. SGS11/135/OHK4/2T/14, by TIP program of the Ministry of Industry and Trade, project No. FR-TI3/477 and by the Ministry of Education, Youth and Sports, institutional research plan MSM6840770040.

6. References

[1] Baldock C *et al* 2010 *Phys. Med. Biol.* **55** R1-63
[2] De Deene Y *et al Phys. Med. Biol.* **47** 3441-63
[3] De Deene Y *et al* 2002 *Phys. Med. Biol.* **47** 2459-70
[4] Sedaghat M *et al* 2011 *Phys. Med. Biol.* **56** 601-25
[5] Venning A J *et al* 2005 *Phys. Med. Biol.* **50** 3875-88
[6] Vávrů K *et al* 2010 *J. Phys. Conf. Ser.* **250** 012099
[7] Chain J N M *et al* 2011 *Macromol. Theory Simul.* **20** 735-51
[8] Bankamp A and Schad LR 2003 *Magn. Reson. Imaging* **21** 929-39
[9] Baras P *et al* 2005 *Med. Phys.* **32** 3339-45
[10] Sandilos P *et al* 2006 *Nucl. Instrum. Meth. A* **569** 572-6

Journal of Physics: Conference Series **444** (2013) 012094 doi:10.1088/1742-6596/444/1/012094

Comparison of the influence of inorganic salts on the NMR dose sensitivity of polyacrylamide-based gel dosimeter

S-I Hayashi[1], H Kawamura[2], S Usui[1] and T Tominaga[1]
[1]Department of Clinical Radiology, Faculty of Health Sciences, Hiroshima International University, Higashi-Hiroshima, Hiroshima, 739-2695, Japan
[2]Department of Radiological Sciences, Ibaraki Prefectural University of Health Sciences, 4669-2 Ami, Ibaraki, 300-0394, Japan

E-mail: rin@hs.hirokoku-u.ac.jp

Abstract. On the NMR dose sensitivities of polyacrylamide-based gel dosimeters irradiated by X-ray, the additive effect of various inorganic salts (electrolytes) is investigated. Among the various combination of cations (Li^+, Na^+, K^+, Mg^{2+}, Ca^{2+}, Sr^{2+}, Ba^{2+}, Zn^{2+} and Al^{3+}) and anions (Cl^-, NO_3^- and SO_4^{2-}), $MgCl_2$ is shown to be the most effective sensitizer. In the result, it is suggested that the extent of the increase of the dose sensitivity may correlate to the hydration power of cations rather than anions. Contrary to the dose sensitivity enhancement, the depression of melting point caused by the additives is also pointed out.

1. Introduction

Recently, several efforts have been made to improve the dose sensitivity of polyacrylamide-type gels [1] which contain *N,N'*-methylene-bis-acrylamide (Bis) as a cross linker [1-3][2-4]. In the gels, crosslinking reactions lead the polymer created by irradiation to precipitation, which contributes to increase the dose response in the measurements of MRI [5, 6], X-ray CT [7], optical CT [8] and ultrasound [9]. However the low solubility of Bis in water (at most 3 wt%) sets a limit on increasing the dose sensitivity. Koeva et al. [1][2] and Jirasek et al. [2][3] have increased the solubility of Bis in the *N*-isopropylacrylamide (NIPAM) -based gel by adding a cosolvent (*e.g*., glycerol and isopropanol). Chain et al. [3][4] have reported that the increase of the concentration of monomer (NIPAM) leads the solubility of Bis to increase even in the gel without a cosolvent. Consequently, they caused the increase of dose sensitivity.

In our recent work, we have investigated the effect of some inorganic salts (electrolytes) on the NMR dose sensitivity of a methacrylic-acid-based gel (MAGAT) dosimeter [10]. In the results, it was found that the addition of inorganic salt, especially $MgCl_2$, obviously increased the dose sensitivity. Moreover as the salt concentration increases, the increase of polymerization rate was also observed via the temperature measurements during irradiation. The results indicated that inorganic salts acted as an accelerator for radiation-induced free-radical polymerization in a methacrylic-acid-based gel. Increased dose sensitivity has been also observed in a polyacrylamide-based gel (PAGAT) and a NIPAM-based gel with $MgCl_2$ under the conventional Bis concentration [11]. Surprisingly, the dose sensitivity in a PAGAT gel with 1.0 M $MgCl_2$ was approximately one order higher than in a basic PAGAT gel.

To gain the deeper insight into the effects of inorganic salts on the NMR dose sensitivity in a PAGAT gel, in the present work, we will further examine the influence of various inorganic salts such as Li, Na, K, Mg, Ca, Sr, Ba, Zn and Al chlorides, nitrates and sulphates.

2. Materials and method

Basic PAGAT gels were prepared in the standard composition as shown in table 1. Various inorganic salts (0.4 and 0.8 M) were added into the basic PAGAT gels, respectively. The resulting solution was subdivided by pouring into test tubes (1.5 cm diameter and 10 cm length). After allowed to room temperature, all samples were stored in a refrigerator at a fixed temperature (4 °C) for one day.

The irradiations were performed using the 10 MV photon beam of a medical linear accelerator (EXL-15SP, Mitsubishi Electric Co.). The test tubes were irradiated in a cubic water tank (30 x 30 x 40 cm^3) at a depth of 10 cm at room temperature. The axis of the test tube was perpendicular to the beam. The SSD was 90 cm, and the field size of 10 x 10 cm^2 at the isocenter was used to guarantee the beam flatness. The doses of 2, 4, 6, 8 and 10 Gy were delivered to each sample with a constant dose rate of 1.69 Gy/min.

Dose readout from the samples was performed using a 1.5 T scanner (EXCELART Vantage, Toshiba Medical Systems Co.) with a quadrature torso SPEEDER coil. A multiple spin-echo sequence was applied using the following parameters, (TR / TE's / FOV / Slice Thickness / Matrix Size) = (4 s / 30, 250 ms / 128 x 224 mm^2 / 5 mm / 128 x 224) with a resulting voxel size of 1.0 x 1.0 x 5.0 mm^3. The transverse relaxation rate R_2 is determined by $R_2 = 1/T_2 = ln[S_1/S_2]/(TE_2 - TE_1)$, where T_2 is the transverse relaxation time, and S_1 and S_2 are the signal intensities at echo times, TE_1 and TE_2, respectively.

Table 1: The composition of basic PAGAT gel and the additives of various inorganic salts. For convenience, the molarity of inorganic salts and THPC per 1000 g of water is represented by M in this article.

Chemical	Concentration
Water	89 wt%
Gelatin	5 wt%
Acrylamide	3 wt%
N,N'-methylene-bis-acrylamide	3 wt%
Tetrakis(hydroxymethyl)phosphonium chloride (THPC)	10 mM
Inorganic salt (electrolyte) (Li, Na, K, Mg, Ca, Sr, Ba, Zn and Al chloride (without Al), nitrate and sulfate	0.4, 0.8 M

3. Results and Discussion

Figures 1 shows the NMR dose responses of the PAGAT gel dosimeters containing various inorganic salts. Some salts not shown in the figures were insoluble in a basic PAGAT gel or inhibited the gelation at room temperature.

Figures 1(a) and 1(b) show the dose responses of the PAGAT containing 0.8 M and 0.4 M of various chlorides, respectively. The results indicate that the enhancement of the sensitivity and the concentration dependence are observed more or less in all gels and $MgCl_2$ is the most effective sensitizer among them. In the previous work, it has been suggested that the enhancement of the dose sensitivity is caused by “gel effect” and correlate with the power of hydration of the metal cation added [10]. The hydration is more effective for cations having a smaller radius and larger charge. In the present results also, the enhancement of the sensitivity is larger for bivalent elements than for univalent ones, and is larger for Mg than for Ca. These results may also support our hypothesis.

In comparison with the sensitivities in PAGAT/$MgCl_2$ (0.4M) and in PAGAT/LiCl (or NaCl, KCl) (0.8M), the former shows larger enhancement of the sensitivity than the latter. The power of hydration of Cl^- is generally smaller than that of cation having smaller radius. (For example, the ratio of the hydrated ion radius and the crystal ion radius of Mg^{2+} and Cl^- are 4.8 and 1.4, respectively [12].) Therefore the influence of cations should be dominant rather than anions.

Figure 1(c) shows the dose responses in the PAGAT containing 0.4 M of various nitrates. Similar tendency on the sensitivity to the gel containing chlorides (figure 1(b)) is observed while no appreciable increase in the sensitivity was observed in univalent salts. In the gel containing nitrates, magnesium salt was again the most effective additive.

Figure 1(d) shows the dose responses of the PAGAT containing 0.4 M of various sulfates. The change in the sensitivity is small but still the enhancement is observed in all gels with each additive. In addition, the R_2 values in un-irradiated gels are a little higher than those in the gels containing chlorides or nitrates. It may be due to the insolubility of the sulfates itself.

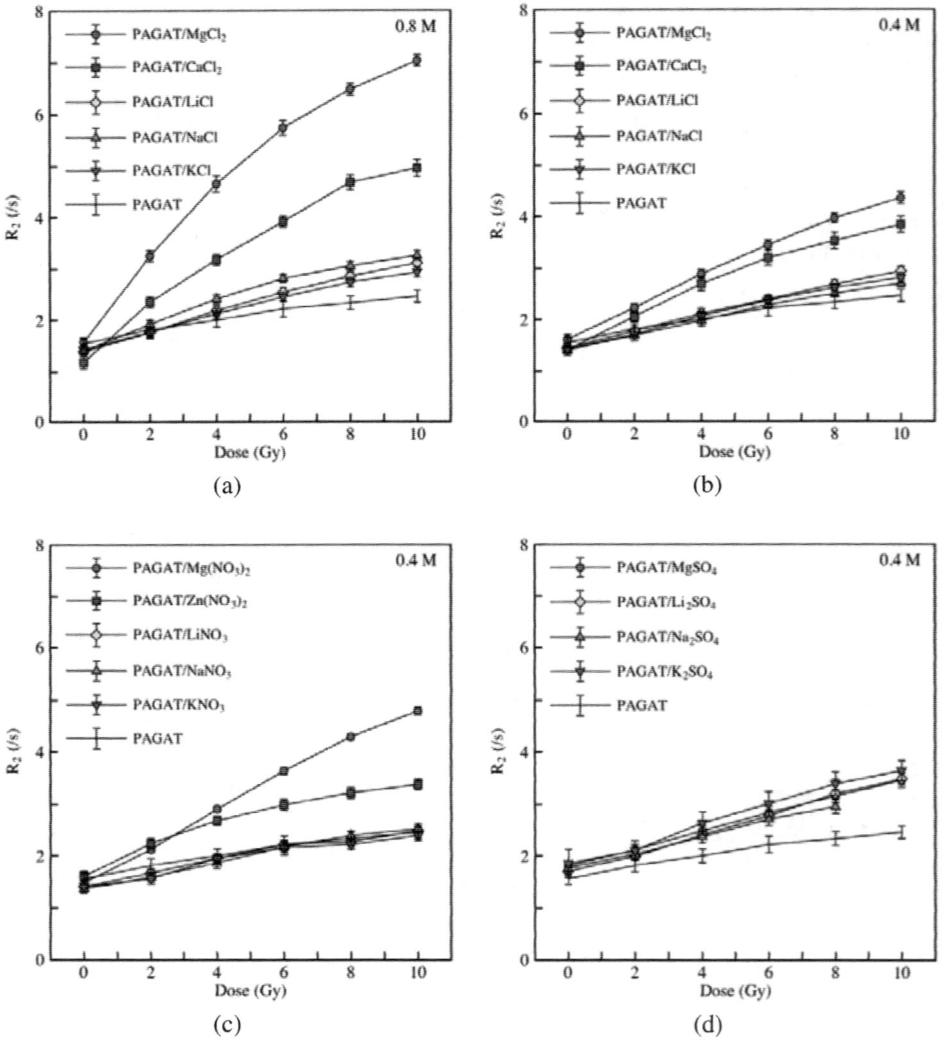

Figure 1: Dose-R_2 responses of PAGAT containing different inorganic salt. (a) : 0.8 M of chlorides, (b) : 0.4 M of chlorides, (c) : 0.4 M of nitrates and (d) : 0.4 M of sulfates. The error bars show the standard deviation in R_2 calculated in a ROI. The data point of PAGAT in all figures comes from same basic PAGAT sample.

The ratio of the hydrated ion radius and the crystal ion radius of NO_3^- and SO_4^{2-} are 1.2 and 1.6, respectively and they are nearly same as Cl^- [12]. Therefore it was expected that the effect is also nearly same as chloride. However, from the above results, they seem to depress the increase of the dose sensitivities except $Mg(NO_3)_2$ while chloride may be the most harmless as a counter ion.

4. Conclusion

In this work, we investigated the influence of various inorganic salts on the NMR dose sensitivities in a polyacrylamide-based gel. In the results, $MgCl_2$ was indicated the most promising sensitizer among the inorganic salts examined in this work. From the viewpoint of the tissue equivalency also, $MgCl_2$ may be more desirable rather than other heavy atoms. However the influence of the depression of melting point was observed in the gel with higher concentration of inorganic salts. In order to apply the gel dosimeter with inorganic salt to clinical use, the optimization of the composition and the use of a cross-linker for gelatin-matrix such as formaldehyde [13] are required. On the other hand, polymer gel dosimeters containing inorganic salt may be also beneficial for use with other imaging modalities such as X-ray and optical CT. The studies on other characteristics of polymer gel containing inorganic salt, such as dependency on dose rate and dose integral, spatial and temporal stability, and water and tissue equivalency, are under way.

5. References

[1] Baldock C *et al* 2010 *Phys. Med. Biol.* **55** R1-63
[2] Koeva V I *et al* 2009 *Phys. Med. Biol.* **54** 2779-90
[3] Jirasek A *et al* 2010 *Phys. Med. Biol.* **55** 5269-81
[4] Chain J N M *et al* 2011 *Phys. Med. Biol.* **56** 2091-102
[5] Lepage M *et al* 2002 *Phys. Med. Biol.* **47** 1881-90
[6] Murry P and Baldock C 2000 *Australas. Phys. Eng. Sci. Med.* **23** 44-51
[7] Trapp J V *et al* 2001 *Phys. Med. Biol.* **46** 2939-2951
[8] Bosi S G *et al* 2009 *Phys. Med. Biol.* **54** 275-83
[9] Mather M L *et al* 2003 *Ultrasonics* **41** 551-9
[10] Hayashi S *et al* 2012 *Radiat. Phys. Chem.* **81** 884-8
[11] Hayashi S *et al* 2012 (In press)
[12] Marcus Y,1977 Introduction to Liquid State Chemistry (Wiley, London)
[13] Pavoni J F 2012 *Med. Phys.* **39** 2877-84

Journal of Physics: Conference Series **444** (2013) 012095 doi:10.1088/1742-6596/444/1/012095

The effect of chemical stability on the NIPAM gel dosimeter using 1H-NMR

You-Ruei Huang[1], Ling-Ling Hsieh[1], Yuan-Jen Chang[23] and Bor-Tsung Hsieh[1]*
[1]Department of Medical Imaging and Radiological Sciences, Central Taiwan University of Science and Technology, Taichung, Taiwan, ROC
[2]Institute of Biomedical Engineering and Material Science, Central Taiwan University of Science and Technology, Taichung, Taiwan, ROC
[3]Department of Management Information Systems, Central Taiwan University of Science and Technology, Taichung, Taiwan, ROC

E-mail: bthsieh@ctust.edu.tw

Abstract. Radiation-induced chemical changes in the N-isopropylacrylamide (NIPAM) gels used in three-dimensional dosimeters were investigated using 1H-NMR in this study. The experimental results show that the signal from C=C bonds of NIPAM and N,N'-Methylenediacrylamide (BIS) are 5.5 and 6.3 ppm, respectively. The double bonds from the NIPAM and BIS disappeared with half-dose (D50) were about 10.90 Gy ± 0.76 Gy and 10.09 Gy ± 0.29 Gy, respectively. This observation demonstrates that the polymerization rate of BIS is faster than that of the NIPAM monomer. The 1H-NMR can indicate the chemical structure changes of the polymer gel dosimeter after irradiation and successfully determine the D50 in the NIPAM gel dosimeter.

1. Introduction

In clinical radiotherapy, polymer gel dosimeter is used to verify three-dimensional (3D) dose distributions. Such common dosimeters as ionization chamber, thermo luminescent diodes, and radiographic film do not allow the measurement of 3D dose distribution due to several spatial laminations [1].

Maryanski et al. proposed the use of a polyacrylamide gel as a potential tool for 3D mapping of dose distribution. Since this initial study, several polymer gels have been proposed, including Bis-acrylamide, Acrylamide, Nitrous oxide And Agarose (BANANA), Bis-acrylamide, Acrylamide, Nitrogen, Gelatin (BANGTM), Polyacrylamide Gel, THPC (PAGAT), and NIPAM gel consisting of gelatin, N-isopropylacrylamide, BIS, and tetrakis (hydroxymethyl) phosphonium chloride (THPC)[2].

Researchers have also investigated several readout tools in an attempt to obtain dose distribution information from polymer gel, including MRI, optical computed tomography (OCT), x-ray computed tomography (CT), and Fourier transform-Raman (FT-Raman) spectroscopy. However, the response of polymer gel dosimeter to radiation has yet to be completely understood. NMR and FT-Raman can quantify the chemical changes in molecules [3–5].

The NIPAM polymer gel consists of NIPAM and BIS dissolved in an aqueous gelatin matrix. The results provide evidence for the polymerization of the monomers that form a NIPAM gel network within the gelatin matrix. In this study, 1H-NMR was used to analyze the technique of correction in characterizing the consumption of monomer and crosslinking in irradiated NIPAM gel. Furthermore, direct evidence of polymer gel formation has been observed in 1H-NMR.

2. Materials and Methods

2.1. Manufacture of the NIPAM gel

The gel used in this study consisted of 5% gelatin (300 Bloom Tape A. Sigma-Aldrich), 5% NIPAM (97%, Sigma-Aldrich), 3% BIS (Merck), 5 mM THPC (80%, Sigma-Aldrich), and 87% deionized water. The method used for the manufacture of the NIPAM gel has been described previously [6]. The gelatin was added to water at room temperature and left to soak for 10 min. The solution was heated and maintained at a temperature of 45 °C. NIPAM and BIS were added and magnetically stirred for 25 min until complete dissolution is achieved. Finally, THPC was added to remove the oxygen. The tube was then covered using a Pyrex screw cap and placed in a refrigerator at 4 °C.

2.2. Irradiation of the NIPAM gel

All samples were irradiated with a Varian 21EX Clinic linear accelerator using 6 MV photons, a dose rate of 400 MU/min, 100 cm surface to source distance (SSD), and field size of 20 cm × 20 cm. At the center of a short side of a customized 30 cm × 30 cm × 4 cm acrylic phantom, a 16 mm diameter hole was punctured to accommodate the Pyrex test tube. The irradiated gels were left in a refrigerator for one day before imaging to allow complete post irradiation polymerization.

2.3. Measurement system

An optical laser scanning system was used to measure the NIPAM gel. An NRC Model 127 He-Ne laser with 20 mW power and 632.8 nm wavelength was used in CT-s1. After a 140 min laser warm up, the laser power deviation was less than 1%. Room temperature was maintained at 22±1°C, and a photodetector was used to scan the gel dosimeter as depicted in a previous research [7].

The NMR measurement was performed using a BRUKER AVANCE 400 SSNMR spectrometer operating at a frequency of 400 MHz for 1H. The 1H-NMR was used to examine the chemical shift of the NIPAM gel and observe the detritions of the NIPAM and BIS after irradiation.

3. Results and Discussion

3.1. Dose response of the NIPAM gel

Fig. 1(a) shows the dose response of the NIPAM gel presented in this study. The dose response is linear from 0 Gy to 15 Gy, with the linearity of 0.997 and sensitivity of 0.023 Gy-1. The dose response curves of the NIPAM gel dosimeter under two distinct photon beam qualities (6 and 10 MV) are obtained. Fig. 1(b) shows the insignificant variation in the dose response within the range of 0 Gy to 15 Gy. The result is similar to those of most published reports on gel dosimeters 【7】

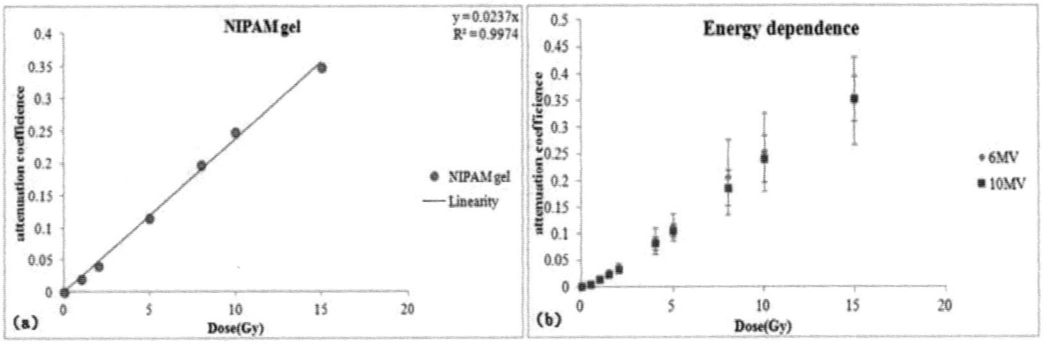

Figure 1: (a) Dose response of the NIPAM gel (b) Energy dependence of the NIPAM gel

3.2. 1H-NMR of the NIPAM gel

Fig. 2 shows the 1H-NMR spectrum of the NIPAM gel. The CH2 functional group of the gelatin is located at 1.04 ppm, and the NH functional group of BIS appears at 8.2 ppm. Bands at 5.5 and 6.3 ppm corresponding to both NIPAM and BIS vinyl group modes, respectively, show clearly decreasing peak intensities as copolymerization and crosslinking increase with absorbed radiation dose. This result indicates that the polymerization induced by NIPAM through irradiation and its effect are achievable using the dose in the current study.

The spectra of 1H-NMR are similar to that of the functional groups polymerized via PAG irradiation, the results of which concur with those of the PAG irradiation performed by Lepage et al. In their work, the C=C bonds of the acrylamide and BIS in PAG are near 5.9 and 6.3 ppm, respectively 【4】. In the current study, the values for NIPAM and BIS are 5.5 and 6.3 ppm, respectively. Considering that the acrylamide has a chemical structure similar to that of NIPAM (Fig. 3), the 1H-NMR spectrum of the NIPAM gel is quite similar to that of PAG (Table 1).

Fig. 4, meanwhile, shows that the C=C intensity gradually decreases as the absorbed dose increases. The result indicates that the D50 of vinyl group to the NIPAM and BIS were about 10.90 Gy ± 0.76 Gy and 10.09 Gy ± 0.29 Gy, respectively. In comparison with PAG gel dosimeter, the D50 of the monomers and crosslinking agent in the PAG gel dosimeter is 7.62 Gy. This difference is possibly attributed to the different monomers and the proportion of ingredients in the gel dosimeter. This observation demonstrates that the polymerization rate of BIS is faster than that of NIPAM. Thus, NIPAM polymerization needs a higher irradiation absorbed dose than BIS.

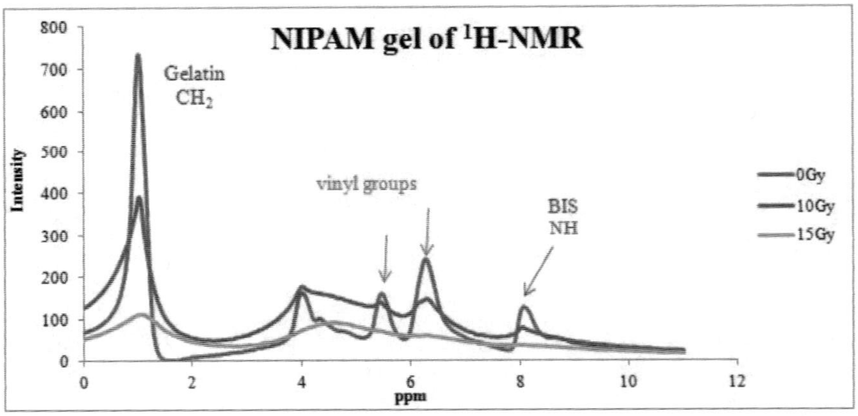

Figure 2: 1H-NMR spectrum of the NIPAM gel

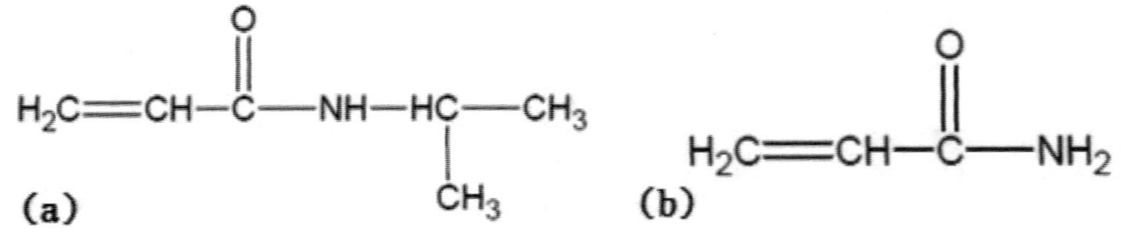

Figure 3: Structural formulas of the (a) NIPAM gel and (b) acrylamide

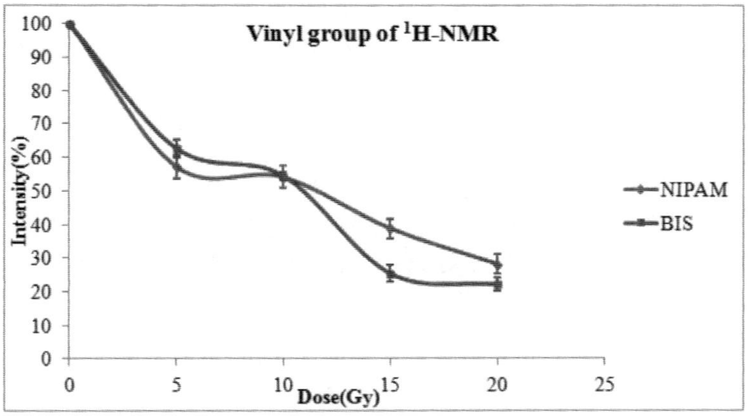

Figure 4: Correlation of the vinyl group of the NIPAM gel in the 1H-NMR spectrum

Table 1: Chemical shift assignment

Chemical Shift (ppm)	Assignment
1.04	CH_2 group
5.5	Vinyl group of NIPAM
6.3	Vinyl group of BIS
8.7	NH group of BIS

4. Conclusion

In this work, we characterized with the chemical changes occurring in a NIPAM gel dosimeter using 1H-NMR.The intensity of the C=C bonds of NIPAM and BIS decrease as the absorbed radiation dose increases during the NIPAM gel dosimeter irradiation. The D50 of NIPAM and BIS were about 10.90 Gy ± 0.76 Gy and 10.09 Gy ± 0.29 Gy, respectively. When the NIPAM gel is irradiated, the amount of monomer and crosslinking decrease, while the amount of polymerization increase. Hence, the decreasing rate of monomer and the increasing rate of polymerization should be investigated further.

5. References

[1] Wagter C D 2004 *J. Phys.: Conf. Ser.* **3** 4-8
[2] Senden R J *et al* 2006. *Phys. Med. Biol.* **51** 3301-14
[3] Lepage M *et al* 2001 *J. Appl. Polym. Sci.* **79** 1572-81
[4] Jirasek A I *et al* J 2001 *Phys. Med. Biol.* **46** 151-65
[5] Hsieh L L *et al* 2012 *PLoS One* **7** 1-7
[6] Chang Y J *et al* 2011 *Nucl. Instrum. Methods Phys. Res. A* **652** 783-5
[7] Hsieh B T *et al* 2011 *J. Radioanal. Nucl. Chem.* **290** 141-8

7th International Conference on 3D Radiation Dosimetry (IC3DDose) IOP Publishing
Journal of Physics: Conference Series **444** (2013) 012096 doi:10.1088/1742-6596/444/1/012096

Monomer consumption in MAGIC-type polymer gels in the Bragg-peak of proton beams observed by volume selective ^{1}H MR-spectroscopy (MRS): proof of principle for high resolution MRS-methodology with a sensitive rf-detector

AI Schmid[1,2], E Laistler[1,2], J Sieg[1,2], B Dymerska[1,2,3], M Wieland[1,2,4], J Naumann[5], O Jaekel[5] and A Berg[1,2]
[1]Center for Medical Physics and Biomedical Eng. Med. University of Vienna, Vienna, Austria
[2]High-Field MR-Centre of Excellence, Medical University of Vienna, Austria
[3]Institute of Solid State Physics, Gdansk University of Technology, Poland
[4]Institut für Radioonkologie, Sozialmedizinisches Zentrum Ost - Donauspital Wien; Austria
[5]Heidelberg Ion Therapy Center (HIT), Heidelberg, Germany

E-Mail: andreas.berg@meduniwien.ac.at

Abstract. Mono-energetic proton and heavy ion beams for tumour therapy feature high dose gradients laterally and at its penetration depth, characterized by the Bragg-peak. The 3-dimensional dosimetry of such Hadron particle beams poses high demands on the spatial resolution of the imaging methodology and linearity of the polymer gel dose response in a wide dose range and at high linear energy transfer (LET). In almost all polymer gels the Bragg-peak dose response is therefore quenched. Volume selective MR-spectroscopy is in principle capable of delivering information on the polymerization process. We here present the MR-methodology to obtain MR-spectroscopic (MRS) data on the monomer consumption at the very small voxel volumes necessary for resolving e.g. the Bragg-peak area. Using additional hardware components, i.e. a strong gradient system and a very sensitive rf-detector at a high field human 7T scanner, MR-microimaging and MRS with 600 μm depth resolution can be implemented at very short measurement time. The vinyl groups of methacrylic acid in a MAGIC-type polymer gel can be resolved by volume selective MRS. The complete monomer consumption in the Bragg-peak due to polymerization is demonstrated selectively in the Bragg-peak indicating one main reason for Bragg-peak quenching in the investigated polymer gel.

1. Purpose/Introduction

The 3D-dosimetry of proton ion beams for tumor therapy with single energy remains difficult due to the fine dose placement in the Bragg-peak. For evaluation beyond planar film dosimetry 3D-arrangements of pin-point ionization-chambers are used as standard for the verification of treatment planning systems [1]. However the spatial resolutions (voxel volume) determined by the ionization chamber distances in these arrangements are significantly lower (10^3) than achievable with polymer gel dosimetric imaging [2, 3]. The MR-Parameter T2 has shown to be most sensitive to the polymerization process in polymer gels subsequent to irradiation [4, 2]. T2-maps can be obtained for resolving the Bragg-peak region [5-8]. Besides from other factors, the spatial resolution in imaging is critical for the accurate determination of dose in the fine Bragg-peak enhancement area for single particle energy [7]. T2-parameter imaging does not give detailed insight in the molecules involved during the polymerization procedure and the amount of monomer consumption. FT-Raman spectroscopy has been applied for the analysis of monomer consumption in acrylamide-Bis-acrylamide polymer gel variants [2]. It has also been demonstrated, that Magnetic Resonance (MR)-spectroscopy and spectroscopic imaging might yield useful information on the relative consumption in the monomer co-monomer mixture in a photon field for poly-acrylamide and methacrylate-co-monomer gels [9]. MR-spectroscopic imaging using a chemical shift selective refocusing pulse might even be used in principle for dosimetric imaging [9].

Heavy ion and proton therapy is characterized by the high ionization density along and especially at the very end of the beam path, i.e. high linear energy transfer dE/dz in the Bragg-peak regime [6] (see figure 1). In addition the high deposited energy per unit mass might result in a complete consumption of the monomer in the very small volume characterized by the Bragg-peak and thus quenching of the dose response in T2-imaging in this region ("Bragg-peak quenching").

Volume selective MR-spectroscopy (MRS), adapted to the very small volumes in the Bragg-peak, might be capable of the detection of this monomer consumption. We therefore investigated, whether it is possible to measure the monomer concentration at the very small voxel volumes within and outside of the Bragg-peak area, using a high field (7T) human MR-scanner equipped with a microscopy insert with a highly sensitive rf-detector.

2. Materials and methods

2.1. Polymer gels

We have manufactured several sets of "normoxic" polymer gels with different additives for sensitivity and dose range modifications. The presented data belongs to a variant using ascorbic acid and copper-sulfate (similar to MAGIC gels [10]) as oxygen scavenger, methacrylic acid MAA (6% w/w) as monomer, gelatin (14% w/w) and water. The MAGIC-variant exhibited a linear dose response in the investigated dose range up to the maximum dose D_{max} = 48 Gy (figure1). The polymer gel is filled into Barex® flacons (o.d = 28 mm) to avoid oxygen penetration from outside. The polymer gels were irradiated with a mono-energetic proton ion beam.

2.2. Irradiation

For reference ^{60}Co-photons were applied to the same type of polymer gels. The T2-response for the polymer gel variant was measured. The specimen at D = 0 and D = 16 Gy were also investigated by MRS. Mono-energetic (E = 59.8 MeV) proton beams for human therapy were applied to the polymer gel at a beam diameter of 6cm at the Heidelberger Ion-beam-Therapy-center (HIT). The penetration depth for this beam amounts to about 28 mm in water. The investigated sample experienced a dose of D = 32 Gy (plateau).

2.3. Magnetic resonance imaging and spectroscopy

Hardware: The high spatial selectivity with small voxels in MRS was achieved on a 7T human ultra-high-field MR-scanner, equipped with a small sized (i.d. 90 mm) strong gradient system (G = 750 mT/m) [11]. A very sensitive solenoid type detector (i.d. 28.5 mm) was designed particularly for the use with the Barex® flacons. Numerical simulations were performed to optimize the geometry of the solenoidal radiofrequency (rf-) detector to improve the homogeneity within the sample volume [12]. A customized housing was designed and built to fit the detector into the gradient insert and allow frequency adjustment and impedance matching from the outside.

Pulse-sequence and protocols: T2-weighted images using Multi-slice multi echo imaging are obtained for reference position (Fig.3a). FOV= $(70\ mm)^2$, voxel size = $(0.546)^2$ x 1 mm^3, Matrix = 128 x 128, TE = 10 ms; 20 echoes. MRS-data was obtained within 2 min at a voxel size of 12 x 12 x 0.6 =

8.6 mm^3 with a depth of 600 μm in the beam direction, necessary to spatially resolve the Bragg-peak area. Volume selection was achieved with a STEAM pulse sequence [13] (TR = 6 s, TE = 15 ms, av:16 , TM < 100 s). The samples were shimmed manually to a water line-width of about 20-40 Hz.

3. Results

The polymer gel variant exhibits a linear dose response to ^{60}Co-photons up to the maximum investigated dose level D_{max} = 48 Gy (figure 1). A high-resolution proton ion beam dose-map of the polymer gel, irradiated with D = 32 Gy (plateau), is shown in figure 2a. The sharp dose edge at the Bragg-peak is clearly delineated.

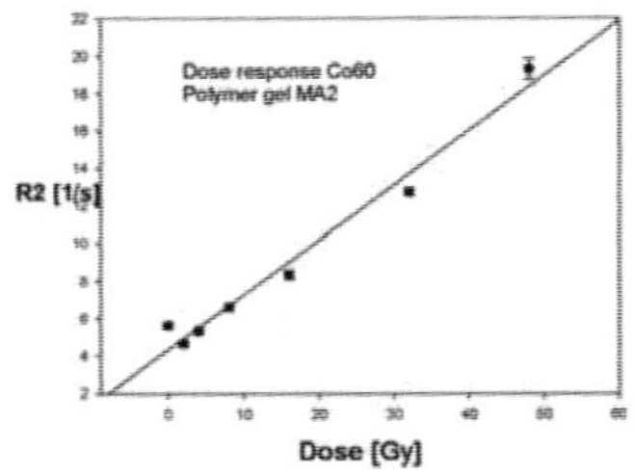

Figure 1: Relaxation rate (R2 = 1/T2) of the polymer gel MA2 to the dose (^{60}Co-photons). The dose response is linear in the investigated range from D = 0 up to 48 Gy.

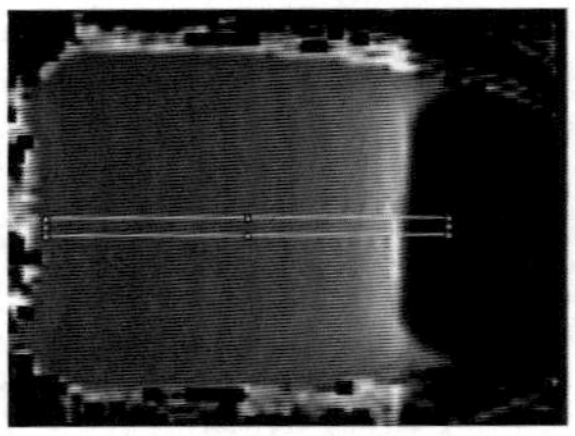

Figure 2a: Dose image calculated from a T2-map of the polymer gel irradiated with a proton beam at D = 32 Gy (plateau). The beam enters at left and penetrates the polymer gel until absorption at the Bragg-peak (bright line at right).

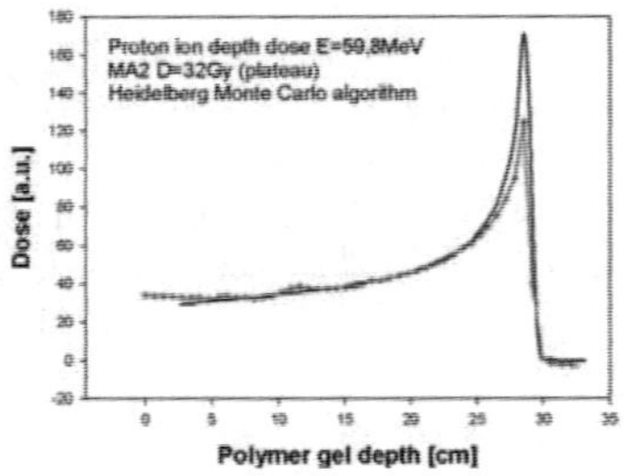

Figure 2b: Dose profile along the indicated path in figure 2a. For reference the result of a Monte-Carlo-simulation for the depth dose path is indicated (normalized to 32 Gy in the plateau and shifted to Bragg-peak coincidence).

Volume selective ^{1}H-spectroscopic data is shown for three different regions of interest in figure 3:

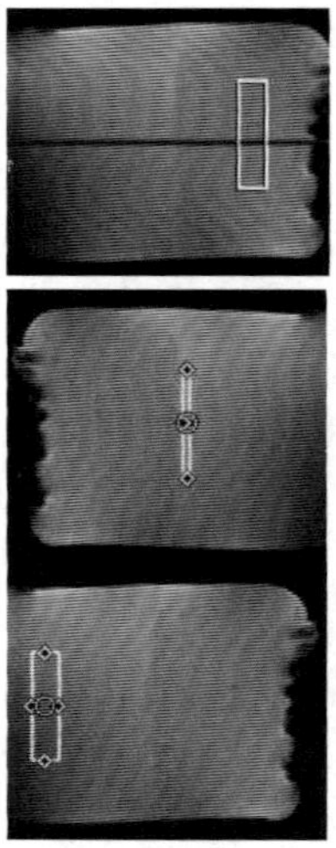

top: position behind the Bragg-peak in area of no dose and no polymerization,

center: small sized volume in the Bragg-peak,

bottom: position of the MRS selected volume in the plateau.

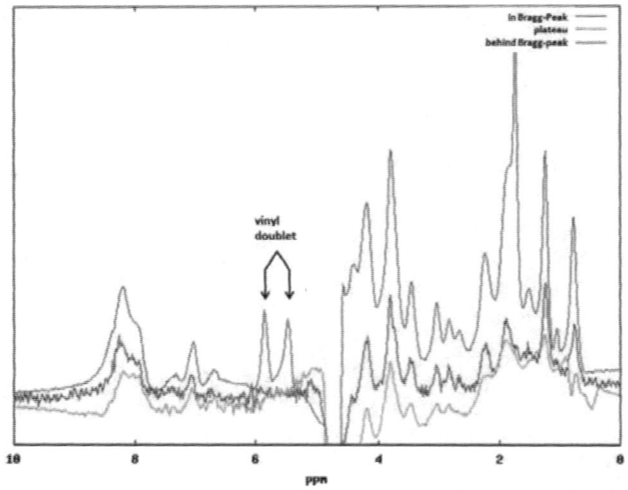

Figure 3a: Positions of the detected volume for ^{1}H -volume selective MR-spectroscopy.
region of the proton beam.

Figure 3b: ^{1}H-MR-spectra from volumes as indicated in figure 3a. Behind the Bragg-peak at D=0Gy the doublet signal of the vinyl protons at 5.4 and 5.8 ppm (arrows) from the monomer is clearly observed. The vinyl doublet disappears in the Bragg-peak (medium spectrum, red) indicative of the monomer consumption.

1. Behind the Bragg-peak (figure 3a, top), i.e. in a region of presumably no dose. The topmost (grey) spectrum in figure 3b reveals the vinyl doublet signal (s = 5.4 and s = 5.8 ppm), connected to the

protons near the double bindings of the carbon atoms in MAA. For D=0 Gy, the doublet-signal is enhanced, indicating high concentration of 6% for MAA in the investigated polymer gel.
2. In the *plateau region* (figure 3a, bottom), the vinyl group doublet is already strongly reduced, indicating monomer consumption (figure 3b, light green spectrum).
3. In the small detected volume *in the Bragg-Peak* (figure 3a, center), the vinyl group has completely disappeared (figure 3b, medium spectrum, dark-red line), indicating that the monomer is completely consumed and the dose response might be suppressed in the Bragg-peak.

4. Discussion/Conclusion

Using additional hardware components, i.e. a strong gradient system and a very sensitive object adapted rf-detector in combination with an ultra-high field human 7 T scanner, volume selective spectroscopy can be implemented even at the very small voxel volumes necessary for investigations of the Bragg-peak. In the MR spectrum, the doublet from protons near the vinyl double bond can easily be detected within 2 min measurement time. The reduction of monomer concentration due to polymerization can be observed. MAGIC polymer gels with a linear dose response far beyond (i.e. D = 48 Gy) a nominal entrance dose ($D_{plateau}$ = 32 Gy) might exhibit significant dose quenching in the Bragg-peak due to complete monomer consumption. Molecule specific information, even in the Bragg-peak in a small area of high LET, as well as highly localized chemical information on the polymerization process in gel dosimetry can be obtained from volume selective spectroscopy. The absolute accuracy and stability of the spectroscopic polymer gel dose response, particularly in the Bragg-peak regime, still has to be investigated for different dose levels. Our findings might represent the basis for future calibration and quantification of spectroscopic data with regard to dose imaging based on MR spectroscopic imaging [9] at very high spatial resolution.

5. Acknowledgements

We thank Dietmar Georg and Uwe Wolff, Medical University of Vienna, Department of Radiation Oncology for the access to the ^{60}Co-Photon beam facility, the introduction to operation of the ^{60}Co-Teratron and ionization chamber reference dosimetry. The project is financially supported by the "Hochschuljubiläumsfonds" of the City of Vienna (HSJ, project nr. H-1208/2003, A. Berg).

6. References

[1] Karger C P *et al* 1999 *Med. Phys.* **26** 2125-32
[2] Baldock C *et al* 2010 *Phys. Med. Biol.* **55** R1-63
[3] Berg A *et al* 2004 *Phys. Med. Biol.* **49** 4087-108
[4] Maryanski M J *et al* 1993 *Magn. Reson. Imaging* **11** 253-8
[5] Ramm U *et al* 2000 *Phys. Med. Biol.* **45** N95-102
[6] Gustavsson H *et al* 2004 *Phys. Med. Biol.* **49** 3847-55
[7] Berg A *et al* 2005 *Biomedizinische Technik* **50** 118-9
[8] Zeidan O A *et al* 2010 *Med. Phys.* **37** 2145-52
[9] Murphy P S *et al* 2000 *Phys. Med. Biol.* **45** 835-45
[10] Fong P M *et al* 2001 *Phys. Med. Biol.* **46** 3105-13
[11] Berg A *et al* 2010 *Proc. ISMRM*, Stockholm, Sweden
[12] Laistler E *et al* 2012 *Proc. ESMRMB*, Lisbon, Portugal
[13] Frahm J *et al* 1989 *Magn. Reson. Med.* **9** 79

A novel method for 4D measurement-guided planned dose perturbation to estimate patient dose/DVH changes due to interplay

B Nelms[1] and V Feygelman[2]
[1]Canis Lupus LLC, Merrimac, Wisconsin 53561, USA
[2]Department of Radiation Oncology, Moffitt Cancer Center, Tampa, Florida 33612, USA

E-mail: alpha@canislupusllc.com

Abstract. As IMRT/VMAT technology continues to evolve, so do the dosimetric QA methods. We present the theoretical framework for the novel planned dose perturbation algorithm. It allows not only to reconstruct the 3D volumetric doe on a *patient* from a measurement in a cylindrical *phantom*, but also to incorporate the effects of the interplay between the intrafractional organ motion and dynamic delivery. Unlike in our previous work, this 4D dose reconstruction does not require the knowledge of the TPS dose for each control point of the plan, making the method much more practical. Motion is viewed as just another source of error, accounted for by perturbing (morphing) the planned dose distribution based on the limited empirical dose from the phantom measurement. The strategy for empirical verification of the algorithm is presented as the necessary next step.

1. Introduction

Historically, the most frequently used method for patient-specific end-to-end testing [1] of inverse-planned delivery was limited comparison of the measured and calculated doses in a *phantom*. While this approach was sufficiently rigorous to ensure safe adoption of IMRT techniques in wide clinical practice, it was pointed out that it lacks correlation to the clinically meaningful metrics [2, 3]. It is now recognized that ideally a deliverable dose to the *patient* should be reconstructed volumetrically based on the measurement and compared to the TPS calculations [3, 4]. But even that may not always be sufficient as the interplay between the dynamic delivery and intrafraction organ motion is likely to alter doses accumulated by different structures. Therefore, a 4D approach to patient dose reconstruction is needed for patient-specific end-to-end testing, particularly in the case of hypofractionated/stereotactic delivery.

A prototype method for measurement-guided VMAT 4D patient dose reconstruction was described by Nelms et al [5]. It is based on 3DVH software package (Sun Nuclear Corp., USA). The basic premise of the approach is that the volumetric dose on the *patient* can be estimated by perturbing the patient TPS dose distribution based on the *phantom* measurements. The software was modified to allow this perturbation to occur at the control point (CP) dose level, thus creating a time-resolved volumetric (4D) dose on a *patient*. While proved accurate [5], this method requires the knowledge of planned 3D dose matrices per CP. Not every TPS exports such information and a substantial amount of data needs to be stored and manipulated, rendering this approach less than ideal in practice. Hence in this paper we set out to demonstrate in theory how measurement-guided 4D dose reconstruction on the patient can be done bypassing the TPS CP doses, by treating the motion as another source of perturbation to the TPS-generated dose that can be quantified from the 3D phantom dose.

2. Method description

2.1. Dose to stationary target in a phantom

A stationary target can be defined by (i.e. discretized into) a collection of small voxel elements in the couch coordinate system (Figure 1). A contoured target volume will contain a finite set of voxel elements in the Cartesian grid, each with a corresponding radius-vector $\mathbf{r}_{ijk} = [x_i, y_j, z_k]$

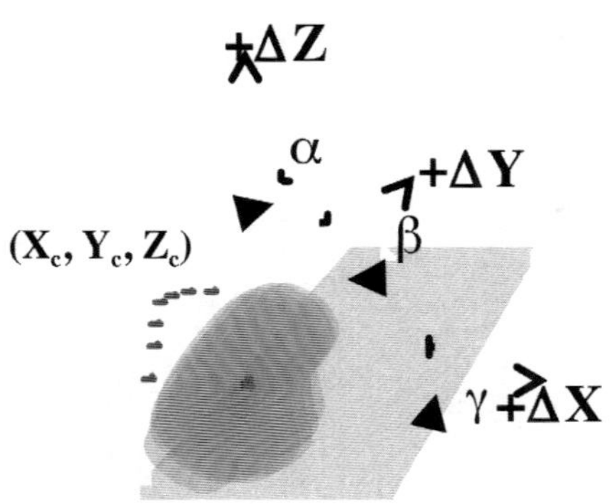

Figure 1: Six degrees of freedom vs. time are used to define the rigid body transformation model for a radiation therapy target in motion. The orthogonal motion axes are parallel to the unified couch coordinate system axes. (ΔX, ΔY, ΔZ) = (0,0,0) represents the centroid of the target volume at t = 0, with that centroid located at point (Xc, Yc, Zc) in the couch coordinate system. Rotations are about this centroid.

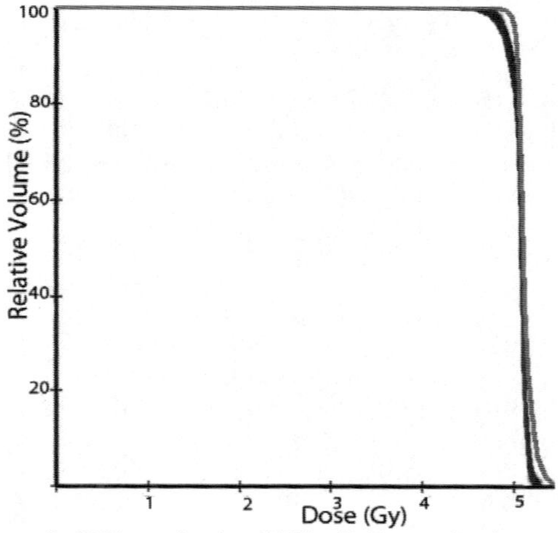

Figure 2: PTV per-fraction DVHs (5 Gy per fraction VMAT plan) simulated for 10 fractions, compared to stationary target DVH. The blue (wider) curves overlap to capture the variation of "Motion DVHs" over 10 fractions, and the red curve is the stationary target DVH.

If continuous, instantaneous dose rate as a function of time in the fixed couch coordinate system is $\bar{d}(x,y,z,t)$, the cumulative dose for a stationary voxel element ***(i,j,k)*** over a single fraction (T_0 to T_f) is given by:

$$D^{\text{Phantom}}_{\text{St, 1 fx}} = \int_{T_0}^{T_f} \bar{d}(x_i, y_j, z_k, t)\, dt$$

And for a course of N fractions, the cumulative dose for a stationary voxel element ***(i,j,k)*** is simply:

$$D^{\text{Phantom}}_{\text{St, N fx}} = N \times \int_{T_0}^{T_f} \bar{d}(x_i, y_j, z_k, t)\, dt$$

2.2. Dose to moving target in a phantom

A target in motion can be approximated by the finite set of voxels that move in time, generally described in the unified couch coordinate system as $r_{ijk}(t) = [x_i(t), y_j(t), z_k(t)]$. For now, target motion is modeled as a rigid body with six degrees of freedom, each of the six a function of time:

$$\{\Delta x(t), \Delta y(t), \Delta z(t), \alpha(t), \beta(t), \gamma(t)\}$$

At any given time ***t*** a moving voxel's location can be mapped to the couch coordinate system (Figure 1):

$$\begin{bmatrix} x_i(t) \\ y_j(t) \\ z_k(t) \\ 1 \end{bmatrix} = \begin{bmatrix} \cos\alpha\cos\beta & \cos\alpha\sin\beta\sin\gamma - \sin\alpha\cos\gamma & \cos\alpha\sin\beta\cos\gamma + \sin\alpha\sin\gamma & \Delta x \\ \sin\alpha\cos\beta & \sin\alpha\sin\beta\sin\gamma + \cos\alpha\cos\gamma & \sin\alpha\sin\beta\cos\gamma - \cos\alpha\sin\gamma & \Delta y \\ -\sin\beta & \cos\beta\sin\gamma & \cos\beta\cos\gamma & \Delta z \\ 0 & 0 & 0 & 1 \end{bmatrix} \begin{bmatrix} x_i \\ y_j \\ z_k \\ 1 \end{bmatrix}$$

The same continuous, instantaneous dose rate as a function of time in the fixed couch coordinate system is used as the source of 4D dose data for all voxels in a moving target. Potential small perturbations in the dose matrices due to patient shape and internal anatomy changes due to motion are ignored. But for a moving target, each voxel's instantaneous location is also a function of time, so per-voxel motion traces are performed to accumulate dose over a full fraction:

$$\mathrm{D}^{\mathrm{Phantom}}_{\mathrm{Mov,1\,fx}} = \int_{T_0}^{T_f} \bar{d}(x_i(t), y_j(t), z_k(t), t)\, dt$$

For a course of N fractions, if the phase of motion is allowed to vary per fraction (with respect to the time frame of the dynamic delivery), the cumulative dose for a stationary voxel element ***(i,j,k)*** is no longer simply multiplicative with respect to the number of fractions. Instead, per-fraction variation due to motion phase must be modeled as:

$$\mathrm{D}^{\mathrm{Phantom}}_{\mathrm{Mov,N\,fx}} = \sum_{fx=1}^{N} \int_{T_0}^{T_f} \bar{d}(x_i(t_{fx}), y_j(t_{fx}), z_k(t_{fx}), t)\, dt$$

where t_{fx} represents an altered phase in the per-voxel motion trace with each fraction "fx". Phases are randomized. If the number of fractions is small (i.e. hypofractionation), many courses of N fractions can be simulated to capture the differences due to small sampling of potential motion phase shifts.

2.3. Accounting for time-discretized 4D dose grids

The six degrees of motion are derived empirically (e.g. from 4D CT) and are entered as values at known relative time points. Assuming there are enough multi-phased, time-resolved data points entered, the motion trajectory can be treated as continuous in time with smooth transitions between the entered data values, each of the six degrees of freedom interpolated linearly with time.

However, the theoretical continuous, instantaneous dose rate as a function of time is not known. Instead, what is known is a time-resolved series of dose grids (i.e. a 4D dose grid set). Each component of the 4D dose series is a 3D grid of dose accumulated over a short period of time ΔT. Those ΔT intervals are determined by the measurement-guided dose reconstruction process and are not all equal. For VMAT plans, ΔT intervals typically range from 0.2-0.4 seconds.

Thus, the theoretical models of dynamic dose must be discretized. For the stationary target, the integral of a full fraction dose to a voxel element becomes a summation:

$$\mathrm{D}^{\mathrm{Phantom}}_{\mathrm{st,1\,fx}} = \sum_{Dose=4D_0}^{4D_f} Dose(x_i, y_j, z_k, \Delta T_i)$$

where $4D_0$ through $4D_f$ represent the range of time-discretized 4D dose grids covering the full fraction delivery. For the moving voxels, instantaneous dose rates per (x,y,z) within the 4D dose grids must be considered in order to model the continuous motion, and the full fraction dose to a moving voxel becomes:

$$D^{\text{Phantom}}_{\text{Mov, 1 fx}} = \sum_{Dose=4D_0}^{4D_f} \int_{\Delta T_i start}^{\Delta T_i end} \left[\frac{Dose(x_i(t), y_j(t), z_k(t), \Delta T_i)}{\Delta T_i} \right] dt$$

This in turn must be discretized in time rather than solved in closed form within each ΔT interval (in order to model motion within each 4D dose time interval). Sub-intervals of ~0.1 seconds are more than sufficient to capture the range of motion.

2.4. *Estimating patient dose/DVH changes due to interplay*

So far we addressed the dose values in a homogeneous cylindrical *phantom*. The dose perturbation method described previously can equally well estimate *patient* dose for a full fraction or any interval thereof, but for the latter the TPS dose per CP has to be known [5]. This is impractical, but fortunately it can be shown that the same perturbation method used to estimate the patient dose from the phantom measurement and dose reconstruction can be readily employed to estimate the impact of motion.

For motion perturbation, each target voxel is still analyzed, but the static patient dose voxels are now modified for motion errors/differences derived from the phantom doses. For each target dose voxel element (***i,j,k***) the estimated patient dose to the moving voxel is well approximated for 1 fraction and N fractions by:

$$D^{\text{Patient}}_{\text{Mov ijk, 1 fx}} = D^{\text{Patient}}_{\text{St ijk, 1 fx}} \times \frac{D^{\text{Phantom}}_{\text{Mov ijk, 1 fx}}}{Dose^{\text{Phantom}}_{\text{St ijk, 1 fx}}}$$

$$D^{\text{Patient}}_{\text{Mov ijk, N fx}} = D^{\text{Patient}}_{\text{St ijk, N fx}} \times \frac{D^{\text{Phantom}}_{\text{Mov ijk, N fx}}}{Dose^{\text{Phantom}}_{\text{St ijk, N fx}}}$$

With patient per voxel doses now approximated for a moving target, all voxels may be binned to create the estimated DVH to the moving target. This "Motion DVH" can directly compared to the stationary target DVH in order to quantify the impact of interplay, in particular to assess if the target dose coverage is robust (or fragile) with respect to the patient-specific target motion.

2.5. *Analysis Example*

Figure 2 illustrates how per-fraction, "Motion DVHs" can be readily compared to the stationary target DVH. Likewise, the summation of each fraction's voxel doses re-binned for the cumulative dose over N fractions reduces to a single DVH-to-DVH comparison.

3. Conclusions

So far only the theoretical framework of the 4D dose perturbation algorithm has been presented. The next necessary step is to verify this formalism empirically. In its most obvious form, such verification would involve measuring the cumulative dose per fraction with any integrating dosimeter in a moving phantom representing a patient, and comparing the results with the planned dose perturbation dose grid derived from the measurement in the stationary cylindrical phantom. In addition, dose rate vs. time values can be obtained from the time-resolved ion chamber measurements [5] and compared to the values predicted by the novel 4D reconstruction algorithm.

4. References

[1] ACR – ASTRO Practice guideline for intensity modulated radiation therapy (IMRT) http://www.acr.org/~/media/ACR/Documents/PGTS/guidelines/IMRT.pdf
[2] Nelms B E *et al* 2011 *Med. Phys.* **38** 1037-44
[3] Zhen H *et al* 2011 *Med. Phys.* **38** 5477-89
[4] van Elmpt W *et al* 2008 *Radiother. Oncol.* **86** 86-92
[5] Nelms B E *et al* 2012 *Med. Phys.* **39** 4228-38

Journal of Physics: Conference Series **444** (2013) 012098 doi:10.1088/1742-6596/444/1/012098

Evaluation of breathing interplay effects during VMAT by using 3D gel measurements

S Ceberg[1], C Ceberg[2], M Falk[3], P Munk af Rosenschöld[3] and SÅJ Bäck[1]
[1]Department of Medical Radiation Physics, Lund University, Skåne University Hospital, Malmö, Sweden
[2]Medical Radiation Physics, Department of Clinical Sciences Lund, Lund University, Lund, Sweden
[3]Radiation Medicine Research Center, Department of Radiation Oncology, Rigshospitalet, Copenhagen, Denmark

E-mail: sofie.ceberg@med.lu.se

Abstract. Respiratory motion during dynamic radiotherapy may affect the absorbed dose distribution both by dose-reducing smoothing and by more complicated interplay effects. In this study we present a novel method to determine the relative importance of these two effects. For the two dynamic deliveries studied in this work, the expected target dose reduction due to the smoothing effect was estimated by measurements convolved by the motion function. Remaining absorbed dose differences were attributed to interplay effects between the motion of the gel phantom and the movement of the modulating MLC leaves during modulated arc radiotherapy. The total dosimetric effect due to breathing motion and dynamic MLC motion during VMAT delivery resulted in an average of about 4% target dose reduction. Comparing with only the smoothing effect, the average difference was decreased to around 1%, and the remaining distribution was attributed to interplay effects. Although the interplay effects were small compared to the smoothing effect, the standard deviations of 1.4-2.3% (1SD) were larger than the narrow distribution for repeated stationary measurement with a standard deviation between 0.5-0.9% (1SD).

1. Introduction

The desire to deliver high absorbed dose to the target volume while minimizing the dose to normal tissues during radiotherapy has resulted in advanced treatment procedures, for instance techniques using intensity modulated beams and arcs [1-3]. The advantages of these techniques are the increased possibility to deliver a high absorbed dose to the target volume while minimizing the dose to normal tissues. However, intra-fractional tumor motion, mostly due to respiration, can be a major challenge to the ambition to deliver the desired dose distributions.

Two types of dosimetric effects due to intra-fractional organ motion has been defined; A) the dose-blurring effect, where the dose delivered to a point in the patient is smeared by the motion of this point in the radiation beam, and B) the interplay effect, where the relative motion between the dynamic MLC leaves and the treatment region may lead to a more complicated dosimetric effect [4].

During a volumetric modulated arc therapy (VMAT) delivery the 3D trajectory of an MLC leaf is a combination of the gantry rotation and the MLC modulation. Investigations of potential breathing interplay effects during this type of treatment delivery by experimental measurements using a high-resolution 3D-detector are highly desirable.

Using gel dosimetry, the absorbed dose can be obtained in the entire irradiated volume with high resolution [5, 6]. For example, a polymer gel dosimeter measures absorbed dose in over 10 000 measurement points in 3D for a target volume less than 5 cm^3. It has been showed that polymer gel is a feasible detector for 3D dose verification of dynamic radiotherapy e.g. [7, 8].

The aim of this study was to investigate any potential breathing interplay effects during VMAT delivery by comparing absorbed dose measurements using 3D polymer gel. The interplay effect was defined as the difference between the absorbed dose volume measured during motion and the volume resulted from a stationary measurement convolved with the motion function. Thus, the blurring effect was subtracted from the measured dose volume during motion, which means that any residual dose differences are due to interplay effects.

2. Material and method

The gel preparation, read-out and data analysis were carried out at Lund Univerity and Skåne University Hospital in Malmö while the VMAT treatment planning and delivery was carried out at Rigshospitalet in Copenhagen.

2.1. Polymer gel phantoms

The normoxic polyacrylamide gel (nPAG) used in this study contained 89% w/w ultra-pure deionised water, 3% w/w acrylamide, 3% w/w N,N'-methylenebisacrylamide, 5% w/w gelatine and 10 mM tetrakis(hydroxymethyl)–phosphonium chloride. Magnetic resonance imaging (MRI) using a 1.5 T unit (Siemens Symphony) was performed about 24 hours post-irradiation. The mixing- and read-out procedures are described elsewhere [8]. MATLAB 7.4.0 was used for image processing and 3D rendering. The R2 data of the irradiated gel phantoms was converted to relative absorbed dose using background subtraction and normalization in a region of homogenous dose [9].

Two batches were prepared; *i*) four identical 500 ml circular gel phantoms (Ø 80 mm) and *ii*) three identical 1.3 liters circular gel phantoms (Ø 10 cm). For each batch, an identical un-irradiated gel phantom was used to acquire a background value. Gel vials irradiated to known doses were used to assure the linearity of the gel dose response.

2.2. Modulated arc treatment plans and delivery

To simulate tumour movement caused by respiratory motion, the gel phantoms were positioned on a programmable motion platform (Standard Imaging, Inc), which was set to carry out sinusoidal motion in the superior–inferior (SI) direction.

Two different RapidArc™ plans were delivered; *i*) A 6 MV lung plan was created in the Eclipse version 8.6 treatment planning system (TPS) using inverse optimisation and calculated with AAA. The plan was delivered using a Novalis TX™ linear accelerator (Varian Medical Systems) with a high definition MLC in one 358-degree arc rotation. The small target had a 4.86 cm^3 volume and was irradiated with a planned target dose of 4 Gy using 790 MU. The collimator rotation was set to 45 degrees. The delivery was carried out both to a stationary gel phantom and to a gel phantom during motion with a peak-to-peak distance of 10 mm during a period of 4 s.

ii) A 18 MV prostate plan was created using a beta version of the pre-clinical RapidArc™ optimizer and calculated with the anisotropic analytical algorithm (AAA). The plan was delivered using a Clinac iX linear accelerator (Varian Medical Systems) in one arc rotation from 210□ to 150□. The planned target dose was 3.3 Gy using 864 MU. The collimator rotation was set to 46 degrees. A dosimetric verification of this delivery using gel dosimetry and Monte Carlo calculation has earlier been published [8]. In this study the delivery was carried out again, both to a stationary gel phantom and to a phantom during motion with a peak-to-peak distance of 20 mm during a period of 5 s. Although the target volume in prostate plans doesn't expect to be affected by a respiratory induced motion, this delivery was useful when investigating interplay effects.

2.3. *Interplay effect definition*

To be able to measure any potential dosimetric differences between the smooting and interplay effect we take advantage of the very high spatial resolution in three dimensions that the polymer gel detector system provides.

In order to evaluate the dosimetric impact of the interplay between the phantom motion and the dynamic treatment delivery, we defined the dosimetric interplay effect as the relative 3D absorbed dose difference between the volumes obtained from the VMAT measurement during phantom motion and the VMAT measurement to a stationary phantom convolved with the motion function of the moving platform. The difference between the measured and convolved gel matrices was calculated as ((“during motion”-“convolved stationary”)/”convolved stationary”)*100. The motion function of the platform was a sinusoidal motion with an amplitude of 10 mm and a frequency of 4 s for the lung plan delivery (*i*) and with an amplitude of 20 mm and a frequency of 5 s for the prostate plan delivery (*ii*). Only the volume within the 90% isodose surface (approximating the target volume) was investigated.

If no relative absorbed dose deviations were detected, there would be no measurable interplay effects between the MLC motion and phantom motion, and the decreased high dose volume measured in the gel in motion would only depend on the smoothing effect.

To estimate the size of a detectable interplay effect, the observed dose difference was compared to the dose difference between two repeated stationary gel measurements for each delivery.

3. Results and discussion

The lung- and prostate VMAT plan deliveries were measured with a stationary gel phantom and with a phantom motion with amplitude of 10 mm and a frequency of 4 s, and amplitude of 20 mm amplitude and 5 s frequency, respectively. The total dosimetric effect of motion was evaluated, i.e. the dose difference between the volume obtained from the measurement during phantom motion and the measurement to a stationary phantom. The isodose volume ≥ 90% was investigated. The mean value and standard deviation of the differences obtained from the measurements were (-3.70±2.56)% with only 26.6% of the voxels within 2% dose difference for the lung plan (figure 1a) and (-4.31±4.80)% with only 32.6% of the voxels within 2% dose difference for the prostate plan (figure 1b).

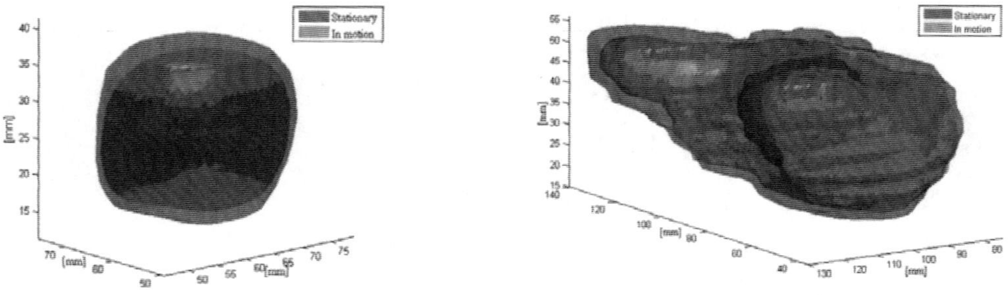

a) b)

Figure 1: The 90% isodose surface overlay of the gel measurements. The red volumes represent the stationary gel measurements and the green volumes represent the gel measurements obtained during motion. The overlay visualizes the relative absorbed dose reduction to target due to respiratory motion.

Investigating the interplay effect for the lung plan, the mean value and standard deviation of the relative dose difference between the measured VMAT delivery during phantom motion and the convolved stationary VMAT delivery resulted in (-1.2 ± 1.4)% with 71.2% of the voxels within 2% dose difference (figure 2a). This result was compared to the dose difference between two repeated stationary convolved gel measurement, which resulted in (0.064±0.51)% with 99.9% of the voxels within 2% dose difference and 94.2% of the voxels within 1% dose difference (figure 2b).

The corresponding interplay result for the prostate VMAT plan delivery was (0.94 ± 2.32)% with 60.0% of the voxels within 2% dose difference. In this case, a repeated stationary measurement using a second gel batch had resulted in a relative dose difference of (0.39±0.92)% [8].

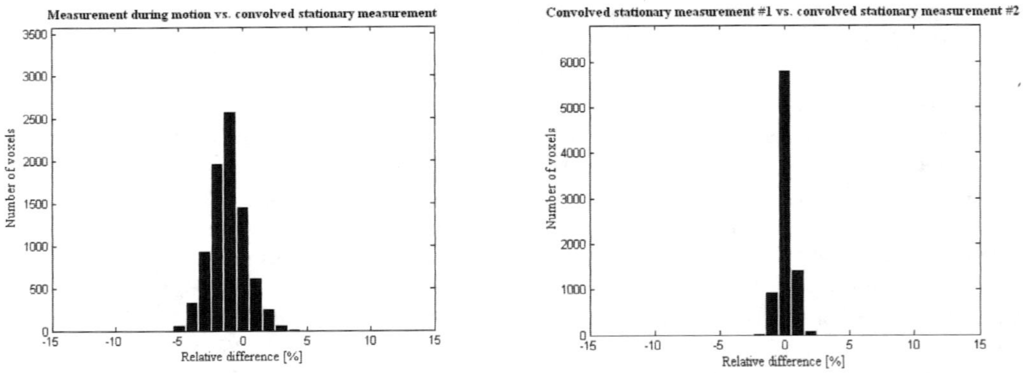

a) b)

Figure 2: The distribution of voxel-by-voxel deviations between the different gel measurement sets within the volumes enclosed by the 90% isodose surface for the lung VMAT plan delivery. The number of voxels is plotted against the difference in relative dose between the two analyzed volumes. Note the different scale of the ordinates. The interplay effect is presented in a) and the difference between repeated measurements in b).

4. Conclusions

The total dosimetric effect due to breathing motion and dynamic MLC motion during VMAT delivery resulted in an average of about 4% target dose reduction. Comparing with the convolved stationary measurement, which includes only the smoothing effect, the average difference was decreased to around 1%, and the remaining distribution was attributed to interplay effects. For repeated stationary measurement, i.e. without interplay effects but including all other measurements uncertainties (e.g. set-up), the differences had a narrow distribution with a standard deviation between 0.5-0.9% (1SD). Thus, the larger standard deviations of 1.4% (1SD) in the lung plan case (with a motion of 1 cm amplitude and 4 s frequency) and 2.3% (1SD) in the prostate plan case (2 cm amplitude and 5 s frequency), were interpreted as interplay effects.

5. References

[1] Mackie T R *et al* 1993 *Med. Phys.* **20** 1709
[2] Otto K *et al* 2008 *Med. Phys.* **35** 310
[3] Yu C X *et al* 1995 *Phys. Med. Biol.* **40** 1435
[4] Jiang B *et al* 2003 *Phys. Med. Biol.* **48** 1773-84
[5] Baldock C *et al* 2010 *Phys. Med. Biol.* **55** R1-63
[6] Gustavsson H *et al* 2004 *Phys. Med. Biol.* **49** 227-41
[7] Ceberg S *et al* 2008 *Phys. Med. Biol.* **53** 387-396
[8] Ceberg S *et al* 2010 *Phys. Med. Biol.* **55** 4885-98
[9] Bjoreland A *et al* 2008 *Acta. Oncol.* **47** 1099–109

The effect of systematic set-up deviations on the absorbed dose distribution for left-sided breast cancer treated with respiratory gating

A Edvardsson and S Ceberg
Department of Medical Radiation Physics, Lund University, Skåne University Hospital, Malmö, Sweden

E-mail: anneli.edvardsson@skane.se

Abstract. The aim of this study was 1) to investigate interfraction set-up uncertainties for patients treated with respiratory gating for left-sided breast cancer, 2) to investigate the effect of the inter-fraction set-up on the absorbed dose-distribution for the target and organs at risk (OARs) and 3) optimize the set-up correction strategy. By acquiring multiple set-up images the systematic set-up deviation was evaluated. The effect of the systematic set-up deviation on the absorbed dose distribution was evaluated by 1) simulation in the treatment planning system and 2) measurements with a biplanar diode array. The set-up deviations could be decreased using a no action level correction strategy. Not using the clinically implemented adaptive maximum likelihood factor for the gating patients resulted in better set-up. When the uncorrected set-up deviations were simulated the average mean absorbed dose was increased from 1.38 to 2.21 Gy for the heart, 4.17 to 8.86 Gy to the left anterior descending coronary artery and 5.80 to 7.64 Gy to the left lung. Respiratory gating can induce systematic set-up deviations which would result in increased mean absorbed dose to the OARs if not corrected for and should therefore be corrected for by an appropriate correction strategy.

1. Introduction

Several studies show a decrease in recurrence using postoperative radiotherapy for breast cancer [1-2]. But studies also show an increased cardiac morbidity and mortality in patients treated with radiotherapy for left-sided breast cancer compared to right-sided [3-4], due to the higher cardiac absorbed dose for the left-sided patients, implying that increased absorbed doses to the heart increase cardiac complications. One way of reducing the absorbed dose to the heart is by using breathing adapted radiotherapy [5]. One example of such technique is free breathing respiratory gating. The patient then breathes freely, however a bit deeper than normal. The beam is only on during end-inspiration, when the treatment volume and heart are furthest apart. Due to this increased spatial distance the absorbed dose to the heart can be decreased [5].

There are several uncertainties in radiotherapy that lead to that the absorbed dose is not delivered where it is intended. One such uncertainty is patient set-up, were the set-up deviation consists of a systematic and a random component [6]. The systematic part of the set-up deviation can be corrected for by using an appropriate correction strategy, such for example the no action level (NAL) protocol [7]. The mean deviation of the first n treatment fractions is calculated and applied as a correction for the remaining treatment fractions. In our clinic n equal to three is used. To avoid overcorrection an adaptive maximum likelihood (AML) factor has been introduced and can be combined with the correction strategy [8-9]. Today an AML factor of 0.75 is used at our clinic.

For the patients treated with respiratory gating at our clinic systematic set-up deviations are induced due to the set-up procedure used. During CT-scanning skin markers are made on the patient during normal free breathing. A treatment plan is made in a CT-set acquired during deep breathing. Before treatment the patient is positioned according to lasers. These laser lines marks the correct patient position during deep breathing. The patient is positioned during normal free breathing and the deep breathing then leads to a deviation in the set-up.

The aim of this study was to investigate the set-up deviations for the patients treated with respiratory gating for left-sided breast cancer and the effect of these set-up deviations on the absorbed dose distribution for the heart, left anterior descending coronary artery (LADCA) and left lung. To evaluate the impact of the systematic set-up deviations on the absorbed dose to the organs at risk (OAR) simulations were made in the treatment planning system (TPS) (Eclipse version 10.0.28, Varian medical systems; CA) and measurements were carried out using the biplanar diode array Delta[4] (ScandiDos, Inc.; Sweden). The effect of two correction strategies, NAL with an AML factor of 0.75 and NAL without AML factor, to correct for the systematic set-up deviations was also investigated.

2. Material and methods

2.1. Acquiring set-up images and calculating set-up deviations

Prospectively, 18 patients treated with respiratory gating for left-sided breast cancer using the real-time positioning management system (RPM, Varian medical systems; CA) were enrolled in this study. The patient´s position at every fraction was verified with two orthogonal kilovolt (kV) set-up images using the on-board imager (OBI, Varian medical systems; CA). All together 352 orthogonal image pairs were acquired. The kV set-up images were manually matched with digitally reconstructed radiographs (DRR) that have been reconstructed from the gated CT-set. The set-up deviations for every treatment fraction were obtained and the overall mean systematic deviation, $m_{overall}$, was calculated as the mean deviation for all patients and all treatment fractions [10]. The NAL correction strategy with AML factor 0.75 and the NAL correction strategy without AML factor were simulated and corrected deviations obtained. The $m_{overall}$ was calculated also for the corrected deviations.

2.2. Simulating the set-up deviations in the TPS

The effect of the systematic set-up deviation on the absorbed dose distribution to the OARs was investigated by shifting the isocenter of the treatment fields in the TPS, corresponding to $m_{overall}$ (table 1). This was made for 19 patients treated with respiratory gating for left-sided breast cancer (not the same patients as above) treated with 50 Gy in 25 fractions.

2.3. Measurement with the Delta[4]

Delta[4] is a biplanar diode array containing two orthogonal detector boards. The center-to-center distance between the diodes is 5 mm in the central 6x6 cm^2 region and 10 mm in the rest of the array. Six treatment plans were delivered to the Delta[4] using the Truebeam (Varian medical systems; CA). Two measurements were carried out for each patient; one without set-up deviations and one where the uncorrected set-up deviations corresponding to the $m_{overall}$ (table 1) were introduced. A 3D gamma evaluation was carried out using the dose deviation and distance to agreement (DTA) criteria 3%/2mm.

3. Results and discussion

With the set-up procedure used at our clinic, systematic set-up deviations are induced using respiratory gating in the anterior and cranial direction (table 1). A NAL correction strategy can be used to correct for these set-up deviations. The best set-up results were achieved using the NAL correction strategy without the AML factor.

Table 1: The overall mean systematic set-up deviation. Positive values in the posterior, caudal and right directions.

	Ant-post (mm)	Cran-caud (mm)	Left-right (mm)
No correction strategy	-6.0	-8.1	0.7
NAL AML =0.75	-1.1	-3.3	1.3
NAL no AML	0.5	-1.0	1.2

The effect of the systematic set-up deviation was evaluated by shifting the $m_{overall}$ in the TPS. This result in an increased average mean absorbed dose to the OARs (table 2). When the set-up deviations for the two correction strategies were simulated smaller difference in the average mean absorbed dose were observed (table 2). In figure 1cumulative histograms of the mean absorbed dose to the OARs for all investigated patients are shown.

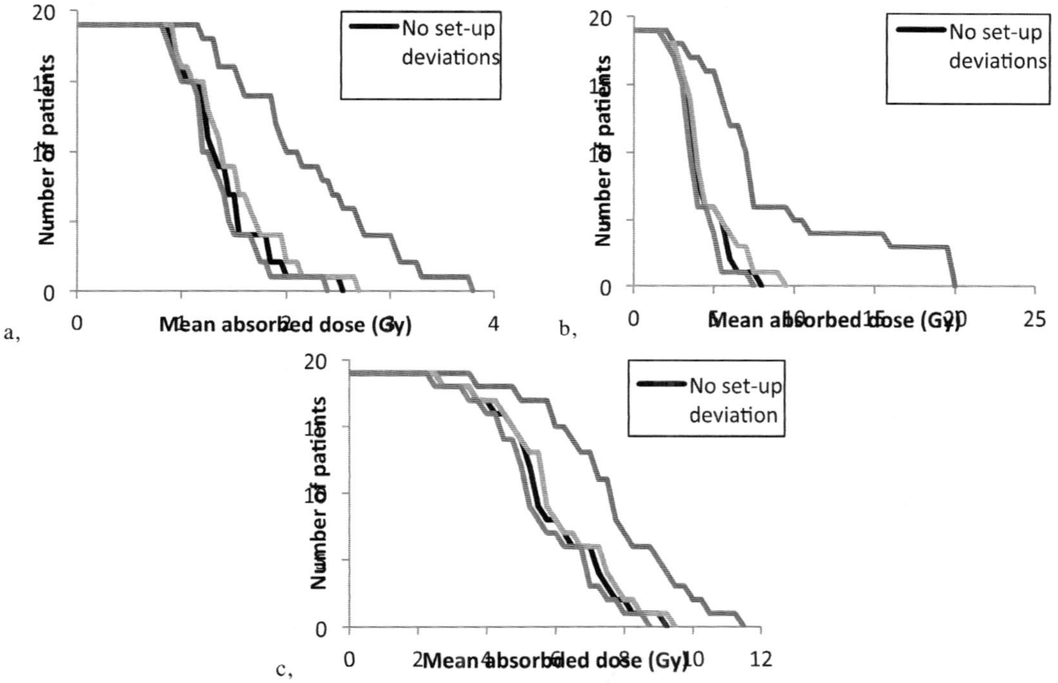

Figure 1: Cumulative histograms for the mean absorbed dose for the heart (a), LADCA (b) and left lung (c).

The gamma evaluation method takes both dosimetric and spatial accuracy into account when comparing two absorbed dose distributions. A value of γ less than one implies that the criterions are fulfilled. When no set-up deviations are introduced, in average 97.3 % of the measuring points have a $\gamma<1$. This value was decreased to 64.2 % when the uncorrected set-up deviations are introduced, implying that the systematic set-up deviations result in a difference between the two absorbed dose distributions. This difference is a shift in the absorbed dose profiles caused by the systematic set-up deviation (figure 2). The measuring points using the Delta[4] were compared to dose profiles acquired from the TPS, both with and without the uncorrected set-up deviations induced. Good agreement between the measurement and the simulation in the TPS were observed (table 2).

4. Conclusion

Using respiratory gating systematic set-up deviations can be introduced. These systematic set-up deviations lead to an increased mean absorbed dose to the heart, LADCA and left lung. This can be

Table 2: Average mean absorbed dose to the OARs. The difference in percent compared to original treatment plan in parentheses.

	Heart (Gy)	LADCA (Gy)	Left lung (Gy)
No set-up deviations simulated	1.38	4.17	5.80
No correction strategy	2.21 (60.2)	8.86 (112.7)	7.64 (31.7)
NAL AML = 0.75	1.48 (6.9)	4.53 (8.8)	6.00 (3.4)
NAL no AML	1.31 (-5.3)	3.86 (-7.5)	5.51 (-5.1)

corrected for using the NAL correction strategy. For such large systematic set-up deviations seen in this study, not using an AML factor results in better set-up than using an AML factor of 0.75.

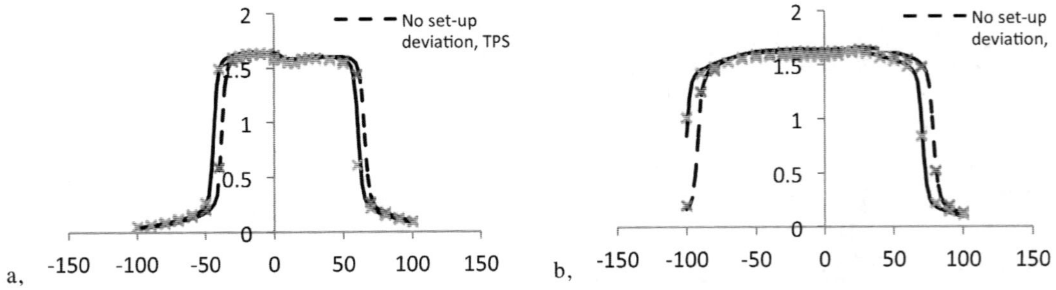

Figure 2: Absorbed dose profile for the detector boards showing the shift in the absorbed dose distribution induced by the systematic set-up deviation in the anterior-posterior direction (a) and the craniocaudal direction (b) for one patient in this study.

5. References

[1] Early Breast Cancer Trialists' Group 2011 *Lancet* **378** 1707-16
[2] Fisher B *et al* 1995 *N. Engl. J. Med.* **333** 1456-61
[3] McGale P *et al* 2011 *Radioth. Oncol.* **100** 167-75
[4] Roychoudhuri R *et al* 2007 *BMC Cancer* **7** 1-5
[5] Korreman S *et al* 2005 *Radiother. Oncol.* **76** 311-18
[6] Hurkmans C *et al* 2001 *Radiother. Oncol.* **58** 105-20
[7] de Boer H C J and Heijmen B J M 2001 *Int. J. Radiat. Oncol. Biol Phys.* **50** 1350-65
[8] Gluhchev G 1998 *Radiother. Oncol.* **48** 79-82
[9] Shalev S and Gluhchev G 1995 *Int. J. Radiat. Oncol. Biol. Phys.* **32** 216
[10] Greener T 2003 *Br. J. Radiol.* 36-43

Journal of Physics: Conference Series **444** (2013) 012100 doi:10.1088/1742-6596/444/1/012100

Towards comprehensive characterization of Cs-137 Seeds using PRESAGE® dosimetry with optical tomography

J Adamson[1], Y Yang[1], L Rankine[1], J Newton[1], J Adamovics[2], O Craciunescu[1] and M Oldham[1]

[1]Dept. of Radiation Oncology, Duke University Medical Center Box 3295, Durham NC 27710

[2]Dept. of Chemistry, Rider University, Lawrenceville, NJ 08648

Email: justus.adamson@duke.edu

Abstract. We describe a method to directly measure the radial dose and anisotropy functions of brachytherapy sources using polyurethane based dosimeters read out with optical CT. We measured the radial dose and anisotropy functions for a Cs-137 source using a PRESAGE® dosimeter (9.5cm diameter, 9.2cm height) with a 0.35cm channel drilled for source placement. The dosimeter was immersed in water and irradiated to 5.3Gy at 1cm. Pre- and post-irradiation optical CT scans were acquired with the Duke Large field of view Optical CT Scanner (DLOS) and dose was reconstructed with 0.5mm isotropic voxel size. The measured radial dose factor matched the published fit to within 3% for radii between 0.5-3.0cm, and the anisotropy function matched to within 4% except for θ near 0° and 180° and radii >3cm. Further improvements in measurement accuracy may be achieved by optimizing dose, using the high dynamic range scanning capability of DLOS, and irradiating multiple dosimeters. Initial simulations indicate an 8 fold increase in dose is possible while still allowing sufficient light transmission during optical CT. A more comprehensive measurement may be achieved by increasing dosimeter size and flipping the source orientation between irradiations.

1. Introduction

Accurate and comprehensive measurement of dose distributions about brachytherapy sources remains a challenging yet relevant problem. Varying dose rates and complex distributions about these sources require a measurement system capable of high spatial accuracy and resolution, and large dynamic range [1]. Due to these challenges, most published characterizations of brachytherapy sources rely heavily on Monte Carlo simulations with experimental verification at select points [2]. 3D dosimetry has potential to provide a more comprehensive verification. A number of studies have investigated gel dosimetry for brachytherapy [3, 4]; sensitivity of the polymerization process to oxygen, energy dependence (for low energy sources), and diffusion of monomers across steep dose gradients close to the source are challenges to achieving accurate results [1]. Another 3D dosimeter, PRESAGE® (Heuris Inc, NJ) [3, 5], is a polyurethane based dosimeter that is solid and transparent. Some of its advantages include being machinable, moldable, insensitive to oxygen exposure, and having minimal signal diffusion, which are important for brachytherapy. PRESAGE® has been shown to be tissue equivalent for Ir-192 and Cs-137. These dosimeters can be analyzed using optical CT, one system of which is the Duke Large field of view Optical CT-Scanner (DLOS) [6]. This scanner has minimal scatter and is capable of reconstructing PRESAGE® distributions with sub-millimeter voxel size.

This study we investigate the feasibility of using PRESAGE®/DLOS to investigate dose distributions about a Cesium-137 source [7]. Specifically we describe initial attempts to directly calculate and verify Monte Carlo based parameters from the formalism developed by the American Association of Physicists in Medicine (AAPM) Task Group 43 (TG-43) [2].

2. Materials & Methods

2.1. Measurement

A cylindrical PRESAGE® dosimeter [8] was obtained (9.5cm diameter) with a 0.35cm diameter channel drilled in the center to midline. An optical CT pre-scan was acquired using DLOS [6] in which 720 projection images were acquired at 0.5° increments in ~15 minutes. The Cs-137 source [7] was inserted into the channel for irradiation and 5.3Gy was delivered at 1cm with the dosimeter placed in water to provide backscatter $(40\ \text{cm})^3$. After irradiation a post optical CT was acquired and the dose distribution reconstructed with $(0.5\ \text{mm})^3$ isotropic voxel size and no median filtering.

2.2. TG-43 Parameters

In TG-43, the dose rate constant, Λ, radial dose function, $g_L(r)$, and anisotropy function, $F(r,\theta)$, are parameters that are specific to each source model. We extracted the radial dose and anisotropy functions from the measured 3D dose distribution. Given the source location and orientation, the radius r and angle θ can be calculated for each voxel of the measured distribution.

We calculate a fit, $\hat{g}_L(r)$, of the radial dose function from the measured distribution which includes all voxels within ±0.1° or ±0.5mm of the transverse plane. A 5-degree polynomial fit is used to model $g_L(r)$, where x-values in the fit are the radius r per voxel, and y-values are:

$$y_i = \frac{I(r_i,\theta_0)}{\bar{I}(r_0,\theta_0)}\frac{G_L(r_0,\theta_0)}{G_L(r_i,\theta_0)} \quad (1)$$

where $G_L(r,\theta)$ is the geometry function, $I(r_i,\theta_0)$ is the intensity accumulated in voxel i of the 3D distribution and $\bar{I}(r_0,\theta_0)$ is the mean intensity for all voxels within ±0.1° and ±0.5mm of θ_0 and r_0.

We calculated the anisotropy function out to the max distance measurable for our dosimeter. At a given r_j and θ_k, we calculate the anisotropy function using voxels in the dosimeter within ±0.5° and ±1.0mm with the equation:

$$\hat{F}(r,\theta) = \sum_{i=1}^{n}\frac{I(r_i,\theta_i)}{G_L(r_i,\theta_i)}\frac{G_L(r_0,\theta_0)}{\bar{I}(r_0,\theta_0)\cdot\hat{g}_L(r_i)} \quad (2)$$

where $\hat{g}_L(r_i)$ is calculated using the polynomial fit described above.

2.3. Data Preparation

Preparatory steps prior to calculating the radial dose and anisotropy functions included defining the region of interest (ROI) within the dosimeter and registering the source location and orientation. The ROI excluded the channel and voxels with zero or negative signal. Finally a morphological erosion algorithm [9] excluded voxels near imperfections and the dosimeter surface. The source location and orientation was determined by iteratively calculating $\hat{g}_L(r)$ and $\hat{F}(r,\theta)$ for different source locations and orientations and minimizing a score function that was the product of their residual errors.

3. Results

Figures 1 and 2 describe the data along the transverse plane defining the radial dose function, and illustrate an advantage of 3D dosimetry. As r increases, so does the sample size of voxels, resulting in a nearly constant standard error. The radial dose function is shown in Figure 3 where data was binned in 1mm increments and error bars represent one standard deviation. The factor matched the published fit to within 3% for radii of 0.5-3.0cm. Beyond 3cm the radial dose function was underestimated by as much as 10%. This underestimation is a topic of current investigation and may be attributed to differences in density or the edge of the dosimeter.

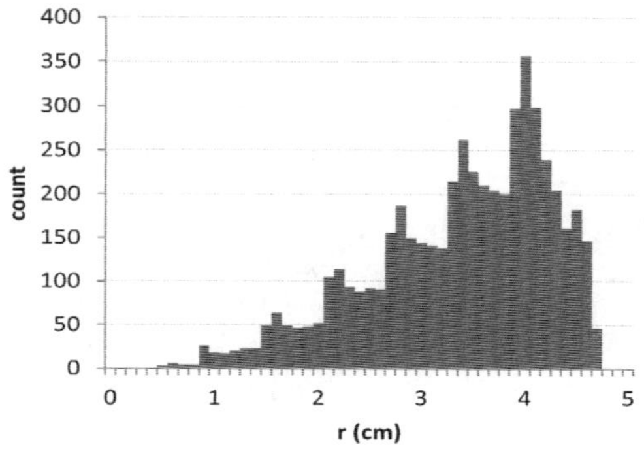

Figure 1: Number of measurement points along the transverse plane with 1mm binning.

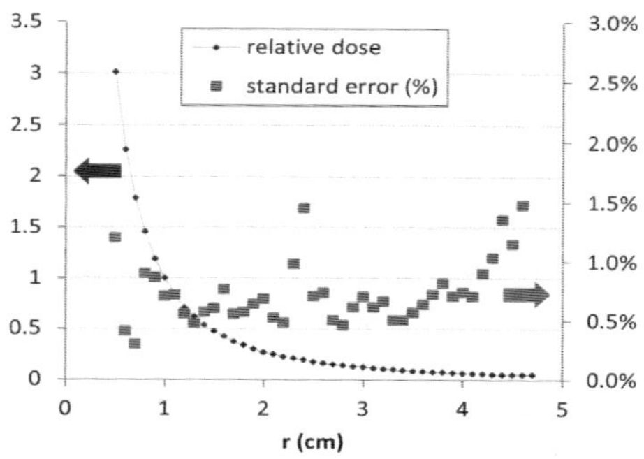

Figure 2: Relative dose & standard error along transverse plane. Diminishing dose is offset by increasing sample size (Fig. 1) resulting in nearly constant standard error.

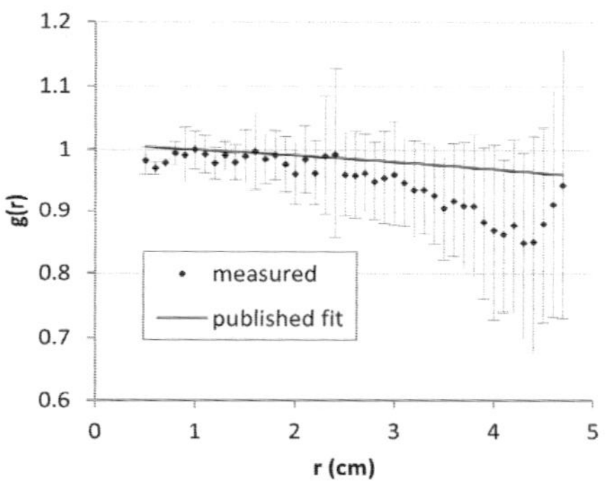

Figure 3: Radial dose function in 1 mm increments measured using PRESAGE® / optical CT and the published fit from Monte Carlo simulations [7].

Table 1 shows the difference from the published anisotropy function. The function matched to within 4% except when θ approached 0° and 180° and radii > 3cm.

4. Discussion

We have investigated feasibility of measuring the radial dose and 2D anisotropy functions for a Cs-137 source using PRESAGE® / Optical CT. One advantage shown here for 3D dosimetry is that low signal at distance is offset by increased number of voxels. Measurement of TG-43 parameters using 3D dosimetry will allow for direct comparison to the factors calculated by Monte Carlo, has potential to allow for investigation of the uncertainty in the various factors. TG-43 states: "sufficient data are not available…to provide a comprehensive uncertainty analysis for anisotropy functions. More research is needed… identifying geometric parameters to which (the anisotropy function) is sensitive" [2]. This is an evident area where source characterization using 3D dosimetry will be of use.

The following options remain to be investigated to optimize source characterization using PRESAGE® / Optical CT:

- Optimizing delivered dose: A tradeoff must be made between achieving acceptable signal to noise ratio distal from the source and achieving sufficient light transmission through the dosimeter during optical CT acquisition. ScanSim is a tool for simulating optical CT imaging [10] which we intend to use to determine the maximum dose that still achieves sufficient light transmission during optical CT with DLOS. Initial estimates indicate an optimal attenuation coefficient of $0.95cm^{-1}$ at 1cm, corresponding to a dose of ~8Gy at 1cm for a high sensitivity dosimeter.

- High Dynamic Range Scanning: DLOS can be operated in a "dual gain" mode where two acquisitions are performed with varied parameters [11]. In this mode a second scan is performed where the light source intensity is increased, which increases sensitivity at high optical densities.
- Increasing dosimeter size: Larger dosimeters have been used in conjunction with DLOS and can be implemented in future work.
- Using multiple irradiations: Systematic errors can be minimized using irradiation of multiple dosimeters. Also the channel in the dosimeter center results in missing data for the anisotropy function which can be overcome by using multiple irradiations with reversed source orientation.
- Determining dose rate constant: This is the only source model specific parameter that was not measured. Each dosimeter is accompanied by cuvettes used to quantify dosimeter sensitivity and in theory could be used to measure the dose rate constant.

Table 1: Measured 2D anisotropy function for a Cs-137 source [7]. Shown is the percent difference from published fit ± standard error (# measurement voxels).

	r (cm)					
theta (°)	0.75	1	1.5	2	3	4
10	-	-	-	-	4.7±4.3% (16)	-
20	-	-	-	-6.6±2.9% (24)	-7.4±1.9% (45)	-
30	-	-	-0.3±2.0% (15)	-3.8±1.8% (58)	0.9±1.3% (86)	-
40	-	-	-1.6±1.2% (21)	-2.5±1.0% (62)	-3.1±0.9% (107)	-0.2±0.8% (266)
50	-	-0.7±1.6% (22)	-0.7±0.7% (29)	-0.3±0.7% (72)	-1.4±0.6% (138)	-4.0±0.7% (312)
60	-	0.7±1.1% (18)	-0.5±0.6% (46)	-1.5±0.5% (89)	-1.7±0.6% (149)	-2.0±0.6% (351)
70	-	1.3±0.8% (21)	-1.1±0.5% (48)	-1.8±0.5% (84)	-1.7±0.5% (179)	-3.5±0.6% (354)
80	1.0±0.8% (12)	0.9±0.6% (30)	-1.3±0.5% (28)	-0.5±0.5% (95)	-0.1±0.4% (142)	0.5±0.6% (382)
90	-	1.8±0.3% (13)	0.2±0.3% (66)	-0.8±0.4% (128)	0.4±0.4% (141)	1.1±0.6% (365)
100	-0.1±0.4% (11)	1.5±0.3% (33)	0.6±0.4% (29)	-0.6±0.4% (87)	-0.4±0.4% (145)	-1.1±0.5% (395)
110	-	2.4±0.4% (24)	1.0±0.5% (47)	-0.4±0.4% (80)	0.0±0.3% (173)	-2.0±0.5% (361)
120	-	1.2±0.7% (17)	0.5±0.4% (39)	-0.6±0.3% (99)	-0.8±0.2% (140)	-2.9±0.4% (351)
130	-	4.1±1.0% (18)	0.7±0.6% (28)	-0.8±0.2% (70)	-1.0±0.2% (155)	-4.1±0.3% (304)
140	-	-	1.8±0.3% (24)	-0.2±0.2% (54)	-4.0±0.3% (113)	-4.9±0.3% (251)
150	-	-	2.5±0.3% (17)	-0.1±0.2% (57)	-3.2±0.6% (84)	-4.7±0.3% (191)
160	-	-	2.6±0.2% (11)	-0.2±0.2% (34)	-2.2±0.2% (55)	-4.9±0.4% (132)
170	-	-	-	-1.6±0.3% (8)	-3.6±0.4% (31)	-4.6±0.6% (72)
172	-	-	-	-1.5±0.3% (12)	-3.9±0.4% (30)	-9.4±0.7% (54)
174	-	-	-	-	-1.8±0.9% (17)	-6.8±1.4% (40)
175	-	-	-	-	-1.7±0.5% (17)	-4.6±1.5% (32)
176	-	-	-	-	-2.6±0.4% (16)	-3.2±1.6% (27)

5. Acknowledgements

This work was supported by NIH R01CA100835.

6. References

[1] Ibbott G and Heard M 2010 *J. Phys.: Conf. Ser.* **250** 012088
[2] Rivard M J *et al* 2004 *Med. Phys.* **31** 633-74
[3] Pierquet M *et al* 2010 *J. Phys. Conf. Ser.* **250** 012091
[4] Fragoso M *et al* 2004 *Phys. Med. Biol.* **49** 5459-74
[5] Gorjiara T *et al* 2011 *Med. Phys.* **38** 2265-74
[6] Thomas A *et al* 2011 *Med. Phys.* **38** 4846-57
[7] Meigooni A S *et al* 2009 *Med. Phys.* **36** 4711-9
[8] Guo P *et al* 2006 *Med. Phys.* **33** 3962-72
[9] Haralick R M *et al* 1987 *IEEE T. Pattern Anal.* **9** 532-50
[10] Oldham M 2010 *J. Phys. Conf. Ser.* **250** 012064
[11] Niebanck M *et al* 2012 *Med. Phys.* **39**

Semi-3D dosimetry of high dose rate brachytherapy using a novel Gafchromic EBT3 film-array water phantom

AL Palmer[1,2] A Nisbet[1,3] and DA Bradley[1]
[1]Department of Physics, Faculty of Engineering and Physical Science, University of Surrey, UK
[2]Medical Physics Department, Queen Alexandra Hospital, Portsmouth Hospitals NHS Trust, Portsmouth, UK
[3]Medical Physics Department, Royal Surrey County Hospital NHS Foundation Trust, Guildford, UK

E-mail: antony.palmer@porthosp.nhs.uk

Abstract. There is a need to modernise clinical brachytherapy dosimetry measurement beyond traditional point dose verification to enable appropriate quality control within 3D treatment environments. This is to keep pace with the 3D clinical and planning approaches which often include significant patient-specific optimisation away from 'standard loading patterns'. A multi-dimension measurement system is required to provide assurance of the complex 3D dose distributions, to verify equipment performance, and to enable quality audits. However, true 3D dose measurements around brachytherapy applicators are often impractical due to their complex shapes and the requirement for close measurement distances. A solution utilising an array of radiochromic film (Gafchromic EBT3) positioned within a water filled phantom is presented. A calibration function for the film has been determined over 0 to 90Gy dose range using three colour channel analysis (FilmQAPro software). Film measurements of the radial dose from a single HDR source agree with TPS and Monte Carlo calculations within 5 % up to 50 mm from the source. Film array measurements of the dose distribution around a cervix applicator agree with TPS calculations generally within 4 mm distance to agreement The feasibility of film array measurements for semi-3D dosimetry in clinical HDR applications is demonstrated.

1. Introduction

Brachytherapy treatments are currently undergoing a period of significant innovation and rapid modernisation, including a fundamental shift away from the use of traditional pre-determined dose distributions and a change from 2D to 3D perspectives of the entire treatment process. It is essential that quality assurance techniques keep pace with developments in brachytherapy treatment planning and delivery to ensure an appropriate level of dosimetric accuracy and quality. Traditional methods of quality control are based solely on isolated source-strength measurements with very little independent measurement of the actual dose delivery: This is inadequate for modern 3D-based brachytherapy which requires multi-dimension verification measurements of the dose delivered when using clinical treatment applicators and the potential of significant patient-specific dose distribution optimisation.

There is however a lack of established methods for quality control measurements of 3D dose distribution delivered by HDR treatment equipment [1]. The measurement challenge in brachytherapy is particularly difficult due to extremely steep dose gradients (6% per mm at the target volume in HDR gynaecology treatments) and orders of magnitude variation in dose deposition across the region of interest. Measurement systems must also be sufficiently flexible to accept a range of HDR applicator designs and provide data within the maximum extent of the applicator dimensions.

Radiochromic film has been used in brachytherapy dosimetry by many investigators over the last decade to make specific measurements or to verify Monte Carlo generated TG43-based source model data [2-4]. However there have been no publications to date utilising the latest Gafchromic EBT3 radiochromic film in brachytherapy applications. There have also been only limited publications on quality control audit methods for brachytherapy dosimetry, all being based on point dose measurements, normally with ion chambers or alanine [5-7]. Further development of practical measurement and analysis techniques is required both for routine in-clinic quality checks of brachytherapy equipment performance and for independent 3D brachytherapy dosimetry audits.

We present a novel film-array test object using EBT3 film and multi-channel dosimetric analysis, suitable for use in quality control or quality audit measurements in HDR brachytherapy. We also evaluate the EBT3 manufacture's useable dose range statement "1 cGy to > 40 Gy", which is particularly relevant for brachytherapy applications [8].

2. Methods and materials

All measurements were performed using Gafchromic EBT3 (ISP Advanced Materials, New Jersey, USA) from a single batch. HDR exposures were performed with the Eckert & Ziegler BEBIG GmbH MultiSource® HDR treatment unit with a Co-60 source. Films were scanned using a 48-bit rgb transmission film scanner (Epson Perfection V750 Pro) at 96 dpi with no colour or sharpness corrections and consistent orientation and time-since-exposure protocol. Images were analysed in three colour channels using FilmQAPro® software (ISP Advanced Materials, New Jersey, USA). Treatment planning system (TPS) dose calculations were made with the Eckert & Ziegler BEBIG GmbH HDR Plus® software, utilizing a TG-43 based dose calculation.

2.1. EBT3 calibration

The dose response of the EBT3 film was evaluated by exposing 5 cm width strips positioned on the central axis in a 10 x 10 cm treatment field from a 6 MV linear accelerator (linac) treatment unit, to doses from 0 to 90.7 Gy. The output of the linac having been measured immediately preceding the EBT3 exposures, with three ionization chambers with calibrations traceable to a primary standard held at the National Physical Laboratory, Teddington, UK.

2.2. Film-array measurements

The film-array water phantom is shown in figure 1. A three-sided Perspex holder was accurately machined to precisely locate parallel EBT3 films in water, eleven central films at 5.0 mm separation, and three outer films at 10.0 mm separation. The holder was positioned centrally within a 50 litre water tank to provide near full scatter conditions. A clamp and stand was used to position the HDR applicator within the film array using alignment guides. Symmetric films at the same distance from the applicator (left and right) were used as consistency checks of the applicator and film alignment.

2.2.1. Single dwell point measurement. The dosimetry system was validated by measuring the dose with radial distance from a single source, repeating the measurement with three separate films, and comparing results to previously published Monte Carlo generated TG43 dose model parameters [9] and treatment planning system predicted dose points.

2.2.2. Clinical treatment applicator measurement. The dose distribution in several 2D planes around a clinical cervix applicator (Eckert & Ziegler BEBIG ring and intrauterine (IU) tube) was measured using the EBT3 film-array phantom: This enabled a semi-3D verification of the intended dose distribution. These images were compared to the isodose distribution calculated by the treatment planning system.

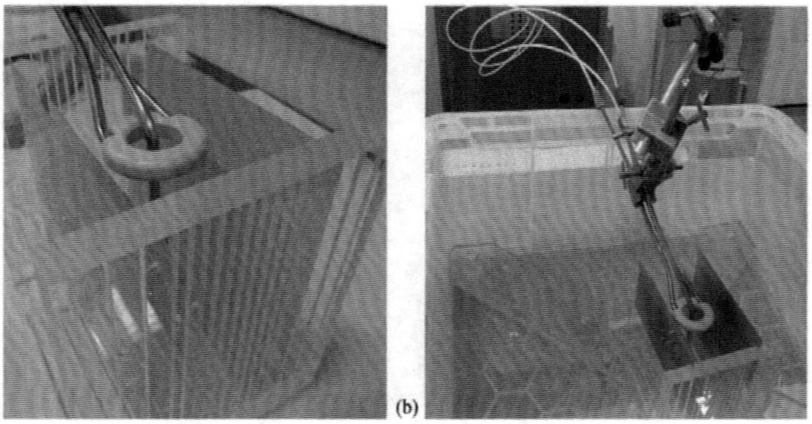

(a) (b)

Figure 1: Film-array holder with four EBT3 films in place around cervix treatment applicator (a) prior to irradiation and (b) in water connected to HDR treatment unit, post irradiation

3. Results

Figure 2 shows the EBT3 colour channel response as a function of dose and the applied cubic calibration fit to the red, green and blue channels over a 0 to 90 Gy range. Figure 3 shows the calibrated film dose response for the red and green channels as a function of radial distance in the case of a single isolated source, compared to TPS and Monte Carlo calculation. Figure 4 shows an isodose comparison of the EBT3 measured and TPS calculated dose distributions in a plane 1 cm above the IU channel of the clinical applicator. In general distance to agreement between the sets of isodose lines is within 4 mm.

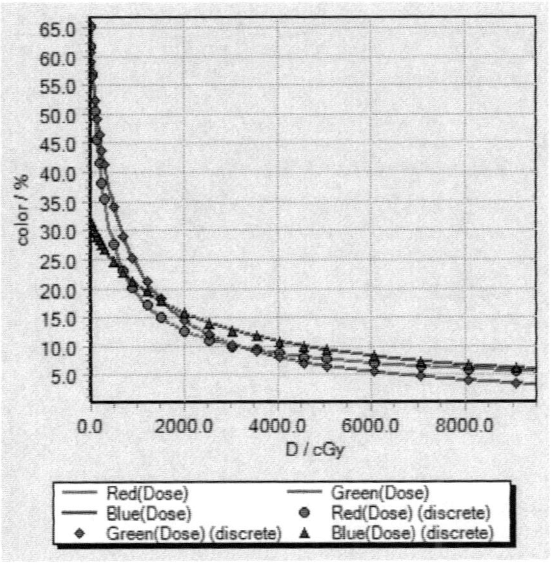

Figure 2: Calibration fit of red, green & blue film channels, over 0 to 90 Gy, using FilmQAPro®

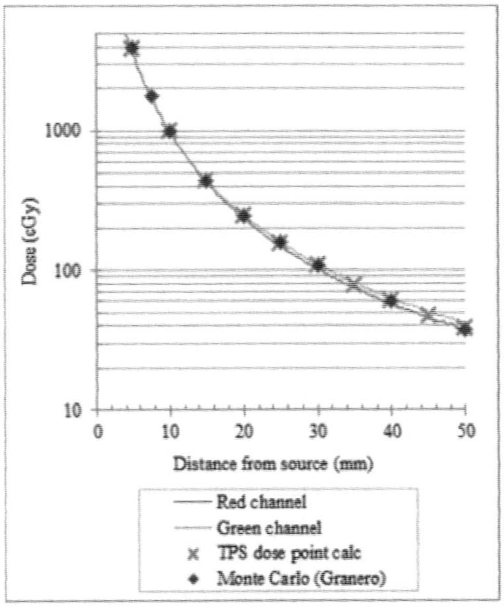

Figure 3: EBT3 measured dose compared to TPS and Monte Carlo from single source

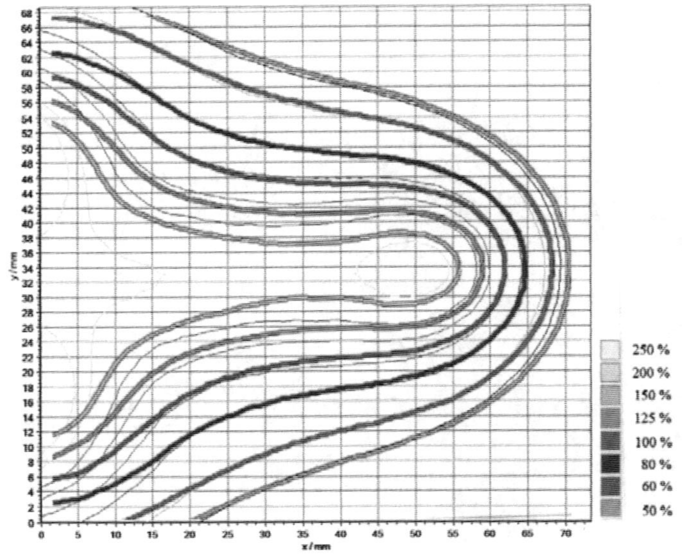

Figure 4: Comparison of EBT3 measured (thin lines) and TPS calculated (thick lines) isodose distribution, 50 to 250 %, for 8.5 Gy prescription dose.

4. Conclusions and Discussion

The use of Gafchromic EBT3 film for brachytherapy applications has been successfully demonstrated, with a calibration curve up to 90 Gy established. The film measurement technique has been successfully validated for the case of a single dwell against TPS and Monte Carlo data. Film measurements in multiple planes around an HDR cervix treatment applicator have shown acceptable agreement with TPS, generally within 4 mm distance to agreement within each plane. Distance to agreement can also be assessed between the multiple film planes (semi-3D analysis).

The EBT3 film-array water phantom has applications as a routine quality control tool, a non-routine commissioning system for new applicators or source models, or as an independent quality audit system, providing semi-3D verification of the actual dose distribution delivered by HDR treatment equipment, in difficult measurement situations.

5. Acknowledgments

The authors gratefully acknowledge Vertec UK for the supply of EBT3 film used in this investigation, and Steve Craig at the University of Surrey for construction of the film array phantom.

6. References

[1] Palmer A *et al* 2012 *J. Contemp. Brachyther.* **4** 81-91
[2] Sureka C S *et al* 2007 *Phys. Med. Biol.* 2007 **52** 525-37
[3] Lliso F *et al* 2011 *J. Contemp. Brachyther.* **3** 32-5
[4] Aldelaijan S *et al* 2011 *Med. Phys.* **38** 6074-83
[5] Elfrink R J M *et al* 2011 *Radiother. Oncol.* 2001 **59** 297-306
[6] Roué A *et al* 2007 *Radiother. Oncol.* **83** 86-93
[7] Carlsson Tedgren A and Grindborg J E 2008 *Radiother. Oncol.* 2008 **86** 126-30
[8] International Specialty Products Advanced Materials (ISP), New Jersey, USA. Gafchromic EBT3 brochure 2011
[9] Granero D *et al* 2007 *Med. Phys.* **34** 3485-8

Journal of Physics: Conference Series **444** (2013) 012102 doi:10.1088/1742-6596/444/1/012102

ROPES eye plaque dosimetry: commissioning and verification of an ophthalmic brachytherapy treatment planning system

J Poder[1], N Annabell[2], M Geso[2], M Alqathami[2] and S Corde[1]
[1] Radiation Oncology Dept, Prince of Wales Hospital, Randwick, NSW, AUS
[2] School of Medical Sciences - RMIT University, Melbourne, VIC, AUS

Email: Joel.Poder@sesiahs.health.nsw.gov.au

Abstract. In this study, the Plaque Simulator™ eye plaque brachytherapy planning system was commissioned for ROPES eye plaques and Amersham Health model 6711 Iodine 125 seeds, using TG43-UI data. The brachytherapy module of the RADCALC® independent checking program was configured to allow verification of the accuracy of the dose calculated by Plaque Simulator™. Central axis depth dose distributions were compared and observed to agree to within 2% for all ROPES plaque models and depths of interest. Experimental measurements were performed with a customized PRESAGE™ 3-D type dosimeter to validate the calculated depth dose distributions. Preliminary results have shown the effect of the stainless steel plaque backing decreases the measured fluorescence intensity by up to 25%, and 40% for the 15 mm and 10 mm diameter ROPES plaques respectively. This effect, once fully quantified should be accounted for in the Plaque Simulator™ eye plaque brachytherapy planning system.

1. Introduction

Choroidal melanoma is the most common primary intraocular malignancy [1] and is commonly treated using plaque brachytherapy, external beam charged particle therapy, surgical resection or enucleation. The Collaborative Ocular Melanoma Study (COMS) trial began in 1986 [2], comparing enucleation against a minimum of 100 Gy I-125 plaque radiation therapy for medium-sized choroidal melanomas. With 12 years follow up, no significant difference in survival was found, establishing eye plaque brachytherapy as an effective treatment modality for choroidal melanoma [3].

Following the publication of AAPM TG-43 report in 1995 providing a dose calculation formalism for Ir-192, I-125 and Pd-103 sources [4], the dose prescription of 100 Gy was lowered to 85 Gy [5]. A TG-43 report update was released in 2004 (TG43-UI [6]) to include new brachytherapy sources and critically reassessed some published brachytherapy dosimetry data, however both the TG-43 and TG43-UI reports assume a full water scattering medium around the sources.

I-125 seeds have been the subject of extensive dosimetric studies using LiF thermoluminescent dosimeters (TLD), diodes, scintillators, Gafchromic film and polymer gels [6]. LiF TLD remains the method of choice recommended in the AAPM TG43-UI report [6], and Monte Carlo codes are often used as benchmarks for any experimental measurements [7].

Whilst the dosimetric studies of model 6711 I-125 seeds are extensive, many of these are limited to the study of the seeds in conjunction with COMS plaques [8]. To date, only one group has published a Monte Carlo study of the dose distributions produced by Radiation Oncology Physics and Engineering Services Australia (ROPES) type eye plaques with model 6711 I-125 seeds [9] and to the best of our knowledge, limited experimental data have been reported.

The Plaque Simulator™ eye plaque brachytherapy planning system (BEBIG, Germany) was recently commissioned in our institution using TG43-UI data for model 6711 Amersham Health (London, UK) I-125 seeds in association with ROPES eye plaques. The brachytherapy module of RADCALC® independent checking program (Lifeline Software, Inc. USA) was used to separately model the ROPES eye plaques loaded with model 6711 I-125 seeds and verify the resulting treatment time calculation accuracy. Central axis eye plaque depth dose distributions, as calculated by Plaque Simulator™ and RADCALC®, are compared for different ROPES eye plaque seed configurations. PRESAGE® is a radiochromic solid dosimeter 3-D dosimeter [10, 11]. Preliminary results obtained using customized PRESAGE™ 3-D dosimeter for central axis fluorescence intensity measurements are also presented with an experimental assessment of the effect of the plaque backing on the distribution.

2. Materials & Methods

2.1. ROPES Eye Plaques loaded with I-125 Seeds (Amersham Health type 6711)

The ROPES eye plaques consist of an acrylic carrier with holes for the I-125 seeds combined with a stainless steel backing shield to place the acrylic insert in. Available plaque diameters are 11, 15 and 18 mm with 4, 9 or 10, and 14 seeds, respectively. The stainless steel backing and the acrylic insert are portion of a spherical shell that offsets the radioactive seeds by approximately 1 mm from the sclera of the eye [9]. The Amersham Health type 6711 I-125 seed is a cylindrically shaped and silver rod on which I-125 has been coated onto the surface. It emits photons with a weighted mean energy of 28.37 keV [7].

2.2. Dose calculations with Plaque Simulator™ and RADCALC®

The Plaque Simulator™ v5.63 software (BEBIG Germany) supports dose calculations for a variety of eye plaques, radioactive nuclides and seed models. The dose calculations are performed three-dimensionally with an algorithm based on the superposition principle, using TG43-UI data and formalism [12].

RADCALC® v6.1 (Lifeline Software, Inc. USA) is a versatile program designed to independently check radiation therapy treatment doses. The brachytherapy module allows dose calculations to points of interest for various brachytherapy sources whose data and model/approximation are user defined. The 3D I-125 seeds geometries were defined in Radcalc® for every ROPES eye plaque type available.

The line-source approximation for the geometry function and 2D anisotropy corrections were used for the I-125 sources in both software systems and lead to single seed dose comparison of difference less than 1% at distances up to 1 cm from the seed centre.

The plaques were loaded with 9.93 U sources and treatment times calculated in Plaque Simulator™ to deliver 85 Gy to a depth of 5 mm. Corresponding treatment times were entered into RADCALC®, taking into account that RADCALC® does not account for the decay of the sources during treatment. Central axis depth dose distributions were then compared for each of the ROPES eye plaques.

2.3. 3-D PRESAGE™ Phantom/Dosimeter and irradiation protocols

Customized 3-D PRESAGE™ phantoms/dosimeters were shaped in hemispheres to match eye ball geometry and fit ROPES eye plaques internal curvature. The following fractional atomic-mass composition optimized water equivalence for low-energy photons with a resulting mass-density of 1.06 g/cm3 (at room temperature) and Zeff=7.46: C (64.086%); N (5.083%); H (9.379%); O (20.922%); Sn (0.011%); Br (0.518%) [13].

Plaque Simulator™ was used to calculate the time needed to deliver a dose of 3 Gy to 5 mm depth. ROPES plaques loaded with model 6711 I-125 seeds were then placed over the hemispherical 3-D PRESAGE™ phantoms/dosimeters, and bound together using film wrap. The arrangement was then submerged in a water tank, in order to provide full scatter conditions, and left for the calculated treatment time. Experiments were repeated for each plaque configuration both with and without the stainless steel backing, in order to quantify its contribution to the central axis fluorescence signal.

Fluorescence intensities obtained with the 3-D PRESAGE^m dosimeters were read using a laser type confocal microscope as described elsewhere [14].

3. Results

3.1. ROPES Eye Plaques Dose Calculation Comparison

Figure 1 presents the calculated percentage difference in the central axis depth dose curves between Plaque Simulator™ and RADCALC® for all ROPES plaque models. Agreement better than 2% between the two programs is observed for depths of 1 – 8 mm, except for the 18 mm diameter eye plaque (14 seed insert) at a depth of 1 mm.

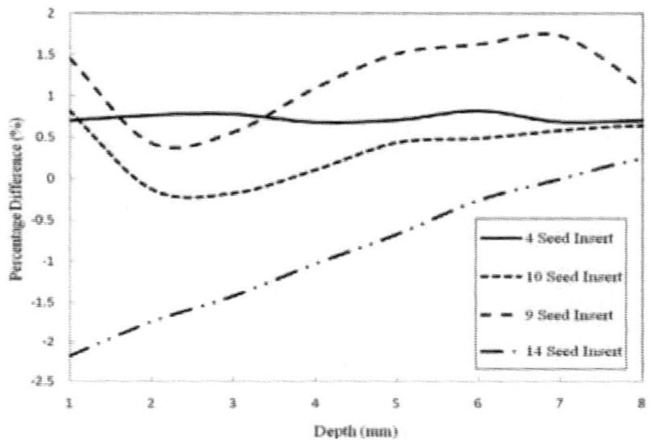

Figure 1: Percentage different between Plaque Simulator™ and RADCALC® for ROPES eye plaque central axis depth dose curves.

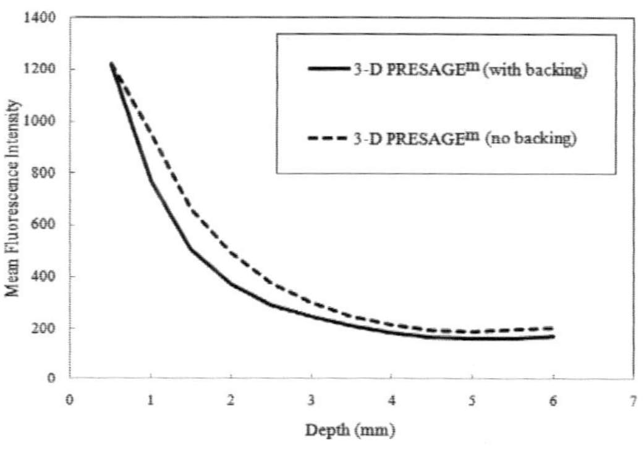

Figure 2: Central axis fluorescence intensity measurements for the 15 mm diameter (10 seeds insert) ROPES plaque.

3.2. 3-D PRESAGE^m Central Axis Measurements

Figure 2 shows the eye plaque central axis fluorescence intensity curves measured by the gel dosimeters for the 15 mm diameter plaque (10 seeds), both with and without the plaque stainless steel backing. The effect of the plaque backing is observed to result in a decrease in signal when compared to the measurements in which the plaque backing was removed, with a maximum difference of -25% occurring at 2 mm depth. The 10 mm plaque (4 seeds) resulted in similar signal decrease, with a maximum of -40% also occurring at 2 mm depth.

4. Discussion and Conclusion

The Plaque Simulator™ eye plaque brachytherapy planning system was commissioned following TG43-UI data and formalism for use with ROPES eye plaques and Amersham Health model 6711 I-125 seeds. The 3D geometries for these plaques loaded with their I-125 seeds were simulated in the RADCALC® independent checking program to allow verification of the doses calculated by Plaque Simulator™. Central axis depth dose comparisons agreed to within 2% for all ROPES plaques seed configurations, except for the 18 mm diameter (14 seeds) eye plaque where a discrepancy of 2.2% is found at 1 mm depth. At this point, there is a superposition of a large number of sources at a short distance, corresponding to a region of large uncertainty in the dosimetric data. Further comparison of the two programs is planned in order to compare off-axis profiles at several depths of interest. Using these off-axis profiles, three-dimensional dose distributions can be obtained. Precise calculations of these dose distributions are necessary in order for Plaque Simulator™ to accurately determine the dose to the sclera of the eye and other organs at risk (lens, macula, etc.).

Preliminary central axis fluorescence intensity curves have been measured using 3-D PRESAGE^m dosimeters. The effect of the eye plaque stainless steel backing on the shape of the curve was investigated. The effect of the stainless steel plaque backing was observed to decrease the measured fluorescence intensity by up to 25% and 40% for the 15 mm and 10 mm diameter ROPES plaques respectively, and was not predicted by any of the dose calculation software systems that use water full scatter conditions. Granero et al. [9] have shown through Monte Carlo simulation, an average decrease in dose due to the stainless steel backing of the 15 mm diameter (4 seeds) eye plaque of 4%.

Future measurements using the 3-D PRESAGE™ dosimeters should be performed with a calibration of the gels to correlate the fluorescence intensity with dose. Also, through read-out with optical CT, three-dimensional dose distributions measured by the gels can be obtained to compare with dose distributions produced by the Plaque Simulator™ and RADCALC® programs. The effect of the plaque backing on the dose distribution can then be fully quantified, and eventually, this effect incorporated into the Plaque Simulator™ planning system. Corrections for inhomogeneities associated with the plaque backing and seed holders are plaque specific, and must be fully quantified for more accurate dosimetry of treatments using ROPES eye plaques.

5. References

[1] Thomson R M and Rogers D W O 2010 *Med. Phys.* **37** 368-76
[2] Collaborative Ocular Melanoma Study Group 1995 Radiation Therapy *COMS Manual of Procedures* (Springfield: National Technical Information Service)
[3] Hawkins B S 2004 *Am. J. Opthalmol.* **138** 936-51
[4] Nath R *et al* 1995 *Med. Phys.* **22** 209-34
[5] Hanson W 1996 *Med. Phys.* **23** 1054
[6] Rivard M *et al* 2004 *Med. Phys.* **31** 633-674
[7] Dolan J *et al* 2006 *Med. Phys.* **33** 4675-84
[8] Chiu-Tsao S T *et al* 1993 *Phys.* **20** 383-89
[9] Granero D *et al* 2004 *Med. Phys.* **31** 3330-36
[10] Brown S *et al* 2008 *Appl. Radiat. Isot.* **66** 1970-74
[11] Gorjiara T *et al* 2011 *Med. Phys.* **38** 2265-74
[12] Astrahan M A *et al* 1990 *Med. Phys.* **17** 1053-57
[13] Alqathami M *et al Radiat. Phys. Chem.* **81** 867-73
[14] Annabell N *et al* 2012 *J. Synchrotron Rad.* **19** 332-9

Journal of Physics: Conference Series **444** (2013) 012103 doi:10.1088/1742-6596/444/1/012103

Application of MAGAT polymer gel dosimetry in breast balloon

N Govi [1,2], P Gueye[1] and S Avery[3]
[1]Hampton University, Hampton, VA
[2]Department of Radiation Oncology, Mary Washington Hospital, VA
[3]Department of Radiation Oncology, University of Pennsylvania, PA

Email: sg.nanda@gmail.com

Abstract. The purpose of this study is to characterize the response of MAGAT normoxic polymer gel for breast brachytherapy applications using two balloon applicators (MammoSite® and Contoura®) and verify the dose distribution with a commercial treatment planning system (BrachyVision® version 8.9.15). We present the fabrication, irradiation and readout of the gel used for the work described herein.

1. Introduction

There are several dosimetry techniques available to determine the radiation dose distribution delivered to the volume of interest during radiation therapy treatments. Each dosimetric tool has a unique advantage and simplicity of determination, along with their own strengths and weaknesses. Gel dosimetry is an inherent dosimetric method used to determine the 3D dose distribution of the absorbed dose in the medium [1, 2]. Different gels have been developed [3] and most recently normoxic polymer gels gains its own traction due to its simplicity of preparation under normal atmospheric condition and its high spatial resolution [4, 5]. A new type of gel composition called MAGAT, that is comprised of methacrylic acid, gelatin, tetrakis (hydroxymethyl) phosphonium chloride (THPC), was found to have desirable characteristics of normoxic polymer gels [6]. In particular, THPC was found to be an aggressive scavenger for dissolved oxygen, and exhibit good temporal stability and spatial resolution with an excellent R2 dose response and sensitivity (from MRI scanning).

2. Methods and Materials

2.1. Fabrication

The preparation of the MAGAT gel (Methaacrylic Gelatin And Tetrakis) in the laboratory used a table-top setup. The gel compositions consisted of a gelatin, methacrylic acid, and THPC. The percentage of the mixtures is summarized in Table 1.

Distilled water (430 g) were placed in a glass container and boiled using a hot plate. About 1/10th of gelatin (40 g) were then added to the boiling water and heated to 48 °C to permit the gelatin to completely dissolve in the water. The mixture was then allowed to cool down, thus providing a clear solution. The temperature of the gel was constantly monitored using a thermometer and when the temperature felt below 35 °C, 30 g of methacrylic acid (~99% titration, Sigma Aldrich) was added and mixed thoroughly. An additional 0.177 g of THPC (80% solution in water, Sigma Aldrich) was mixed to the solution which was again cooled for another 20 minutes. This clear gel solution was poured in six glass vials and into a larger (500 ml and 750 ml) glass phantom. Two glass containers, one with the single lumen (MammoSite®) and the other with the mutli-lumen (Contura®) balloon were placed inside the gel container and inflated with a saline solution (35 ml) and a contrast agent (5 ml). The entire setup was placed properly at the center of the container. Finally, the bottles were wrapped with an aluminum foil and stored in a refrigerator with a set temperature of 4°C. They were allowed to cool for 24 hours.

Table 1: Chemical composition of MAGAT polymer gel

1	Gelatin	8 % (40 g)
2	Methacrylic acid	6 % (30 g)
3	THPC (2mM)	0.177 g
4	Water	86% (430g)

2.2. Irradiation

The irradiation of the MAGAT gel was performed 24 hours after the preparation. Three sets of gels phantoms were prepared for this study: (i) calibration vials (ii) single lumen balloon catheter (MammoSite®) and (iii) Multi lumen balloon catheter (Contura®).

A Varian 2300ix linear accelerator was used to acquire the calibration data using the six vials. They were placed inside a perplex water phantom and positioned at the center of the field at 10 cm depth (FS= 10x10 cm^2, Figure 1(a). The theoretical dose requested varied from 0 (un-irradiated) to 7 Gy using 6 MV photon beams. The R2 dose response of the calibration vials are calculated based on the pixel value of the scanned MRI image is shown in Figure 1b.

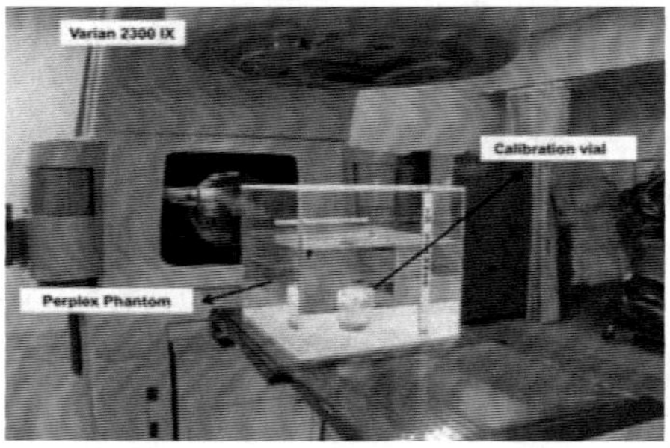

(a)

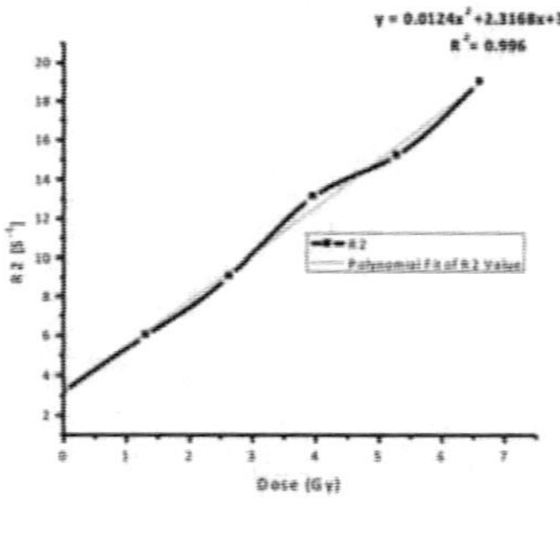

(b)

Figure 1: (a) Calibration of vials using 6 MV X-ray beams (b) dose response of the MAGAT polymer gel for the calibration vials.

Two bottles of gel, one for the MammoSite® balloon and one for the Contura® balloon, were irradiated using the VariSource HDR brachytherapy afterloader machine (Figure 2a). A treatment plan was created for the two balloons using the Brachyvision® software version 8.9.15. The radiation dose of 340 cGy was prescribed at PTV (planning target volume), 1 cm from the surface of the balloon using a single dwell position. After irradiation, the gel containers were stored in a cool place at 18-20°C for the post polymerization and the gel to get stabilized.

2.3. Gel Readout

The MAGAT polymer gels used for this study were scanned using a 1.5 Tesla MRI scanner (Siemens Esperee Magnetron). The images were acquired under a spin echo sequence with a CPMG radio frequency pulse scheme. A total of 864 images were obtained with 32 echo images having the following parameter values: echo time spacing ΔTE = 22.4 ms, threshold ratio TR = 9970 ms and pixel size of 1 mm x 1mm with a slice thickness of 2 mm. Twenty seven sets of images were obtained and the scanned images were analyzed using a Matlab® code which uses the maximum likelihood estimator to estimate the R2 values .The decay constant R2 values (longitudinal relaxation rate=1/T2) were subsequently calculated from the MRI data using a maximum likelihood method from the exponential decay equation. The isodose distribution for the Mammosite® balloon using MAGAT gel is shown in figure 2b.

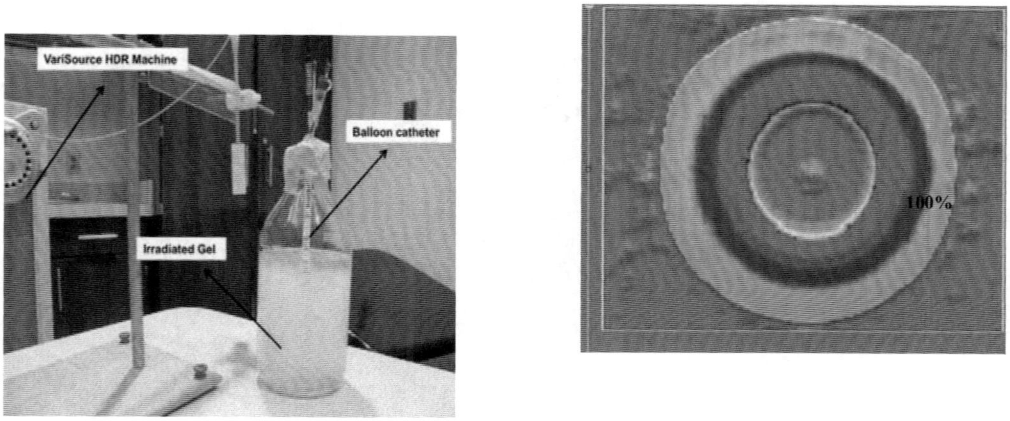

(a) (b)

Figure 2: (a) Balloon catheter irradiation (Ir-192 radioactive source) (b) Isodose distribution for MammoSite® balloon catheter (100% - 340 cGy)

3. Results and Discussion

The dose maps of the MammoSite® and Contura® balloons obtained from the MRI scanned data were compared to the treatment planning system: the gel dosimetry results provided are in very good agreement, less than 5%, with the TPS.

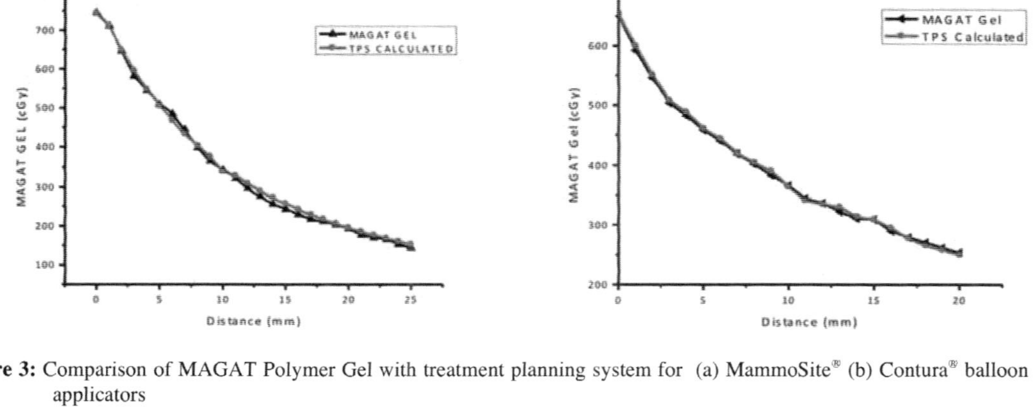

Figure 3: Comparison of MAGAT Polymer Gel with treatment planning system for (a) MammoSite® (b) Contura® balloon applicators

The MAGAT polymer gel was successfully used as a preliminary investigation for balloon brachytherapy applications to evaluate the 3D dose distribution of balloon catheters. It has been demonstrated that the MAGAT polymer gel can be used as a reliable tool for three dimensional dose verification of balloon brachytherapy techniques [7-9].

4. References

[1] Baldock C *et al* 2010 *Phys. Med. Biol.* **55** R1-63
[2] Brown S *et al* 2008 *Appl. Radiat. Isot.* **66** 1970-74
[3] Gustavsson H *et al* 2004 *Phys. Med. Biol.* **49** 227-41
[4] De Deene Y *et al* 2002 *Phys. Med. Biol.* **47** 3441-63
[5] De Deene Y *et al* 2002 *Phys. Med. Biol.* **47** 2459-2470
[6] Hurley *C et al* 2005 *Science Direct* 443-456
[7] Ibbott G *et al* 1997 *Int. J. Rad. Onc. Biol. Phys.* **38** 1097-1103
[8] Avery S *et al* 2010 *J. Phys.: Conf. Ser.* **250** 012033
[9] Watanabe Y *et al* 2005 *J Appl Clin Med Phys* **6** 133-42

Polymer gel dosimetry of an electron beam in the presence of a magnetic field

J Vandecasteele[1] and Y De Deene[1,2]
[1]Department for Radiation Oncology and Experimental Cancer Research, Ghent University, De Pintelaan 185, 9000 Gent, Belgium
[2]Institute of Medical Physics, School of Physics, University of Sydney, Sydney NSW, Australia

E-mail: Jan.Vandecasteele@UGent.be

Abstract. The effect of a strong external magnetic field on 4 MeV electron beam was measured with polymer gel dosimetry. The measured entrance dose distribution was compared with a calculated fluence map. The magnetic field was created by use of two permanent Neodymium (NdFeB) magnets that were positioned perpendicular to the electron beam. The magnetic field between the magnets was measured with Hall sensors. Based on the magnetic field measurement and the law of Biot-Savart, the magnetic field distribution was extrapolated. Electron trajectories were calculated using a relativistic Lorentz force operator. Although the simplified computational model that was applied, the shape and position of the calculated entrance fluence map are found to be in good agreement with the measured dose distribution in the first layer of the phantom. In combination with the development of low density polymer gel dosimeters, these preliminary results show the potential of 3D gel dosimetry in MRI-linac applications.

1. Introduction

In recent times, MRI scanners were combined with high energy x-ray treatment devices creating so called MRIGRT [1-3]. This combination results in the presence of a strong magnetic field in a region where charged particles are created, thus affecting the dose distribution inside a patient. A preliminary study was set-up to investigate the potential use of polymer gel dosimeters (both unit density and lung equivalent gel dosimeters [4]) in the presence of a strong magnetic field. A simulation program was developed to predict the charged particles trajectories in vacuum in the presence of an electromagnetic field.

2. Materials and methods

2.1. Gel fabrication

The PAGAT dosimeter used in this study, is composed of gelatin (6% w/w), acrylamide (3% w/w), N,N'-methylene-bis-acrylamide (3% w/w) and 5 mM Bis[tetrakis(hydroxymethyl)phosphonium] sulphate (THPS) as antioxidant. The polymer gel was fabricated according to a procedure as described elsewhere [5]. The gel was at 32 °C when it was poured in 21 test tubes and two Barex™ containers (20 x 20 x 6 cm^3).

2.2. Irradiation

Two PAGAT gel dosimeters were irradiated with a 4 MeV electron beam (4 x 1 cm² at isocentre) from a clinical linear accelerator (Elekta Synergy). The first phantom was irradiated in the presence of a magnetic field (figure 1b and 1c). A dedicated magnet holder was constructed to position the NdFeB magnets 3.3 cm from each other. In between the magnets sufficient space was left to allow the electron beam to pass (figure 1c). The second PAGAT gel dosimeter was irradiated at the same distance from the source, but without the presences of the magnets.

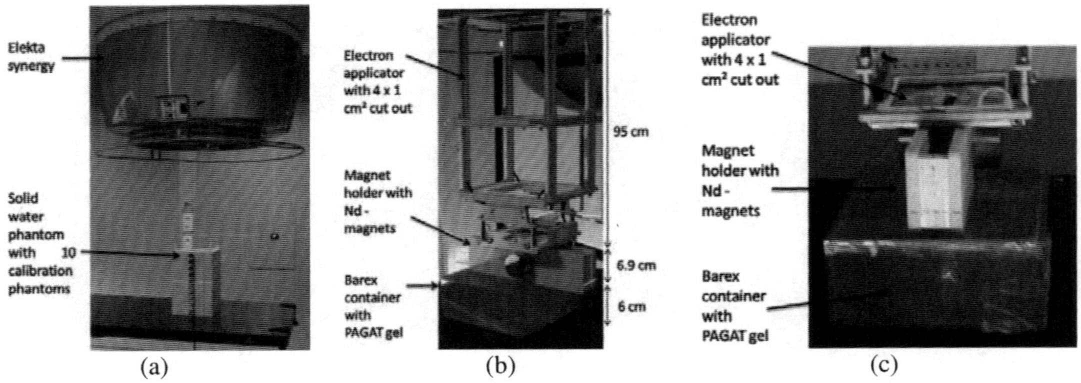

Figure 1: A set of calibration phantoms was irradiated with a 6 MV photon beam using a dedicated solid water phantom (a). Set-up of the irradiation of a gel phantom with an 4 MeV electron beam in the presence of 2 magnets (b). An appropriate cut-out was placed in the electron applicator assuring that the 4 x 1 cm² beam passes through the space between the 2 magnets (c).

A set of 8 calibration phantoms was irradiated in a solid water phantom with a 6 MV photon beam with field size 10 x 10 cm² and SSD 90 cm (figure 1a). In this phantom, 10 calibration phantoms can be positioned along the depth of the irradiation beam. An ion chamber measurement was performed at each position in the solid water phantom and compared to in-house beam data for the same irradiation setup in water. Finally, five calibration phantoms were irradiated with a 4 MeV electron beam under reference conditions so that 1 MU equals 1 cGy. This allowed us to compare the response of the gel to a photon beam exposure versus the response of the gel to an electron beam exposure.

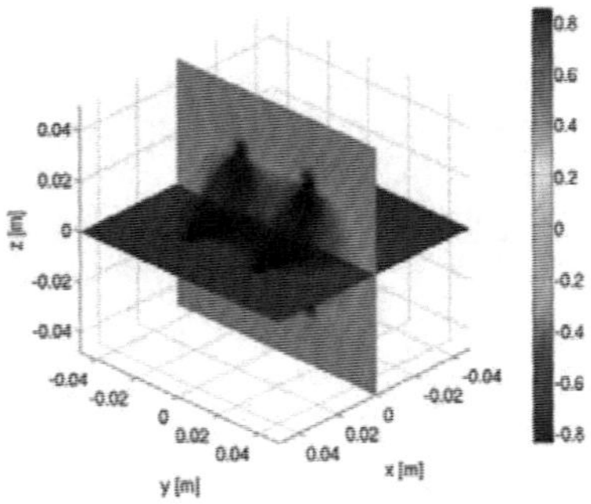

Figure 2: The calculated magnetic field distribution expressed in Tesla around the magnets. The magnets are represented in the figure by two round discs.

2.3. Computer simulations

The magnetic field generated by the two NdFeB magnets was calculated using the law of Biot-Savart (figure 2). The magnitude of the magnetic field was independently calibrated by Hall sensor probe measurements of the magnetic field generated by the two magnets. The magnetic field strength amounts to 0.49 Tesla in between the two magnets. The electron beam trajectories were predicted by a relativistic calculation of the Lorentz force exerted on the charged particles in the presence of the electromagnetic field. Because no in-house data on the energy spectrum of the 4 MeV electron beam was available, literature values were used [6]. No particle interactions with matter were included in the our simulations.

2.4. MRI read-out

Quantitative NMR spin-spin relaxation rate (R_2) maps were recorded using a multiple spin-echo sequence with 32 spin-echoes with a 1.5 Tesla MRI scanner (Siemens Avanto). The phantoms were separately scanned in the body coil along with the calibration phantoms using following imaging parameters: TR: 3000 ms:, TE: 40 – 1280 ms, voxel size: 1 × 1 × 3 mm³, number of acquisitions: 4 and bandwidth: 130 Hz/pixel. The centre of the recorded slice was positioned at 8 mm from the top of the phantom (including 4 mm Barex™). A fitting of the intensity of 31 equidistant consecutive base images to a monoexponential decay using χ^2-minimalisation was performed to obtain R_2 maps [7]. R_2 maps were calibrated to dose maps using the dose-R_2 relationship extracted from the calibration phantoms.

3. Results and discussion

3.1. Calibration

The dose-R_2 response of the calibration phantoms irradiated with a 6 MV photon beam (figure 3, blue markers) matches the dose-R_2 response of those irradiated with the 4 MeV electron beam (figure 3, red markers). The novel solid water phantom is an easy, reproducible and fast set-up to irradiate a set of calibration vials to a known dose. However, because the 10 calibration vials are positioned underneath each other, the effect of the accumulating borosilicate glass wall will affect the dose deposition in the gel. An ion chamber measurement is necessary to evaluate the dose at each position along the depth of the solid water phantom. The dose reduction caused by the borosilicate glass of the calibration vials amounts to 0.44% per millimetre glass.

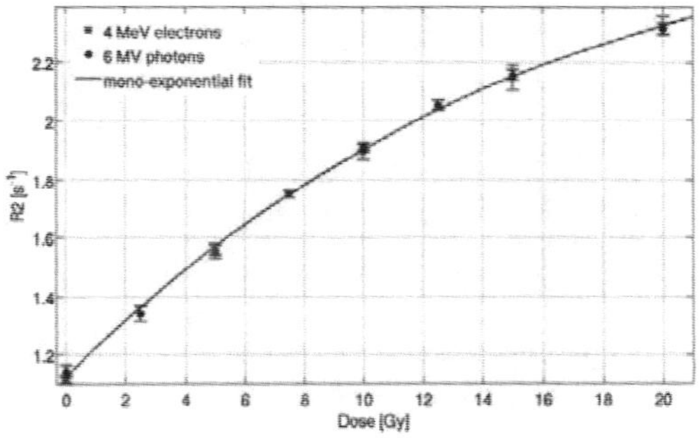

Figure 3: The dose response of calibration phantoms irradiated with 6 MV photon beam as compared to the dose response of calibration phantoms irradiated to a 4 MeV electron beam.

3.2. Electron dose measurements and simulations

In figure 4, nineteen electron trajectories are shown corresponding to the nineteen energy bins used in the simulations. A total of 1000 electron trajectories were simulated. In figure 5, a comparison between the simulated dose distribution and the gel measured dose distribution is displayed. The simulations were rescaled to the measured dose values in the gel. Because no interaction with matter is included in the simulation, actual dose maps were not calculated. Figure 5a and 5b display the simulated entrance fluence maps. Figure 5c and 5d display the gel measured entrance dose maps. A good agreement with the simulations is found.

(a) (b)

Figure 4: Nineteen electron trajectories are shown corresponding to the nineteen energy bins used in the simulations without an external magnetic field (a) and with and external magnetic field applied (b).

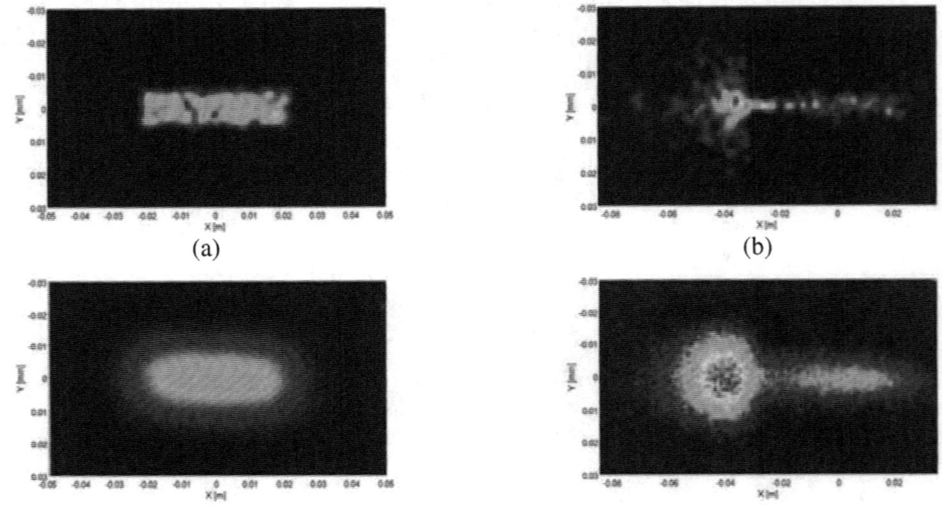

(c) (d)

Figure 5: Results of the simulation of a 4 x 1 cm² 4 MeV electron beam without (a) and with (b) an external magnetic field applied. In c and d the polymer gel dosimetry measurements are shown.

4. Conclusions

Simulations and gel measurements are in good agreement. In future work, lung equivalent gel dosimeters will be used in a strong magnetic field to investigate the influence of the applied magnetic field on the 3D dose distribution and Monte-Carlo simulations will be applied to incorporate the particle interactions in the phantom.

5. Acknowledgements

This research is supported by the Institute for the Promotion of Innovation though Science and Technology in Flanders (IWT -Vlaanderen).

6. References

[1] Raaijmakers A J E *et al* 2007 *Phys. Med. Biol.* **52** 7045-54
[2] Aubin J *et al* 2010 *Phys. Med. Biol.* **55** 4861-9
[3] Raaymakers B W *et al* 2007 *Phys. Med. Biol.* **49** 4109-118
[4] De Deene Y *et al* 2006 *Med. Phys.* **33**, 2586-97.

Journal of Physics: Conference Series **444** (2013) 012104 doi:10.1088/1742-6596/444/1/012104

[5] De Deene Y *et al* 2006 *Phys. Med. Biol.* **51** 653-73
[6] Kovar I *et al* 1983 *Phys. Med. Biol.* **28** 1441-6
[7] De Deene Y *et al* 1998 *Signal Process* **70** 85-101

The effect of mixed dopants on the stability of Fricke gel dosimeters

K Penev and K Mequanint
Department of Chemical and Biochemical Engineering, University of Western Ontario, London, Ontario, N6A 5B9, Canada

E-mail: kmequani@uwo.ca

Abstract. Auto-oxidation and fast diffusion in Fricke gels are major drawbacks to wide-spread application of these gels in 3D dosimetry. Aiming to limit both processes, we used mixed dopants: the ferric-specific ligand xylenol orange with a ferrous-specific ligand (1,10-phenanthroline) and/or a bi-functional cross-linking agent (glyoxal). Markedly improved auto-oxidation stability was observed in the xylenol orange and phenanthroline doped gel at the expense of increased background absorbance and faster diffusion. Addition of glyoxal limited the diffusion rate and led to a partial bleaching of the gel. It is conceivable that these two new compositions may find useful practical application.

1. Introduction

The ferrous sulfate – gelatin 3D dosimeter (Fricke gel) has historically suffered from two major drawbacks: relatively fast diffusion of the radiation product (ferric iron, Fe^{3+}) following irradiation [1-3] and significant auto-oxidation of the ferrous iron (Fe^{2+}) during storage [4]. More recently, as fast optical scanning has become available, the diffusion rate in Fricke – xylenol orange gels, following short irradiation protocols, has been deemed acceptable [5]. Arguably, therefore, the auto-oxidation becomes a more important factor that limits both the useful dose range and shelf life of Fricke gels.

It has previously been shown that introduction of ligands that bind strongly to Fe^{2+}, like 1,10-phenanthroline, alone or in combination with formaldehyde, in a Fricke gel leads to spontaneous reduction of the product following irradiation [6]. The effect was attributed to the increased redox potential of the iron-phenanthroline complex. Conversely, xylenol orange, which is the current dopant of choice in Fricke gels, does not bind to Fe^{2+}, but forms a strong complex with the Fe^{3+}, thus decreasing the redox potential of the system. Therefore, the addition of xylenol orange likely favors auto-oxidation of the ferrous iron. In our study, we tested the properties of Fricke gels doped with xylenol orange, phenanthroline and/or glyoxal as a cross-linking agent. To our best knowledge, and from two previous reviews of the system [6, 7], such use of xylenol orange with ferrous-specific ligands and cross-linkers has not been investigated.

2. Materials and methods

Stock solutions were prepared using analytical grade reagents (supplied by Sigma-Aldrich): H_2SO_4, $Fe(NH_4)_2(SO_4)_2 \cdot 6H_2O$, 1,10-phenanthroline (Pn), xylenol orange disodium salt (XO) from Fluka, NaCl, and glyoxal (Gx). For each gel (25.0 g), the required amounts of the stock solutions were added together and adjusted to 12.5 g with de-ionized water, followed by 12.5 g of warm (50 – 60 °C) 10% solution of type A porcine gelatin (300 bloom from Sigma-Aldrich), which was prepared at 80 °C. The warm gel was poured into poly(methyl methacrylate) cuvettes (10 mm path length, square cross-section), closed with polypropylene cuvette caps and let to solidify in a refrigerator at 10 °C. All gels contained 5% gelatin; and the following compositions were tested (concentrations in mM): **FXG** (H_2SO_4/NaCl/Fe^{2+}/XO = 50/1.0/0.1/0.1), **FXGxG** (H_2SO_4/NaCl/Fe^{2+}/XO/Gx = 50/1.0/0.1/0.1/10), **FXPnG** (H_2SO_4/NaCl/Fe^{2+}/XO/Pn = 25/1.0/0.1/0.1/0.3), **FXPnGxG** (H_2SO_4/NaCl/Fe^{2+}/XO/Pn/Gx = 25/1.0/0.1/0.1/0.3/10).

Irradiations were performed using a Co-60 source (Eldorado 6, Atomic Energy of Canada Ltd.) at the London Regional Cancer Program (London, Canada). For sensitivity and stability determination the cuvettes were irradiated in full. For diffusion measurement, the top halves of the cuvettes were shielded with a lead block and a dose of 10 Gy was delivered to the bottom halves. The cuvettes were irradiated at 22 °C, transported at 4 °C for a period of 1.5 hours after irradiation, and scanned at 22 °C. Optical absorbance readings of the fully irradiated cuvettes were performed at 585 nm with a GENESYS 10S UV-Vis spectrophotometer (Thermo Scientific). The half-irradiated gels were imaged using a Vista10 conebeam optical CT scanner (Modus Medical Devices Inc., London, Canada), with an amber LED source, filtered at 590±2 nm. Time course images were taken regularly for the following 4.5 hours. For long term storage, the gels were kept in a cool environment (at 10 °C).

The light transmittance profiles of the diffusion cuvettes were obtained with image processing software (ImageJ, version 1.44p). The data was fitted to an approximation of the complementary error function in order to determine the diffusion parameter (n, mm^2), whose increase over time (t, h) gives the diffusion coefficient (D, $mm^2 \cdot h^{-1}$) [8]:

$$D = \left(\frac{\sqrt[3]{4} - 1}{4 \cdot \ln 2} \right) \frac{n}{t} = 0.212 \frac{n}{t} \quad (1)$$

3. Results

3.1. Dose response and chemical stability

Figure 1 shows the dose response in the tested gels immediately after irradiation and, in the same gels, eight days later. All compositions, except FXPnG, show significant change in their absorbance.

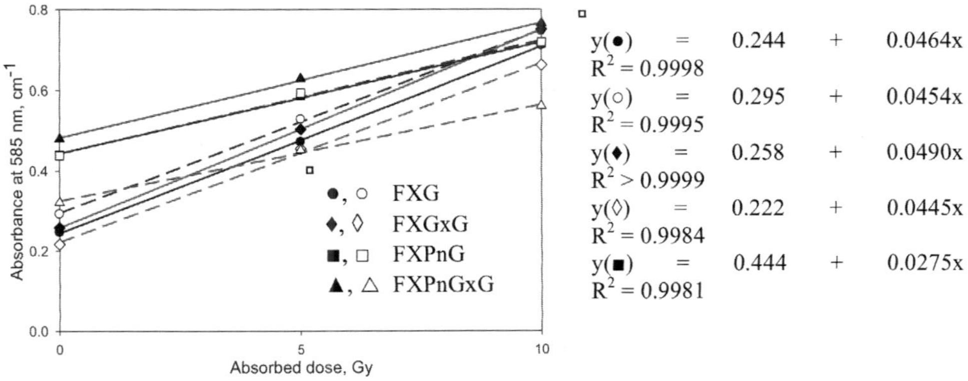

Figure 1: Dose response: 3 hours after irradiation (solid lines, filled symbols) and 8 days after irradiation (dashed lines, open symbols).

Figure 2 presents the effect of long-term storage in a cool environment (10 °C) on the background absorbance of the gels (2a), and the initial dose response (2b). Not given in the figure is the linearity of the dose response: all four compositions gave linear response ($R^2 > 0.995$) for the first 21 days after irradiation (22 days after preparation), and $R^2 > 0.989$ after 50 days (51 days after preparation).

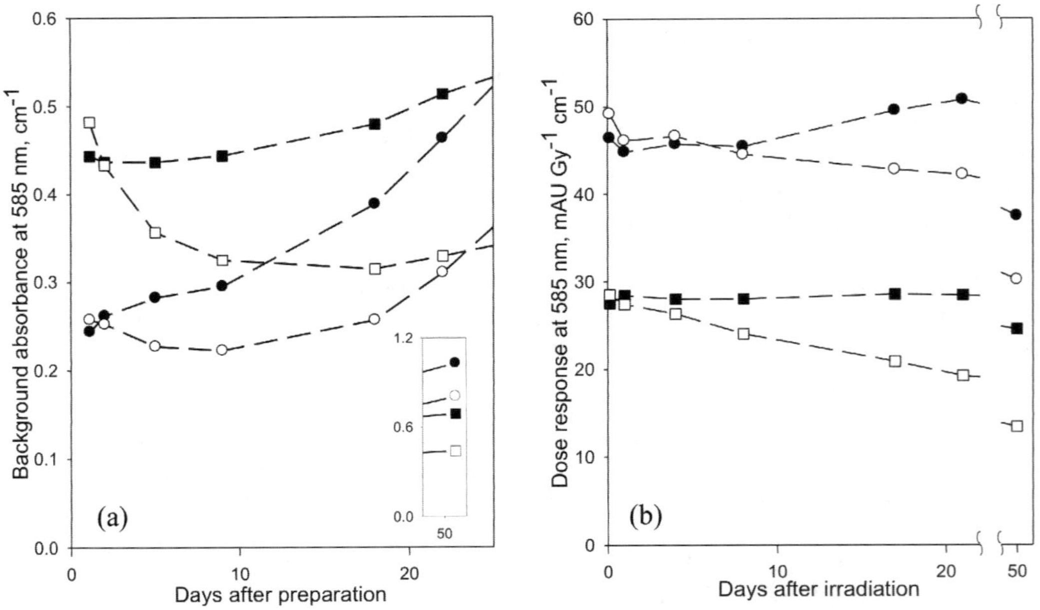

Figure 2: Chemical stability during storage at 10 °C. 2a: background absorbance, 2b: change of the initial dose response (the gels were not re-irradiated).
Legend: □ – FXG, □ – FXGxG, □ – FXPnG, □ – FXPnGxG, mAU = milli (Absorbance Units).

Addition of phenanthroline led to an approximately 40% decreased sensitivity and increased background absorbance. Nevertheless, for FXPnG, both characteristics of the gel remained very stable: there was minimal apparent auto-oxidation for the first week of storage and practically constant dose response for at least three weeks. The presence of glyoxal had minimal initial effect on both background absorbance and sensitivity. However, partial bleaching of the gels was seen in the following days. Auto-oxidation was eventually apparent in approximately a week for FXGxG and three weeks for FXPnGxG. The initial dose response decreases monotonically in those gels.

3.2. Diffusion coefficients and summary of results

The change of the diffusion parameter (*n*) versus time is presented in figure 3.

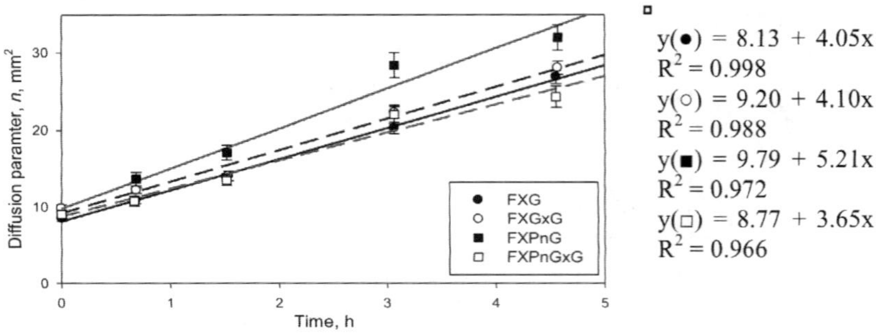

Figure 3: Diffusion parameter (*n*) at 22 °C, starting approximately 1.5 hours after irradiation.

The diffusion coefficients, estimated from equation (1) and the linear fits from figure 3, are given in table 1, together with a summary of the other characteristics of the gels.

Table 1: Summary of the results.

Gel code	Bkgrd. abs. mAU·cm^{-1}	Dose response mAU·Gy^{-1}·cm^{-1}	Diff. coeff. mm^2·h^{-1}	Auto-oxidation during storage	Dose response after irradiation
FXG	224	46.4	0.86 ± 0.02	fast	unstable
FXGxG	258	49.0	0.87 ± 0.06	bleaching at first	decreasing
FXPnG	444	27.5	1.10 ± 0.11	slow	stable
FXPnGxG	482	28.5	0.77 ± 0.08	bleaching at first	decreasing

4. Discussion and conclusions

The compositions of the tested gels were chosen on the basis of screening experiments and a review of the published literature [1-3, 6, 7]. Glyoxal, which is a di-aldehyde, was preferred as a cross-linker over the mono-functional and less stable formaldehyde. The sulfuric acid concentration was based on previous investigations [7] and preliminary testing of the stability of the ferrous-phenanthroline complex. Sodium chloride was used as an additive that may minimize the effect of organic impurities.

As stated in the beginning, given the advances in fast optical scanning, the auto-oxidation in Fricke gels may be considered one of their main disadvantages. We have shown that addition of phenanthroline and xylenol orange, i.e. FXPnG, yields a very chemically stable gel. Compared to FXG, the radiation sensitivity of the gel is decreased but still adequate for clinical applications. However, the initial background absorbance is almost doubled, and the diffusion rate is increased by 30%. In order to decrease the initial background absorbance, we suggest that the gelatin solution be pre-treated, as suggested in the literature [3].

Introduction of glyoxal in FXPnGxG led to a drastic decrease of the diffusion rate without affecting the sensitivity of gel initially. Instead of auto-oxidation, bleaching of the gel was observed during storage; in a few days the background of the gel stabilized and in several weeks auto-oxidation was seen. That effect may be useful for clinical applications but requires further research. FXGxG showed nearly identical diffusion rate as FXG, indicating that the cross-linking ability of glyoxal may be affected by pH. During storage FXGxG behaved in a similar manner as FXPnGxG, but on a shorter time-scale.

Overall, both FXPnG and FXPnGxG show interesting properties that may be advantageous in clinical settings.

5. Acknowledgements

We would like to acknowledge the help of Mr. Thomas Gungor, Ms. Jennifer Dietrich and Dr. Kevin Jordan, as well as the support from the Ontario Centres of Excellence, Modus Medical Devices Inc. (London, Canada) and the London Regional Cancer Program (London, Canada).

6. References

[1] Harris P J *et al* 1996 *Phys. Med. Biol.* **41** 1745-53
[2] Baldock C *et al* 2001 *Australas. Phys. Eng. Sci. Med.* **24** 19–30
[3] Schreiner LJ 2004 Review of Fricke gel dosimeters *J. Phys. Conf. Ser.* **3** 9
[4] Jordan K and Battista J 2004 *J. Phys.: Conf. Ser.* **3** 232
[5] Jordan K 2010 *J. Phys.: Conf. Ser.* **250** 012043
[6] Rae W I D *et al* 1996 *Med. Phys.* **23** 15
[7] Baldock C 2006 *J. Phys.: Conf. Ser.* **56** 14
[8] Kron T *et al* 1997 *Magn. Reson. Imaging* **15** 211

Journal of Physics: Conference Series **444** (2013) 012106 doi:10.1088/1742-6596/444/1/012106

Preliminary dosimetry investigation of Tc-99m diagnostic radionuclide by NIPAM gel dosimeter

You-Ruei Huang[1], Yuan-Jen Chang[2,3], Ling- Ling Hsieh[1], Bi- Wei Yu[1], Chien-Hau Chu[4] and Bor-Tsung Hsieh[1]

[1]Department of Medical Imaging and Radiological Sciences, Central Taiwan University of Science and Technology, Taichung, Taiwan, ROC

[2]Institute of Biomedical Engineering and Material Science, Central Taiwan University of Science and Technology, Taichung, Taiwan, ROC

[3]Department of Management Information Systems, Central Taiwan University of Science and Technology, Taichung, Taiwan, ROC

[4]Health Physics Division, Institute of Nuclear Energy Research, Taoyuan, Taiwan, ROC

E-mail: bthsieh@ctust.edu.tw

Abstract. The N-isopropylacrylamide (NIPAM) gel dosimeter was investigated as a suitable material for measuring absorbed doses from radionuclide sources. In this study, NIPAM gel dosimeter was used to evaluate the dose distributions of the Tc-99m radionuclide in NIPAM gel. The accumulated radioactivity range of the Tc-99m NIPAM gel is from approximately 0 MBq to 13.6 MBq (about 0.37 mCi). The NIPAM gel dosimeter with high stability and high-dose linear and non-energy dependent properties can provide various radiopharmaceutical activity intensities in the conduct of dose assessment in nuclear medicine, thereby producing the most promising dose verification tools.

1. Introduction

Treatments involving the use of ionizing radiation, such as radiotherapy and targeted radionuclide therapy (TRT), play an essential role in medicine and are fast becoming common with the development of new therapies and technologies [1]. TRT involves the intravenous or oral administration of a radionuclide-labeled pharmaceutical that is appropriate for a type of cancer. Along with these treatments is the need for dosimeter materials that can absorb radiation doses to ensure the health and safety of the patient [2]. Such common dosimeters as ionization chamber, thermo luminescent diodes, and radiographic film do not allow the measurement of three-dimensional (3D) dose distribution due to several spatial laminations [3]. Polymer gel dosimeter provides true 3D absorbed dose information and is very suitable as a dose distribution tool for radionuclide dosimeters [4].

In recent years, polymer gel dosimeter is used for radionuclides in brachytherapy and TRT for the purpose of obtaining dose distribution. Dosimetry for radionuclide therapy and diagnostic studies currently relies on imaging to map the distribution of the radiopharmaceutical in the body. Absorbed radiation dose from radionuclide is more difficult to measure than that from an external beam. Wuu et al., for example, used Re-188 source inserted into BANG gel to measure dose distribution [5], while Kelly et al. used MRI to evaluate dose uniformity with Tc-99m source in PAG gel dosimeter [6].

The current study investigates the polymer gel dosimeters of the Tc-99m radionuclide in conjunction with the N-isopropylacrylamide (NIPAM) gel for the production of nuclear medicine. The radiation doses from different activities of the radionuclide are measured by analyzing the radiation-induced change in the gels with optical computed tomography (OCT).

2. Materials and Methods

2.1. NIPAM gel preparation and Optical Computed Tomography measurement

The NIPAM gels were manufactured following the method described by Hsieh et al. [7]. A 4 wt% gelatin (300 Bloom Tape A. Sigma-Aldrich) was added to 88 wt% deionized water and stirred for 10 min at room temperature of 22 °C. The gelatin solution was heated to 45 °C with a hot plate until it became clear and transparent. With continuous stirring, 4 wt% NIPAM (97%, Sigma-Aldrich) and 4 wt% BIS (Merck) were poured into the gelatin solution and dissolved for 20 min. The solution was left to stir for 5 min, after which 3320 MBq of the Tc-99m, was added to the Pyrex test tube via a syringe. Next, 10 mM THPC (80%, Sigma-Aldrich) was added to the solution, which was continuously stirred for 2 min before the NIPAM gel was added to the Tc-ppm in the Pyrex test tube via a syringe. After removing the empty syringes, the tubes were shaken and stored behind a lead shield.

After 12 T1/ to 15 T1/2 decay time, the irradiated tubes were analyzed via an OCT using Chang et al.'s protocol [8]; Afterwards, an NRC Model 127 He-Ne laser with 20 mW power and 632.8 nm wavelength was used in CT-s1. After a 140 min laser warm up, the laser power deviation was less than 1%. Room temperature was maintained at 22±1 °C, and a photodetector was used to scan the gel dosimeter as depicted in a previous research.

2.2. External beam irradiation

In the present study, a Varian 21EX Clinic linear accelerator was used as a high energy irradiated apparatus. The setup condition of linear accelerator included gantry at 0°, energy of 6 MV, dose rate of 400 MU/min, and field size of 20 cm × 20 cm. Another medium energy X-ray apparatus used was the Pantak HF 420C. The condition of Pantak HF 420C included an energy range of 80 kVP to 250 kVP, and source-axis distance (SAD) of 150 cm. Furthermore, the absorbed dose calibrations were performed using Kodak X-O mat V film.

3. Results and Discussion

3.1. Dose response of the NIPAM gel

Fig. 1(a) shows the dose response of the NIPAM gel presented in this study. The dose response is linear from 0 Gy to 3 Gy, with linearity of 0.999 and sensitivity of 0.009 Gy-1. Fig. 1(b) shows the dose response curve for different beam energies of 100, 150, and 250kV. No significant changes in the dose responses at 0.1 and 0.5 Gy appear in the NIPAM gel.

3.2. Dose verification of the radiochromic film

Fig. 2(a) shows the energy dependence of the V film. When its irradiated dose is above 2.25 Gy, energy dependence is significant. Hence, the radioactivity of Tc-99m is approximately 1600 MBq (about 2.2 Gy). The result avoids the energy dependence that influences the accuracy of the measurement dose.

3.3. Dose verification of Tc-99m NIPAM gel dosimeter

Based on the results, the NIPAM gel dosimeter has the potential to measure absorbed dose from the Tc-99m radionuclide. When using the V film as a dose verification dosimetry, the corresponding dose of the Tc-99m ranges up to 2.2 Gy. The V film is the most commonly used radiochromic film in clinical practice. Its energy range is from 8 MV to 18 MV. The V film dose verification range is 2.25 Gy, and the accumulated radioactivity range of the Tc-99m NIPAM gel is approximately 0 MBq to 13.6 MBq (about 0.37 mCi) (Fig. 2(b)).

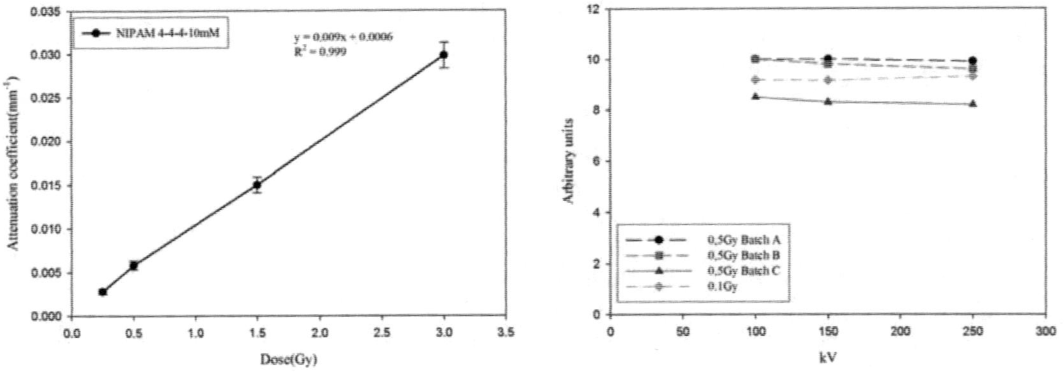

(a) (b)

Figure 1: (a) Dose response and (b) Energy dependence of the NIPAM gel

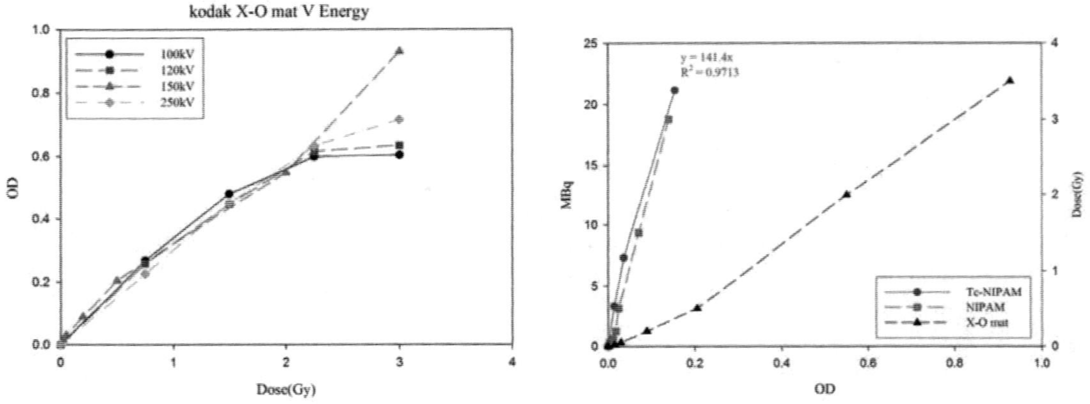

(a) (b)

Figure 1: (a) Dose response and (b) Energy dependence of the NIPAM gel

4. Conclusion

This study suggests that the radiolabeled NIPAM gels are suitable as dosimeters in radionuclide studies. The NIPAM gel dosimeter provides a variety of radiopharmaceutical activities of dose assessments in nuclear medicine that, in turn, lead to the development of the most promising dose verification tools.

References

[1] Gear J I *et al* 2006 *Phys. Med. Biol.* **51** 3503-16
[2] Schwarcke M *et al* 2010 *J. Phys.: Conf. Ser.* **250** 012082
[3] Oldham M *et al* 2003 *Med. Phys.* **30** 623-34
[4] Baldock C *et al* 2010 *Phys. Med. Biol.* **55** R1-63
[5] Wuu C S 2003 *Med. Phys.* **30** 132-7
[6] Braun K *et al* 2009 *J. Phys.: Conf. Ser.* **164** 012050
[7] Hsieh B T *et al* 2011 *J. Radioanal. Nucl. Chem.* **290** 141-8
[8] Chang Y J *et al* 2011 *Nucl. Instrum. Methods Phys. Res. A* **652** 783-5

Journal of Physics: Conference Series **444** (2013) 012107 doi:10.1088/1742-6596/444/1/012107

Deformable gel dosimetry II: experimental validation of DIR-based dose-warping

UJ Yeo[1], ML Taylor[1,2], JR Supple[1], RL Smith[1,3], T Kron[1,2] and RD Franich[1]
[1]School of Applied Sciences and Health Innovations Research Institute, RMIT University, Melbourne, VIC, Australia
[2]Physical Sciences, Peter MacCallum Cancer Centre, Melbourne, VIC, Australia
[3]William Buckland Radiotherapy Centre, The Alfred Hospital, Melbourne, VIC, Australia

E-mail: rick.franich@rmit.edu.au

Abstract. Algorithms exist for the deformation of radiotherapy doses based on patient image sets, though these are sometimes contentious because not all such image calculations are constrained by appropriate physical laws. By use of a deformable dosimetric gel phantom, 'DEFGEL', we demonstrate a full 3D experimental validation of a range of dose deformation algorithms publicly available. Spatial accuracy in low contrast areas was assessed using "ghost" fiducial markers (digitally removed from CT images prior to registration) implanted in the phantom. The accuracy with which the different algorithms deform dose was evaluated by comparing doses measured with the deformable phantom to warped planned doses, *via* 3D g-analysis. Mean spatial errors ranged from 1.9 mm with a g_{3D} passing ratio of 95.8 % for the *original Horn and Schunck* algorithm to 3.9 mm with a g_{3D} passing ratio of 39.9 % for the *modified demons* algorithm.

1. Introduction

There is an increasing clinical awareness of and interest in anatomic deformation and the dosimetric consequences thereof. Understanding the cumulative dose distribution in deforming anatomy may be accomplished by deformable image registration (DIR) based 'dose-warping' [1], whereby the dose distribution is morphed according to the geometric changes of anatomy evident in the patient images. The warped distribution can then be added to previous fractions in order to interpret the accumulated dose [2, 3]. What is *not* well known is the degree to which this approach is acceptable. This work quantitatively evaluates the accuracy of DIR and presents an experimental validation of dose-warping, achieved using a fully three-dimensional, deformable, integrating dosimeter.

2. Materials and Methods

We previously reported on a deformable, tissue-equivalent, dose-sensitive gel phantom dubbed 'DEFGEL' [4] based on normoxic polyacrylamide gel [5]. The gel was poured into a thin latex membrane and moulded into a cylindrical shape of 46 mm diameter. To investigate displacement of the internal structure after deformation, sixteen Aluminium fiducial markers (FMs) were implanted into the DEFGEL during the gel setting phase. The DEFGEL was then deformed – in this case, by applying a bilateral compression of equal displacement from both sides.

Eleven different DIR algorithms available in the open source MATLAB toolkit DIRART [6] were assessed in terms of spatial accuracy by taking CT images of the DEFGEL in both the deformed (compressed) and undeformed states. The high contrast FMs were digitally removed from the images prior to performing DIR so as not to contribute to the calculation and lead to a biased accuracy assessment, and later used to evaluate accuracy in uniform intensity regions. The resultant deformation vector field (DVF) was then applied to the original image of the deformed DEFGEL (containing the FMs). The locations of the FMs in this *calculated* image were then compared to the known locations from the original image of the DEFGEL in its undeformed state.

To investigate the validity of DIR-based dose-warping, a four dynamic-arc stereotactic treatment adapted from a patient plan was delivered to the DEFGEL in the deformed state. The DEFGEL returned to its undeformed state and then the measured dose was read out *via* optical CT imaging [7-9]. The planned dose distribution (from TPS) was warped by applying the DVF obtained from DIR. The measured and warped doses were then quantitatively compared *via* 3D γ-analysis using criteria of 3 % dose difference and 3 mm distance to agreement (DTA).

3. Results and Discussion

The first assessment of the 11 DIR algorithms was to test whether the general shape of the boundary or contour of the DEFGEL phantom could reasonably be calculated. As shown in figure 1, not all algorithms tested were able to replicate the DEFGEL contour. With the exception of the *double force demons*, it was observed that the *optical flow* category and *demons* category algorithms performed reasonably well. Algorithms A-G are represented by a single calculated image as these results were visually indistinguishable. Only these seven algorithms able to reasonably register the DEFGEL contour were assessed further for accuracy within the uniform intensity region (i.e. A-G).

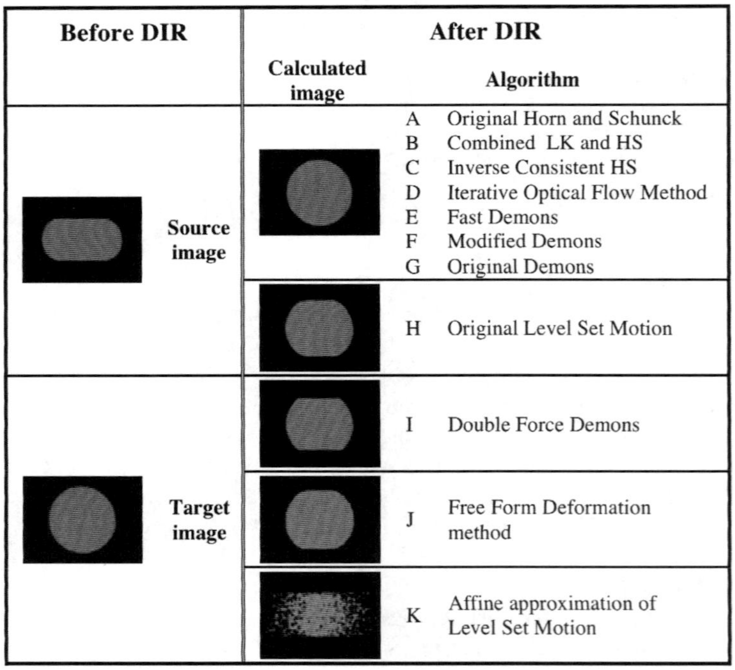

Figure 1: The calculated non-rigid registration of the source image to the target image for 11 DIR algorithms. The seven successful algorithms (A-G) are represented by a single calculated image.

Table 1: The mean error, $\bar{\Delta}$, standard deviation, σ_Δ and maximum error, Δ_{max} (values in parentheses are the maximum error as a percentage of the maximum deformation), of calculated FM positions.

Algorithm	$\bar{\Delta}$ (mm)	σ_Δ (mm)	Δ_{Max} (mm)
Original Horn and Schunck	1.9	0.9	4.3 (38%)
Modified Demons	3.9	1.1	5.6 (50%)

Table 1 shows the mean error, $\bar{\Delta}$, standard deviation, σ_Δ, and maximum error, Δ_{max}, of the 16 calculated FM positions for the best performing algorithm, *original Horn and Schunck*, and the worst performing algorithm, *modified demons*. The error is defined as the magnitude of the vector pointing from the actual position (determined from the original CT image) to the calculated position.

Figure 2 shows a comparison of calculated and measured doses delivered to undeformed and deformed DEFGELs. Transverse plane gamma maps (at D_{max}) are shown with 3 % / 3 mm criteria. The gamma map for the undeformed case i.e. comparison of planned and measured dose distributions in the absence of deformation, indicates the achievable agreement.

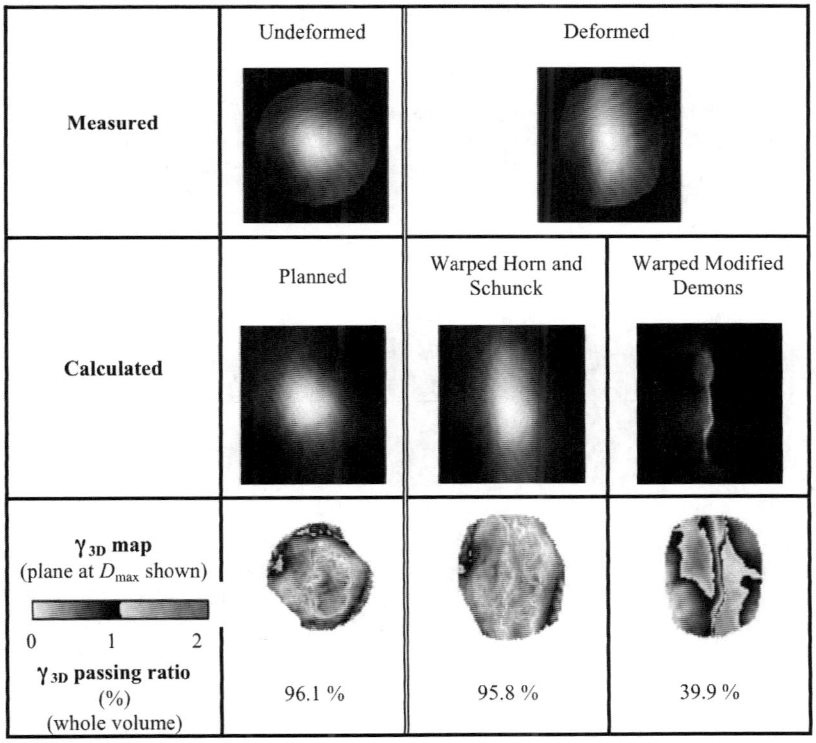

Figure 2: A comparison of calculated and measured doses delivered to undeformed and deformed DEFGELs. Transverse plane gamma maps (at D_{max}) are shown with 3 % / 3 mm criteria.

It is clear from the results for the *modified demons* algorithm that although a given algorithm may be able to accurately register the *contour* of an object, the accuracy *within* the uniform intensity region of such an object may be quite poor.

While the mean errors are small for both algorithms, the relatively large standard deviations and maximum errors indicate varying accuracy throughout the phantom. Although the maximum errors for both algorithms are larger than the DTA criterion, this does not necessarily mean those regions would fail the γ-analysis. If these large errors were in an area of uniform dose they would likely pass as the dose at the reference point would still closely match that of the evaluated point. This is supported by the fact that most voxels failing are in high dose gradient regions. The γ_{3D} passing ratio of 95.8 % for the *original Horn and Schunck* algorithm is exceptionally good, being just 0.3 % less than the achievable accuracy indicated by the undeformed case (γ_{3D} = 96.1 %). While the maximum error for the *modified demons* algorithm was only slightly larger than that for the *original Horn and Schunck*, the mean error was larger than the DTA criterion, leading to a much lower γ_{3D} passing ratio of 39.9 %.

4. Conclusions

Using a tissue equivalent, mass and density preserving deformable gel (DEFGEL) phantom implanted with high contrast fiducial markers, we were able to demonstrate a quantitative evaluation process to systematically investigate the accuracy of DIR algorithms. Using a method of mathematically erasing the markers prior to registration, we were able to assess the accuracy of algorithms within areas of near-uniform intensity, rather than only at known landmarks. This eliminates bias introduced when using intensity-based algorithms. We have also demonstrated that dose-warping using deformation vector fields obtained *via* deformable image registration can accurately represent the true (measured) dose in a deformed medium.

5. Acknowledgements

This work is supported by an RMIT University Research & Innovation Emerging Researcher Industry Award (Dr R Franich, Dr Taylor). Mr Smith is supported by Cancer Australia Grant 616614, funded by the Radiation Oncology Section, Australian Government Department of Health and Ageing.

6. References

[1] Brock K K *et al* 2003 *Med. Phys.* **30** 290-5
[2] Guerrero T *et al* 2005 *Radiat. Prot. Dosimet.* **115** 497-502
[3] Schaly B *et al* 2004 *Phys. Med. Biol.* **49** 791-805
[4] Yeo U J *et al* 2012 *Med. Phys.* **39** 2203-13
[5] Baldock C *et al* 2010 *Phys. Med. Biol.* **55** R1-63
[6] Yang D *et al* 2011 *Med. Phys.* **38** 67-77
[7] Islam K *et al* 2003 *Med. Phys.* **30** 2159
[8] Olding T *et al* 2010 *Phys. Med. Biol.* **55** 2819-40
[9] Olding T and Schreiner LJ 2011 *Phys. Med. Biol.* **56** 1259-79

Dosimetric effects of a high-density spinal implant

T Kairn[1,2], SB Crowe[2], J Kenny[1,3], J Mitchell[1], M Burke[1], D Schlect[1] and JV Trapp[2]
[1]Premion, The Wesley Medical Centre, Suite 1, 40 Chasely St, Auchenflower, Qld 4066, Australia
[2]Science and Engineering Faculty, Queensland University of Technology, GPO Box 2434,Brisbane Qld 4001, Australia
[3]Australian Clinical Dosimetry Service, ARPANSA, Yallambie Vic, Australia

E-mail: kairn@physics.org

Abstract. In this study, a treatment plan for a spinal lesion, with all beams transmitted though a titanium vertebral reconstruction implant, was used to investigate the potential effect of a high-density implant on a three-dimensional dose distribution for a radiotherapy treatment. The BEAMnrc/DOSXYZnrc and MCDTK Monte Carlo codes were used to simulate the treatment using both a simplified, recltilinear model and a detailed model incorporating the full complexity of the patient anatomy and treatment plan. The resulting Monte Carlo dose distributions showed that the commercial treatment planning system failed to accurately predict both the depletion of dose downstream of the implant and the increase in scattered dose adjacent to the implant. Overall, the dosimetric effect of the implant was underestimated by the commercial treatment planning system and overestimated by the simplified Monte Carlo model. The value of performing detailed Monte Carlo calculations, using the full patient and treatment geometry, was demonstrated.

1. Introduction

High-density, metallic implants are often necessary to for the survival or quality of life of radiotherapy patients. However, these implants can cause problems for successful radiotherapy treatment planning, when they are located within or close to targeted lesions and organs. In addition to producing star and shadow artefacts in CT images, the presence of high-density and high-atomic-number implants can also result in inaccurate density assignment and underestimation of dosimetric effects, by conventional treatment planning systems.

Monte Carlo simulations have been shown to provide valuable reference data, for establishing the accuracy of treatment planning dose calculation algorithms, in the presence of heterogeneous tissue [1]. Previous Monte Carlo simulation studies of the dosimetric effects of high-density implants have relied on the use of simplified anatomical models (usually homogeneous water) and simplified implant models (usually cylinders, cubes or planes) [2, 3]. It has been argued, however, that the results of simple phantom studies may exaggerate the effects of density heterogeneities on dose, since patient treatments use complex treatment plans, with multiple gantry angles, delivered to complex anatomies where variable thicknesses of air and bone intersect the radiation beams [4].

The current study examines one clinical case, where a radiotherapy treatment was planned for delivery through a titanium vertebral implant, and evaluates the reliability of a simplified Monte Carlo model of this complex treatment, in comparison with a Monte Carlo simulation that includes the full complexity of the patient anatomy and treatment geometry.

2. Methods

2.1. Treatment plan

A sample treatment plan was devised using CT data and prescription information from a recent patient treatment. The treatment involved the surgical excision of the laminae from seven vertebrae in the cervical-thoracic vertebrae [5], and the implantation of a Vertex Max reconstruction system (Medtronic, Sydney, Australia). This titanium implant consisted of two rods which were attached, via two parallel rows of screws, to the affected vertebrae.

The radiotherapy treatment was planned using the Eclipse treatment planning system (Varian Medical Systems, Palo Alto, USA), version 8.6, and dose was calculated using the AAA convolution-superposition algorithm. In order to provide an apparently homogeneous dose to the targeted tumour, the sample treatment plan used in this study used a forward-planned (field-in-field) IMRT technique. All beams were transmitted through the implant before reaching the targeted lesion. The titanium implant was expected to have a substantial effect on the dose to the targeted region of the spinal cord.

2.2. Monte Carlo simulations

The treatment was planned for delivery using a 10 MV photon beam from a Varian iX linear accelerator (Varian Medical Systems, Palo Alto, USA) with a Millennium 120-leaf MLC. A Monte Carlo linear accelerator model with output matched to the clinical beam was designed and commissioned to match clinical beam data, using BEAMnrc [6].

Using the DOSXYZnrc Monte Carlo user code, rectilinear phantom model was designed to approximate the major features of the implant in the patient's neck. Fourteen $1 \times 1 \times 0.8$ cm^3 blocks of titanium, with density 4.54 g/cm3 were positioned in two parallel rows at 2 cm depth in a $10 \times 10 \times 29$ cm^3 volume of water, representing the titanium screw heads that comprised the most substantial parts of the implant. In each row, adjacent titanium blocks are separated by 0.6 cm or 1.0 cm, mimicking the pattern of screw positions in the patient's implant.

The simple, rectilinear model included neither the remaining components of the implant (narrow screw shafts, rods joining the screws and base plate fixing the implant to the skull), nor the air in the trachea, nor the bones of the spine, shoulders or skull. Additionally, only one 19.3×3.0 cm^2 rectangular 10 MV photon field from the planned treatment was applied to the simplified model.

In order to produce a more-detailed Monte Carlo simulation, including the full complexity of the treatment to the patient's heterogeneous anatomy, treatment planning data including the CT images, treatment plan, structures (including the contours of low-density shadow artefacts produced by the high-density implant) and dose distribution were exported from the treatment planning system in DICOMRT format and used as inputs for MCDTK (the Monte Carlo DICOM ToolKit) [7, 8], which produced BEAMnrc and DOSXYZnrc input files based on treatment plan information, and which produced a voxelised DOSXYZnrc 'egsphant' phantom model from the CT data. To produce an egsphant file that was not affected by CT scanning artefacts, the MCDTK code was adapted to reassign the materials and densities of the voxels enclosed by the artefact contours exported from the treatment planning system, so that they matched the surrounding tissues.

A Monte Carlo simulation of the complete radiotherapy treatment was completed, and MCDTK was used to convert the resulting Monte Carlo simulation output into a three dimensional array of absolute dose values, taking into account the number of monitor units per beam as well as the effect of backscatter from the secondary collimators into the linear accelerator's monitor chamber [9, 10].

3. Results and discussion

Figures 1(a), (b) and (c) show the results of evaluating dose in the simple, rectilinear phantom, using the treatment planning system and the Monte Carlo calculations. In figure 1(b), profiles through the implant indicate that the treatment planning system's calculation predicted an increase in absorbed dose within the implant components, which had an apparently minimal effect on the dose between the implant components. By contrast, the Monte Carlo dose distribution shows a decrease in dose within the implant components, and a dramatic increase in scattered dose adjacent to implant components, which is compounded for the section where the model screw heads are placed closer together. Additionally, the Monte Carlo data show that the attenuation of the beam by the implant results in an obvious depletion of dose 5 cm downstream, while the treatment planning system's calculations suggest that the implant has a reduced effect at this distance. In the dose shadow of the implant components, the treatment planning system and Monte Carlo calculations differ by up to 9.5%. This is similar to results obtained in previous simple-phantom studies where Monte Carlo calculations were compared with treatment planning dose predictions generated using CADPLAN [2] and Pinnacle [3].

Figure 2 shows the results of dose calculations made for the full patient treatment, using the conventional treatment planning system and Monte Carlo, and provides immediate visual confirmation of the effects predicted by the simplified model simulation: isodoses obtained from the Monte Carlo data describe regions of high dose between the implant screws as well as regions of depleted dose on the anterior (downstream) side of the implant, while these features are barely evident in the isodoses obtained from the treatment planning system's dose calculation.

There are differences between the results of the detailed patient treatment simulation and the results obtained using the simplified model. The rectilinear model predicts regions of under-dosage of up to 9.5% at a depth 5 cm downstream of the implant, while the patient treatment simulation shows regions of up to 5% under-dosage, at depths 2-3 cm downstream of the implant. This difference in the level of under-dosage seems to result from the use of multiple fields delivered from three gantry angles, in the clinical treatment plan, so that regions under-dosed by one beam are compensated by another. The regions of 5% under-dose remain concerning, however, since they are located within the planned treatment volume (PTV). The variation in dose across the PTV is predicted by the planning system as 10%, but the detailed Monte Carlo model shows that this value increases to 16%, when the dosimetric effects of the implant are accurately evaluated.

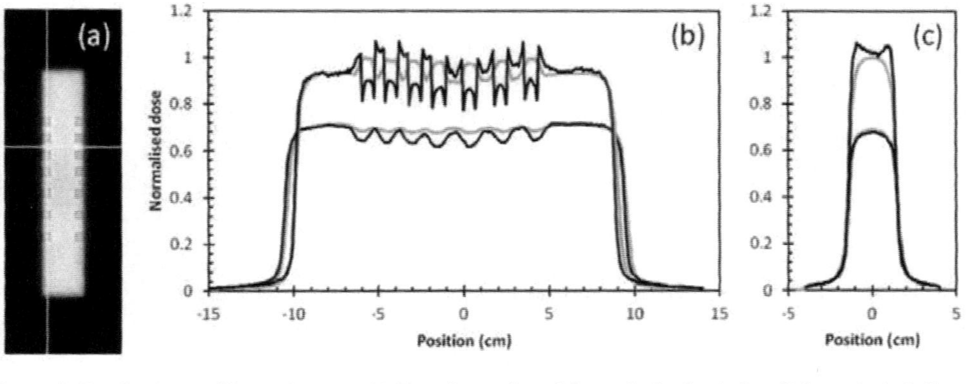

Figure 1: Results for rectilinear phantom: (a) Dose image from Monte Carlo simulation (lighter pixels indicate higher dose, darker pixels indicate lower dose), showing the locations of the profiles shown in (b), as a vertical white line, and (c), as a horizontal white line. Profiles in (b) and (c) are from the treatment planning system's dose calculation (grey lines) and from Monte Carlo simulations (black lines). In profiles (b) and (c), the upper pair of curves represent dose calculated at the depth of the implant and the lower pair of curves represent dose calculated at a depth 5 cm downstream of the implant.

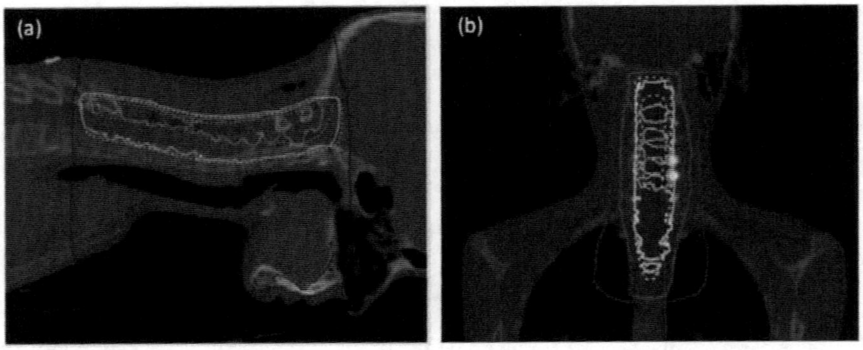

Figure 2: Sagittal (a) and coronal (b) planes showing isodose lines generated from treatment planning dose data (dotted lines) and from Monte Carlo simulations (solid lines), for the full patient treatment. Lines represent doses of 45 Gy (red), 43 Gy (orange), 35 Gy (yellow), 25 Gy (green) and 10 Gy (blue).

4. Conclusion

The Monte Carlo simulations done using both a simplified and detailed model of a radiotherapy treatment planned for a patient with a high-density implant showed that the implant had a profound effect on local and downstream dose, and that this effect was substantially underestimated by the treatment planning system. The simplified model of the patient geometry, irradiated by one radiotherapy beam, provided useful qualitative information. However, the use of this simple model, alone, might lead to an overestimation of the dosimetric effect of the implant on the whole patient treatment. The detailed Monte Carlo simulation of the entire treatment provided information in regions of specific clinical interest, such as the heterogeneity of dose to the PTV, which was not accurately predicted by the planning system and could not be obtained from the simplified Monte Carlo model.

This study confirms the value of performing additional Monte Carlo calculations, preferably using detailed patient models, in cases where treating through high-density implants cannot be avoided.

5. Acknowledgments

This study was supported by the Australian Research Council, the Wesley Research Institute, Premion and the QUT, through linkage grant number LP110100401. Computational resources used in this work were provided by the HPC and Research Support Unit, QUT.

6. References

[1] Papanikolaou N and Stathakis A 2009 *Med. Phys.* **36** 4765-75
[2] Ding G X and Yu C W 2001 *Int. J. Radiat. Oncol. Biol. Phys.* **51** 1167-75
[3] Palleri F *et al* 2008 *Nucl. Instrum. Methods Phys. Res. B* **266** 5001-6
[4] Seco J and Evans PM 2006 *Med. Phys.* **33** 540–552
[5] Sinha A K *et al* 1997 *Can. J. Surg.* **40** 218-26
[6] Rogers D W O *et al* 1995 *Med. Phys.* **22** 503-24
[7] Crowe S B *et al* 2012 *IFMBE Proc.* **39** 1803-6
[8] Crowe S B *et al* 2012 *IFMBE Proc.* **39** 1807-10
[9] Popescu I A *et al* 2005 *Phys. Med. Biol.* **50** 3375-92
[10] Kairn T *et al* 2009 *Australas. Phys. Eng. Sci. Med.* **32** 129-35

The study of N-isopropylacrylamide gel dosimeter doped iodinated contrast agents

YJ Chang[1,2], LL Hsieh[3], MH Liu[3], JS Liu[4] and BT Hsieh[3]
[1] Institute of Biomedical Engineering and Material Science, Central Taiwan University of Science and Technology, Taichung, Taiwan, ROC
[2] Department of Management Information Systems, Central Taiwan University of Science and Technology, Taichung, Taiwan, ROC
[3] Department of Medical Imaging and Radiological Science, Central Taiwan University of Science and Technology, Taichung, Taiwan, ROC
[4] Tungs' Taichung MetroHarbor Hospital, Department of Diagnostic Radiology, Taichung, Taiwan, ROC

E-mail: bthsieh@ctust.edu.tw

Abstract. Low toxicity of N-isopropylacrylamide (NIPAM) dosimeter was doped with clinical iodinated contrast medium agents(Iobitridol (Xenetix® 350) and organically bound iodine (Conray® 60) as radiation sensitizers; The suitable gel dosimeter preparation formula in this research was 5 w/w% gelatin, 5 w/w% N-isopropylacrylamide, 3 w/w% N,N-methylene-bis-acrylamide, and 5 mM Tetrakis phosphonium chloride. The spiral CT was irradiator, and 120 kVp was the operating tube voltage. The maximum radiation dose was 0.6 Gy, and optical CT was the gel measurement device used. The results showed SERs with the addition of radiosensitizers were 10.70 (Xenetix® 350) and 9.67 (Conray® 60), respectively. Thus, the polymerized gel dosimeter could be used in the efficacy evaluation of low-energy and low-radiation dose.

1. Introduction

Gel dosimeters were previously used for dose verification in radiotherapy. However, they have low sensitivity to low radiation dose. In 2005, Corde et al [1] boosted the radiation sensitivity of Fricke gel dosimeters by adding high-z number materials (iodine and barium) and giving 33.5 and 80 keV X-ray radiation using a single-energy synchronized accelerator. In the following year, Boudou et al [2] boosted the radiation sensitivity of polymeric gel dosimeters by adding iodine-containing developers and giving 50 keV single-energy X-ray radiation. In clinical practice, doctors usually add contrast agents into low-energy radiators to enhance regional image contrast ratio and obtain images with higher diagnostic value. This research aims to add iodinated contrast agents into a low toxicity gel dosimeter(NIPAM), use a low-energy X-ray machine, give low-dose radiation, and observe the CTDIw (up to 0.6Gy) of the gel dosimeter.

2. Materials and method

2.1. NIPAM gel preparation

In this study, gel was prepared from gelatin, NIPAM, BIS, and THPC using two formulas: (1) 6 w/w%, 5 w/w%, 2.5 w/w%, and 5 mM, respectively, and (2) 5 w/w%, 5 w/w%, 3 w/w%, and 5 mM, respectively [3]. Linac (Varian) was used to radiate until 5 Gy linearity and sensitivity were achieved. Briefly, gelatin was dissolved in deionized water and heated to 45 °C. When the water was already transparent, NIPAM and BIS were added with continuous stirring for 25 min. Finally, THPC was added. After 2 min, the mixture was poured into a Pyrex spiral test tube. In this study, radiosensitizers

Iobitridol (Xenetix® 350) and Iothalamate (Conray® 60) were added after dissolving NIPAM and BIS in water solution. As a result, two phase symmetries of the additives were produced: NIPAMXenetix® 350 and NIPAMConray® 60. For equal iodine equivalent, the addition doses are shown in Table 1.

Table 1. Addition dosages of contrast agents to NIPAM dosimeter

Contrast medium agents	Concentration in this product	Contrast agent doped amounts in 13mL NIPAM dosimeter
Xenetix® 350(NIPAMXenetix® 350)	350mg/mL	1.30mL
Conray® 60(NIPAMConray® 60)	282mg/mL	1.61mL

2.2. Irradiation gel dosimeter

The doped NIPAM dosimeters were irradiated up to 0.6 Gy(CTDIw) by spiral computed tomography (Hi-speed, GE). The indoor temperature of the radiation chamber was 22 °C, the output tube voltage was 120 kVp, the tube current was 660 mAs, the CTDIw doses for the reference were 0.1, 0.2, 0.3, 0.4, and 0.6 Gy, and the radiation thickness was 10 mm. Polystyrene was used as prosthesis in this process. Additionally, a medium-energy X-ray machine (Pantak HF 420C, National Standards Lab) was used. QRT filter disk assembly was used to simulate the quality of CT beam (error within 5% compared with IEC61267). CAFCHROMIC® XR CT2 and NIPAMXenetix® 350 were radiated with 120 kV. The beam center distance was 150 cm. Doses of 0.1, 0.15, 0.2, 0.25, 0.05, 0.1, 0.15, 0.2, 0.25, and 0.3 Gy were given separately. Sensitivity comparisons were made regarding the radiation results against the gel dosimeter.

2.3. Radiated gel measurement using Optical Computed Tomography

After post-irradiated 24hrs, NIPAM gels were measured by optical computed tomography (a single 632nm He-Ne laser beam). It was evaluated of gel at different optical densities was detected before/after irradiation.

$$\text{Optical density(OD)} = \log \frac{I_{in}}{I_{out}} \qquad (1)$$

I_{in}: Decrease of 632nm laser light intensity in NIPAM gel.
I_{out}: Decrease of 632nm laser light intensity in irradiated NIPAM gel

2.4. Quantification of radiation sensitivity

This study is mainly targeted at low-energy and low-dose gel dosimeter reactions. Therefore, quantifying the reaction results is helpful for verifying the test results. The radiation sensitizer of sensitivity enhancement ratio (SER) is defined as the ratio between the slope of doped radiation sensitizer to NIPAM gel and without doped radiation sensitizerto NIPAM gels:

$$SER = \frac{slope^{doped}}{slope^{non-doped}} \qquad .(2)$$

3. Results and discussion

3.1. Optimal gel formulation

Different gel formulas also generate different reaction results. Studies show that the appropriate BIS content is 2.5 w/w% to 3 w/w% [3]. The gelatin content in this study was decreased from 6 w/w% to 5 w/w% [3]. As shown in Figure 1, the linearity (R2) and sensitivity (slope) were 0.9954 (0.1293 Gy^{-1}) and 0.9843 (0.0693 Gy^{-1}), respectively. The results showed that the polymerization reaction increased with decreasing gelatin content. Because gelatin consumes long-lived radicals, higher gelatin concentrations were relatively lower polymerization rates [3, 4]. The formula 5 w/w% gelatin, 5 w/w% NIPAM, 3 w/w% BIS, and 5 mM THPC was adopted because it enables the gel dosimeter to have higher linearity and sensitivity.

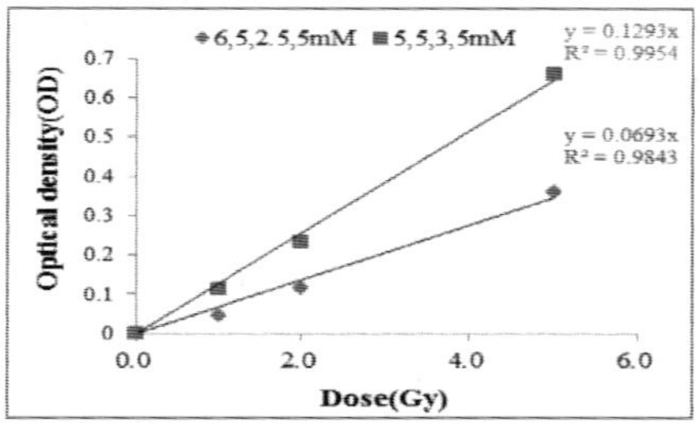

Figure 1. R-square value R^2 and slope of the two formulas of gel dosimeters radiated with 6 MV photon

3.2. Measurement results using Optical Computed Tomography

Figure 2(a) showed when the samples were irradiated with X-ray photons whose most probable energy is close to the K-edge absorption of iodine (~33 keV) [5]. Comparing of different energy to irradiate NIPAMXenetix$^{\text{® 350}}$ and NIPAM$^{\text{Conray® 60}}$, has great linearity of 0.995 and highest sensitivity of 0.428 Gy-1. Figure 2 (b) showed NIPAMXenetix® 350 had highest slice width response profile than NIPAM$^{\text{Conray® 60}}$ even non-doped to NIPAM gel (purple). According to literature, polymer gel can be used as an alternative dosimeter to TLD for the determination of Slice width dose profile (SWDP) [6]. However, there are also some problems in application of gel dosimeter for CTDI measurements such as their low sensitivity [7] In this study, NIPAM gel doped Iodine to sensitivity enhancement that can be more easier to determination of SWDP.

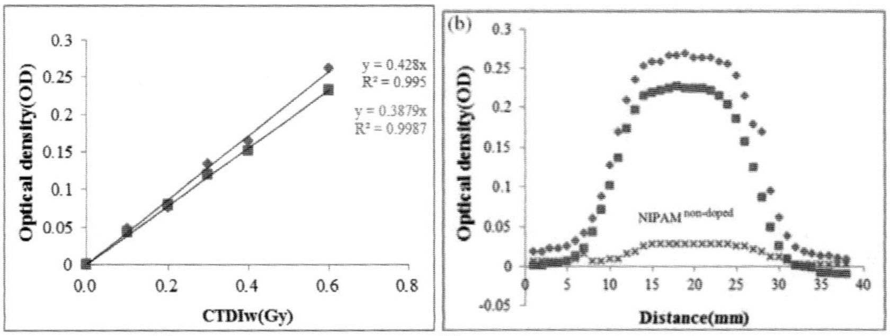

(a) (b)

Figure 2. (a) Scatter plots drawn according to the data generated by optical SPECT scanning NIPAMXenetix$^{\text{® 350}}$ (blue) and NIPAM$^{\text{Conray® 60}}$ (red) radiated with 120 kVp; (b) Scatter plots drawn according to the optical CT data when X-ray energy is 120 kVp and maximum CTDIw dose is 0.6 Gy

3.3. Sensitivity enhancement ratio (SER)

Table 2 showed the SER of NIPAM$^{\text{Xenetix® 350}}$was higher than NIPAM$^{\text{Conray® 60}}$. For kilovoltage energy X-rays, increase is produced by photoelectrons and Auger electrons after photoelectric effect on the heavy atoms [8]. This effect can be produced more free radical to increase polymerization for NIPAM gel. However this enhancement effect of the probability relative reduction with increasing energy of X-ray.

3.4. Comparison sensitivity of radiated radiochromic film and NIPAM gel

Chemical film radiation is used in alignments such as verifications of beam evenness, photon flatness, beam width, and so on. Changes in the ash medium of chemical films are used to determine the radiation reaction. The enhancement results for NIPAM$^{\text{Xenetix® 350}}$ with radiochromic film of the maximum difference to 3.54 at the dose is 0.3 Gy (Figure 3). In this study, the optimal formulation only doped to 1.30mL in 13 mL NIPAM gels. To increase doped concentration of radiation sensitizers to polymer gel dosimeter can be more enhanced sensitivity of radiation doses [5]. In this study, the content of Xenetix® 350 added is 1.30 mL. The gap could be narrowed down when the volume of the radiosensitizer added is increased.

Table 2. SER values of different radiosensitizers in NIPAM gels.

Irradiated gels	Gel response of sensitivity	Correlation coefficient	SER
$NIPAM^{non\text{-}doped}$	0.04	0.91	1.00
$NIPAM^{Xenetix® 350}$	0.43	0.99	10.70
$NIPAM^{Conray® 60}$	0.39	0.99	9.67

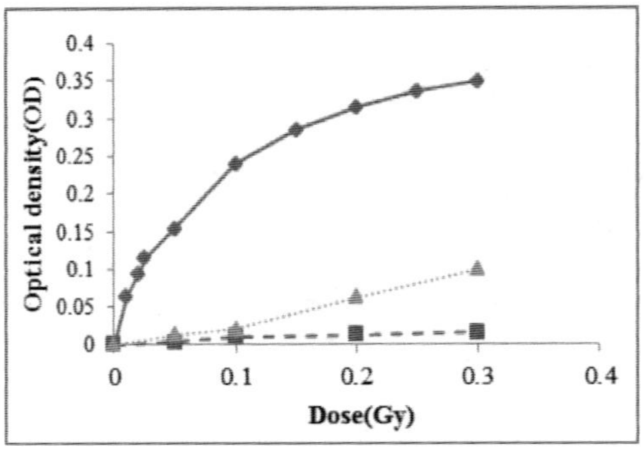

Figure 3. OD reactions of v CAFCHROMIC® XR CT2 and NIPAM after being radiated with 120 kV a medium-energy X-ray machine (Pantak HF 420C, National Standards Lab)

4. Conclusion

In the past, polymeric gel dosimeters are used for dose verification in high-energy radiation therapy plans. Our results suggested that radiosensitizers are helpful in the feasibility of low-energy and low-dose radiation verifications using NIPAM gel dosimeter.

5. References

[1] Corde S and Adam J F 2005 *Radiat. Prot. Dosimetry* **117** 425-31
[2] Boudou C *et al* 2006 *J. Phys.: Conf. Ser.* **56** 145-8
[3] Chang Y J *et al* 2011 *Nucl. Instrum. Methods Phys. Res. A* **652** 783-5
[4] McAuley K 2006 *J. Phys.: Conf. Ser.* **56** 35-44
[5] Meesat R *et al* 2009 *Phys. Med. Biol.* **54** 5909-17
[6] Hill B *et al* 2008 *Physica Medica* **24** 149-158
[7] Karimiafshar L *et al* 2009 *IFMBE Proc.* **25** 541-4
[8] Gastaldo J and Caroline B 2008 *Eur. J. Radiol.* **68** 118-20

Radiological characteristics of charged particle interactions in the first clay-nanoparticle dichromate gel dosimeter

ML Taylor[1,2†], T Maeyama[3], N Fukunishi[3], KL Ishikawa[4,5,6], K Fukasaku[4,7], T Furuta[9], S Takagi[4,8], S Noda[4,9], R Himeno[4,9] and S Fukuda[10]

[1] RMIT University/School of Applied Sciences, Melbourne, Australia
[2] Peter MacCallum Cancer Centre/Physical Sciences, Melbourne, Australia
[3] RIKEN/Nishina Centre for Accelerator-Based Science, Wako, Japan
[4] RIKEN/Computational Science Research Program, Wako, Japan
[5] The University of Tokyo/Department of Nuclear Engineering and Management, Graduate School of Engineering, Tokyo, Japan
[6] The University of Tokyo/Photon Science Centre, Graduate School of Engineering, Tokyo, Japan
[7] Himon'ya Hospital, Department of Neurosurgery, Tokyo, Japan
[8] The University of Tokyo/Department of Mechanical Engineering, Graduate School of Engineering, Tokyo, Japan
[9] RIKEN/Advanced Centre for Computing and Communication, Wako, Japan
[10] National Institute of Radiological Sciences / Research Centre for Charged Particle Therapy, Chiba, Japan

E-mail: michael.taylor@rmit.edu.au

Abstract. The incorporation of clay nanoparticles into gel dosimeters shows promise for significant diffusion reduction – but to what extent does the presence of the nano-clay influence charged particle interactions and, in particular, what is the impact on water equivalence? In this work, we quantify the radiological characteristics of electron, proton and carbon ion interactions in the RIKEN dichromate nanoclay gel and specifically evaluate the water equivalence over a broad energy range. Results indicate that the radiological properties are sufficiently representative of tissues that this low-diffusion gel could readily be used for validation of complex dose distributions. Electron and proton ranges are within 1 % of those in water. Mean effective atomic numbers for electron interactions in the range 10 keV – 10 GeV are within 1 % of those of water which, coupled with the similar mass density, ultimately means the overall impact on dose distributions is not great. The range of C^{6+} ions in the nanoclay gel is closer to that of water (< 4 %) than a common polymer gel dosimeter (< 7 %), though experimentally measured R_1 values indicate an over-response at low doses.

1. Introduction

Contemporary radiotherapy involves increasingly small fields and often complex dose distributions [1]and, within this context, gel dosimetry possesses many advantages over other dosimetric methods (1). Polymer gels have reduced diffusion when compared to ferrous-sulphate (Fricke) based formulations, but exhibit greater artefacts (due mostly to Mie scattering) when imaged with modern optical readout techniques compared to ferrous-sulphate gels, which are amenable to spectrophotometric approaches

(2). A novel alternative currently being explored is dichromate gel dosimetry, with a clay nanoparticle additive as a diffusion suppressant (3). This, however, raises the immediate question as to the radiological influence of the nano-clay and, particularly, the extent of deviation from water equivalence. In this work, we investigate the fundamental characteristics of charged particle (electron, proton and Carbon ions) interactions and quantify any differences to water and tissue.

2. Methods

A range of calculations and comparisons have been made for electron, proton and Carbon ion interactions. For electron interactions: (i) Mean energy-dependent effective atomic numbers have been calculated *via* the Taylor approach (4-7); (ii) Collisional interaction cross sections have been presented as a function of kinetic energy using the standard Bethe-Bloch (8-10) approach recommended by the ICRU (11); (iii) Radiative interaction cross sections have been determined using the combined method of Seltzer and Berger (12, 13); (iv) Continuous slowing down approximation (CSDA) ranges have also been calculated. For proton interactions: (i) electronic and (ii) nuclear stopping powers have been determined, and (iii) projected proton ranges have been computed according to Ziegler (14, 15). Equivalent calculations have also been undertaken for Carbon ions. Example dose distributions have also been computed. In each case, comparisons are made to water and ICRU-defined tissues. Experimental measurements for C^{6+} ion beams (Heavy Ion Medical Accelerator, NIRS Japan) are also shown, with readout undertaken *via* MRI (1.5T Philips Achieva).

3. Results

The mean effective atomic number for electron interactions in the nanoclay gel over the range 10 keV – 10 GeV are presented in Table 1, along with other media (relevant to medical physics) for comparison. Stopping powers are shown in Figure 1(a) for electron interactions, with the ratio to water presented as an insert. The continuous slowing down approximation ranges for electrons in the nanoclay gel are presented in Figure 1(b); plotted as dashed lines against the right axis is the ratio of ranges in gel to water and tissue. A depth-dose distribution of 1 MeV electrons in the nanoclay gel (not shown) indicates a shift in d_{max} of approximately 0.2 mm relative to water. Figure 1(c) shows that the electronic stopping power for proton interactions is lower than that for water (by a maximum of 5 %) while nuclear stopping powers, which comprise a small fraction of the total stopping power, are similar. Proton ranges are within several percent of those in water; see Figure(d). Figure 1(e) indicates that the range of 270 MeV/u C^{6+} ions in nanoclay gel is within 6 mm of that of water, compared to 10 mm for a common polymer gel formulation (BANG-3). Both measured and calculated data are shown. The R_1 values indicate an over-response to the high LET radiation in low-dose regions.

Table 1. Mean effective atomic numbers for the partial and total electron interaction processes; the spread of values over the 10 keV – 10 GeV range is indicated by the standard deviation. Along with the nanoclay gel, data for water, tissues and other dosimeters are provided for comparison.

$\overline{Z}_{eff}$	Radiative	σ	Collisional	σ	Total	σ
Nanoclay gel	4.605	0.113	3.659	0.462	3.732	0.534
Water	4.569	0.112	3.623	0.465	3.696	0.534
Tissue (16)	4.613	0.108	3.688	0.416	3.772	0.502
Bone (16)	5.757	0.156	4.273	0.231	4.552	0.562
BANG gel (17)	4.598	0.104	3.709	0.418	3.784	0.493
TLD-100 (4)	6.653	0.066	6.200	0.204	6.230	0.237
R (nanoclay/water)	1.008	-	1.010	-	1.010	-

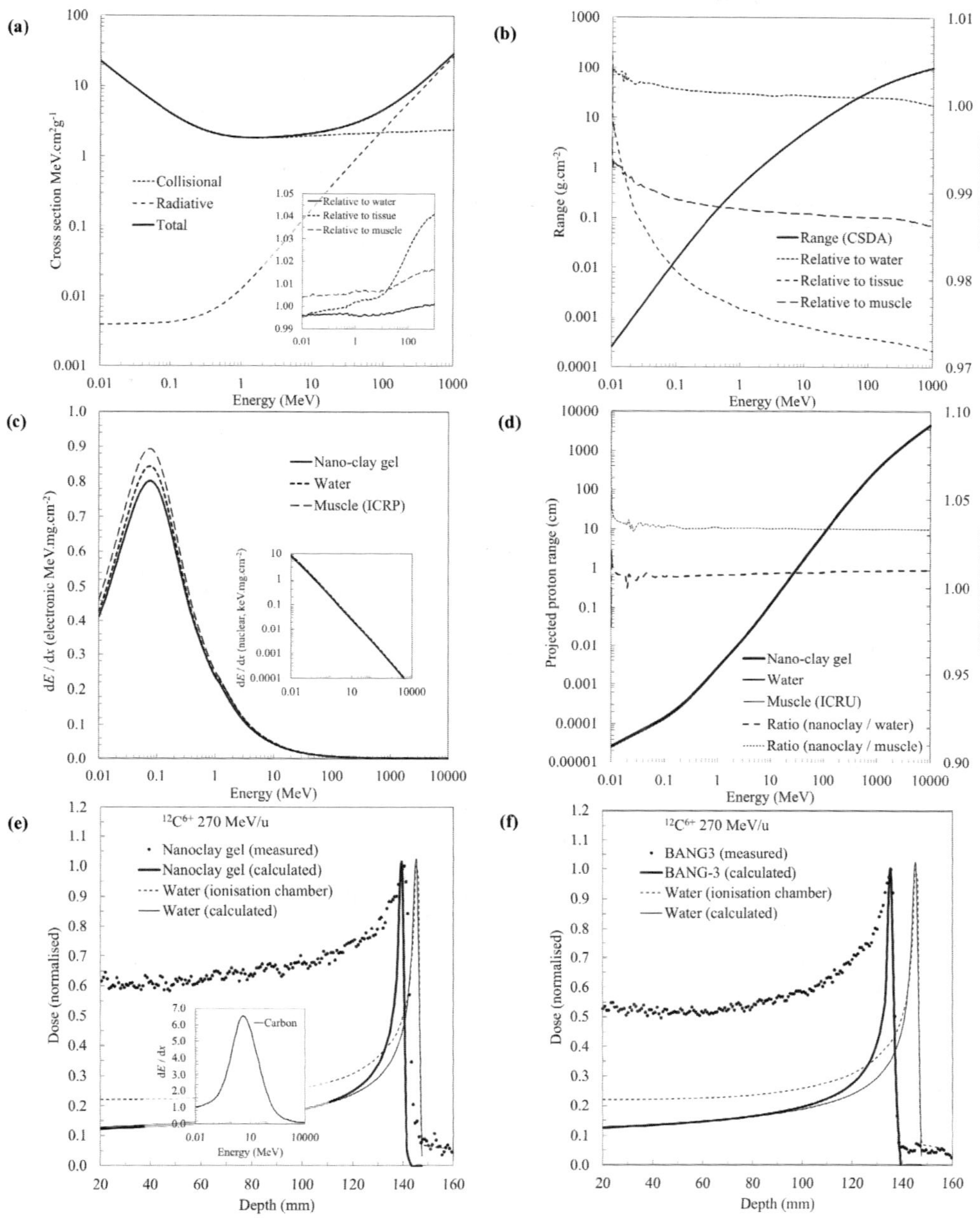

Figure 1 (a) Collisional, radiative and total electron interaction cross sections in the nanoclay gel; data relative to water and human tissues are provided as an insert. **(b)** The continuous slowing down approximation range of electrons (left axis); the dashed lines indicate the ratio of ranges in gel to those in water and tissue (right axis). **(c)** Proton electronic stopping powers plotted alongside those for water and muscle tissue; the insert indicates the nuclear stopping power. **(d)** Proton range as a function of energy (left axis; note these effectively overlap); the dashed lines indicate the ratio of ranges in gel to those in water and tissue (right axis). **(e)** The depth dose profile corresponding to 270 MeV/u C^{6+} ions in the nanoclay gel, both measured and calculated. For comparison, an equivalent profile in water (measured with an ionisation chamber) is also provided, demonstrating a 6 mm difference in range. The insert indicates the total (electronic and nuclear) stopping power for Carbon ions in the nanoclay gel. **(f)** For comparison to the nanoclay gel, we have also measured (and calculated) depth dose curves in BANG gel. The discrepancy compared to water is more pronounced in this case (10 mm).

4. Conclusions
The nanoclay gel exhibits broadly similar radiological properties to water, tissue and other dosimeters over the keV – GeV range for electron, proton and Carbon interactions. The addition of clay nanoparticles raises the effective atomic number only at the sub-percent level and the mass density is not significantly different to water; consequently, resultant dose distributions are not greatly impacted. By comparison, the difference in effective atomic number for the BANG gel formulation is greater relative to water (due mostly to collisional interactions). Ultimately, the results suggest that the formulation could be readily employed for validation of phantom-mapped patient plans.

5. Acknowledgments
Research support: (i) Research and Development of the Next-Generation Integrated Simulation of Living Matter (MEXT Next-Generation Supercomputer Project, Japan), (ii) RIKEN President's Discretionary Fund (Strategic Programs for R&D), (iii) Postdoctoral researcher program of RIKEN & MEXT (JSPS) KAKENHI (24760725). We thank the HIMAC staff for providing beamtime.

6. References
[1] Taylor M *et al* 2011 *Acta. Oncol.* **50** 483-508
[2] Baldock C *et al* 2010 *Phys. Med. Biol.* **55** R1-63
[3] Maeyama T *et al* 2012 *IFMBE Proc.* **39** 1183-6
[4] Taylor M 2011 *Nucl. Instrum. Methods Phys. Res. B* **269** 770-3
[5] Taylor M *et al* 2009 *Radiat. Res.* **171** 123-6
[6] Taylor M *et al* 2012 *Med. Phys.* 2012 **39** 1769-78
[7] Taylor M *et al* 2008 *Australas. Phys. Eng. Sci. Med.* **31** 131-8
[8] Bethe H 1930 *Ann Physik* **5** 325
[9] Bethe H 1932 *Zeitschrift für Physik* **76** 293-9
[10] Bloch F 1933 *Annalen der Physik* **408** 285-320
[11] ICRU 1984 ICRU Report No. 37 (Washington, DC)
[12] Seltzer S and Berger M 1985 *Nucl. Instr. Meth. Phys. Res. B* **12** 95-134
[13] Seltzer S and Berger M 1986 *Data Nucl Data Tables* **35** 345-418
[14] Ziegler J and Biersack J 1985 The stopping and range of ions in matter (Pergamon Press)
[15] Ziegler J *et al* 2010 *Nucl. Instr. Meth. Phys. Res. B* **268** 1818-23
[16] ICRU 1989 Report No 44 (Bethesda, MD)
[17] Zeidan O *et al* 2010 *Med. Phys.* **37** 2145-52

Institute of Physics
Dirac House, Temple Back
Bristol BS1 6BE UK

CURRAN ASSOCIATES INC
proceedings.com